Artificial Intelligence and Industry 4.0

Intelligent Data Centric Systems

Artificial Intelligence and Industry 4.0

Edited by

Aboul Ella Hassanien

Professor of Information Technology, Faculty of Computer and
Information, Cairo University, Cairo, Egypt

Jyotir Moy Chatterjee

Assistant Professor, Information Technology department, Lord Buddha
Education Foundation (Asia Pacific University), Kathmandu, Nepal

Vishal Jain

Associate Professor, Department of Computer Science and Engineering,
School of Engineering and Technology, Sharda University,
Greater Noida, U. P. India

Series Editor

Fatos Xhafa

Universitat Politècnica de Catalunya, Spain

ACADEMIC PRESS

An imprint of Elsevier

Academic Press is an imprint of Elsevier
125 London Wall, London EC2Y 5AS, United Kingdom
525 B Street, Suite 1650, San Diego, CA 92101, United States
50 Hampshire Street, 5th Floor, Cambridge, MA 02139, United States
The Boulevard, Langford Lane, Kidlington, Oxford OX5 1GB, United Kingdom

ISBN 978-0-323-88468-6

For information on all Academic Press publications
visit our website at https://www.elsevier.com/books-and-journals

Publisher: Mara Conner
Editorial Project Manager: Emily Thomson
Production Project Manager: Kamesh Ramajogi
Cover Designer: Christian J. Bilbow

Typeset by STRAIVE, India

Contents

PART 2 Role of artificial intelligence in industry 4.0

CHAPTER 7   **Society 5.0: Effective technology for a smart society** **175**
M. Hanefi Calp and Resul Bütüner

CHAPTER 8   **Big data analytics for strategic and operational decisions** **195**
Brahim Jabir and Noureddine Falih

Contributors

Manish Mohan Baral
Department of Operations, GITAM School of Business, GITAM (Deemed to be University), Visakhapatnam, Andhra Pradesh, India

Smita Vinit Bhoir
Department of Computer Engineering, K.J. Somaiya College of Engineering, Ramrao Adik Institute of Technology, Mumbai, Maharashtra, India

Resul Bütüner
Department of Computer, Ankara Beypazarı Fatih Vocational and Technical Anatolian High School, Ankara, Turkey

M. Hanefi Calp
Department of Management Information Systems, Faculty of Economics & Administrative Sciences, Ankara Hacı Bayram Veli University, Ankara, Turkey

Elias Carrum
Mexican Materials Research Corporation, Saltillo, Coahuila, Mexico

Arvind Chel
Department of Mechanical Engineering, MGM's JNEC Research Centre, MGM University, Aurangabad, Maharashtra, India

Venkataiah Chittipaka
School of Management Studies, Indira Gandhi National Open University, Delhi, India

Jagjit Singh Dhatterwal
Department of Artificial Intelligence & Data Science Koneru Lakshmaiah Education Foundation, Vaddeswaram, AP, India

Noureddine Falih
LIMATI Laboratory, Sultan Moulay Slimane University, Beni Mellal, Morocco

Mustansar Hussain
Department of Chemistry and EVSC, New Jersey Institute of Technology, Newark, NJ, United States

Brahim Jabir
LIMATI Laboratory, Sultan Moulay Slimane University, Beni Mellal, Morocco

Vitthal Jumbad
Jawaharlal Nehru Engineering College Research Centre, MGM University, Aurangabad, Maharashtra, India

Kuldeep Singh Kaswan
School of Computing Science and Engineering, Galgotias University, Greater Noida, Uttar Pradesh, India

Geetanjali Kaushik
Department of Civil Engineering, Hi-Tech Institute of Technology, Aurangabad, Maharashtra, India

Sumit Koul
Department of Mathematics and Scientific Computing, NIT Hamirpur, Hamirpur, Himachal Pradesh, India

Sanjay Kumar
Galgotias College of Engineering and Technology, Greater Noida, Uttar Pradesh, India

Ibtisam Yakub Mogul
ATHE Coordinator Computing, University of Bolton Academic Centre RAK, UAE

Marlyn Montalvo-Martel
Mexican Materials Research Corporation, Saltillo, Coahuila, Mexico

Subhodeep Mukherjee
Department of Operations, GITAM School of Business, GITAM (Deemed to be University), Visakhapatnam, Andhra Pradesh, India

Anant Nemade
Jawaharlal Nehru Engineering College Research Centre, MGM University, Aurangabad, Maharashtra, India

Alberto Ochoa-Zezzatti
Autonomous University of Juarez City, Juarez City, Chihuahua, Mexico

Surya Kant Pal
Department of Mathematics, School of Basic Sciences and Research, Sharda University, Greater Noida, Uttar Pradesh, India

Amit Pandey
Computer Science Department, College of Informatics, Bule Hora University, Bule Hora, Ethiopia

Sunita R. Patil
Department of Computer Engineering, K.J. Somaiya Institute of Engineering and Information Technology, Mumbai, Maharashtra, India

Pedro Perez
Mexican Materials Research Corporation, Saltillo, Coahuila, Mexico

Sudhir Rana
College of Healthcare Management and Economics, Gulf Medical University,
Ajman, United Arab Emirates

Martín Montes Rivera
Aguascalientes Polytechnique, Aguascalientes, México

Sebastián Pérez Serna
Aguascalientes Polytechnique, Aguascalientes, México

Samir Telang
Jawaharlal Nehru Engineering College Research Centre, MGM University,
Aurangabad, Maharashtra, India

Artificial intelligence and Industry 4.0 with healthcare

Influence and implementation of Industry 4.0 in health care

Sumit Koul

Department of Mathematics and Scientific Computing, NIT Hamirpur, Hamirpur, Himachal Pradesh, India

1.1 Introduction

With automation and globalization, the manufacturing industry, which originated in the 18th century, is progressing at a rapid rate. Before the industrial revolution, production was completed manually. Nowadays, much manual labor has been replaced by automated machinery. Over the years, industry has progressed through various stages of change, and we are currently experiencing the Fourth Industrial Revolution. Industry 4.0 touches many sectors, one of the largest of which is the healthcare sector. The healthcare sector is growing at a rate of around 15% annually. Advanced and innovative health systems have improved, and continue to improve, the health and livelihood of humankind.

Big data is a set of large and complex data consisting of both structured and unstructured information. Algorithms are used to manage this data wisely, providing the right data to the right user at the right time. Big data is typically described using the five Vs: volume, velocity, variety, veracity, and value. Handling big data has progressively become easier over time as related technology has developed and advanced. Blockchain is a type of new database in which data are saved in blocks. The basic unit of blockchain is known as a block, which contains various elements of the data, including timestamp, cryptographic hash of the previous block, and transaction data. Blockchain is a decentralized distributed database that is used to feed and maintain continuously growing information. Once the data are entered in a block it cannot be further modified. There are different types of blockchain technologies, for example, bitcoin [1]. To secure data in a blockchain, the data are entered cryptographically. The user uses a private key that helps others to validate the authenticity of data using the public key. The Internet of Things (IoT) is another category of network interrelated with big data. The concept of IoT was introduced in 1999 by technologist Kevin Ashton who described it as a technology that connected several devices with the help of radio frequency identification (RFID) tags for supply chain management. IoT is relevant because of the growth of transportable items,

Artificial Intelligence and Industry 4.0. https://doi.org/10.1016/B978-0-323-88468-6.00002-4

communication, cloud computing, and data analytics. The IoT does not only include computers; it encompasses computers, cars, smart phones, household items, cameras, wearable health devices, and more. Every device connected to the IoT generates and exchanges data that can be analyzed and used for numerous applications in a wide variety of fields, including health care.

As Industry 4.0 is in the early stages, there is a gap in the literature regarding its application in the healthcare sector. In addition, the healthcare sector is unable to use some of the advanced technology from Industry 4.0 due to cost as well as lack of skilled operators. There is scant research on blockchain technology in health care, machine learning for detecting and treating disease, and storing, securing, and managing patient data. This chapter aims to fill this gap.

Section 1.2 of this chapter reviews and tracks the transformation of Industry 1.0 (the start of the industrial revolution) to Industry 4.0, while Section 1.3 discusses the evolution of Healthcare 1.0 to Healthcare 4.0. Section 1.4 provides a brief literature review and Section 1.5 discusses big data in health care. Sections 1.6 and 1.7 discuss the use of the IoT and blockchain technology in health care, respectively. Section 1.8 presents applications of Healthcare 4.0 and Section 1.9 presents three case studies highlighting the use of Healthcare 4.0 in monitoring and predicting different diseases. Section 1.10 summarizes the chapter, Section 1.11 discusses future scope, and finally, Section 1.12 highlights recent innovations and research on Industry 4.0 in the healthcare sector.

1.2 Industry 1.0 to 4.0

The industrial revolution occurred in the late 18th century to about the middle of the 19th century. The first phase of industrial revolution that occurred in this period is known as Industry 1.0, which heralded a transition from human labor to machinery and thus increased productivity. The iron and textile industries played an integral role in this first phase of the industrial revolution, which saw the advent of water and steam-powered machines. The steam engine, considered an "atmospheric engine," was invented by Thomas Newcomen in 1712. Coal was used to fuel engines made of iron. Rural industries transformed into cities and craftsmen became waged laborers. All these changes led to economic development in Great Britain. They also affected population growth; mortality rates for infants and children greatly decreased during this time.

Industry 1.0 opened the door for Industry 2.0, which began in America in the 19th century and lasted to about the beginning of the 20th century. This period saw the discovery of electricity and inventions like the light bulb, phonograph, and steamboat. Other inventions discovered during this time include the wireless telegraph and electric railways, elevators, and motors. The first mechanical computer was invented by Charles Babbage during this period. Industry 2.0 led to development in social and economic life and fields like communication, business, and transportation. Likewise, there were great changes in the healthcare sector, such as the introduction of antibiotics and health insurance, and education; more than 80% of children attended school during this period. Due to the invention of the harvesting machine, the agriculture sector also developed during this period. Production in the steel, iron, and oil industries also increased as more types of machinery were introduced. By 1910, coal-fired, steam-generating plants and hydroelectric power stations were established.

Industry 3.0 occurred in the last few decades of the 20th century and was brought about by the introduction of electronic devices like computers. The development of semiconductors, personal computing, and mainframe computing took place during this time. Analog devices switched to digital technology. Robots, programmable logic controllers, and integrated circuits began to be used to perform manual tasks. Before this time, computers were mostly speculative, unpredictable, and extremely large compared to the power they could provide. Industry 3.0 was the direct result of major developments in computers and information technology and was categorized by digitalization and integration through the incorporation of electronics with automated technology.

Currently, we are experiencing the "Fourth Industrial Revolution," a term coined by Klaus Schwab, the Founder and Executive Chairman of the World Economic Forum. Also known as Industry 4.0, this revolution has brought about many changes, mainly related to the Internet and telecommunications. These include big data analytics, virtual systems, cloud computing, robotics, IoT, AI, 3D printing, and more. Ref. [2] describes Healthcare 4.0 and the different technologies that benefit humanity. Table 1.1 presents a timeline and description of the four phases of the industrial revolution.

Industry 4.0 can be classified as being integrated either horizontally, vertically, or via engineering. Horizontal integration means that a business should compete and cooperate with other similar businesses to achieve an efficient production system. Vertical integration involves connecting all business units and processes within an organization. It ensures high flexibility and helps to configure production lines in

Table 1.1 Industry 1.0 to 4.0.

	Date	Description	Application
Industry 1.0	18th century	Water and steam-powered machines	Rural industries transformed into cities and craftsmen became waged laborers
Industry 2.0	Mid-19th century	Electricity	Development in social and economic life and fields like communication, business, and transportation
Industry 3.0	20th century	Computers	The usage of electronics with Information Technology to advance the dynamics of production
Industry 4.0	21st century	Internet of Things (IoT)	Big data analytics and robotics

an easy manner. Engineering integration gives information about the creation of product-centric value, customer satisfaction, increasing the quality of the product, technology for recycling the product, production planning, and maintenance. The characteristics of the Fourth Industrial Revolution are data simulation, IoT, improved realism, cyber systems, additive production, mechanical robots, blockchain, and so on. Industry 4.0 helps international companies improve their competitiveness in the world market. Industry 4.0 productions are less expensive and more efficient. Ref. [3] identified barriers that can hinder the execution of Healthcare 4.0 in the real world and suggested some solutions. Ref. [4] emphasized seven different methods to secure large amounts of electronically stored patient data. Ref. [5] examined the role of data science in Industry 4.0. Fig. 1.1 shows the architecture of the industry in chronological order.

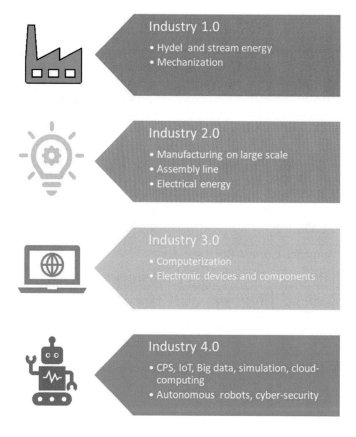

FIG. 1.1

Chronological approach to industry.

1.3 **Healthcare 1.0 to 4.0**

Healthcare 1.0 introduced solutions for public health problems. These solutions were evidence-based and thus effective. Healthcare 2.0 saw the establishment of bigger hospitals and better medical education, with doctors training in specialties so they could provide better-quality treatment. This time period also saw the rise of therapeutics and increased manufacturing and appropriate usage of antibiotics. Healthcare 3.0 heralded the arrival of digitization, which led to the quick manufacture of medicines, among other advancements. In addition, the storage of patient data became easier and more efficient during this phase.

Healthcare 4.0 is the current stage of revolution in this sector, further enhanced by new technologies and thus developing at a rapid pace. It is patient-focused as opposed to hospital-centric. Artificial intelligence (AI) and the IoT, cornerstones of Healthcare 4.0, help both patients and healthcare providers manage health and quality of life. With the help of AI, doctors can more easily predict and diagnose disease. AI also helps with extracting, processing, and delivering data that are collected via sensors and securely stored in the cloud. With the help of the IoT, patients can monitor their health status regularly. Innovations in health care have numerous beneficial effects. For example, treatments in rural areas have become more advanced. Treatment costs have decreased and analytic technologies are able to detect insurance fraud and bogus claims. Blockchain technology is being used to manage big data and big data analytics can help providers predict disease at earlier stages. Electronic health records (EHRs) are simpler and more secure. Healthcare 4.0 has many advantages. As such, the authors of Ref. [6] propose a problem-oriented methodology to prioritize the integration of Industry 4.0 technologies in hospitals. Fig. 1.2 explains the various generations of healthcare systems and the evolution of Healthcare 1.0 to Healthcare 4.0.

1.4 **Literature review**

Ref. [7] propose a blockchain system called GuardHealth for data privacy, preservation, and sharing. The system is designed to manage confidentiality, authentication, preservation, and sharing of sensitive information. In Ref. [8], the authors emphasize Healthcare 4.0 as well as its importance in connecting with smart devices to improve health. The authors also describe research challenges as well as opportunities in this sector. In Ref. [9], the authors discuss the impact of Industry 4.0 in health care. They refer to "smart hospitals," in which healthcare facilities are aimed at optimizing resources and processes via the use of technology. The authors also refer to "smart patients," who are patients that use Internet-connected devices to monitor their health. In their analysis, the authors classified the scientific contributions to health care into eight categories, depending on the type of enabling technology to which they refer. These categories are cyber physical systems (CPS), the IoT, edge/cloud computing, addictive manufacturing, robotics, AI, virtual reality, and augmented

FIG. 1.2

Hierarchy of generations in the healthcare system.

reality, which integrate computation, communication, and physical processes. In Ref. [10], the researchers examine fog computing, cloud computing, and the IoT. The authors' three-layer fog computing technique allows for real-time data collection, processing, and transmission, thus leading to better diagnosis and treatment. Ref. [11] conducted a review of the applications of Industry 4.0 in the medical field and identified 19 important applications that enhance diagnostics, telemedicine, patient-doctor relationships, and more.

Industry 4.0 consists of mechanization procedures, production components, and intelligent systems. This contains digital deployment, automation robots, sensing elements, cyber safety, etc. It enables hospitals to share their information digitally which enhance patient's treatment well in time and also other factors like cost and time management are involved in it. In Ref. [12], the authors focus on the security of patient data. They propose a biometric-based authentication scheme to ensure secure access of patient electronic health records (EHRs) from any location. In Ref. [13], the authors explain the revolution of health care as a result of the Fourth Industrial Revolution. They proclaim the IoT and the Internet of Services (IoS) as some of the most important advances in Industry 4.0. The IoT includes vehicles, home appliances, and other gadgets that connect through the Internet via the IoS. These technologies, including biosensors, artificial organs, and smart devices, help patients monitor their health. In addition, virtualization technologies help patients consult with their healthcare providers at any time from any place. Value chain organizations help facilitate usefulness and productivity in the face of budget constraints. Healthcare 3.0 was hospital-centric, whereas Healthcare 4.0 is patient-focused. The principal designs of Industry 4.0 are interoperability, virtualization, decentralization, real-time capability, service-oriented, and modularity. Interoperability simply means the flow of information at all levels. A Cyber-physical system is one of the main services which combine the reading and recording of the data in a meaningful way. Virtualization creates the virtual copy of every real thing. Smart factories produce a virtual copy of the needed medical things and help to monitor these using Cyber-Physical systems. Virtualization permits personalization in the healthcare sector and precise medicine for patients, workers, and caretakers. Improving health services through cyber-physical system adherence: The technologies like 5G, IoT, Narrow Band IoT, cloud computing, big data, network slices, and cryptography for security in real-time CPS. The decentralization of healthcare is a challenging fact for underdeveloped nations. The quality of the smart devices, gadgets related to healthcare such as wearable, bio-actuators, etc. is not guaranteed. Mobile Edge Cloud computing is a technology that helps in the decentralization process in the healthcare sector. In Ref. [14], the authors study the important fundamental principles in medical services and recommended some useful technologies to enhance said sector. In Ref. [15], the authors explain the lack of policies in the healthcare system of the past. They emphasize shifting to a new healthcare system that is highly equipped with different technologies to overcome these gaps. In Ref. [16], the authors discuss several issues related to telesurgery and its associated challenges in the new era of health care. In Ref. [17], the author has explained the blockchain technology in healthcare sector such as to safe guard the patient's record, supply chain management system, etc. Table 1.2 shows the comparison in the healthcare sector by different authors. The principal designs of Industry 4.0 are interoperability, virtualization, decentralization, real-time capability, service-oriented, and modularity. Interoperability simply means the flow of information at all levels. A Cyber-physical system is one of the main services which combine the reading and recording of the data in a meaningful way. Virtualization creates the virtual copy of every real thing. Smart factories produce a virtual copy of the needed medical things and help to monitor these using Cyber-Physical systems. Virtualization permits personalization in the healthcare sector and precise medicine for patients, workers, and caretakers. Improving health services through cyber-physical system adherence: The technologies like 5G, IoT, Narrow Band IoT, cloud computing, big data, network slices, and cryptography for security in real-time CPS. The decentralization of healthcare is a challenging fact for underdeveloped nations. The quality of the smart devices, gadgets related to healthcare such as

Table 1.2 Comparison of different healthcare technologies by various authors.

References	Purpose	Description
[2]	New technology to enhance health care	How the adoption and integration of Industry 4.0 technologies in the health domain is changing the way to provide traditional services and products
[14]	Implementation of technology in health care	Introduction to Healthcare 4.0
[18]	Blockchain technology in health care	Blockchain technology to improve the interoperability of patient health information between healthcare organizations while maintaining the privacy and security of data
[3]	Barriers to the execution of Healthcare 4.0	Fifteen barriers that can affect the adoption of Healthcare 4.0 in the Indian healthcare sector
[12]	Patient data security	A biometric-based authentication scheme to ensure secure access to patient EHRs from any location
[10]	Fog computing in health care	Three-layer fog computing to enhance efficiency and reliability of patient data interpretation by medical practitioners
[8]	Revolution of Healthcare 1.0 to Healthcare 4.0	Evolution of Healthcare 4.0 and associated challenges and research opportunities
[19]	Blockchain technology in health care	Blockchain technology for managing patient data
[7]	What's new in Healthcare 4.0	Blockchain technology and machine learning algorithms
[20]	Edge-based approach called BodyEdge	Applications of big data and AI

wearable, bio-actuators, etc. is not guaranteed. Mobile Edge Cloud computing is a technology that helps in the decentralization process in the healthcare sector.

1.5 Big data in health care

The healthcare sector is developing rapidly and thus a massive amount of data is being collected in hospitals as well as private clinics. This data is known as big data, due to both its nature and size. It has become a vital element in every field of health care. The digitization of healthcare information is increasing up to 48% per year. In Ref. [21], the authors discuss the 10 Vs of healthcare big data: Volume, Velocity, Variety, Veracity, Validity, Viability, Volatility, Vulnerability, Visualization, and Value. The volume of data in the healthcare sector is increasing daily. Patient details, clinical notes, medical records, scanning reports, insurance data, radiology images, 3D images, and so on are some of the types of data in the healthcare sector. This data can be text-based, image-based, or video-based. Previously, health data was handled manually. Nowadays, this data is handled both manually and technologically via smart devices and the Internet. The velocity of data is also changing. In the past, data was handled manually and was more manageable. There are three varieties of healthcare data: structured data, semistructured data, and unstructured data. The veracity of information refers to the quality of data. Data quality measures should be integrated, trustworthy, complete, noise-free, and unbiased. Validity refers to the correctness and accuracy of the data. Viability means finding the appropriate information from a large sample. Volatility refers to the period when the data is relevant. The vulnerability of data is related to the safety as well as the confidentiality of the data. Proper visualization of information helps to garner valuable insights from the data. The value of data is determined via analysis and governance.

As healthcare data is growing exponentially, it is essential to manage all this data and determine what is the relevant information for the desired purpose. Big data analytics is very helpful in this regard. It helps doctors to carry out proper investigations and provide better treatment. There are four categories of big data analytics: descriptive analytics, diagnostic analytics, predictive analytics, and prescriptive analytics. In health care, diagnostic analytics uses the results of descriptive analytics to determine the cause of disease. Predictive analytics is used to predict disease. It combines the technology of machine learning, data mining, statistical modeling, and AI. Predictive analytics can be used to make patients aware of their conditions so they can take any necessary precautions. Prescriptive analytics combines both descriptive analysis and predictive analytics to determine what to do with the predicted model. Prescriptive analytics allows for faster diagnosis and treatment. This helps the healthcare provider determine proper and accurate treatment in less time.

Healthcare big data has many advantages. It has transformed the healthcare system to a patient-oriented system and thus has reduced patient costs. It has also reduced hospital readmissions. In the past, the chances that patients would need multiple follow-up visits were very high. Now, with the help of big data analytics, medical experts can analyze patients' medical history and identify those individuals who need further care and those who do not, thus preventing unnecessary doctor visits and hospital admissions. Another advantage of big data analytics is workforce optimization. It can help predict how many patients will come in on a particular day so that the organization can determine just how many staff are needed. Big data analytics also helps with real-time alerts. A patient's wearable device can alert their physician to any emergencies in real time. Big data analytics is integral to maintaining and securing patient EHRs. Blockchain technology allows management and exchange of healthcare information at any time anywhere in the world as long as the patient gives their permission. Big data is useful in hospital network analysis, controlling public health research, medical practice efficiency, implementing safety measures, increasing patient engagement, and much more. It can be used to predict and treat numerous disease conditions, including cancer, diabetes, drug abuse, heart disease, and so on.

Big data technology has a significant impact on modern health care. Ref. [20] has shown several applications of big data and AI in enhancing Industry 4.0.

1.6 Internet of Things in the healthcare sector

Ref. [22] discusses the use of wearable devices to improve human health. The IoT is a smart technology that is used to connect devices using sensors, microcontrollers, and transceivers via the Internet, enabling communication and interaction among people anywhere at any time in the world. There are several types of IoT sensor-based devices that connect to the Internet, such as heart rate monitors, continuous glucose monitors (CGMs), sleep tracking devices, and more. All of these devices contribute to creating a personalized healthcare system that takes into account the unique behavioral, cultural, and biological characteristics of each patient. By using the digital identity of each patient, the IoT ensures the personalization of healthcare services. Remote health monitoring is one of the specialties of IoT. It helps patients monitor their health from home, thus reducing the pressure on doctors, other medical staff, and administration. It also saves time. It is a great advantage to people living in rural areas, elderly patients, and people who may be too busy to seek out regular health care. The use of IoT in health care has brought about drastic changes in chronic disease management, rehabilitation, and ambient-assisted living. There are many examples of applications of the IoT in health care. For example, patients suffering from Parkinson's disease can use IoT devices to monitor gait and detect tremors. ECG sensors can predict cardiac events by measuring the rate of heart activity. Another example is the SPHERE wearable device, which is a sensor node worn on the wrist that is tailored for low-maintenance residential health and behavior monitoring. It is designed to help elderly people remain safe and comfortable at home. Other IoT devices for health care include sensors to monitor pulse, respiratory rate, body temperature, and blood pressure.

The requirements of the IoT are privacy, reliability, confirmation, ease of use, novelty of information, endorsement, resiliency, fault acceptance, self-curing, protection, interoperability, and seclusion. With the introduction of the IoT in the healthcare industry, the efficiency of health care increases, the cost of treatment reduces, and more focus is placed on patients and providing them effective and proper treatment. In today's era industry 4.0 is not only showing its importance in healthcare but also in other sectors such as agriculture Ref. [23] described machine learning as well as deep learning applications in the agriculture sector.

1.7 Blockchain technology in the healthcare sector

In Ref. [1], the authors propose a peer-to-peer electronic cash system using bitcoin cryptocurrency technology (blockchain). Ref. [24] coined the concept of blockchain and investigates the suitability of blockchain-based solutions for governance challenges in genomic data sharing. Blockchain has a wide variety of applications in the healthcare sector. This technology helps to protect and safeguard patient information using cryptographic technology. A blockchain is an advanced data structure where data are stored in blocks. The characteristics of blockchain technology are decentralization, traceability, transparency, and immutability. Decentralization of blockchain means there is no single authority for any kind of data transactions that occur within the system. Immutability means the persistence of the blockchain ledger; data cannot be altered or changed once it is entered and saved. Traceability refers to tracing the data with verifiable addresses and timestamps. Transparency means that the members who are involved in the network system can see all the information stored in it. Each block in a blockchain consists of data, a timestamp, hashes of the present block, and a hash of the previous block. In Ref. [19], the authors discuss many ways to enhance the medical sector using blockchain. The authors developed an algorithm to enhance the capability of medical works in accessing patient information. Ref. [18] describes different types of blockchain technology, such as public blockchain, private blockchain, and hybrid blockchain. In the case of a public blockchain, there is open access to the system and anyone can participate in the blockchain. In private blockchain technology, the participant needs permission to join the network system and transactions can be viewed and accessed by authorized users. In a hybrid blockchain, the data can be kept both public and private. Blockchain can confer patient-related benefits as well as organization-related benefits. It keeps patient data safe and accessible and allows organizations to use data to manage medical insurance and track clinical trials along the pharmaceutical supply chain. Although there are many benefits to blockchain technology, there are also many challenges. These are related to scalability, authorization and security issues, energy usage, lack of operational speediness, and lack of technical skills on the part of users.

Smart contracts are a technology that works with blockchain technology. They are computer programs or transaction protocols that automatically execute, control, or document legally relevant events and actions according to the terms of a contract or an agreement. A registrar contract is a global contract that maps the participant identification string with a public key known as the Ethereum address identity. Here, strings are used more than the cryptographic public key identity that uses the existing ID form. A registrar contract also maps with a special contract called a summary contract. The next structure is the patient-provider relationship (PPR) contract. In the PPR contract, there are two nodes. The first node stores and manages the healthcare data for the second node. PPR identifies the data held by the care provider through the assortment of data pointers and the accessibility of the permission. Next is the summary contract, which helps participants track their medical data. The summary contract provides notifications to participants and the relationship is stored in a status variable. Providers can create the relationship in the summary contract of the patient and create a new relationship by updating the data. It is the decision of patients whether they want to accept, reject, or delete a relationship according to the previous history. The last structure of the smart contract is the system node description. The system node can integrate with EHR infrastructure. The software components implemented for the design of a system node are the backend library, Ethereum client, database gatekeeper, and EHR manager. Patient nodes in our system contain the same basic components as providers. The implementation of patient nodes can be executed on a local personal computer or even a mobile smartphone via the Internet. By using the nodes in the summary contract, any missing data can be recovered or retrieved from the network at any time from anywhere. These databases can work as input summation of a patient's health. In Ref. [25], the authors discuss the potential impacts of blockchain technology in development and realization of real-world cyber-physical production systems (CPPSs). Ref. [26] provides a systemic review of blockchain technology in the healthcare sector. In Ref. [27], the authors consider challenges and future perspectives of blockchain technology in medicine and health care.

1.8 Applications in healthcare 4.0

In Ref. [28], the authors propose an edge-based architecture to support applications for Healthcare 4.0. Ref. [29] discusses enhancing smart farms as well as the pharmaceutical industry to increase their reliability and efficiency. Ref. [30] addresses security issues in telesurgery and Ref. [31] focuses on securing information as well as predicting future events in health care via technology such as deep learning blockchain. Fig. 1.3 presents applications in Healthcare 4.0.

1.8.1 Observing physiological and pathological signals

Wearables and sensing devices organized by wireless sensor networks (WSNs) and wireless body area networks (WBANs) work with data from the cloud and fog computing and use big data analytics to support health-monitoring applications. The information collected from these devices provide details about patient health status. The new cloud technologies being developed may eventually replace the hospital information system. Patient monitoring systems consist of three components: sensing and collecting data about patient movements and physical changes, communication software and hardware to relay the data, and analysis of the collected data. There are several types of sensors that monitor physical, mental, and physiological health. With the help of the IoT, the data collected can be transferred to a patient's or doctor's smart device via Bluetooth, and the data can then be stored in the cloud.

1.8.2 Self-management, wellness monitoring, and prevention

Healthcare 4.0 mainly supports patient self-management. Big data analytics helps in preventing disease. Researchers are always investigating new technologies that will advance the healthcare sector and make it more intelligent. Algorithms in big data analytics identify risk factors to help prevent and manage disease.

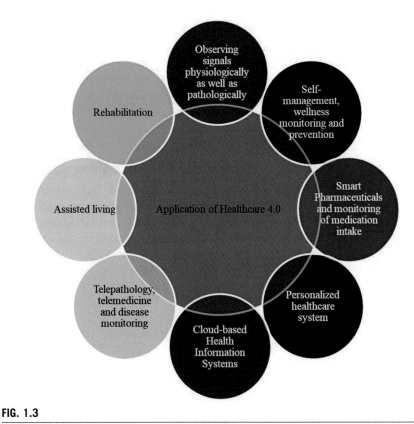

FIG. 1.3

Applications of Healthcare 4.0.

1.8.3 **Smart pharmaceuticals and monitoring of medication intake**

A common issue among the elderly is their propensity to forget to take their medications. This problem can be solved by monitoring their intake of medicine with the help of technology. Several apps are available that alert users to take their medicine at the correct time. Smart pharmaceuticals are those that can be ordered electronically and delivered by mail.

1.8.4 **Personalized healthcare system**

A personalized healthcare system is a patient-oriented system. Patients can wear sensor-based devices to gather information about their health from a variety of sources. Some smart health devices are doctor-approved and proven to improve patient health.

1.8.5 Cloud-based health information systems

The data in the healthcare sector is increasing exponentially and thus it is very important to manage all this data efficiently. Cloud-based architecture helps in the data collection process. There are several types of data types in the healthcare sector, including patient information, treatment information, administrative information, scanning reports, medical images, and so on. Cloud storage is one of the safest ways to secure big data without losing it.

1.8.6 Telepathology, telemedicine, and disease monitoring

Nowadays, telemedicine, telepathology, and disease monitoring have become more advanced with the help of information technology. Smart devices via the IoT can be used to detect conditions and diseases such as cancer, cardiovascular disease, diabetes, Parkinson's, Alzheimer's, and more. Teleconsultation in surgery allows remote consultation via video conferencing. Surgeries can be recorded and the recording sent to any number of experts for feedback.

1.8.7 Assisted living

Disease incidence and hospital visits increase as the population ages. Systems that integrate apps for in-home health monitoring are emerging to realize enhanced living environments (ELEs) for eldercare. Video conferencing, telecommunication, robotics, and wearable devices all play a role in this regard. WBANs may be used along with sensors for creating ambient-assisted living, which uses AI techniques to enable elderly people to live independently for as long as possible.

1.8.8 Rehabilitation

One of the main applications of Healthcare 4.0 is in home-based rehabilitation. It saves costs and results in better quality of treatment. WBANs are used in this case to detect a patient's physical movements via sensors and virtual instruments that provide real-time feedback to the patient and their medical provider.

1.9 Case studies

The case studies presented in this section demonstrate how Healthcare 4.0 technologies are being used to improve patient care for various conditions.

1. Case study 1: Diabetes.

   There are two types of diabetes mellitus: type 1 or juvenile diabetes, and type 2 diabetes. Both types are chronic and require ongoing treatment in consultation with a medical professional. Diabetes can increase the risk of heart attack, kidney failure, eye disease, and more. Conventional treatment in the time before

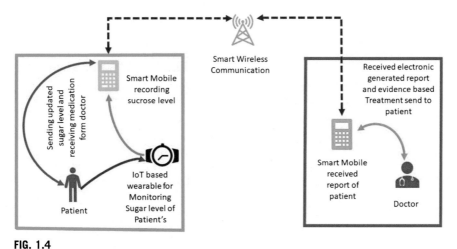

FIG. 1.4

Glucose-monitoring device for diabetics.

Healthcare 4.0 consisted of in-person appointments with a provider and prescription drugs to manage sugar levels. As such, diabetics typically need to visit their providers on a regular basis, which can be costly and time-consuming. Today, diabetics have access to many different technologies for monitoring their blood sugar, heart rate, oxygen level, pulse rate, and blood pressure at home on their own. This allows the patient to detect any anomalies and communicate them to their doctor remotely via the IoT. The doctor can then suggest further treatments if applicable, as well as monitor the patient's info on a regular basis (e.g., weekly, monthly, etc.). Fig. 1.4 shows a wearable continuous glucose monitoring (CGM) device, which the patient wears on their wrist. It measures glucose levels via the patient's sweat and sends the data to a mobile device (e.g., smartphone). This data can be downloaded via the IoT and viewed by the patient's healthcare provider. Prior to the development of CGMs, diabetics needed to check their blood sugar via a finger prick. CGMs are less invasive and more efficient and thus a good example of innovation in Healthcare 4.0.

2. Case study 2: COVID-19.

At the start of the COVID-19 pandemic, patients could not be diagnosed with the virus without visiting a diagnostic center or lab. However, this was not advisable because of the risk of spreading the virus to others. As such, Healthcare 4.0 was tasked with finding a remote and safe technique for COVID testing. Consequently, at-home testing kits were developed. An at-home test contains a sterilized nasal swab, extraction tube, test card, biohazard bag, and instruction manual (see Fig. 1.5). Before self-administering the test, the individual must download an app on their smartphone for capturing the bar code reading. When the testing is complete, the person receives their result on their mobile device. These rapid tests save time and money and allow the person to stay at home, thus

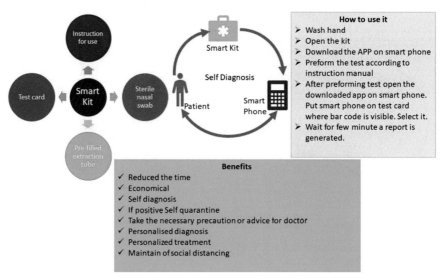

FIG. 1.5

COVID-19 at-home testing kit.

avoiding exposing others to the virus. At-home rapid tests are another innovation of Healthcare 4.0.

3. Case study 3: Heart disease.

In this case study, machine learning algorithms such as linear regression, support vector machine, decision tree classifier, logistic regression, k-nearest neighbor, and Gaussian naive Bayes are used to predict the risk of heart attack [32]. In the past, to detect and diagnose disease, the medical practitioner needed to have all the data related to the patient's previous history as well as information on similar case studies, data that may not have been available initially. This delayed information led to delayed diagnosis. Today, however, with the help of machine learning, a patient's full history and data from similar cases are just a click away. In this study, various machine learning algorithms were used to detect heart disease. Table 1.3 presents the various techniques with mean square error to determine the optimal model for determining risk of heart attack.

Table 1.3 Comparison of various algorithms in machine learning.

S. No.	Different algorithms	R^2 score	Mean squared error
1	Support vector machine	0.5348	0.1147
2	Logistic regression	0.5348	0.1147
3	Linear regression	0.5320	0.1154
4	Decision tree classifier	0.4684	0.1311
5	Gaussian naive Bayes	0.3355	0.1639
6	k-Nearest neighbor	0.2690	0.1803

Table 1.3 shows that support vector machine and logistic regression are the best algorithms for predicting the risk of heart attack.

1.10 Conclusion

This chapter presented a comprehensive overview of the evolution of health care in the context of Industry 4.0. It tracked and described the evolution of Industry 1.0 to Industry 4.0 as well as the evolution of Healthcare 1.0 to Healthcare 4.0. The Fourth Industrial Revolution in the healthcare sector promises to transform the health of patients by providing more accurate and personalized service. Technological advances such as the IoT, blockchain technology, big data, and AI have also increased productivity and efficiency in health care. The quote "Prevention is better than cure" is very meaningful in the case of big data analytics. Predictive data analysis can predict and detect disease at earlier stages. Blockchain technology protects patient data collected via wearable sensors and devices connected to the IoT. This data can be transmitted in real time to medical providers for analysis. The possibilities of Healthcare 4.0 are seemingly endless and the implementation of the innovations discussed in this chapter have been and will continue to be a boon to the field of modern medicine.

1.11 The recent innovations and research solutions of Industry 4.0 in healthcare sector

(i) To collect and interpret the data of developed countries to analyze the work done in the medical sector regarding the development of health value chain 4.0. This research work analyses the patients' data regarding their symptoms, diagnosis and continuous treatment. This analysis will contribute in the efficient and quick working using health chain 4.0.

(ii) The verification and assessment of parameters related to the expansion of various products by some corporations in the medical health sector need to be taken into consideration in the time to come. This can lead to the growth in economical sector as well as increase the production of the related companies also.

(iii) A survey in the hospital need to be done by collecting the data of all patients that includes the case study for analyzing the specific characteristics in the 4.0 industry.

(iv) Identify how smart hospitals are empowering employees, while increasing new technologies annually in the control of patient data.

References

[1] S. Nakamoto, Bitcoin: A Peer-to Peer Electronic Cash System, vol. 9, 2008. www.Bitcoin.org.

[2] G. Aceto, V. Persico, A. Pescapé, Industry 4.0 and health: Internet of Things, Big Data, and cloud computing for Healthcare 4.0, J. Ind. Inf. Integr. 18 (2020), https://doi.org/10.1016/j.jii.2020.100129, 100129.

[3] P. Ajmera, V. Jain, Modelling the barriers of Health 4.0–the fourth healthcare industrial revolution in India by TISM, Oper. Manag. Res. 12 (2019) 129–145.

[4] J.J. Hathaliyaa, S. Tanwar, R. Evans, Securing electronic healthcare records: a mobile-based biometric authentication approach, J. Inf. Sec. Appl. 53 (2020), 102528.

[5] S. Sajid, A. Haleem, S. Bahl, M. Javaid, T. Goyal, M. Mittal, Data science applications for predictive maintenance and materials science in context to Industry 4.0, Mater. Today: Proc. 45 (2021) 4898–4905, https://doi.org/10.1016/j.matpr.2021.01.357.

[6] G.L. Tortorella, F.S. Fogliatto, M. Vijaya Sunder, A.M.C. Vergara, R. Vassolo, Assessment and prioritisation of Healthcare 4.0 implementation in hospitals using quality function deployment, Int. J. Prod. Res. (2021), https://doi.org/10.1080/00207543.2021.1912429.

[7] Z. Wang, N. Luo, P. Zhou, GuardHealth: blockchain empowered secure data management and graph convolutional network enabled anomaly detection in smart healthcare, J. Parallel Distrib. Comput. 142 (2020) 1–12, https://doi.org/10.1016/j.jpdc.2020.03.004.

[8] J. Li, P. Carayon, Health Care 4.0: a vision for smart and connected health care, IISE Trans. Healthc. Systems Eng. (2021), https://doi.org/10.1080/24725579.2021.1884627.

[9] L. Cassettari, C. Patrone, S. Saccaro, Industry 4.0 and its applications in the healthcare Sector: a systematic review, in: XXIV Summer School "Francesco Turco"-Industrial System Engineering, 2019, pp. 136–142.

[10] A. Kumari, S. Tanwar, S. Tyagi, N. Kumar, Fog computing for Healthcare 4.0 environment: opportunities and challenges, Comput. Electr. Eng. 72 (2018) 1–13, https://doi.org/10.1016/j.compeleceng.2018.08.015.

[11] M. Javaid, A. Haleem, Industry 4.0 applications in medical field: a brief review, Curr. Med. Res. Practice 9 (2019) 102–109, https://doi.org/10.1016/j.cmrp.2019.04.001.

[12] J.J. Hathaliya, S. Tanwar, S. Tyagi, N. Kumar, Securing electronics healthcare records in Healthcare 4.0: a biometric-based approach, Comput. Electr. Eng. 76 (2019) 398–410, https://doi.org/10.1016/j.compeleceng.2019.04.017.

[13] A.C.B. Monterio, R.P. Franca, V.V. Estrela, A. Khelassi, N. Razmjooy, Y. Iano, Health 4.0: applications, management, technologies and review, Med. Technol. J. 2 (2018) 262–272.

[14] J. Chanchaichujit, A. Tan, F. Meng, S. Eaimkhong, An introduction to Healthcare 4.0, in: Healthcare 4.0, Palgrave Pivot, Singapore, 2019, pp. 1–15, https://doi.org/10.1007/978-981-13-8114-0_1.

[15] D. Sharma, G.S. Aujla, R. Bajaj, Evolution from ancient medication to human-centered Healthcare 4.0: a review on health care recommender systems, Int. J. Commun. Syst. (2019), https://doi.org/10.1002/dac.4058.

[16] R. Gupta, S. Tanwar, S. Tyagi, N. Kumar, Tactile-internet-based telesurgery system for Healthcare 4.0: an architecture, research challenges, and future directions, IEEE Netw. 33 (6) (2019) 22–29, https://doi.org/10.1109/MNET.001.1900063.

[17] S Koul, T Krishna, Introduction to Blockchain Technology and Its Role in the Healthcare Sector, in: Ambikapathy, R Shobana, Logavani, Dharmasa (Eds.), Reinvention of Health Applications with IoT: Challenges and Solutions, 1st, CRC Press, Boca Raton, 2022, pp. 55–80. https://doi.org/10.1201/9781003166511.

[18] I. Abu-elezz, A. Hassan, A. Nazeemudeen, M. Househ, The benefits and threats of blockchain technology in healthcare: a scoping review, Int. J. Med. Inform. 142 (2020) 1–9, https://doi.org/10.1016/j.ijmedinf.2020.104246.

[19] S. Tanwar, K. Parekh, R. Evans, Blockchain-based electronic healthcare record system for Healthcare 4.0 applications, J. Inf. Sec. Appl. 50 (2020), https://doi.org/10.1016/j.jisa.2019.102407, 102407.

[20] S.K. Jagatheesaperumal, M. Rahouti, K. Ahmad, A. Al-Fuqaha, M. Guizani, The duo of artificial intelligence and Big Data for Industry 4.0: review of applications, techniques, challenges, and future research directions, IEEE Internet Things J. (2020). https://arxiv.org/abs/2104.02425.

[21] P.K.D. Pramanik, S. Pal, M. Mukhopadhyay, Healthcare big data: a comprehensive overview, in: N. Bouchemal (Ed.), Intelligent Systems for Healthcare Management and Delivery, IGI Global, 2018, pp. 72–100.

[22] S. Pirbhulal, O.W. Samuel, W. Wu, A.K. Sangaiah, G. Li, A joint resource-aware and medical data security framework for wearable healthcare systems, Futur. Gener. Comput. Syst. 95 (2019) 382–391.

[23] S. Koul, Machine learning and deep learning in agriculture, in: Patel, et al. (Eds.), Smart Agriculture: Emerging Pedagogies of Deep Learning Machine Learning and Internet of Things, CRC Press, 2021, pp. 1–19.

[24] M. Shabani, Blockchain-based platforms for genomic data sharing: a de-centralized approach in response to the governance problems? J. Am. Med. Inform. Assoc. 26 (2019) 76–80, https://doi.org/10.1093/jamia/ocy149.

[25] J. Lee, M. Azamfar, J. Singh, A blockchain enabled cyber-physical system architecture for Industry 4.0 manufacturing systems, Manuf. Lett. 20 (2019) 34–39.

[26] C.C. Agbo, Q.H. Mahmoud, J.M. Eklund, Blockchain technology in healthcare: a systematic review, Healthcare 7 (2019) 56, https://doi.org/10.3390/healthcare7020056.

[27] A.A. Siyal, A.Z. Junejo, M. Zawish, K. Ahmed, A. Khalil, G. Soursou, Applications of blockchain technology in medicine and healthcare: challenges and future perspectives, Cryptography 3 (1) (2019). https://www.mdpi.com/2410-387X/3/1/3.

[28] P. Pace, G. Aloi, R. Gravina, G. Caliciuri, G. Fortino, A. Liotta, An edge-based architecture to support efficient applications for healthcare industry 4.0, IEEE Trans. Ind. Inf. 15 (1) (2019) 481–489, https://doi.org/10.1109/TII.2018.2843169.

[29] J. Wan, S. Tang, D. Li, M. Imran, C. Zhang, C. Liu, Z. Pang, Reconfigurable smart factory for drug packing in healthcare Industry 4.0, IEEE Trans. Ind. Inf. 15 (1) (2019) 507–516, https://doi.org/10.1109/TII.2018.2843811.

[30] R. Gupta, S. Tanwar, S. Tyagi, N. Kumar, M.S. Obaidat, B. Sadoun, HaBiTs: blockchain-based telesurgery framework for healthcare 4.0, in: 2019 International Conference on Computer, Information and Telecommunication Systems (CITS), 2019, pp. 1–5, https://doi.org/10.1109/CITS.2019.8862127.

[31] P. Bhattacharya, S. Tanwar, U. Bodke, S. Tyagi, N. Kumar, BinDaaS: blockchain-based deep-learning as-a-service in healthcare 4.0 applications, IEEE Trans. Netw. Sci. Eng. 8 (2021) 1242–1255, https:/doi.org/10.1109/TNSE.2019.2961932.

[32] A. Janosi, W. Steinbrunn, M. Pfisterer, R. Detrano, UCI Machine Learning Repository, University of California, School of Information and Computer Science, Irvine, CA, 1988. http://archive.ics.uci.edu/ml.

Impact of artificial intelligence in the healthcare sector

Subhodeep Mukherjee[a], Venkataiah Chittipaka[b], Manish Mohan Baral[a], Surya Kant Pal[c], and Sudhir Rana[d]

[a]*Department of Operations, GITAM School of Business, GITAM (Deemed to be University), Visakhapatnam, Andhra Pradesh, India* [b]*School of Management Studies, Indira Gandhi National Open University, Delhi, India* [c]*Department of Mathematics, School of Basic Sciences and Research, Sharda University, Greater Noida, Uttar Pradesh, India* [d]*College of Healthcare Management and Economics, Gulf Medical University, Ajman, United Arab Emirates*

2.1 Introduction

The ascent of digital innovation fueled by artificial intelligence (AI) has become a significant driver for change across different businesses [1]. In 2016, Gartner's report found that 9% of firms conveyed AI innovation, which by 2019 expanded to 25% [2]. Today, AI remains an essential innovation for organizations [3]. Due to progression of systems administration and expanded information preparation, AI has become a vital part of computerized change [4]. India's healthcare industry encompasses emergency clinics, drugs, diagnostics, medical equipment and supplies, clinical protection, and telemedicine. Partners in the reception and execution of AI in medical services include professionals, engineers, examination and industry bodies, government, and speculators [5]. Microsoft and Google have likewise met to deal with various activities to help construct an AI foundation for the nation [6] via a pilot project with hospitals in India. The increasing use of AI in medical services can be viewed as an assortment of advances in innovation [7]. For example, Practo, a doctor consultation and appointment booking app for patients, has been chipping away at robotizing understanding associations with AI utilization [8].

Digital advancements in health care incorporate characteristic language preparation, specialists, AI, master frameworks, chatbots, and voice acknowledgement. However, there are challenges in building up these arrangements, such as the absence of thorough, delegate, interoperable, and clean information. To address this issue, the Electronic Health Record (HER) Standards created by the Ministry of Health and Family Welfare in 2016 provide guidelines for standardization and homogeneity, interoperability in capture, storage, transmission, and use of healthcare information across various health IT systems [5].

Artificial Intelligence and Industry 4.0. https://doi.org/10.1016/B978-0-323-88468-6.00001-2

The research presented in this chapter studies the impact of AI in India's healthcare systems and reviews the status of its implementation in India's hospitals. For this study, we identified technological-organizational-environmental (TOE) factors via a literature review. The four factors from the technological perspective (TP) are: (1) cost-effectiveness (CE), (2) relative advantage (RA), (3) security and privacy concerns (SPC), and (4) complexity (COMP). The three factors from the organizational perspective (OP) are: (1) human resource readiness (HRR), (2) top management support (TMS), and (3) organizational readiness (ORR). Finally, the three factors from the environmental perspective (EP) are: (1) competitive pressure (CP), (2) support from technology vendors (STV), and (3) environmental uncertainty (EU). Artificial intelligence adoption (AIA) is a dependent variable in this study with four indicators: AIA1, AIA2, AIA3, and AIA4.

Our study attempts to answer the following questions:

1. What is the impact of AI in Indian healthcare systems?
2. What are the TOE factors in the adoption of AI in India's hospitals?

2.2 Literature review
2.2.1 AI in health care

In health care, AI addresses the issue of large volumes of medical data, which can be difficult to manage and analyze, thus potentially undermining evidence-based practice [9]. This is known as "channel disappointment," where the principal issue isn't an excessive amount of data, but rather how such data is investigated [10]. Insufficient data recovery frameworks cause problems in purpose-care settings, such as trouble distinguishing all pertinent proof in various data assets and the absence of fundamental healthcare data education [11]. For example, analysts have utilized keen calculations to extricate data from radiology reports in an archive traversing different organizations [12]. They reported that the methodology "gives a compelling programmed technique to comment on and separate clinically critical data from an enormous assortment of free-text radiology reports."

Expressive AI is the most broadly utilized in medical care innovation today and holds the most promise [13]. It evaluates events in real time and uses the information to recognize patterns and minor changes that may be helpful for clinical experts. For example, expressive AI can distinguish designs in bone breaks (e.g., wrist fractures) and skin injuries [14]. Prescient AI processes data to make predictions about what's to come [15]. Clinical experts utilize simulated intelligence to obtain knowledge and proactively propose activities. AI can play a critical role in proactive medical care innovations. Prescient AI can perform some of the clinician's roles, thus replacing human work [16]. Prescriptive AI assists proactive AI and can distinguish subtle patterns that healthcare providers may not recognize and suggest potential medicines [17]. This dynamic capacity makes prescriptive AI the most fascinating and disputable use case in the near term [18]. AI can help specialists obtain, analyze, and apply clinical information in association with doctors, medical caretakers, and scientists, thus improving clinician productivity and quality of care [19].

2.2.2 Utilization of AI in health care in India

The utilization of AI in health care in India is expanding with new businesses and information and communications technology (ICT). Organizations are using AI to address medical services difficulties in the country. Such difficulties include the lopsided proportion of specialists to patients, helping specialists become more productive, providing customized and high-quality medical services to country zones, and preparing specialists and attendants in complex methodology [20]. Organizations offer a scope of arrangements including computerization of clinical data, mechanized examination of clinical trials, identification and screening of disease, wearable sensor-based clinical devices, tolerant administration frameworks, proactive medical care, and infection anticipation [21]. In India, assistive AI has the most potential for development. Simultaneously, advancements that can supplant specialists have minimal odds of succeeding, one reason being an irreconcilable situation among the clinical establishment [22]. The healthcare industry in India is comprised of various segments [23]. Through an audit of organizations creating AI answers for wellbeing, health professionals utilizing AI, and analysts investigating the capability of AI in health care, it was discovered that AI is being used in an assortment of ways across various sectors, as we discuss in the sections that follow.

2.2.2.1 Hospitals

Hospitals in India include government hospitals, district hospitals, healthcare centers, general emergency clinics, mid- and top-level private medical clinics, and nursing homes. An example of AI utilization in hospitals is that of the Manipal Hospitals group, which has adopted IBM's Watson for Oncology to help specialists diagnose and treat seven kinds of malignancies [24]. In this case, AI is used to examine patient information, which proves and improves the nature of the data, thus increasing understanding and trust among patients and providers. Patients are aware that their data is being examined and are asked for their permission. The technology works by assembling a board of disease specialists who form proposals for detailed patient profiles. "These suggestions address the most realistic estimations of these specialists, upheld by clinical writing and individual experience. I.B.M. has never permitted an independent investigation of Watson for Oncology [11]. No, follow up is done to assess whether its suggestions help patients."

2.2.2.2 Pharmaceuticals

The pharmaceutical sector in India incorporates manufacturing, extracting, handling, sanitizing, and bundling of synthetic materials for human and animal medicine [25]. Tekkeşin [26] reviewed the use of prescriptive AI in this area. The most well-known utilization of AI in drugs is in medication revelation, where AI is used to review all accessible writing on a specific atom for a medication (e.g., directed atom revelation), something that is impossible to do manually [27].

2.2.2.3 Diagnostics

Diagnostics includes organizations and research facilities that offer logical or analytic administrations. Notwithstanding more influential organizations, such as Google and IBM, India is host to new businesses representing considerable authority in outfitting AI to analyze illness. According to our review of this sector, diagnostics in India utilizes prescient AI. For example, qure.ai utilizes learning innovation to help diagnose disease. It also uses medical imaging to suggest customized therapy plans [19]. Orbuculum uses AI to predict disease, for example, malignancy, diabetes, neurological problems, and cardiovascular sicknesses, through genomic information [28]. In India, AI is being employed through chatbots, such as Wysa, a technology that helps people manage stress and improve mental health. An individual can speak anonymously with an AI-powered framework (chatbot) that offers a sympathetic ear and recommends counseling professionals [29]. These chatbots are not intended to provide a diagnosis or deal with more significant issues (these are moved to specialists).

2.2.2.4 Medical equipment and supplies

The medical supplies sector produces clinical tools and equipment, such as dental, muscular, and ophthalmologic devices, as well as research facility instruments, and more. AI is also being used in this area. For example, Niramai is a software-based medical device used to detect breast cancer at an early stage. The device filters the chest territory like a camera and uses cloud-computing to analyze the obtained images for early indications of breast cancer [30]. The company ten3T Healthcare developed a wearable device for heart patients that monitors the patient's vital signs and communicates this information through the cloud for examination by specialists. IntelliSpace Consultative Critical Care (ICCC) by Phillips is another innovation in healthcare AI. It enables the creation of tele-ICU facilities that use remote monitoring and consultation to support high-quality patient care [31].

2.2.2.5 Health insurance

Healthcare coverage and clinical repayment offices cover a person's healthcare costs. AI is also being used in this area in India [29]. AI can computerize insurance claims by breaking down large amounts of data in a short time, thus decreasing processing time, calculating costs, and improving client experience. Bajaj Allianz General Insurance utilizes BOING, a chatbot that responds to client questions about healthcare coverage. ICICI Lombard utilizes its chatbot MyRA to sell protection strategies. HDFC Life's chatbot SPOK automatically reads, understands, categorizes, prioritizes, and responds to customer emails.

2.2.2.6 Telemedicine

Telemedicine can help address the difficulties of providing medical care to provincial and far-off territories [21]. It uses electronic interchanges and programming to offer clinical assistance to patients distantly [32]. It is often utilized for follow-up visits, monitoring chronic conditions, drug the board, expert conferences, and other clinical activities that can be provided through secure video channels [7]. In India, telemedicine AI can help address the difficulties of providing medical care to provincial and remote regions, as well as provide online education to clinicians and prepare them for board certification. The company SigTuple can examine blood slides and create a pathology report without help from a pathologist [33]. This technology can be used in remote locations for a fraction of what it typically costs [34]. The Philips Innovation Campus (PIC) in Bengaluru tackles innovation to make medical care affordable and available.

2.3 Research framework and development of hypotheses

Depietro et al. [35] developed a multipoint view of the TOE model, which is an authoritative-level model. This model contends that the reception variables of innovation influence the association remotely and inside with three points of view: mechanical, hierarchical, and ecological [36]. In the TOE model, the innovation viewpoint considers the internal and external assets needed to adopt innovation in an organization [37]. The TOE framework has been used in earlier research and technology adoption [38–60].

2.3.1 Technological perspective (TP)

2.3.1.1 Cost-effectiveness (CE)

CE refers to the costs of adopting and implementing the latest technology [61]. Organizations need to save costs in the implementation process. CE negatively influences its adoption in a business intelligence system [53]. AI in health care will reduce costs significantly over time. The expenses incurred while adopting the latest technology include infrastructure, software, implementation, employee training, and consultancy support costs.

H1 Cost-effectiveness will positively impact the AIA.

2.3.1.2 Relative advantage (RA)

RA refers to the benefits of the latest technology in the adoption process [62]. If the technology can provide RA, it is more likely to be adopted by the organization [45,46]. AI offers many advantages to the healthcare sector. It optimizes cost, time, profitability, and operational excellence.

H2 Relative advantage will positively impact the AIA.

2.3.1.3 Security and privacy concerns (SPC)

SPC refers to the security factors that firms must consider before adopting innovative technologies [60,63,64]. Previous studies have shown that safety for technology adoption is a significant concern [59,60,65]. AI can be used to identify, track, and monitor people across multiple devices, whether they are at work, at home, or in public. This means that even if your data is anonymized once included in a large data set, AI can de-anonymize it using inferences from other devices.

H3 Security and privacy concerns will positively impact the AIA.

2.3.1.4 Complexity (COMP)

Rogers [62] Defined complexity as "the degree to which an innovation is perceived as relatively difficult to understand and use." COMP has been found to be negatively correlated with the rate of technology adoption [66]. COMP is an innovation that is relatively difficult to utilize and understand. Due to a lack of skills and knowledge, few firms consider adopting complex innovations.

H4 Complexity will positively impact the AIA.

2.3.2 Organizational perspective (OP)

2.3.2.1 HR readiness (HRR)

The human resources department of an organization must prepare the business to adopt the latest technology. It should train employees to utilize the technology [41,67–69]. HRR also considers the available budget and resources as well as employee skills when considering adopting new technology [70]. A firm manager's job is to motivate, influence, and coordinate with employees and service providers. They need to ensure that the employees learn all the AI technicities and implement them properly.

H5 HR Readiness will positively impact the AIA.

2.3.2.2 Top management support (TMS)

Any project or technology adoption will not succeed [38-40,71] without support of management, who must understand the requirements and advantages of the adopted technology [42]. Hospital management needs to understand the benefits of AI. Management also provides the money for the technology as well as creates an organizational strategy for implementing it. TMS helps in adopting the latest technology by motivating and supporting employees.

H6 Top management support will positively impact the AIA.

2.3.2.3 Organizational readiness (ORR)

ORR refers to an organization's capabilities in the adoption process [38,72–74]. These capabilities may be financial or technological. Organizations must consider their budget and how much money they can spend for the adoption process [55,69]. The physical assets were available in the firms required for technology adoption. TC of the firms is measured by the availability of computer hardware, data, and networking.

H7 Organizational readiness will positively impact the AIA.

2.3.3 Environmental perspective (EP)

2.3.3.1 Competitive pressure (CP)

Various organizations adopt the latest technology to create an edge in the market and differentiate from competitors [41,44,75]. The organization needs to adopt the latest technology to survive in the market [76,77]. Competitive pressure means the firms' pressure from their next competitors in the market in the same industry. Competitive pressure is defined as the extent of the competitive atmosphere within the industry in which the companies operate. This pressure, in turn, helps firms accept the latest technologies.

H8 Competitive pressure will positively impact the AIA.

2.3.3.2 Support from technology vendors (STV)

Support from vendors is significant for the adoption of the latest technology [54,78,79]. Organizations need to discuss the overall process and create a standard minimum adoption process that will benefit both the organization and the vendors [35,59,60,80]. Support is provided by either the vendors or the firms' suppliers. Without the help of the suppliers, technology adoption will not succeed.

H9 Support from technology vendors will positively impact the AIA.

2.3.3.3 Environmental uncertainty (EU)

The relations between the organizational structures and organizational innovation in environments with high uncertainty have a positive impact and require accurate information to respond to environmental requirements [41,52,67,79]. EU refers to the problems faced by the organizations in the supply chain areas like vendor management or local issues. there is a positive relationship between a company's structure relationship and its innovation. Uncertainty leads to the unavailability of the information, and accurate information is required to respond as conditions necessitate.

H10 Environmental uncertainty will positively impact the AIA.

Table 2.1 Demographics of survey respondents.

SI. no.	Characteristics	Percentage
A	Gender	
1	Male	63
2	Female	37
B	Respondents current position	
1	Doctors	42
2	Nursing staff	37
3	Medical officers	21

2.4 Research methodology

We used a questionnaire to gather information from various hospitals in India. We selected respondents via a stratified random sampling method. We used a seven-point Likert scale ranging from "strongly disagree" to "strongly agree" to examine responses. We distributed 461 questionnaires and received 323 completed surveys. To check the biasness of the questionnaire, we used the Harman single-value test.

2.4.1 Demographics of the respondents

Data was generated using the survey method. A cross-sectional plan includes testing and looking at individuals from diverse segments. This methodology allows the specialist to gather the necessary information simultaneously. Table 2.1 lists the demographics of the respondents.

2.5 Data analysis
2.5.1 Reliability and validity
2.5.1.1 Cronbach's alpha
Reliabilty test was performed using Cronbach's alpha for all the costructs. The recommended value is greater than 0.70, which was achieved by all the factors [81].

2.5.1.2 Composite reliability
Composite reliability (CR) was measured for all factors [82]. All the CR values shown in Table 2.4 for the three constructs are within the acceptable threshold (i.e., >0.7), which means the CR values are reliable [83].

2.5.2 Harman test
We used the Harman single-value test via IBM's SPSS 20.0 software to check the biasness of the data collected. Here, all the factors are taken in a single factor using the principal components analysis method. The recommended value should be less than 50% [84]. All three constructs (TP, OP, and EP) achieved values less than 50%, as shown in Table 2.2.

Table 2.2 Parameters of factor analysis, common method bias, and Cronbach's alpha values.

Constructs	Latent variable	Indicators	Harman test (%)	Cronbach's alpha (α)	Composite reliability (CR)	KMO value	Total variance explained (%)	Rotated component matrix
TP	CE	1CE	31.49	0.834	0.887	0.825	69.56	0.869
		2CE						0.759
		3CE						0.770
		4CE						0.851
	RA	RA1		0.846	0.895			0.767
		RA2						0.860
		RA3						0.896
		RA4						0.771
	SPC	SPC1		0.73	0.848			0.876
		SPC2						0.844
		SPC3						0.691
	COMP	COMP1		0.847	0.801			0.885
		COMP2						0.936
		COMP3						0.796
OP	HRR	HRR1	21.72	0.726	0.845	0.699	67.95	0.830
		HRR2						0.783
		HRR3						0.798
	TMS	TMS1		0.728	0.844			0.812
		TMS2						0.877
		TMS3						0.714
	ORR	ORR1		0.762	0.864			0.743
		ORR2						0.882
		ORR3						0.842

Continued

Table 2.2 Parameters of factor analysis, common method bias, and Cronbach's alpha values—*Cont'd*

Constructs	Latent variable	Indicators	Harman test (%)	Cronbach's alpha (α)	Composite reliability (CR)	KMO value	Total variance explained (%)	Rotated component matrix
EP	CP	CP1	47.20	0.884	0.91	0.889	73.61	0.891
		CP2						0.820
		CP3						0.806
		CP4						0.868
	STV	STV1		0.875	0.907			0.868
		STV2						0.908
		STV3						0.882
		STV4						0.697
	EU	EU1		0.872	0.905			0.828
		EU2						0.876
		EU3						0.885
		EU4						0.765

2.5.3 **Exploratory factor analysis (EFA)**

Social scientists frequently use factor analysis to ensure that the variables used to measure a specific concept are measuring the intended idea. We performed exploratory factor analysis (EFA) to group the factors into components. Again, we used the SPSS 20.0 software. In research, the Kaiser-Meyer-Olkin (KMO) test is used to determine the sampling adequacy of data used for factor analysis. The value of the KMO test should be greater than 0.60, which is the recommended threshold level [85]. The sum of variances for all the main components is the total difference. The variance fraction of a significant component is the ratio between the variance of the main component and the total variance. We used principal axis factoring for extraction. The rotated component matrix is the output of main component analysis, sometimes called loadings. It contains estimates of the correlations between individual variables and the components estimated.

In this research, there are 14 variables for TP constructs, 9 variables for OP constructs, and 12 variables for EP constructs. Table 2.2 lists the parameters of factor analysis, common method bias, and Cronbach's alpha values.

2.5.4 **Construct validity (CV)**

CV is "the extent to which a test measures or claims to measure what it says" [86]. The validity of the structure is usually checked by comparing the test with other tests measuring similar qualities to determine the strong correlation between the two measures. Build validity means whether the construction is measured adequately by a scale or test. An example is measuring human intelligence, emotional level, skill, or ability. The average variance extracted (AVE) is the quantity of variance a construct captures due to a measuring error. AVE measures the variation level captured by a build compared to the measurement failure level. Values greater than 0.7 are outstanding, whereas a variance level of 0.5 is acceptable. The extracted average variance is calculated as follows: total multiple squared correlations plus the total sum of each variable and then separated according to the number of factors in that variable. Discriminant validity tests are not related to concepts or measures that should not be linked. A successful assessment of discriminant validity shows that a concept test is not significantly correlated with other tests to measure concepts theoretically. To make it clear that the instrument is associated with measures of the same idea, it is essential to show that it is not related to standards of different concepts. For discriminant validity to be established, one must demonstrate that unrelated measures do not connect. Table 2.3 presents the discriminant validity matrix and AVE for TP constructs.

Table 2.4 presents the discriminant validity matrix and AVE for OP constructs.
Table 2.5 presents the discriminant validity matrix and AVE for EP constructs.

Table 2.3 Discriminant validity matrix and AVE for TP constructs.

	AVE	Variance extracted between factors			
		RA	**CE**	**COMP**	**SPC**
RA	0.823	1			
CE	0.813	0.669	1		
COMP	0.873	0.72	0.653	1	
SPC	0.803	0.662	0.711	0.704	1

Note: CE, *cost-effectiveness;* COMP, *complexity;* RA, *relative advantage;* SPC, *security and privacy concerns.*

Table 2.4 Discriminant validity matrix and AVE for OP constructs.

	AVE	Variance extracted between factors		
		ORR	**HRR**	**TMS**
ORR	0.823	1		
HRR	0.804	0.661	1	
TMS	0.801	0.659	0.643	1

Note: HRR, *HR readiness;* ORR, *organizational readiness;* TMS, *top management support.*

Table 2.5 Discriminant validity matrix and AVE for EP constructs.

	AVE	Variance extracted between factors		
		STV	**EU**	**CP**
STV	0.839	1		
EU	0.839	0.703	1	
CP	0.846	0.71	0.71	1

Note: CP, *competitive pressure;* EU, *environmental uncertainty;* STV, *support from technology vendors.*

2.5.5 Structural equation modeling

We used structural equation modeling (SEM) to test our hypotheses and develop three models for the three constructs [87–89]. We used AMOS 22.0 software for this research. Table 2.6 lists all the model parameters for the three constructs. All three constructs satisfied all the model parameters. Fig. 2.1 presents the final measurement model for the TP constructs. The independent and dependent variables along with their indicators are CE (1CE, 2CE, 3CE, and 4CE), RA (RA1, RA2, RA3, and RA4), SPC (SPC1, SPC2, and SPC3), and COMP (COMP1, COMP2, and

Table 2.6 Model parameters for TP, OP, and EP constructs along with their benchmarks.

Goodness-of-fit indices	Default model for technological perspective	Default model for organizational perspective	Default model for environmental perspective	Benchmark
Absolute goodness-of-fit measure				
χ^2/df (CMIN/DF)	2.359	2.922	2.415	Lower limit:1.0 Upper limit 2.0/3.0 or 5.0
Goodness of fit index	0.910	0.925	0.915	>0.90
Incremental fit measure				
Comparative fit index	0.937	0.916	0.954	≥ 0.90
Incremental fit index	0.938	0.917	0.955	≥ 0.90
Tucker-Lewis's coefficient	0.923	0.911	0.944	≥ 0.90
Absolute badness of fit measure				
Root mean square error of approximation	0.065	0.077	0.066	Within 0.08

COMP3). T/he dependent variable is AIA, which has four indicators: AIA1, AIA2, AIA3, and AIA4. The value of R^2 for the TP construct is 60%, R^2 for the OP construct is 64%, and R^2 for the EP construct is 68%.

Fig. 2.2 presents the final measurement model for the OP constructs. The independent and dependent variables and their indicators are: HRR (HRR1, HRR2, and HRR3), TMS (TMS1, TMS2, and TMS3), and ORR (ORR1, ORR2, and ORR3). The dependent variable is AIA, which has four indicators: AIA1, AIA2, AIA3, and AIA4.

Fig. 2.3 presents the final measurement model for the EP constructs. The independent and dependent variables along with their indicators are: CP (CP1, CP2, CP3, and CP4), STV (STV1, STV2, STV3, and STV4), and EU (EU1, EU2, EU3, and EU4). The dependent variable is AIA, which has four indicators: AIA1, AIA2, AIA3, and AIA4.

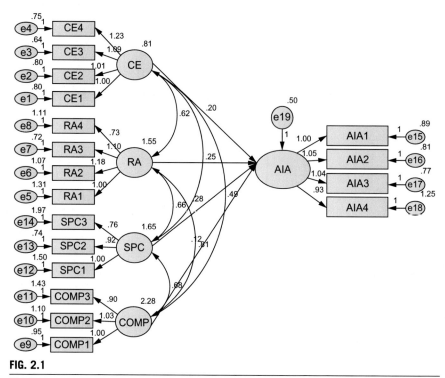

FIG. 2.1

Final measurement model for TP constructs. Note: *AIA*, artificial intelligence adoption; *CE*, cost-effectiveness; *COMP*, complexity; *RA*, relative advantage; *SPC*, security and privacy concerns.

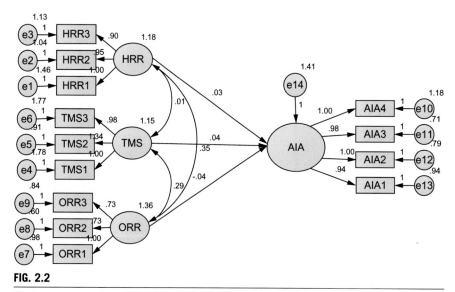

FIG. 2.2

Final measurement model for OP constructs. Note: *AIA*, artificial intelligence adoption; *HRR*, HR readiness; *ORR*, organizational readiness; *TMS*, top management support.

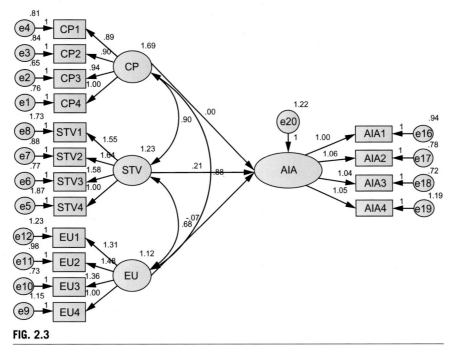

FIG. 2.3

Final measurement model for EP constructs. Note: *AIA*, artificial intelligence adoption; *CP*, competitive pressure; *EU*, environmental uncertainty; *STV*, support from technology vendors.

2.6 Discussion

The proposed model is based on TOE frameworks, consisting of technological, organizational, and environmental constructs. The research presented in this chapter studies the adoption of AI in healthcare sectors and reviews the status of AI adoption in different healthcare sectors.

2.6.1 Technological perspective (TP)

TP helps to identify the technological factors that can impact the adoption of any technological innovations. The Cronbach's alpha and composite reliability values for the construct TP were greater than 0.7, which is the recommended level for the four factors. The KMO test value of the construct is 0.825, which is also greater than the recommended level of 0.6 and allows the data for factor analysis. The total variance explained is 69.56%, and in the rotated component matrix, the variables were grouped under four groups. Only the loadings greater than |0.40| are considered in this research because these are typically high and hence are more significant [90].

For further analysis in this research, we utilized four components with fourteen indicators. The CE component showed a positive influence in the study and thus Hypothesis 1 could not be rejected. Earlier studies of AI in different sectors supported this as well [6,91]. AI will reduce labor costs as well as the time it takes to complete repetitive jobs. The loadings of the four indicators (1CE, 2CE, 3CE, and 4CE) are 0.869, 0.759, 0.770, and 0.851, respectively. The RA component showed a positive influence on AI and thus the Hypothesis 2 could not be rejected. Earlier studies showed the same results [6,92]. The loadings of the four indicators (RA1, RA2, RA3, and RA4) are 0.767, 0.860, 0.896, and 0.771, respectively. The SPC component showed a positive influence on AI and thus Hypothesis 3 could not be rejected. Earlier studies showed the same results [42,93,94]. The loadings of the three indicators (SPC1, SPC2, and SPC3) are 0.876, 0.844, and 0.691, respectively. The COMP component showed a positive influence on AI and thus Hypothesis 4 could not be rejected. Earlier studies showed the same results [40,66]. Prior research found that COMP negatively influenced the adoption of the latest technology [40,66]. The loadings of the three indicators (COMP1, COMP2, and COMP3) are 0.885, 0.936, and 0.796, respectively. Therefore H1, H2, H3, and H4 could not be rejected.

In this research, we explain and validate the components using the SEM approach, which is the most appropriate method to prove validity. This technique has not been used in any prior research.

2.6.2 Organizational perspectives (OP)

The Cronbach's alpha and composite reliability values for the construct OP were greater than 0.7, which is the recommended level [81,88,95] for the three factors. The KMO test value of the construct is 0.699, which is also greater than the recommended level of 0.6 [90], which allows the data for factor analysis. The total variance explained was 67.95%, and in the rotated component matrix, the variables were grouped under three categories. Only the loadings greater than |0.40| are considered in this research because these are typically high and hence are more significant [90].

For further analysis in this research, three components with nine indicators were utilized. The HRR component showed a positive influence on AI and thus Hypothesis 5 could not be rejected. Earlier studies showed the same results [41,67,68]. The loadings of the three indicators (HRR1, HRR2, and HRR3) are 0.830, 0.783, and 0.798, respectively. The TMS component showed a positive influence on AI and thus Hypothesis 6 could not be rejected. Earlier studies showed the same results [44,45,80,96]. The loadings of the three indicators (TMS1, TMS2, and TMS3) are 0.812, 0.877, and 0.714, respectively. The ORR component showed a positive influence on AI and thus Hypothesis 7 could not be rejected. Earlier studies showed the same results [41,67,69,92,97]. The loadings of the three indicators (ORR1, ORR2, and ORR3) are 0.743, 0.882, and 0.842, respectively. Therefore H5, H6, and H7 could not be rejected.

In this research, we explain and validate the components using the SEM approach, which is the most appropriate method to prove validity. This technique has not been used in any prior research.

2.6.3 **Environmental perspectives (EP)**

The Cronbach's alpha and composite reliability values for the construct EP were greater than 0.7, which is the recommended level for the three factors. The KMO test value of the construct is 0.889, which is also greater than the recommended level of 0.6 [90], which allows the data for factor analysis. The total variance explained was 73.61%, and in the rotated component matrix, the variables were grouped under three categories. Only the loadings greater than $|0.40|$ are considered in this research because these are typically high and hence are more significant [90].

For further analysis in this research, 4 components with 12 indicators were utilized. The CP component showed a positive influence on AI and thus Hypothesis 8 could not be rejected. Earlier studies showed the same results [38-41,43,67,92,98]. The loadings of the four indicators (CP1, CP2, CP3, and CP4) are 0.891, 0.820, 0.806, and 0.868, respectively. The STV component showed a positive influence on AI and thus Hypothesis 9 could not be rejected. Earlier studies showed the same results [43,80,96]. The loadings of the four indicators (STV1, STV2, STV3, and STV4) are 0.868, 0.908, 0.882, and 0.697, respectively. The EU component showed a positive influence on AI and thus Hypothesis 10 could not be rejected. Earlier studies showed the same results [38,48-50,73,99]. The loadings of the four indicators (EU1, EU2, EU3, and EU4) are 0.828, 0.876, 0.885, and 0.765, respectively.

In this research, we explain and validate the components using the SEM approach, which is the most appropriate method to prove validity. This technique has not been used in any prior research.

2.7 **Conclusion**

This study identified the influences of AI adoption in Indian hospitals. AI can change the face of healthcare industries, as it has many applications in numerous areas of health care, including surgery, diagnosis, clinical decision-making, and virtual assistance through applications like chatbots. We undertook a literature review in this chapter to identify TOE factors and developed a questionnaire for survey-based research in Indian hospitals. Our target population was hospital employees. We used EFA and SEM to analyze the collected data. Results show the developed model was well suited and the proposed hypotheses were accepted.

2.8 **Limitations and future scope of the study**

One of the limitations of this study is that the survey was carried out only in private hospitals in India and not in government hospitals. Thus, there is a chance that we may get different results if we conduct the survey in government hospitals or private hospitals. A future study can include a survey at both government and private hospitals and comparing the data. This study can be further extended to other countries, for example, developed countries that have already adopted AI in their healthcare systems. This study can also be further extended to other sectors like the automobile, textile, logistics, supply chain, and retail industries.

Appendix A. Measurement items

Constructs	Variables	Labels	Measures	Sources
Technological perspective	Relative advantage (RA)	RA 1	Artificial intelligence adoption provides timely decision-making information	Senyo et al.
		RA 2	Artificial Intelligence implementation increases today's business profitability	Senyo et al. [54]
		RA 3	Artificial Intelligence helps organizations to communicate better with their partner	Added by the authors
		RA 4	Our productivity increase with the usage of artificial intelligence services increases	Ramaswamy et al. [79]
	Security and privacy concerns (SPC)	SPC 1	Artificial Intelligence provides privacy for its users	Puklavec et al. [53]
		SPC 2	The data movement from one system to another is safe and secured.	Added by the authors
		SPC 3	In sharing sensitive information with our suppliers, I would find artificial Intelligence safe	Added by the authors

Constructs	Variables	Labels	Measures	Sources
	Complexity (COM)	COM 1	Employees find the skills needed for using these technologies too complex	Priyadarshinee et al. [52]
		COM 2	It will be challenging to integrate artificial Intelligence into the existing process	Priyadarshinee et al. [52]
		COM 3	My healthcare firms consider artificial intelligence development to be a complex process	Priyadarshinee et al. [52]
	Cost-effectiveness (CE)	1 CE	The artificial intelligence cost is much higher than the benefits	Makena [100]
		2 CE	Maintenance costs and artificial intelligence support for the company are extremely high	Makena [100]
		3 CE	Integrating Artificial Intelligence into the current information management system will be prohibitively costly	Makena [100]
		CE 4	The budget of IT in healthcare firms is low	Added by the authors
Organizational perspective	Top management support (TMS)	TMS 1	Artificial Intelligence is enthusiastically adopted in healthcare firms by senior management	Nam et al. [73]
		TMS 2	Adequate artificial intelligence resources have been allocated by top management	Nam et al. [73]
		TMS 3	Our top management actively supports employees in their day-to-day tasks to use new technologies	Nam et al. [73]
	HR readiness (HRR)	HRR 1	In our organization, we still do not have the technical abilities to use artificial Intelligence	Priyadarshinee et al. [52]
		HRR 2	Efficient IT training is required	Priyadarshinee et al. [52]
		HRR 3	The staff can share their knowledge with each other	Priyadarshinee et al. [52]

Constructs	Variables	Labels	Measures	Sources
	Organizational readiness (ORR)	ORR 1	The staff has technological capabilities to solve problems	Priyadarshinee et al. [52]
		ORR 2	The staff can quickly learn new technologies	Priyadarshinee et al. [52]
		ORR 3	The staff usually come up with new hotel ideas	Added by the authors
Environmental perspective	Competitive pressure (CP)	CP 1	Artificial Intelligence enables healthcare firms to deal more effectively with political uncertainty	Mahakittikun et al. [48]
		CP 2	Artificial Intelligence enables healthcare firms to deal more effectively with demand uncertainty	Mahakittikun et al. [48]
		CP 3	Artificial Intelligence enables my healthcare firms to deal more effectively with food insecurity	Added by the authors
		CP 4	Artificial Intelligence enables healthcare firms to cope more effectively with competition uncertainty	Mahakittikun et al. [48]
	Support from technology vendors (STV)	STV 1	Vendors of the healthcare firms will support the adoption of artificial intelligent	Mahakittikun et al. [48]
		STV 2	Vendors of the healthcare firms will does support the adoption of artificial intelligent	Mahakittikun et al. [48]
		STV 3	Artificial intelligent will help in creating good communication between the healthcare firms and its vendors	Added by the authors
		STV 4	It offers better visibility of the market	Lin [75]
	Environmental uncertainty	EU 1	Artificial intelligent enables the healthcare firms to deal more effectively with political uncertainty	Lin [75]

Constructs	Variables	Labels	Measures	Sources
		EU 2	Artificial intelligent enables the healthcare firms to deal more effectively with demand uncertainty	Lin [75]
		EU 3	Artificial intelligent enables my healthcare firms to deal more effectively with market insecurity	Lin [75]
		EU 4	Artificial intelligent enables the healthcare firms to cope more effectively with competition uncertainty	Lin [75]
Artificial intelligence adoption	Artificial intelligence adoption (AIA)	AIA 1	Healthcare firms like the idea of using artificial intelligence	Added by the authors
		AIA 2	The productivity of our employees increases (would increase)	Added by the authors
		AIA 3	Artificial intelligence is intended for our healthcare firms in the future	Added by the authors
		AIA 4	I predict that in the future, this healthcare firm will use artificial intelligence	Added by the authors

Appendix B. Questionnaire

Part I: Organizations and employee details

    Q1. Year of establishment: _____.
    Q2. Gender:
        (a) Male
        (b) Female
    Q3. Respondents current position:
        (a) Doctors
        (b) Nursing Staff
        (c) Medical Officers

Part II: To what extent do you agree with the following statements?
    Likert scale: (Strongly agree = 7; Agree = 6; Slightly agree = 5; Neutral = 4; Slightly disagree = 3; Disagree = 2; and Strongly disagree = 1).

Q4: Cost-effectiveness	Strongly disagree	Disagree	Slightly disagree	Neutral	Slightly agree	Agree	Strongly agree
The artificial intelligence cost is much higher than the benefits							
Maintenance costs and artificial intelligence support for the company are extremely high							
Integrating Artificial Intelligence into the current information management system will be prohibitively costly							
The budget of IT in healthcare firms is low							

Q5: Relative advantage	Strongly disagree	Disagree	Slightly disagree	Neutral	Slightly agree	Agree	Strongly agree
Artificial Intelligence adoption provides timely decision-making information							
Artificial Intelligence implementation increases today's business profitability							
Artificial Intelligence helps organizations to communicate better with their partner							
Our productivity increase with the usage of artificial intelligence services increases							

Q6: Security and privacy concerns	Strongly disagree	Disagree	Slightly disagree	Neutral	Slightly agree	Agree	Strongly agree
Artificial Intelligence provides privacy for its users							
The data movement from one system to another is safe and secured							
In sharing sensitive information with our suppliers, I would find artificial Intelligence safe							

Q7: Complexity	Strongly disagree	Disagree	Slightly disagree	Neutral	Slightly agree	Agree	Strongly agree
Employees find the skills needed for using these technologies too complex							
It will be challenging to integrate artificial Intelligence into the existing process							
My healthcare firms consider artificial intelligence development to be a complex process							

Q8: Top management support	Strongly disagree	Disagree	Slightly disagree	Neutral	Slightly agree	Agree	Strongly agree
Artificial Intelligence is enthusiastically adopted in healthcare firms by senior management							
Adequate artificial intelligence resources have been allocated by top management							
Our top management actively supports employees in their day-to-day tasks to use new technologies							

Q9: HR readiness	Strongly disagree	Disagree	Slightly disagree	Neutral	Slightly agree	Agree	Strongly agree
In our organization, we still do not have the technical abilities to use Artificial Intelligence							
Efficient IT training is required							
The staff can share their knowledge with each other							

	Strongly disagree	Disagree	Slightly disagree	Neutral	Slightly agree	Agree	Strongly agree
Q10: Organizational readiness							
The staff has technological capabilities to solve problems							
The staff can quickly learn new technologies							
The staff usually come up with new hotel ideas							
Q11: Competitive pressure							
Artificial Intelligence enables healthcare firms to deal more effectively with political uncertainty							
Artificial Intelligence enables healthcare firms to deal more effectively with demand uncertainty							
Artificial Intelligence enables my healthcare firms to deal more effectively with food insecurity							
Artificial Intelligence enables healthcare firms to cope more effectively with competition uncertainty							
Q12: Support from technology vendors							
Vendors of the healthcare firms will support the adoption of artificial intelligent							
Vendors of the healthcare firms will do support the adoption of artificial intelligent							
Artificial intelligence will help in creating good communication between healthcare firms and their vendors							
It offers better visibility of the market							

Q13: Environmental uncertainty	Strongly disagree	Disagree	Slightly disagree	Neutral	Slightly agree	Agree	Strongly agree
Artificial intelligence enables healthcare firms to deal more effectively with political uncertainty							
Artificial intelligence enables healthcare firms to deal more effectively with demand uncertainty							
Artificial intelligence enables my healthcare firms to deal more effectively with market insecurity							
Artificial intelligence enables healthcare firms to cope more effectively with competition uncertainty							

Q14: Artificial intelligence adoption	Strongly disagree	Disagree	Slightly disagree	Neutral	Slightly agree	Agree	Strongly agree
Healthcare firms like the idea of using artificial Intelligence							
The productivity of our employees increases (would increase)							
Artificial Intelligence is intended for our healthcare firms in the future							
I predict that in the future, this healthcare firm will use artificial intelligence							

References

[1] F.K. Badi, K.A. Al, F.J. Alhosani, A. Stachowicz-Stanusch, N. Shehzad, A.M.A.N.N. Wolfgang, Challenges of AI adoption in the UAE healthcare, Vision (2021), https://doi.org/10.1177/0972262920988398. February. Sage Publications India Pvt. Ltd.

[2] P. Kumar, Y.K. Dwivedi, A. Anand, Responsible artificial intelligence (AI) for value formation and market performance in healthcare: the mediating role of Patient's cognitive engagement, Inf. Syst. Front. (2021), https://doi.org/10.1007/s10796-021-10136-6.

[3] A. Abubakar, E.B. Mohammed, H. Rezapouraghdam, S.B. Yildiz, Applying artificial intelligence technique to predict knowledge hiding behavior, Int. J. Inf. Manag. 49 (2019) 45–57, https://doi.org/10.1016/j.ijinfomgt.2019.02.006. Elsevier.

[4] M. Al Mohammed, H. Rashid, Comprehensive review on the challenges that impact artificial intelligence applications in the public sector, in: Proceedings of the International Conference on Industrial Engineering and Operations Management, no. August, 2020, pp. 2078–2087.

[5] M. Bothra, A. Mahajan, Mining artificial intelligence in oncology: tata memorial hospital journey, Cancer Res. Stat. Treat. 3 (3) (2020) 622, https://doi.org/10.4103/CRST.CRST_59_20. Medknow Publications and Media Pvt. Ltd.

[6] E.T. Albert, AI in talent acquisition: a review of AI-applications used in recruitment and selection, Strateg. HR Rev. 18 (5) (2019) 215–221, https://doi.org/10.1108/shr-04-2019-0024. Emerald.

[7] M.E. Matheny, D. Whicher, S.T. Israni, Artificial intelligence in health care: a report from the National Academy of medicine, JAMA (2020), https://doi.org/10.1001/jama.2019.21579. American Medical Association.

[8] B. Mesko, The role of artificial intelligence in precision medicine, Expert Rev. Precis. Med. Drug Dev. (2017), https://doi.org/10.1080/23808993.2017.1380516. Taylor and Francis Ltd.

[9] G. Baryannis, S. Validi, S. Dani, G. Antoniou, Supply chain risk management and artificial intelligence: state of the art and future research directions, Int. J. Prod. Res. 57 (7) (2019) 2179–2202, https://doi.org/10.1080/00207543.2018.1530476.

[10] R.E. Bawack, S.F. Wamba, K.D.A. Carillo, A framework for understanding artificial intelligence research: insights from practice, J. Enterp. Inf. Manag. 34 (2) (2021) 645–678, https://doi.org/10.1108/JEIM-07-2020-0284.

[11] B. Blobel, P. Ruotsalainen, M. Brochhausen, F. Oemig, G.A. Uribe, Autonomous systems and artificial intelligence in healthcare transformation to 5P medicine-ethical challenges, Technol. Forecast. Soc. Change (2020), https://doi.org/10.3233/SHTI200330.

[12] S. Bag, J.H.C. Pretorius, S. Gupta, Y.K. Dwivedi, Role of institutional pressures and resources in the adoption of big data analytics powered artificial intelligence, sustainable manufacturing practices and circular economy capabilities, Technol. Forecast. Soc. Chang. 163 (2021) 120420, https://doi.org/10.1016/j.techfore.2020.120420. Elsevier Inc.

[13] L. Pumplun, P. Buxmann, Intelligent Systems and Hospitals: Joint Forces in the Name of Health? Accessed June 16, 2021, https://doi.org/10.30844/wi_2020_f8-pumplun.

[14] N. Mohammadzadeh, R. Safdari, Artificial intelligence tools in health information management, Int. J. Hosp. Res. 2012 (2012). www.ijhr.tums.ac.ir.

[15] way, ES Taie, Artificial Intelligence as an Innovative Approach for Investment in the future of Healthcare in Egypt, 2021, Researchgate.Net. Accessed June 20 https://www.researchgate.net/profile/Eman_Taie/publication/342358151_Artificial_intelligence_as_an_innovative_approach_for_investment_in_the_future_of_healthcare_in_Egypt/

links/5fb45e3445851518fdb07c7c/Artificial-intelligence-as-an-innovative-approach-for.

[16] G. Kaur, H. Gupta, "Application of Artificial Intelligence in Medical Field-A Review." Ij-Tejas.Com, 2021, Accessed 20 June http://www.ij-tejas.com/paper/vol5Issue2/7.pdf.

[17] O. Er, A.Ç. Tanrikulu, A. Abakay, Use of artificial intelligence techniques for diagnosis of malignant pleural mesothelioma, Dicle Med. J./Dicle Tip Dergisi 42 (1) (2015) 5–11, https://doi.org/10.5798/diclemedj.0921.2015.01.0520.

[18] D. Thesmar, D. Sraer, L. Pinheiro, N. Dadson, R. Veliche, P. Greenberg, Combining the power of artificial intelligence with the richness of healthcare claims data: opportunities and challenges, Pharmacoeconomics 37 (6) (2019) 745–752, https://doi.org/10.1007/s40273-019-00777-6. Springer International Publishing.

[19] O. Iliashenko, Z. Bikkulova, A. Dubgorn, Opportunities and challenges of artificial intelligence in healthcare, E3S Web Conf. 110 (2019) 201–206, https://doi.org/10.1051/e3sconf/201911002028. (Icemci). 02028. EDP Sciences.

[20] V. Jain, N. Singh, S. Pradhan, P. Gupta, Factors influencing AI implementation decision in Indian healthcare industry: a qualitative inquiry, in: IFIP Advances in Information and Communication Technology, vol. 617, Springer Science and Business Media Deutschland GmbH, 2020, pp. 635–640, https://doi.org/10.1007/978-3-030-64849-7_56.

[21] A. Mahajan, T. Vaidya, A. Gupta, S. Rane, S. Gupta, Artificial intelligence in healthcare in developing nations: the beginning of a transformative journey, Cancer Res. Stat. Treat. 2 (2) (2019) 182, https://doi.org/10.4103/crst.crst_50_19. Medknow.

[22] S. Reddy, J. Fox, M.P. Purohit, Artificial intelligence-enabled healthcare delivery, J. R. Soc. Med. 112 (1) (2019) 22–28, https://doi.org/10.1177/0141076818815510.

[23] S. Secinaro, D. Calandra, A. Secinaro, V. Muthurangu, P. Biancone, The role of artificial intelligence in healthcare: a structured literature review, BMC Med. Inform. Decis. Mak. 21 (1) (2021) 1–23, https://doi.org/10.1186/s12911-021-01488-9. BioMed Central Ltd.

[24] F. Jiang, Y. Jiang, H. Zhi, Y. Dong, H. Li, S. Ma, Y. Wang, Q. Dong, H. Shen, Y. Wang, Artificial intelligence in healthcare: past, present and future, Stroke Vasc. Neurol. (2017), https://doi.org/10.1136/svn-2017-000101. BMJ Publishing Group.

[25] M. Chen, M. Decary, Artificial intelligence in healthcare: an essential guide for health leaders, Health Manage. Forum 33 (1) (2020) 10–18, https://doi.org/10.1177/0840470419873123. SAGE Publications Inc.:.

[26] A.İ. Tekkeşin, Review 8, 2019, https://doi.org/10.14744/AnatolJCardiol.2019.28661.

[27] R. Malik, S. Pande, Nishi, Artificial intelligence and machine learning to assist climate change monitoring, J. Artif. Intell. Syst. 1 (1) (2020) 168–190, https://doi.org/10.33969/ais.2020.21011.

[28] N. Noorbakhsh-Sabet, R. Zand, Y. Zhang, V. Abedi, Artificial intelligence transforms the future of health care, Am. J. Med. 132 (7) (2019) 795–801, https://doi.org/10.1016/j.amjmed.2019.01.017. Elsevier Inc.

[29] M.H. Stanfill, D.T. Marc, Health information management: implications of artificial intelligence on healthcare data and information management, Yearb. Med. Inform. (2019), https://doi.org/10.1055/s-0039-1677913. NLM (Medline).

[30] R. Pinninti, S. Rajappa, Leptomeningeal metastasis from extracranial solid tumors, Cancer Res. Stat. Treat. 3 (Suppl. 1) (2020) S65–S70, https://doi.org/10.4103/CRST.CRST.

[31] J. Guo, B. Li, The application of medical artificial intelligence Technology in Rural Areas of developing countries, Health Equity 2 (1) (2018) 174–181, https://doi.org/10.1089/heq.2018.0037.

[32] T.Q. Sun, R. Medaglia, Mapping the challenges of artificial intelligence in the public sector: evidence from public healthcare, Gov. Inf. Q. 36 (2) (2019) 368–383, https://doi.org/10.1016/j.giq.2018.09.008. Elsevier.

[33] S.F.S. Alhashmi, S.A. Salloum, C. Mhamdi, Implementing artificial intelligence in the United Arab Emirates healthcare sector: an extended technology acceptance model, Int. J. Inf. Technol. Lang. Stud 3 (3) (2019) 27–42.

[34] S.N. Yoon, D.H. Lee, Artificial intelligence and robots in healthcare: what are the success factors for technology-based service encounters? Int. J. Healthc. Manag. 12 (3) (2019) 218–225, https://doi.org/10.1080/20479700.2018.1498220. Taylor & Francis.

[35] R. Depietro, E. Wiarda, M. Fleischer, The context for change: organization, technology and environment, in: The Processes of Technological Innovation, vol. 199, Elsevier, 1990, pp. 151–175.

[36] N. Al-Qirim, The adoption of ECommerce communications and applications Technologies in Small Businesses in New Zealand, Electron. Commer. Res. Appl. 6 (4) (2007) 462–473, https://doi.org/10.1016/j.elerap.2007.02.012. Elsevier:.

[37] S. Khemthong, L.M. Roberts, Adoption of internet and web technology for hotel marketing: a study of hotels in Thailand, J. Bus. Syst. Govern. Ethics 1 (2) (2006), https://doi.org/10.15209/jbsge.v1i2.74. Victoria University.

[38] S. Al-Isma'Ili, M. Li, J. Shen, Q. He, Cloud computing adoption determinants: an analysis of Australian SMEs, in: Pacific Asia Conference on Information Systems, PACIS 2016—Proceedings, 2016.

[39] N. Alkhater, R. Walters, G. Wills, An empirical study of factors influencing cloud adoption among private sector Organisations, Telematics Inform. 35 (1) (2018) 38–54, https://doi.org/10.1016/j.tele.2017.09.017.

[40] O. Alsetoohy, B. Ayoun, S. Arous, F. Megahed, G. Nabil, Intelligent agent technology: what affects its adoption in hotel food supply chain management? J. Hosp. Tour. Technol. 10 (3) (2019) 317–341, https://doi.org/10.1108/JHTT-01-2018-0005.

[41] Y. Alshamaila, S. Papagiannidis, F. Li, Cloud computing adoption by SMEs in the north east of England: a multi-perspective framework, J. Enterp. Inf. Manag. 26 (3) (2013) 250–275, https://doi.org/10.1108/17410391311325225.

[42] M.M. Baral, A. Verma, Cloud computing adoption for healthcare: an empirical study using SEM approach, FIIB Bus. Rev. (2021), https://doi.org/10.1177/23197145211012505. May. SAGE PublicationsSage India: New Delhi, India, 231971452110125.

[43] S. Fosso Wamba, M.M. Queiroz, L. Trinchera, Dynamics between blockchain adoption determinants and supply chain performance: an empirical investigation, Int. J. Prod. Econ. 229 (2020) 107791, https://doi.org/10.1016/j.ijpe.2020.107791. Elsevier B.V.

[44] E. Gökalp, M.O. Gökalp, S. Çoban, Blockchain-based supply chain management: understanding the determinants of adoption in the context of organizations, Inf. Syst. Manag. (2020) 1–22, https://doi.org/10.1080/10580530.2020.1812014. Taylor & Francis.

[45] B. Haryanto, A. Gandhi, Y.G. Sucahyo, The determinant factors in utilizing electronic signature using the TAM and TOE framework, in: 2020 5th International Conference on Informatics and Computing, ICIC 2020, Institute of Electrical and Electronics Engineers Inc., 2020, https://doi.org/10.1109/ICIC50835.2020.9288623.

[46] K.K. Hiran, A. Henten, An integrated TOE-DoI framework for cloud computing adoption in higher education: the case of sub-Saharan Africa, Ethiopia, Adv. Intell. Syst. Comput. 1053 (2020) 1281–1290. Springer https://doi.org/10.1007/978-981-15-0751-9_117.

[47] A. Kumar, B. Krishnamoorthy, Business analytics adoption in firms: a qualitative study elaborating TOE framework in India, Int. J. Global Bus. Competitiveness 15 (2) (2020) 80–93, https://doi.org/10.1007/s42943-020-00013-5. Springer Singapore.

[48] T. Mahakittikun, S. Suntrayuth, V. Bhatiasevi, The impact of technological-organizational-environmental (TOE) factors on firm performance: Merchant's perspective of Mobile payment from Thailand's retail and service firms, J. Asia Bus. Stud. (2020), https://doi.org/10.1108/JABS-01-2020-0012. Emerald Group Holdings Ltd.

[49] P. Maroufkhani, W.K.W. Ismail, M. Ghobakhloo, Big data analytics adoption model for small and medium enterprises, J. Sci. Technol. Policy Manag. 11 (2) (2020) 171–201, https://doi.org/10.1108/JSTPM-02-2020-0018.

[50] M. Narmetta, S. Krishnan, Competitiveness, change readiness, and ICT development: an empirical investigation of TOE framework for poverty alleviation, in: IFIP Advances in Information and Communication Technology, vol. 618, Springer Science and Business Media Deutschland GmbH, 2020, pp. 638–649, https://doi.org/10.1007/978-3-030-64861-9_55.

[51] P. Priyadarshinee, R.D. Raut, M.K. Jha, B.B. Gardas, Understanding and predicting the determinants of cloud computing adoption: a two staged hybrid SEM—neural networks approach, Comput. Hum. Behav. 76 (2017) 341–362, https://doi.org/10.1016/j.chb.2017.07.027. Elsevier B.V.

[52] P. Priyadarshinee, R.D. Raut, M.K. Jha, S.S. Kamble, A cloud computing adoption in Indian SMEs: scale development and validation approach, J. High Technol. Managem. Res. 28 (2) (2017) 221–245, https://doi.org/10.1016/j.hitech.2017.10.010. Elsevier.

[53] B. Puklavec, T. Oliveira, A. Popovič, Understanding the determinants of business intelligence system adoption stages an empirical study of SMEs, Ind. Manag. Data Syst. 118 (1) (2018) 236–261, https://doi.org/10.1108/IMDS-05-2017-0170. Emerald Group Publishing Ltd.

[54] P.K. Senyo, J. Effah, E. Addae, Preliminary insight into cloud computing adoption in a developing country, J. Enterp. Inf. Manag. 29 (4) (2016) 505–524, https://doi.org/10.1108/JEIM-09-2014-0094.

[55] F. Shahzad, G.Y. Xiu, I. Khan, M. Shahbaz, M.U. Riaz, A. Abbas, The moderating role of intrinsic motivation in cloud computing adoption in online education in a developing country: a structural equation model, Asia Pac. Educ. Rev. 21 (1) (2020) 121–141, https://doi.org/10.1007/s12564-019-09611-2. Springer Netherlands.

[56] M. Skafi, M.M. Yunis, A. Zekri, Factors influencing SMEs' adoption of cloud computing services in Lebanon: an empirical analysis using TOE and contextual theory, IEEE Access 8 (2020) 79169–79181, https://doi.org/10.1109/ACCESS.2020.2987331. Institute of Electrical and Electronics Engineers Inc.

[57] A.A. Tashkandi, I. Al-Jabri, Cloud computing adoption by higher education institutions in Saudi Arabia: Analysis based on TOE, in: 2015 International Conference on Cloud Computing, ICCC 2015, IEEE, 2015, pp. 1–8, https://doi.org/10.1109/CLOUDCOMP.2015.7149634.

[58] B. Umam, A.K. Darmawan, A. Anwari, I. Santosa, M. Walid, A.N. Hidayanto, Mobile-based smart regency adoption with TOE framework: an empirical inquiry from Madura Island districts, in: ICICoS 2020 - Proceeding: 4th International Conference

on Informatics and Computational Sciences, 2020, https://doi.org/10.1109/ICI-CoS51170.2020.9299025. Institute of Electrical and Electronics Engineers Inc.

[59] L.-W. Wong, G.W.-H. Tan, V.-H. Lee, K.-B. Ooi, A. Sohal, Unearthing the determinants of blockchain adoption in supply chain management, Int. J. Protein Res. 58 (7) (2020) 2100–2123, https://doi.org/10.1080/00207543.2020.1730463. Taylor and Francis Ltd.

[60] L.W. Wong, L.Y. Leong, J.J. Hew, G.W.H. Tan, K.B. Ooi, Time to seize the digital evolution: adoption of blockchain in operations and supply chain management among Malaysian SMEs, Int. J. Inf. Manag. 52 (2020) 101997, https://doi.org/10.1016/j.ijinfomgt.2019.08.005. Elsevier Ltd.

[61] G. Premkumar, M. Roberts, Adoption of new information technologies in rural small businesses, Omega 27 (4) (1999) 467–484, https://doi.org/10.1016/S0305-0483(98)00071-1. Pergamon.

[62] E.M. Rogers, Diffusion of innovations: Modifications of a model for telecommunications, in: Die Diffusion von Innovationen in Der Telekommunikation, Springer, Berlin, Heidelberg, 1995, pp. 25–38, https://doi.org/10.1007/978-3-642-79868-9_2.

[63] A.-M. Stjepić, M.P. Bach, V.B. Vukšić, Exploring risks in the adoption of business intelligence in SMEs using the TOE framework, J. Risk Financ. Manag. 14 (2) (2021) 58, https://doi.org/10.3390/jrfm14020058. MDPI AG.

[64] S.F. Wamba, M.M. Queiroz, Blockchain in the operations and supply chain management: benefits, challenges and future research opportunities, Int. J. Inf. Manag. (2020), https://doi.org/10.1016/j.ijinfomgt.2019.102064. Elsevier Ltd.

[65] M. Kouhizadeh, S. Saberi, J. Sarkis, Blockchain technology and the sustainable supply chain: theoretically exploring adoption barriers, Int. J. Prod. Econ. 231 (2021) 107831, https://doi.org/10.1016/j.ijpe.2020.107831. Elsevier B.V.

[66] H. Ahmadi, M. Nilashi, L. Shahmoradi, O. Ibrahim, Hospital information system adoption: expert perspectives on an adoption framework for Malaysian public hospitals, Comput. Hum. Behav. 67 (2017) 161–189, https://doi.org/10.1016/j.chb.2016.10.023. Elsevier Ltd.

[67] Y. Alshamaila, S. Papagiannidis, F. Li, A. Agostino, K.S. Søilen, B. Gerritsen, J.N. Makena, P. Gupta, A. Seetharaman, J.R. Raj, The usage and adoption of cloud computing by small and medium businesses, Int. J. Comput. Appl. Technol. Res. 33 (5) (2013) 861–874, https://doi.org/10.7753/ijcatr0205.1003. Elsevier Ltd.

[68] M. Amini, A. Bakri, Cloud computing adoption by SMEs in the Malaysia: a multiperspective framework based on DOI theory and TOE framework, J. Inf. Technol. Inf. Syst. Res. (JITISR) 9 (2) (2015) 121–135.

[69] F. Badr, A.-A. Nasser, M.S. Jawad, Factors of cloud computing adoption by small and medium size enterprises (SMEs), Int. J. Innov. Res. Electr. Electron. Instrum. Control. Eng. 7 (1) (2019) 2321–5526.

[70] A. Wulandari, B. Suryawardani, D. Marcelino, Social media technology adoption for improving MSMEs performance in Bandung: A technology-organization-environment (TOE) framework, in: 2020 8th International Conference on Cyber and IT Service Management, CITSM 2020, Institute of Electrical and Electronics Engineers Inc, 2020, https://doi.org/10.1109/CITSM50537.2020.9268803.

[71] F. Alharbi, A. Atkins, C. Stanier, Understanding the determinants of cloud computing adoption in Saudi healthcare organisations, Complex Intell. Syst. 2 (3) (2016) 155–171, https://doi.org/10.1007/s40747-016-0021-9. Springer Berlin Heidelberg.

[72] S. Kamble, A. Gunasekaran, H. Arha, Understanding the blockchain technology adoption in supply chains-Indian context, Int. J. Prod. Res. 57 (7) (2019) 2009–2033, https://doi.org/10.1080/00207543.2018.1518610. Taylor and Francis Ltd.

[73] K. Nam, C.S. Dutt, P. Chathoth, A. Daghfous, M. Sajid Khan, The adoption of artificial intelligence and robotics in the hotel industry: prospects and challenges, Electron. Mark. (2020), https://doi.org/10.1007/s12525-020-00442-3.

[74] A. Pateli, N. Mylonas, A. Spyrou, Organizational adoption of social media in the hospitality industry: an integrated approach based on DIT and TOE frameworks, Sustainability 12 (17) (2020) 7132, https://doi.org/10.3390/su12177132. MDPI AG.

[75] H.F. Lin, Understanding the determinants of electronic supply chain management system adoption: using the technology-organization-environment framework, Technol. Forecast. Soc. Chang. 86 (2014) 80–92, https://doi.org/10.1016/j.techfore.2013.09.001. Elsevier Inc.

[76] I. Mrhaouarh, C. Okar, A. Namir, N. Chafiq, Cloud computing adoption in developing countries: a systematic literature review, in: 2018 IEEE International Conference on Technology Management, Operations and Decisions, ICTMOD 2018. IEEE, 2018, pp. 73–79, https://doi.org/10.1109/ITMC.2018.8691295.

[77] A.H. Ngah, Y. Zainuddin, R. Thurasamy, Applying the TOE framework in the halal warehouse adoption study, J. Islamic Account. Bus. Res. 8 (2) (2017) 161–181, https://doi.org/10.1108/JIABR-04-2014-0014. Emerald Group Publishing Ltd.

[78] C. Low, Y. Chen, W. Mingchang, Understanding the determinants of cloud computing adoption, Ind. Manag. Data Syst. 111 (2011), https://doi.org/10.1108/02635571111161262.

[79] R. Ramaswamy, H. Gangwar, D. Hema, Journal of Enterprise information management understanding determinants of cloud computing adoption using an integrated TAM-TOE model, J. Enterp. Inf. Manag. 28 (1) (2015) 107–130.

[80] A.A. Ergado, A. Desta, H. Mehta, Determining the barriers contributing to ICT implementation by using technology-organization-environment framework in Ethiopian higher educational institutions, Educ. Inf. Technol. (2021) 1–19, https://doi.org/10.1007/s10639-020-10397-9. January. Springer.

[81] J.C. Nunnally, Psychometric Theory 3E, Tata McGraw-hill Education, 1994.

[82] J. Henseler, C.M. Ringle, R.R. Sinkovics, The use of partial least squares path modeling in International marketing, Adv. Int. Mark. 20 (2009) 277–319, https://doi.org/10.1108/S1474-7979(2009)0000020014.

[83] J.F. Hair, M. Sarstedt, T.M. Pieper, C.M. Ringle, The use of partial least squares structural equation modeling in strategic management research: a review of past practices and recommendations for future applications, Long Range Plann. 45 (5–6) (2012) 320–340, https://doi.org/10.1016/j.lrp.2012.09.008. Pergamon.

[84] N.P. Podsakoff, Common method biases in behavioral research: a critical review of the literature and recommended remedies, J. Appl. Psychol. 885 (879) (2003) (10-1037).

[85] J.F. Hair, M. Sarstedt, C.M. Ringle, J.A. Mena, An assessment of the use of partial least squares structural equation modeling in marketing research, J. Acad. Market Sci. 40 (3) (2012) 414–433, https://doi.org/10.1007/s11747-011-0261-6. Springer.

[86] S. Mukherjee, V. Chittipaka, Analysing the adoption of intelligent agent technology in food supply chain management: an empirical evidence, FIIB Bus. Rev. (2021). https://doi.org/10.1177/23197145211059243.

[87] B.M. Byrne, in: B.M. Byrne (Ed.), Structural Equation Modeling with EQS and EQS/ WINDOWS: Basic Concepts …, Google Books., 1994. https://books.google.co.in/ books?hl=en&lr=&id=dBR0mU6W8YAC&oi=fnd&pg=PR7&dq=byrne+struc tural+equation+modeling&ots=J0j8Oyegn5&sig =QYQr128fX4j0g7E0H8DrUvltMmo&redir_esc=y#v=onepage& q=byrnestructuralequationmodeling&f=false.

[88] B.M. Byrne, Structural Equation Modeling with AMOS: Basic Concepts, Applications, and Programming (Multivariate Applications Series), vol. 396, Taylor & Francis Group, 2010, p. 7384.

[89] B.M. Byrne, Structural Equation Modeling with Mplus, 2013, https://doi.org/10.4324/ 9780203807644.

[90] J.F. Hair, W.C. Black, B.J. Babin, R.E. Anderson, R.L. Tatham, Multivariate Data Analysis, seventh ed., Prentice Hall, Upper Saddle River, NJ, 2010.

[91] R. Pillai, B. Sivathanu, Adoption of artificial intelligence (AI) for talent acquisition in IT/ITeS organizations, Benchmarking 27 (9) (2020) 2599–2629, https://doi.org/ 10.1108/BIJ-04-2020-0186. JAI Press.

[92] F. Cruz-Jesus, A. Pinheiro, T. Oliveira, Understanding CRM adoption stages: Empirical analysis building on the TOE framework, Comput. Ind. 109 (2019) 1–13, https://doi. org/10.1016/j.compind.2019.03.007. Elsevier B.V.

[93] P. Helo, Y. Hao, Artificial intelligence in operations management and supply chain management: an exploratory case study, Prod. Plan. Control (2021), https://doi.org/ 10.1080/09537287.2021.1882690. Taylor & Francis.

[94] S.K. Paul, S. Riaz, S. Das, Organizational Adoption of Artificial Intelligence in Supply Chain Risk Management. IFIP Advances in Information and Communication Technol- ogy, vol. 617, Springer International Publishing, 2020, https://doi.org/10.1007/978-3- 030-64849-7_2.

[95] R.B. Kline, Principles and Practice of Structural Equation Modeling, Guilford Publica- tions, 2015.

[96] E. Gide, R. Sandu, A study to explore the key factors impacting on cloud based service adoption in Indian SMEs, in: Proceedings—12th IEEE International Conference on E-Business Engineering, ICEBE 2015. IEEE, 2015, pp. 387–392, https://doi.org/ 10.1109/ICEBE.2015.72.

[97] T. Clohessy, T. Acton, N. Rogers, Blockchain adoption: technological, Organisational and environmental considerations, in: Business Transformation through Blockchain, 2019, pp. 47–76, https://doi.org/10.1007/978-3-319-98911-2_2.

[98] S.S. Abed, Social commerce adoption using TOE framework: an empirical investiga- tion of Saudi Arabian SMEs, Int. J. Inf. Manag. 53 (2020), https://doi.org/10.1016/j.ijin- fomgt.2020.102118. Elsevier Ltd: 102118.

[99] F.A. Nuskiya, Factors influencing cloud computing adoption by SMEs in eastern region of Sri Lanka, J. Inf. Syst. Inf. Technol. (JISIT) 2 (1) (2017) 2478. 0677 http://ir.lib.seu. ac.lk/bitstream/handle/123456789/3008/Paper2.pdf?sequence=1&isAllowed=y.

[100] J.N. Makena, Factors that affect cloud computing adoption by small and medium Enter- prises in Kenya, Int. J. Comput. Appl. Technol. Res. 2 (5) (2013) 517–521, https://doi. org/10.7753/ijcatr0205.1003.

Role of artificial intelligence in industry 4.0

Embedded system for model characterization developing intelligent controllers in industry 4.0

3

Martín Montes Rivera[a], Alberto Ochoa-Zezzatti[b], and Sebastián Pérez Serna[a]
[a]*Aguascalientes Polytechnique, Aguascalientes, México,* [b]*Autonomous University of Juarez City, Juarez City, Chihuahua, Mexico*

3.1 Introduction

Industry 4.0, also known as the Fourth Industrial Revolution, is a concept that originated in Germany in 2011 to address future challenges in industrial environments. This index considers the mechanical loom in 1784 as Industry 1.0, electricity and mass production in 1870 as Industry 2.0, and programable logic controllers and information technology systems in 1960 as Industry 3.0 [1–3].

Industry 4.0 brings technologies like Machine-to-Machine (M2M) communications, the Internet of Things (IoT), Cyber-Physical Systems (CPSs), Big Data Analytics (BDA), intelligent robots, cobots, Artificial Intelligence (AI), Virtual Reality (VR), Augmented Reality (AR), and intelligent controllers, among other innovations [1,4,5].

Integration of technologies, levels of connectivity, understandability, performance, decision making, monitoring, flexibility, and control are the main advantages of Industry 4.0 [4,6]. Nevertheless, these technologies demand more specialized workers because robots with AI trend other job positions [1,4,7]. On the other hand, technologies like AI and robotics have boosted human capabilities, technologies, productivity, and efficiency in different scenarios, including control processes [7,8].

The control field in Industry 4.0 creates a twin of an object or a business process for managing it. An automated system is a tool for optimization because the digital twin allows offline optimization, which improves the efficiency of the process. Thus, modeling and characterization become a fundamental part of the control processes to generate digital twins or mathematical models [9].

The more classic approach for modeling processes based on systems theory and systems analysis describes systems with equations and parameters associated with physical and chemical properties [9].

Artificial Intelligence and Industry 4.0. https://doi.org/10.1016/B978-0-323-88468-6.00004-8

Determining the mathematical model for predicting the behavior of an industrial process implies a high level of knowledge in physics, mathematics, system analysis, simulation, parameterization, and validation. In particular, parametrization must be regularly updated since some components are subject to wear and tear [10].

Alternatively, AI techniques have tackled the problem of predicting a system's behavior by using algorithms that learn from their mistakes and adjust themselves to be more precise and efficient in different scenarios [11].

Nowadays, AI applications include supporting control theory methodologies and replacing classic techniques with alternatives that do not require a mathematical model or determining it automatically, as shown in [11–17].

Table 3.1 lists works that apply AI techniques to characterize and control processes autonomously from 1999 to 2021, like in the application we describe in this chapter.

Despite that AI algorithms solve problems of classical control theory, industrial applications still misuse them. Conversely, other more straightforward techniques are used but sometimes wrongly tuned [26].

Proportional Integral Derivative (PID) controllers are among the most popular industry controllers; they are used in 95% of industrial applications. The popularity of PID is related to its ability to compensate processes and the simplicity of its tuning methods. Nevertheless, PID controllers have severe limitations simultaneously controlling uncertainties, disturbances, overshoot, energy consumption, and error by using only three degrees of freedom. Moreover, in several cases, PID gains are wrongly selected, causing dangerous behaviors and high-power consumption [13,26].

This chapter proposes an intelligent control assistant for Industry 4.0 capable of characterizing and controlling systems automatically. In addition, we implement the first stage for generating the Transfer Function (TF)—frequency-domain description well documented in control theory to predict linear systems' behavior [27].

TFs have been optimized before with stochastic search algorithms, but the cost functions used rapidly increase the problem dimensionality as the order of the system increases.

The previously proposed cost functions with this dimensionality problem identify poles and zeros in the TF structure, requiring one parameter per pole and one parameter per zero, increasing two parameters per order in the system ($2n$), like in the works in [24,25,28].

However, control theory has simplified the characterization problem with general formulas for TFs that correspond with specific orders and parameters determined through time and frequency response analysis. This approach reduces the increase of dimensionality per order to one parameter (n).

Our proposal's novelty is that we use the control theory formulas for TFs and extract the time response parameters, which implemented in the cost function maintain a low dimensionality in the search space. Then, we optimize the cost function with a genetic algorithm since it is a well-documented algorithm for numerical optimization previously used in TFs, as shown in the works in [24,25,28].

Table 3.1 Description of recent works applying AI techniques to control design and characterization of systems.

Name	Year	Description
Characterization of neural networks automatically mapped on automotive-grade microcontrollers [18]	2021	A neural network system to control and update the intrusion detection sensors and battery capacity in the automotive sector, the proposed controller, gives a mean absolute error of 0.0434
Design a robust intelligent controller for rigid robotic manipulator system having two links and payloads [19]	2020	A fuzzy PID control scheme with GA to control a robotic arm, the proposed controller, reduces the system error by 20%
Appropriate feature set and window parameters selection for efficient motion intent characterization towards intelligently smart EMG-PR system [20]	2020	A pattern recognition (PR) algorithm integrates prostheses for people with amputated limbs from electromagnetic signals, obtaining a minimum decoding error less than 10% in the optimal window parameters of 250 ms/100 ms
Optimal pitch angle control for wind turbine using intelligent controller [21]	2020	Initially, a PI controller for the optimum angle of inclination of wind turbines, but using a neural network type NARMA-L2 power generation was optimized by more than 20%
Realization of the sensorless permanent magnet synchronous motor drive control system with an intelligent controller [22]	2020	It consists of sensorless control of a permanent magnet using a PID controller based on a neural network (RBFNN). Its error is approximately zero
IIR system identification using cat swarm optimization [23]	2011	This work presents an approach for characterizing transfer functions using cat swarm optimization, particle swarm optimization, and genetic algorithms. This last one has a root mean square error (RMS) of 0.004922 units
Genetic algorithm-based identification of transfer function parameters for a rectangular flexible plate system [24]	2010	A control system based on genetic algorithms to control and characterize the vibrations of a plate. Said model generates the best mean square with an error of 0.00088 after 144 generations
Applying genetic algorithm to modeling nonlinear transfer functions [25]	1999	A genetic algorithm technique to nonlinear transfer function approximation to replace the classic Chebyshev. The GA method gives a maximum error of 1.2–1.5 times less than the conventional method

The TF generation in this work is offline, allowing IoT-enabled devices to transmit the input-output records and then process them on a server. Moreover, we maintain a low-dimensionality problem by mixing control theory concepts with the proposed cost functions.

After the first stage, we will use the obtained TF for generating the correct tuned controller for the corresponding applications.

The rest of the chapter is structured as follows. The section on "Theoretical Framework" includes the concepts and definitions used in work and the "Methods" section consists of the proposed methodology for implementing stage one of the intelligent control assistant. "Results and Discussion" evaluates and analyzes the proposed methodology and "Conclusions" shows the level of accomplishment in the proposed goal supported by the results and the future work of this research.

3.2 Theoretical framework

3.2.1 Processing unit Raspberry PI

The acquisition system we use is the Raspberry computer board (see Fig. 3.1) developed by the Raspberry Pi Foundation (Raspberry.org). It works with different operative systems. One of the most common is the Raspian version of Ubuntu Linux, which we use in this proposal [29,30].

FIG. 3.1

Processing system RaspberryPi.

FIG. 3.2

Middle of the RaspberryPi board.

Linux is a multitasking operating system like Windows. However, Linux differs from Windows because it is an open-source system. The RaspberryPi and Linux are the first options in several applications for embedded systems due to benefits that include many compatible devices, such as sensors and drivers, and the possibilities of commercializing new technologies based on open-source applications codes [29].

The RaspberryPi features a Broadcom BCM2835 SoC (system-on-chip) processor for multimedia processing that allows integration of most systems and components into a single component in the middle of the board (Fig. 3.2), including memory, the graphics processing unit, and the audio and communications hardware [30].

There are different ports on the RaspberryPi. The most relevant to use are listed below as detailed in [31]:

- The Universal Serial Bus (USB) port allows connecting the keyboard, mouse, and any other device with a similar communication protocol.
- The Ethernet port, also known as the network port, allows access to the Internet or communications with other devices with the RJ45 connector.
- The Audio-Visual (AV) jack allows the transmission of audio and video signals.
- The power supply micro-USB port.
- The HDMI connector transmits combined signals of audio and video to screen devices.
- The General-Purpose Input/Output (GPIO) is a port with a series of metal pins that allow connection to electronic devices like light-emitting diodes (LEDs), buttons, potentiometers, and sensors of all kinds for signal acquisition and process control.

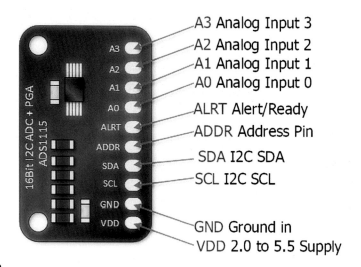

A3 Analog Input 3
A2 Analog Input 2
A1 Analog Input 1
A0 Analog Input 0

ALRT Alert/Ready
ADDR Address Pin

SDA I2C SDA
SCL I2C SCL

GND Ground in
VDD 2.0 to 5.5 Supply

FIG. 3.3

ADC converter ADS1115.

3.2.2 Analogic digital converter ADS1115

The ADS1115 device (Fig. 3.3) is the Analogic Digital Converter (ADC) module for the RaspberryPi. This device has a four-channel connection, which provides a high-resolution analog-to-digital conversion. In addition, this converter operates with signals ranging from 2v to 5v with excellent compatibility with other devices. The RaspberryPi communicates with the ADS1115 using a 2-wire protocol I2C bus that provides up to 16 single-ended channels or eight differential channels. Another essential part of the ADS1115 is its sample capture speed of 860 samples per second. The ADS1115 also includes a Programmable Gain Amplifier (PGA) that transforms inputs from ±256 mV to ±6.144 V [32,33].

Table 3.2 shows a list of the features included in the ADS1115 ADC converter and the units for each feature, including resolution, samples per second, power supply, consumption in the low-current configuration, details of I2C pin address, and PGA amplification level, among others.

3.2.3 Genetic algorithms

Genetic Algorithms (GAs) are programmable methods for solving problems based on the principles of natural selection proposed by Charles Darwin, focusing on the survival of the fittest [34].

The evolutionary algorithms optimize the candidate solutions using maximization and minimization of the fitness function. This optimization begins by measuring the fitness of a randomly generated population of candidate solutions with the cost function. These solutions mentioned previously are known as chromosomes. Each chromosome is an arrangement of genes. Genes improve based on the fitness

Table 3.2 ADS1115 features [33].

Parameter	Value	Units
Resolution	16 bits	Bits
Samples per second	8–860 samples	Samples
Power supply	2.0VDC to 5.5VDC	VDC
Low-current consumption	150	µA
I2C	4	Pin-selectable addresses
PGA	x16	Amplifier gain
Low-drift voltage reference	✓	N/A
Oscillator	✓	N/A
Four single-ended or two differential inputs	✓	N/A
Programmable comparator	✓	N/A

function and the main operators in evolutive algorithms. Thus, for each iteration or generation, the GA improves the average fitness of the population. There are three main operators in GAs. The first is the selection function, which selects the chromosomes that have a better fit. These selected chromosomes generate the next generation and the algorithm eliminates those with the worst fitness. The second function is the crossover function, which generates offspring with the chromosomes obtained in the selection function. A crossover point is selected and these crossover sites mix the selected chromosomes to create new ones. The third function is the mutation operator, which changes the chromosomes in random ranges in small percentages, diversifying and changing the population. Fig. 3.4 shows the previous algorithm [34–36].

GAs cover a wide field of study called evolutionary computation. Evolutionary computation mimics biological processes of reproduction through the selection of the fittest. The processes of these algorithms must have randomness as well as natural biological selection. These optimization methods allow establishing a level of control and randomness. GAs are efficient at finding solutions without using complex mathematics. While more information is available, a better solution will be provided [35,36].

This work uses the GA implemented on a computer that works as a server that receives the step response data acquired by the Raspberry Pi. The sections that follow describe the general implementation of the operators in GAs used in this work on the server computer.

3.2.3.1 Population
The population generation we use in this work includes two possibilities for encoding chromosomes. The first one uses two genes for the first-order systems that determine the gain and time constant parameters required in this system. The second situation

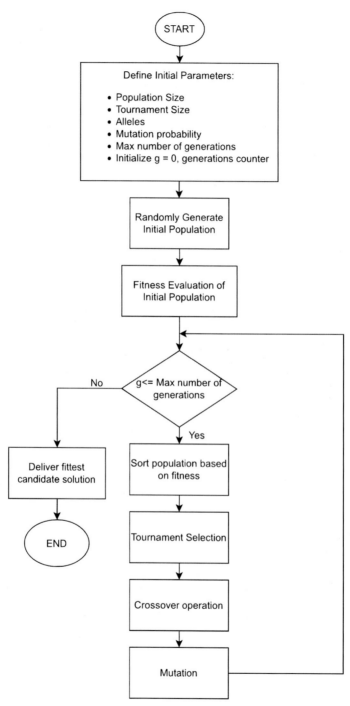

FIG. 3.4

Flow diagram of a Genetic Algorithm.

uses three genes for second-order systems, including gain, natural frequency, and damping factor, required in this type of system.

All the genes used in this work are fixed-size binary arrays, best known as binary encoding, the most common representation used in GAs. Thus the number of alleles is the possible numbers expressed with that array [37].

In decoding, we start by converting the binary number to a decimal. Then we convert this number into a floating-point number by dividing the result of the conversion by a gain parameter P, which allows us to modify the position of the point in the conversion, as shown in Eq. (3.1).

$$\text{Floating} = \frac{\text{Decimal(binary)}}{P} \qquad (3.1)$$

In this work, we maintain $P = 100$ and gradually increase the length of the binary array, which increases the number of alleles and the maximum value of the gene to avoid overflow in the obtained candidate solutions.

3.2.3.2 Selection

The selection method used in this work is tournament selection, which controls the pressure of selection with the size of the tournament [35].

We selected this method since it is a good alternative against noisy data, which is the case in the characterization of systems. It commonly uses an ADC with noisy signals to acquire the systems' response [35,38]. The algorithm used for the tournament selection is the one described in [35].

3.2.3.3 Crossover operation

The crossover operation used in this work follows the binary schema with one crossover point described in [35], which selects a random position across the binary array called the crossover point. After that, each parent splitting it into two sections in the crossover point creates two new elements in the offspring with the parents' split sections interchange.

3.2.3.4 Mutation

We follow the uniform mutation schema for binary encoding in GA, as described in [35], which toggles the value of the bit if a random value in ranges [0,1] is over the probability of mutation pm, which represents the probability of mutations in ranges [0,1].

The mutation rate is a problem-dependent parameter. However, several works recommend maintaining pm less than 10% or 0.1 because greater values affect the parents' inheritance, as described in [35,39].

3.2.4 Transfer function

The TF represents a dynamic linear system with invariance in time infinite dimensions. It is a rational function of complex variables used to describe high-order systems and infinite-dimensional systems governed by partial differential equations [40].

A short description of the TF consists of a linear system of inputs/outputs. An efficient method to obtain solutions to these functions is to mix them with block diagrams [40]. As mentioned, the TF is a linear time-invariant (LTI), described as the Laplace quotient transforms the output and the input [41].

The TF allows finding the poles and zeros of the function that describe the region of convergence, which places the limits where the function exists and where it does not. The main idea of the TF is to obtain the response of the system in frequency [40].

Eq. (3.2) describes how to obtain the TF represented with the function $H(s)$ in the frequency domain (s), with the function $Y(s)$ as the system's output and the function $X(s)$ as its input.

$$H(s) = \frac{Y(s)}{X(s)} \tag{3.2}$$

3.2.4.1 First-order systems

First-order systems have one pole and a general description without delay response, as shown in Eq. (3.3) [41].

$$H(s) = \frac{K}{\tau s + 1} \tag{3.3}$$

The parameter K is the system gain, which modifies the output multiplying the input, that is, $K = \frac{\Delta y}{\Delta u}$, the time constant (τ) measures the settling time (t_s) given by $t_s = 5\tau$.

3.2.4.2 Second-order systems

These systems have two poles instead of one, and their generic TF $H(s)$ in Eq. (3.4) depends on three parameters: K, ω_n, and ξ [41].

$$H(s) = \frac{K\omega_n^2}{s^2 + 2\xi\omega_n s + \omega_n^2} \tag{3.4}$$

As in the first-order systems, K is the system gain with $K = \frac{\Delta y}{\Delta u}$ [41].

The parameter ξ is the damping factor. Three possibilities exist for damping in these systems: underdamped $(\xi < 1)$, overdamped $(\xi > 1)$, and critically damped $(\xi = 0)$ [41].

The parameter ω_n is the natural frequency or eigenfrequency, the frequency at which the system vibrates freely after a disturbance [42].

The frequency modeling consists of evaluating the system's response to sinus and exponential signals in a system. Then the system's expressions are in terms of frequency. In general terms, the variable s refers to the frequency domain that replaces the variable t in the time domain. The TF represents a change from the time domain to the frequency domain [43].

3.3 **Methods**

3.3.1 **General description of the identification process**

The TF identification process has five stages of implementation: industrial processes, generation of testing signals, reading of testing signals, characterization, and sending TF results. Each stage occurs in a corresponding place or a device, as shown in Fig. 3.5.

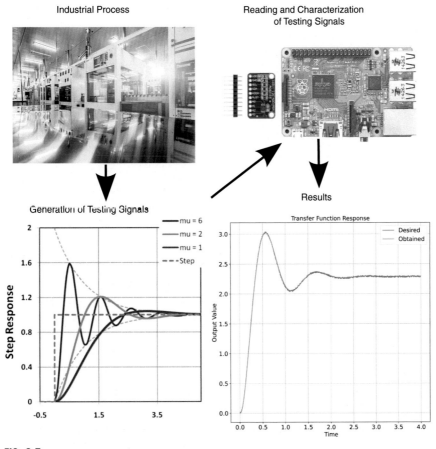

FIG. 3.5

Stages of TF identification process.

3.3.2 Time analysis for the transfer functions

Time analysis for first- and second-order systems involves obtaining the following parameters:

- The system's gain, determined with the start and end times in the step response
- The overshoot identified with maximum system output value of step response
- Rise and settling times for identifying the time constants
- Start and minimum input values representing the starting conditions of the system's input

Let $y(T) = y(1T) + y(2T) + \ldots + y(kT)$ be the output times series of the step response of the system, with k as the number of samples of the output and T the sampling period. Then, the time parameters used in the cost functions are given by:

(1) Starting value $y(1T)$ is the first sample of the output
(2) The ending value $y(kT)$ is the final sample of the output
(3) The maximum output is $y(MT) = \max(y(1T), y(2T), \ldots, y(kT))$, where M is the corresponding sampling with the maximum value of $y(T)$
(4) The minimum output is $y(mT) = \min(y(1T), y(2T), \ldots, y(kT))$, where m is the corresponding sampling with the minimum value of $y(T)$
(5) The rising time given by $yt_r = rT$ with r as the rising sample, achieved when $y(iT) \geq 0.9y(kT)$ the first time, using $i = 1, 2, \ldots, k$ as the exploring index for testing the inequality
(6) The settling time $yt_s = ST$ with 2% of error considers the last sample S that $\frac{y(iT) - y(kT)}{y(kT)} \geq 0.02$ is true with i as the exploring index
(7) The overshoot is $yM_p = \frac{y(MT)}{y(kT)} - 1$
(8) The overshoot time given by $yt_p = pT$ with p as the sample at which $y(iT) = yM_p \cdot y(kT)$ with i as the exploring index

3.3.3 Fitness function for characterizing TFs

The Mean Absolute Error (MAE) has been used to optimize TFs and is a standard metric used when optimizing cost functions in evolutive algorithms like [35]. Thus, we use MAE in the main components of our proposed cost functions, which depend on two components. The first one C_1 calculates MAE between the desired output and the candidate solution response. The second one C_2 calculates the MAE between the desired time analysis parameters and the obtained ones with the candidate solution. However, the evaluation of these parameters depends on the order of the system.

3.3.3.1 First-order systems fitness function

For first-order systems, the (C_1) component calculates MAE between the desired output $y(iT)$ and the obtained output $\widehat{y}(iT)$ across $i = 1, \ldots, k$ samples, as in Eq. (3.5).

$$C_1 = \frac{1}{k}|y(iT) - \widehat{y}(iT)| \tag{3.5}$$

where $\widehat{y} = f(H(s), K, \tau)$ is the step response of TF generated with K and τ, the corresponding parameters of first-order TF substituted with genes of the candidate solution.

The second component C_2 in first-order systems calculates the MAE between the desired time analysis parameters and the obtained ones.

For the first-order systems, we use the desired time properties in the array $yTR_n = [y(MT), y(mT), y(1T), y(kT), yt_r, yt_s]$ and the obtained ones in the array $\widehat{y}TR_n = [\widehat{y}(MT), \widehat{y}(mT), \widehat{y}(1T), \widehat{y}(kT), \widehat{y}t_r, \widehat{y}t_s]$ because these properties are required for determining the first-order general TF. Thus, Eq. (3.6) shows the second component.

$$C_2 = \frac{1}{N}\sum_{j}^{N} g_j |yTR_j - \widehat{y}TR_j| \tag{3.6}$$

where g_j is the corresponding gain in ranges $[0, 1]$ for prioritizing the time properties, and $j = 1, \ldots, N$ with N is the number of time properties used.

Eq. (3.7) is the characterization cost function for first-order systems, where we divide by 2 because the MAE between step responses and desired time characteristics are equally prioritized. Since we are characterizing first-order systems, C_1, C_2, and f depend on K and τ.

$$f(K, \tau) = \frac{1}{2}[C_1(K, \tau) + C_2(K, \tau)] \tag{3.7}$$

3.3.3.2 Second-order systems fitness function

For second-order systems, the C_1 component again estimates the error between the system's step response and the one obtained with gene substitution. However, the obtained output is $\widehat{y} = f(H(s), K, \xi, \omega_n)$.

The second component C_2 determined via Eq. (3.6) calculates the MAE between the desired time analysis parameters and the obtained parameters. Nevertheless, for the second-order systems, we use the desired time properties in the array $yTR_n = [y(MT), y(mT), y(1T), y(kT), yt_r, yt_s, yM_p, yt_p]$ and the obtained properties in the array $\widehat{y}TR_n = [\widehat{y}(MT), \widehat{y}(mT), \widehat{y}(1T), \widehat{y}(kT), \widehat{y}t_r, \widehat{y}t_s, \widehat{y}M_p, \widehat{y}t_p]$. These properties are required to determine the second-order general TF.

The characterization TF for second-order systems' cost function is given in Eq. (3.8), dividing again by 2 for equal importance to the MAE step response and the MAE time desired characteristics. However, C_1, C_2, and f depend on K, ξ, and ω_n, because we are optimizing a TF for second-order systems.

$$f(K, \xi, \omega_n) = \frac{1}{2}[C_1(K, \xi, \omega_n) + C_2(K, \xi, \omega_n)] \qquad (3.8)$$

3.4 Results and discussion

3.4.1 Design of experiment

We tested our proposed methodology in five stable TFs obtaining random noise in each representation reading the ADS1115 signal responses of the Raspberry PI. However, we reduced the error by implementing a lowpass Infinite Impulse Response (IIR) fifth-order filter to reduce the noise perceived in the time response acquired for each TF. The discrete TF of the IIR is shown in Eq. (3.9).

$$H(z) = \frac{\sum_{k=0}^{N} b_k z^{-k}}{\sum_{k=0}^{M} a_k z^{-k}} \qquad (3.9)$$

The parameters $N = M = 5$ are the order of the filter. The coefficients depend on cutting frequency. We use a cutting frequency of 10 Hz for a sampling frequency of 250 Hz, obtaining the coefficients in Eqs. (3.10), (3.11).

$$a = \begin{bmatrix} 1.00000000 \\ -4.18730005 \\ 7.06972275 \\ -6.00995815 \\ 2.5704293 \\ -0.44220918 \end{bmatrix} \qquad (3.10)$$

$$b = \begin{bmatrix} 2.1396152e-05 \\ 1.0698076e-04 \\ 2.1396152e-04 \\ 2.1396152e-04 \\ 1.0698076e-04 \\ 2.1396152e-05 \end{bmatrix} \qquad (3.11)$$

We calculate MAE, Mean Absolute Percentage Error (MAPE), Root Mean Square Error (RMSE), and the R^2 for each output signal with noise and after filtering. These metrics allow us to compare our results with other works in the literature and support that the obtained TFs explain the behavior we acquire with the ADS1115 [23,35,44].

The first three TFs describe a first-order system, and the last two describe a second-order system. The proposed systems were defined with different values for the TFs parameters. In the case of first-order systems, the different levels were the gain (K) and the time constant (τ). In second-order systems, the different levels were natural frequency ω_n, damping factor (ξ), and gain (K).

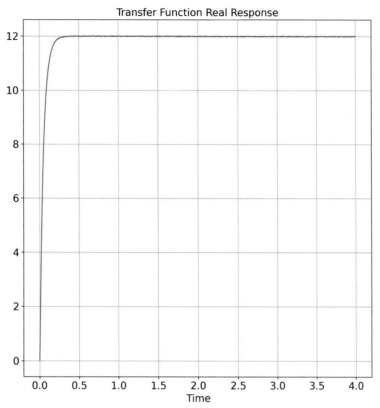

FIG. 3.6

First-order step response to the transfer function in Eq. (3.12).

The TFs for first-order systems are described as in Eqs. (3.12)–(3.14) and in Figs. 3.6–3.8.

The TF in Eq. (3.12) is a first-order system proposed for the conducted experiment with $\tau=0.05$ and gain $K=12$.

$$H(s) = \frac{12}{0.05s + 1} \tag{3.12}$$

Fig. 3.6 shows the time response of TF in Eq. (3.12). Its settling time (5τ) is 0.25 s, and its settling value is 12 according to its gain value.

The TF in Eq. (3.13) is a first-order system proposed for the conducted experiment with $\tau=0.1$ and gain $K=2$.

$$H(s) = \frac{2}{0.1s + 1} \tag{3.13}$$

Fig. 3.7 shows the time response of TF in Eq. (3.13) Its settling time is 0.5 s, given by 5τ and its settling value is 2 according to its gain value.

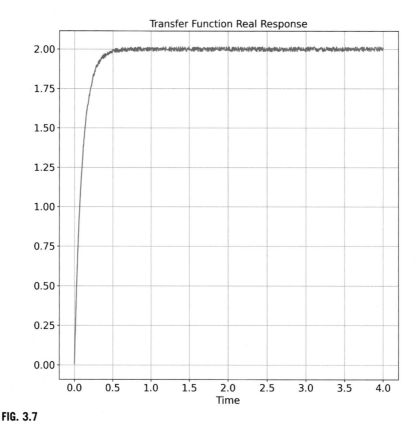

FIG. 3.7

First-order step response to the transfer function in Eq. (3.13).

The TF in Eq. (3.14) is a first-order system proposed for the conducted experiment with $\tau=0.15$ and gain $K=0.5$.

$$H(s) = \frac{0.5}{0.15s + 1} \tag{3.14}$$

Fig. 3.8 shows the time response of TF in Eq. (3.14). Its settling time (5τ) is 0.75 s, and its settling value is 0.5 according to its gain value.

The TFs for second-order systems are as shown in Eqs. (3.15), (3.16) and in Figs. 3.9 and 3.10.

The TF in Eq. (3.15) is a second-order system proposed for the conducted experiment with $K\omega_n^2 = 80$, $\omega_n^2 = 35$, and $2\xi\omega_n=4$.

$$H(s) = \frac{80}{s^2 + 4s + 35} \tag{3.15}$$

Fig. 3.9 shows an underdamped time response for the TF in Eq. (3.15) produced because the natural frequency is $\omega_n = \sqrt{35} \approx 5.916$ and the damping factor

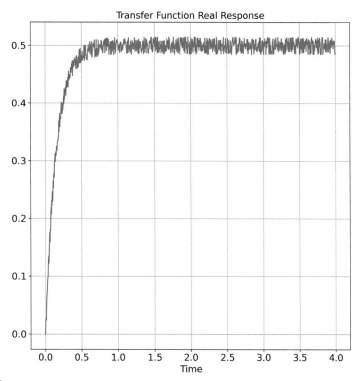

FIG. 3.8

First-order step response to the transfer function in Eq. (3.14).

$\xi = \frac{4}{2\omega_n} \approx \frac{4}{2(5.916)} \approx 0.338$ (obtained from the second coefficient in the denominator and Eq. 3.4). The settling value is 2.285 since the gain $K = \frac{80}{35} \approx 2.285$ (obtained from the numerator in Eqs. 3.15, 3.4).

The TF in Eq. (3.16) is a second-order system proposed for the conducted experiment with $K\omega_n{}^2 = 50$, $\omega_n{}^2 = 50$, and $2\xi\omega_n = 15$.

$$H(s) = \frac{50}{s^2 + 15s + 50} \tag{3.16}$$

Fig. 3.10 shows an overdamped time response for the TF in Eq. (3.16) produced because the natural frequency is $\omega_n = \sqrt{50} \approx 7.071$ and the damping factor $\xi = \frac{15}{2\omega_n} \approx \frac{15}{2(7.071)} \approx 1.061$ (obtained from the second coefficient in the denominator and Eq. 3.4). The settling value is 1.0 since the gain $K = \frac{80}{80} = 1$ (obtained from the numerator in Eqs. 3.15, 3.4).

Identifying the best parameters for GAs in the characterization of 1st and 2nd order systems implies building a representative dataset and splitting it for cross-validation with a technique like K-fold Cross-Validation to support the selected

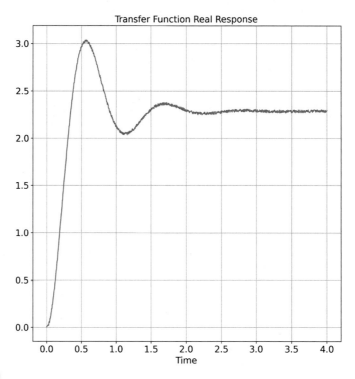

FIG. 3.9

Second-order step response to the transfer function in Eq. (3.15).

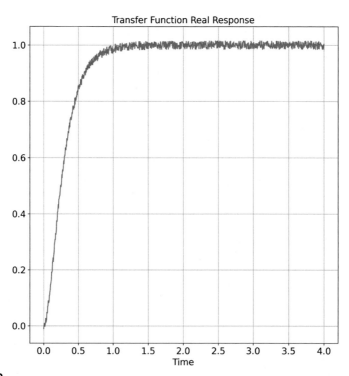

FIG. 3.10

Second-order step response to the transfer function in Eq. (3.16).

parameters suppressing the experiment's variance shown in [44,45]. However, this is not the goal of this work.

3.4.2 Results

3.4.2.1 Configuration of parameters in the optimization algorithm

The main criterion for proposing these techniques is the time required for optimization, since in Industry 4.0, intelligent controllers should optimize the models and controls in competitive times to those required in conventional approaches.

We set the number of chromosomes (Nchromosomes) to 100 in first- and second-order systems because this allows us to obtain the fitness values of the initial population in about 3 s. Then, we set the number of generations (Ngenerations) to 2000 because this maintains an entire optimization with the GA in less than 3 min or 180 s.

We identify the number of bits per gene based on producing candidate solutions that can reach the values we used in the proposed models for the experiment. In industrial implementation, the users must gradually increase the number of bits to find candidate solutions that produce good-quality genes.

Next, we identify the probability of mutation and the tournament size parameters by gradually increasing the other parameters to find mean cost values less than 0.1 for the TFs. We tested the tournament size (tournament_size) parameter with steps of one unit from 5 to 100 elements. Similarly, we tested the probability of mutation (*pm*) with steps of 0.01 from 0 to 0.10.

This approach allows use of the tournament size and *pm* to control the convergence speed for obtaining solutions as fast as possible and in less than the maximum established time of 3 min.

The parameters used in GAs for optimizing the first-order systems fitness function are as follows:

- Nchromosomes $=100$
- Ngenerations $=2000$
- Bits per gene $=14$
- tournament_size $=5$
- *pm* $=0.08$

The parameters used in GAs for optimizing second-order systems fitness function are as follows:

- Nchromosomes $=100$
- Ngenerations $=2000$
- Bits per gene $=10$
- tournament_size $=5$
- *pm* $=0.08$

The proposed parameters are not the best for each solution but maintain diversity for solving the TFs and obtain candidate solutions with the desired fitness value we set as 0.1.

However, the user could change them in more complex systems, accepting the increase in time for optimization.

On the other hand, reaching the cost value we set allows the user to define the expected quality and reach it in the least possible time found by setting different input parameters for the algorithm.

3.4.2.2 First-order transfer function

Using the configuration with the selected GA parameters, we required 0.00867 s for declaring the GA initial parameters, 1.97013 s to generate and evaluate the initial population, an average time per generation of 0.08616 s, and a total time of 174.29880 s for finding the first-order system in Eq. (3.12).

The GA found the best candidate solution at 608 generations. The training behavior for the TF in Eq. (3.12) is shown in Fig. 3.11 for the generations of improvement, with red showing the worst cost and blue showing the best cost across iterations. On the other hand, we set 2000 generations as an initial parameter for finding TFs with more dispersed parameters.

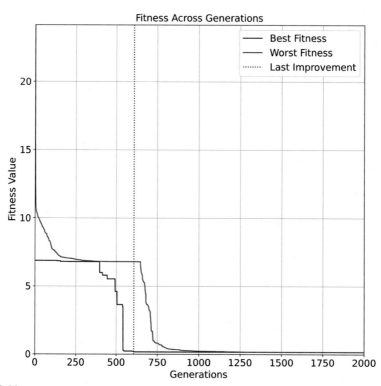

FIG. 3.11

Training response for first-order transfer function in Eq. (3.12).

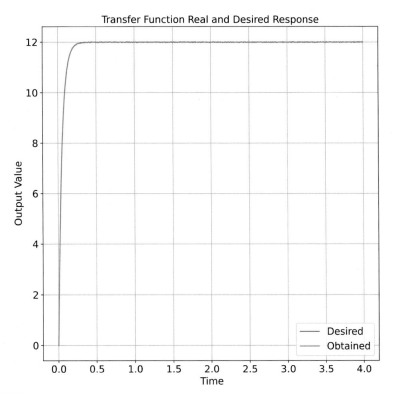

FIG. 3.12

Desired-obtained comparison response for first-order transfer function in Eq. (3.12).

The best TF after 608 generations is the same proposed TF. Fig. 3.12 compares the desired response (blue) with noise and the obtained response (orange) for TF in Eq. (3.12). The differences between the system and the obtained model are due to the inclusion of the noise.

Eq. (3.17) is the TF obtained with the GA for the TF in Eq. (3.12), which exhibits minimal discrepancies with the time parameters of the system response produced by the noise, as shown in Table 3.3, supporting the minimal differences perceived in Fig. 3.12. Moreover, Eqs. (3.17), (3.12) are the same.

$$H(s) = \frac{12}{0.05s + 1} \tag{3.17}$$

The units in Table 3.3 yt_r and yt_s are seconds, but the rest of the variables depend on the characterized problem. However, in this case, they are dimensionless because we did not define any specific problem.

We found that the proposed algorithm obtains the error metrics for the filtered signal of MAE$=4.50975$, MAPE$=0.19117\%$, RMSE$=0.02003$, $R^2=0.99959$,

Table 3.3 Comparison among expected and obtained values in transfer function of Eq. (3.12).

Time-response parameters	Expected values	Obtained values	Absolute error	Relative error (%)
$y(MT)$	12.0043	12.0000	0.0043	0.0355
$y(mT)$	0.0000	0.0000	0.0000	0.0000
$y(1T)$	0.0000	0.0000	0.0000	0.0000
$y(kT)$	12.0005	12.0000	0.0005	0.0041
yt_r	0.1201	0.1161	0.0040	3.3333
yt_s	0.1481	0.1481	0.0000	0.0000
K	12.0000	12.0000	0.0000	0.0000
τ	0.0500	0.0500	0.0000	0.0000

and the signal without the filter MAE$=7.55744$, MAPE$=0.165972\%$, RMSE$=0.00875$, and $R^2=0.99992$. However, the proposed algorithm obtains a TF that generates a time-response parameters error less than 4% compared with the expected TF with noise (Table 3.3). However, we obtained the same TF proposed.

Using the configuration with the selected GA parameters, we required 0.00760 s for declaring the GA initial parameters, 2.11557 s to generate and evaluate the initial population, an average time per generation of 0.08930 s, and a total time of 180.72331 s for finding the first-order system in Eq. (3.13).

The training behavior for the TF in Eq. (3.13) is shown in Fig. 3.13, with red showing the worst cost and blue showing the best cost across iterations. Fig. 3.13 shows that we found the best candidate solution at 1246 generations. On the other hand, we set 2000 generations as an initial parameter for finding TFs with more dispersed parameters.

The best TF after 1246 generations is the same proposed TF. Fig. 3.14 compares the desired response (blue) and the obtained response (orange) for TF in Eq. (3.13). Again, the reading noise produces the differences between the system and the obtained model.

Eq. (3.18) and Table 3.4 are the parameters of the TF solution in Eq. (3.13), which exhibit minimal discrepancies with the system response, supporting the minimal differences perceived in Fig. 3.14. Moreover, the noise produces variations in the time parameters because the TF in Eq. (3.18) obtained with the GA is the same as shown in Eq. (3.13).

$$H(s) = \frac{2}{0.1s + 1} \tag{3.18}$$

The units in Table 3.4 yt_r and yt_s are seconds, but the rest of the variables depend on the characterized problem. However, in this case, they are dimensionless because we did not define any specific problem.

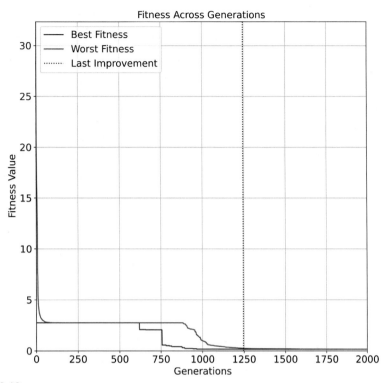

FIG. 3.13

Training response for first-order transfer function in Eq. (3.13).

We found that the proposed algorithm obtains the error metrics for the filtered signal of MAE$=1.73289$, MAPE$=0.20405\%$, RMSE$=0.00222$, $R^2=0.99990$, and the signal without the filter MAE$=7.37430$, MAPE$=0.50085\%$, RMSE$=0.00849$, and $R^2=0.99854$. However, the proposed algorithm obtains a TF that generates a time-response parameters error less than 2% compared with the expected TF with noise (Table 3.4). However, we obtained the same TF proposed.

Using the configuration with the selected GA parameters, we required 0.00077 s for declaring the GA initial parameters, 1.62185 s to generate and evaluate the initial population, an average time per generation of 0.08947 s, and a total time of 180.56262 s for finding the first-order system in Eq. (3.14).

The training behavior for the TF in Eq. (3.14) is shown in Fig. 3.15, with red showing the worst cost and blue showing the best cost across iterations. Fig. 3.15 shows that we found the best candidate solution at 1396 generations. On the other hand, the worst TF still improves after 2000 generations.

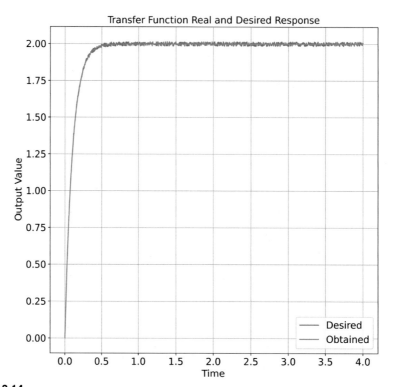

FIG. 3.14

Desired-obtained comparison response for first-order transfer function in Eq. (3.13).

Table 3.4 Comparison among expected and obtained values in transfer function of Eq. (3.13).

Time-response parameters	Expected values	Obtained values	Absolute error	Relative error (%)
$y(MT)$	2.0070	2.0000	0.0070	0.3507
$y(mT)$	0.0000	0.0000	0.0000	0.0000
$y(1T)$	0.0000	0.0000	0.0000	0.0000
$y(kT)$	2.0061	2.0000	0.0061	0.3034
yt_r	0.2362	0.2322	0.0040	1.7094
yt_s	0.3003	0.2963	0.0040	1.3423
K	2.0000	2.0000	0.0000	0.0000
τ	0.1000	0.1000	0.0000	0.0000

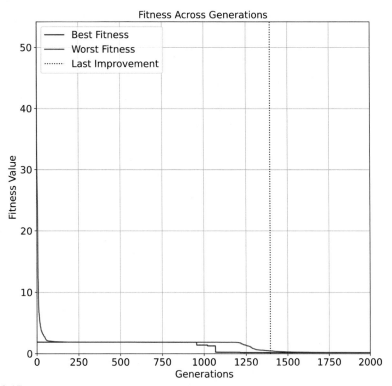

FIG. 3.15

Training response for first-order transfer function in the Eq. (3.14).

The best TF we obtained after 1396 generations varies in the time constant of the proposed TF. Fig. 3.16 compares the desired response (blue) and the obtained response (orange) for TF in Eq. (3.14). However, the differences between the system and the obtained model are minimal and produced by the noise.

Eq. (3.19) and Table 3.5 are the parameters of the TF solution in Eq. (3.14), which exhibit minimal discrepancies with the system response produced by the noise. Therefore, the minimal differences perceived in Fig. 3.16 supports that the noise produces the variation in the time constant.

$$H(s) = \frac{0.5}{0.13s + 1} \qquad (3.19)$$

The units in Table 3.5 yt_r and yt_s are seconds, but the rest of the variables depend on the characterized problem. However, in this case, they are dimensionless because we did not define any specific problem.

We found that the proposed algorithm obtains the error metrics for the filtered signal of MAE $= 4.04726$, MAPE $= 1.27533\%$, RMSE $= 0.00742$, $R^2 = 0.98772$, and the signal without the filter MAE $= 8.88712$, MAPE $= 2.02320\%$,

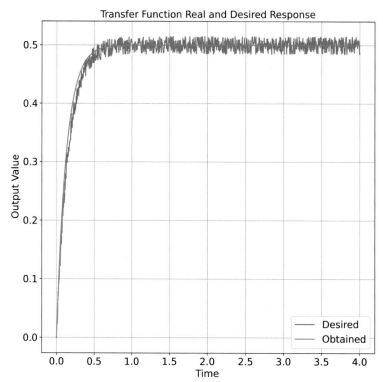

FIG. 3.16

Desired-obtained comparison response for first-order transfer function in Eq. (3.14).

Table 3.5 Comparison among expected and obtained values in transfer function of Eq. (3.12).

Time-response parameters	Expected values	Obtained values	Absolute error	Relative error (%)
$y(MT)$	0.5060	0.5000	0.0060	1.1856
$y(mT)$	0.0000	0.0000	0.0000	0.0000
$y(1T)$	0.0000	0.0000	0.0000	0.0000
$y(kT)$	0.4877	0.5000	0.0123	2.5169
yt_r	0.3203	0.3003	0.0200	6.2500
yt_s	0.3764	0.3884	0.0120	3.1915
K	0.5000	0.5000	0.0000	0.0000
τ	0.1500	0.1300	0.0200	13.3333

RMSE$=0.01119$, and $R^2=0.97249$. Despite that, we obtain a significant *error* of 13.3333% for the τ parameter (Table 3.5). The $y(kT)$, yt_r, and yt_s relative errors are high because their expected value is near zero, which leads to higher relative errors.

3.4.2.3 Second-order transfer function

Using the configuration with the selected GA parameters, we required 0.00455 s for declaring the GA initial parameters, 2.93409 s to generate and evaluate the initial population, an average time per generation of 0.09265 s, and a total time of 188.23864 s for finding the first order system in the Eq. (3.15).

The training behavior for the TF in the Eq. (3.15) is in Fig. 3.17, with red showing the worst cost and blue for the best cost across iterations. Fig. 3.17 shows that we found the best candidate solution at 592 generations, and the worst TF converged to this solution after 2000 generations.

The comparison between the desired response (blue) and the obtained response (orange) is shown in Fig. 3.18 for the TF shown in Eq. (3.15). Again, the differences between the system and the obtained model are minimal and produced by the noise.

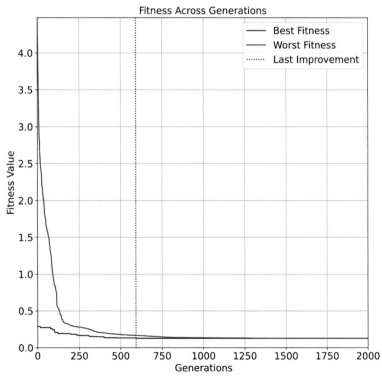

FIG. 3.17

Training response for second-order transfer function in Eq. (3.15).

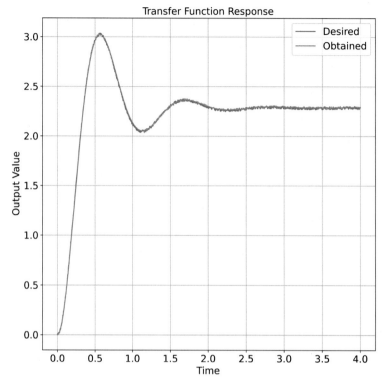

FIG. 3.18

Desired-obtained comparison response for first-order transfer function in Eq. (3.15).

Eq. (3.20) and Table 3.6 are the parameters of the TF solution in Eq. (3.15), which exhibit minimal discrepancies with the system response, supporting the minimal differences perceived in Fig. 3.18.

$$H(s) = \frac{82.44}{s^2 + 4.08s + 36} \qquad (3.20)$$

The units in Table 3.6 yt_r and yt_s are seconds, but the rest of the variables depend on the characterized problem. However, in this case, they are dimensionless because we did not define any specific problem.

We found that the proposed algorithm obtains a time-response with an error less than 1% (Table 3.6). The proposed algorithm obtains the error metrics for the filtered signal of MAE = 5.55396, MAPE = 0.75597%, RMSE = 0.00686, $R^2 = 0.99980$, and the signal without the filter MAE = 8.80183, MAPE = 1.01674%, RMSE = 0.01069, and $R^2 = 0.99951$.

Using the configuration with the selected GA parameters, we required 0.01301 s for declaring the GA initial parameters, 3.29715 s to generate and evaluate the initial

Table 3.6 Comparison among expected and obtained values in transfer function of Eq. (3.15).

Time-response parameters	Expected values	Obtained values	Absolute error	Relative error (%)
$y(MT)$	3.0288	3.0254	0.0033	0.1105
$y(mT)$	0.0000	0.0000	0.0000	0.0000
$y(1T)$	0.0000	0.0000	0.0000	0.0000
$y(kT)$	2.2925	2.2908	0.0017	0.0761
yt_r	0.3163	0.3163	0.0000	0.0000
yM_p	32.1147	32.0693	0.0454	0.1414
yt_p	0.5686	0.5686	0.0000	0.0000
yt_s	1.3413	1.3413	0.0000	0.0000
K	2.2900	2.2800	0.0100	0.4367
ω_n	6.0000	5.9800	0.0200	0.3333
ξ	0.3400	0.3400	0.0000	0.0000

population, an average time per generation of 0.09322 s, and a total time of 189.75016 s for finding the first-order system in Eq. (3.16).

The training behavior for the TF in Eq. (3.16) is shown in Fig. 3.19, with red showing the worst cost and blue showing the best cost across iterations. Fig. 3.19 shows that we found the best candidate solution at 1390 generations, and the worst TF converged to this solution after 2000 generations.

The comparison between the desired response (blue) and the obtained response (orange) is shown in Fig. 3.20 for the TF shown in Eq. (3.16). Again, the differences between the system and the obtained model are minimal and produced by the noise.

Table 3.7 shows the parameters of the TF solution in Eq. (3.16), which exhibit minimal discrepancies with the system response, supporting the minimal differences perceived in Fig. 3.20.

$$H(s) = \frac{44.93}{s^2 + 13...21s + 44.49} \tag{3.21}$$

The units in Table 3.7 yt_r and yt_s are seconds, but the rest of the variables depend on the characterized problem. However, in this case, they are dimensionless because we did not define any specific problem. We found that the proposed algorithm obtains a time-response relative error of 0.5251% (Table 3.6).

We found that the proposed algorithm obtains the error metrics for the filtered signal of MAE$=1.96264$, MAPE$=0.08332\%$ RMSE$=0.00250$, $R^2=0.99846$, and the signal without the filter MAE$=7.31315$, MAPE$=2.07987\%$ RMSE$=0.00851$, $R^2=0.99822$. Despite that, we obtain a significant error of 10.6312% for the yt_p parameter (Table 3.5). The yt_p, ω_n, ξ, relative errors are high because the significant noise in this signal modifies the original parameters. However, the comparison between the obtained TF and the systems signal is minimal.

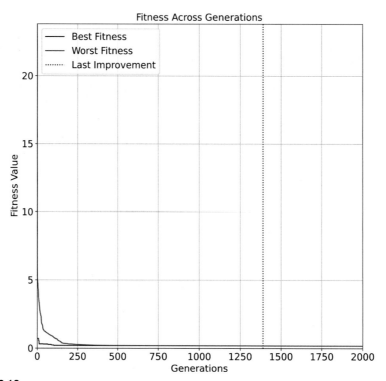

FIG. 3.19

Training response for second-order transfer function in Eq. (3.16).

3.4.3 Discussion

The proposed cost function successfully identifies all the attempted TFs with a maximum mean percentage error of 1.27533%, even in noise, a situation that other works in the literature we found did not explore. However, the error levels are competitive with other works in the literature for characterizing TFs with Gas, but we take a step forward verifying the results in acquired signals. Moreover, four of our TFs obtained an R^2 greater than 0.99, and one obtained 0.98. Thus, the obtained TFs are promising results for first- and second-order characterizations, successfully explaining the characterized signals' behavior. Despite this, the cases with more significant errors occur in measurements with more significant noise. After comparing the system and model responses, we obtain minimal variations even with the affectations produced with the ADC reading noise. Our proposal produced candidate solutions using the TF model, which has several methods and metrics for analysis in control theory. This model allows for a more reliable structure than the others proposed in the literature using black-box structures like artificial neural networks, which produce solutions incompatible with control theory analysis techniques. Moreover, our solution obtains

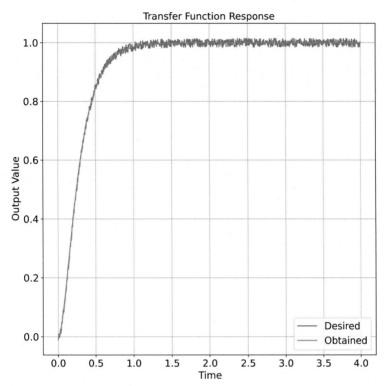

FIG. 3.20

Desired-obtained comparison response for first-order transfer function in Eq. (3.16).

Table 3.7 Time parameters comparison between expected and obtained values in transfer function of Eq. (3.16).

Time-response parameters	Expected values	Obtained values	Absolute error	Relative error (%)
$y(MT)$	1.0053	1.0000	0.0053	0.5251
$y(mT)$	0.0203	0.0202	0.0001	0.2718
$y(1T)$	0.0203	0.0202	0.0001	0.2718
$y(kT)$	1.0027	1.0000	0.0027	0.2719
yt_r	0.5966	0.5966	0.0000	0.0000
yM_p	0.2546	0.0000	0.2546	0.0000
yt_p	3.6156	4.0000	0.3844	10.6312
yt_s	0.7487	0.7407	0.0080	1.0695
K	1.0000	1.0100	0.0100	1.0000
ω_n	7.0711	6.6700	0.4011	5.6719
ξ	1.0607	0.9900	0.0707	6.6619

competitive results with the implementations described in the system characterization works of Table 3.1 and only requires the system's step response as an input.

3.4.3.1 Integration of the proposed algorithm into industry 4.0

Based on our results, the controller we proposed could become part of Industry 4.0 processes.

Every time a new machine or system requires maintaining specific temperature values, position or speed of conveyors, fluids flow, among other variables in Industry 4.0, they will require a controller.

The used controller can be an active device that will send the obtained data to a server after applying to the plant input signals like in this work. The server infers a model representation of the system, then obtains a suitable quality controller and sends it back to the controlling system.

The obtained candidate control has desired characteristics in the industrial process like saving energy, avoiding overshoots, resistance to disturbances, and other characteristics identified with a multicriterial cost function, like those used in this identification process.

3.5 Conclusions

In this chapter, we presented an assistant for designing intelligent controllers for Industry 4.0. We described the first stage of characterizing linear systems with TFs for first- and second-order systems.

The novel approach followed in this work for identifying TFs mixes the general descriptions and the time parameters used for identification in control theory. Our proposed approach uses two components in the cost function. The first component evaluates the MAE between the real output signal and the obtained one, and the second one considers the MAE among the desired time properties.

We successfully characterized the first- and second-order TFs by reducing the error among each output sample and the error in the desired time proprieties, obtaining an R^2 greater than 0.99 and one equal to 0.98. Nevertheless, despite the lower levels of error between the system and the obtained response, the cost function affects the perception of the quality of the candidate solution in the presence of the ADC reading noise. Thus the process for measuring the time-response parameters must include a filter for reducing these errors.

3.5.1 Future work

We have found that the cost functions proposed in this work allow obtaining reasonable solutions, as shown in Section 3.4.2 of this chapter. However, we found that using a GA takes several iterations in a simple problem with three dimensions. Thus, we will try different algorithms in the future to improve optimization speed.

References

[1] Z. Grodek-Szostak, L.O. Siguencia, A. Szelag-Sikora, G. Marzano, The impact of industry 4.0 on the labor market, in: 2020 61st Int. Sci. Conf. Inf. Technol. Manag. Sci. Riga Tech. Univ. ITMS 2020—Proc, Institute of Electrical and Electronics Engineers Inc, 2020, https://doi.org/10.1109/ITMS51158.2020.9259295.

[2] C. Emeric, D. Geoffroy, D. Paul-Eric, Development of a new robotic programming support system for operators, Procedia Manuf. 51 (2020) 73–80, https://doi.org/10.1016/j.promfg.2020.10.012.

[3] L.M. Villar, E. Oliva-Lopez, O. Luis-Pineda, A. Benešová, J. Tupa, J.A. Garza-Reyes, Fostering economic growth, social inclusion & sustainability in industry 4.0: a systemic approach, Procedia Manuf. 51 (2020) 1755–1762, https://doi.org/10.1016/j.promfg.2020.10.244.

[4] H. Fatorachian, H. Kazemi, Impact of industry 4.0 on supply chain performance, Prod. Plan. Control 32 (2021) 63–81, https://doi.org/10.1080/09537287.2020.1712487.

[5] E. Oztemel, S. Gursev, Literature review of industry 4.0 and related technologies, J. Intell. Manuf. 31 (2020) 127–182, https://doi.org/10.1007/s10845-018-1433-8.

[6] M. Hernandez-de-Menendez, R. Morales-Menendez, C.A. Escobar, M. McGovern, Competencies for industry 4.0, Int. J. Interact. Des. Manuf. 14 (2020) 1511–1524, https://doi.org/10.1007/s12008-020-00716-2.

[7] C. Webster, S. Ivanov, Robotics, artificial intelligence, and the evolving nature of work, in: G. Babu, P. Justin (Eds.), Digit. Transform. Bus. Soc. Theory Cases, Springer, 2019, pp. 127–143, https://doi.org/10.1007/978-3-030-08277-2_8.

[8] D. Goodley, D. Cameron, K. Liddiard, B. Parry, K. Runswick-Cole, B. Whitburn, et al., Rebooting inclusive education? New technologies and disabled people, Can. J. Disabil. Stud. 9 (2020), https://doi.org/10.15353/cjds.v9i5.707.

[9] N.A. Spirin, V.Y. Rybolovlev, V.V. Lavrov, I.A. Gurin, D.A. Schnayder, A.V. Krasnobaev, Scientific problems in creating intelligent control systems for technological processes in pyrometallurgy based on industry 4.0 concept, Metallurgist 64 (2020) 574–580, https://doi.org/10.1007/s11015-020-01029-1.

[10] K. Velten, Mathematical Modeling and Simulation: Introduction for Scientists and Engineers, Wiley, 2009, https://doi.org/10.1002/9783527627608.

[11] R. Ilieva, K. Anguelov, Y. Nikolov, Mathematical algorithms for artificial intelligence, in: AIP Conf. Proc, vol. 2172, American Institute of Physics Inc, 2019, p. 110015, https://doi.org/10.1063/1.5133618.

[12] M. Montes Rivera, M. Paz Ramos, M.J. Orozco, Automatic generator of decoupling blocks using genetic programming, in: New Trends Networking, Comput. E-learning, Syst. Sci. Eng, Springer, 2015, pp. 281–290, https://doi.org/10.1007/978-3-319-06764-3.

[13] J. Hernandez-Barragan, J.D. Rios, A.Y. Alanis, C. Lopez-Franco, J. Gomez-Avila, N. Arana-Daniel, Adaptive single neuron anti-windup PID controller based on the extended Kalman filter algorithm, Electronics 9 (2020) 636, https://doi.org/10.3390/electronics9040636.

[14] M.M. Ferdaus, M. Pratama, S.G. Anavatti, M.A. Garratt, E. Lughofer, PAC: a novel self-adaptive neuro-fuzzy controller for micro aerial vehicles, Inf. Sci. 512 (2020) 481–505, https://doi.org/10.1016/j.ins.2019.10.001.

[15] C. Hua, S. Wu, X. Guan, Stabilization of t-s fuzzy system with time delay under sampled-data control using a new looped-functional, IEEE Trans. Fuzzy Syst. 28 (2020) 400–407, https://doi.org/10.1109/TFUZZ.2019.2906040.

[16] S.M. Udrescu, M. Tegmark, AI Feynman: A physics-inspired method for symbolic regression, Sci. Adv. 6 (2020), https://doi.org/10.1126/sciadv.aay2631, eaay2631.

[17] Z. Pahlavan Yali, M. Hossein Fatemi, Symbolic regression via genetic programming model for prediction of adsorption efficiency of some pesticides on MWCNT/PbO 2 nanocomposite, Iran Chem. Soc. Anal. Bioanal. Chem. Res. 8 (2020) 65–77, https://doi.org/10.22036/ABCR.2018.152505.1263.

[18] G. Crocioni, G. Gruosso, D. Pau, D. Denaro, L. Zambrano, G. di Giore, Characterization of Neural Networks Automatically Mapped on Automotive-grade Microcontrollers, ArXiv, 2021.

[19] A. Saxena, J. Kumar, V.K. Deolia, Design a robust intelligent controller for rigid robotic manipulator system having two links and payloads, in: 2020 Int. Conf. Power Electron. IoT Appl. Renew. Energy its Control. PARC 2020, Institute of Electrical and Electronics Engineers Inc, 2020, pp. 159–163, https://doi.org/10.1109/PARC49193.2020.236581.

[20] M.G. Asogbon, O.W. Samuel, Y. Jiang, L. Wang, Y. Geng, A.K. Sangaiah, et al., Appropriate feature set and window parameters selection for efficient motion intent characterization towards intelligently smart emg-pr system, Symmetry (Basel) 12 (2020) 1–20, https://doi.org/10.3390/sym12101710.

[21] M.S. Abdul-Ruhman, M.N. Hawas, Optimal pitch angle control for wind turbine using intelligent controller, in: IOP Conf. Ser. Mater. Sci. Eng, vol. 745, Institute of Physics Publishing, 2020, p. 012017, https://doi.org/10.1088/1757-899X/745/1/012017.

[22] H.K. Hoai, S.C. Chen, H. Than, Realization of the sensorless permanent magnet synchronous motor drive control system with an intelligent controller, Electron 9 (2020) 365, https://doi.org/10.3390/electronics9020365.

[23] G. Panda, P.M. Pradhan, B. Majhi, IIR system identification using cat swarm optimization, Expert Syst. Appl. 38 (2011) 12671–12683, https://doi.org/10.1016/J.ESWA.2011.04.054.

[24] A.R. Tavakolpour, I.Z. Mat Darus, O. Tokhi, M. Mailah, Genetic algorithm-based identification of transfer function parameters for a rectangular flexible plate system, Eng. Appl. Artif. Intel. 23 (2010) 1388–1397, https://doi.org/10.1016/j.engappai.2010.01.005.

[25] S.L. Loyka, Applying genetic algorithm to modeling nonlinear transfer functions, in: 4th Int. Conf. Telecommun. Mod. Satell. Cable Broadcast. Serv. ITELSIKS 1999—Proc, vol. 1, IEEE Computer Society, 1999, pp. 247–250, https://doi.org/10.1109/TELSKS.1999.804737.

[26] Z.S. Hou, Z. Wang, From model-based control to data-driven control: survey, classification and perspective, Inf. Sci. 235 (2013) 3–35, https://doi.org/10.1016/j.ins.2012.07.014.

[27] B. Yang, C.A. Tan, Transfer functions of one-dimensional distributed parameter systems, J. Appl. Mech. Trans. ASME 59 (1992) 1009–1014, https://doi.org/10.1115/1.2894015.

[28] T. He, L. Hong, A. Kaufman, H. Pfister, Generation of transfer functions with stochastic search techniques, in: Proc. IEEE Vis. Conf, IEEE, 1996, pp. 227–234, https://doi.org/10.1109/visual.1996.568113.

[29] J.C. Shovic, Raspberry Pi IoT Projects, Apress, 2016, https://doi.org/10.1007/978-1-4842-1377-3.

[30] E. Upton, G. Halfacree, Raspberry Pi® User Guide, John Wiley & Sons, Ltd, Chichester, 2016, https://doi.org/10.1002/9781119415572.

[31] G. Halfacree, The Official Raspberry Pi Beginner's Guide How to Use Your New Computer, Raspberry Pi Press, 2019.

[32] R.M. Antosia, Voltmeter design based on ADS1115 and arduino uno for DC resistivity measurement, JTERA (J. Teknol. Rekayasa) 5 (2020) 73–80, https://doi.org/10.31544/jtera.v5.i1.2019.73-80.

[33] T. Instruments, ADS111x Ultra-Small, Low-Power, I2C-Compatible, 860-SPS, 16-Bit ADCs With Internal Reference, Oscillator, and Programmable Comparator datasheet (Rev. D) | Enhanced Reader, 2018, pp. 1–54. moz-extension://15da3b32-413b-4251-ad74-6e63e54d6073/enhanced-reader.html?openApp&pdf＝https%3A%2F%2Fwww.ti.com%2Flit%2Fds%2Fsymlink%2Fads1115.pdf%3Fts%3D1611929357734 (Accessed 30 January 2021).

[34] R. Pal, S. Yadav, R. Karnwal, Aarti, EEWC: energy-efficient weighted clustering method based on genetic algorithm for HWSNs, Complex Intell. Syst. 6 (2020) 391–400,- https://doi.org/10.1007/s40747-020-00137-4.

[35] T. Weise, Global Optimization Algorithms-Theory and Application, second ed., Institute of Applied Optimization, 2009.

[36] M. Abbasi, M. Rafiee, M.R. Khosravi, A. Jolfaei, V.G. Menon, J.M. Koushyar, An efficient parallel genetic algorithm solution for vehicle routing problem in cloud implementation of the intelligent transportation systems, J. Cloud Comput. 9 (2020) 6, https://doi.org/10.1186/s13677-020-0157-4.

[37] A. Kumar, Encoding schemes in genetic algorithm, Int. J. Adv. Res. IT Eng. 2 (2013) 1–7.

[38] M. Montes Rivera, A. Padilla Díaz, J.C. Ponce Gallegos, J. Canul-Reich, A. Ochoa Zezzatti, M.A. Meza de Luna, L. Martínez-Villaseñor, I. Batyrshin, A. Marín-Hernández, Performance of human proposed equations, genetic programming equations, and artificial neural networks in a real-time color labeling assistant for the colorblind, in: Adv. Soft Comput., vol. 11835, Springer International Publishing, Cham, 2019, pp. 557–578, https://doi.org/10.1007/978-3-030-33749-0.

[39] L. Chambers, The Practical Handbook of Genetic Algorithms, second ed., Chapman and Hall/CRC, New York, 2000, https://doi.org/10.1201/9781420035568.

[40] K. Johan Åström, R.M. Murray, Feedback Systems: An Introduction for Scientists and Engineers., 2010, https://doi.org/10.5860/choice.46-2107.

[41] B. Girod, R. Rabenstein, A. Stenger, Signals and Systems, Wiley, 2001.

[42] P. Deepak Mane, A.A. Yadav, A.M. Pol, V.A. Kumbhar, Comparative analysis of natural frequency for cantilever beam through analytical and software approach, Int. Res. J. Eng. Technol. 5 (2018) 656–671.

[43] N. Zhuo-Yun, Z. Yi-Min, W. Qing-Guo, L. Rui-Juan, X. Lei-Jun, F.-O. PID, Controller design for time-delay systems based on modified Bode's ideal transfer function, IEEE Access 8 (2020) 103500–103510, https://doi.org/10.1109/ACCESS.2020.2996265.

[44] E. Olvera-Gonzalez, M.M. Rivera, N. Escalante-Garcia, E. Flores-Gallegos, Modeling energy led light consumption based on an artificial intelligent method applied to closed plant production system, Appl. Sci. 11 (2021) 2735, https://doi.org/10.3390/app11062735.

[45] D. Berrar, Cross-validation, in: Encycl. Bioinforma. Comput. Biol. ABC Bioinforma, vols. 1–3, Elsevier, 2018, pp. 542–545, https://doi.org/10.1016/B978-0-12-809633-8.20349-X.

Industry 4.0 multiagent system-based knowledge representation through blockchain

Kuldeep Singh Kaswan[a], **Jagjit Singh Dhatterwal**[b], **Sanjay Kumar**[c], **and Amit Pandey**[d]

[a]*School of Computing Science and Engineering, Galgotias University, Greater Noida, Uttar Pradesh, India,* [b]*Department of Artificial Intelligence & Data Science Koneru Lakshmaiah Education Foundation, Vaddeswaram, AP, India,* [c]*Galgotias College of Engineering and Technology, Greater Noida, Uttar Pradesh, India,* [d]*Computer Science Department, College of Informatics, Bule Hora University, Bule Hora, Ethiopia*

4.1 Introduction

Technology that will cause a shift in manufacturing process effectiveness, reliability, and control, known as Industry 4.0, is one of the greatest industrial developments to date [1]. One such technology is ambient intelligence (AmI), which is beginning to be used in smart devices to gather data in a smart, automated fashion for decision-making and developing consumer products. Thus, companies can make strategic and operational decisions in real time, encouraging consumers to become more interested in the manufacture of tailored products. To accomplish this, we recommend that industries utilize blockchain and multiagent systems (MAS). The goal is to build a paradigm to support decision-making processes, on which entity the current dependencies can be solved. Our model has two elements (blockchain and MAS) that allow the relationships, logic, and understanding of an object to be represented.

4.2 Literature survey

Büth et al. [2] developed a hybrid modeling technique applying agent-based simulation to an industrial setting and discrete event simulations. The discrete section contains components like time and occurrences, passive objects, and causing periodicity, while an agent-based section aids based on a particular scenario for modeling the surroundings, active entities, and particular triggering.

Artificial Intelligence and Industry 4.0. https://doi.org/10.1016/B978-0-323-88468-6.00009-7

Musli et al. [3] analyzed and contrasted possible ways to realize self-adaptation capabilities in cyber-physical production systems (CPPS) based on architectures, multiagent bases, and self-organizing. Their comprehensive analysis shows that MAS is subject to at least three cyber-physical system (CPS) technology layers.

Dorri et al. [4] described several MAS application areas, one of which is a smart network, which shows MAS to be a feasible approach for modeling improved reliability.

Rocha et al. [5] proposed a multiagent-based flexible edge computing architecture to enable agencies to coordinate execution, evaluate quality problems appropriately, and triggering a rehabilitation mechanisms and strategies if appropriate.

Suganuma et al. [6], with the aid of a multiagent's architecture, addressed the problem of establishing mobility between edges and cloud computing. This method should lead to better environmental flexibility and user-oriented, targeted elements such as "loading and job processing" between the boundary and the cloud.

Filz et al. [7] evaluated a matrix-structured production system using an agent-based simulation for material supply chains. Factors of simulations such as necessary advanced techniques per station, the transportation of manufactured goods, autonomous guided automobiles, or supply locations needed to be shown resulting to a simulated of unpredictable customized for resources. The authors underline the agent-based simulations to achieve adaptability and dynamic. A component for the authors proposed analysis of the generated information to improve the design process for the materials.

Cruz Salazar et al. [8] reviewed how MAS tendencies may be used to "enable CPPS migrations." In the case of fixed programming language, the authors expected that MAS would be quickly applicable.

Karnouskos et al. [9] identified four principal obstacles for the use of MAS inside the framework of CPPS as "structures, interfaces, statistics and distributed intelligence." In addition, because of the cyber-physical connection of CPS organizations and their penetration of the various levels of this pyramid, they emphasized the need to overthink the previous traditional hierarchy in MAS.

According to Gorodetsky et al. [10], MAS trends to be a natural element of CPS for operational administration. They perceived a clear link between the creation of CP-MAS that may bear out tasks within the CPS independently, supporting the integration of business twins, and digital revolution objects for simulation studies or reconfiguration in the smart city environment with the connectedness of the numerous mechanisms.

With respect to the digital twin idea, Pires et al. [11] compared the corresponding capabilities for the modeling, data collection, analysis, and processes phases necessary which assign the "agent-based simulation" inside the interpretation stage. In this respect, any logic is named for the implementation of a potential technology platform. The authors also see the benefits of applying digital "twins" to design and manufacturing management, performance monitoring, preventative analysis and regular maintenance, and product development, as well as to "real-time surveillance," actual information, computing positioning of different operational circumstances, and social benefit.

Pantoja et al. [12] concentrate on use of MAS in all-embedded computation and technologies rather than a fully intelligent plant. MAS is seen as a way of creating and managing environmental knowledge in open, IoT-based ecosystems.

4.3 What is industry 4.0

Digital transformation is the key to the 21st century, with continuous advancement in different areas of technology that affects the development and distribution of products [13]. Three industrial revolutions have already taken place and subsequent revolutions have led to fundamental changes in industrial processes. For example, the Third Industrial Revolution, which began in the 1970s, focused on growing robotics and the use of electronics and technologies to increase regulation of industrial production processes.

Newer developments such as the IoT, connectivity, big data, cloud computing and embedded systems, and Web mobility [14] are key innovations in what is being called Industry 4.0, a concept first introduced in Germany in 2011 [15]. The main concept behind Industry 4.0 is to create new values for the sector by designing new business models and tackling different social challenges. Industry 4.0 has shaped customer attitudes regarding the development, personalization, and delivery of goods [16].

Smart plants are based on advanced digital infrastructure manufacturing technologies and must be capable of collecting, producing, and propagating intelligence [17]. Industry 4.0 is based on the idea of smart factories and the future of all manufacturing processes. Manufacturing systems are transitioning from independent to interactive and, accessible to dynamic supply chain. One of the concepts of Industry 4.0 is the cyber-physical system (CPS), which controls advanced technologies between their physical and computational characteristics. A social internet design is established when CPS and a consumer element are linked. This method covers service development and efficient customer functions and introduces an entirely new parameter: the customer. The incorporation of the customer component into this architecture enables personalized delivery of goods and services to businesses to respond to consumer needs.

4.3.1 Virtual Sensors in Intelligent industry

There is an ongoing requirement to gain knowledge to attain the capability of a SME (small to medium enterprise). The pyramids of information and accessibility, called intelligence pyramids, illustrate this method adequately. Ackoff divides the way information is generated into four categories: data, intelligence, wisdom, and real understanding. The process of generating data is begun by the internal quantity of data that a firm can supply for its customers, and it is necessary to look at external sources to reinforce their consumer approximations. External data may be used to optimize procedures and goods as well as to warn companies that they are searching for their customers or other external considerations that can aid decision-making [18].

The value of sensory technologies for the SME and in the adoption and progress of Industry 4.0 is unclear. These physical instruments are also a way to capture internal environmental data. However, sensors can also serve as windows to the outer world, which cannot be overlooked by organizations in our view.

The construction process model shares the virtual role and a detailed analysis of various influences that can be inspired by earlier work [19]. It provides a responsibility of the project of an online community's sensor using data from online social networking platforms to help SMEs and organizations fully comprehend compliance and selection characteristics.

By integrating alternative sources with industry operations, companies may identify and evaluate the appreciations of their consumers by producing their products and/or services wisely and can function as an "external windows."

4.3.2 Intelligent model virtual sensor description

Fig. 4.1 shows our proposal for a standalone virtual sensor, a bridge between industrial and foreign sources of information. The external data collected allow an industrial information layer to increase the "intelligence" connected to smart processes that may be accessed via Internet or other channels of marketing, thus maintaining a continual transmission of communication.

The nodes of the sensor guarantee that more than information-gathering operations may be carried out. To comprehend and establish correlations and links

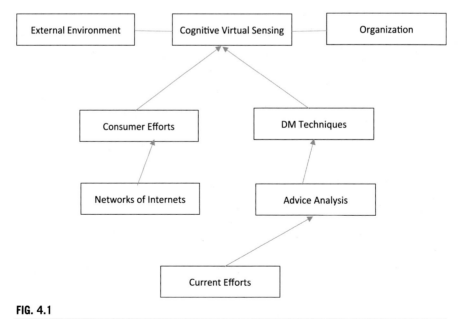

FIG. 4.1

Structure of intelligent virtual sensor model.

between them, the different data types gathered by the sensors might be received from numerous sources. The intellectual feature of sensors will address certain questions for the company that may help to produce the goods that customers want. These questions address demographic properties, the clustering of consumers based on custom metrics, customer preferences and opinions, and other types of input. For example: "What social groups are involved in those goods?" "What age demographic buys more products within a population group?" "What sort of items have consumers purchased?" "Why are profits up or down?" "What do our goods mean or what do customers say?" "What sort of products do our consumers want?"

Clustering capability is one of the components of smart virtual sensors. A variety of actions (represented by modules), adaptable to the demands of industry processes and the goals of an organization, can be taken by a customer or a consumer community.

We selected a range of "default" modules to create a layer of information capable of enhancing Industry 4.0. The modules are as follows [20].

4.3.2.1 Clustering

This module is responsible for clusters of customers with unique corporate features to connect separate customer groups with various goods.

4.3.2.2 Opinion analysis

This is the external input that can lead to manufacturing and decision-making processes. To consider consumer views and interests about certain goods or services, this module is responsible for producing information focused on an individual or community of customers. This information is very useful to companies because it can strengthen the connection between them and their customers' demands. Such information could be used to truly know customers and to account for their views in the decision-making process. The integration of this generated information into decision-making processes makes it easier for a community of customers to develop customized goods that match their needs and, in particular, for a company to produce and execute products that exceed customer standards, thus improving the relationship between the company and its customers.

4.3.2.3 Trends

This module forecasts emerging developments to allow companies to become proactive in the market and fulfill their customer expectations in a limited period. The estimation of patterns and consumer needs includes the uncertainty element, since outcomes cannot be tested, but the quantity and accuracy of previous information may be minimized. To achieve comprehensive prediction, several sources of information must be created and examined, such as knowledge sharing and competitive analysis. Consumer information is derived from an analysis performed in the perception analysis module, and business knowledge is gained from continuous market analysis and from industry stakeholders. The forecast process has a major effect on companies and can be attributed to success or loss because it impacts the success of goods, which I might quantify by the wealth of customers, for example, and the number of purchases, which encourages organizations to play an active position in the market rather than respond to changes. One explanation for this paradigm is that organizations should individualize their goods in compliance with their customers' desires. This can only be achieved after obtaining data from customers' information processes. This knowledge layer is targeted at promoting policy-making, allowing firms to take more choice and be engaged in generating their items in accordance with their device's (or market) demands.

4.3.3 Industry 4.0 vertical integration

Industry 4.0 functionality allows the development of smart, adaptable, and variable factories that use smart virtual sensors. Image signal processing may become an active component of IT, that can be used in analysis and database management to combine the information into a digital IT solution. This IT solution can fulfil the demand by manipulating the information in such a manner that may be part of the business decision.

4.4 What is blockchain

Blockchain is a shared, verified, and unchangeable ledger in which information may be accumulated in sequences. These pieces are then appended to the block list and replayed throughout the dispersed network. Several connections to themselves and the preceding block are used in each block, thus the blockchain name. The references are based on the implementation of cryptographic hatch functions for each block and guarantee the integrity of the block due to its one-way operation. Blockchain uses SHA-256 hatch functions for its original version [21]. These hash functions also include a member-speaking agreement protocol, which makes the entire system stable but highly inefficient, also known as mining. Advanced hardware, called mining platforms, are equipped for the handling of transactions. These mining platforms speed up the multiple hash calculation that each block, known as a proof of work (PoW), would be called upon to validate and receive the respective incentives.

Everybody may modify their blockchain copy, but it is very difficult to get someone to work on it. The rule is that the longest chain is often followed and is worked on by more people. If the distributed blockchain is to be changed, one requires more computing resources than the other nodes. The blockchain attacker must monitor at least more than 50% of the machine mining force so that a longer period can be generated.

In recent years, several blockchain ventures have implemented smart contracts, far from only including a registry of transactions. These can be called spread programs running on a node network. The nodes have no self-confidence or confidence in one another. The immutability of the blockchain brings confidence and renders smart contracts incredibly vulnerable to exploitation. In the case of parties deciding to trade money, the use of smart contracts is best suited if certain conditions at the moment of the exchange are verifiable [22].

An architecturally designed approach consists of a network of collaborating node systems that store transaction data and processors that authenticate the incorporation of the data. It uses a confidential evidence system where all transactions in the blockchain have been previously verified and deposited on numerous decentralized nodes worldwide. This is a trustworthy framework for public accounting. Any trusted authorities or intermediate private entities, including banks or brokerages in a commercial facility, engaging in each transaction is excluded.

4.5 **Blockchain as the industry engine 4.0**

Bitcoin's debut introduced blockchain technology to the world. To secure and broadly disseminate a database, blockchain technology integrates networking among peers, an encryption algorithm, and a decentration mechanism, without the requirement for a reliable intermediary [23]. Anyone can link and verify the data in a blockchain, but no one can change it. Blockchain runs on a secure information state network, providing an environment in which it is easy to share values securely between two or more individuals who may not be familiar with one another. Components of blockchain include smart agreements and digital contracts, which are used to guarantee the fulfillment of contractual terms before transmitting a value. This minimizes the amount of human participation required for contract formation, enforcement, and execution. The rules can be generated in these forms of contract in an easy and fast way.

The value of goods tailormade to fulfill market needs is increasing with Industry 4.0. This demand for high-value products places new requirements on the supply chain, giving companies a better view of suppliers and making them smarter, more open, and effective. Using the blockchain approach, clear and unchanging archives of the supply chain can be maintained, and data exchange between vendors and businesses becomes easier. Blockchain also makes it easy for a company to evaluate potential suppliers and track goods and monitor their development across various points in the supply process. Blockchain addresses data safety concerns as well, which are paramount in Industry 4.0, by storing all data safely, securely, and permanently.

The based on distributed cryptocurrency structure offers a pair-to-peer network to reduce disruption because there is no core issue where failures would occur. It is also evident that the ideas associated with blockchain, build importance for the integration of I4.0. Key principles such as honesty, openness, protection and data durability makes blockchain more important in Industry 4.0.

Smart contracts are automated agreements that guarantee that contract conditions are fulfilled prior to sharing any value, minimizing human interference in the formation, execution, and compliance of a contract. Such contracts allow rules to be generated more easily and simply, as the contracts are described in the block source code. PoW is a prerequisite to overcome an advanced and often named mathematical programming problem to avoid attacks and other network abuses when carrying out blockchain transactions. Data for all recent network operations are stored in each block, holding all transactions in blockchain records [24].

Everybody can install data and check data in a blockchain, but no one can modify the data. Blockchain operates on a decentralized system and there must be a consensus in this network on the data situation, providing an atmosphere where two or more bodies can freely share their values efficiently without being aware of each other.

There are three types of blockchain: public, proprietary, and hybrid. A public blockchain allows anyone to participate without restriction. A private blockchain is one in which only authorized users can access the data. A hybrid blockchain is a combination of both public and private blockchain solutions, which ensures that each transaction remains secret but is also monitored by an unchangeable record [25].

Industry 4.0 is generating integrated logistics requirements, allowing industry to provide manufacturers with a smart, open, and productive view while addressing customer needs more easily. By implementing blockchain, open and immutable data from the supply chain can be retained and the exchange of data between manufacturers and companies becomes easier and enables the registration and progress of goods across various points of the supply network to be monitored. Industry 4.0 needs to defend against misuse of knowledge, for example, about new technologies and intellectual property. Blockchain guarantees the required degree of protection because the stored data is permanent. The decentralized and distributed blockchain structure allows the establishment of a peer-to-peer network that mitigates the risks of vulnerabilities because there can be no primary fault. With all this, blockchain technology can be used to facilitate the development, with accountability, confidence, and dignity, of autonomous organizations that can be independent and share value between departmental organizations and international entities.

4.6 Blockchain-based conventional networks

Blockchain technology will turn many existing conventional networks into more stable, distributive, open, and collaborative systems. It also has potential for a much greater effect, that is, transforming the way businesses can share value without relying on a central infrastructure. The Blockchain Platform for Industrial Internet of Things (BPIIoT) is one such program.

This technology is aimed at strengthening cloud-based architectural operational capabilities or cloud-based manufacturing (CBM). CBM is a form of service providing consumers with the ability to select and manufacture resources. BPIIoT (Blockchain-Based Platform for Industrial IoT) consists of a smart agreement, based on a blockchain network, between healthcare professionals and production resources. It enhances machinery output, servicing, and monitoring on demand.

Another platform of a distributed manufacturing model focused on multiparticipations blockchain, called e-chain, seeks to address challenges such as lack of flexibility and collaboration to enhance value between organizations. This system consists of several smaller and more independent blockchains that constitute a joint distribution channel. Any blockchain holds its own past records in this network, and the digitalization of properties takes place here. In addition, the structure of this network allows participants to perform multiple transactions [26].

Another element of this concept is the value distribution channel. This channel ensures supervised, authorized, and accessible resources and values among blockchains in the real world that may be transferred to other cryptographic techniques.

Blockchain, which utilizes supply chain management to enhance product traceability. The aim of this platform is to provide a trustworthy traceability platform that involves many organizations, focused on a globally dispersed private blockchain. Blockchain provides straightforward traceability data that is tamper-resistant, increases the volume of information, and automates regulatory checking.

4.7 **Decision making concept**

Decisions are focused upon the ability to accomplish previously fixed targets, which are largely evidenced by their commitment to individual and collective interests and can be separated into subdecisions and tasks. A certain outcome shall be attained after a decision is taken, and it would be expected to decide whether the purpose has been accomplished.

The assessment process is a daunting activity to conduct, as it uses the outcomes to give a positive or negative connotation to the decisions before they are taken during the decision-making process. There is an ambiguity about the feasibility of a decision, as the decision process will be affected by the factors used in the process itself. However, there are some factors that make the process more successful, such as access to full and reliable information. The outcome and the planned performance depend on the operation or condition of the process in progress [27]. The decision-making technique is a very complicated and challenging job since each aspect has a different meaning from the final decision. The process of decision-making must be included. This significance can be illustrated by considering several factors such as decision-making categories, prior experience, and views of individual circumstances and personal interests [28].

4.8 **What is ambient intelligence**

A hypothetical outline of AmI, published in 2001 by the Information Society Technologies Advisory Group (ISTAG) of the European Union, emphasizes greater user-friendliness, more efficient services support, user-empowerment, and support for human interactions. AmI technology encompasses many aspects of our existence, from furnishings to vehicles and textiles [29].

To accomplish the intelligence process, expertise needs to be given throughout the world (in smart households, clever cities or even clever plants), and without artificial intelligence, that idea is not practicable and possibilities economic opportunities that AmI has been described as jobs for the next generation of businessmen and industrialists.

It is difficult for professionals to recognize that their capacity must be coupled with other technologies (computerization, connection, AI, digital intelligence, language generation, image processing, smart robotics).

4.8.1 Industry as a smart environment

A smart environment is an ecosystem of communicating objects that have the capacity to coordinate, provide services, and control complicated information (e.g., sensors, actuators, information devices, and other network-connected devices). The physical smart environment is intelligent in design and benefits from the interface of various devices and computer systems, aimed at enhancing services to people. When we start looking at businesses as smart environments, the vision is still true.

To create a genuine smart environment, like AmI, three technological fields must converge: omnipresent computing, intelligent systems, and context awareness. The smooth incorporation of smart devices into daily life must be easy, natural, and nonintrusive. Flexible computations enable a smooth interaction between ambient systems and their users. Intelligent apps include tools like data mining, statistical analysis, machine learning, and optimization methodologies. Contextual knowledge requires environmental adaptation and is crucial in the perception and management of contexts for a technology that changes demands (uses sensors).

4.8.2 Sensors

Sensors are essential in fulfilling the needs of smart environments, particularly in the sense of environmental perception. AmI or smart environments also come together with intelligent and environmentally integrated sensors designed to monitor or quantify movement, light, temperature, moisture, and other circumstances. The position of sensing devices in an EMS is important for setting time limits for location and object connections, assisting in meaning determination, identifying behaviors, experiences, and relationships, and recognizing behavioral trends.

4.8.3 Decision-making in ambient intelligence

As previously mentioned, the idea of AmI was presented by ISTAG in 2001 based on the theory that human beings would be surrounded at some stage by intelligent, computational, and networking interfaces [30].

The shear force on the demands of individuals with the goal of creating user-friendly and accessible services to meet customers and people by applying stress and pressure in the field of ubiquitous computing [31], ubiquitous, constructive, ambient, embedded machinery and telecommunication technologies.

AmI manifests in daily life in commonplace things like clothing, furnishings, and cars [32]. Input/output technologies and intelligent applications, such as smart devices, electrosensor systems, micro-electromechanical structures, technologies incorporated, and pervasive connectivity are features of AmI.

AmI can generate big data that may be used in the challenging decision-making process. Making a decision based on big data is always challenging. As such, AmI is used to make decisions that are tailored to the factors of a given environment. Decisions are dependent on the data that comes from various sources within the AmI, such as a computer network or IoT sensors, which is then processed to generate concrete information and environmental awareness, thereby allowing a smoother decision-making mechanism.

4.8.4 Better decision-making in industry

Industry 4.0 ensures that circumstances among physical and digital systems are approximated, synchronized, and specified to increase efficiency of internal and external environments in an industrial ecosystem. Technologies like sensors, equipment, and IT structures enable an organization, both internally and externally to become more aware of what happens in market systems and to build possibilities for improved decision-making [33]. These developments in organizations are not only mirrored in technology, but also reform fragments such as transformation from centralized systems to data generation and decision-making decentralization [34].

With numerous sources of data, such as manufacturing equipment and consumer service systems allowing for improved decision making with access to more knowledge quantity and quality [35].

Decisions may be made in many different situations, whether they be political, logistical, or organizational, and are relevant to many stages of the manufacturing process and even business decisions [36] Organizations make decisions to enhance their procedures, increase their versatility, improve productivity, customize their products, minimize costs, and simplify processes [37].

In the future, decision-making mechanisms will be a central factor in enterprise with increasing numbers of individuals engaging in the process, transforming how people view and communicate with the various sectors [38].

4.9 **Multiagent systems**

MAS networks are smart systems in an unregulated world [39]. Considering the wide variety of problems these systems can solve, this infrastructure has been investigated to administer broadly distributed systems in industrial applications [40].

An agent is any person that detects and acts on its environment, constantly executing its task, with strong autonomy and coexisting with other individuals and processes in an evolving environment [41]. An MAS is a network of agents that function in a variety of operational and development contexts for resolving issues that cannot be addressed by one agent [42].

An MAS consists of several agents who interact through exchange of information and bargaining in a complicated, dynamic, and usually integrated environment to achieve a certain purpose. Those systems have a selection of main feature elements: sovereignty (the capacity of agents to carry out their own operations), complexity, and adaptation (environmentally compatible agents) [43].

The center of an MAS is its relationships since it displays all its key elements and facilitates the collaboration and organization of activities. The management of these operations is important to ensure, in the sense of decentralization functions utilizing a variety of delivery bodies, that the MAS offers an alternative to the creation of decision-making systems [44].

Therefore, stability, scalability, autonomy, and reduction of problem complexity are the key advantages of an MAS. The features of these processes have been placed in a position where dynamics, instability, and distribution are especially beneficial.

4.10 Blockchain for the representation of knowledge from multiagent systems

Organizations have defined a series of business processes between them, usually in a supply chain, in each manufacturing setting. The supply chain may experience increased demand and a stronger need for intelligent communication between entities. There are a variety of dependencies between one or more individuals in the supply chain that share value among them, making it difficult to grasp the transactions.

Our methodology attempts to address dependence and enhance decision-making in a world created by a variety of different stakeholders and diverse priorities. Our strategy, presented in Fig. 4.3, starts by suggesting that we have a network of entities with several connections between them, in which the collection of such links create a repeating mechanism, producing connections, thought, and information within a network of entities. Therefore, two elements (MAS and blockchain) form our solution.

An agent is something perceived by the sensors and effectors of its environment. An MAS is a community of agents who cooperate together to tackle issues that a single agent cannot handle alone. These bare is a special kind of intelligent system where autonomous agents exist in an unregulated environment or are constantly aware of the situation which makes this approach mostly suitable for the modeling of multiagent distributed problems, so one component of us relies on the basic activities of an entity are accomplished in this element. The interaction layer and the logic layer form two layers.

The interaction layer refers to the procedures defined for coordination, dependence, ties, and other forms of relationship between multiple agencies. This layer is the input point for the operations.

4.10.1 Multiagent systems in industry

MAS, particularly those relating to object modeling, methodologies, and organizations, have many industrial applications. They can be used in manufacturing connectivity, supplier management, production scheduling, planning and holonic manufacturing systems, and a range of sophisticated systems to handle technological issues, which can all offer a greater competitive edge: mobility, scalability, reconfigurability and efficiency.

MAS are also valuable in engineering, health, and industry. They are commonly used in manufacturing, modeling, intelligence, and communications [45]. The early 1990s saw the first industrial ventures for MAS in the automotive and transportations sectors. For example, Metamorphic, a multiagency integration architecture, and Production 2000+, a 5-year, multimedia device built in Daimler Crystal's factory with a view to boosting cylinder head production. Another example is an agent-based scheme that requires ships to carry out transcontinental oil transportation in a very large fleet of crude carriers to be carried out in real time [46].

MAS have applications in production forecasting, supply chain, distribution, power and smart grids, houses, domestic automation, and much more. In the supply chain, a prototype has been developed to assess the possibility of establishing a multilevel MAS decision-making mechanism [47]. The areas in which this system is used must be addressed to competition in the industrial sector, while in real-life applications of I4.0 are powered by internet-based networks, clever manufacturing, and service advancements.

Organizations in diverse sectors have created a process alliance that usually works in a supply chain to bring essential benefits for their companies and to adapt to an increasingly challenging customer. In the era of Industry 4.0, businesses must respond more rapidly to customized customer requirements. This requirement will create more demands in the supply chain and make it necessary for producers and customers to communicate.

It becomes harder to make fast and simple decisions on which other organization to rely for both resources and/or goods. Several dependencies between two or more firms will often be set up exchanging their values in a supply chain, since the processes that create this activity can link and because a broad notion of all interactions from a large supply chain is difficult to obtain [48]. The biggest challenge is that there is no substantial way to know which other companies to rely on in a world in which enterprises must make quick and accurate decisions, causing a role in which multiple deficits will contribute to manufacturing processes.

Our methodology is intended to resolve dependence and enhance the mechanism of decision-making in a multientity context with various reasons and targets. Our proposal aims to create a model enabling any business to choose more intelligently who should rely on its production parameters to improve decision-making.

4.10.2 Model presentation

Fig. 4.2 presents our framework, which aims to be a route of access for the depiction of interentities in the industrial environment, to help identify whether the other route is selected to communicate.

To reach their market objectives, blockchain starts to include a collection of industries as a network of organizations with partnerships that reflects the various processes generated by the industries. A repetitive mechanism exists with any individual within this network in which connections are formed first with another entity and then a justification process is carried out using the data obtained from the interaction in order to use something in future procedures. This logic is expressed in the model by two distinct elements: blockchain and MAS, to represent information, thought, and interactions.

We adopt private blockchain in this model to allow access to data, regardless of its form. Using a blockchain means that the data remain real, clear, permanent, and secure. The blockchain consists of a public and private profile of the organization that guarantees that an entity has network information, information about other entities, and its own profile. This data would be made more reliable over time and by the experiences of the company [49].

A private profile requires access to authorization such that the information is only available to the person it refers to. In terms of the public image, the information about the company is largely available and the following variables are presented in the network and stored [50]:

* **Inputs**—These reflect the specifications of the organization, namely, whether it is essential to other network entities in order, for example, raw materials, to satisfy its business processes.

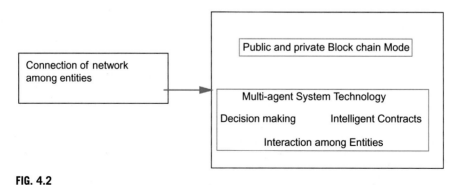

FIG. 4.2

Integrated structure of MAS-based blockchain.

- **Outputs**—This is what the network and its organizations must sell. Ultimately, this reflects another entity's feedback and may differ from a good to a service.
- **Credibility**—This refers to the connectivity and stability of each organization. Even the individual they refer to modify that value by just viewing the value of the biography so that other people may see our reputations.

The logic module and requirements module comprise the logic layer. The decision module helps the entity to get feedback while picking the preferred entity. The smart contract module contains several rules to ensure safe trading with the preferred party and that the circumstances are consistent with the preferred party.

As the interactive layer of the MAS produces data that generates or updates information at the blockchain, the principal elements of the proposed model remain in constant connection. The data obtained is used by the MAS logic layer to generate smart contracts and enhance decision-making. We think this model will resolve dynamic dependency between companies and provide a solution to allow companies to better determine which organization chooses to respond to their corporate requirements.

4.11 **Comparison between the existing model and our proposed model**

If the objective is to make strategic decisions based on simulations, one must consider how complicated and realistic the computation is. An MAS may show complicated situations, but it naturally demands the knowledge engineer to deeply examine the surroundings and to begin producing potential distinct agents. The qualitative researcher and the stakeholders in question must in this situation agree on the specifics of such an MAS simulation and make certain generalized assumptions about an agent's behavior without sacrificing believability or real-world depiction of simulated results. Table 4.1 lists the roles and tasks before evaluation of Industry 4.0, and Table 4.2 lists the agents and their behaviors after the evolution of Industry 4.0.

Table 4.1 Roles and tasks before evaluation of industry 4.0.

Sr. no.	Roles	Description	Tasks
1	Production incharge	Specifies the coordination and frameworks of staff. It might fluctuate in certain personnel in manufacturing	Send configuration information, collect manufacturing material from the workpiece, cut work, grinders work, color scheme, assemble work

Continued

Table 4.1 Roles and tasks before evaluation of industry 4.0—cont'd

Sr. no.	Roles	Description	Tasks
2	Order configuration	Specifies the ordering system for the configuration and deduction of the materials required for the suitable range	Item selection, commodity ordering, transmit information settings
3	Customer roles	Representing consumer groups that acquire various types of items	Product configuration, ordering of product
4	Material management	The kinds of materials required to manufacture bicycle frameworks are represented	Collect materials, cutting services, grinder jobs, coloring jobs, assembling work
5	Driver-less transport system or automated guided vehicle	It signifies a transportation in the intelligent plant transporting the required stuff from and to various warehouses and equipment	Collect the products for manufacturing tasks, cutting work, molding work, repainting work, housekeeping duties
6	Autonomous storage system	The technological logistical system independently symbolizes the storing of products in storage technologies	Collect resources for manufacturing, store the complete outcome
7	Cutting machine	Identifies the machine(s) and the slicing tasks performed	Slicing tasks
8	Grinding machine	Describes the machine(s) for sharpening and work performed	Sharpening tasks
9	Painting machine	Describes the machine(s) for repainting and the color work performed	Coloring tasks
10	Assembly machine	The assembling facility and the installation works carried out	Installation tasks
11	Final product	Specifies the finished product at the end of a framing	Manufacturing configuration, ordering, storing finishing products
12	Truck	Indicates the vehicles for the conveyance to the intelligent plant of original material	Material delivery
13	Shipping truck	It is the transporter used to deliver finished goods from the intelligent plant to the consumer	Commodities delivery

Table 4.2 Agents and their behavior after the evaluation of industry 4.0.

No.	Agent	Belief	Desire	Intention		
				Initialization of agent behavior	Action performed by agent	Unnecessary task removal by agent
1	Main	The actual status based on a transition from state	Simulations of the whole procedure	Initializing collaboration of the various agents	–	–
2	Order defined by choice of configurator variables: woodColor, woodNoColor, aluColor, aluNoColor, pvcColor or pcvNoColor	Technique of simulating of client order	Randomized picking of variables listed in schedule assumptions	–	–	
3	Material defined by choice of variables: isWood, isAlu, isPVC, isGlass, isAttechment	Manufacturing simulations of wooden aluminum PVC profiles, interfaces, economic assistance material or plastic packaging for materials profile pictures, listings, glass board	Randomized picking of parameters listed in section assumptions	–	–	
4	Automated guided vehicle	Automatic storage definition—Differences among restricted or available statuses	Multimodal transport simulator from inside plant	Agent retains a valuable service for transportation simulations	In the manufacturing transportation	Agent releases resource → end of transport

Continued

Table 4.2 Agents and their behavior after the evaluation of industry 4.0—cont'd

No.	Agent	Belief	Desire	Intention			
				Initialization of agent behavior	Action performed by agent	Unnecessary task removal by agent	
5	Cutting machine	ResourcePool definitions—Differentiate among restricted or free status	Cutting process simulation	Agent reserves free resource → start of cutting process	Agent reserves resource for defined time interval → cutting time	Agent releases resource → end of cutting process	
6	Grinding machine	ResourcePool definitions—Differentiate among restricted or free status	Grinding process simulation	Agent reserves free resource → start of grinding process	Agent reserves resource for defined time interval → grinding time	Agent releases resource → end of grinding process	
7	Painting machine	ResourcePool definitions—Differentiate among restricted or free status	Painting process simulation	Agent reserves free resource → start of painting process	Agent reserves resource for defined time interval → painting time	Agent releases resource → end of the painting process	
8	Assembly machine	ResourcePool definitions—Differentiate among restricted or free status	Assembly process simulation	Agent reserves free resource → start of assembly process	Agent reserves resource for defined time interval → assembly time	Agent releases resource → end of the assembly process	
9	Truck	Agent loading and unloading time intervals	Simulation of material unloading of the transportation vehicle	Start material delivery	Material unloading	Stop material delivery	
10	Shipping truck	Agent loading and unloading time intervals	product delivery simulation	Start product delivery	Truck loading	End of product delivery	

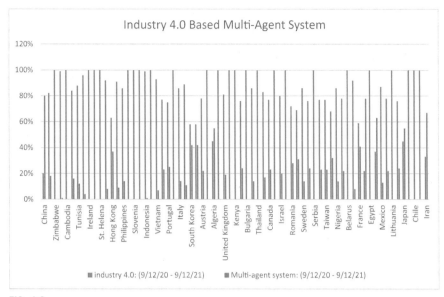

FIG. 4.3

Industry 4.0-based multiagent system.

4.12 Country-wise comparison of multiagent-based industry 4.0

Fig. 4.3 presents a country-wise comparison of multiagent-based Industry 4.0.

Fig. 4.3 shows China, Japan, and Iran country use industry 4.0 maximum dependent on multiagent system but rest of the countries approaches to achieving this technology. Most of the countries have benefits to adopt to this technology to produced maximum commodities in required time. And, organization earn a lot of profits for this adoption's minimum wastage of raw material.

4.13 Conclusion and future work

This chapter discussed a model approach to enhance interaction in an Industry 4.0 environment and to increase cooperation among industries and organizations. With the launch of Industry 4.0, industries began developing their manufacturing technology, profiting from the quantity of data created and digitizing production pipelines, and enhancing their technology and industrial processes. The innovations and developments brought about by Industry 4.0 urge industries to adapt rapidly to these changes so that they may continue to achieve commercial profits and stay competitive. Industry 4.0 provides new ways for industries to create value while controlling their production more effectively, producing better products faster, and inventing goods that match customer demands and expectations. In addition to addressing the essential demands of customers, industries will have to tackle new problems arising from ever-growing trends. As shown from the literature analysis, one of these problems is the development of CPS. In Industry 4.0, collaboration implies that resources, expertise, and information can be shared, and that industries can readily and effectively communicate, so they can continue to prosper during this period of the Fourth Industrial Revolution.

The future work in this field will aid decision-making processes by using MAS. To improve the definition of all actors and structure explicit activities that will lead to effective decisions, an MAS requires additional developments. In the future, creating decision-making skills within MAS is an important objective. A framework or algorithm such as the Markov decision-making procedure or the fluid inference system are opportunities to be explored to understand better the influence that they will have on MAS models. By defining the activities of the decision-making agent (DMA), the MAS progresses to full state and pushes forward the general model.

References

[1] H.S. Kang, J.Y. Lee, S. Choi, H. Kim, J.H. Park, J.Y. Son, B.H. Kim, S.D. Noh, Smart manufacturing: past research, present findings, and future directions, Int. J. Precis. Eng. Manuf. 3 (1) (2016) 111–128.

[2] L. Büth, N. Broderius, C. Herrmann, S. Thiede, Introducing agent-based simulation of manufacturing systems to industrial discrete-event simulation tools, in: Proceedings of the 2017 IEEE 15th International Conference on Industrial Informatics (INDIN), Emden, Germany, 24–26 July 2017, IEEE, Piscataway, NJ, 2017, pp. 1141–1146.

[3] A. Musli, J. Musli, D. Weyns, T. Bures, H. Muccini, M. Sharaf, Patterns for self-adaptation in cyber–physical systems, in: S. Biffl, A. Lüder, D. Gerhard (Eds.), Multi-Disciplinary Engineering for Cyber–Physical Production Systems, Springer International Publishing, Cham, Switzerland, 2017, pp. 331–368.

[4] A. Dorri, S.S. Kanhere, R. Jurdak, Multi-agent systems: a survey, IEEE Access 6 (2018) 28573–28593.

[5] A.D. Rocha, R. Silva Peres, J. Barata, J. Barbosa, P. Leitao, Improvement of multistage quality control through the integration of decision modeling and cyber–physical production systems, in: Proceedings of the 2018 International Conference on Intelligent Systems (IS), Funchal-Madeira, Portugal, 25–27 September 2018, IEEE, Piscataway, NJ, 2018, pp. 479–484.

[6] T. Suganuma, T. Oide, S. Kitagami, K. Sugawara, N. Shiratori, Multiagent-based flexible edge computing architecture for IoT, IEEE Netw. 32 (2018) 16–23 (CrossRef).

[7] M.A. Filz, C. Herrmann, S. Thiede, Simulation-based data analysis to support the planning of flexible manufacturing systems, in: M. Putz, A. Schlegel (Eds.), Simulation in Produktion und Logistik 2019, Wissenschaftliche Scripten, Auerbach, Germany, 2019, pp. 413–422.

[8] L.A. Cruz Salazar, D. Ryashentseva, A. Lüder, B. Vogel-Heuser, Cyber–physical production systems architecture based on multi-agent's design pattern—comparison of selected approaches mapping four agent patterns, Int. J. Adv. Manuf. Technol. 105 (2019) 4005–4034.

[9] S. Karnouskos, L. Ribeiro, P. Leitao, A. Luder, B. Vogel-Heuser, Key directions for industrial agent based cyber–physical production systems, in: Proceedings of the 2019 IEEE International Conference on Industrial Cyber Physical Systems (ICPS), Taipei, Taiwan, 6–9 May 2019, IEEE, Piscataway, NJ, 2019, pp. 17–22.

[10] V.I. Gorodetsky, S.S. Kozhevnikov, D. Novichkov, P.O. Skobelev, The framework for designing autonomous cyber–physical multi-agent systems for adaptive resource management, in: Industrial Applications of Holonic and Multi-Agent Systems, Springer Nature Switzerland AG, 2019, pp. 52–64. Piscataway, NJ.

[11] F. Pires, A. Cachada, J. Barbosa, A.P. Moreira, P. Leitao, Digital twin in industry 4.0: technologies, applications and challenges, in: Proceedings of the 2019 IEEE 17th International Conference on Industrial Informatics (INDIN), Helsinki-Espoo, Finland, 22–25 July 2019, IEEE, Piscataway, NJ, 2019, pp. 721–726.

[12] C.E. Pantoja, J. Viterbo, A.E.F. Seghrouchni, From thing to smart thing: towards an architecture for agent-based AmI systems, in: Agents and Multi-agent Systems, 2019.

[13] Roland Berger Strategy Consultants, M. Blanchet, T. Rinn, G. de Thieulloy, G. von Thaden, Industry 4.0. The New Industrial Revolution. How Europe Will Succeed, Semantic Scholar, 2014.

[14] S. Wang, J. Wan, D. Li, C. Zhang, Implementing smart factory of industry 4.0: an outlook, Int. J. Distrib. Sens. Netw. 12 (1) (2016) 3159805.

[15] J. Qin, Y. Liu, R. Grosvenor, A categorical framework of manufacturing for industry 4.0 and beyond, in: Procedia CIRP52, 2016, pp. 173–178.

[16] J. Lee, H.A. Kao, S. Yang, Service innovation and smart analytics for industry 4.0 and big data environment, in: Procedia CIRP16, 2014, pp. 3–8.

[17] F. Longo, L. Nicoletti, A. Padovano, Smart operators in industry 4.0: a human centered approach to enhance operators' capabilities and competencies within the new smart factory context, Comput. Ind. Eng. 113 (2017) 144–159.

[18] A. Bahga, V.K. Madisetti, Blockchain platform for industrial internet of things, J. Softw. Eng. Appl. 9 (2016) 533–546.

[19] K.S. Kaswan, J.S. Dhatterwal, Intelligent agent based case base reasoning systems build knowledge representation in Covid-19 analysis of recovery infectious patients, in book entitled, in: Application of AI in COVID 19″ accepted in Springer series: Medical Virology: From Pathogenesis to Disease Control, July 2020, https://doi.org/10.1007/978-981-15-7317-0. ISBN No. 978-981-15-7317-0 (e-Book), 978-981-15-7316-3 (Hard Book).

[20] C.E. Pantoja, J. Viterbo, A.E.F. Seghrouchni, From Thing to Smart Thing: Towards an Architecture for Agent-Based AmI Systems, in: G. Jezic, Y.H.J. Chen-Burger, M. Kusek, R. Šperka, R.J. Howlett, L.C. Jain (Eds.), Agents and Multi-Agent Systems, Springer, Singapore, 2020, pp. 57–67.

[21] F. Shrouf, J. Ordieres, G. Miragliotta, Smart factories in industry 4.0: a review of the concept and of energy management approached in production based on the internet of things paradigm, in: IEEE International Conference on Industrial Engineering and Engineering Management, January 2015, 2014, pp. 697–701.

[22] Deloitte, Industry 4.0. Challenges and Solutions for the Digital Transformation and Use of Exponential Technologies, Deloitte, 2015, pp. 1–30.

[23] K.S. Kaswan, J.S. Dhatterwal, Blockchain of IoT based earthquake alarming system in smart cities, book entitled, in: Integration and Implementation of the Internet of Things Through Cloud Computing, IGI Global, June 2021. Published in. ISBN 13: 9781799869818, ISBN10: 1799869814, EISBN14: 9781799869832.

[24] K. Rabah, Overview of blockchain as the engine of the 4th industrial revolution, Mara Res. J. Bus. Manag. 1 (1) (2016) 125–135.

[25] K.S. Kaswan, J.S. Dhatterwal, Smart grid using IoT, book entitled, in: Integration and Implementation of the Internet of Things Through Cloud Computing, IGI Global, June 2021. Published in. ISBN 13: 9781799869818, ISBN10: 1799869814, EISBN14: 9781799869832.

[26] R. Santos, G. Marreiros, C. Ramos, J. Bulas-Cruz, Argumentative agents for ambient intelligence ubiquitous environments, in: Proceedings of Artificial Intelligence Techniques for Ambient Intelligence. 18th European Conference on Artificial Intelligence, ECAI 2008, 2008.

[27] J.W. Dean, M.P. Sharfman, Does decision process matter? A study of strategic decision-making effectiveness, Acad. Manag. J. 39 (2) (1996) 368–396.

[28] S. Srinivasan, J. Singh, V. Kumar, Multi-agent based decision support system using data mining and case based reasoning, Int. J. Comput. Sci. Issues 8 (4) (2011). No 2. ISSN (Online): 1694–0814.

[29] J. Lee, B. Bagheri, H.A. Kao, A cyber-physical systems architecture for industry 4.0-based manufacturing systems, Manuf. Lett. 3 (2015) 18–23.

[30] I.C. Veronica, G. Mirela, B.D. Maria, Modern approaches in the context of ambient intelligence, Ann. Univ. Oradea Econ. Sci. Ser. 18 (4) (2009) 963–968.

[31] D.J. Cook, J.C. Augusto, V.R. Jakkula, Ambient intelligence: technologies, applications, and opportunities, Pervasive Mob. Comput. 5 (4) (2009) 277–298.

[32] M. Marques, C. Agostinho, G. Zacharewicz, R. Jardim-Goncalves, Decentralized decision support for intelligent manufacturing in industry 4.0, J. Ambient Intell. Smart Environ. 9 (3) (2017) 299–313.

[33] M. Rußmann, M. Lorenz, P. Gerbert, M. Waldner, J. Justus, P. Engel, M. Harnisch, Industry 4.0. The Future of Productivity and Growth in Manufacturing, Boston Consulting, April 2015, pp. 1–5.

[34] T. Dory, P. Waldbuesser, Connected cognitive entity management: new challenges for executive decision-making, in: Proceedings of 6th IEEE Conference on Cognitive Info communications, CogInfoCom 2015, 2016, pp. 235–240.

[35] G. Marreiros, R. Santos, C. Ramos, J. Neves, P. Novais, J. Machado, J. BulasCruz, Ambient intelligence in emotion based ubiquitous decision making, in: Proceedings of the International Joint Conference on Artificial Intelligence (IJCAI 2007)—2nd Workshop on Artificial Intelligence Techniques for Ambient Intelligence (AITAm I 2007), 2007, pp. 86–91.

[36] K. Upasani, M. Bakshi, V. Pandhare, B.K. Lad, Distributed maintenance planning in manufacturing industries, Comput. Ind. Eng. 108 (2017) 1–14.

[37] M. Oprea, Applications of Multi-Agent Systems, Information Technology, Kluwer Academic Publishers, Boston Dordrecht London, 2004, pp. 239–270.

[38] A. Rai, R.J. Kannan, Membrane computing based scalable distributed learning and collaborative decision making for cyber physical systems, in: 2017 IEEE 26th International Conference on Enabling Technologies: Infrastructure for Collaborative Enterprises (WETICE), 2017, pp. 24–27.

[39] Y.S. Eddy, S. Xun, Y.S.F. Eddy, S. Member, H.B. Gooi, S. Member, Multi agent system for distributed management of microgrids, IEEE Trans. Power Syst. 30 (1) (2014) 24–34.

[40] M. Glavic, Agents and Multi-Agent Systems: A Short Introduction for Power Engineers, University of Liege Electrical Engineering and Computer Science Department, Tech. Rep, 2006, pp. 1–21.

[41] J.Y. Zhao, Y.J. Wang, X. Xi, Simulation of steel production logistics system based on multi-agents, Int. J. Simul. Model. 16 (1) (2017) 167–175.

[42] M.K. Adeyeri, K. Mpofu, T. Adenuga Olukorede, Integration of agent technology into manufacturing enterprise: a review and platform for industry 4.0, in: Proceedings of 5th International Conference on Industrial Engineering and Operations Management, IEOM 2015, 2015.

[43] A. Aldea, R. Bañares-Alc'antara, L. Jim'enez, A. Moreno, J. Mart'ınez, D. Ria͠no, The scope of application of multi-agent systems in the process industry: three case studies, Expert Syst. Appl. 26 (1 SPEC.ISS) (2004) 39–47.

[44] J. Himoff, P. Skobelev, M. Wooldridge, MAGENTA technology: multi-agent systems for industrial logistics, in: Proceedings of the Fourth International Joint Conference on Autonomous Agents and Multiagent Systems, AAMAS 2005, February 2016, 2005, pp. 60–66.

[45] P. Leitão, V. Marík, P. Vrba, Past, present, and future of industrial agent applications, IEEE Trans. Ind. Inform. 9 (4) (2013) 2360–2372.

[46] J.E. Hernandez, J. Mula, R. Poler, A.C. Lyons, Collaborative planning in multitier supply chains supported by a negotiation-based mechanism and multi-agent system, Group Decis. Negot. 23 (2) (2014) 235–269.

[47] G. Marreiros, R. Santos, C. Ramos, J. Neves, J. Bulas-Cruz, ABS4GD: a multiagent system that simulates group decision processes considering emotional and argumentative aspects, in: AAAI Spring Symposium Series, 2008, pp. 88–95.

[48] S.A. Abeyratne, R.P. Monfared, Blockchain ready manufacturing supply chain using distributed ledger, Int. J. Res. Eng. Technol. 05 (09) (2016) 1–10.

[49] G. Marreiros, R. Santos, C. Ramos, J. Neves, Context-aware emotion-based model for group decision making, IEEE Intell. Syst. Mag. 25 (2) (2010) 31–39.

[50] Z. Xiong, Y. Zhang, D. Niyato, P. Wang, Z. Han, When mobile blockchain meets edge computing: challenges and applications, IEEE Commun. Mag. (2017) 1–17.

Artificial Intelligence: A tool to resolve thermal behavior issues in disc braking systems

Anant Nemade[a], Samir Telang[a], Vitthal Jumbad[a], Arvind Chel[b], Geetanjali Kaushik[c], and Mustansar Hussain[d]

[a]*Jawaharlal Nehru Engineering College Research Centre, MGM University, Aurangabad, Maharashtra, India,* [b]*Department of Mechanical Engineering, MGM's JNEC Research Centre, MGM University, Aurangabad, Maharashtra, India,* [c]*Department of Civil Engineering, Hi-Tech Institute of Technology, Aurangabad, Maharashtra, India,* [d]*Department of Chemistry and EVSC, New Jersey Institute of Technology, Newark, NJ, United States*

5.1 Introduction

Artificial Intelligence (AI) coordinates with computer science and is concerned with developing computer-operated machines that can complete tasks that typically require human intelligence. With the use and explosion of available data of computing capacity, the world is making rapid developments in AI. The Responsible AI for Social Empowerment (RAISE) 2020 summit virtually hosted by India brought the issues around artificial intelligence to the center of policy discussions. AI-driven solutions can be applied to a variety of fields like health, transport, agriculture, manufacturing, human safety, education, and training. Countries across the world are making efforts to become a part of an AI-led digital economy that is expected to contribute around $15.8 trillion to the global economy in the next 10 years [1]. The RAISE 2020 summit brought together global experts to create a roadmap for responsible AI. They formed an action plan that can help to build replicable models with a strong foundation of ethics built in. More than 38,700 people from 125 countries participated in this summit [2].

5.2 Artificial Intelligence in the automobile sector

AI in automobiles is used to replicate the decision/thinking ability of humans by using precollected data and algorithms [3]. AI finds wide application in the automotive sector such as in supply chain, production, engine diagnosis, postproduction, and

117

so on [4]. At present, AI also has applications in speech, image and video recognition, natural language processing, autonomous objects, conversational agents, prescriptive modeling, augmented modeling, smart automation, advanced simulation, complex analytics, and predictions [5]. Driver assistance, autonomous driving, and driver risk assessment systems are emerging areas in the automobile sector in which AI is being implemented. AI is also presently used for predictive maintenance and in the automotive insurance industry [6]. AI-powered robots are being used to manufacture automobile parts without error [7]. AI helps to develop cars using exoskeletons, which will improve plant efficiency. Companies like Hyundai Motor Company are training exoskeletons for electric charging stations and the sale of vehicles. Exoskeletons will reduce human effort by 70% as well as naturally decrease fatigue. The US-based company OTTO Motors introduced an AI-powered material handling system in their plant [8]. OTTO is also engaged in a self-driving vehicle project based on AI. Material handling systems improve plant efficiency and reduce accidents [9]. Tesla launched its Model S car with autopilot capability, and continuous improvements are being carried out to improve the accuracy and safety of this self-driving car with the help of AI [10]. Companies like Nexyad are developing software to alert drivers to potential risks. This software will compute risk more than 20 times per second and alert the driver of potential accidents ahead of time [11]. In most vehicle service stations, predictive maintenance data is being used for automotive services based on AI. The data suggests the maintenance schedule for the vehicle and determines which parts are due to be replaced (e.g., filters, spark plugs, etc.) [12]. AI will be a market force for a new automobile revolution. Figures released by different experts show expected growth in automotive industries within the next 10 years will be 60% globally [13]. The market share of AI will be on the higher side as more and more automobile companies introduce AI in their forthcoming models [14].

Automation in vehicles is classified as follows: 0 for no automation, 1 for driver assistance, 2 for partial automation, 3 for conditional automation, 4 for high automation, and 5 for full automation. More emphasis is being placed on higher-quality vehicles with better safety systems [15]. This will increase vehicle reliability and customer confidence. In crucial situations, a driver's decision can save their life of the lives of others. However, if the car's safety systems are not working as intended/expected, risk of accident, injury, or even death increases regardless of the driver's actions [16]. This may happen during any mechanical malfunction of different parts of automobile. While overtaking the vehicle if the proper power will not be delivered by internal combustion engine then also accidental risk increases, as in running engine too many thermal processes are taking place [17]. Therefore, the engine control unit (ECU) should alert the driver well in advance. Similarly, if the car's active safety brakes do not operate properly, the driver's actions (e.g., stepping on the brakes) will do nothing to decrease the risk of accident [18]. Thus, all the working parts of an automobile must be functioning properly before introducing or using AI vehicle technology [18]. Today, automobiles use internal combustion (IC) engines, but in the future this may not be the case. It is possible that traditional IC engines will be replaced by unconventional sources of power and today's automobile engines will go the way of the railway steam engine [19]. Automobile original equipment manufacturers (OEMs) are taking a keen interest in AI and many of them are already collaborating with different technology providers [20]. For example, BMW has already collaborated with the Chinese technology company Tencent for developing semiautomated or level 3 cars for the Chinese market. Bosch and Mercedes Benz are also working with tech companies to develop level 4 and level 5 autonomous vehicles. To maximize share in the future market, many car manufacturers are collaborating with AI technology-based companies. In the next 4–5 years, we are likely to see many AI-based vehicles on the road. Technology companies are at the forefront in this scenario, as they are sharing their AI experiences with the automobile industry [21]. Recently, Apple, a big brand in mobile technology, announced that they will introduce their first car in 2024. This is a perfect example of diversified manufacturing, as Apple was previously not a player in the field of automobile manufacturing. Now, however, due to increasing possibilities and advances in AI technology, Apple is entering into this new market [22].

To accelerate AI innovations, many technology companies are following an open-platform approach, which will create a competitive market in AI implementation [23]. For example, the technology company Baidu collaborated with more than 120 OEMs, suppliers, and chipmakers to launch Apollo, an open-source, autonomous driving platform. Some OEMs have allowed third-party developers access to their vehicle data and controls and some of this AI technology is already in use [24]. China's Didi Chuxing ride-hailing service uses augmented reality (AR), including computer vision positioning and 3D face-recognition technologies, to enable services [25]. General Motors has begun to apply machine learning to their design services. Their Project Dreamcatcher is a generative design system that the company used to design a seatbelt bracket. The Dreamcatcher computer-aided design system produced 150 bracket designs that were 40% lighter and 20% stronger than conventional brackets. Continental Engineering Services developed Advanced Driver Assistance Systems that can generate 5000 miles of vehicle test data per hour as compared to 6500 miles of vehicle test data per month [18]. Audi uses AI to find faults in sheet metal via high-resolution cameras for quality control. A cloud-based image classification tool is being used by General Motors to check the predictive maintenance in robots; there are more than 6000 robots in the automotive industry [19]. Volkswagen is using machine learning to forecast the expected sales of 250 different car models in more than 110 countries. The company also initiated a modern showroom using AI and virtual reality technology to fulfill customer needs all over the world [20]. A voice assistance and hybrid navigation system is being implemented in Toyota cars based on a data communication module. The hybrid navigation system automatically suggests an alternate driving route [26]. Micheline installed a predictive diagnostic system for tires before their installation. Mercedes Benz is using AI in mobility services to track the delivery of parcels [27].

Autonomous driving vehicles equipped with AI technology are the future of the automotive industry. by using the AI, but at present driver monitoring system is providing great assistance to vehicles. Driver monitoring system continuously observes the eye gaze, eye openness and head position of the driver, with this observation eyesight positioning can provide alert about distracted driving and focus on road instructions [28]. In addition, eye openness and eyelid position can give information about the drowsiness of the driver and accordingly feedback can be given. A driver's sitting position is continuously monitored by sensors and accordingly the air bags are being exposed. The main difference in applying AI to machines is that the machine must learn how to accomplish tasks rather than being programmed to do so [29]. Big data is required for AI to perform even a single task. AI can make decisions only if logical permutations and combinations are fed into the system as data. Machine learning depends on neural networks. During machine learning or training, a neural network processes data regarding the steps to be carried out during a particular process [30]. One can correlate the function of a neural network with the neurons in the human brain; neural networks are made up of interconnected algorithms that exchange data about a specific task or tasks to be carried out [31]. There are many hurdles to implementing AI, such as the high cost of the technology, the expertise required to operate the technology, the difficulty of integrating AI projects with existing systems and processes, the immaturity of AI technology, and resistance to its implementation[32]. Once AI technology has matured, it will be easier to apply it across a wide variety of sectors and industries. It will save time and costs and lead to more effective and productive processes [33].

5.3 AI in automobile disc braking systems

Automobile brakes have a long history. They were initially introduced in simple form to stop the motion of a wheel. Over time, braking systems have improved significantly [34]. Table 5.1 is a list of different types of brakes and their description.

As shown in Fig. 5.1, disc brakes are used in almost all vehicles due to their quick response and reliability [35]. These brakes are mostly oil-pressure operated (hydraulic); when applying the brake pedal, oil creates pressure on both sides of the caliper in which friction pads are fitted [36]. This action reduces the speed of the vehicle and ultimately stops it [37]. This system only works if the mechanical linkages are in proper working condition [38]. In almost every critical road situation, a driver will apply the brakes to avoid an accident. Brake manufacturers give assurance that the brakes will be applied as you push the pedal [39]. This mechanical action of braking creates too many phenomena that are hurdles to effective braking. There is always the big question that if brakes are working as intended by the manufacturer, then why is the number of accidents increasing [40]?

Disc brakes are mostly fitted on the front side of vehicles and drum brakes are fitted on the rear side of vehicles, as shown in Fig. 5.2 [41]. Disc brakes are more efficient and quicker in action than drum brakes, therefore, for modern high-speed vehicles, disc brakes are fitted to the front wheels. Fig. 5.3 shows the different parts of a disc brake. Naturally, more efficient and effective disc brakes will generate more frictional heat than drum brakes and therefore they experience more thermal problems [42,43].

An annual report released by the central government of India shows the statistics of road accidents in 13 states where most accidents occur (see Fig. 5.4) [44–46].

Table 5.1 Types of automobile brakes.

Sr. no.	Types of brakes	Description
01	Wooden block brakes	First used in horse-drawn carriages and steam engines in the period 1795–1800. Wooden blocks were used to stop rotating wheels
02	Mechanical-type drum brakes	The first application of these types of brakes dates to 1900. They consisted of rope wrapped around the rotating drum
03	Internal-type shoe brakes (drum brakes)	Brake shoes were located on the internal side of the drum and stopped the wheel by creating friction between the drum and expanding shoes
04	Hydraulic brakes	Introduced in 1917, hydraulic brakes operated by hydraulic pressure with the help of brake pedal
05	Disc brakes	Disc brakes became mainstream in the 1950s but were first patented in 1902 by William Lanchester. These brakes are more effective and responded faster than previous versions
06	Antilock brake system (ABS)	ABS is a recent technology, rather than an improvement in braking systems, wherein brakes are applied in an on-and-off nature. Continuous hydraulic pressure during braking locks the wheels and thus skidding may take place. To avoid this, an ABS automatically applies and releases hydraulic pressure intermittently so that locking of the wheels can be avoided
07	Electric brake force distribution (EDB) system	Not all brakes should be applied hard to stop the vehicle, as the load on the wheels are different. According to the weight on the wheel, the braking force needed will vary. An EDB system distributes force as per the need
08	Traction control system (TCS), dynamic rear proportionating (DRP), electronic stability control (ESC)	TCS controls the track of moving vehicle, whereas DRP maintains the balance between the front and rear wheels. ESC acts to stabilize the moving vehicle and prevent it from rolling during braking at high speed

FIG. 5.1

Disc brakes and calipers of a four-wheel automobile.

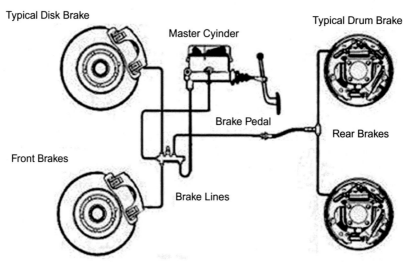

Typical Automotive Braking System

FIG. 5.2

Braking system of a four-wheel automobile.

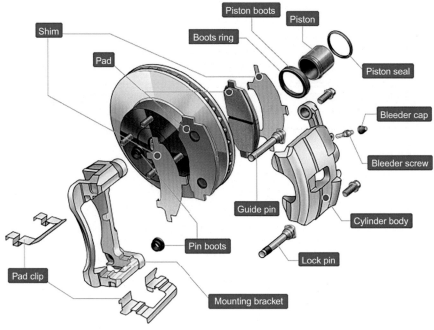

FIG. 5.3

Disc brake assembly of a four-wheel automobile.

84%
DEATHS IN 13 STATES, UP ON TOP

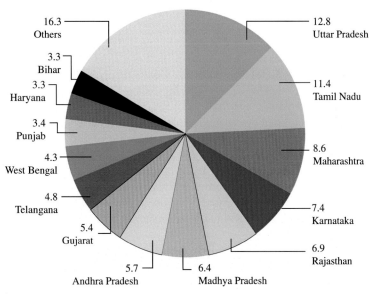

16.3
Others

3.3
Bihar

3.3
Haryana

3.4
Punjab

4.3
West Bengal

4.8
Telangana

5.4
Gujarat

5.7
Andhra Pradesh

6.4
Madhya Pradesh

12.8
Uttar Pradesh

11.4
Tamil Nadu

8.6
Maharashtra

7.4
Karnataka

6.9
Rajasthan

FIG. 5.4

Accidents and deaths statistics from the Government of India 2018–19.

There are numerous reasons for road accidents such as road conditions, driver error, aggressive driver behavior, vehicle condition, improper driving training, overloading, and so on. However, these reasons are beyond the control of automotive manufacturers [47]. Automotive manufacturers cannot guarantee that the safety features of a vehicle will prevent all accidents from occurring. If the braking system of a vehicle works properly and efficiently, most, but not all, accidents can be avoided [48].

5.4 Notable brake complaints observed at servicing centers

We analyzed complaints regarding braking systems and grouped them into different categories. Different car manufacturers use different braking systems, but the related problems are nearly the same. Following are the notable complaints collected from a survey carried out at different service stations of different car manufacturers.

1. Brake pad wear: Almost all car dealers reported this complaint. Brake wear is a direct result of frequent application of brakes during traffic [49]. The number of vehicles on the road is increasing daily and exceeding road capacity, thus leading to increased traffic and more wear of brake pads [50].
2. Noise: In most cases, brake noise is due to lack of maintenance. During regular scheduled servicing, customers are usually asked to replace their vehicle's brake pads; however, many customers do not do this [51]. Modern brake pads have a wear indicator metallic strip attached to them. After sufficient wear, the metal strip comes into contact with the brake disc/drum and begins making noise. This is one potential reason for the noise. Another reason may be that the anticattle clips and springs that hold the pads in place in the anchor bracket [52] are creating vibrations during initial engagement of the pads. Yet another reason for brake noise may be stones or other hard materials trapped between the pad and disc [53].
3. Pedal pulsation: Vehicles fitted with disc brakes encounter this type of problem. Pedal pulsation is caused because of rotor thickness variation or lateral runout. This type of complaint may be due to manufacturer error in some vehicles [54]. Generally, it is ignored by drivers because the noise is only heard when brakes are applied. If not rectified in due course, however, the noise will become more notable and persistent [55].
4. Uneven wear of discs (grinding): In most of the complaints received regarding slow responsive brakes, it was observed that the disc brake had grinding marks on the braking area. This is because of the use of friction pads for longer than the recommended period or stacking of hard material between the pads and disc [56].
5. Brake pulling: This is a result of poor maintenance. Sometimes it can be fixed by adjustment, but in other cases, it is an assembly that shows erratic behavior due to wear in multiple parts. In this case replacement is the only solution [57].

6. ABS and non-ABS problems: People who drive vehicles without ABS habitually apply brakes nearer to objects [58]. When they drive a vehicle fitted with ABS, their general complaint is nonresponsive brakes. They observe that stopping distance is more in ABS-fitted vehicles [59].
7. Dusting tire disc: Brakes are working properly, but the driver notices black dust on the outer surface of the tire rim [60]. This is due to wearing of the friction pads and dust deposition due to heat formation and oil [61].
8. Nonresponsive brakes at high speed: This problem, in which the brakes do not respond when a driver tries to stop a vehicle at high speed, is directly related to accidents. Despite continuous hard braking, the vehicles in question do not respond as quickly at high speeds as they do at low speeds [62]. This is a problem of modern cars that go much faster than their predecessors as well as driver error; it is usually not a problem with the braking system [63]. Still the similar type of problems is hitting day by day is increasing. This is the worry area for dealers. Problem arises due to rise in temperature during long braking [64].
9. Nonavailability of spare braking system parts: Sometimes spare parts are not available and therefore repairs can be expensive [65]. For example, one of the leading car manufacturers in India fits their vehicles with Bosch systems, which are hard to procure parts for. Thus, they suggest total replacement of the system with a TVS (Brake Manufacturing Company in India) system [66].
10. Thermal conning of brake discs: When a vehicle is stopped by applying a long brake at high speed the brakes heat up. This leads to thermal conning of the brake disc [67], that is, bending of the disc. This thermal conning leads to uneven wear of friction pads, which reduces the contact surface of pads with the disc, resulting in reduced breaking efficiency [68].
11. Steering vibration or wobbling: This problem is observed in many cases, but sometimes drivers fail to report this during servicing because they consider it to be a road surface problem [69]. This may happen due to uneven wear of the rotor surface, which leads to uneven contact of the friction pads with the rotor, creating vibrations in steering or wobbling in wheels [70].

This collected data and the proposed technical solutions can be used with AI to develop an automatic monitoring system to make braking systems more reliable and safe [71]. This has already been done to monitor tire pressure via sensors that automatically indicate tire air pressure. Another monitoring system of this type is being used to automatically repair tire punctures [72].

5.5 Literature survey

Table 5.2 lists some previous research in the field of automotive AI.

Table 5.2 Lists of some previous research in the field of automotive AI.

Sr. no.	Authors	Paper title	Contribution
01	Deng [73]	Agglomeration of technology innovation network of new energy automobile industry based on IoT and artificial intelligence	This paper gives insight into AI development in automobile industries. Intelligent technologies are changing the driving habits of people. Perception, communication technology, and embedded systems will strongly support future vehicles. This paper supports the early detection of thermal problems in braking systems of vehicles with the help of AI
02	Canyi et al. [74]	Research on application of artificial intelligence method in automobile engine fault diagnosis	The authors suggest using a BBM data collector to diagnose car engines with prefed data available in the collector with the help of AI. The fuel requirement diagnosis will help to reduce consumption. This type of model will help to predict vehicle braking problems
03	Zhao [75]	The Development Jilin Province's Automobile Industry Artificial Intelligence and LC Economy	This paper discusses the application of AI in the automobile industry to reduce carbon pollution
04	Rouf et al. [76]	AI in Mechanical Engineering	In Mechanical Engineering application of AI has increased the scope for intelligent machine manufacturing in all fields of upcoming autonomous vehicles. In automobile sector application will bust-up the new era of AI autonomous vehicles
05	Hofmann et al. [77]	AI and Data Science in the Automotive Industry	This paper examines the role of automatic optimization as a key technology in combination with data analytics. It presents examples of the way that data science and machine learning are currently being used in the automotive industry

Table 5.2 Lists of some previous research in the field of automotive AI –cont'd

Sr. no.	Authors	Paper title	Contribution
06	Oh and Kang [78]	Object Detection and Classification by Decision-Level Fusion for Intelligent Vehicle Systems	The authors propose a new object-detection and classification method using decision-level fusion to detect and classify three object classes: cars, pedestrians, and cyclists
07	Chen et al. [79]	DeepDriving: Learning Affordance for Direct Prediction in Autonomous Driving	The authors train a deep Convolutional Neural Network using recording from 12 h of human driving in a video game and show that their model can work well to drive a car in a very diverse set of virtual environments
08	Girshik et al. [80]	Rich feature hierarchies for accurate object detection and semantic segmentation	The authors propose a simple and scalable detection algorithm that improves mean average precision (mAP)
09	Aly [81]	Real time Detection of Lane Markers in Urban Streets	The authors propose a real-time approach to lane marker detection in urban streets
10	Pomerleau and Alvinn [82]	ALVINN: An Autonomous Land Vehicle in a Neural Network	The authors propose a three-layer back-propagation network designed for the task of road following

5.6 Application of AI to resolve thermal problems during long braking

The major issue of brake failure lies within the change in mechanical phenomenon during continuous braking of high-speed vehicles [83]. Friction between brake discs and friction pads creates heat that ranges from room temperature to a maximum of 1100°C [84]. During this rise in temperature, the friction pads lose their bonding capacity and thus the brakes become ineffective. Fig. 5.5 is a graph showing how temperature affects the coefficient of friction of braking material [85].

It has been observed that for normal operating temperatures, the pad wear per unit energy absorbed is independent of speed. It is only influenced by brake pressure and other driving behavior [17]. A driver who applies the brakes more aggressively will have more wear than a driver who always smoothly increases the brake pressure to its steady-state level. The influence of brake pressure in pad wear is easy to model even

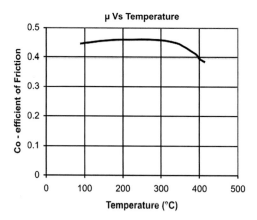

FIG. 5.5

Graph of temperature vs coefficient of friction in braking systems.

though it requires a significant amount of testing and verification [86]. The wear of braking pad material is exponentially dependent on increase in temperature, typically above 200–250°C. Fig. 5.6 is a graph of wear factor [87]. Below the critical temperature, binder plays a minor role in wear resistance. Above critical temperature, the wear rate is strongly influenced by the thermal decomposition of the resin [88]. In some studies, wear rate was found to decrease with increasing load. This may occur because of greater exposure to the contact region and an increase in pressure [18]. Also, in this research it has been noted that wear rate decreases with an increase in velocity. There is no correlation between the wear rate and μ, though they are both affected by change in temperature [89].

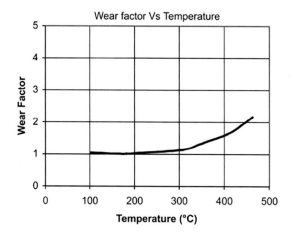

FIG. 5.6

Graph of temperature vs wear factor.

In a braking system, particularly one that uses disc brakes, heat transfer takes place via convection. Air circulating between the vanes of two chicks (two face plates of brake disc on which friction pads slides are called chicks) of disc is the most prominent area of heat transfer [90]. Air circulation is quite good in a car traveling about 60–70 km/h [91]. However, if the vehicle's speed increases beyond this range, the air gets trapped in the vanes and external air force creates locking at the edge of the disc by forming an air circulation block, which reduces the convection capacity of brakes at high speed [92]. The heat generated during braking must be transmitted to some other parts from the interface of the disc and friction pads. If heat accumulates, the friction material starts losing its coefficient of friction [93]. In this case, AI can play a crucial role in detecting the heat generated and can be used to apply artificial cooling accordingly so that the coefficient of friction material retains its value throughout the braking period and thus effective brakes can be applied [94].

Fig. 5.7 shows the setup of our experiment investigating brake fade due to heat generation in brake discs to generate data for AI. This setup enables us to measure the temperature generated during braking by using a rubbing thermostat [95].

FIG. 5.7

Experimental setup.

The experiment uses a hydraulic braking system to apply the brakes, pressure measuring gauge installed in line to measure the braking pressure when the brakes are applied. [96]. An electric motor is used to drive the brake assembly with inertia load. The brake disc can easily be installed on setup. To create this realistic experiment, the brake disc was cooled with forced air [97].

For our experiment, the initial temperature of brake disc was set to room temperature and, with the help of an electric motor, the brake disc was rotated at 2540 RPM, measured by tachometer [98]. The inertia weight was 42 kg [99]. After rotating the brake disc, sufficient time was given to stabilize the system. Then, the brakes were applied at regular intervals with varying time [100]. Stopping time of the disc was noted after each brake [101]. The braking time was increased after each braking and the temperature and pressure applied to the brake was recorded. The thermostat can be set at any position (e.g., t1, t2, or t3), as shown in Fig. 5.8 [102].

Experimental setup is designed for fitting different friction pads [24]. We used an off-market friction pad for our experiment, as shown in Fig. 5.9.

The data generated from our experiment can be used for the implementation of AI in thermal performance of brakes [103]. We also used ANSYS software to simulate the same braking situation [104].

Using the finite element thermal analysis method, we performed steady-state thermal analysis with some boundary conditions on brake disc pad assembly [105]. The heat flux of 57,317.72 W/m^2 was applied on the surface of the disc where pads are in contact with it. The film coefficient for convection was taken as 106 W/m^{2}K and the coefficient of conductivity was 51 W/mK. Brake pressure of 2.70 MPa was applied on the pads [106]. Finite temperature analysis of the disc brake during braking was done using the software ANSYS 15.0 [107]. Fig. 5.10 shows the temperature distribution between the interface of the braking disc and friction pads at different braking times. Finite element analysis considers the whole body as an object, therefore the temperature difference may occur in both actual and finite analysis [108].

Analysis shows that heat generation is more at the interface of friction pads and disc area in contacts, leading to nonoperation of brakes by reducing the coefficient of friction (μ) and fast wear of the braking pads, known as brake fade [109]. However,

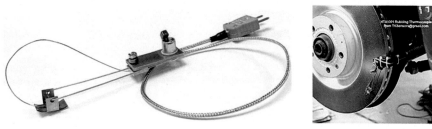

FIG. 5.8

Rubbing K-type thermocouple and locations on brake disc.

FIG. 5.9

Friction pad of disc brake.

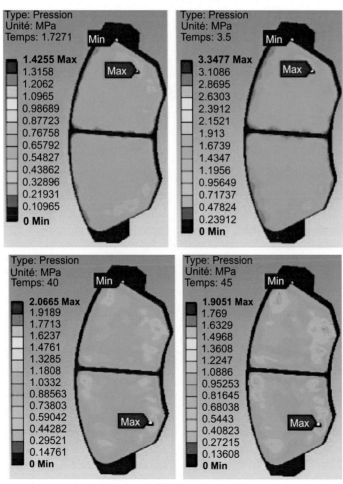

FIG. 5.10

Results of temperature distribution at different time durations in simulation.

(Continued)

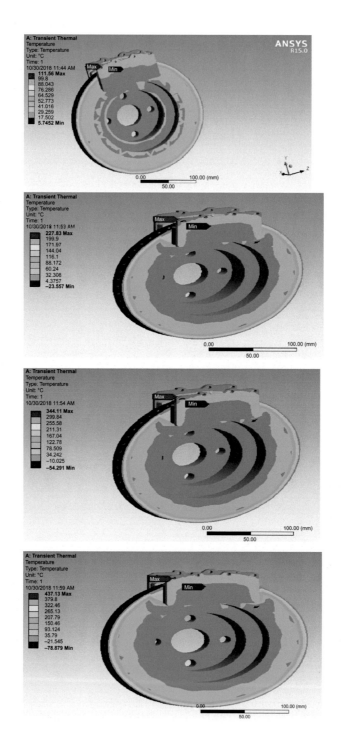

FIG. 5.10—cont'd

change in (μ) due to rise in temperature depends upon the frictional material [110]. Fig. 5.10 also shows the total deformation at braking with variation of the coefficient of friction. It is observed that there is no significant variation with an increase of the coefficient of friction [111]. Temperature increase weakens the bonding between friction pad molecules; many off-market friction pads have a shorter lifetime than expected and the percentage of accidents is increasing [112].

The automobile industry needs reliable components with longer life spans [113].

5.7 Results and conclusion

AI has great potential for application in the automobile industry. There is enough preexisting data to feed AI innovations, and future experiments will generate even more useful data to develop AI algorithms for vehicles [114]. Repetitive tasks requiring human intelligence can be assigned to machines powered by AI [63]. Big data collected from vehicles will serve as an input to AI algorithms [25].

Sensors connected to braking systems will provide suggestions for corrective action and monitor AI-operated braking systems. AI assisted vehicles will detect thermal problems of braking systems and automatically by taking corrective action will resolve the issues related to thermal brake fed [115]. During the initial braking process, the coefficient of friction increases as the process of braking continues. At initial temperature up to 145°C [22], the coefficient of friction shows improvement; however, beyond this temperature, the coefficient of friction decreases and brakes became inactive at the friction area [116]. The material begins to lose its bonding properties and thus wear begins to take place [117]. During braking, maximum time drag braking is required to stop the vehicle [118]. This continuous friction increases the temperature at peak level and damages the surface of friction pads [119].

In the case of disc braking systems, we compared the experimental temperature increase in drag braking with the simulated temperature increase using ANSYS 15.0 software and compared the results [120]. Both the experimental and simulated data show slight variations in temperature, as measured by a rubbing K-type thermocouple [121].

Table 5.3 shows that there is no considerable difference in temperature increase in either the experiment or the simulation. Therefore, Ansys simulation data can be generated for AI and used to develop an algorithm for solving the brake fade problem during drag braking [122]. This process can be carried out using any suitable software presently used for error detection and then correlating those results for AI developments [123].

In a mechanical system, it is very difficult to monitor the thermal changes taking place during long braking, but using modern techniques and equipment like a rubbing thermocouple [124] will generate correct and time-dependent data for AI application [125]. Therefore, software with validation from experimentation can solve the problem of generating data for AI [53]. AI can then give the instruction to the vehicle to cool the disc brakes so that the braking coefficient of friction pads will not change

Table 5.3 Comparison of temperatures obtained via experiment vs temperatures obtained via simulation.

Sr no.	Measured temperature rise in °C (experimentally)	Measured temperature rise in °C (by simulation)	Difference in temperature rise
01	109	111	2
02	119	122	3
03	148	149	1
04	185	187	2
05	199	200	1
06	212	214	2
07	220	223	3
08	224	227	3
09	342	344	2
10	434	438	4

with an increase in temperature [126]. These sorts of systems are essential for driverless cars with the finest correction factors from the experimental and previous data collection [127]. AI has great potential in the automobile industry to reduce accidents and increase human safety [128]. Too many controls will lead to confusion while driving, hence it is better to allot some functions to AI systems that will automatically take care of certain things. [129].

In this proposed model, we conducted an experiment and a simulation. Other researchers have also used simulation software for developing AI systems, but their data was based on theoretical assumptions [130] and AI systems using theoretical data have the potential to fail [131]. In our model, there is only a slight difference between the experimental data and simulated data. Only collected data from the several experiences will not serve to develop model for AI but it needs experimental and simulation support for better implementation. Automobile industry will require strong implementation of AI for the better future [132].

5.8 Future scope

Experiments conducted to resolve the intricate problems associated with automobile braking systems generate accurate thermal data that can be fed into an AI system to operate braking systems with more precision. The automobile industry has already started developing driverless cars with the help of AI. To accomplish this, big data regarding the operation of vehicles is necessary. This methodology can be applied for any type of problem associated with automobiles, such as fuel supply, engine diagnostics, overloading, component failure, engine temperature, and so on [133]. There must be a correlation between experimental data and automation for the successful implementation of AI in the automotive industry [132].

References

[1] Report on RAISE, Indiaai.gov.in, AI for all, An Overview, 2020.

[2] C-DAC, A Make in India Force That's Powering India's AI Revolution. https://indiaai.gov.in/article/c-dac-a-make-in-india-force-that-s-powering-india-s-ai-revolution.

[3] S. Prasetya, et al., Artificial intelligence for smart electric vehicle braking system, J. Mech. Eng. Res. Dev. 1024-1752, 46 (6) (2020) 106–112.

[4] Accelerating Automotive, AI Transformation. www.capgemini.com.

[5] Artificial Intelligence, Reshaping the Automotive Industry. www.futurebridge.com.

[6] Artificial Intelligence in Cars Powers an AI Revolution in Automobile Industry. www.builtin.com.

[7] Making Automotive Industries Smarter With AI. www.CIO.com.

[8] D. Olive, et al., Artificial Intelligence Is a Key to Automated Driving Equipping Cars With It Is a Bit Like Teaching Students. www.bosch-mobility.com.

[9] Safety Requirements, Braking for Autonomous Vehicles, Report by SAE. https://www.sae.org/news/2018/03/soft-tests-for-autonomous-vehicles.

[10] S. Krasnowshtanov, et al., Dependence of braking vehicle characteristics on environmental condition, Adv. Res. 158 (2018), https://doi.org/10.2991/avent-18.2018.39.

[11] A. Singh, The Future of Artificial Intelligence in the Automotive Industry, artificialintelligence.oodles.io, 2020.

[12] Artificial Intelligence in Automotive Market, 2020. www.gminsights.com.

[13] H. Chae, C.M. Kang, B. Kim, J. Kim, C.C. Chung, J.W. Choi, Autonomous braking system via deep reinforcement learning, in: 2017 IEEE 20th International Conference on Intelligent Transportation Systems (ITSC), 2017, pp. 1–6, https://doi.org/10.1109/ITSC.2017.8317839.

[14] P. Pavual Arockiyaraj, P.K. Mani, An autonomous braking system of cars using artificial neural network, Int. J. Circuit Theory Appl. 9 (9) (2016) 3665–3670.

[15] S. Sen, Artificial intelligence in automobile: an overview, Int. J. Innov. Res. Sci. Eng. Technol. 2319-8753, (2018). www.ijirset.com.

[16] Ministry of Road Transport, Government of India, MORTH Report, 2019. morth.nic.in.

[17] P. Zang, et al., Fade Behaviour of Copper-Based Brake Pad During Cyclic Emergency Braking at High Speed and Overload Condition, Elsevier, 2019, https://doi.org/10.1016/j.wear.2019.01.126.

[18] V.T. Rao, V.S.R. Ramasubramanian, K.N. Seetharamu, Analysis of temperature field in brake disc for fade assessment, in: Warmer and Stofffüber Tragung, vol. 24, 1989, pp. 9–17, https://doi.org/10.1007/BF01599500.

[19] U.S. Hong, et al., Wear mechanism of multiphase friction materials with different phenolic resin matrices, Wear 266 (7-8) (2009) 739–744, https://doi.org/10.1016/j.wear.2008.08.008.

[20] K.W. Liew, U. Nirmal, Frictional performance evaluation of newly designed brake pad materials, Mater. Des. 48 (2013) 25–33, https://doi.org/10.1016/j.matdes.2012.07.055.

[21] A.J. Day, Book on Braking of Road Vehicles, vol. 02, Butterworth-Heinemann Publ. (imprint of Elsevier), 2014, pp. 25–26.

[22] R. Yun, P. Filip, Y. Lu, Performance and evaluation of eco-friendly brake friction materials, Tribol. Int. 43 (11) (2010) 2010–2019, https://doi.org/10.1016/j.triboint.2010.05.001.

[23] V. Matějka, et al., On the running-in of brake pads and discs for dyno bench tests, Catal. Today 209 (2017) 170–175, https://doi.org/10.1016/j.triboint.2017.06.008.

[24] R.C. Dante, Handbook of Friction Materials and Their Application, Elsevier, 2016, pp. 258–261, ISBN: 978-0-08-100620-7.

[25] D.W. Dockery, C.A. Pope III, Health effects of fine particulate air pollution: lines that connect, J. Air Waste Manage. Assoc. 28 (2006) 24–26, https://doi.org/10.1080/10473289.2006.10464485.

[26] R.M. Harrison, A.M. Jones, J. Gietl, J. Yin, D.C. Green, Estimation of the contributions of brake dust, tire wear, and resuspension to non- exhaust traffic particles derived from atmospheric measurements, Environ. Sci. Technol. vol. 46 (2012) 6523–6529, https://doi.org/10.1021/es300894r.

[27] G. Valota, S. De Luca, A. Söderberg, Using Finite Element Analysis to Simulate the Wear in Disc Brakes During a Dyno Bench Test Cycle, Eurobrake, Dredsen, Germany, 2017, pp. 19–21. Issue-01.

[28] J. Wahlström, Towards a cellular automaton to simulate friction, wear, and particle emission of disc brakes, Wear 313 (2014) 75–78, https://doi.org/10.1016/j.wear.2014.02.014.

[29] Road Accidents in India, Reiterates Commitment to Reduce the Number of Road Accidents and Fatalities by 50 Percent by 2020 Press Information Bureau Government of India Ministry of Road Transport & Highways 09-June-2016 19:21 IST, 2015.

[30] G. Valota , S. De Luca, A. Söderberg, Using finite element analysis to simulate the wear in disc brakes during a dyno bench test cycle, in: Eurobrake, Dredsen, Germany, Issue 01, 2017, pp. 19–21.

[31] Report for first half of, Encouraging—Road Accidents Drop by 3 % and Fatalities Drop by 4.75 % Between January to July 2017 Press Information Bureau Government of India Ministry of Road Transport & Highways, 2017.

[32] Report of Washington State Department of transportation USA, Experimental Study of Automotive Brake System Temperatures, 1997. Jully-1997, WA-RD-434-1.

[33] S. Senthil Gavaskar, et al., Failure analysis in light commercial vehicles by using SQC tools, Int. J. Eng. Res. Technol. 3 (11) (2014). Nov 2014.

[34] P. Biradar, et al., Automatic brake failure detection with auxiliary braking system, Int. J. Sci. Adv. Res. Technol. 2 (3) (2016). March -2016.

[35] R. Jagdeeshwaran, et al., Brake fault diagnosis using clonal selection classical algorithm—a statistical learning approach, J. Eng. Sci. Technol. (2015) 14–23.

[36] S.D. Oduro, Brake failure and its effect on road traffic accidents in Kumasi metropolis Ghana, Int. J. Sci. Technol. (2012). Sep. 2012.

[37] A. Choudhary, et al., Automatic brake failure indicator, Int. J. Eng. Sci. Res. Technol. (2016) 4877–4882. April 2016.

[38] P. Gregory, et al., An investigation on falure of automotive components in cars, Int. Res. J. Eng. Technol. 04 (06) (2017). June-2017.

[39] S.R. Abhang, et al., Design and analysis of disk brake, Int. J. Eng. Innov. Technol. 8 (4) (2014) 165–167. Feb-2014.

[40] R.H. Rana, et al., Optimum design of braking system for a formula 3 race cars with numeric computations and thermal analysis, Int. Res. J. Eng. Technol. 4 (2) (2017). Feb- 2017.

[41] U. S. Department of Transport, "Examination of ABS-Related Driver Behavioral" NHTSA Light Vehicle Antilock Brake System Research Program-2001, 2001.

[42] S.C. Mudd, Technology for Motor Mechanics 2, second ed., Gibrine Publishing Company, Maryland, 1972, pp. 213–241.

[43] S.J. Zammit, Motor Vehicle Engineering Science for Technicians, Longman Group UK Limited, London, 1987, p. 64.

[44] W. Odero, M. Khayesi, P.M. Heda, Road traffic injuries in Kenya: magnitude, cause and status of intervention, Inj. Control. Saf. Promot. 10 (2008) 53–61. 2003.

[45] D.J. Leeming, R. Hartley, Heavy Vehicle Technology, second ed., Hutchinson & Co. Publishing Ltd, 2001.

[46] M. Peden, R. Scurfield, D. Sleet, D. Mohan, A.A. Hyder, C. Mathers, World Report on Road Traffic Injury Prevention, World Health Organization, Geneva, 2004, pp. 1784–1790.

[47] A.K. Agnihotri, H.S. Joshi, Pattern of road traffic injuries: one year hospital-based study in Western Nepal, Int. J. Inj. Control Saf. Promot. 13 (2) (2006) 128–130.

[48] P. Banthia, B. Koirala, A. Rauniyar, D. Chaudhary, T. Kharel, S.B. Khadka, An epidemiological study of road traffic accident cases attending emergency department of teaching hospital, J. Nepal Med. Assoc. 45 (162) (2006) 238–243.

[49] World Health Organization, World Report on Road Traffic Injury Prevention: Summary, World Health Organization, Geneva, 2004.

[50] P. Owasu-Anah, et al., Survey of the couses of brake failure in commercial mini-buses in Kumasi, Ghana, Res. J. Appl. Sci. Eng. Technol. 7 (23) (2014). June 20, 2014.

[51] F.K. Afukaar, W. Agyemang, W. Ackaah, I. Mosi, Road Traffic Crashes in Ghana, Statistics 2007, 2008. Consultancy Service Report for National Road Safety Commission of Ghana.

[52] A.D. McPhee, D.A. Johnson, Experimental heat transfer and flow analysis of a vented brake rotor, Int. J. Therm. Sci. 47 (4) (2007) 458–467.

[53] H. Heinz, Vehicle and Engine Technology, second ed., Butterworth–Heinemann publications, Nurumberg, 2001.

[54] V. Surblys, et al., Research of the brake testing efficiency, Elsevier, Procedia Eng. 134 (2016) (2015) 452–458.

[55] N. Podaprigora, et al., Method of assessing the influence of operational factors on brake system efficiency in investigating traffic accidents, Elsevier, Procedia Eng. 20 (2017) (2016) 516–522.

[56] S.A.A. Albatlan, Automotive brake pipes characteristics and their effects on brake performance, Ain Shams Eng. J. 2012 (3) (2012) 279–287.

[57] Darredy Ramana Reddy, Development of a new bio-degradable friction material for brake pads from palm kernel shell, Int. J. Sci. Eng. Res. 11 (1) (2017).

[58] R. Maske, et al., Automatic brake failure indicator and braking system, Int. J. Adv. Res. Innov. Ideas Educ. 3 (2017). ISSN(O)-2395-4396.

[59] M.B. Swarup, B. Hari Prasad, FMEA-based failure analysis of brake-by-wire automotive safety-critical system, Int. J. Adv. Res. Comput. Commun. Eng. 3 (5) (2014) 6652–6655.

[60] P. Sinha, Architectural design and reliability analysis of a fail-operational brake-by-wire system from ISO 26262 perspectives, Reliab. Eng. Syst. Saf. 96 (10) (2011) 1349–1359. https://doi.org/10.1016/j.ress.2011.03.013.

[61] F. Talathi, et al., Analysis of Heat Conduction in a Disk Brake System, Springer, 2009, https://doi.org/10.1007/s00231-009-0476-y.

[62] A.M. Puncioiu, et al., Analysis of heat conduction in a drum brake system of the wheeled armored personnel carriers, IOP Conf. Ser.: Mater. Sci. Eng. 95 (2015), https://doi.org/10.1088/1757-899X/95/1/012039, 012039.

[63] E. Jeong, O. Cheol, Evaluating the Effectiveness of Active Vehicle Safety Systems, Elsevier, 2017, https://doi.org/10.1016/j.aap.2017.01.015.

[64] P.D. Neis, et al., Towards a better understanding of the structures existing on the surface of brake pads, Tribol. Int. (2017), https://doi.org/10.1016/j.triboint.2016.09.033.

[65] Q. Jian, et al., Numerical and Experimental Analysis of Transient Temperature Field of Ventilated Disc Brake Under the Condition of Hard Braking, Elsevier, 2017, https://doi.org/10.1016/j.ijthermalsci.2017.08.013.

[66] S. Lakkam, et al., A study of heat transfer on front and back vented brake disc affecting vibration, Eng. J. (2017), https://doi.org/10.4186/ej.2017.21.1.169. http://www.engj.org/.

[67] A.A. Alnaqi, et al., Reduced scale thermal characterization of automotive disc brake, Appl. Therm. Eng. (2018), https://doi.org/10.1016/j.applthermaleng.2014.10.001.

[68] İ. Mutlu, et al., The effects of porosity in friction performance of brake pad using waste tire dust, SciElo (2015), https://doi.org/10.1590/0104-1428.1860.

[69] A. Adamowicz, Effect of convective cooling on temperature and thermal stresses in disk during repeated intermittent braking, J. Frict. Wear 37 (2) (2015) 107–112, https://doi.org/10.3103/S1068366616020021. 2016.

[70] M. Timur, et al., Heat Transfer of Brake Pad Used in the Autos After Friction and Examination of Thermal Tension Analysis, 2014, https://doi.org/10.5755/j01.mech.20.1.6595.

[71] S.A.A. Albatlan, Study Effect of Pads shapes on Temperature Distribution for Disc Brake Contact Surface, 2013, e-ISSN: 2278-067X, p-ISSN: 2278-800X www.ijerd.com.

[72] B. Ali, et al., Thermomechanical modelling of disc brake contact phenomena, FME Trans. 41 (2013) 59–65. 60 vol. 41, no 1, 2013.

[73] Y. Deng, Agglomeration of technology innovation network of new energy automobile industry based on IoT and AI, J. Ambient. Intell. Humaniz. Comput. (2021), https://doi.org/10.1007/s12652-021-03102-2.

[74] D. Canyi, W. Li, Y. Rong, F. Li, et al., Research on Application of Artificial Intelligence Method in Automobile Engine Fault Diagnosis, IOP Publishing, 2021, https://doi.org/10.1088/2631-8695/ac01ad.

[75] X. Zhao, The Development Jilin Province's Automobile Industry Artificial Intelligence and LC Economy, Springer, 2021, pp. 11–13.

[76] S. Rouf, M. Ali, A. Hussain, AI in Mechanical Engineering, IEEE, 2018.

[77] F. Hofmann, F. Neukart, T. Back, AI and data science in the automotive industry, 2017. arXiv preprint arXiv:1709.01989.

[78] S.I. Oh, H.B. Kang, Object detection and classification by decision-level fusion for intelligent vehicle system, Sensors 17 (1) (2017) 207.

[79] C. Chen, A. Seff, A. Kornhauser, et al., Deep Driving: Learning Affordance for Direct Prediction in Autonomous Driving, IEEE, 2015, pp. 2722–2730.

[80] R. Girshik, et al., Rich Feature Hierarchies for Accurate Object Detection and Semantic Segmentation, IEEE, CVPR, 2014.

[81] M. Aly, Real Time Detections of Lane Markers in Urban Streets, IEEE, 2008, pp. 07–12.

[82] D.A. Pomerleau, Alvinn, An Autonomous Land Vehicle in a Neural Network, DTIC Document, 2007.

[83] A. Vdovin, A coupled approach for vehicle brake cooling performance simulations, Elsevier, Int. J. Therm. Sci. 132 (2019) 257–266.

[84] M. Mathissen, et al., A novel real-world braking cycle for studying brake wear particle emissions, Wear (2018), https://doi.org/10.1016/j.wear.2018.07.020.

[85] E. Marin, et al., Diagnostic Spectroscopic Tools for Worn Brake Pad Materials: A Case Study, Journal Homepage, 2019. www.elsevier.com/locate/wear.

[86] L. Wei, Y.S. Choy, C.S. Cheung, A Study of Brake Contact Pairs Under Different Friction Conditions With Respect to Characteristics of Brake Pad Surfaces, Elsevier, 2019, https://doi.org/10.1016/j.triboint.2019.05.016.

[87] H. Lan, C. Guo, M. Han, Y.H. Song, S.W. Jia, Thermal-mechanical coupling finite element simulation and experimental study of disc brake friction pair, Int. Conf. Phys. Math. Stat. Model. Simul. (2018) 24–29. ISBN 978-1-60595-558-2 10.12783/dtetr/pmsms2018/24892.

[88] B. Ali, B. Mostefa, Thermomechanical modelling of disc brake contact phenomena, FME Trans. 4 (2012) 59–65, https://doi.org/10.1016/j.ijthrmalsci.2012.05.006.

[89] A. Loizou, S. QiH, A.J. Day, A fundamental study on the heat partition ratio of vehicle disc brake, J. Heat Transf. 135 (12) (2013), https://doi.org/10.1115/1.4024840. 121302-1-8.

[90] D.L. Singaravelu, F.P. VijayR, Influence of various cashew friction dusts on the fade and recovery characteristics of non-asbestos copper free brake friction composites, Wear (2019) 1129–1141, https://doi.org/10.1016/j.wear.2018.12.036.

[91] A. Afzal, M.A. Mujeebu, Thermo-mechanical and structural performance of automobile disc brakes: a review of numerical and experimental studies, Arch. Comput. Methods Eng. (2018) 36–45, https://doi.org/10.1007/s11831-018-9279-y.

[92] C.H. Galindo-Lopez, M. Tirovic, Understanding and improving the convective cooling of brake disc with radial vanes, J. Automob. Eng. D (2008), https://doi.org/10.1243/09544070JAUTO594.

[93] B. Gadhimi, F. Kowsary, M. Khorami, Thermal analysis of locomotive wheel mounted brake disc, J. Appl. Therm. Eng. (51) (2012) 948–952, https://doi.org/10.1016/j.applethermaleng.2012.10.051. Elsevier.

[94] M. Pevec, I. Potrc, G. Bombek, D. Vranesevic, Prediction of the cooling factors of a vehicle brake disc and its influence on the results of the thermal numerical simulation, Int. J. Automot. Technol. 13 (5) (2012) 725–733, https://doi.org/10.1007/s12239-012-0071-y.

[95] K. Stevens, M. Tirovic, Heat dissipation from the stationary brake disc, Proc. Inst. Mech. Eng. C J. Mech. Eng. Sci. 232 (9) (2017) 1703–1733, https://doi.org/10.1177/0954406217707983.

[96] W.T. Yan, S. Wu, G. Feng, Y. Xie, Role of vane configuration on the heat dissipation performance of ventilated brake discs, Appl. Therm. Eng. (2018), https://doi.org/10.1016/j.applthermaleng.2018.03.002.

[97] G.J. Balotine, P.D. Neis, F.N. Ferreira, Analysis of the influence of temperature on the friction coefficient of friction material, in: ABCM Symposium Series in Mechatronics, vol. 4, 2010, pp. 898–906. DOI: https//: analysis Corregidor Jean_01 06 09.doc.

[98] A. Belhocine, C.D. Cho, M. Nouby, Y.B. Yi, A.R. Abu Backer, Thermal analysis of both ventilated and full disc brake rotors with frictional heat generation, Appl. Comput. Mech. (2014) 5–24. DOI: http://hdl.handle.net/11025/11668.

[99] A.D. McPhee, D.A. Johnson, Experimental heat transfer and flow analysis of a vented brake rotor, Int. J. Therm. Sci. 47 (2008) 458–467, https://doi.org/10.1016/j.ijthermalsci.2007.03.006.

[100] P. Verma, L. Menapace, A. Bonfanti, R. Ciudin, S. Gialanella, G. Straffelini, Braking pad disc system: wear mechanisms and wear fragments, Wear (2015) 251–258, https://doi.org/10.1016/j.wear.2014.11.019.

[101] V. Matejka, I. Metinoz, J. Wahistrom, M. Alemani, G. Perricone, On the running in of brake pads and discs for dyno bench tests, Tribol. Int. (2017) 01–15, https://doi.org/10.1016/j. triboint.2017.06.008.

[102] X.D. Nong, Y.L. Jiang, M. Fang, L. Yu, C.Y. Liu, Numerical analysis of novel SiC$_{3D}$/AI alloy co-continuous composites ventilated brake disc, Int. J. Heat Mass Transf. 108 (2016) 1374–1382, https://doi.org/10.1016/j.ijthermalsci.2016.11.108. Elsevier.

[103] G. Straffelini, L. Maines, The relationship between the wear of semi metallic friction materials and pearlitic cast iron in dry sliding, Wear 307 (1-2) (2013) 75–80, https://doi.org/10.1016/j.wear.2013.08.020.

[104] E. Palmer, R. Mishra, J. Fieldhouse, An optimization study of a multiple row pin vented brake disc to promote brake cooling using computational fluid dynamics, Proc. Inst. Mech. Eng. D: J. Automob. Eng. 223 (7) (2009) 865–875, https://doi.org/10.1243/09544070jauto 1053.

[105] M. Duzgun, Investigation of thermo structural behaviours of different ventilation application, J. Mech. Sci. Technol. 26 (1) (2012) 235–240, https://doi.org/10.1007/s12206-011-0921-y. Springer.

[106] T.P. Newcomb, R.T. Spurr, Friction materials for brakes, Tribology 4 (2) (1971) 75–81, https://doi.org/10.1016/0041-2678(71)90135-7.

[107] S.A.A. Albatlan, Study effect of pads shapes on temperature distribution for disc brake contact surface, Int. J. Eng. Res. Dev. 8 (9) (2013) 62–67. e-ISSN 2278-067X, p-ISSN 2278-800X www.ijerd.com.

[108] A. Adamowicz, P. Grzes, Analysis of disc brake temperature distribution during single braking under non-axisymmetric load, Appl. Therm. Eng. 31 (6–7) (2011) 1003–1009, https://doi.org/10.1016/j.applthermaleng.2010.12.016. Elsevier.

[109] T.J. Mackin, S.C. Noe, K.J. Ball, B.C. Bedell, D.P. Bim-Merle, M.C. Bingaman, R.S. Zimmerman, Thermal cracking in disc brakes, Eng. Fail. Anal. 9 (1) (2002) 63–76, https://doi.org/10.1016/s1350-6307(00)00037-6.

[110] A. Adamowicz, Effect of convective cooling on temperature and thermal stresses in disc during repeated intermittent braking, J. Frict. Wear 1068-3666, 37 (2) (2016) 107–112, https://doi.org/10.3103/S 1068366616020021.

[111] Y.M. Huang, S.H. Chen, Analytical study of design parameters on cooling performance of brake disc, SAE Int. (2006), https://doi.org/10.4271/2006-01-0692. ISBN 0-7680-1631-2.

[112] D.D. Karan, Thermo-mechanical performance of automotive disc brakes, Mater. Today: Proc. 5 (1) (2018) 1864–1871, https://doi.org/10.1016/j.matpr.2017.11.287. Elsevier.

[113] H.B. Yan, Q.C. Zhang, T.J. Lu, Heat transfer enhancement by X-type lattice in ventilated brake disc, Int. J. Therm. Sci. 107 (2016) 39–55, https://doi.org/10.1016/j.ijthermalsci.2016.03.026.

[114] H. Keller, Brake Disc, Particularly an Internally Ventilated Brake Disc, U.S. Patent (7967115), 2011.

[115] B.D. Garg, S.H. Cadle, P.A. Mulawa, P. Groblicki, C. Laroo, G.A. Parr, Brake wear particulate matter emission, Environ. Sci. Technol. 34 (21) (2000) 4463–4469, https://doi.org/10.1021/es001108h.

[116] M. Timur, H. Kuşçu, Heat Transfer of brake pad used in the Autos after friction and examination of the thermal tension analysis, Mechanica 20 (1) (2014) 17–23, https://doi.org/10.5755/j01.mech.20.1.6595.

[117] H. Hagino, M. Oyama, S. Sasaki, Laboratory testing of airborne brake wear particle emissions using a dynamometer system under urban city driving cycles, Springer J. 131 (2016) 269–278, https://doi.org/10.1016/j.atmosenv.2016.02.014.

[118] İ. Mutlu, İ. Sugözü, A. Keskin, The effect of porosity in friction performance of brake pad using waste tyre dust, J. Polym. 25 (5) (2015) 440–446, https://doi.org/10.1590/0104-1428.1860.

[119] M. Kumar, J. Bijwe, Studies on reduced scale tribometer to investigate the effects of metal additives on friction coefficient- temperature sensitivity in brake materials, Wear 269 (2010) 838–846, https://doi.org/10.1016/j.triboint.2009.12.062.

[120] F. Talati, S. Jalalifar, Analysis of heat conduction in a disc brake system, Heat Mass Transf. 45 (2009) 1047–1059, https://doi.org/10.1007/s00231-009-0476-y. Springer.

[121] M. Mathissen, J. Grochowicz, C. Schmidt, R. Vogt, F.H. Farwick zum Hagen, T. Grabiec, T. Grigoratos, A novel real-world braking cycle for studying brake wear particle emissions, Wear (2018), https://doi.org/10.1016/j.wear.2018.07.020.

[122] M. Eriksson, S. Jacobson, Tribological surface of organic brake pads, Tribol. Int. 33 (12) (2000) 817–827, https://doi.org/10.1016/S0301-679X(00)00127-4.

[123] D.R. Reddy, B. Balunaik, Development of a composite material from agro waste for wear resistance application, Mater. Sci. Forum 773–774 (2013) 319–324, https://doi.org/10.4028/www.scientific.net/msf.773-774.319.

[124] H.A. Kemmer, Investigation of the Friction Behaviour of Automotive Brakes Through Experiments and Tribological Modelling, PhD Thesis, University of Paderborn, 2002.

[125] A. Nemade, S. Telang, A. Chel, Thermal investigation of disc brake fade during long braking, Souvenir of COPEN-11, IIT Indore, in: International Conference on Precision, Meso, Micro and Nano Engineering, 2019, pp. 70–71.

[126] N. Podoprigora, V. Dobromirov, A. Pushkarev, V. Lozhkin, Methods of assessing the influence of operational factors on brake system efficiency in investigating traffic accidents, Transp. Res. Procedia 20 (2017) 516–522, https://doi.org/10.1016/j.trpro.2017.01.084.

[127] M.B. Swarup, B. Hari Prasad, FMEA-based failure analysis of brake-by-wire automotive safety-critical system, Int. J. Adv. Res. Comput. Commun. Eng. 3 (5) (2014) 6652–6655.

[128] Q. Jian, Y. Shui, Numerical and experimental analysis of transient temperature field of ventilated disc brake under the condition of hard braking, Int. J. Therm. Sci. 122 (2017) 115–123, https://doi.org/10.1016/j.ijthermalsci.2017.08.013.

[129] A. Belhocine, M. Bouchetara, Temperature and thermal stresses of vehicles gray cast brake, J. Appl. Res. Technol. 11 (5) (2013) 674–682, https://doi.org/10.1016/s1665-6423(13)71575-x.

[130] X. Ji, L. He, W. Wang, A. Meng, J. Zhong, J. Song, Hyperbolic distribution of all-wheel independent braking force for cornering vehicle, IFAC Proc. Vol. 46 (21) (2013) 422–427, https://doi.org/10.3182/20130904-4-jp-2042.00101.

[131] G.P. Ostermeyer, J. Merlis, Modelling the friction boundary layer of an entire brake pad with an abstract cellular automaton, Lubricants 6 (2) (2018) 41–44, https://doi.org/10.3390/lubricants6020044.

[132] B. Tang, J.L. Mo, Y.K. Wu, X. Quan, M.H. Zhu, Z.R. Zhou, Effect of the friction block shape of railway brakes on the vibration and noise under dry and wet conditions, Tribol. Trans. (2019) 1–42, https://doi.org/10.1080/10402004.2018.1545954.

[133] P.C. Verma, M. Alemani, S. Gialanella, L. Lutterotti, U. Olofsson, G. Straffelini, Wear debris from brake system materials: a multi-analytical characterization approach, Tribol. Int. 94 (2016) 249–259, https://doi.org/10.1016/j.triboint.2015.08.011.

Proposal of a smart framework for a transportation system in a smart city

6

Marlyn Montalvo-Martel[a], Alberto Ochoa-Zezzatti[b], Elias Carrum[a], and Pedro Perez[a]

[a]*Mexican Materials Research Corporation, Saltillo, Coahuila, Mexico,* [b]*Autonomous University of Juarez City, Juarez City, Chihuahua, Mexico*

6.1 Introduction

Worldwide urbanization is creating profound changes in cities and these cities must develop strategies and plans to adapt to these changes and guarantee sustainability. In recent years, the Smart City (SC) concept has become popular due to the interest in achieving sustainability goals with the help of "intelligence." This definition was elaborated for the first time in the 1990s [1] as an attempt at applying new technologies to city planning processes.

Most studies on this subject agree that for a city to become "smart," it must provide mobility, society, quality of life, environment, economy, and government. These factors are considered the pillars of a smart city, and each one of them takes into account culture and wellness, government transparency, sustainable constructions, infrastructure, technologies, and so on (Fig. 6.1). Mobility is one of the most difficult issues to deal with in large metropolitan areas. It involves environmental and economic aspects and requires both technology and people. Smart mobility is influenced by Information and Communication Technologies (ICTs), which are used to improve traffic flows and to collect citizen points of view on the quality of local public transportation services [2].

According to Ref. [3], a city's mobility generally aims to optimize the logistics systems within the urban area, taking into account the costs and benefits for the public and private sectors. According to the above, the public sector tries to avoid traffic congestion and environmental problems. Therefore, globally optimized logistics systems offer advantages.

From a logistical point of view, the fast growth of urbanization rates and the potential change in the demand for goods in urban areas justify the need to develop new urban logistics systems that guarantee efficient urban mobility [4].

Artificial Intelligence and Industry 4.0. https://doi.org/10.1016/B978-0-323-88468-6.00007-3

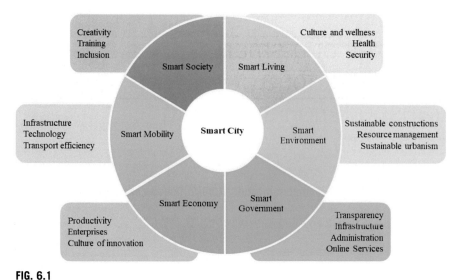

FIG. 6.1

Pillars of a Smart City.

Ref. [5] defines city logistics (urban mobility) as "a system that improves its transport activities (…), taking into account the environmental impact of vehicles, traffic jams, and energy consumption within the economy."

Global urbanization, especially in developed countries such as England and the Netherlands, indicates that it is time to establish regulated policies and urban designs according to the needs of the population to create smart and sustainable cities. One way to do this is to provide public transportation services that contribute to reducing private vehicle use, fuel costs, tickets, and transit taxes, as well as recovering spaces lost to road infrastructure for pedestrian use [6].

In Mexico, efforts are being made to diversify mobility and implement regulatory measures to reduce the use of cars and improve public transportation [6]. A study carried out by the Instituto Mexicano para la Competitividad (IMCO) compared urban mobility options and the ability to offer accessible public transportation in the country's most populous and industrialized cities. Only the cities of Saltillo, the Valley of Mexico, and Guadalajara were found to be capable of providing adequate public transportation [6].

According to Ref. [6], Saltillo is working to establish itself as a Smart City via the actions implemented by the municipal authority headed by Eng. Manolo Jiménez Salinas. Examples of these actions are the participation of business organizations, citizen councils, and committees, the use of new technologies such as LED lighting and solar panels, the use of methane gas, and communication tools to form support groups in which citizens participate with authorities to maintain public safety.

The municipality of Saltillo, which is the capital of Coahuila, is one of the main metropolitan areas of Mexico with the largest population (see Fig. 6.2).

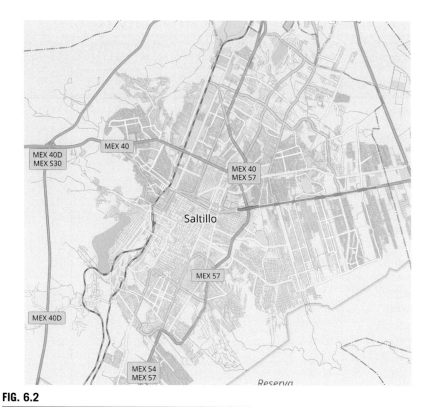

FIG. 6.2

Map of Saltillo metropolitan area [7].

This municipality is characterized as an industrial city. Table 6.1 breaks down the number of homes that have a car or truck in the municipalities that make up Coahuila. Using this table, it is possible to calculate the coverage density of private and public transportation.

Note that 61% of inhabited private homes in Saltillo have their own motor vehicles and thus 39% of homes do not have private transportation (Fig. 6.3). Based on the information provided in Table 6.1 and taking into account that the average number of people living in a household is 3.7 [9], the number of people requiring access to

Table 6.1 Access to private transportation in Coahuila [8].

Municipality or delegation	Inhabited private homes that have a car or truck	Total inhabited private homes	Density %
Abasolo	214	362	59
Acuna	20,392	36,749	55
Allende	4412	6322	70

Continued

Table 6.1 Access to private transportation in Coahuila [8]—cont'd

Municipality or delegation	Inhabited private homes that have a car or truck	Total inhabited private homes	Density %
Arteaga	3077	5901	52
Candela	287	529	54
Castanos	4213	6856	61
Cuatro Cienegas	1975	3605	55
Escobedo	490	810	60
Francisco I. Madero	6461	13,976	46
Frontera	11,402	19,466	59
General Cepeda	1256	3102	40
Guerrero	385	619	62
Hidalgo	290	500	58
Jimenez	1644	2666	62
Juarez	316	448	71
Lamadrid	274	549	50
Matamoros	11,561	26,128	44
Monclova	39,195	58,196	67
Morelos	1628	2239	73
Muzquiz	10,660	17,884	60
Nadadores	1081	1778	61
Nava	4932	7163	69
Ocampo	1784	2809	64
Parras	5496	11,869	46
Piedras Negras	25,929	41,040	63
Progreso	556	967	57
Ramos Arizpe	11,287	20,293	56
Sabinas	10,811	17,071	63
Sacramento	343	639	54
Saltillo	113,945	187,764	61
San Buenaventura	4022	6224	65
San Juan de Sabinas	7655	11,926	64
San Pedro	10,540	25,313	42
Sierra Mojada	1201	1627	74
Torreon	92,794	172,680	54
Viesca	1921	5269	36
Villa Unión	1204	1788	67
Zaragoza	2507	3608	69

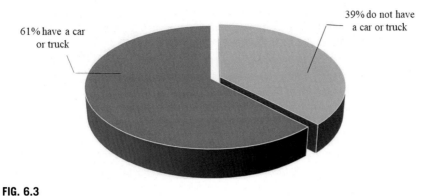

61% have a car or truck

39% do not have a car or truck

FIG. 6.3

Potential demand for public transportation.

public transportation is 694,727. According to the 2010 population census, Saltillo had 269,681 registered motor vehicles in circulation. Of these, 2287 were public passenger trucks (see Fig. 6.4). This provides public transportation authorities with an indicator to determine whether the public vehicle fleet should be increased to satisfy the demand for public transportation.

According to Ref. [7], in Saltillo, there are approximately 58 routes for public transportation, in which some official stops are determined. Despite the establishment of official stops, public transport trucks stop each time a user requests a stop regardless of where the truck is as long as it remains on the established route. This causes assigned stop times not to be respected and schedules to be delayed. In

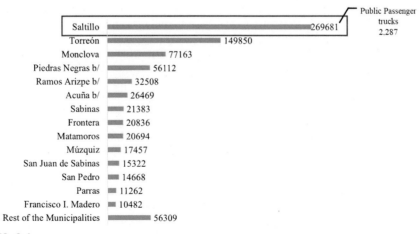

FIG. 6.4

Motor vehicles in circulation registered by municipalities [8].

addition, it causes traffic congestion since repeated, unscheduled stops obstruct the path of other vehicles. Repeated stopping and starting also results in higher fuel costs and increased emissions of CO, CO2, HC, and NOx [10–12].

As such, an algorithm is needed to explore the available public space and develop solutions to these issues. One potential algorithm is the Spider Monkey Optimization (SMO) meta-heuristic, which is an algorithm that mimics the actions of spider monkeys. Animals that follow this social behavior initially work in a large class ruled by an adult female who divides the class into smaller subclasses, if necessary, based on the need for foraging.

Considering the study by Nayak et al. [13], SMO is one of the most recent (and most popular) stochastic optimization techniques, which is mathematically modeled by [14]. Due to its fast convergence and effectiveness, SMO has attracted the attention of many researchers searching for solutions to complex issues. We use this algorithm in our study to determine and establish the best truck stop locations on an urban route.

6.2 Literary review

In the category of nature-inspired methods, there is a meta-heuristic approach used to fix optimization issues known as Swarm Intelligence (SI). It refers to the collective behaviors of social individuals with each other and with their environment to implement their social learning skills to overcome obstacles. These behaviors have been studied by many researchers, and algorithms have been created to solve nonconvex, nonlinear, or mixed optimization problems in numerous fields of science and engineering. Earlier studies [15–18] have revealed that algorithms based on SI have great potential to find solutions to real-life optimization problems, solutions that are hard to find using deterministic procedures [19]. In addition to SMO, some techniques that have been developed in current years include Particle Swarm Optimization (PSO) [16], Artificial Bee Colony Optimization (ABC) [20], Ant Colony Optimization (ACO) [15], and Bacterial Foraging Optimization (BFO) [21].

SMO has multiple applications in dissimilar areas, particularly real-valued optimization tasks, within a brief timber [22]. It has been used for:

1. designing electromagnetic antenna arrays [23]
2. Code Division Multiple Access (CDMA) multiuser detection [24]
3. optical power flow, pattern synthesis of sparse linear array and antenna arrays [25]
4. automatic generation of control [26]
5. rule mining for diabetes classifications [27]
6. designing optimal fuzzy rule base [28]
7. improving quality and diversity of particles and distributing them in particle filters to provide a robust object tracking framework [29]
8. numerical classification [30]

9. the economic dispatch problem [31]
10. optimizing models of a multi-reservoir system [32]
11. energy-efficient clustering for Wireless Sensor Networks (WSNs) [33]

SMO has also been used by [34], who presented it for the synthesis of sparse linear arrays. Several investigations have adapted SMO for optimization problems in various fields. Many of these studies used SMO to explain global optimization problems [35], improved the algorithm's local search capability [36], or proposed fitness-based position update tactics for spider monkeys [37].

On the other hand, to understand the grouping and regrouping properties that are used by the SMO, it's important to take into account that the grouping approach uses the whole population created in the optimization method and the regrouping procedure can be assumed as the continuation of the process of evolution [38]. Every SMO application listed above is a floating-point numeral optimization where mathematical processes are not hard to apply for dealing inside features and characters in SMO. Procedures that adopt grouping and regrouping techniques are effective for different numerical optimizations [39–41].

In this case, connection grouping and regrouping are indirectly assumed by SMO to master the results. The multi-population method is effective for discrete optimization (e.g., transportation problems) and is the motivation for [42] to develop an optimal approach to the Traveling Salesman Problem (TSP). In this study, the authors propose discrete SMO (DSMO) to solve the TSP. In this case, TSP solutions are represented by each spider monkey and an optimal solution to the TSP is found using the interactions between them. Results demonstrate the effectiveness of the DSMO algorithm.

In their research, Ref. [43] utilized an adapted SMO with Multilayer Perceptron (MLP) to resolve a classification problem in five diverse data groups. The original SMO with MLP was used to determine the best grouping outcome. The SMO was then improved by other meta-heuristic procedures, such as Differential Evolution (DE) and Gray Wolf Optimization (GWO), to train the MLP [44] suggest a different SMO taking into account an exponential adaptive approach for step size to accelerate the convergence ability of the swarm. In fact, Ref. [45] compared exponential SMO, chaotic SMO, and sigmoidal SMO and found the latter gave the best results.

Ref. [46] also modified SMO to solve a multi-objective resource allocation problem using the Taylor series model. They called the new optimization model Taylor-Spider Monkey Optimization (TaySMO). Another study by Ref. [47] used SMO to solve a multi-objective problem of improving network lifetime.

The social behavior of spider monkeys inspired [48] to carry out a stochastic optimization technique that mimics the searching behavior of these monkeys. Through trials on test problems and real-world problems, the authors found that, for almost all problems, trustworthiness (achievement rate), efficacy (average number of utility assessments), and accuracy (objective utility) of the SMO technique are greater than that of ABC, PSO, and DE. Therefore, it can be established that SMO is a competitive applicant in the field of swarm-based optimization techniques.

Ref. [49] applied SMO to minimize travel time and compared it to the ABC algorithm. SMO performed better than ABC due to its dispersed, stochastic, and self-organizing characteristics, which make it appropriate for the nature of traffic networks. However, the ABC method is faster in terms of execution time.

Using SMO to solve discrete-type optimization tasks has its own complications. Normal SMO workers cannot be applied for the TSP or related issues. New workers must be considered, or SMO must be improved for new solicitations. Hence, developing an instrument to solve transportation problems using SMO is the principal inspiration of this study. This chapter presents our proposal to use DSMO to minimize waiting times at truck stops.

6.3 Spider monkey optimization

6.3.1 Structure of fission–fusion culture

Fission–fusion culture was first described by Hans Kummer in his 1971 book exploring the social system of Hamadryas baboons [50]. The struggle for food among inhabitants of the group caused the group to split up to search for food (fission). Once food was found, the group came back together (fusion).

6.3.2 Spider monkey conduct

Spider monkeys inhabit rainforests in South and Central America, reaching almost to Mexico [51]. They are some of the smartest monkeys in the world and are known for the way they hang on their tails, which makes them look like spiders [52]. Spider monkeys always prefer living in a "parent group," which is a unit group. They are divided or combined depending on the insufficiency or accessibility of food. They communicate with one another through positions, motions, and screams.

6.3.3 Social conduct

Spider monkeys live in groups of 40–50. They hunt in small groups and diverse directions by day and night and share their bounties. The female leader decides where to hunt. If the group does not discover enough food, the leader splits the group into smaller ones, and they search on their own.

Communication. Unlike other animals, spider monkeys communicate through postures and attitudes. When they are separated by long distances, they communicate by chattering or shouting. Every monkey has its own particular sound, so it is easy for others to identify it.

Fig. 6.5 shows spider monkeys' hunting behavior.

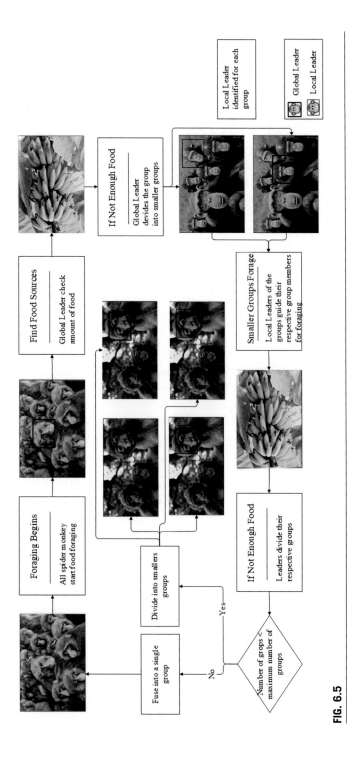

FIG. 6.5

Diagram of Spider Monkeys hunting behavior.

6.3.4 Spider monkey optimization process

SMO is a meta-heuristic approach that takes into account spider monkeys' intellectual hunting behavior. Spider monkey behavior is related to fission–fusion society. The cornerstone of this technique is group social associations in which the leader makes the decision of scattering or combining. The head of the whole group is called the global leader and the heads of the small groups are called local leaders. However, it is important to point out that the SMO technique does not improve a solution, because it is supported by the scarce food phenomenon and is a nongenerational algorithm. Because SMO is a technique based on SI, every small group has a minimum number of monkeys. Consequently, when additional fission results in one group having less than the minimum amount of monkeys, fusion occurs. The SMO algorithm has six stages: (1) local leader, (2) global leader, (3) local leader learning, (4) global leader learning, (5) local leader decision, and (6) global leader decision.

We describe each of these stages in the sections that follow.

Stage 0. In this initialization stage, SMO generates an initial spider monkey swarm (N) fitted to a uniform distribution, where SM_i represents the ith spider monkey in the group. Each SM_i is prepared as Eq. (6.1):

$$SM_{ij} = SM_{minj} + U(0,1) \times \left(SM_{maxj} - SM_{minj}\right) \tag{6.1}$$

where SM_{maxj} and SM_{minj} are the search space's upper and lower limits, respectively, in $j - $ th dimension, and $U(0,1)$ is a uniformly distributed arbitrary number.

Stage 1: Local leader. In this stage, each spider monkey has the opportunity to change position depending on the knowledge of its local group members and local leader. They consider every spider monkey's fitness value at its new location and if it is fitter at the new position than at its previous position, it updates. The equation for the updated position is:

$$SM_{newij} = SM_{ij} + U(0,1) \times \left(LL_{kj} - SM_{ij}\right) + U(-1,1) \times \left(SM_{rj} - SM_{ij}\right) \tag{6.2}$$

where SM_{ij} is the $j - $ th dimension of ith SM, LL_{kj} represents the $j - $ th length of the local leader of the $k - $ th group, and SM_{rj} is the $j - $ th length of a spider monkey forms selected arbitrarily the $k - $ th group such that $U(-1,1)$ is a homogeneously distributed arbitrary number in the interval $(-1,1)$ and $r \neq i$.

As a result, it's understood by Eq. (6.2) that the spider monkey, who will update its position, is attracted to wherever the local leader is to keep its tenacity or confidence. The last element helps to add variations in the hunt procedure, which could benefit to keep the procedure's stochastic nature to avoid premature stagnation.

Stage 2: Global leader. Once the local leader stage is completed, the procedure takes the step to the global leader stage. To update solutions, a probabilistic selection is used, which is a fitness function. From the objective function f_i, the fitness $f_i t_i$ can be found by Eqs. (6.3) and (6.4).

$$f_i t_i = \frac{1}{1 + f_i}, if \ f_i \geq 0 \tag{6.3}$$

$$f_i t_i = 1 + abs(\ f_i), if \ f_i < 0 \tag{6.4}$$

The probability selection p_i is estimated by taking into account the roulette wheel selection, and if f_it_i is the fitness of ith, the probability of being selected is calculated by:

$$p_i = \frac{f_it_i}{\sum\limits_{i=1}^{n} f_it_i} \qquad (6.5)$$

or

$$p_i = 0.9 \times \frac{f_it_i}{\max f_it_i} \qquad (6.6)$$

To update position, a spider monkey uses its determination, the skill of the global leader, and the experience of neighboring spider monkeys. Here, the position update equation is:

$$SM(new)_{ij} = SM_{ij} + U(0,1) \times \left(GL_j - SM_{ij}\right) + U(-1,1) \times \left(SM_{rj} - SM_{ij}\right) \qquad (6.7)$$

where GL_j is the location of the global leader in the jth iteration. This position-update equation is separated into three sections. Section one shows the parent spider monkey's perseverance, section two demonstrates the parent spider monkey's attraction to the global leader, and finally, section three is used to maintain the procedure's stochastic conduct. In Eq. (6.7), factor number two improves the exploitation of the identified hunt area, while factor number three assists the hunt procedure to evade a rash convergence or to decrease the opportunity of being trapped in a local optimum.

Stage 3: Global leader learning. Here, the proposed technique determines the best solution. As a result, the obtained spider monkey is the swarm's global leader. Moreover, it is possible to know the specific global leader position, which, if it is updated, increases the Global Limit Count (GLC) by 1; otherwise it is 0. Finally, it is necessary to compare the GLC, the global leader, and the Global Leader Limit (GLL).

Stage 4: Local leader learning. In this segment, the local leader location is updated through a greedy selection between group members. Note that if the local leader does not improve its location, then the local leader counter, known as the Local Limit Count (LLC), is increased by 1; otherwise the LLC is 0. This procedure is implemented for each group to discover its related local leader. The LLC is a counter that increases until it touches a fixed edge called the Local Leader Limit (LLL).

Stage 5: Local leader decision. Before this stage, the global leader and the local leaders were identified. If a local leader does not update to a specific limit (LLL), every group member updates their locations using two main methods: random initialization or global leader's experience using Eq. (6.8), which is calculated with a probability p_r known as the perturbation rate.

$$SM(new)_{ij} = SM_{ij} + U(0,1) \times \left(GL_j - SM_{ij}\right) + U(0,1) \times \left(SM_{rj} - SM_{ij}\right) \qquad (6.8)$$

Starting from this equation, these group solutions are refused from the existing local leader, and solutions are drawn to the global leader to modify the current hunt positions and directions. Moreover, according to p_r, some solution dimensions are arbitrarily established to present some disruption in the solutions' current positions. Now, the LLL is the variable that verifies that the local leader doesn't get caught in a local minimum. It is usually determined as $D \times N$, where N is the total number of spider monkeys and D is the dimension.

Stage 6: Global leader decision. As in the local leader choice stage, the global leader fuses groups into a single group or splits the group into smaller groups if the global leader does not get updated to a specific limit (GLC). At this point, GLL is the variable that checks for any early convergence and fluctuates from $N/2$ to $2 \times N$. GLC is 0 and the group numbers are compared to maximum groups if GLC is greater than GLL. The global leader splits the groups if the groups' current number is less than the predefined groups' maximum number; otherwise, it combines them to form a single or parent group.

Fig. 6.6 describes this fission–fusion process in 10 steps.

Step 1: The population, GLL, LLL, and perturbation rate *pr* are initialized.

Step 2: The population is evaluated.

Step 3: Global and local leaders are identified.

Step 4: The position is updated by the local leader stage.

Step 5: The position is updated by the global leader stage.

Step 6: Learning through the global leader learning stage.

Step 7: Learning through the local leader learning stage.

Step 8: The position is updated by the local leader decision stage.

Step 9: Fission or fusion are decided by the use of the global leader choice stage.

Step 10: If the end condition is satisfied, the global leader location is declared as the optimal result; otherwise go to step 4.

Fig. 6.7 shows the SMO pseudocode, which takes into account the six stages of the algorithm and the 10 steps of the methodology.

6.3.5 Application methodology SMO

SMO strikes an improved balance between exploration and exploitation as seeking the optimum. The local leader stage examines the hunt area since in this stage every group member updates their places with great disturbance in the dimensions. Meanwhile, the global leader stage encourages exploitation; the best contenders have extra opportunities to upgrade places. Due to this property, the SMO is the best candidate among search-based optimization procedures. SMO can be used for inactivity assessment as well. The local and global leader learning stages are implemented to assess whether the exploration process has deteriorated. When local leaders go inactive, the global leader makes the decision to continue the search for food. The local leader choice stage makes an extra examination and the decision based on fission or fusion is made in the global leader choice stage. Consequently, SMO exploration and exploitation are much more composed if convergence speed is maintained.

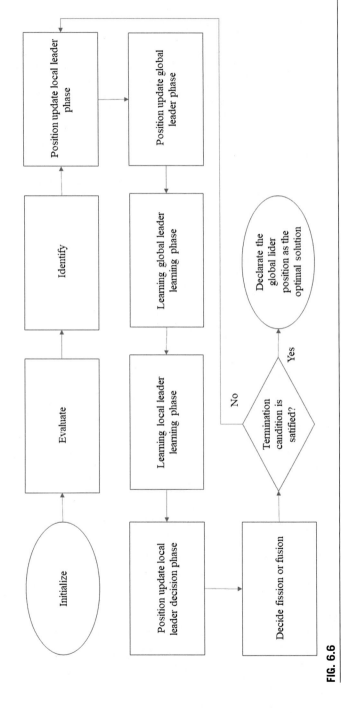

FIG. 6.6

Spider Monkey Optimization methodology.

Step 1 **Initialize** population, local leader limit, global leader limit and pertubation rate pr;

Step 2 **Evaluate** the population;

Step 3 **Identify** global and local laders;

Step 4 **Position update** by local leader phase;

    for each member $SM_i \in k^{th}$ group **do**

      for each $j \in \{1, ..., D\}$ **do**

      if $U(0,1) \geq pr$ **then**

$$SM\,new_{ij} = SM_{ij} + U(0,1) * (LL_{kj} - SM_{ij}) + U(-1,1) * (SM_{rj} - SM_{ij})$$

      else

        $SM\,new_{ij} = SM_{ij}$

      end if

      end for

    end for

Step 5 **Position update** by global leader phase;

    count = 0;

    while count $<$ *group size* **do**

      for each member $SM_i \in$ group **do**

        if $U(0,1) < prob_i$ **then**

        count = count + 1

        Randomly select $j \in \{1, ..., D\}$

        Randomly select $SM_r \in$ group s.t.r$\neq$i

        $SM\,new_{ij} = SM_{ij} + U(0,1) * (GL_j - SM_{ij}) + U(-1,1) * (SM_{rj} - SM_{ij})$

        end if

      end for

    end whille

Step 6 **Learning** through global leader learning phase;

Step 7 **Learning** through local leader learning phase;

Step 8 **Position update** by local leader decisión phase;

    if LocalLimitCount $>$ *LocalLeaderLimit* **then**

      LocalLimitCount = 0

    for each $j \in \{1, ..., D\}$ **do**

      if $U(0,1) \geq pr$ **then**

$$SM\,new_{ij} = SM_{ij} + U(0,1) * (SM_{maxj} - SM_{minj})$$

      else

$$SM\,new_{ij} = SM_{ij} + U(0,1) * (GL_j - SM_{ij}) + U(0,1) * (SM_{rj} - LL_{kj})$$

      end if

    end for

    end if

Step 9 **Decide fission or fusión** using global leader decisión phase;

    if GlobalLimitCount $>$ Global*LeaderLimit* **then**

      GlobalLimitCount = 0

    if Number og groups $<$ MG **then**

      Divide the swarms into groups

    else

      Combine all the groups to make a single group

    end if

      update Local Leaders positions

    end if

Step 10 If termination condition is satisfied stop and declare the global leader position as the optimal solution else go to step 4

FIG. 6.7

Spider Monkey Optimization pseudocode [19].

6.3.5.1 SMO parameters

SMO is composed of four control variables:

1. local leader limit (LLL)
2. global leader limit (GLL)
3. maximum group numbers (MGN)
4. perturbation rate p_r

If the local group leader does not update in a certain number of repetitions, the LLL has to be redirected. When the GLL value is reached without being updated, the group is divided into smaller subgroups by the global leader. In the case of the p_r and MGN, the first is determined by a function of the size of the current perturbation and the second is the function of a maximum number of groups in the population.

According to Ref. [19], the recommended parameter configuration (SMO algorithm for numerical optimization) is:

- $MGN = \frac{n}{10}$ (the minimum number of spider monkeys in a group should be 10)
- Global leader limit $\in \left[\frac{n}{2}, 2 \times N\right]$
- Local leader limit is set to $D \times N$
- $p_r \in [0.1, 0.8]$

6.4 Practical application

As previously discussed, the municipality of Saltillo, which is the capital of Coahuila, is one of the main metropolitan areas in Mexico with the largest population. It is characterized as an industrial city. As such, we use Route 2A in Saltillo in our case study of public transportation (see Fig. 6.8).

Route 2A has a total distance of 21,113 m (21.113 km), a trip of about 1:40 h. On this route, 21 trucks operate and leave every 8 min [53].

For the efficient management of the resources of a public transportation system, it is essential to design correctly itineraries, schedules, frequencies, and fleet of trucks [54]. In addition, when implementing a public transportation system in a city, or modifying an existing one, it is important to establish the location of truck stops. Rational distribution of stops is an advantage not only for the transportation system but also for traffic in general. The number of stops is significantly influenced by the demand for transportation in the different areas of the city.

Demand can change unexpectedly or randomly; however, since the demand for transportation depends on economic activities that have a high degree of routine and repetition, there may be a certain tendency to show a more or less stable cyclical behavior [55].

To identify mobility trends of transportation, it is necessary to know the distribution of the population by socioeconomic levels: high, medium-high, medium-low, and low.

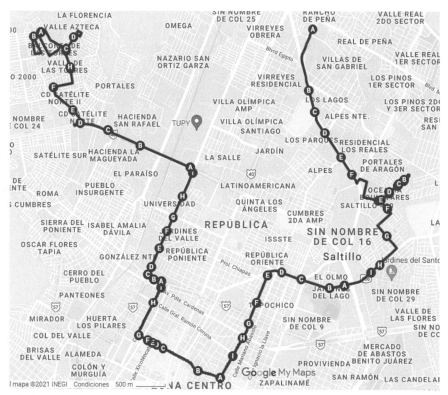

FIG. 6.8

Route 2A map.

Fig. 6.9 shows that the population with a high socioeconomic level is located mostly towards the north of the municipality. Here is where there are the largest number of cars per home. The populations at the lowest socioeconomic levels are those that use public transportation. Therefore, it is in the areas of medium-low and low socioeconomic levels where the greatest number of truck stops should be located.

Other influencing factors are the location of the areas with the corridors that have the greatest social activity and the main work destinations, as well as the residential areas because they show the origins of the trips. Fig. 6.10 shows the secondary zone of the urban area according to the land uses of the Urban Development Master Plan in force in Saltillo. Visually identifying destinations allows us to understand the urban structure of the origins and destinations of the trips that are carried out within the municipality.

Urban density is an important factor in understanding how cities work. Low-density cities put the sustainability of public transportation systems at risk due to lack of sufficient demand from users because they are dispersed throughout the urban

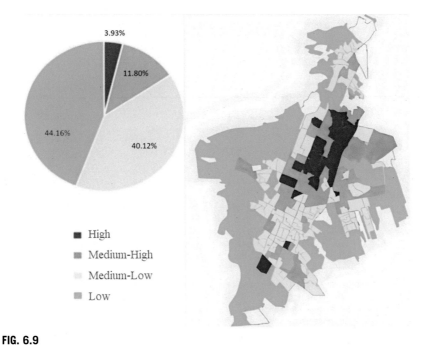

FIG. 6.9

Distribution of the population by socioeconomic level [56].

territory. According to Ref. [57], Saltillo has an urban density of 4217 inhabitants per km^2 (42.17 Hab/ha), while its metropolitan area has 4290 inhabitants per km^2 (42.9 Hab/ha). Fig. 6.11 displays the urban density map, which is the number of people per hectare who live in the urban area of the municipality. Note that the largest number of people per hectare or square kilometer is to the south of the municipality.

Each stop means delays for passengers on board. The greatest influences in determining passenger travel time are disembarkation and embarkation times along with the capacity of the truck. Reducing the number of stops saves time, but it tends to increase the loading and unloading times of the truck.

According to Article 52 of Chapter 3 of the Sustainable Mobility Law for the State of Coahuila de Zaragoza [58], the public collective transportation service must have "(…) a capacity that may not be less than twenty-two seated passengers, in which up to twenty percent of passengers in addition to the number of seats available to the unit may be admitted (…)."

According to field research, the urban public transportation fleet in Saltillo is heterogeneous. One of the trucks used is a Volkswagen Volksbus 9.160 OD model. Its capacity is 34 seated passengers [59], therefore, according to the Sustainable Mobility Law of the State of Coahuila de Zaragoza, the truck has a total capacity of 40 passengers.

FIG. 6.10

Urban structure [56].

6.4.1 Initial conditions

Tables 6.2 and 6.3 show the general initial conditions of the system. The *Node* column indicates origin (i) and destination ($i + 1$), the second column indicates the number of people known as the demand (X_1), the third column indicates distance between two public transportation truck stops in meters (X_2), and the fourth column indicates time in minutes (X_3) it takes for the truck to go from one stop to the next.

In addition, we consider the constant variable capacity (X_4), which, as we previously explained, is 40 passengers. We added a restriction in the model so that the distance between the truck stops does not exceed 500 m because this is the maximum distance between two consecutive truck stops existing on Route 2A.

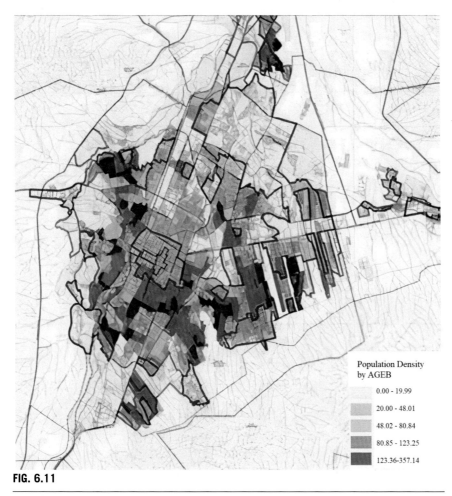

FIG. 6.11

Density of the municipal seat, 2010 [56].

Eq. (6.9) shows the model of the transportation route, which uses the SMO algorithm to minimize waiting time.

$$W_T = \sqrt{\frac{D_i * (X_{i+1} - X_i)^2}{\sum Di}} + B_C \qquad (6.9)$$

where W_T is the waiting time to be minimized;
    D_i is the demand (amount of people) in the node i;
    X_i is the node in the position i;
    X_{i+1} is the node in the position $i + 1$; and.
    B_C is the constant truck capacity.

Table 6.2 General initial conditions (Part I).

Node		Node i demand—X_1	Distance—X_2	Time—X_3
i	$i+1$	(Amount of people)	(m)	(min)
0	1	9	998	1.00
1	2	6	355	0.90
2	3	11	377	0.94
3	4	2	451	0.99
4	5	3	322	0.95
5	6	12	844	2.00
6	7	2	213	0.34
7	8	5	189	0.90
8	9	4	121	0.37
9	10	1	266	1.00
10	11	3	177	0.78
11	12	6	198	0.39
12	13	4	171	0.64
13	14	9	200	0.54
14	15	3	599	0.99
15	16	12	644	2.00
16	17	4	181	0.81
17	18	2	481	1.00
18	19	7	244	0.94
19	20	10	479	0.96
20	21	6	330	0.58
21	22	12	266	1.00
22	23	6	456	1.00
23	24	4	345	0.23
24	25	3	251	0.35
25	26	12	321	0.82
26	27	3	393	0.17
27	28	5	398	0.65

6.4.2 SMO control parameters

The following SMO control parameters have been established by researchers.

Population size or swarm, $N = 50$.

Dimension of problem, $D = 4$

Minimum number of individuals in a group (10) and maximum group number (MGN),

$$MGN = \frac{N}{10} = 5$$

Table 6.3 General initial conditions (Part II).

Node		Node i demand—X_1	Distance—X_2	Time—X_3
i	$i+1$	(Amount of people)	(m)	(min)
28	29	3	668	3.00
29	30	4	124	0.30
30	31	6	68	0.20
31	32	12	81	0.31
32	33	11	163	0.22
33	34	7	594	1.00
34	35	8	261	0.69
35	36	1	125	0.37
36	37	7	120	0.19
37	38	11	152	0.65
38	39	6	121	0.10
39	40	5	325	0.73
40	41	2	292	0.74
41	42	8	232	0.63
42	43	12	357	0.78
43	44	9	408	0.93
44	45	6	127	0.55
45	46	8	879	1.00
46	47	8	590	1.00
47	48	9	695	2.00
48	49	8	262	1.00
49	50	7	499	2.00
50	51	12	557	2.00
51	52	8	24	0.13
52	53	8	269	0.45
53	54	9	574	1.00
54	55	11	126	0.15
55	56	5	1512	6.00
56	57	9	638	2.00

Global leader limit, $GLL \in \left\{\frac{N}{2}, 2N\right\} \Rightarrow GLL \in \{25, 100\}$

$$\text{Let } GLL = 75$$

Local leader limit, $LLL = D \times N = 4 \times 50 = 200$

Perturbation rate, $p_r \in [0.1, 0.8]$

$$\text{Let } p_r = 0.8$$

6.5 Discussion of results

Tables 6.4 and 6.5 show the average of the values that correspond to each of the monkeys and the function evaluated in each of the positions of these monkeys.

Table 6.5 (Part II) contains the values that correspond to the monkey with the best evaluation, according to the method proposed by [19]. It is possible to visualize the values that correspond to the fitness function with the best performance (which are shown in a bold format in Table 6.5), that is, the shortest waiting time for the user is 18 min ($f(x)$), the demand (X_1) is 97 people, the average distance between two consecutive stops (X_2) is 489 m, and the mean arrival time at a stop (X_3) is 4.13 min with a truck capacity (X_4) of 40 passengers.

Fitness values are estimated by means of the fitness function equation (f_it_i):

$$f_it_i = \frac{1}{1+f_i}, if \ f_i \geq 0 \tag{6.10}$$

$$f_it_i = 1 + abs(f_i), if \ f_i < 0 \tag{6.11}$$

The highest fitness value is 0.986, which is equal to monkey 28, who became the global leader (this value is in bold in Table 6.6). As things stand, there is just one group, so the 28th spider monkey is both the global and local leader (Table 6.6).

Table 6.4 SMO results (Part I).

SM	Mean demand X_1	Mean distance X_2	Mean time X_3	Capacity X_4	$f(X_1,X_2,X_3,X_4)$
1	66	316	0.68	40	16
2	81	428	3.85	40	18
3	82	456	2.35	40	20
4	83	221	0.58	40	19
5	60	282	0.8	40	16
6	78	362	2.04	40	20
7	67	396	2.6	40	18
8	89	417	3.87	40	17
9	65	367	2.53	40	16
10	83	489	3.59	40	20
11	75	321	1.92	40	19
12	73	434	3.36	40	18
13	83	357	2.36	40	16
14	92	289	1.18	40	15
15	81	412	2.9	40	17
16	85	266	1.01	40	16
17	70	298	2.7	40	15
18	85	439	3.94	40	18
19	69	421	2.92	40	18
20	84	458	3.18	40	19

Table 6.5 SMO results (Part II).

SM	Mean demand X_1	Mean distance X_2	Mean time X_3	Capacity X_4	$f(X_1,X_2,X_3,X_4)$
21	78	489	3.82	40	20
22	71	258	0.34	40	17
23	73	410	4.72	40	20
24	65	311	2.35	40	18
25	65	130	0.24	40	15
26	84	301	0.92	40	17
27	82	512	4.76	40	19
28	**97**	**489**	**4.13**	**40**	**18**
29	88	460	4.16	40	17
30	82	307	2.61	40	19
31	82	178	0.58	40	16
32	84	232	0.57	40	16
33	75	491	4.6	40	20
34	80	341	2.36	40	18
35	80	324	1.69	40	16
36	85	243	0.25	40	14
37	67	378	1.41	40	17
38	60	432	3.09	40	18
39	78	436	3.12	40	19
40	77	478	4.07	40	20
41	89	449	3.5	40	18
42	82	332	0.9	40	16
43	67	421	3.23	40	19
44	78	289	0.68	40	17
45	72	230	0.49	40	15
46	72	421	3.12	40	18
47	67	202	0.2	40	16
48	86	303	0.89	40	15
49	79	365	2.63	40	19
50	70	390	1.28	40	18

Global Leader Learning Stage. In this stage, the swarm's global leader is chosen and a comparison with the fitting results is made. If the global leader updates its position, the GLC is 0; otherwise, it is 1. The fitting of the 28th spider monkey is the greatest in the updated swarm, which is why it is the universal leader. The GLC is 0 as long as the global leader has been improved.

Local Leader Learning Stage. In this stage, the groups' local leaders are selected in this stage and a comparison with the fitting results is made. If the local leader improves their position, the LLC is 0; otherwise, it is 1. In this case, the 28th spider

Table 6.6 Fitness values.

SM	Fitness	SM	Fitness
1	0.919	26	0.181
2	0.178	27	0.095
3	0.079	**28**	**0.986**
4	0.790	29	0.873
5	0.729	30	0.703
6	0.933	31	0.546
7	0.956	32	0.093
8	0.200	33	0.192
9	0.520	34	0.099
10	0.212	35	0.068
11	0.933	36	0.091
12	0.921	37	0.151
13	0.640	38	0.241
14	0.434	39	0.124
15	0.302	40	0.157
16	0.565	41	0.757
17	0.066	42	0.926
18	0.744	43	0.872
19	0.032	44	0.452
20	0.418	45	0.971
21	0.285	46	0.872
22	0.935	47	0.976
23	0.781	48	0.712
24	0.299	49	0.277
25	0.271	50	0.562

monkey is both the global and local leader since there is only one group. The GLC is 0 as long as the global leader has been improved.

Tables 6.7 and 6.8 show the nodes proposed by the SMO algorithm to reduce user waiting times. The estimated waiting time is 18 min.

To test the effectiveness of the proposed model solved by the SMO algorithm, we compared the waiting times achieved with other models, including Genetic Algorithm (GA), PSO, and Simulated Annealing (SA). Table 6.9 presents the results of this comparison. The SMO algorithm shows the best performance.

AG parameter settings:

- Crossover probability $= 0.8$
- Mutation probability $= 0.5$
- Swarm size $= 100$

Table 6.7 Route proposed by the SMO algorithm (Part I).

Node		Node *i* demand	Distance	Time
i	*i* + 1	(Amount of people)	(m)	(min)
0	1	9	950	4.71
1	2	6	370	3.99
2	3	11	385	5.01
3	4	2	454	4.42
4	5	3	342	4.08
5	6	12	481	4.55
6	7	2	280	4.12
7	8	5	202	4.33
8	9	4	171	4.56
9	10	1	290	4.69
10	11	3	198	3.88
11	12	6	206	3.32
12	13	4	189	4.43
13	14	9	289	4.03
14	15	3	450	4.29
15	16	12	578	3.99
16	17	4	302	4.66
17	18	2	481	4.34
18	19	7	351	4.02
19	20	10	479	3.87
20	21	6	370	5.14
21	22	12	302	4.44
22	23	6	457	4.72
23	24	4	445	4.15
24	25	3	432	4.55
25	26	12	423	4.15
26	27	3	432	4.51
27	28	5	425	4.29

PSO parameter settings:

- Inertia weight decreases linearly from 0.9 to 0.4
- Acceleration coefficients ($c_1 = 0.05$, $c_2 = 0.07$)
- Swarm size $= 100$

SA parameter settings:

- Temperature $= 100°C$
- Swarm size $= 100$

Table 6.8 Route proposed by the SMO algorithm (Part II).

Node		Node *i* demand	Distance	Time
i	*i* + 1	(Amount of people)	(m)	(min)
28	29	3	589	3.35
29	30	4	354	4.54
30	31	6	232	5.14
31	32	12	180	2.29
32	33	11	290	3.95
33	34	7	458	3.57
34	35	8	365	3.63
35	36	1	324	4.35
36	37	7	267	4.09
37	38	11	243	3.74
38	39	6	232	4.52
39	40	5	375	3.57
40	41	2	356	3.77
41	42	8	312	3.99
42	43	12	432	4.91
43	44	9	498	4.30
44	45	6	253	4.52
45	46	8	781	3.88
46	47	8	550	4.69
47	48	9	589	4.04
48	49	8	382	3.44
49	50	7	481	3.93
50	51	12	489	4.17
51	52	8	100	4.81
52	53	8	378	4.58
53	54	9	459	3.82
54	55	11	278	4.41
55	56	5	1302	4.02
56	57	9	638	4.24

Table 6.9 Comparison of waiting times achieved by different optimization algorithms.

	Algorithms			
	SMO	GA	PSO	SA
Waiting time	18 min	26 min	33 min	39 min

Table 6.10 Comparison of computational time in seconds among different optimization algorithms.

	Mean computational time (s)			
	Small	**Medium**	**Large**	**Very large**
	1–9	**9–18**	**18–27**	**27–36**
SMO		9.315		
GA			19.552	
PSO		12.789		
SA				29.771

We also compared the computational times of the four algorithms (Table 6.10). The results confirm that the SMO algorithm has the shortest computational time of 9315 s.

Regarding the convergence of the different algorithms, Fig. 6.12 shows that none of the methods converged in the 50 running generations due to the great variability of

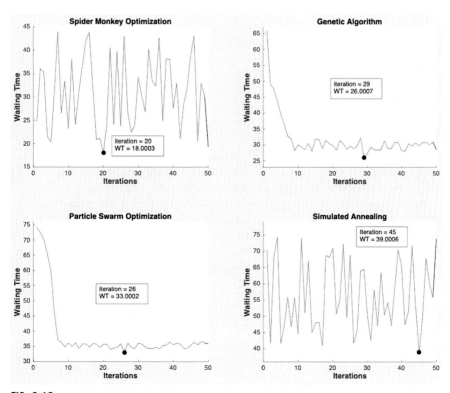

FIG. 6.12

Comparison of convergence among different optimization algorithms.

the waiting time in the plots. Therefore, a sub-optimal value of user waiting time was obtained, which was thrown by the spider monkey in iteration 20. Therefore it is recommended to increase the number of iterations in the models to ensure their convergence.

6.6 Conclusions

In this chapter, we presented a novel technique based on Swarm Intelligence (SI) called Spider Monkey Optimization (SMO) to model a public transportation system in Saltillo, the capital of Coahuila, Mexico. The SMO algorithm is inspired by the social behavior of spider monkeys and is used by many researchers to solve complex logistical problems. We chose Saltillo for our experiment because it is an urban area that has experienced rapid expansion of industrial development and economic activity. As such, we deemed it necessary to define a public transportation route taking into account current transportation constraints such as number of truck stops, distance between stops, total transportation time, truck capacity, and passenger demand.

The SMO algorithm is a useful tool that has shown successful results in logistic and transport applications. In SMO, examination of the area is done in the global and local leader stages, while search for solutions is done in the global and local leader decision stages. Whenever there is a large number of user-dependent considerations in an SMO algorithm, it is beneficial to conduct additional investigations. To achieve convergence, the number of iterations run by the model should be increased. The regulation of self-adaptive consideration can benefit the strength improvement and the procedure reliability.

The SMO algorithm we applied to Route 2A in Saltillo helped reduce expected passenger waiting time to an average of 18 min, with an average distance between stops of 489 m, and truck capacity of 40 passengers. We compared these results with those of other algorithms and the SMO algorithm showed the best performance. Waiting times determined by Genetic Algorithm (GA), Particle Swarm Optimization (PSO), and Simulated Annealing (SA) were 26, 33, and 39 min, respectively.

In future works, it will be possible to include the number of urban trucks, loaded weight, and vehicular float number in the SMO model. A proposal for new truck stop locations can also be made to fulfill the distance between stops proposed by the SMO algorithm. These new conditions will provide a robust and effective model that can be applied to other public transportation routes in Saltillo.

References

[1] S. Alawadhi, A. Aldama-Nalda, H. Chourabi, J.R. Gil-Garcia, S. Leung, S. Mellouli, et al., Building understanding of smart city initiatives, in: H.J. Scholl, M. Janssen, M. A. Wimmer, C.E. Moe, L.S. Flak (Eds.), Electronic Government, vol. 7443, Springer Berlin Heidelberg, Berlin, Heidelberg, 2012, pp. 40–53, https://doi.org/10.1007/978-3-642-33489-4_4.

[2] C. Benevolo, R. Dameri, B. D'Auria, Smart Mobility in Smart City. Action Taxonomy, ICT Intensity and Public Benefits, vol. 11, 2016, pp. 13–28, https://doi.org/10.1007/978-3-319-23784-8_2.

[3] J.D.J.Á. Montero, A.E. Sarmiento, La logística urbana, la ciudad logística y el ordenamiento territorial logístico, RETO 4 (2016) 21–40.

[4] E. Fatnassi, O. Chebbi, J. Chaouachi, Viability of implementing smart mobility tool in the case of Tunis City, in: K. Saeed, W. Homenda (Eds.), Computer Information Systems and Industrial Management, vol. 9339, Springer International Publishing, Cham, 2015, pp. 339–350, https://doi.org/10.1007/978-3-319-24369-6_28.

[5] E. Taniguchi, R.G. Thompson, Modeling City Logistics, 1790, Transportation Research Record, 2002, pp. 45–51.

[6] A. Duque-García, Trasporte y movilidad eficiente en el ámbito nacional e internacional ¿Por qué la ciudad inteligente es necesaria?, vol. 13, 2019, pp. 10–13.

[7] A. Vázquez, Rutas de Camiones de Saltillo y su Área Metropolitana—rutadirecta.com, rutadirecta, 2009. https://saltillo.rutadirecta.com/.

[8] Instituto Nacional de Estadística y Geografía, Censo de Población y Vivienda 2010, INEGI, 2016. https://www.inegi.org.mx/programas/ccpv/2010/.

[9] Instituto Nacional de Estadística y Geografía, Principales resultadosde la Encuesta Intercensal 2015, Coahuila de Zaragoza, 2016.

[10] J.-Q. Li, S.D. Gupta, L. Zhang, K. Zhou, W.-B. Zhang, Evaluate bus emissions generated near far-side and near-side stops and potential reductions by ITS: an empirical study, Transp. Res. Part D: Transp. Environ. 17 (2012) 73–77, https://doi.org/10.1016/j.trd.2011.09.012.

[11] B. Narotzki, A.Z. Reznick, T. Mitki, D. Aizenbud, Y. Levy, Bus Stop Air Quality: An Empirical Analysis of Exposure to Particulate Matter at Bus Stop Shelters, 2014, https://doi.org/10.1089/rej.2013.1540.

[12] Y. Pan, S. Chen, T. Li, S. Niu, K. Tang, Exploring spatial variation of the bus stop influence zone with multi-source data: a case study in Zhenjiang, China, J. Transp. Geogr. 76 (2019) 166–177, https://doi.org/10.1016/j.jtrangeo.2019.03.012.

[13] J. Nayak, K. Vakula, P. Dinesh, B. Naik, Spider monkey optimisation: state of the art and advances, Int. J. Swarm Intell. Res. 4 (2019) 175–198.

[14] J.C. Bansal, H. Sharma, S.S. Jadon, M. Clerc, Spider monkey optimization algorithm for numerical optimization, Memet. Comput. 6 (2014) 31–47, https://doi.org/10.1007/s12293-013-0128-0.

[15] M. Dorigo, M. Birattari, T. Stützle, Ant colony optimization, IEEE Comput. Intell. Mag. 1 (2006) 28–39, https://doi.org/10.1109/MCI.2006.329691.

[16] J. Kennedy, R. Eberhart, Particle swarm optimization, in: Proceedings of ICNN'95—International Conference on Neural Networks, vol. 4, 1995, pp. 1942–1948, https://doi.org/10.1109/ICNN.1995.488968.

[17] K. Price, R.M. Storn, J.A. Lampinen, Differential Evolution: A Practical Approach to Global Optimization, Springer-Verlag, Berlin Heidelberg, 2005, https://doi.org/10.1007/3-540-31306-0.

[18] J. Vesterstrom, R. Thomsen, A comparative study of differential evolution, particle swarm optimization, and evolutionary algorithms on numerical benchmark problems, in: Proceedings of the 2004 Congress on Evolutionary Computation (IEEE Cat. No.04TH8753), vol. 2, 2004, pp. 1980–1987, https://doi.org/10.1109/CEC.2004.1331139.

[19] H. Sharma, G. Hazrati, J.C. Bansal, Spider monkey optimization algorithm, in: J.C. Bansal, P.K. Singh, N.R. Pal (Eds.), Evolutionary and Swarm Intelligence Algorithms, Springer International Publishing, Cham, 2019, pp. 43–59, https://doi.org/10.1007/978-3-319-91341-4_4.

[20] D. Karaboga, An idea based on honey bee swarm for numerical optimization, technical report—TR06, Erciyes University, 2005.

[21] K.M. Passino, Bacterial foraging optimization, Int. J. Swarm Intell. Res. 1 (2010) 1–16, https://doi.org/10.4018/jsir.2010010101.

[22] V. Agrawal, R. Rastogi, D.C. Tiwari, Spider monkey optimization: a survey, Int. J. Syst. Assur. Eng. Manag. 9 (2018) 929–941.

[23] A.A. Al-Azza, A.A. Al-Jodah, F.J. Harackiewicz, Spider monkey optimization: a novel technique for antenna optimization, IEEE Antennas Wirel. Propag. Lett. 15 (2016) 1016–1019, https://doi.org/10.1109/LAWP.2015.2490103.

[24] A. Kaur, Comparison analysis of CDMA multiuser detection using PSO and SMO, Int. J. Comput. Appl. 133 (2016) 47–50, https://doi.org/10.5120/ijca2016907739.

[25] U. Singh, R. Salgotra, M. Rattan, A novel binary spider monkey optimization algorithm for thinning of concentric circular antenna arrays, IETE J. Res. 62 (2016) 736–744, https://doi.org/10.1080/03772063.2015.1135086.

[26] A. Sharma, H. Sharma, A. Bhargava, N. Sharma, Power law based local search in spider monkey optimization for lower order system modeling, Int. J. Solids Struct. 48 (2017) 150–160, https://doi.org/10.1080/00207721.2016.1165895.

[27] R. Cheruku, D.R. Edla, V. Kuppili, SM-RuleMiner: spider monkey based rule miner using novel fitness function for diabetes classification, Comput. Biol. Med. 81 (2017) 79–92, https://doi.org/10.1016/j.compbiomed.2016.12.009.

[28] J. Dhar, S. Arora, Designing fuzzy rule base using spider monkey optimization algorithm in cooperative framework, Future Comput. Inform. J. 2 (2017) 31–38, https://doi.org/10.1016/j.fcij.2017.04.004.

[29] R. Rohilla, V. Sikri, R. Kapoor, Spider monkey optimization assisted particle filter for robust object tracking, IET Comput. Vis. 11 (2017) 207–219, https://doi.org/10.1049/iet-cvi.2016.0201.

[30] B. Omkar, D. Preet, D. Swarada, D. Poonam, Dengue fever classification using SMO optimization algorithm, Int. Res. J. Eng. Technol. 04 (2017) 1683–1686.

[31] A.F. Ali, An improved spider monkey optimization for solving a convex economic dispatch problem, in: Nature-Inspired Computing and Optimization, Springer, 2017, pp. 425–448.

[32] M. Ehteram, H. Karami, S. Farzin, Reducing irrigation deficiencies based optimizing model for multi-reservoir systems utilizing spider monkey algorithm, Water Resour. Manag. 32 (2018) 2315–2334, https://doi.org/10.1007/s11269-018-1931-7.

[33] N. Mittal, U. Singh, R. Salgotra, B. Sohi, A boolean spider monkey optimization based energy efficient clustering approach for WSNs, Wirel. Netw. 24 (2018) 2093–2109, https://doi.org/10.1007/s11276-017-1459-4.

[34] H. Wu, Y. Yan, C. Liu, J. Zhang, Pattern synthesis of sparse linear arrays using spider monkey optimization, IEICE Trans. Commun. E100.B (2017) 426–432, https://doi.org/10.1587/transcom.2016EBP3203.

[35] P.R. Singh, M.A. Elaziz, S. Xiong, Modified spider monkey optimization based on Nelder–Mead method for global optimization, Expert Syst. Appl. 110 (2018) 264–289, https://doi.org/10.1016/j.eswa.2018.05.040.

[36] S. Kumar, V. Sharma, R. Kumari, Self-adaptive spider monkey optimization algorithm for engineering optimization problems, Int. J. Inf. Commun. Comput. Technol. II (2015) 96–107.

[37] S. Kumar, R. Kumari, V.K. Sharma, Fitness based position update in spider monkey optimization algorithm, Procedia Comput. Sci. 62 (2015) 442–449, https://doi.org/10.1016/j.procs.2015.08.504.

[38] M.N. Omidvar, X. Li, Y. Mei, X. Yao, Cooperative co-evolution with differential grouping for large scale optimization, IEEE Trans. Evol. Comput. 18 (2014) 378–393, https://doi.org/10.1109/TEVC.2013.2281543.

[39] M.A. Al-Betar, M.A. Awadallah, I. Abu Doush, A.I. Hammouri, M. Mafarja, Z.A.A. Alyasseri, Island flower pollination algorithm for global optimization, J. Supercomput. 75 (2019) 5280–5323, https://doi.org/10.1007/s11227-019-02776-y.

[40] F.B. Özsoydan, A. Baykasoğlu, A multi-population firefly algorithm for dynamic optimization problems, in: 2015 IEEE International Conference on Evolving and Adaptive Intelligent Systems (EAIS), 2015, https://doi.org/10.1109/EAIS.2015.7368777.

[41] F.B. Özsoydan, A. Baykasoğlu, Quantum firefly swarms for multimodal dynamic optimization problems, Expert Syst. Appl. 115 (2019) 189–199, https://doi.org/10.1016/j.eswa.2018.08.007.

[42] M.A.H. Akhand, S.I. Ayon, S.A. Shahriyar, N. Siddique, H. Adeli, Discrete spider monkey optimization for travelling salesman problem, Appl. Soft Comput. 86 (2020), https://doi.org/10.1016/j.asoc.2019.105887, 105887.

[43] M. Diallo, X. Shengwu, B. Singh, Improved spider monkey optimization algorithm to train MLP for data classification, 3C Tecnol. (2019) 142–165, https://doi.org/10.17993/3ctecno.2019.specialissue2.142-165.

[44] A. Sharma, N. Sharma, H. Sharma, B.J. Chand, Exponential adaptive strategy in spider monkey optimization algorithm, in: A.K. Nagar, K. Deep, J.C. Bansal, K.N. Das (Eds.), Soft Computing for Problem Solving 2019, Springer, Singapore, 2020, pp. 1–15, https://doi.org/10.1007/978-981-15-3287-0_1.

[45] B. Sharma, V.K. Sharma, S. Kumar, Comparative analysis of selected variant of spider monkey optimization algorithm, in: H. Sharma, K. Govindan, R.C. Poonia, S. Kumar, W.M. El-Medany (Eds.), Advances in Computing and Intelligent Systems, Springer, Singapore, 2020, pp. 365–372, https://doi.org/10.1007/978-981-15-0222-4_33.

[46] R. Menon, A. Kulkarni, D. Singh, M. Venkatesan, Hybrid multi-objective optimization algorithm using Taylor series model and Spider Monkey Optimization, Int. J. Numer. Methods Eng. 122 (2021) 2478–2497, https://doi.org/10.1002/nme.6628.

[47] D. Samiayya, A. Ramalingam, Multi-Objective Spider Monkey Optimization for Energy Efficient Clustering and Routing in Wireless Sensor Networks, Research Square, 2021.

[48] J. Chand Bansal, H. Sharma, S. Singh Jadon, M. Clerc, Spider monkey optimization algorithm for numerical optimization, Memet. Comput. 6 (2014) 31–47, https://doi.org/10.1007/s12293-013-0128-0.

[49] S.E. Ezekwere, V.I.E. Anireh, M. Daniel, Application of the spider monkey optimization algorithm in a class of traffic delay problem, SSRG Int. J. Comput. Sci. Eng. 7 (2) (2020) 48–56.

[50] I.D. Couzin, M.E. Laidre, Fission–fusion populations, Curr. Biol. 19 (2009) R633–R635, https://doi.org/10.1016/j.cub.2009.05.034.

[51] Spider Monkeys | National Geographic, 2020. https://www.nationalgeographic.com/animals/mammals/group/spider-monkeys/. (Accessed 26 December 2020).

[52] Spider Monkey—Facts, Diet & Habitat Information, 2020. https://animalcorner.org/animals/spider-monkey/. (Accessed 26 December 2020).

[53] Instituto Municipal del Transporte, Ley de Transparencia y Acceso a la Información. Artículo 28: Ruta 2A, Municipio de Saltillo, 2018. https://saltillo.gob.mx/transparencia-imt/imt-articulo-28/.

[54] A. Ibeas, L. Dell'Olio, B. Alonso, O. Sainz, Optimizing bus stop spacing in urban areas, Transp. Res. E: Logist. Transp. Rev. 46 (2010) 446–458, https://doi.org/10.1016/j.tre.2009.11.001.

[55] V.M. Islas-Rivera, C. Rivera-Trujillo, V.G. Torres, Estudio de la demanda del transporte, Instituto Mexicano Del Transporte, 2002.

[56] Instituto Municipal de Planificación, Sistema de Indicadores de Movilidad Urbana Saltillo, vol. 2015, 2015. http://implansaltillo.mx/publicaciones-4/sistema-de-indicadores-de-movilidad-urbana-saltillo-2015.

[57] Instituto Municipal de Planificación, Propuesta Ordenamiento del Sistema de Transporte Público de la Zona Conurbada de Saltillo, 3, IMPLAN, 2015.

[58] Congreso del Estado de Coahuila, Ley de Transporte y Movilidad Sustentable para el Estado de Coahuila de Zaragoza, 2018, p. 976.

[59] MAN Truck and Bus México, Autobús Volkswagen urbano minibús 9.160 OD, MAN Truck and Bus, 2017. http://www.mantruckandbus.com.mx.

Society 5.0: Effective technology for a smart society

M. Hanefi Calp[a] and Resul Bütüner[b]

[a]*Department of Management Information Systems, Faculty of Economics & Administrative Sciences, Ankara Hacı Bayram Veli University, Ankara, Turkey,* [b]*Department of Computer, Ankara Beypazarı Fatih Vocational and Technical Anatolian High School, Ankara, Turkey*

7.1 Introduction

The technological and social behavior of private or public institutions and even societies are developing with the rapid spread of digitalization. These digital innovations include artificial intelligence (AI), autonomous robots, augmented and virtual reality, cyber-physical systems, cloud computing, the Internet of Things (IoT), and similar technologies. Production systems are changing via these technologies and thus radically transforming societies. This transition to a digital society has brought about high technology costs (technology procurement, use, etc.) as well as production costs. Regardless, the contribution of technology to institutions and societies is undeniable. In addition, the foundations of human-centered societies are laid with these technologies, and information societies are formed. At this point, the concept of Society 5.0 comes to the fore [1–4]. Countries worldwide are faced with many challenges, such as aging populations, declining birth rates, aging and outdated technical or technological infrastructure, and so on. Several steps have been taken to eliminate or at least minimize such difficulties. Society 5.0, an initiative first launched in Japan, aims to create sustainable smart societies that put people at the center, recognizing their right to comfortable and safe lives [5–7].

Society 5.0 represents the combination of the innovation economy and job creation capabilities to tackle social problems. In addition, it is designed to meet certain needs of society with technological support [8]. Faster progress can be achieved in almost every area of daily life with the steps and developments taking place in science and technology. However, the competitiveness of countries, institutions, and organizations is increasing [6]. In addition, some actions that take place in the life cycle of the individual evolve into a new form with the destructive effect of technology. It can be said that AI technologies developed to facilitate the happiness and lives of people are effective in every field to meet the goals of Society 5.0. Considering the

Artificial Intelligence and Industry 4.0. https://doi.org/10.1016/B978-0-323-88468-6.00006-1

transformative effect of technology on culture, AI technologies can be given as an example of changing cultural phenomena [9].

Society 5.0 is transforming individual lives as well as industries. In this context, Society 5.0 focuses on cities and regions, energy, disaster prevention and mitigation, health care, agriculture and food, logistics, manufacturing and services, finance, utilities, and cooperation. In addition, Society 5.0 shows a transition and development from a uniform society to individual abilities, from ecological restrictions to society in harmony with nature, from productivity to value creation, from social inequality to a society of opportunities, and from concerns to peace ([10] b).

Society 5.0 was proposed in the 5th Science and Technology Basic Plan as a future society that Japan wishes to realize [11]. In this context, Society 5.0 goals are very important in terms of creating a "super-smart society" thanks to the capabilities of AI systems of smart devices that manage and direct human behavior. It is seen that humanoid robots, developed with AI technologies in the period from 2016 to the present day, are very effective in a super-smart society [9,12].

In this study, we reveal the importance of the concept of Society 5.0, which focuses on individuals, their lives, and their relationship with AI. Specifically, we discuss developments in the field, keeping up with these developments, raising awareness, producing solutions against the aging world population, ensuring the virtual and real-world work together, benefiting from current technologies, and maximizing the integration of society with AI. In addition, this study addresses creating the foundations of a more livable world by minimizing negative effects such as unemployment in society and eliminating some occupations.

In Section 7.2, we present a literature review. In Section 7.3, we discuss and describe the four industrial revolutions the world has experienced up to this point. In Section 7.4, we define the concept of Society 5.0, its relationship with AI, its goals and innovations, and challenges to its implementation. We conclude the chapter with recommendations and future scope.

7.2 Literature review

There are many studies in the literature on the subject of study. For example, Saracel and Aksoy discussed the development stages of societies up to the period of Society 5.0 and the goals and contributions of Society 5.0. In particular, they demonstrated the development of individuals, institutions, and societies that can establish, operate and use Society 5.0 systems in the face of COVID19, which is a major epidemic all over the world. In addition, they emphasized that compliance with Society 5.0 at the institutional level plays an important role in the policies of states toward digital technologies. Thus, Society 5.0 requires societies that can install and operate e-based systems [13].

Salimova et al. examined the challenges of Industry 4.0 and Society 5.0 and the opportunities these concepts bring. They defined Society 5.0 as the humanization of industrial production beyond the technological, organizational, and economic transformation based on the latest scientific and technical developments. In addition, they identified the difficulties and opportunities in the process of implementing the concepts of Industry 4.0 and Society 5.0 in Russia. For this, the authors developed a questionnaire and received support from representatives of the Russian business world (mainly industrial enterprises) as participants. According to the survey results, these (the difficulties and opportunities in the process of implementing the concepts of Industry 4.0 and Society 5.0)should not only be identified as the priority of technological development at the national level but also mechanisms should be developed to motivate the business world in the field of socially oriented technologies. In addition, the results of the study revealed that the economy of the Russian Federation is facing the same problems and constraints as most countries in the world in their transition to a digital society [14].

Akin et al. evaluated the concept of Society 5.0 and discussed the integration of Society 5.0 technologies into the daily lives of people and society. The authors analyzed previous studies on Society 5.0 and found that researchers in Japan, Turkey, and Indonesia conducted the most studies in this field [15].

Kocaman et al. examined the transformation process from Education 1.0 to 4.0 by addressing the issues of digitalization in education during the transition to Society 5.0 and talked about applications related to digital transformation. By examining the studies and practices within the scope of digital transformation at formal and informal levels in Turkey, the authors revealed developments in education and discussed their contributions to digital transformation. They emphasized that in recent times when digitalization was prioritized, it was guiding in terms of examining the tools and methods in the digitalization process in education, bringing together the practices and examples related to the subject, and realizing the goals of digital transformation in education in Turkey [16].

A study by Avsar examined both Industry 4.0 and Society 5.0 and proposed the introduction of Industry 4.0, which is a new production model perspective, to all areas of society by integrating with Society 5.0 as a goal for Turkey. In addition, the researchers emphasized that the success rate could increase when all segments of the society were prepared for this process. Finally, they suggested that education should be updated in accordance with this new industry and society approach [17].

Korkusuz et al. aimed to determine the metaphorical perceptions developed by undergraduate and high school students about the concepts of Society 5.0 and Industry 4.0. Using phenomenology, one of the qualitative research methods, the authors analyzed students' sentences regarding the concepts of "Industry 4.0" and "Society 5.0" using the content method. The study sample consists of students in Computer Engineering and Computer Education and Instructional Technologies Education at Balıkesir University, Turkey, and students from different high schools [18].

Cark aimed to reveal the effects of digital transformation on the workforce and professions with a general perspective for Turkey. Therefore, the author reviewed published studies on the effects of digital transformation on the workforce and future professions. It was concluded that advanced technologies associated with Industry 4.0 are rapidly transforming working life [19].

Savaneviciene et al. aimed to reveal the individual innovativeness of different generations (Baby Boomers, Generation X, Generation Y, and Generation Z) in the context of Society 5.0 in Lithuania. The study focused on the concepts of generation, intergenerational diversity, generational differences, and the theoretical aspects of individual innovation. Three methods were used for data analysis: hierarchical clustering analysis, multidimensional scaling, and regression for categorical data. Finally, the study examined the contribution of different generations to the creation of Society 5.0 in Lithuania ([6]).

In this study, it emerged with the aim of creating a "Bravely Defying the Future" culture, which was envisaged as the next stage following the hunter-gatherer society, agricultural society, industrial society, and information society, which was put forward as an inclusive goal by the Japanese Business Federation in 2016. The study examined the usual and possible outcomes of Society 5.0 through the transition periods of societies, narrative traditions, and belief ecologies. It also evaluated the cultural changes brought about by social robots in human life in the context of human-robot integration. The study suggested that the future culture would be conceived in the context of human-robot symbolism in the process of creating a super-intelligent society of Society 5.0 [9].

A study by Un discussed the concepts of digitalization and Society 5.0 and examined the effects of information and communication technologies and the socio-cultural and ethical effects of these technologies on elderly care workers and the elderly in need of care. In particular, the issue that AI robots could take part in the fields of spiritual social services and care beyond body care and cleaning was handled with concern. As a result, an evaluation was made on the possible effects of digitalization on the health and care sector [20].

Turkeli conducted research on elderly care management in the period of Society 5.0. It was revealed that the demographic characteristics of Turkey have changed significantly in the last 50 years and that this change would continue in the coming years, changing Turkey from a country with a young population to an aging society. It was emphasized that Society 5.0 was a human-centered and technology-based approach. In Society 5.0, sensors, devices, robots, and the Internet play an important role in human life. Considering the aging of Turkey's population and the position of robots, the study attempted to answer the question: "Can robots replace humans for the care of the elderly in the age of Society 5.0?" ([21]).

Gokten examined the possible effects of technological advances on cost and cost control. The author's study suggested that cost and management accounting practices will change rapidly in the near future. The author stated that with the increase in indirect expenses, these practices have turned into an activity-oriented cost-based strategic management tool. It is expected that business processes will be carried out through cyber-physical systems in the future. In addition, factories will constitute the most important tangible assets of digitally integrated value chain elements [22].

A study by Yildirim discusses cyber-physical systems that include software, hardware, and biological factors. The study also discusses the concepts of AI, Big Data, cyber security, cloud computing, and digital twin, which are the main elements in the design of cyber-physical systems. The author emphasized that systems that not only provide communication between machines but also keep people in the loop are being developed and that the developing technologies, as well as software and hardware, are also biological factors. The author also gave examples of cyber-physical systems in health care, which is one of the sectors most affected by Society 5.0 [23].

Buyukgoze and Dereli examined questions about Society 5.0 and digital health. They defined the concept of Society 5.0, which emerged as a result of industrial revolutions, as the use of technology for the benefit of humanity. They emphasized that the concept of Society 5.0 includes sustainable development goals. They also examined questions about Society 5.0 and digital health, such as what was done in the field of digital health to prolong healthy life span, how digital health applications make our lives easier, and how digital health applications and home care applications are carried out in rapidly aging societies [24].

Buyukbingol examined Society 5.0 in the context of the idea of value-oriented technology. The author used qualitative and theoretical methods to scan the literature data related to Society 5.0. The study revealed that "technological man" would need a value production in terms of social meaning and that increasing economic welfare alone would not be sufficient in terms of maintaining the social whole. It was concluded that the production of these two values together was important for both individual and social happiness [25].

A study was conducted by Aktug and Sevinc to determine the socioeconomic impact of robotic systems, which are becoming increasingly widespread in the globalizing world. The authors examined the effects of the use of smart and robotic systems on the economy in digital transformation and change in industrial, socioeconomic, and sociocultural fields. The study emphasized that the systems and transformations in question provide a great economic revolution [26].

Fukuda conducted research on the transformation of the Science, Technology, and Innovation (STI) ecosystem into Society 5.0. The study examined a forward-looking model of STI activities in Society 5.0 from an ecosystem perspective. The study and statistical analysis of STI policies were conducted in comparison with Germany and the United States to define the historical development of Japan's ecosystem in these three areas (STI). Major socioeconomic risks in Japan's STI ecosystem were classified as capital, labor, and spatial risks. Increasing system resilience by creating value for society was demonstrated in Society 5.0 as crucial to reducing risk in these three areas and stimulating productivity and growth [27].

Researchers have studied the Japanese Society 5.0 framework to create super-intelligent societies. As in other countries, there are several problems that threaten Japan, including declining birth rates, aging, and lack of competition. Based on a community-centered approach, Society 5.0 aims to leverage technological advances to solve these problems. Another of its aims is to contribute to the development and growth of the country in all respects and to establish a better world order where no one is excluded from the technological advances of the current society. The United Nations Sustainable Development Goals (SDGs) were developed to achieve this goal. The SDGs evaluate the use of modern technology and propose the best strategies and tools to guarantee sustainability within the framework of a new society that demands constant renewal [28].

Piorunkiewicz and Zdonek aimed to identify keywords (priority terms) and efforts to expose digital vulnerabilities for their inclusion in Industry 4.0 and Society 5.0 processes. The researchers used visualization techniques, correlation analysis, text mining, and multidimensional cluster analysis to analyze a dataset of 288 digital products and services in Europe. The research focused on texts promoting digital open-data products and services, the most popular of which were apps, websites, and platforms. Its purpose was to provide real-time information on public issues in areas such as transport, education, culture and sport, economy, finance, and health [29].

Akkaya et al. defined Society 5.0 as a human-centered social understanding and the period in which the technological opportunities offered by Industry 4.0 will serve the welfare of people. However, they stated that Society 5.0 is resistant to social change in many respects. The integration and uncertainty of both individuals and organizations into current social life practices hinders the transformation of Society 5.0. With the digital solutions and applications related to the COVID-19 pandemic, there has been a radical change in life practices and thus the Society 5.0 transformation process has accelerated. The researchers suggested that decision-makers and administrators should initiate change, especially in education and other fields, and new practices and methods that serve human welfare should contribute to social transformation. In this context, the process was analyzed and discussed and suggestions were made within the scope of the study [30].

7.3 Industrial revolutions

Society 5.0 is explained as bringing developing technology and the digitalizing world to the service of society [24]. Since the Industrial Revolution in the 18th century, three more industrial revolutions have taken place [31,32]. Table 7.1 shows the historical development process of the various industrial revolutions.

In the sections that follow, we define and describe the four industrial revolutions that have thus far taken place.

7.3.1 Industry 1.0

The Industrial Revolution began in England in the middle of the 18th century. It turned agricultural societies into industrial societies. Developments in this period moved most employees from the fields to the factories. This period is known for the use of water and steam power in machinery and transportation [13,24].

Production areas suitable for today's factory structure were created with the development of steam-powered manufacturing machines. Mass-produced factory products began to be delivered to countries that lagged in production when steam technology began to be used in ships and trains. Railway networks began to expand

Table 7.1 Historical development of industrial revolutions

Society name	Description of process	Period	Trigger event	Explanation
Industry 1.0 (Society 1.0)	Hunter-Gatherer Society	Until ~13,000 BC	Natural Life	Living that the Society in harmony with nature, performing hunting activities
Industry 2.0 (Society 2.0)	Agricultural Society	Until the 18th century	Improvement in Irrigation Techniques	Beginning of agricultural activities
Industry 3.0 (Society 3.0)	Industrial Society	Late 18th–20th century, beginning of the century	Inventions in Steam Engines	Industrialization and mass production
Industry 4.0 (Society 4.0)	Information Society	Late 20th-21st century, beginning of the century	Computer Technology	Effective use of information and communication technologies and information sharing begins
Industry 5.0 (Society 5.0)	Super-Intelligent Society	...	Industry 4.0	Benefiting from Industry 4.0 according to the interests of the society, producing new sustainable solutions in the protection of the environment

all over the world with the use of steam power in railways. Moreover, humanity overcame the phenomenon of distance and began to create a wider market with inventions such as steamships, fossil fuel cars, and the telegraph. The invention of steam machines is very important due to the power and performance they produce even in places that require the most advanced technology today [33].

7.3.2 Industry 2.0

This revolution began with cheap steel production methods and spread with electrical and chemical techniques. It continued with the use of electricity in factories and cities in 1882 [24]. Electric power and the development of automatic processes are the key elements of the second revolution. Innovations in this period continued to progress with the emergence of higher quality of life. The critical point of the period is the use of assembly lines in production to meet increasing demand. "total quality management" and "scientific management" principles that emerged in this period are among the management practices still used today [13].

In this period, the production of synthetic chemicals was realized along with the invention of safflower and even the production of artificial fertilizers. At the end of this period, synthetic textile products started to be produced. The most important point of this second phase of the industrial revolution is that it enabled almost all industries except agriculture to be mechanized. Manufacturing of fabricated shoes and ready-made clothing also emerged in this period. Developing technology provided consumers with updated and more qualified products [33,34].

7.3.3 Industry 3.0

This revolution emerged because of programmable machines, with the use of digital technologies in production. The third industrial revolution can be described as the informatics revolution in which computers and the Internet advanced rapidly [24]. Automation devices needed to be programmed with the transition to mass production, and factory machines started to work under computer control. More serial and efficient production was started by using these computers with their own memory. Thus, labor costs were reduced. As a result, the industry had to focus on problems such as overcapacity [13].

In this period, developments and needs in terms of information and production were solved with programmable logical control (PLC) systems. Productivity increased by using PLC-supported automated production. In previous industrial revolutions, organized structures were established with the mechanical power of production systems and later with larger energy systems. The most important development of the third industrial revolution is that the machines started to take place in the control and partial management of parts of the production processes, and the era of intelligent machines based on artificial intelligence [33].

7.3.4 Industry 4.0

The concept of Industry 4.0 was first introduced in Germany by business, policy, and academic representatives. Industry 4.0 is a revolution in which the Internet of Services (IoS), IoT, cyber-physical systems, and smart factories are the main concepts. In addition to these are smart products, machine-to-machine communication, Big Data, and cloud technologies. Industry 4.0 is characterized as having a "high level of complexity" and "collective use of products and production processes with a communication network" [2].

Smart factories come to the fore with the use of physical systems such as machines and robotics, which are cyber-physical systems controlled by automation systems based on machine learning algorithms. Flexible production in line with the demands of customers is the biggest advantage of this innovation. The third

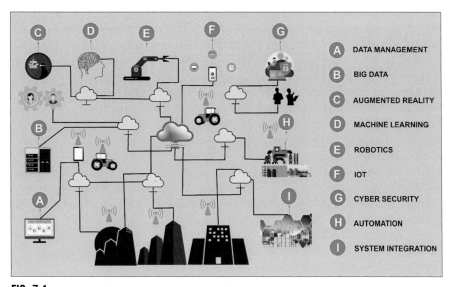

FIG. 7.1

An example for Industry 4.0.

Source: own work.

industrial revolution enabled the transition of production to an automated system with the use of electronics, thus increasing production speed. It evolved with a parabolic tempo, not a linear one, unlike previous industrial revolutions. [33,35].

As shown in Fig. 7.1, companies can produce smarter, faster, higher quality, and competitive products by using Industry 4.0 technologies such as data management, IoT, cloud technologies, cyber-physical systems, robots, augmented reality, machine learning, AI, smart factories, smart products, and Big Data.

7.4 The concept of Society 5.0 and artificial intelligence
7.4.1 The transition process from Industry 4.0 to Society 5.0

It has become easier to obtain and share information with Industry 4.0. As a result, the amount of information produced is also increasing considerably. There are some developments in the industry that have occurred with the rapid increase of knowledge. The use of information and communication technologies in business processes is among them. This has led to less utilization of manual power in the industry, with machines taking an active role and the digitalization of the industry [22]. Industry 4.0 and technological transformations do not only apply to the industry and production system; they also have effects on social systems [4].

The digital world is a phenomenon for the benefit of society when talking about a digitalizing industry. The concept of Society 5.0 has come to the fore with this understanding. While expressing this concept, it is stated that the developments in technology are not actually a threat to society. On the contrary, they are an auxiliary element that meets the needs of society and provides convenience in daily life [36]. The concept of a smart society has emerged with the increasing use of intelligent systems such as AI. In this context, human and technology elements were used together in Society 5.0. Technology has been more involved in hard work, and people have been more involved in monitoring, control, and evaluation processes [22,36]. Industry 4.0 focuses on production in technology, but Society 5.0 focuses on increasing the welfare of people via the innovations of Industry 4.0 and social responsibility, technology, and sustainability in improving the quality of life [13].

The keywords of Society 5.0 are agility, adaptability, mobility, and agency. Agility, adaptability, and action are crucial and require the implementation of Industry 4.0 (and therefore digital transformation) using additional techniques that consume few resources for production [37,38]. As shown in Table 7.2, in the evolution process from Industry 4.0 to Society 5.0, intelligent and mass production takes place under the criteria of the economic situation, target group, the center of attraction, safety, environment, and social problems in general.

In short, Society 5.0 aims to integrate technological elements into every process of society. Thus, a society that is aware of technological achievements, not afraid of technology, and lives with technology is formed [22].

7.4.2 The concept of Society 5.0

Japan introduced the concept of Society 5.0 within the scope of the 5th Science and Technology Basic Plan after Germany introduced the concept of Industry 4.0. Japan revealed the vision of a super-smart society in the face of various challenges such as rapid aging of the population, air pollution, and natural disasters [39,40]. The characteristics of a "super-smart society" targeted by Society 5.0 were stated by Harayama [41] as follows:

• A society that enables individuals with diverse needs to provide necessary products and services in the amount and time they need.

Table 7.2 The transition process from Industry 4.0 to Society 5.0

Criterion	Problem (Industry 4.0)	Solution (Society 5.0)
Economic situation	Getting rid of economic value creation	Problem-solving and value creation
Target group	Getting rid of individuality	Variation
Center of attraction	Getting rid of inequality	Localization
Safety	Relief from anxiety	Flexibility
Environment	Freedom from resource and environmental constraints	Sustainability and environmental compatibility

- A society where all people can get high-quality services.
- A society in which people can live in comfort and prosperity that allows for a variety of differences, such as age, gender, region, or language.

The purpose of Society 5.0 is to help scientists balance economic progress with social problem solving by imitating the structures and processes that can be described in biological evolution [42]. In other words, it is to perfectly adapt production processes to environmental and human needs, constantly upgrading services, process data, and products with smart systems and associated substructure [13].

Society 5.0 is a lifestyle that can provide the necessary goods and services to those in need at the right time and in the right amount, can fully respond to a wide variety of social needs, can provide easy and high-quality services to all kinds of people, and can eliminate differences in age, gender, religion, and language; it is defined as a society that can offer opportunities [43]. However, Society 5.0 is also defined as a human-centered society that balances the resolution of social problems with a system that greatly integrates cyberspace and physical space [44].

Society 5.0 aims to ensure the industrial competitiveness of those who implement it by using sustainability, comprehensive, and efficiency and therefore the power of intelligence and knowledge. It develops economically and aims to protect the world with development and gains and to create a smart society that creates value that can be used for the good and benefit of humanity. Society 5.0 can be defined as a "smart society" in which physical space and cyberspace are strongly integrated. This new society, called the smart society, appears to be a desire for balance in the search for optimization of the previous four societies (hunter-gatherer society, agricultural society, industrial society, and information society). It was mostly created by using Industry 4.0 concepts, although it includes elements from almost all the previous societies [37,38,45,46].

7.4.3 The factors that reveal Society 5.0

Societies, no matter how well off economically, cannot be considered advanced when they degenerate in scientific, artistic, social, political, legal, ethical, esthetic, and ecological aspects. All technological and scientific developments have negative consequences as well as positive contributions to human life. The essence of sophistication lies in the ability to cope with these problems and manage them well.

According to a report by Keidanren [10], the Japanese Federation of Businessmen, which introduced Society 5.0 to the world, identified three reasons for the establishment of a "super-smart society" or "smart society."

- The world is facing a great wave of change brought about by technological innovations, while it is rapidly developing with digital technologies such as AI, IoT, robotics, and biotechnology. These innovations will trigger a revolutionary wave of change not only technologically but also socially. In addition to large companies and entrepreneurs (Google, Amazon, Facebook,

Apple, etc.), countries can also take measures, initiate projects, and develop strategies (e.g., German Industry 4.0, China 2025, France 2020 plan).

- The center of gravity of the world economy is shifting from the West to Asia. In addition to the emergence of China as a superpower, the rapid growth of India and the member states of the Association of Southeast Asian Nations show that the economic and geopolitical indicators are changing rapidly. In addition, it shows that the population in the West is gradually decreasing and aging in some countries when evaluated in terms of demographic variables. However, global population growth is very high in the world. Therefore, the rapid change in population dynamics threatens sustainability with the emergence of new social problems and requires the production of new economic and geographical policies.

- The United Nations SDGs was adopted in 2015 due to the increase in social problems such as ecological and inequality at various levels, climate change and environmental pollution, and the worsening of global problems. It should be noted that the concept of Society 5.0 and the SDGs overlap and support each other. In addition, investments in social, ecology, and governance are also increasing in the financial sector. This tendency is becoming increasingly effective in the view and common perception that not only dealing with the economy does not provide political and economic stability, therefore taking measures that may be in the interest and benefit of all humanity is as important as profitability and positively affects the economy [37].

Society 5.0 refers to six main pillars: infrastructure, technological innovation, finance, healthcare, logistics, and AI. Technology and innovation should be used to help and advance society, not to replace the role of humans [47]. Society 5.0, considered the next stage of Industry 4.0, envisages the efficient use of the relationship between humans and robots. Society 5.0 has determined various visions for itself to produce solutions to environmental, human, and economic problems. The realization of these visions will only be possible if all the foundations of technological and scientific advances are processed in economic and social life in a planned manner [48]. Fig. 7.2 shows the general structure of Society 5.0.

Within the general structure of Society 5.0, technologies such as AI, 5G networks, Big Data, robots, and IoT have the potential to produce solutions to social problems. Therefore, Society 5.0 should be investigated as one of today's most important concepts.

7.4.4 The goals of Society 5.0

Society 5.0 was originally designed to provide economic benefits for individuals in general. Its purpose is to provide those in need with goods and services at the right time and in the right amount, thus facilitating human wellbeing. It is hoped that the implementation of Society 5.0 practices in Japan will benefit the whole nation, beyond facilitating/improving the lives of its individuals. The aim of Society 5.0

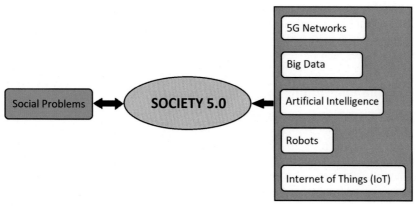

FIG. 7.2

The general structure of Society 5.0.

Source: own work.

is to meet the needs of a wide variety of social infrastructure to ensure that all types of people have easy access to high-quality services to live safe and comfortable lives. Society 5.0 technologies are being used to solve problems related to aging societies, declining populations, and major natural disasters [5,49].

The Society 5.0 philosophy in Japan suggests the implementation of some goals such as [24,48]:

- To develop solutions against the aging world population
- To ensure that the virtual and real-world work together
- To benefit from the IoT by considering the interests of society
- To develop solutions against natural disasters and environmental pollution

The understanding that puts people at the center aims at a life where the needs of the society are met at a sufficient level and quality service is provided [40]. The goals of Society 5.0 are:

- *Increasing the power of individuals by carrying out individual reform*

Every individual, regardless of race, gender, age, and so on, has the ability to live a comfortable and healthy life safely and to realize their own lifestyle.

- *Providing new values by reforming companies*

It is the realization of the new society and economy by promoting productivity enhancement through digitization and reforming business models and supporting innovation and globalization.

- *Creating a better future by solving social problems*

It is an effort to realize a rich and powerful future in order to solve many problems such as a declining population, rapidly aging societies, and natural disasters. It is to contribute to solving problems on a global scale through the spread of new businesses, services, and processes worldwide [49].

7.4.5 The innovations provided by Society 5.0

Society 5.0 contributes to innovation by analyzing preconditions, proposing solutions for accelerating innovations, and realizing the invention–innovation–diffusion process. Basically, the activities of organizations are focused on technological innovations, which academics conceptualize as "the development of new services and products that meet the needs of the organizations to solve the problems in the internal and external environment" [50]. Society 5.0 proposes a model of a "super-smart society" in which the relationship of humans with machines and robots considers the impact of digitalization from a demographic, economic, and sociological perspective. Society 5.0, in which the IoT is used extensively, has a great contribution to every field where people live and operate. The most important of these are the contributions to home life. In addition, Society 5.0 makes important contributions to almost every level of education and even to the most effective and efficient way of continuous education. Students with access to the Internet can easily and quickly receive all kinds of information, exchange information quickly with various social media and communication channels, and work faster using a learning management system without having to buy books or other school equipment [13].

In addition, Society 5.0 makes significant contributions to solving vital problems encountered in real life, such as providing early detection and prevention of natural disasters via unmanned aerial vehicles and air surveillance radars, providing a high level of security with sensors and AI-based tools, facilitating patient-doctor machine cooperation, monitoring health status (heart rate, steps, calories, etc.) using wearable technologies (wristbands, rings, etc.), realizing financial transactions using blockchain technology, data tracking via IoT technology (weather forecast, water, and humidity data), and providing savings in all areas with sensors and IoT technologies [24,51,52].

Society 5.0 provides AI, physical and cyberspace, IoT, and Big Data innovations. Much more information is obtained from sensors in cyberspace than in physical space. These Big Data are analyzed with AI, and the results are presented to the public and used for the benefit of society. In other words, objects and systems communicate with each other in cyberspace and the results obtained by AI are used for society's benefit. All these innovations offered by Society 5.0 aim to transform society from a technology-centered world order to a human-oriented order via innovative and human-based solutions to social problems [53].

7.4.6 The relationship between Society 5.0 and artificial intelligence

AI refers to programmed machines that exhibit features of the human mind such as learning and problem solving [16,54–56]. Here, the topics of the hunter-gatherer period left their place for other subjects with the development of tools. Several reasons, such as catching game animals, not being developed enough to kill them in a single shot, inability to neutralize them, hardly neutralizing animals, and inability to explain various natural phenomena with the available technologies, are some of the main factors that shaped the technological developments in this period. Technological developments that made the invention of agriculture possible have played a role in changing human life in many areas. The world is moving with Society 5.0 toward a "radical change period" in which the process of creating knowledge and value under the domination of digital technologies changes daily, and social and industrial structures emerge under the rule of smart devices. In this context, AI technologies have begun to be integrated into human life at a time when the world has become a global village as a result of developments in information and communication technologies. In addition, AI technologies are integrated into human life, enabling the future society to transform in the axis of smart and digital devices [57].

The Society 5.0 revolution has reached predictable dimensions. It is causing and will continue to change and transform human life. Technological inventions and tools play an important role in change and transformation [9].

Society 5.0 is defined as an intelligent society that includes standardized processes to evaluate human demands and meet their needs using AI technology. Therefore, a digital knowledge-based society must work for environmental, economic, and social sustainability and efficiency. Human and social resources are at the center of developments in smart societies and smart cities, which require innovative methods and techniques for predictive and adaptive processes. Here, it is aimed to design a knowledge-based economy with a digital infrastructure that can work together and provide dynamic real-time interactions among various smart city subsystems. Society 5.0, on the other hand, is accumulating data in cyberspace with AI. New data is produced by AI, taking into account human-machine cooperation and interaction. Thus, it aims to balance economic progress and social problem solving by developing similar to biological evolution [42]. In summary, it is the aim of Society 5.0 to perfect the harmony of human needs and production processes and to continuously upgrade process data, products, and services with smart systems and related infrastructures [13].

In line with these purposes, technological sustainability, integration, and transparency have been determined as basic principles [18]. Society 5.0 will not only make the new information obtained from information societies available to humans, but it will also share this information with AI robots. AI-based technological assets and people will be able to obtain information outside of the economy/work area, in every frame of social life, and will engage in mutual information sharing and cooperation. The scope and extent of the digitalization of information will extend to all areas, ranging from legal areas to social systems. In short, the digitalization and transformation process will progress continuously and rapidly in every field [58].

7.4.7 **The obstacles to Society 5.0**

Implementing Society 5.0 goals in real life cause a social transformation. However, some potential obstacles are encountered in the said transformation process. The obstacles to overcome in order to implement Society 5.0 are:

- socio-political obstacles
- obstacles in the legal system
- technological obstacles
- lack of qualified human resources
- social resistance

To overcome socio-political obstacles, it is recommended to implement national strategic moves with the support of industry and academia, provide state support, establish an IoT platform, and establish a think tank.

To overcome obstacles in the legal system, it is recommended to establish rules to direct data use and implementation, support regulations and system reform, and supervise the legislative process on copyright-intellectual property rights.

To overcome technological obstacles, it is recommended to support technologies such as AI, cyber security, and autonomous robots, and to make some improvements related to innovation in science and technology.

To overcome the lack of qualified human resources, it is recommended to organize training for the dynamic participation of all citizens, support the provision of personnel for cyber security, data science, and international standardization, and encourage women's participation in order to reveal talents.

To overcome social resistance, it is recommended to create social consensus by distributing the national vision among all stakeholders and addressing ethical and social implications from human-machine relationships to philosophical issues [48,53].

7.5 **Conclusions**

In this study, we examined in detail the Japanese concept of Society 5.0 and its relationship with AI. First, we described all the industrial revolutions that led up to this period. Second, and were explained, then the birth, definition, goals, obstacles of the Society 5.0, the innovations it provides, and its relationship with artificial intelligence are examined. The "super-smart society" model put forward by the Japanese aims to enable everyone to live a satisfying and prosperous life by using both humans and AI. In addition, this model aims to create a more livable world by minimizing negative effects such as unemployment. It is a very important approach in terms of providing for the needs of all individuals in a timely fashion.

The main purpose of Society 5.0 is to produce solutions to combat the aging world population, to ensure that the virtual and real-world work together, to benefit from the IoT in a way that contributes to society, and to generate ideas for mitigating environmental pollution and natural disasters. In addition, it was observed that with the transfer of business processes to the digital environment, AI and robot technology improved and the need for manual power decreased. In short, it can be said that digital development and transformation create new employment opportunities and eliminates many business areas. However, AI focuses on facilitating human life, ensuring efficiency in production, and surviving in a competitive environment.

7.6 Recommendations and future scope

As for the proposals and future perspective of the study, we recommend working harder to ensure sustainability in social, environmental, and economic terms of the digitally effective, efficient, and knowledge-based society. At this point, the focus should be on adapting to new technologies, increasing competencies, analytical thinking, continuous improvement and learning, and creative and innovative approaches. In addition, current AI technologies that will meet the possible future needs of society and make their lives easier should be developed and technical training should be implemented.

Future studies should examine new technologies, identify priority areas, and analyze the contributions of technological innovation in all aspects of society.

References

[1] E. Çay, A. Yıkmış, ve Sola Özgüç, C., Özel eğitimde teknoloji kullanımına ilişkin özel eğitim öğretmenlerinin deneyim ve görüşleri, Eğitimde Nitel Araştırmalar Dergisi—J. Qual. Res. Educ. 8 (2) (2020) 629–648, https://doi.org/10.14689/issn.2148-624.1.8c.2s.9m.

[2] M. Hermann, T. Pentek, B. Otto, Design principles for industrie 4.0 scenarios, in: 2016 49th Hawaii International Conference on System Sciences (HICSS), IEEE, 2016, January, pp. 3928–3937.

[3] İ. Özen, Teknoloji Muhasebesi, Electron. Turkish Stud. 15 (6) (2020).

[4] E. Saribay, O.E. Erkan, A research on the industry 4.0 perception of university students, Kongre Başkanı 77 (2019).

[5] M.E. Gladden, Who will be the members of society 5.0? Towards an anthropology of technologically posthumanized future societies, Soc. Sci. 8 (5) (2019) 148.

[6] A. Savanevičienė, G. Statnickė, S. Vaitkevičius, Individual innovativeness of different generations in the context of the forthcoming Society 5.0 in Lithuania, Eng. Econ. 30 (2) (2019) 211–222.

[7] Y. Shiroishi, K. Uchiyama, N. Suzuki, Society 5.0: for human security and well-being, Computer 51 (7) (2018) 91–95, https://doi.org/10.1109/MC.2018.3011041.

[8] C. Holroyd, Technological innovation and building a 'super smart' Society: Japan's vision of Society 5.0, J. Asian Public Policy (2020) 1–14.

[9] R. Kocak, Fifth industrial revolution: society 5.0 and artificial intelligence culture, Int. J. Folk. Stud. S.5 (2020). s.1-17.

[10] Keidanren, Society 5.0—CoCreating the Future. Keidanren Policy & Action, 2018. https://www.keidanren.or.jp/en/policy/2018/095_proposal.pdf. Access Date: 17.06.2021.

[11] Government of Japan, The 5th Science and Technology Basic Plan, January 12, 2016, pp. 1–23. https://www8.cao.go.jp/cstp/kihonkeikaku/5basicplan_en.pdf. A.D: 18.12.2015.

[12] A. Deguchi, et al., From smart city to society 5.0, in: Society 5.0, Edit. Hitachi UTokyo Laboratory (H-UTokyo Lab.) Springer, Singapore, 2020, pp. 43–65.

[13] N. Saracel, I. Aksoy, Society 5.0: super smart society, Soc. Sci. Res. J. 9 (2) (2020) 26–34.

[14] T. Salimova, N. Vukovic, N. Guskova, I. Krakovskaya, Industry 4.0 and society 5.0: challenges and opportunities, the case study of Russia, IPSI BgD Trans. Internet Res. 17 (1) (2021). 4-+.

[15] N. Akin, E.M. Akyol, O. Dalkilic, An evaluation of the concept of society 5.0 in the light of academic publications, Atatürk Univ. J. Econ. Adm. Sci. 35 (2) (2021) 577–593.

[16] A.K.K. Karoglu, K.B. Cetinkaya, E. Cimsir, Digital transformation in education in Turkey in the process of society 5.0, J. Univ. Res. 3 (3) (2020) 147–158. December 2020.

[17] I. Avsar, Industry 4.0 Interpretation: Society 5.0. AL-FARABI 2nd International Congress on Social Sciences, 2018, pp. 510–514. Gaziantep http://www.iksadfua rkongre.org/.

[18] M.E. Korkusuz, G. Durak, N.K. Ari, Metaphoric perceptions of undergraduate and high school students about the concepts of industry 4.0 and society 5.0, Necatibey Fac. Educ. Electron. J. Sci. & Math. Educ. 14 (2) (2020) 1504–1527.

[19] O. Cark, Dijital Dönüşümün İşgücü ve Meslekler Üzerindeki Etkileri, Int. J. Entrep. Manag. Inq. (2020) 19–34.

[20] S. Un, Elderly care with information and communication technologies in "society 5.0", HAK-İŞ Int. J. Labor Soc. (2020) 313–330.

[21] E. Turkeli, Toplum 5.0 Döneminde Yaşlı Bakım Yönetimi. Endüstri 5.0—Dijital Toplum (s. 153–174), içinde Ekin Yayınevi, 2021.

[22] P.O. Gokten, Manufacturing in the dark: scope of the cost in the new age, J. World Account. Sci. 20 (4) (2018) 880–897.

[23] T. Yildirim, Cyber-physical systems: from the information society to the superintelligent society, in: 12th International Conference on Electrical and Electronics Engineering (ELECO), IEEE, Bursa, 2020, pp. 266–269.

[24] S. Buyukgoze, E. Dereli, Toplum 5.0 ve Dijital Sağlık. IV, in: International Scientific and Vocational Studies Congress—Science and Health, Bilmes Sh, Ankara, 2019, pp. 276–281. 7-10 November.

[25] A. Buyukbingol, Society 5.0 in the context of the value-oriented technology idea, in: Innovation and Global Issues Congress, InGlobe Academy, Antalya, 2020.

[26] S. Aktug, M. Sevinc, Society 5.0: the economic revolution of intelligent systems and robots, in: International Academic Studies Congress 2021 Spring, Roting Academy, Mersin, 2021, pp. 139–146.

[27] K. Fukuda, Science, technology and innovation ecosystem transformation toward society 5.0, Int. J Prod. Econ. 220 (2020), 107460.

[28] C. Narvaez Rojas, G.A. Alomia Peñafiel, D.F. Loaiza Buitrago, C.A. Tavera Romero, Society 5.0: a Japanese concept for a superintelligent society, Sustainability 13 (12) (2021) 6567.

[29] A. Sołtysik-Piorunkiewicz, I. Zdonek, How society 5.0 and industry 4.0 ideas shape the open data performance expectancy, Sustainability 13 (2) (2021) 917.

[30] B. Akkaya, A. Gunsel, I. Yikilmaz, Digital management towards society 5.0: a review of the framework for kurt lewin theory during COVID-19 pandemic. Emerging challenges, solutions, and best practices for digital enterprise, Transformation (2021) 120–137.

[31] N. Mete, İşletmelerde kurumsallaşma ve inovasyon arasındaki ilişkinin incelenmesi üzerine bir araştırma, Master's thesis, Pamukkale University Institute of Social Sciences, 2019. Access Adress http://acikerisim.pau.edu.tr:8080/xmlui/handle/11499/26637.

[32] M. Onder, Endüstri 4.0 Devrimi ve Haritacılık Mesleğine Yansımaları, TMMOB Harita ve Kadastro Mühendisleri Odası 17 (2019) 25–27.

[33] E.C. Inan, Endüstri 4.0 vizyonunun üretim süreçlerinde getireceği verimlilik, master's thesis, Istanbul Kultur University/Graduate Education Institute/Department of Business Administration/Department of Business Administration), 2019.

[34] P. Kennedy, Büyük Güçlerin Yükseliş ve Çöküşleri, İş Bankası Yayınları, Ankara, 2005.

[35] U. Dombrowski, T. Wagner, Mental strain as field of action in the 4th industrial revolution, Proc. CIRP 17 (2014) 100–105.

[36] H. Develi, "Endüstri 4.0'dan Toplum 5.0'a", Dünya Gazetesi, 2 Kasım 2017, 2017, Access Adress https://www.dunya.com/kose-yazisi/endustri-40dan-toplum-50a/389146. Access Date: 12. 08. 2018.

[37] Z. Eren Ugurlu, Toplum 5.0 ve Dijital Dünyada Toplumsal Dönüşüm ve Eğitim 5.0, Dijital Dönüşüm ve Süreçler & Digital Transformation and Processes, İstanbul Gelişim Üniversitesi Yayınları, 2020. 2020.

[38] Keidanren (Japan Business Federation), Healthcare in Society 5.0, 2018, (March 20, 2018) Access Adress: https://www.keidanren.or.jp/en/policy/2018/021_overview.pdf.

[39] T. Bulut, Sanayi 4.0 mı yoksa Toplum 5.0 mı? 2017, Access Adress http://www.sanayigazetesi.com.tr/sanayi-40-mi-yoksa-toplum-50-mimakale,1307.html. 25.02.2018.

[40] B.D. Nazlican, O. Mecik, Türkiye'de Endüstri 4.0'ın İşgücü Piyasasına Etkileri: Firma Beklentileri. Süleyman Demirel Üniversitesi İktisadi ve İdari Bilimler Fakültesi Dergisi, 23(Endüstri 4.0 ve Örgütsel Değişim Özel Sayısı), 2018, pp. 1581–1606.

[41] Y. Harayama, Society 5.0: Aiming for a New Human-Centered Society. Collaborative Creation Through Global R&D Open Innovation for Creating the Future, 2017. Volume 66 Number 6 558–559 August 2017.

[42] R. Foresti, S. Rossi, M. Magnani, C.G. Lo Bianco, N. Delmonto, Smart society and artificial intelligence: big data scheduling and the global standard method applied to smart maintenance, Engineering (2019) 1–12.

[43] R. Carraz, Y. Harayama, Japan's Innovation Systems at the Crossroads: Society 5.0, Digital Asia, 2018, pp. 33–45.

[44] K. Matsuda, S. Uesugi, K. Naruse, M. Morita, Technologies of production with Society 5.0, in: 2019 6th International Conference on Behavioral, Economic and Socio-Cultural Computing (BESC), IEEE, 2019, October, pp. 1–4.

[45] M. Fukuyama, Society 5.0: aiming for a new human-centered society, Jpn. Spotlight 27 (2018) 47–50. Access Address https://www.jef.or.jp/journal/pdf/220th_Special_Article_02.pdf.

[46] B. Salgues, Society 5.0: Industry of the Future, Technologies, Methods, first ed., ISTE Ltd and John Wiley & Sons, Inc, New York, 2018.

[47] L. Ellitan, L. Anatan, Achieving business continuity in industrial 4.0 and society 5.0, Int. J. Trend Sci. Res. Dev. (2019) 235–239.

[48] C. Sahin, Ülkelerin endüstri 4.0 düzeylerinin COPRAS yöntemi ile analizi: g-20 ülkeleri ve Türkiye, Master's Thesis, Bartin University, Institute of Social Sciences, 2019.

[49] Keidanren (Japan Business Federation), Toward Realization of the New Economy and Society. Reform of the Economy and Society by the Deepening of "Society 5.0", 2016, Retrieved April 16, 2018 from Access Adress: http://www.keidanren.or.jp/en/policy/2016/029_outline.pdf.

[50] V. Potočan, M. Mulej, Z. Nedelko, Toplum 5.0: balancing of industry 4.0, economic advancement and social problems, Kybernetes 50 (3) (2020) 794–811, https://doi.org/10.1108/K-12-2019-0858.

[51] C.E. Erkılıç, A. Yalçın, Evaluation of the wearable technology market within the scope of digital health technologies, Gazi İktisat ve İşletme Dergisi 6 (3) (2020) 310–323.

[52] C. Wise, This Digital Pill Wants to Make Following Your Prescription Easier, 2018, Mayıs 23 https://www.pbs.org/newshour/science/following-aprescription-is-hard-this-digital-pill-wants-to-help. Access Date: 22.06.2021.

[53] D.N. Celep, Toplum 5.0: İnsan Merkezli Toplum, Türk Eğitim Derneği, TED Batman Koleji, 2020. https://www.tedbatman.k12.tr/wp-content/uploads/2020/05/Toplum-5.0-%C4%B0nsan-Merkezli-Toplum-.pdf.

[54] M.H. Calp, Medical diagnosis with a novel SVM-CoDOA based hybrid approach, Brain 9 (4) (2018) 6–16.

[55] M.H. Calp, Evaluation of multidisciplinary effects of artificial intelligence with optimization perspective, Brain 10 (1) (2019) 20–29.

[56] M.H. Calp, The role of artificial intelligence within the scope of digital transformation in enterprises, in: G. Ekren, A. Erkollar, B. Oberer (Eds.), Advanced MIS and Digital Transformation for Increased Creativity and Innovation in Business, IGI Global, 2020, pp. 122–146. http://doi:10.4018/978-1-5225-9550-2.ch006.

[57] Director General for Science, Technology and Innovation Cabinet Office, Cross-ministerial Strategic Innovation Promotion Program (SIP) Energy Systems of an IoE Society Research and Development Plan, July 25, 2019, pp. 3–17. Access Adress https://www8.cao.go.jp/cstp/english/08_ioe_rdplan.pdf. A.D:11.10.2020.

[58] S. Silkin-Un, "Toplum 5.0"da bilgi ve iletişim teknolojileri ile yaşlı bakımı, HAK-İŞ Uluslararası Emek ve Toplum Dergisi 9 (24) (2020) 313–330.

Big data analytics for strategic and operational decisions

8

Brahim Jabir and Noureddine Falih

LIMATI Laboratory, Sultan Moulay Slimane University, Beni Mellal, Morocco

8.1 Introduction

Companies compete to reach commercial goals and take advantage of all available means to achieve profits. Recently, corporations have discovered that data is the most important type of wealth and its optimal exploitation can help organizations withstand competition. A company is distinguished from others by the way it collects, processes, purifies, and analyzes data to extract value that it can exploit in its interest. However, this exploitation is difficult when the data is too large. Over time, companies have collected big and important data that they exploit to reach business goals. This is where the concept of Big data comes from. Big data is a group of very large and complex data packages that are difficult to deal with using traditional database management systems (DBMS) in various processes of storage, search, representation, and analysis. Three characteristics describe Big data. The first is Volume, which is the size of data collected from several sources [1]. The second characteristic is Variety, which refers to the data's structure and type. There are two varieties of data: structured and unstructured. The third characteristic is Velocity, which refers to the speed in collecting and analyzing data in real time to make appropriate decisions [2].

There are several stages of working with Big data: organizational, strategical, and procedural stages. Over the years, many programs and applications have been developed to help collect and analyze data to reach correct decisions. This has contributed to the emergence of Big data Analytics as an essential strategy that enables the analysis of extensive data [3]. The aim is to collect and analyze customer data obtained from multiple sources such as social networking sites, digital images, sensors, and so on to make predictions [4].

It is not easy to deal with Big data using traditional techniques, thus the development of online database management methods, such as Apache's Cassandra, Hadoop, and Spark software [5]. These methods provided effective ways to exploit data better. However, the results achieved by these techniques still need to be improved to make better use of the data regardless of its size and type and in real

Artificial Intelligence and Industry 4.0. https://doi.org/10.1016/B978-0-323-88468-6.00008-5

time [6]. This chapter defines the advantages of Big data Analytics and how it is used to analyze Big data. It discusses its limits and challenges and presents a case study that sheds light on the proposed solutions to these problems. The chapter concludes with a discussion about these solutions and their benefits.

8.2 Overview

The question of the digital transformation of companies is certainly not new, but with the concept of digital intelligence, we are now going further. The new economic players of the Web and leaders in their markets, Uber and Airbnb, demonstrate this quite well: Big data analytics is the keystone of success. In other words, investing in your own digital intelligence is now the sine qua non for success [7,8].

Digital intelligence has become the absolute weapon in the field of economic warfare, a way for businesses to distance themselves from their competitors and become the absolute leaders in their respective markets. Digital intelligence has only existed conceptually since 2015 [9]. Fig. 8.1 presents the digital transformation of companies to inform strategic and operational decisions for smart enterprise.

In the early days of the Internet, digital information was very thin. These were mainly website addresses and user login credentials. With the development of the Web and digital technologies [10], we have gone to one, then two, and then three successive evolutionary paradigms, to arrive today at a real explosion of data.

According to the International Data Corporation (IDC), each individual generates 1.7 megabytes of data per second. These massive data are the result of cross-channel and mobility, centered on the individual and his behavior in terms of socialization,

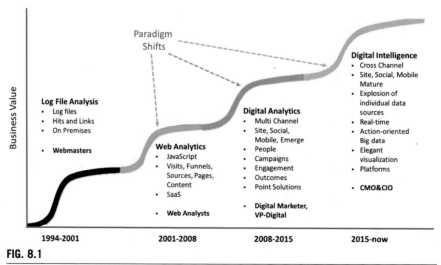

FIG. 8.1

The digital transformation of companies.

interactions, consumption or day-to-day management, personal information, facial recognition, bank details, vacation location, intimate conversations, consumption habits, tastes, hobbies, family situation, place of residence, favorite restaurants, etc. [11]. These data allow companies to benefit from a global vision of their consumers, to have a perfect knowledge of the market, to make a detailed analysis of consumption habits, to better understand changing trends for extremely precise targeting of messages, this is done after collecting, analyzing and extracting meanings from the Big data [12].

The recent momentum of Internet of Things (IoT) technology, which wirelessly interconnects all kinds of objects and devices at all stages of the cyber-physical management cycle, produces large amounts of accessible real-time data on agricultural processes as well as across the entire supply chain. In this regard, operations and transactions are very important sources of process-mediated data. In addition, sensory devices and robots produce non-traditional data such as images, videos, and other types of machine-generated data, while social media is also an important source of human-made data. As a result, to these large amounts of heterogeneous data, access to explicit information and decision-making capabilities is provided at a level that was not previously possible. The key success factor in creating value from this data is Big data Analytics [13]

Big data is changing the scope and organization of agriculture, as it is used to provide predictive information about agricultural operations, to make operational decisions in real time, and to redefine processes [14]. Important issues, such as improving efficiency, food safety and security, and climate change and sustainability, need to be addressed by Big data applications [15].

According to a study by IBM and Harvard Business Review Analytics, 70% of the world's largest companies out of 600 surveyed have experienced a digital transformation in the field of data investing where decisions are made based on knowledge extracted from data [16]. This of course requires a profound change in the way organizations operate. They need to become much more agile, operate iteratively, take more risk, and instill a data-driven corporate culture for all their employees. This is a radical but necessary shift in the transition we are experiencing today due to the digitization of society at all levels.

Of course, no one will achieve a complete transformation overnight. But this inventory shows a changing economic world in which the big issues are out of place. It is up to companies and their leaders to adapt so as not to be left behind. Many solutions already exist for this, especially in the field of marketing "automation."

8.2.1 Business analytics

Business analytics is a set of concepts that emerged from business intelligence. It is an approach that brings together various disciplines to work together to extract values from data by analyzing it using well-studied plans and strategies [17]. Enterprises use this strategy to reach business aims by identifying new opportunities, discovering more features, changing and developing specific systems, understanding customer behavior, and anticipating problems before they occur [18]. These business analytics techniques combine machine learning, deep learning, data science, data management, statistical methods, and many other frameworks [19].

8.2.2 Type of analytics

To understand the concept of "analytics," we must present its three types and understand the way each type works and its purpose [2]. We start with the first type, which is descriptive analytics [20], also called reporting. It answers the questions of "what happened?" and "what is happening?." It is performed mainly on the past data of the business indices in order to understand current trends. For example: producing a report that deals with demographic characteristics of customers, areas of interest or consumer behavior in accordance with sales indices or financial data. Often the analysis can also draw conclusions from data derived from social networks. For example: for which posts did the customers mark "I liked," deriving data from the accounts they are following on the "Instagram" network or from the posts they themselves upload on any other network. It also facilitates understanding of why and how some events happen and explains why some results occur. For example, descriptive analytics can be used to determine the totality of a company's deals or identify customers who signed up for a service before a specific period. This type of analytics uses several techniques such as data mining and data aggregation.

The second type is predictive analytics [21] which answers the questions of "what will happen?" and "for what reason will it happen?" It is based on inquiry to support modeling, and presentation requires some realistic methods that can analyze and explain current and chronic events to give experiences and expectations about unknown future events. This type of analytics is used to predict and intercept problems before they occur, obtain new experiences, and increase efficiency. For example, using predictive analytics, a company can identify the risks of losing liquidity from speculating on a new project.

The last type is prescriptive analytics, this analysis combines the information found from the previous types of analyzes and creates an action plan for the organization to deal with the issue or decision. Most data-driven companies use prescription analysis because forecasting and descriptive analysis are not sufficient to improve data performance. It answers the questions of "what should we do?" and "why should we do it?." Companies use this type of analytics to set new measures of progress, discover opportunities [22].

8.2.3 Big data analytics

Big data has been defined extensively in the literature, many researchers define Big data as data that is "too big," "too hard," and "too fast" [23–25]. These phrases refer to the three characteristics of Big data discussed previously: Volume, Variety, and Velocity (see Fig. 8.2).

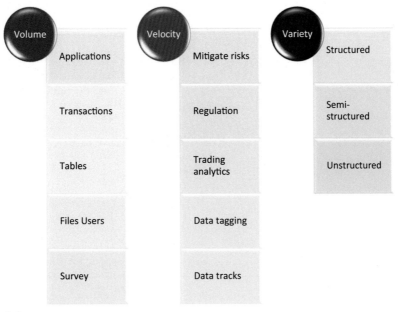

FIG. 8.2

The three Vs of Big data.

Big data is huge data that varies in size, shape, and type and is difficult to analyze and manage using traditional means. As previously discussed, companies are trying to exploit this data for their benefit by compiling, purifying, filtering, analyzing, and discovering the relationships between their processes. The methods used vary according to the goal and the company's ability.

Big data Analytics is defined in scientific research as the advanced analytical methods used to process Big data to extract value from it (see Fig. 8.3) [26]. It provides solutions for companies to exploit the capabilities provided by data to gain insight into and guidance for enhancing decisions. Big data Analytics combines a group of methods and techniques based on algorithms, predictive models, data mining, and various aids to extract information from Big data [27]. Its main objective is to search for the value of Big data that can be used to produce insights and make the right decisions. However, there are challenges in analyzing Big data, such as the diversity of the data in terms of source, shape, and size, rapid dispersal of information, information quality, fast data retrieval, information tagging, and information storage.

FIG. 8.3

Big data Analytics process.

FIG. 8.4

Decision in enterprise areas.

These means of dealing with data contributed to the emergence of the so-called intelligent enterprise, where Mellote stressed that smart organizations are those that exploit the power of technologies to meet the challenges they face in the various stages of production and rely mainly on the knowledge extracted from the processed data [28,29]. Another set of interpretations consistent with the first definition defines a smart enterprise as a system that integrates all modern technological developments in analytics to obtain, process, and analyze information [30]. The third group focuses on insights, predictions, and foresight and emphasizes that smart enterprises are organizations that can transform data into predictions and then actions [31]. Through all these definitions, it becomes clear to us that an intelligent enterprise is an organization that relies on information to build a path that can improve performance and increase productivity. This decision depends on several means and techniques that allow discovering the capabilities offered by this information that enable it to influence all areas of the organization shown in Fig. 8.4 [32].

Big data Analytics consists of a set of abstract layers that represent the levels of dealing with Big data [33]. As shown in Fig. 8.5, these layers consist of blocks and stacks linked to each other, and often the distributed file system layer is the core of the Big data-centric application.

8.2.4 Enterprise Big data sources

Data sources differ according to the company and its field of operation. The Human Resources Department, the information cell, usually supervises the goal of data collection, the process of data collection, and the analysis and processing of the data

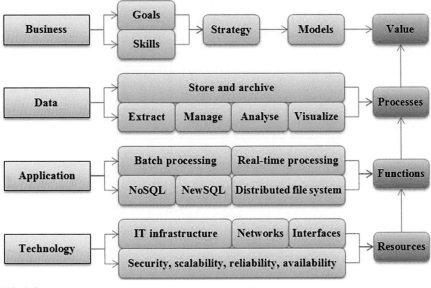

FIG. 8.5

Big data Analytics components.

collected. This information is stored in massive databases or is extracted from other sources and worked on in real time using several techniques. This data is subject to purification, filtering, and processing to discover the links between them and determine the best way use it. Table 8.1 lists sources of Big data [34].

8.3 Opportunities

Big data Analytics is a critical strategy that develops business and provides services that enhance competition. A smart enterprise skilled in analyzing Big data can exploit analytics to make proper and correct business decisions. Big data Analytics guarantees many benefits to an organization [35,36]. Some of the most important of these benefits are orientation to market sentiment and customer base segmentation, knowledge of competitors details, understanding customer trends and behavior, better planning and advance prediction of problems, increasing production, classifying the causes of turnover, and much more [37].

Big data Analytics can be used enterprise for a variety of purposes.

- **Email marketing campaigns**. A marketing automation tool effectively allows segmentation of a database according to certain predefined criteria (e.g., sending an email when an Internet user has not validated their shopping cart), so that in just a few minutes it will be able to personalize emails, send newsletters, and conduct A/B testing.

Table 8.1 Big data sources

Information source	Description	Examples
Local databases	A database that contains a set of data about customers, employees, competitors, and products, such as personal details, methods of progress, profits, and so on. This data can be collected via traditional or advanced technological means	Data: experiences, objectives, pay, execution, rating, comportment, age, gender, prices, etc. Databases: Oracle, Informix, SAP-Sybase
Customer survey data	Recorded data about customers obtained via various techniques, such as surveys. This data includes customer tendencies, customer satisfaction, needs and requests, and so on	Classify customers and determine their degree of dedication, inclinations, satisfaction, purchases, and potential for additional business purchases
Operational performance data	Collected data related to the effectiveness of the institution. These data are related to the assessment of the productive management of the business	Complaints resolved, calls left, time spent on certain tasks, number of new customers, etc.
Employee survey data	Scope of attitude data recorded in surveys to be analyzed and identify all information relating to the employee (attitudes, commitment information, and rules, dealt with suppliers)	Employee commitment, work level, performance, comportment, fulfillment, impression of equity, anxiety, suggestions, etc.
Sales performance data	Various data related to business, such as transactions and income. This data is considered among the important for companies to reach their goals	Deals of the month, achieved goals, profit percentages, turnover, new purchases, income achieved, top of the line, item attributes, etc.
Social media data	Textual data extracted from social media or company websites about customer behavior, customer satisfaction, customer needs for a particular product, and so on	IRaMuTeQ (an open-source license with an API to interact with R language), Lexico3, Hyperbase
Recorded data from sensors	Sensors record data such as employee attendance, customer traffic and behavior, resource usage and send it to a database for analysis	IoT sensor network

- **Content creation**. This tool allows creation of all types of content (call-to-action buttons, personalized content, forms, etc.) to meet user expectations.
- **Managing leads**. Through the segmentation of databases, automation can achieve what is called nurturing, that is, the procedure aimed at strengthening the bond between a company and its prospects.
- **Social media marketing**. A good marketing tool is capable of monitoring social networks and should allow automation of publications and campaigns on these media.
- **Analysis**. A powerful tool must be able to analyze what a company has done to determine whether it was beneficial or not. The results should help to set up a strategy for the continuous improvement of web services.

Professionals have a wide variety of Big data solutions to choose from. Some of these solutions for marketing automation include:

ActiveCampaign: This is one of the fastest growing marketing solutions providers today. It is a cloud software platform for small-to-mid-sized businesses that allows automation of email marketing campaigns, content and social media management, lead nurturing, and analysis.

HubSpot: Designed by Brian Halligan and Dharmesh Shah and launched in 2006, this solution is often considered the benchmark in the field. It is designed to increase the effectiveness of online marketing campaigns through the personalization of site content to each visitor and the constant acquisition of new prospects. A blog platform with a high-performance SEO tool is available as well as a module to automate marketing campaigns on social networks where the company is present. A CRM accompanies everything.

InfusionSoft: This solution is mainly intended for small businesses. It automates lead management, social media marketing strategy, content publication, and email marketing campaigns. More than 300 integrations into third-party software are possible and InfusionSoft has the advantage of being very intuitive and therefore very easy to learn. This tool is thus particularly renowned for "building" marketing campaigns accessible even to beginners.

Marketo: Recently bought by Adobe to boost the functionality of its own products, Marketo is a mid-range tool primarily intended for large companies. It is very intuitive and therefore relatively easy to learn. This solution can be integrated into many third-party platforms, which also explains its success on a global scale.

Ontraport: Admittedly, this is not the best-known marketing automation tool on the market, but like ActiveCampaign, it deserves mention for its excellent value for the money. Aimed mainly at small businesses, this solution offers a wide variety of functionalities since it can automate marketing campaigns as well as sales force. It includes many features specific to e-commerce, making it a highly competitive tool.

Recently, Ontraport updated its graphical interface to respond to critics who found the dashboard unappealing.

Regarding Big data Analytics, a large group of companies has produced systems to deal with Big data. Table 8.2 shows the most mainstream analytic systems currently known.

Table 8.2 Details of Big data Analytics systems

Software system	Details
APACHE HADOOP	Apache Hadoop is a robust and open-source system that interacts with data distributed on a group of devices. It uses simple programming models for this purpose, programmed to extend the range of individual servers to a large number of connected servers, each of which provides the advantages of local storage and computing. Its components make it capable of analyzing data more quickly and accurately: HDFS, MapReduce, and YARN [38]
APACHE ZOOKEEPER	Apache Zookeeper is among the most powerful frameworks, providing a common space between all servers, enabling the unification of communications between distributed systems. This distribution enables the system with all its connected parts to unify work, address problems, and work on large data, which speeds up the process of data analysis. In addition, the method of its construction enables the treatment of defects quickly [39]
APACHE STORM	Apache Storm is among the powerful systems based on the Stream Processing System, which provides a shared and distributed processing system in real time for data in a flexible and fast manner. Its construction method provides low latency to ensure fast and efficient processing. It relies on JSON as an interaction protocol and provides collection, filtering, and reading and writing services to and from many disparate sources [40]
KAFKA	Kafka is a mega project that provides real-time data pipelines and streaming applications, providing, distributing, and iterating services. It also supports the parallel loading of data from a range of sources, including the data consumed in real time. After filtering and purifying data, it is saved in any NoSQL system. The data can also be stored in databases that are not connected to the Internet. Kafka is the most important system adopted in Big data Analytics [41]
SPARK	Apache Spark is an open-source system that handles smart data. It is characterized by speed and accuracy in data analysis, and it deals well with images and text, as well as supports data flow, this makes it a powerful system that processes huge data in real time [42]
CASSANDRA	Cassandra is a NoSQL family distributed database. Its strength is that it can collect many different types of data from multiple sources. The collected data is duplicated to multiple instances. This technology helps to avoid damages and data loss [43]

8.4 Challenges

Whatever happens, companies and their leaders will have to focus their efforts on seven key points to adapt: leadership, corporate culture, brand, innovation, quality, and operational and technological challenges. The latter is the subject of this chapter. Technical challenges in Big data are due to the data's large quantity, diversity, varying uses, and speed [44]. Because of the complexity of the data, the lack of experts in the field, the prohibitive costs of systems, and the lack of comprehensive

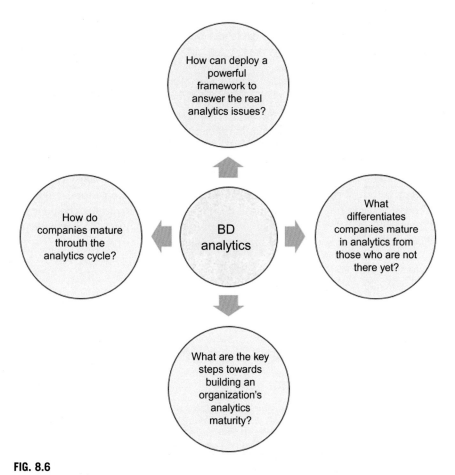

FIG. 8.6

Big data Analytics issues.

frameworks, finding ways to analyze Big data effectively is still the subject of expert research. The operational and technological issues revolve around the control of information with its processing and analysis operations, and how to follow technological and legislative developments, which directly influence the markets.

A review of the possible solutions and the results provided by these methods, highlights the problems of collecting and analyzing Big data (see Fig. 8.6). In the next section, we propose solutions for dealing with Big data to obtain accurate results.

8.5 Proposed solution

In this part of the study, we present some alternative solutions to the problems of collecting and analyzing data more quickly and more accurately despite the different data sources. The first solution depends on a new framework that combines its components with all the basic layers of decision-making. The second solution depends on deep learning, which has become a unique and important way to analyze Big data.

Solution 1:

In the corporate world, decision-making depends on the type and quality of the data collected as well as the intervening people, business planning, and business objectives, all of which gives us the impression that the technical aspect is not the only part of the data analysis solution, but rather that the solution is a multidimensional framework that is easy to handle and understand [45]. The interaction between these dimensions produces data that can be used to make the right decision, which will later turn into a procedure in the company's interest. We have proposed a framework (Fig. 8.7) that brings together all the measurements to be adopted and related to its dimensions, and offers a hierarchical decomposition of its various components and dimensions which are connected as shown in (Fig. 8.7), the main components of this framework are summarized in the following list:

- actors
- communication
- enterprise content
- data sources
- internal hierarchical arrangements
- strategies and culture
- external powers
- interaction models
- computing infrastructure

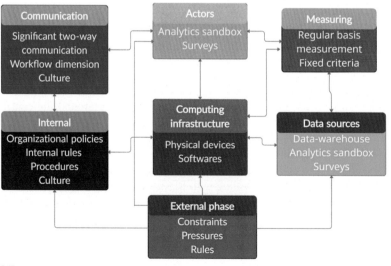

FIG. 8.7

Proposed analytics framework.

Solution 2: Deep learning

The second solution proposed is related to the integration of deep learning as one of the means that can help collect and analyze data, and can be adapted, as we like, according to the quality of the data, to extract values with or without intervention [46]. Deep learning can be an alternative solution to common problems encountered in Big data analysis (tagging, indexing, retrieving, reporting, etc.). It is called deep learning because it is a network that is made up of tens or even hundreds of "layers" of neurons, we are talking about algorithms capable of mimicking the actions of the human brain thanks to artificial neural networks and capable of analyze a huge structured and unstructured data. Here we propose two structures for analyzing Big data: Convolutional Neural Networks (CNNs) and Deep Belief Networks (DBNs).

A CNN is a suitable solution for data collection and analysis because it is consistent and consists of several layers, starting with a convolution layer characterized by a hierarchical shape unlike others. This layer creates feature maps from the collected data, to pass to another layer. The other connected layers are Maxpooling and Dense [47,48]. After setting up the model and adjusting its parameters to increase its accuracy (width and depth, layer arrangement method, number learning steps, activation function, GPU runtime environment, CPU, etc.), the outputs are stored in a comprehensive memory. The simplest solution for using this model on our data is the application of the transfer learning technique, this technique designates all the methods which make it possible to transfer the knowledge acquired from the resolution of given problems to deal with another problem(*Transfer learning refer to the situation where what has been learned in one setting is exploited to improve generalization in another setting*). It means the application of knowledge on our own data to extract meanings, this exploited knowledge, is already obtained by scientists after learning a basic network on data larger than ours. This training is reused to create models during which data is collected and prepared to predict the results reached, and these results are exploited for decision-making.

A DBN is a model consisting of input, output, and hidden layers. This model deals with supervised and unsupervised methods, making it powerful in dealing with all types of data. A DBN is described as a composition of many restricted Boltzmann machines (RBM) which consists of two directly connected layers, and it has nodes in its two layers [49]. Each layer contains nodes that are linked to the nodes of the other layer and are not linked to nodes of the same layer. We can utilize this network separation to deal with Big data analytics proficiently. A Graphical Processing Unit (GPU)-based compositional model is introduced for this technique to handle the enormous measure of information favoring Big data. This strategy helps to deal with a gigantic measure of data with less time to extract value and make a decision [50].

Fig. 8.8 is an example of a deep learning model for analyzing text data coupled with the vector representation of word-embedding words and the use of pretrained words from the word embedding algorithms (GloVe). The gloVe is an unsupervised learning algorithm for obtaining vector representations of words developed by MIT. The proposed model shown in Fig. 8.8 consists of a set of deep learning layers for each layer and its operation. The first allows us to establish a link between words and the layer known in deep learning:

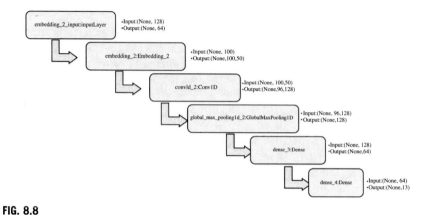

FIG. 8.8

Proposed deep learning model.

- convolution layer
- max pooling layer
- dense layer
- output layer

8.6 Case study

Our case study involves determining the degree of satisfaction of the customers of an agricultural company that sells irrigation products. We use a technique that integrates the solutions we mentioned in the previous section, as this combination of techniques enables data analysis and extraction of values. Fig. 8.9 is the model that incorporates these solutions. The first proposed solution helps us understand data and the work methods, thus what makes this a robust model is its hierarchical decomposition of these different components. Likewise, the chance of separating an unpredictable

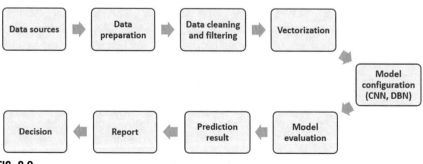

FIG. 8.9

Architecture of the system.

process, framework, or device into its parts offers the likelihood to study and understand them and afterward incorporate the outcomes attempting to comprehend the total working framework. The interaction between this model's different components presents the objective set of Big data Analytics with these three types. It can offer dashboards and reporting; it can make insights and predict future performance. We adopt deep learning to establish the model (Fig. 8.9), which introduces data from various sources, preprocesses it, and applies the cleaning operations. It only applies simple filters. Then the data is vectorized by word-embedding, term frequency-inverse document frequency (TF-IDF) (only for machine learning algorithms), and one-hot-encoding representations. Once the data is vectorized, it enters our models (DBN and CNN). After performing the processing operations using deep learning, we obtain the prediction results that the company will use.

The data we collected is data written by users of one of the electronic portals of an agricultural company that sells materials used in irrigation. This portal allows consumers to express their opinions and reactions to the products provided by the company. We combined this data with the data we collected from Facebook using the site's Graph API over a period of 3 months. This allowed us to select 10,000 comments that mention the name of the company and express certain sentiments. We also captured more than tens of posts from the agriculture company's website. The size of this collected data is estimated to be tens of gigabytes.

The processing of the text (comments and posts from Facebook and the company website) is a critical phase in the sentiment analysis process. The objective is to extract variables (in the form of words or sequences of words) to use them in the classification. The quality of processing therefore clearly has a major impact on the performance of classification models and the results obtained at the end of the process. We begin this step by cleaning up and normalizing the text: removing signs, symbols, repeated letters, stop words or words that do not provide any information on the subject studied. The next task is the "tokenization" operation by which the text of the comment is divided into lexical units (tokens). In a text in modern or dialectal Arabic used in Morocco, these units are more complex since they often consist of more than one word, hence the importance of the task of remove suffixes or stemming. To develop and apply a rooting on the comments collected, we were inspired by the light10 stemmer, which proposes the elimination of the most common prefixes and suffixes of a token in modern Arabic language, on which we add an extension for Moroccan dialect Arabic.

8.7 Results and discussion

To classify Facebook and website comments, we applied the preceding algorithm based on deep learning. The deep learning model that we proposed was trained on the collected data comprised of two classes. Each class expresses a certain feeling of the customer (satisfaction, no satisfaction). We trained the new model on 2000 epochs. We then obtained the results shown in Fig. 8.10.

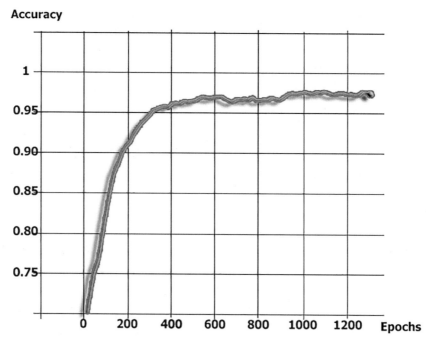

FIG. 8.10

The training accuracy.

The graph in Fig. 8.10 shows the ability of the model to generalize, where a score of 97% is reached after training (the score is given by the metrics Precision). This means that this model can classify words collected from a company's website and from Facebook, and express the most frequently used word, whether to discover the degree of satisfaction of site visitors (consumers) with the services provided by the company. Data is collected from the site using word-embedding weighting and one-hot-encoding and from Facebook using the Facebook Graph API. This operation of text classification arranged collected text based on 10 previous classes. This allowed us to identify the most frequently used words and determine the general orientation of the consumer in this space, which is an example of the exploitation of information to know the consumer's feelings (satisfied or not). Fig. 8.11 shows

Satisfaction_analyzer | (' The product is extremelty good ' |) |

The product is extremely good- - - - - - - - - - -{ ' satisfaction ': 0.8735,   ' no satisfaction ' : 0.1727}

FIG. 8.11

The results of the model on a test post.

Table 8.3 Comparison of the proposed model with existing models

Proposed solution	Existing models [44]
It is open to all possibilities of development and can be applied to various decision-making problems in the company	Few of the experienced personnel can deal with these systems
It does not require much knowledge of how algorithms work	Converting data into knowledge requires an understanding of technology and an understanding of business processes
Accurate results in data classification	There is no great credibility in choosing the right information
It gives instant results in a quick time	Models are expensive and complex
It can work on data of various shapes and sizes	
Values related to corporate decision can be extracted	
It can be updated and upgraded	

the model result on a sentence and how the model classified it in the correct class. This sentence is specific to a customer who is satisfied with the product.

The satisfaction score is higher than others (0.8735 vs 0.1727), which means that the author of this comment is satisfied with the product. The model based on deep learning has done the job well and can be adopted as a solution for sentiment analysis. To show the added value of these proposed solutions, we compare them with the existing solutions through Table 8.3.

From the results of this case study, it becomes clear that making the appropriate decision is difficult and requires concerted human and technological efforts. However, deep learning is part of the solution that must be better exploited to collect and analyze data on the horizon of decision-making for a better future for companies. These decisions are derived from data analysis. For example, according to the case study, if most consumers are not satisfied with a particular product, the company must replace it or improve it.

8.8 Conclusion

Big data Analytics is a set of analyses conducted on large and complex data that smart organizations use to drive and plan current and future business. The data is the secret of this process. Using it correctly is what distinguishes one company from another. However, most companies find it arduous to analyze and exploit the data due to its large size, diversity of form, and assortment of sources. This chapter provided a comprehensive briefing on this concept, associated problems, and existing solutions. We proposed a framework that supports Big data analysis that can help an organization reach its desired goals. The chapter also presented deep learning and demonstrated its ability to collect and analyze Big data. Deep learning is an important solution for reliable Big data Analytics. We also presented a case study in which we created a deep learning model and collected and analyzed consumer data to discover customer satisfaction. In future work, we will propose other deep learning models that combine CNN and DBN and train them on many types of data to obtain results that are more accurate and adopt them in decision-making.

References

[1] C.C. Qi, Big Data management in the mining industry, Int. J. Miner. Metall. Mater. 27 (2) (2020) 131–139.

[2] B. Jabir, N. Falih, K. Rahmani, HR analytics a roadmap for decision making: case study, Indones. J. Electr. Eng. Comput. Sci. 15 (2) (2019) 979–990.

[3] F. Ciampi, S. Demi, A. Magrini, G. Marzi, A. Papa, Exploring the impact of Big Data analytics capabilities on business model innovation: the mediating role of entrepreneurial orientation, J. Bus. Res. 123 (2021) 1–13.

[4] D. Buhalis, K. Volchek, Bridging marketing theory and Big Data analytics: the taxonomy of marketing attribution, Int. J. Inf. Manag. 56 (2021), 102253.

[5] C.C. Wei, T.H. Chou, Typhoon quantitative rainfall prediction from big data analytics by using the apache hadoop spark parallel computing framework, Atmosphere 11 (8) (2020) 870.

[6] W. Yu, G. Zhao, Q. Liu, Y. Song, Role of big data analytics capability in developing integrated hospital supply chains and operational flexibility: an organizational information processing theory perspective, Technol. Forecast. Soc. Chang. 163 (2021), 120417.

[7] D. Guttentag, Progress on Airbnb: a literature review, J. Hosp. Tour. Technol. 10 (4) (2019) 814–844. https://doi.org/10.1108/JHTT-08-2018-0075.

[8] J.M. Jordan, Challenges to large-scale digital organization: the case of Uber, J. Organ. Des. 6 (1) (2017) 1–12.

[9] G. Gopal, C. Suter-Crazzolara, L. Toldo, W. Eberhardt, Digital transformation in healthcare–architectures of present and future information technologies, Clin. Chem. Lab. Med. 57 (3) (2019) 328–335.

[10] T. Hooley, J. Hutchinson, A.G. Watts, Careering Through the Web: The Potential of Web 2.0 and 3.0 Technologies for Career Development and Career Support Services, UKCES, 2011.

[11] F. Shahzad, J.K. Khattak, M.J. Khattak, F. Shahzad, Impact of consumer socialization on soft drink consumption and mediating role of consumer generational behavior, Br. Food J. 117 (3) (2015) 1205–1222. https://doi.org/10.1108/BFJ-08-2013-0219.

[12] S. Ramgovind, M.M. Eloff, E. Smith, The management of security in cloud computing, in: 2010 Information Security for South Africa, IEEE, 2010, August, pp. 1–7.

[13] J. Chen, Q. Jiang, Y. Wang, J. Tang, Study of data analysis model based on big data technology, in: 2016 IEEE International Conference on Big Data Analysis (ICBDA), IEEE, 2016, March, pp. 1–6.

[14] J.M. Talavera, L.E. Tobón, J.A. Gómez, M.A. Culman, J.M. Aranda, D.T. Parra, L.E. Garreta, Review of IoT applications in agro-industrial and environmental fields, Comput. Electron. Agric. 142 (2017) 283–297.

[15] R.H. Ip, L.M. Ang, K.P. Seng, J.C. Broster, J.E. Pratley, Big data and machine learning for crop protection, Comput. Electron. Agric. 151 (2018) 376–383.

[16] P. Jain, Concept note on HR analytics, Int. J. Res. Anal. Rev. 7 (2) (2020).

[17] P. Mikalef, I. Pappas, J. Krogstie, P.A. Pavlou, (Eds.),, Big Data and Business Analytics: A Research Agenda for Realizing Business Value, Elsevier, 2020.

[18] Z. Wang, N. Wang, X. Su, S. Ge, An empirical study on business analytics affordances enhancing the management of cloud computing data security, Int. J. Inf. Manag. 50 (2020) 387–394.

[19] M. Kraus, S. Feuerriegel, A. Oztekin, Deep learning in business analytics and operations research: models, applications and managerial implications, Eur. J. Oper. Res. 281 (3) (2020) 628–641.

[20] A. Abbasi, W. Li, V. Benjamin, S. Hu, H. Chen, Descriptive analytics: examining expert hackers in web forums, in: 2014 IEEE Joint Intelligence and Security Informatics Conference, IEEE, 2014, September, pp. 56–63.

[21] S.B. Golas, M. Nikolova-Simons, R. Palacholla, J. op den Buijs, G. Garberg, A. Orenstein, J. Kvedar, Predictive analytics and tailored interventions improve clinical outcomes in older adults: a randomized controlled trial, npj Digit. Med. 4 (1) (2021) 1–10.

[22] D. Bertsimas, N. Kallus, From predictive to prescriptive analytics, Manag. Sci. 66 (3) (2020) 1025–1044.

[23] V. Chang, An ethical framework for Big Data and smart cities, Technol. Forecast. Soc. Chang. 165 (2021), 120559.

[24] Z.S. Ageed, S.R. Zeebaree, M.M. Sadeeq, S.F. Kak, H.S. Yahia, M.R. Mahmood, I.M. Ibrahim, Comprehensive survey of Big Data mining approaches in cloud systems, Qubahan Acad. J. 1 (2) (2021) 29–38.

[25] F. Cappa, R. Oriani, E. Peruffo, I. McCarthy, Big Data for creating and capturing value in the digitalized environment: unpacking the effects of volume, variety, and veracity on firm performance, J. Prod. Innov. Manag. 38 (1) (2021) 49–67.

[26] S. Bag, L.C. Wood, L. Xu, P. Dhamija, Y. Kayikci, Big Data analytics as an operational excellence approach to enhance sustainable supply chain performance, Resour. Conserv. Recycl. 153 (2020), 104559.

[27] S. Khanra, A. Dhir, M. Mäntymäki, Big data analytics and enterprises: a bibliometric synthesis of the literature, Enterp. Inf. Syst. 14 (6) (2020) 737–768.

[28] J. Springett, U.S. Patent Application No. 11/613,847, 2007.

[29] N. Kruschwitz, First look: the second annual new intelligent enterprise survey, MIT Sloan Manag. Rev. 52 (4) (2011) 87.

[30] J. Ranjan, C. Foropon, Big Data analytics in building the competitive intelligence of organizations, Int. J. Inf. Manag. 56 (2021), 102231.

[31] H.P. Lu, C.I. Weng, Smart manufacturing technology, market maturity analysis and technology roadmap in the computer and electronic product manufacturing industry, Technol. Forecast. Soc. Chang. 133 (2018) 85–94.

[32] C.R. Greer, S.A. Youngblood, D.A. Gray, Human resource management outsourcing: the make or buy decision, Acad. Manag. Perspect. 13 (3) (1999) 85–96.

[33] A. Oussous, F.Z. Benjelloun, A.A. Lahcen, S. Belfkih, Big Data technologies: a survey, J. King Saud Univ.-Comput. Inform. Sci. 30 (4) (2018) 431–448.

[34] M.H. Salas-Olmedo, B. Moya-Gómez, J.C. García-Palomares, J. Gutiérrez, Tourists' digital footprint in cities: comparing big data sources, Tour. Manag. 66 (2018) 13–25.

[35] J.T. Avella, M. Kebritchi, S.G. Nunn, T. Kanai, Learning analytics methods, benefits, and challenges in higher education: a systematic literature review, Online Learn. 20 (2) (2016) 13–29.

[36] K.V.N. Rajesh, Big data analytics: applications and benefits, IUP J. Inform. Technol. 9 (4) (2013).

[37] J.A. Delgado, N.M. Short Jr., D.P. Roberts, B. Vandenberg, Big data analysis for sustainable agriculture on a geospatial cloud framework, Front. Sustain. Food Syst. 3 (2019) 54.

[38] P. Zikopoulos, C. Eaton, Understanding Big Data: Analytics for Enterprise Class Hadoop and Streaming Data, McGraw-Hill Osborne Media, 2011.

[39] S. Haloi, Apache ZooKeeper Essentials, Packt Publishing Ltd., 2015.

[40] J.S. van der Veen, et al., Dynamically scaling apache storm for the analysis of streaming data, in: Big Data Computing Service and Applications (BigDataService), 2015 IEEE First International Conference on, IEEE, 2015, March, pp. 154–161.

[41] R. Ranjan, Streaming big data processing in datacenter clouds, IEEE Cloud Comput. 1 (1) (2014) 78–83.

[42] R. Xin, et al., Graysort on Apache Spark by Databricks, GraySort Competition, 2014.

[43] A. Chebotko, et al., A big data modelling methodology for apache cassandra, in: Big Data (BigData Congress), 2015 IEEE International Congress on (pp. 238–245), IEEE, 2015, June.

[44] D.P. Acharjya, K. Ahmed, A survey on Big Data analytics: challenges, open research issues and tools, Int. J. Adv. Comput. Sci. Appl. 7 (2) (2016) 511–518.

[45] M. Janssen, H. van der Voort, A. Wahyudi, Factors influencing big data decision-making quality, J. Bus. Res. 70 (2017) 338–345.

[46] M.M. Najafabadi, F. Villanustre, T.M. Khoshgoftaar, N. Seliya, R. Wald, E. Muharemagic, Deep learning applications and challenges in Big Data analytics, J. Big Data 2 (1) (2015) 1–21.

[47] B. Jabir, N. Falih, K. Rahmani, Accuracy and efficiency comparison of object detection open-source models, Int. J. Online & Biomed. Eng. 17 (5) (2021).

[48] B. Jabir, N. Falih, A. Sarih, A. Tannouche, A strategic analytics using convolutional neural networks for weed identification in sugar beet fields, Agris On-Line Pap. Econ. Inform. 1 (March) (2021) 49–57.

[49] S. Hu, Y. Xiang, D. Huo, S. Jawad, J. Liu, An improved deep belief network based hybrid forecasting method for wind power, Energy 224 (2021), 120185.

[50] Y. Zhang, P. Geng, C.B. Sivaparthipan, B.A. Muthu, Big Data and artificial intelligence based early risk warning system of fire hazard for smart cities, Sustainable Energy Technol. Assess. 45 (2021), 100986.

Person-based automation with artificial intelligence Chatbots: A driving force of Industry 4.0

Smita Vinit Bhoir[a], Sunita R. Patil[b], and Ibtisam Yakub Mogul[c]

[a]*Department of Computer Engineering, K.J. Somaiya College of Engineering, Ramrao Adik Institute of Technology, Mumbai, Maharashtra, India,* [b]*Department of Computer Engineering, K.J. Somaiya Institute of Engineering and Information Technology, Mumbai, Maharashtra, India,* [c]*ATHE Coordinator Computing, University of Bolton Academic Centre RAK, UAE*

9.1 Introduction

9.1.1 Chatbots

A chatbot is an artificial intelligence (AI) agent. It is used to provide human interaction with a computer system. Interaction is provided using natural languages. A variety of recently used mobile messaging apps, blogs, and smartphones serve the same purpose [1]. We have been following AI very closely for many years, and one of the most promising expressions of interaction between humans and machines is the chatbot [2]. Although theoretically a chatbot can be advanced into something even more sophisticated, at its core it is still a client-side computer that responds to input from a Web interface [3].

9.1.2 Role of Chatbots in Industry 4.0

Traditionally, manufacturing and industrial practices function and contribute to industry via the use of automation, but with the changes and trends in Industry 4.0, industry is also being improved via the advancement of data exchange [4]. Fig. 9.1 shows some of the latest ideas in both the present as well as the future. It also presents some of the technological tools being used today such as autonomous robots, Big Data applications, cybersecurity systems, system integration, and the Internet of Things (IoT). The main goal of Industry 4.0 is to create factories managed by intelligence. Industry 4.0 is one of the most significant milestones in the idea of "Industry 4.0" standards [5]. The point is not to replace humans with computers, but rather to establish a new form of interaction between the two [6]. An efficient interface should be developed to provide human-based automation. A real-time connector

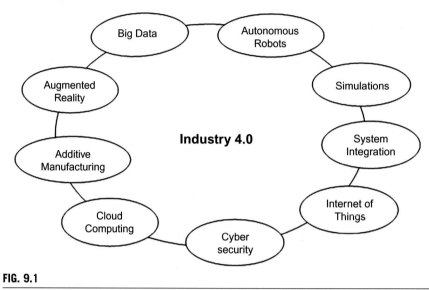

FIG. 9.1

Role of Industry 4.0.

Credit: W. Bodrow, Vision of Industry 4.0. Iwama. 2016, pp. 55–58. https://doi.org/10.2991/iwama-16.2016.10.

can act as a bond between people and machines that gives the machines access to information [7].

According to Industry 4.0 standards, person-based automation is required to be available; however, in many systems, it seems like manual jobs typically performed by humans have been replaced by automation. In automation, humans must communicate with and through automated systems while avoiding interaction problems, especially when it comes to human-automation interactions that pose danger to people. The future of work is likely to be a combination of both human and AI working towards the same common objectives. The question now is what we can do to satisfy consumer demands that are human-centric.

In engineering and industry, some chatbots are machines that can independently learn and make decisions on their own. They use the data they receive to make decisions that the engineers need. Humans need to take a bigger role in the workplace, as machines are used as a physical representation of thoughts and not necessarily the idea behind them. With assistance from the chatbot assistant, technologists make the most effective use of their expertise. Thus, because of the Industry 4.0 strategy, a chatbot is the best interface candidate, as shown in Fig. 9.2.

9.1.3 Importance of Chatbots

Chatbot apps are technologies that assist in communication by streamlining communications between individuals and providers. At the same time, they give businesses a new way to interact with customers, by reducing the traditional cost of customer

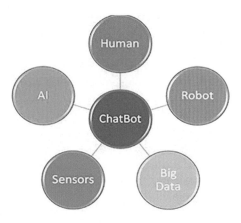

FIG. 9.2

Chatbots' new role.

Credit: D.K.-S. Management and undefined 2019, How Chatbots Influence Marketing. content.sciendo.com,
https://doi.org/10.2478/manment-2019-0015.

care and in turn strengthening the process of customer interaction and operational performance [8]. A chatbot needs to be able to perform both these tasks efficiently, such as answering questions and chatting with a user. Human support is key here. Human involvement is essential in the setup, training, and optimization of chatbot systems, regardless of the platform. It seems that chatbots are a sort of semi-intelligent guidance tool that is supposed to offer users an interactive chance to respond to their questions [9]. A chatbot that is easy to use will greatly increase productivity. The chatbot has become very popular in various businesses because it has satisfied a need for customer service without hiring additional staff [10].

Some of the benefits of chatbots include:

- Ease of use
  Chatbots are designed to be user-friendly to interact with people.
- 24/7 availability
  Chatbots are automated, hence they are available 24/7 to satisfy customer requirements.
- Personalized experience
  Chatbots provide a personalized experience to users. In many e-commerce industries, chatbots act as one-to-one marketing agents.
- Continuous improvement
  With AI technologies, as per the current demand of users, chatbots are continuously improving.

9.1.4 Key concepts

We discuss some key concepts related to chatbots in the sections that follow.

9.1.4.1 Pattern matching
In chatbots, a request-response model is used. According to the pattern-matching technique, the response is generated for a given input [7]. The downside to this technique is that the answers are probable and monotonous. In this model, the history of responses is not stored, which leads to repetitive conversations [11].

9.1.4.2 Artificial Intelligence Markup Language (AIML)
Artificial Intelligence Markup Language (AIML) was developed for chatbots and used between 1995 and 2000. It uses tag-based markup language like XML script. AIML is based on pattern-matching techniques (das Graças Bruno Marietto et al. [8]). The following is an example of AIML script:

```
<?xml version = "1.0" encoding = "UTF-8"?>
<aiml version = "1.0.1" encoding = "UTF-8"?>
<category>
<pattern> HELLO ALICE </pattern>

<template>

        Hello User

</template>

</category>
</aiml>
```

where,
    < aiml >—defines the beginning and end of an AIML document.
    < category >—defines the unit of knowledge in the bot's knowledge base.
    <pattern>—defines the pattern to match what a user may input to a bot.
    < template >—defines the response of a bot to the user's input.

9.1.4.3 Latent semantic analysis (LSA)
The creation of chatbots can be done using Latent Semantic Analysis (LSA) infusion with AIML [12]. It uses vector representation to discover similarities between words. The major differences between AIML and LSA are that LSA is used to generate responses for template-based requests, whereas AIML is used to generate responses to all unanswered requests..

9.1.4.4 Chat script
The chat script [13] is an open-source expert framework. It includes topical rules to find the best match to the input query. It is also case-sensitive, expanding the potential answers that can be given based on the intended emotion to the same user feedback, as the upper case is usually used to denote focus in conversations.

9.1.4.5 Natural language unit (NLU)

Natural language unit (NLU) analyzes natural language spoken. Its understanding is vital to analyze user input given in natural languages and, accordingly, a chatbot can generate the response. Most methods of NLP are based on machine learning (ML) [10].

9.1.5 **Popular Chatbots**

In the following sections, we describe some well-known chatbots.

1. *ELIZA*

The ground-breaking chatbot ELIZA was developed at MIT in 1966. This chatbot operates on pattern matching. It takes input from the user and searches via pattern matching for a valid response in a document [14].

2. *Endurance*

The goal of the Endurance chatbot is to "recognize deviations in conversational branches that may imply an immediate recollection issue" using NLP [15]. It is a companion bot for dementia patients. Because Endurance's software is cloud-based, doctors and family members can review the logs of their communication with health-care providers to identify possible memory loss and communication barriers that can indicate increased severity of disease. The chatbot's conversational software is an open-source project, which means that the code that makes up the software base for the bot can be contributed to by anyone. Despite the fact the project is in its early stages, it has already shown immense potential to help researchers and medical staff better understand how the brain is affected by Alzheimer's disease.

3. Disney chatbot

Chatbots might be more popular in the customer care sector, but that hasn't prevented global television conglomerate Disney from utilizing the tool to reach younger viewers [16], as it did with a chatbot that starred a character from the 2016 animated family crime caper Zootopia, as shown in Fig. 9.3.

4. MedWhat

For patients, the goal of the MedWhat chatbot is to diagnose a condition or disease as quickly and as accurately as possible. For physicians, the goal is to make diagnosis and treatment as transparent as possible. Its adaptive business model is controlled by the intelligent ML technique that presents questions to humans and offers more proximate answers to those who do the best at that answer. A screenshot of the MedWhat app is shown in Fig. 9.4.

9.2 **NBC Chatbot**

These days, reading the news over morning coffee is as much about finding out whether we should be hunkering down in the basement planning for impending nuclear destruction as it is about keeping up with the day's headlines [17]. In this

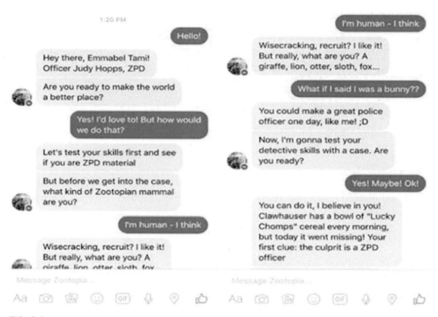

FIG. 9.3

Disney chatbot.

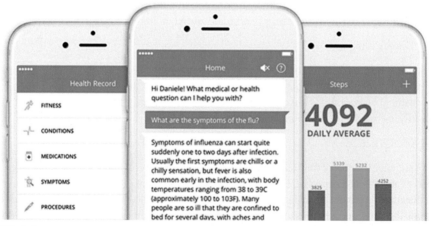

FIG. 9.4

MedWhat chatbot.

Credit: K. Ramesh, S. Ravishankaran, A. Joshi, K. Chandrasekaran, A survey of design techniques for conversational agents, Commun. Comput. Inf. Sci. 750 (2017) 336–350, https://doi.org/ 10.1007/978-981-10-6544-6_31.

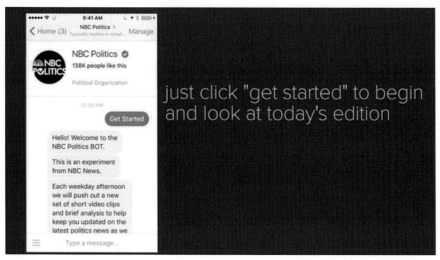

FIG. 9.5

NBC chatbot.

Credit: K. Ramesh, S. Ravishankaran, A. Joshi, K. Chandrasekaran, A survey of design techniques for conversational agents, Commun. Comput. Inf. Sci. 750 (2017) 336–350, https://doi.org/ 10.1007/978-981-10-6544-6_31.

era, information is flooding in from such a vast array of sources that it is nearly impossible to process it all and separate the signals from the noise. This is why different news outlets have incorporated custom bot technology into their Facebook pages to quickly process information and respond to readers with a clear and concise summary. Fig. 9.5 is a screenshot of the NBC chatbot.

9.3 A.L.I.C.E

A.L.I.C.E. stands for Artificial Linguistic Internet Computer Entity. It was introduced, redesigned, and launched as an Internet entity in 1995. Fig. 9.6 shows an analysis of the usability of the chatbot's aesthetics. ALICE, which stands for Sociality, Liar, Emotivity, Cognitive Inhibition, and Disinhibition, relies on an old codebase, but still offers a remarkable conversational experience. A.L.I.C.E., like many of her contemporary peers, struggles with the limitations of certain questions and sometimes returns a mix of unintended responses and comments. However, chatbots are much than they were just 5 years ago. Thanks to Dr. Walsch, these programs can now interact with people in a far more natural manner. In Spike Jonze's 2013 science fiction romance, Her, the character Ava is a bot, which served as a model for the female artificial intelligence..

9.3.1 Chapter organization

This chapter is organized into seven sections. Section 9.1 introduces chatbots and describes their role in Industry 4.0, identifies key concepts, and describes some of the most popular chatbots in use today. Section 9.2 focuses on the history of chatbots,

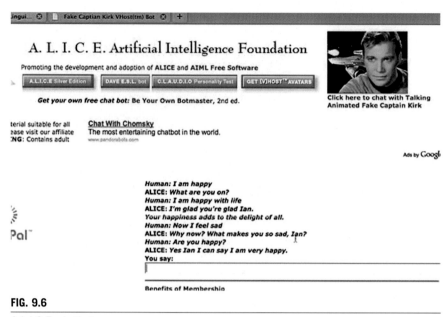

FIG. 9.6

A.L.I.C.E. chatbot.

AI and ML, chatbot models, challenges in implementing chatbots, and a literature survey of various ML techniques used to analyze user sentiments and their classification. Section 9.3 covers the general architecture of chatbots with user intention analysis and provides details about the proposed model of Hybrid Intention-based Multiclass Sentiment Analysis (HI-MCM). Section 9.4 discusses the applications of chatbots in various domains, and Section 9.5 highlights the scope of research work in chatbots. Section 9.6 presents and explains the investigational results. Finally, Section 9.7 concludes with a discussion of the need for chatbots, their importance, and how ML approaches are useful in building efficient chatbots.

9.4 Related work

9.4.1 History of Chatbots

In 1950, Alan Turing introduced "The Turing Test" at the same time he also introduced the concept of the chatbot. As previously mentioned, the first known chatbot ELIZA was developed in 1966. It played the role of a psychotherapist to return answers to user queries. It was developed using pattern matching, but its ability to speak was poor. In 1972, PARRY was introduced to overcome the limitations of ELIZA [10]. In 1995, the A.L.I.C.E. chatbot was created, which used AIML [10] with the underlying intelligence. Developers today can use AIML to construct the building blocks of chatbots [9]. Recently, chatbots have been integrated with messaging systems on major websites such as eBay, SmarterChild [14], Swelly [18], Endurance [12], Casper [13], Disney, Google Assistant [19], Marvel, Lyft, and so on.

9.4.2 **Machine learning and artificial intelligence algorithms in Chatbots**

Because of AI, intelligent agents or chatbots have become part of daily life [20]. Social technologies available in the marketplace have always been a way for people to solve existing problems. One of the ways technology has been able to offer solutions to current problems is through chatbots. A chatbot is a soft-spoken AI code. It makes it easier to communicate with a user in a natural language. To provide this service, it uses websites, message applications, mobile apps, and telephone. Chatbots are used in marketing and advertising for many brands. Today, most people who need interact with a client prefer to use a chatbot because it is quicker and easier than communicating with a human representative. Bots are used to generate conversations by utilizing market information and ML.

AI can be used in building a chatbot by playing a significant role in the bot's design. Algorithms increase the efficiency of chatbots and offer many other advantages, including making the user feel as if they are communicating with an actual human. The computer system understands the client's request and proposes a solution. AI is widely used in companies to help with customer service inquiries [21].

Chatbots use AI algorithms to process questions. First, the algorithm checks the last transaction of the chat the customer initiated and then functions to solve the present problem. AI, ML, and NLP all contribute to providing user-friendly solutions [17].

The dialogue between the AI and the ML bot results in a good response once the conversation has been understood. The first thing that a chatbot should do is correctly understand the question. The use of natural language (such as "Buy" or "Raise a Question") is the main use among some of the clients or customers. NLP chatbots were developed to read and understand crude language.

An NLP chatbot can, in some situations, dictate a more sympathetic way of interacting with its users. These NLP systems take the user's problem, tokenize it, and create a formal description of that problem using ML techniques. This NLP system trains a chatbot with many examples, and it provides a better and more useful output to users. It requires training in many different scenarios. Chatbot trainers should conduct themselves in an appropriate formal manner so that the users stay engaged in the conversation. There is no one way to ask a question. If a user asks certain questions in a specific fashion, the chatbot is expected to answer in the same manner. NLP programs can be used to make chatbots that understand human interactions and respond in line with the user's demeanor and attitude. People can still distinguish different tone types, and in some cases, in-group language, but not completely.

NLP is made up of three core areas, which we discuss in the sections that follow.

1. Natural language unit (NLU)

The languages that we use to communicate with others in our daily lives are very flexible. Before we start scaling, we must conduct logical operations ensuring that the computer's regular algorithms comprehend our daily language.

2. Natural language generation (NLG)

As the accountant for the company in question uses the process, the accounting system is automatically audited and processed. The NLG unit generates natural languages.

3. Natural language interaction (NLI)

After completing the NLU and NLG, based on the client's requirements, their requirements, interaction training is provided for bots through the NLI..

Scripted chatbots are built with the programming language and the AI chatbots are run depending on the information they receive from the users.

4. Scripted chatbots (scripting)

If someone asks for something, scripted chatbots are not able to respond, but they can catch the general theme of the question.

5. Artificial Intelligence chatbots

AI chatbots are developed with NLP in mind. Intelligent agents can learn from their past interactions and give appropriate responses. In the present age, these chatbots communicate with customers through voice or text.

Even though the words used in ordinary language are different from one another, we understand each other because we communicate with variations in phrases and expressions. As a challenge, we must deal with synonyms, slang, spelling, antinomy, accent, punctuation, and non-ASCII characters. If someone is going to be asked a question, they are guaranteed to be able to respond if they want to. Technicians must use NLP to help make the system work by teaching it all the factors that might be involved and sending it in a nonstop response to the user's query. Thus, the most difficult thing for an AI is to apply the differentiating factors of each individual's voice, while still sharing a common voice. Although it is a preprogrammed application, a chatbot can still produce an answer for the present matter [22]. If we decided to limit ourselves to preprogrammed software, then we cannot meet our requirements. To combat this challenge, NLP applications are the most effective response.

9.4.3 Models of Chatbots

Chatbots that provide information fall into the category of Information Service, as do chatbots that assist people with obtaining services or information. Generalized (or prioritized) conversation models are divided into two main types: generative (or prioritized) models and selective models [18,23,24]. Note that hybrids are also a possibility. A properly trained model considers the previous lines of dialogue and predicts the context leading to the next line.

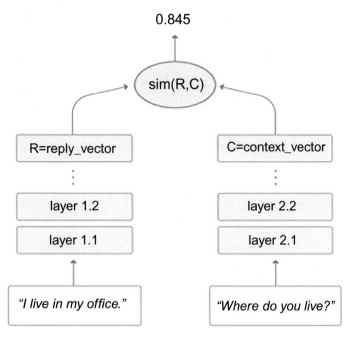

FIG. 9.7

The selective model: Illustrative example.

Credit: M. das Graças Bruno Marietto et al., Artificial Intelligence Markup Language: A Brief Tutorial. Accessed January 25, 2021 [Online]. Available: https://arxiv.org/abs/1307.3091.

9.4.3.1 The selective model

Instead of estimating probability p(reply | context; w), selective models learn by applying similarity functions. Some models use a pool of possible variables for responses, and others use the mean of a predefined set of samples as a response value (see Fig. 9.7). The intuition is that the neural network takes the present context and the candidate response as inputs and returns to the program whether it is appropriate to continue the conversation.

In the case of the selective network, the context and the reply are ranked through the first tower, and then the context flows through the second tower for decisions. Each tower you build may have any architecture you want. The tower takes in its input and embeds it in the semantic vector space (vectors R and C in the illustration). Next, cosine similarity between the context vector and reply vector is computed, resulting in the context vector.

$$C^{\wedge}T*R/\left(\|C\|*\|R\|\right) \tag{9.1}$$

At inference time, we can calculate the similarity between the context and all possible answers and choose the answer based on the greatest similarity. To train the model, we use triplet loss. Triplet loss is defined on triplets (ctx, $reply_{correct}$, $reply_{wrong}$) and is equal to:

$$L = \max \left(0, \sin \left(ctx, reply_{wrong}\right) - \sin \left(ctx, reply_{correct}\right) + \acute{a}\right) -> \min \qquad (9.2)$$

Triplet loss for selective models is very similar to max-margin loss in support vector machines (SVMs). A negative sample is given by $reply_{wrong}$ and a positive sample by $reply_{correct}$ (3,4) and in the simplest case, it is a random reply from the pool of answers. Thus, by minimizing the loss we can learn the discrete values of similarity function for ranking, where absolute values are not informative. After reading all the messages, we will rate each answer. The one with the highest score will be chosen.

9.4.3.2 The generative model

Based on previous user communication as well as the present message, our predictive model predicts that a similar response will be given. Stemming from its massive computational prowess, the generative adaptive mode relies on mathematical formulae comprising complex ML models that are very expensive to construct, train, and test. To better understand, see Table 9.1, which has a sample of question-answer dialogue between two persons.

Table 9.1 Context and reply for generative models.

Context		reply
Hi! < eos>	**Hi there. <eos>**	**How are you? <eos>**
Hi there. <eos>	How are you? <eos>	I am fine.And how are you? <eos>
How are you? <eos>	I am fine.Andhow are you? <eos>	Me too fine.<eos>

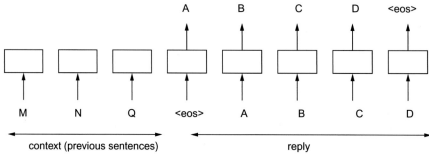

FIG. 9.8

The generative model using seq2seq framework.

Table 9.2 Selective vs. generative models.

Selective models	Generative models
They cannot cover all dialogue topics	They can generate an arbitrary answer (more like general AI)
They are inconsistent in grammar generation, e.g., incorrect speaker gender	They can generate an answer to incorrect grammar form, e.g., correct speaker gender
There exist predefined answers, so there is less chance of grammar mistakes in answers generated	They can generate an answer with incorrect grammar or syntax
They are less prone to general answer problems	They are prone to general answer problems
In a selective model, you can customize your answers	Difficult to impose properties on model replies

For modeling the dialogue, a sequence-to-sequence (seq2seq) method that emerged in the neural machine translation area is employed and is successfully adapted to the dialogue problems. In Fig. 9.8, the architecture consists of a top-down recursive neural network (RNN) followed by a second RNN using the same parameters as the previous RNN. The left one (corresponding to M-N-Q tokens) is called the *encoder*, while the right one (corresponding to $<\text{eos}> - \text{A-B-C-D}$ tokens) is called the *decoder*.

9.4.4 Selective vs. generative models

The choice to use the selective or generative model depends on the need of the client as shown in Table 9.2.

9.4.5 Challenges in Chatbot implementation

Using AI chatbots to improve customer satisfaction is now much easier than it had been in the past. However, integrating AI chatbots comes with several unique challenges.

- **Recognizing the intention of the user**

The major issue for any chatbot is to grasp the user's intention and decode the message's and query's hidden intent. The greater a chatbot's intelligence, the greater its capacity to read consumer conversations and reveal their intentions. Modern chatbots' interpretation of the message is a significant value proposition. As a result, chatbot development companies are now placing greater emphasis on creating algorithms that are proactive, flexible, and more capable of processing user communications. Given that robots may only respond in a relatively specified fashion, many chatbots may have difficulty recognizing the query's intent when presented with queries that have a diverse sentence structure, syntax, or word sequence. Requesting the user to post their question in basic terms will help solve this problem because it will be processed more quickly. This challenge can now be creatively addressed with modern chatbot algorithms. Today's chatbots, to better understand a customer's inquiry, compare the inquiry to previous requests and look for versions that fit the current situation. With this, it is possible to get a little bit closer to understanding the customer's intent, even if the chatbot still cannot decipher the inquiry.

- **Creating a Chatbot at an Affordable Price**

Because of the increased expense, chatbots are often overlooked by small and medium-sized businesses. Many people around the world still view chatbots as a high-cost investment that only major corporations can afford. However, this notion is quickly fading as an increasing number of startups and small businesses turn to chatbots for intelligent business communication. Even still, using a chatbot requires a significant financial commitment. Developing and implementing chatbots can be expensive, therefore we propose some solutions to this problem. One option is to buy a ready-to-use product right out of the box. The simplest and most cost-effective approach to designing and implementing chatbots is to use a chatbot solution. In exchange for the lower cost, you must accept less customizability of such solutions. Another low-cost option is to create a chatbot on a self-service platform using a basic drag-and-drop framework. Chatbots, like webpages, can be developed in a modular fashion using a variety of themes. As a last resort, a specialist can be employed to design the full chatbot software from scratch and include all the bespoke qualities and aspects that are appropriate for the company's targeted audiences.

- **Chatbot Security**

Some Internet chatbots resemble spam, so visitors avoid engaging with them. Therefore the security elements of the chatbot should be prioritized over everything else. To keep your data safe, make sure your chatbots are using the HTTPS protocol. Ensure that the chatbot security is beta tested before it goes live for the whole audience. Another thing to keep an eye on is whether the chatbot is answering inquiries that could lead to a security breach. Showing users how and where the chatbot stores and uses their data can help to build trust.

- **Chatbot Testing**

Finally, when the chatbot is out of the development lab, it needs to be tested before being implemented. Chatbot testing is complex. As NLP capability is increasingly improving, chatbots are now frequently updated. Therefore the testing mechanism should be used for every update to check the effects of each value addition. Among the chatbot testing methods, the first focuses on automated testing through a variety of automation testing platforms like testYourBot, Bot Testing, Zypnos, Dimon, and others. These automation testing platforms can evaluate test results and test script code across a variety of test cases. The second testing method is manual testing of conversational logic. Such tests are generally carried out within a group of testers.

The users check the bot across a variety of use cases and contexts and evaluate the results.

- **Building a more human chatbot**

Last, but not least, the difficulty is to create a chatbot that gives the user on the other end the impression of a human chat. To be effective, the chatbot must be culturally sensitive, responsive to greetings and personal gestures, and should put individuals at ease during the conversation. A chatbot's personality should appeal to the intended audience to be successful. To create the best AI chatbot requires investment in time and training. This way, it can easily identify the correct sentiments and emotions of a human voice and respond in the right tone.

The major challenge is to understand user intent and create a human chatbot. The proposed model is implemented to understand user intentions.

9.4.6 Machine learning in Chatbots for sentiment analysis

Recently, various ML and NLP techniques are being used to understand and classify text entered in conversational chatbots. Table 9.3 presents a literature survey of different research papers on sentiment analysis and text classification techniques.

Text mining refers to the task of retrieving or extracting the top quality (relevant) information from text, whereas text classification is a task of text mining and the act of separating a set of text documents into two or more classes where every text document can be assigned to one or more classes. The process of text classification is described in this chapter and various classification methods like Naive Bayes, K-nearest neighbor (KNN), support vector machine, decision tree, and neural network are discussed. Compared to other classifiers, SVM shows good performance in accuracy, speed of learning, classification speed, and tolerance to irrelevant examples that create noise. However, it is hard to say that any one technique is better than the others, as performance depends on the criteria and requirements of the particular organization. There is a need for an efficient algorithm to classify text for conversational chatbots.

9.5 Proposed work
9.5.1 The general architecture of Chatbots

The related work and literature indicate that there is a need for chatbots that can understand the different aspects of conversation and send replies accordingly to the end user. In the future, chatbots are going to become an integral part of conversations or conversational agents with the help of technology [30]. Thus, let us understand how a chatbot works. Fig. 9.9 shows the general architecture of a chatbot [31].

A chatbot has two separate assignments:

- analyzing the user request
- returning the response

Table 9.3 Literature review of sentiment analysis and text classification.

Sr. No.	Paper title	Performance analysis	Application	Algorithms
1	Classifying Positive or Negative Text Using Features Based on Opinion Words and Term Frequency—Inverse Document Frequency [13]	Accuracy = 0.87	E-commerce	KNN, SVM
2	Characterizing human opinion in social networks using machine learning algorithm [19]	Accuracy = 0.85	Popularity prediction (self-driven car)	Random forest, AdaBoost, SVM, and Logistic Regression Classifiers
3	Improved Random Forest for Classification [23]	Speed = faster Overall recall = ~0.89	Medical science Industry	Improved Random Forest
4	Sentiment analysis of movie reviews using machine learning techniques, [15]	Accuracy = 0.81 Mean absolute error = ~0.21	Popularity prediction (Movie rating)	Naive Bayes, Random forest, KNN
5	Rating prediction by exploring user's preference and sentiment [16]	Root Mean square Error (RME) = ~1.2 Mean Absolute Error (MAE) = ~ 0.9	Recommender System (User sentiment)	Matrix factorized framework LDA + word2vec
6	Application of machine learning techniques to sentiment analysis [17]	Faster Scalable	User sentiment	Naive Bayes Decision Tree
7	A survey on Twitter data analysis techniques to extract public opinion [25]	Improved accuracy	User sentiment	Hybrid model (SVM + Naive Bayes)
8	Dual Sentiment Analysis: Considering Two Sides of One Review [20]	Accuracy = 0.73	User sentiment	DSA3
9	A Joint Segmentation and Classification Framework for Sentence Level Sentiment Classification [26]	Accuracy = 0.85	Collective social action	A Joint Segmentation and Classification Framework for Sentence Level Sentiment Classification

#	Title	Result	Application	Method
10	Online Stock Forum Sentiment Analysis [21]	Improved accuracy	Stock market	GARCH-based SVM
11	Sentiment Analysis for Social Networking Messages Sentiment Detection	Improved accuracy and precision	Business sentiment	Feature Relation Network using chi-square
12	SentiView: Sentiment Analysis and Visualization for Internet Popular Topics [22]	Scalable to any social media	Public sentiment	SVM
13	Your age is no secret: Inferring microbloggers' ages via content and interaction analysis [24]	Accuracy = 0.81	User profile	SVM
14	Classification of Twitter Accounts into Automated Agents and Human Users [27]	Accuracy = 0.83	User Classification	Random Forest
15	Aspect based sentiment analysis in social media with classifier ensembles [28]	Accuracy = 0.9	User sentiment	SVM, Naive Bayes, Maximum Entropy, and Ensemble classifier
16	Aspect Level Sentiment Analysis using machine learning [29]	Accuracy = 0.8	E-commerce	SVM

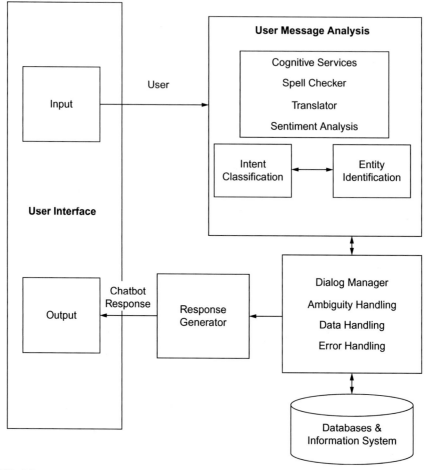

FIG. 9.9

The architecture of a chatbot.

Credit: M. das Graças Bruno Marietto et al., Artificial Intelligence Markup Language: A Brief Tutorial. Accessed January 25, 2021 [Online]. Available: https://arxiv.org/abs/1307.3091.

9.5.1.1 Analyzing user request

This is the first mission to be performed by a chatbot. To extract exactly what the user wants, the system analyzes the request from the user and identifies what the user is trying to define in terms of the entry point. To know a response to the query, a chatbot must know what information is critical to the response it can provide, and it must also not be overloaded via the long request it receives. It should be possible to extract relevant data and entities from the query, and the bot should have enough time to deal with the request before responding.

- **Returning the Response**

In theory, the chatbot is supposed to do what the users want and, if it cannot provide that, the chatbot should provide the next best thing. The other answer may be:

- A predefined and generic text
- A text extracted from a knowledge base that comprises numerous responses
- A piece of information, which is provided in context, and is based on the user's data
- Data stored in enterprise systems
- A query of "ambiguous" that the chatbot can interpret with a clear understanding of the user's purpose

Many approaches and tools can be used to build a chatbot. Combining various types of AI, such as ML, NLP, and semantic (or noun) comps can be the best methods for achieving the desired results.

9.5.2 **Proposed model**

The previous section highlighted the importance of analyzing user intention for text-based inputs through preprocessing of data and classification of text. The classifier must classify text accurately [32]. Also, one of the major drawbacks of existing chatbots is there is no such system to understand the aspect of the given input text. In such cases, aspect-based sentiment analysis (ABSA) plays an important role and can be useful over web data analysis. Its two major tasks are aspect-based feature extraction and feature classification. Here, the Hybrid Intention-based Multiclass Classification Model (HI-MCM) is proposed to improve classification accuracy and improve user intention analysis. Fig. 9.10 shows our proposed model.

The proposed framework performs the classification of user sentiments over web data in four main phases: data collection, preprocessing, feature extraction based on aspects to identify implicit and explicit aspects, and feature classification into multiple classes. The sentiment classification based on aspects or user consists of a rule-based approach and principal component analysis (PCA), which are used for detection of sentiment words and the selection of their features, respectively. Finally, their classification is done using HI-MCM.

1. Data Collection

The data present over the web is inconsistent and diverse, hence it is difficult to obtain the data related to the topic of interest to generate a response. Depending upon the data collected, the classification parameters can be decided. It is difficult to collect statistics on students on Twitter about their educational experiences since the terminology they use on social media platforms is inconsistent and diverse. By utilizing Twitter APIs, we were able to locate student records. To begin, we used the

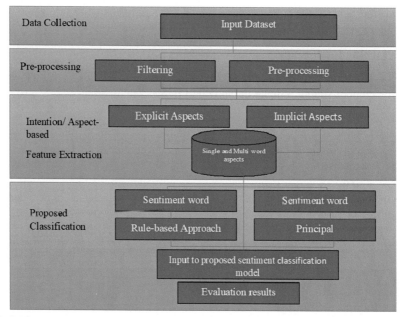

FIG. 9.10

The Hybrid Intention-based Multiclass Classification Model (HI-MCM).

Twitter hashtag #engineeringProblems to find tweets containing the keywords engineer, students, class, assignment, professor, and so on. We also discovered other popular hashtags, including #engineeringlife, #engineeringstudents, #exam, and #examstress. We collected around 25,000 tweets from different users. We only considered tweets that were relevant to the learning experiences and ignored all other tweets that were irrelevant. Using this data, we created categories of issues that students encountered during their time in school (Figs. 9.11 and 9.12).

2. Preprocessing

The preprocessing step is essential to remove unwanted data, noise and outliers. Preprocessing is an essential step before training a model.

Fig. 9.13 illustrates the preprocessing of the tweets. To convey a deeper meaning, Twitter users employ a limited set of symbols. For example, the hashtag is represented by #, the user account is represented by #, and the retweet is represented by #. Users also employ repeated letters to communicate their thoughts or draw attention to a specific word in their sentences. Users use stopwords, nonletter symbols, and punctuations, which create confusion in the reader's mind. As a result, there is a need to preprocess the data before model training.

We did the following to preprocess the data:

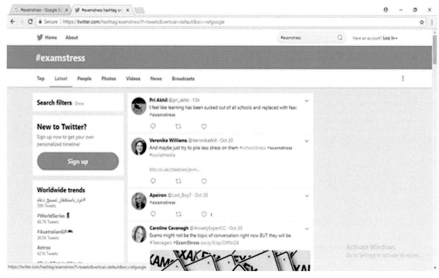

FIG. 9.11

Data collection from Twitter.

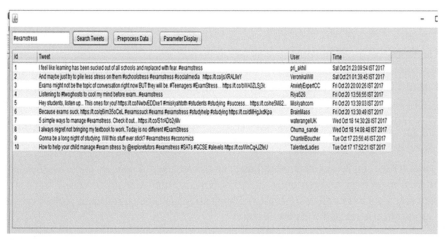

FIG. 9.12

Data collection.

- Deleted all the hashtags from the posts. Only the # symbol was removed from frequently appearing hashtags, and the hashtag text was left intact.
- Substituted the word "negtoken" for the negative phrases nothing, never, can't, andn't.
- Removed all nonletter symbols, and punctuation from the text.

FIG. 9.13

Data preprocessing.

- Deleted any letters that reappeared in words. If more than two similar terms are found, we replaced them all with a single letter. When it comes to words like "too" and "muuchh," we changed them to "more."

3. Facet-based feature extraction

The facet-based feature extraction needs to be performed to find implicit or unambiguous aspects of sentences [33]. Association Rule Mining (ARM) along with the Part of Speech (POS) pattern is used to identify the unambiguous single and multiword facet. To identify implicit facet the relationship between the opinion and facet are considered.

- Unambiguous feature extraction

Using ARM, single and multiword facet can be extracted. It is used to define the significant aspects of a given objective. By analyzing the data for repeated "if/then" patterns, association rules are formed. To classify the most significant partnerships, supporting criteria and trust are used.

- Implicit facet extraction

The dependency parsers are used to capture the grammar-related relations, which helps to identify implicit facets and their relationships with opinions given.

4. Proposed Hybrid Intention-based multiclass Sentiment classification model (HI-MCM)

The proposed HI-MCM sentiment classification is divided into two phases: feature detection and selection, and feature classification. A rule-based approach uses PCA for senti-word detection and feature selection respectively. After this, classification is done with the help of the HI-MCM.

- Rule-based method

The rule-based sentiment word detector can be used for sentiment word detection as shown by the proposed pseudo-code using the rule-based method

```
For each sentence in the input text
Read the aspect in the sentence
IF the aspect match THEN
Get sentimentword
IF sentimentword distance less than or equal to limit value
Add the sentimentword into Result
ELSE
Remove sentimentword
END IF
ELSE Display no sentiment word
END IF
END FOR

FOR each sentimentword in the Result
Compute the sentimentwordvalue
Display sentimentwordvalue
END FOR
```

- Sentiment word feature selection

For sentiment feature extraction, the multisentiment analysis model is proposed. It aims to classify each one of the tweets into the following classes: "love," "happy," "enjoyment," "neutral," "hate," "sad," and "anger." These training datasets can be given as inputs for their classification using various ML techniques. We can extract the following set of features:

- sentiment-related features
- punctuation features
- syntactic features
- semantic features
- unigram features
- top words
- pattern-related features

- Baseline Classification methods

The Naive Bayes classifier is an easy-to-use classifier model that is also successful. It is more easily built than other classifiers because it directs a system to sort the system into different classes or categories. However, it has two major drawbacks in that it implies that feature words are separate and unrelated to each other. In addition, its precision is low when compared with other ML classifiers. The SVM classifier, on the other hand, has several advantages over other ML classifiers, such as its tolerance for noisy data, speed, and accuracy. However, declaring a single classifier or standard model as efficient and most suitable for the problem is incorrect. Consequently, SVM is integrated with Naive Bayes classifier in the suggested system to minimize disadvantages and improve classifier performance to address the classification issue. The combination classifier algorithm, also known as the hybrid classifier algorithm, is described in this chapter. The hybrid model starts with the dataset being run through a Naive Bayes classifier before being handed off to an SVM for classification.

In their paper (Ingole et al. [34]), Ingole et al. demonstrated the pros and cons of using SVM and Naive Bayes classifiers individually. The proposed classifier discussed in Section 9.6 enhances the performance of the classifier, and the experimental results showed that the HI-MCM can achieve up to 97% classification accuracy.

9.6 Applications of Chatbots

In a perfect world, chatbots make marketers' work easier and increase sales because of the great interaction with the customer. Chatbots can be used as marketing gadgets, logos for computer labs, and reputation management. Previously, customers used to wait for a long time for customer service representatives to contact them. Now, because of chatbots, this is no longer an issue. Chatbots can also provide customers with solutions that they want or expect. These chatbots work 24/7, thus making it easier for people to use them anywhere, anytime, all year long. Chatbots are used in various domains, as described in the sections that follow.

- **Environments for Education**

The increased demand for education created competition at the higher level in the field [35]. Ineffective learning systems and environments lead to high dropout numbers and these students face tremendous challenges in the education sector. It has been observed that as dropout student's count is rising, it leads to reduced support from the instructor. Some e-learning activities can be sponsored by Personal Assistance Learning chatbots by preserving data by repeating old lessons that are missed by students. In the review, students are encouraged because chatbots can answer questions about the educational content. The chatbots can be used to address administration-related problems in education and obtain more details about the same. Studies [12] have shown that the participation of the number of students for registration to various courses offered by universities has increased, as it becomes easier with the help of chatbots.

- **Chatbots in Customer service**

It is a very beneficial idea to have customer care system through which various services are provided by the company [36]. A chatbot can provide uncomplicated navigation of a website and assist by guiding users. AI chatbots also assist customers in ordering and asking for the required items.

- **Chatbots in News**

Chatbots are useful for accessing news and information in a personalized pattern. Instead of surfing the Internet, a person can interact with a chatbot to get the news that's relevant to them.

- **Chatbots in Medicine**

The use of chatbots built by AI can help to organize visits with a doctor. They can also be helpful for ordering prescriptions and checking the prices of medications. Several studies have shown the importance of using chatbots in medicine [37,38].

- **Chatbots in In-app support**

These kinds of chatbots are more welcoming to customers. This is the 24/7 support that a system needs. These chatbots can be used to send notifications to the customer while they are searching for something else.

- **Chatbots in Travel and Tourism**

Some websites help tourists find information about their planned vacation itinerary, (location, route, scenery, rates, etc.) Chatbots can suggest nearby shops, hotels, and tour programs [39].

- **Chatbots in Booking Tickets**

A chatbot can help a person buy tickets to movies, plays, museums, events, flights, and so on.

9.7 Weaknesses of Chatbots and scope for improvement

Every technological system has advantages and disadvantages, and this is certainly the case with chatbots. For both providers and consumers, data protection is a major concern [36]. Companies are accountable for the proper security and management of customer data, especially if they have a stand-alone chatbot program. Data protection and privacy [37] need to be ensured for each user and client application. A chatbot may fail if it is not able to understand the intention of the user. A frustrating conversation may lead to the spoiling of customer relationships. For chatbot providers, toxic content [38] can be a significant disadvantage. Toxic material is called copyright infringement. In addition, chatbots can be hacked, thus potentially exposing customer data. It is important to detect deception in some applications where chatbots are used. The signs of deception depend on the skills of the chatbot. Fraud detection becomes difficult due to the adaptive communication skills of chatbots. Some additional reasons for the failures of chatbots are long answers where one or two of the essential information consists of hiding sentences among several others that can lead to the discouragement of users. Short and direct texts, therefore, are favored [39].

User spelling errors can also fail to provide the classification. The ongoing research on chatbots has tremendous scope for minimizing all the weaknesses and ensuring the highest level of satisfaction to customers.

9.8 Results and discussion

In this chapter, we focused on the implementation of a classification model in chatbots to classify engineering student's study-related problems. Students use social networking platforms to share their views, ideas, and problems. The primary challenge, however, is to analyze this anonymous data. It requires human comprehension due to its size and scope. Big Data such as this cannot be handled manually. Thus, there is a need for automated data processing techniques. The existing literature surveyed shows the pros and cons of SVM and Naïve Bayes classifiers. We proposed a hybrid model that uses the advantageous features of the Naïve Bayes classifier and SVM to examine a large amount of informal data.

In our experiment, we took engineering students' Twitter posts as the inputs. We conducted a qualitative analysis of about 34,000 tweets posted by engineering students about their college life and found that the students face lots of problems. We collected tweets over 6 months duration as a data collection step. We analyzed sampled data and the five categories are found in the Qualitative Study, which explains the problems faced by the student during their learning process. The preprocessed Twitter data was given as an input to our HI-MCM, which we used for classification.

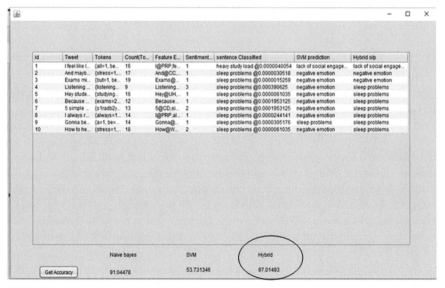

FIG. 9.14

Tweets search.

Table 9.4 Comparative analysis of the proposed model with existing models.

Sr. No.	Technique used	Accuracy
1.	Naïve Bayes	91.045
2.	SVM	53.731
3.	Proposed HI-MCM model	97.014

To show the efficiency of the proposed work, we used the HI-MCM to classify tweets reflecting students' problems. The experimental results show that our model reduced classification time and improved accuracy compared to the other classifiers (see Fig. 9.14).

After an extensive survey of engineering students' live and analyzing their problems, we identified five main categories of classification: Heavy Study Load, Lack of Social Engagement, Negative Emotion, Sleep Problems, and Diversity Issues like Cultural conflicts.

Table 9.4 presents a comparison of the experimental results of the proposed HI-MCM with existing Naïve Bayes and SVM classification techniques. The experimental results show the improved performance of the proposed system.

9.9 Conclusion and future work

The experimental results showed the accuracy of each classification algorithm. The Naïve Bayes classifier attained 91% accuracy, the SVM attained 53% accuracy, and the hybrid model attained the highest accuracy of 97%. The proposed work considered the analysis of user intentions and their aspects. Our HI-MCM showed improved accuracy.

The detailed discussion and results provided in this chapter highlight the need for personalized automation in the era of Industry 4.0. A chatbot is an AI program that simulates natural language conversation with a user via messaging apps, mobile apps, or phones. The important role of chatbots in Industry 4.0 highlights the demand for AI, ML, and NLP-based bots. The request-response model of chatbots with the user intention analysis design indicates that if chatbots can analyze and classify user intentions properly, then expected and appropriate user-friendly responses can be generated.

To improve chatbot result accuracy and for stronger analysis of user intentions, we proposed the HI-MCM Sentiment Analysis Classification Model.

We hope the study presented in this chapter will encourage chatbot researchers to develop chatbots that generate user-friendly responses and overcome the weaknesses of currently available chatbots.

References

[1] T.P. Nagarhalli, V. Vaze, N.K. Rana, A review of current trends in the development of Chatbot systems, in: 2020 6th Int. Conf. Adv. Comput. Commun. Syst. ICACCS 2020, 2020, pp. 706–710, https://doi.org/10.1109/ICACCS48705.2020.9074420.

[2] P. Smutny, P. Schreiberova, Chatbots for learning: a review of educational chatbots for the Facebook Messenger, Comput. Educ. 151 (June 2019) (2020) 103862, https://doi.org/10.1016/j.compedu.2020.103862.

[3] R. Khan, A. Das, Introduction to chatbots, in: Build Better Chatbots, Springer, 2014, pp. 1–11.

[4] W. Bodrow, "Vision of Industry 4.0," Iwama, pp. 55–58, 2016, https://doi.org/10.2991/iwama-16.2016.10.

[5] T. Kaufmann, Geschäftsmodelle in Industrie 4.0 und dem Internet der Dinge, 2016, (Accessed January 25, 2021) [Online]. Available www.localmotors.com.

[6] D. Kaczorowska-Spychalska, How Chatbots Influence Marketing, content.sciendo.com, 2019, https://doi.org/10.2478/manment-2019-0015.

[7] M. das G. Bruno Marietto, et al., Artificial intelligence markup language: a brief tutorial, Int. J. Comput. Sci. Eng. Surv. 4 (3) (2013) 1–20, https://doi.org/10.5121/ijcses.2013.4301.

[8] M. das Graças Bruno Marietto, et al., Artificial Intelligence Markup Language: A Brief Tutorial, (Accessed January 25, 2021. [Online]. Available https://arxiv.org/abs/1307.3091.

[9] N. Akma, M. Hafiz, A. Zainal, M. Fairuz, Z. Adnan, Review of chatbots design techniques, Int. J. Comput. Appl. 181 (8) (Aug. 2018) 7–10, https://doi.org/10.5120/ijca2018917606.

[10] S. Jung, Semantic Vector Learning for Natural Language Understanding, Elsevier, 2019. (Accessed January 25, 2021) [Online]. Available https://www.sciencedirect.com/science/article/pii/S0885230817303595.

[11] K. Ramesh, S. Ravishankaran, A. Joshi, K. Chandrasekaran, A survey of design techniques for conversational agents, Commun. Comput. Inf. Sci. 750 (2017) 336–350, https://doi.org/10.1007/978-981-10-6544-6_31.

[12] C.P.-N.H. Chi, et al., Intelligent Assistants in Higher-Education Environments: The FIT-EBot, a Chatbot for Administrative and Learning Support Toward a Living Lab for Business Development View Project Overview of Regression Model for Software Project Effort Estimation View project Intelligent Assistants in Higher-Education Environments: The FIT-EBot, a Chatbot for Administrative and Learning Support ACM Reference Format, dl.acm.org, Dec. 2018, pp. 69–76, https://doi.org/10.1145/3287921.3287937.

[13] S. Tongman, N. Wattanakitrungroj, Classifying positive or negative text using features based on opinion words and term frequency—inverse document frequency, in: ICAICTA 2018 - 5th Int. Conf. Adv. Informatics Concepts Theory Appl, 2018, pp. 159–164, https://doi.org/10.1109/ICAICTA.2018.8541274.

[14] E. Adamopoulou, L. Moussiades, An Overview of Chatbot Technology, vol. 584, IFIP. Springer International Publishing, 2020.

[15] S. Vinit Bhoir, An efficient fake news detector, in: 2020 International Conference on Computer Communication and Informatics (ICCCI), 2020, pp. 1–9.

[16] X. Ma, X. Lei, G. Zhao, X. Qian, Rating prediction by exploring user's preference and sentiment, Multimed. Tools Appl. 77 (6) (2018) 6425–6444, https://doi.org/10.1007/s11042-017-4550-z.

[17] M. Ashok, S. Rajanna, P.V. Joshi, S.S. Kamath, A personalized recommender system using Machine Learning based Sentiment Analysis over social data, in: 2016 IEEE

Students' Conf. Electr. Electron. Comput. Sci. SCEECS 2016, no. March 2016, 2016, https://doi.org/10.1109/SCEECS.2016.7509354.

[18] P. Suta, X. Lan, B. Wu, P. Mongkolnam, J.H. Chan, An overview of machine learning in chatbots, Int. J. Mech. Eng. Robot. Res. 9 (4) (2020) 502–510, https://doi.org/10.18178/ijmerr.9.4.502-510.

[19] V. Lavanya, S. Shetty, Characterizing Human Opinion in Social Network Using Machine Learning Algorithms, 2017, (Accessed June 21, 2021) [Online]. Available https://www.researchgate.net/profile/Savita_Shetty/publication/327074066_Characterizing_Human_Opinion_in_Social_Network_Using_Machine_Learning_Algorithms/links/5f2ba018a6fdcccc43ac8792/Characterizing-Human-Opinion-in-Social-Network-Using-Machine-Learning-Algorithms.pdf.

[20] R. Xia, F. Xu, C. Zong, Q. Li, Y. Qi, T. Li, Dual sentiment analysis: considering two sides of one review, IEEE Trans. Knowl. Data Eng. 27 (8) (2015) 2120–2133, https://doi.org/10.1109/TKDE.2015.2407371.

[21] S. Madhusudhanan, M. Moorthi, A survey on sentiment analysis, Indian J. Comput. Sci. Eng. 9 (2) (2018) 69–80, https://doi.org/10.21817/indjcse/2018/v9i2/180902030.

[22] J. Ge, M.A. Vazquez, U. Gretzel, Sentiment analysis: a review, Adv. Soc. Media Travel. Tour. Hosp. New Perspect. Pract. Cases 5 (4) (2017) 243–261, https://doi.org/10.4324/9781315565736.

[23] A. Paul, D. Mukherjee, P. Das, Improved random forest for classification, IEEE Trans. Image Process. (2018). *ieeexplore.ieee.org*, Accessed: Jun. 21, 2021. [Online]. Available https://ieeexplore.ieee.org/abstract/document/8357563/.

[24] M. Kataria, S.M. Shah, Extended comprehensive sentiment analysis for informal opinion text, Int. J. Engine Res. V4 (05) (2015) 481–484, https://doi.org/10.17577/ijertv4is050625.

[25] S. Judith Sherin Tilsha, M.S. Shobha, A survey on twitter data analysis techniques to extract public opinion, IJARCSE 5 (11) (2015) 536–540.

[26] D. Tang, B. Qin, F. Wei, L. Dong, T. Liu, M. Zhou, A joint segmentation and classification framework for sentence level sentiment classification, IEEE Trans. Audio Speech Lang. Process. 23 (11) (2015) 1750–1761, https://doi.org/10.1109/TASLP.2015.2449071.

[27] Z. Gilani, E. Kochmar, J. Crowcroft, Classification of twitter accounts into automated agents and human users, in: Proc. 2017 IEEE/ACM Int. Conf. Adv. Soc. Networks Anal. Mining, ASONAM 2017, 2017, pp. 489–496, https://doi.org/10.1145/3110025.3110091.

[28] I. Perikos, I. Hatzilygeroudis, Aspect based sentiment analysis in social media with classifier ensembles, in: Proc. 16th IEEE/ACIS Int. Conf. Comput. Inf. Sci. ICIS 2017, 2017, pp. 273–278, https://doi.org/10.1109/ICIS.2017.7960005.

[29] D. Shubham, P. Mithil, M. Shobharani, S. Sumathy, Aspect level sentiment analysis using machine learning, IOP Conf. Ser. Mater. Sci. Eng. 263 (4) (2017), https://doi.org/10.1088/1757-899X/263/4/042009.

[30] A.T. Neumann, et al., Chatbots as a tool to scale mentoring processes: individually supporting self-study in higher education, Front. Artif. Intell. 4 (May) (2021) 1–7, https://doi.org/10.3389/frai.2021.668220.

[31] M. Woschank, E. Rauch, H. Zsifkovits, A review of further directions for artificial intelligence, machine learning, and deep learning in smart logistics, Sustainability 12 (9) (2020), https://doi.org/10.3390/su12093760.

[32] S.V. Bhoir, An Efficient FAKE NEWS DETECTOR, ieeexplore.ieee.org, 2020. (Accessed January 25, 2021) [Online]. Available https://ieeexplore.ieee.org/abstract/document/9104177/.

[33] N. Zainuddin, A. Selamat, Hybrid Sentiment Classification on Twitter Aspect-Based Sentiment Analysis, Springer, 2018. (Accessed January 25, 2021) [Online]. Available https://link.springer.com/article/10.1007/s10489-017-1098-6.

[34] P. Ingole, S. Bhoir, A.V. Vidhate, Hybrid Model for Text Classification, ieeexplore.ieee. org, 2018. (Accessed January 25, 2021) [Online]. Available https://ieeexplore.ieee.org/abstract/document/8474738/.

[35] W. Huang, K. Foon Hew, D. Gonda, K.F. Hew, D.E. Gonda, Designing and Evaluating Three Chatbot-Enhanced Activities for a Flipped Graduate Course, researchgate.net, 2019, https://doi.org/10.18178/ijmerr.8.5.813-818.

[36] D. Zumstein, S. Hundertmark, Chatbots-an interactive technology for personalized communication, transactions and services 56 publications 158 citations see profile, IADIS Int. J. WWW/Internet 15 (1) (2021) 96–109. 203AD (Accessed: January 25, 2021) [Online]. Available https://www.researchgate.net/publication/322855718.

[37] G. Neff, P. Nagy, Talking to Bots: Symbiotic Agency and the Case of Tay, 2016, (Accessed January 25, 2021) [Online]. Available http://ijoc.org.

[38] M. Coblenz, J. Aldrich, B.A. Myers, J. Sunshine, Interdisciplinary Programming Language Design, Oct. 2018, pp. 133–146, https://doi.org/10.1145/3276954.3276965.

[39] R.M. Schuetzler, G.M. Grimes, J.S. Giboney, The effect of conversational agent skill on user behavior during deception, Comput. Hum. Behav. 97 (2019) 250–259, https://doi.org/10.1016/j.chb.2019.03.033.

Index

Note: Page numbers followed by *f* indicate figures and *t* indicate tables.

Printed in the United States
by Baker & Taylor Publisher Services

Contents

Figures

Tables

Examples

Technical notes

Preface

This book is a third edition of our work which was originally published by Cambridge University Press in 2003 under the title Benefit-Cost Analysis. *The title of the second edition of the book, published by Routledge, was amended to reflect more closely common usage but the subject matter remained unchanged. The book is intended for those with a basic understanding of micro-economics who wish to learn how to conduct a social cost-benefit analysis. We use the term social benefit-cost analysis to refer to the appraisal of a private or public-sector project from the viewpoint of the public interest, broadly defined. As suggested above the terms cost-benefit analysis and benefit-cost analysis (with or without the social prefix) are used interchangeably in professional practice. In this third edition of the book we have expanded our discussion of some aspects of cost-benefit analysis, introduced some issues of recent concern, and included additional case studies. However the emphasis on the practical application of economic principles in a spreadsheet framework remains unchanged.*

A social cost-benefit analysis of a proposed publicly or privately funded project, or public policy change, may be commissioned by a municipal, state or federal government, by a government aid agency, such as USAID, or by an international agency such as the World Bank, the IMF, the Asian Development Bank, the UN, EU or OECD. Proponents of a private project which has significant impacts on the broader community may also commission an economic analysis of this type to support an application for approval to proceed with their project. Sometimes the scope of the required analysis is broader than the measurement of economic benefits and costs: an evaluation of the distribution of costs and benefits may be called for; an impact analysis may be required to determine the effects of the project on employment and economic growth; an environmental impact statement may be commissioned; and a social impact analysis dealing with factors such as crime and family cohesion may be sought. This book concerns itself mainly with the economic benefits and costs of projects, although it does touch on the questions of income distribution and economic impact. The main questions addressed are: do the benefits of the project exceed the costs, no matter how widely costs and benefits are spread? and which group or groups of individuals benefit and which bear the costs?

Cost-benefit analysis relies mainly on microeconomic theory, although some understanding of macroeconomics is also useful. The person whose background should be sufficient to allow them to benefit from this book is someone who studied a principles of economics subject as part of an economics, commerce, arts, science or engineering degree; a person with an undergraduate economics training will find the organizational principles set out in the book to be innovative and of considerable practical use. We develop many of the microeconomic principles used in cost-benefit analysis in the course of our presentation and a student with little background in economics, but a special interest in project appraisal, may find the book to be a useful review of relevant sections of basic microeconomic theory.

The book has several unique features: the close integration of spreadsheet analysis with analytical principles is a feature of some financial appraisal texts, but is unusual in a book dealing with social cost-benefit analysis; the particular layout of the spreadsheet is unique in offering an invaluable cross-check on the accuracy of the economic appraisal; and the book is structured in a way that allows readers to choose the level of analysis which is relevant to their own purposes.

The book emphasises practical application. It develops a spreadsheet-based template which is recommended for use in conducting a social cost-benefit analysis and which, as noted above, provides a check on the accuracy of the analysis. The application of the template is illustrated by a series of case studies, including a full-scale study of a social cost-benefit analysis of a proposed private investment project in a developing economy. This case study, together with reference to the necessary economic principles, is developed stage by stage in Chapters 4–6, after the basic methods of project appraisal have been outlined in Chapters 2–4. At the completion of Chapters 1–6 the reader should be capable of undertaking a spreadsheet-based cost-benefit analysis of a project which is of a small enough scale as to have no effect on market prices and where non-marketed effects are not an issue.

The case study (International Cloth Products) is a hypothetical example of a typical industrial project with public interest implications. This type of project was chosen as an illustration because of its simplicity – the student can readily comprehend the establishment and furnishing of a factory, the purchase of raw materials, the employment of a labour force, and the sale of product – so that the focus of attention is on the principles of cost-benefit analysis and their application in the spreadsheet framework. At the end of each chapter the principles discussed in the chapter are applied to further develop the cost-benefit analysis of the ICP project. In addition, other case studies relating to agriculture, traffic regulation and outdoor recreation are discussed in the body of the text. Projects similar to ICP are included as assignments in Appendix 1, but this appendix also includes applications of cost-benefit analysis to a wider range of investments, such as provision of outdoor recreation, health, education, crime prevention and emissions abatement as such projects are of particular concern in advanced economies.

Chapters 7–10 deal with issues which are more technical than those covered in Chapters 1–6: the analysis of consumer and producer surplus; non-market valuation techniques; uncertainty, information and risk; and foreign exchange market imperfections. By way of illustration these issues are introduced into the ICP case study spreadsheets. Many projects, however, can be appraised without consideration of these more technical matters, which is why they are deferred to the latter part of the book.

The three remaining chapters of the book consider the evaluation of the income distribution effects of a project, estimating its economic impact, and, finally, writing the report of the cost-benefit analysis. A sample report on the cost-benefit analysis of the ICP project, drawing on the principles established in Chapters 1–6, is presented in Chapter 13.

Acknowledgements

We have benefited considerably from teaching cost-benefit analysis in various universities and other organizations over the years and we would like to thank our students for their contribution to our understanding and presentation of the concepts which are the subject of this book. In particular we would like to thank Angela McIntosh for permission to include her case study in Chapter 13. We would also like to thank the School of Economics, University of Queensland, where we have conducted most of our teaching, for the opportunity to present our ideas about cost-benefit analysis in a stimulating and diverse environment. We would like to thank the School's administrative staff for excellent technical support over many years. We have received very helpful comments from Dr David Dorenfeld and from four anonymous reviewers which have informed preparation of this third edition and would like to thank them, together with all our colleagues who have contributed to our work.

Note to the instructor

The book is intended as the required text for one or two courses in applied cost-benefit analysis. It provides a framework involving practical application and leading to the acquisition of a valuable set of project appraisal skills. It can be supplemented by a range of other readings chosen to reflect the emphasis preferred by the Instructor. Since cost-benefit analysis can be applied to a very wide range of resource allocation problems there is an almost endless array of related publications which the student might find useful. We do not attempt to provide an extensive bibliography and suggest the use of a search engine to identify relevant material. Instead we list a few publications we have found useful in the past or that we refer to in the text. The text includes exercises and a selection of major cost-benefit analysis case studies, which can be assigned for credit, is presented in Appendix 1.

A one-semester undergraduate or postgraduate course can be based on Chapters 1–6 of the book, or Chapters 1–7 if issues of consumer and producer surplus are to be included. Students with a working knowledge of investment and discounted cash flow analysis could skim Chapters 2 and 3, and Chapters 8 and 9 dealing with non-market valuation and risk analysis could then be added to the course. Chapter 10 will be required reading for those intending to work on developing-country problems, but probably not otherwise. Chapters 11 and 12 deal with peripheral issues: the distribution of benefits and costs, and the impact of a project on the economy. The latter chapter also provides a useful introduction to macroeconomics for students who have not taken a principles course. Chapter 13, dealing with the way in which the results of a cost-benefit analysis should be reported, can be recommended as reading and serve as a model for the presentation of a cost-benefit analysis major assignment.

While the text can be supplemented with other reading, including reference to chapters in other cost-benefit texts which cover some issues in more detail, we have, as noted above, confined our suggestions for further reading to basic references, or to articles that we have drawn material from, and have not tried to incorporate the many available texts on cost-benefit analysis which use of a search engine would yield. In our course we recommend to our students some articles or book chapters but the choice is very much a matter of individual taste. Some classes might benefit from a set of lectures on the basic microeconomic principles upon which cost-benefit analysis draws, together with reference to a text in microeconomics or public finance.

The text is supported by a Support Material website (www.routledge.com/9781032320755), which can be accessed by students, with a section restricted to Instructors. Instructors can access PowerPoint presentations for each chapter and solutions to exercises and case studies. Instructors can download all the spreadsheets for the examples in the text, together with solution spreadsheets for the case study assignments in Appendix 1, from the website.

End-of-chapter exercises are provided in the text to reinforce the student's understanding of the material, and additional exercises are provided on the website. However, we believe that undertaking a case study assignment is an important part of learning how to conduct a cost-benefit analysis. The case studies provided in the book vary considerably in their level of difficulty. At the simpler level, it is possible to use the National Fruit Growers (NFG) or the International Cloth Products (ICP) projects as case study assignments even though the spreadsheet solutions are provided in the text. If the assignment is posed as undertaking a sensitivity analysis of the net present value of the NFG or ICP project to changes in one or two critical variables the student needs to construct a spreadsheet, based on our template, consisting of formulae. She has the values provided in the solution spreadsheets in the text as a guide, but this simply mimics what happens in laboratory sessions where students can ask the instructor to check their spreadsheets for accuracy. Constructing a spreadsheet based on formulae, for use in a sensitivity analysis, is a test of the student's level of understanding of the material in the text.

Appendix 1 provides a series of case study assignments of varying levels of difficulty and solutions are available on the Instructors' Website. Students will generally need help in completing these assignments and in our course we include tutorial sessions in the computer laboratory.

Our teaching of the course is based on two hours of lectures and class discussion per week plus a one-hour computer lab session. To start with we use the lab session to make sure everyone is comfortable with using spreadsheets and can access the various financial subroutines required in cost-benefit analysis. We then spend some time developing the cost-benefit analysis of the NFG case study project as an example of the practical application of the approach. After 6–7 weeks of lectures and lab work students are generally ready to undertake a major cost-benefit analysis assignment at a level of complexity similar to the case studies included in Appendix 1, which are presented roughly in order of difficulty. In the second part of the semester we use the class and lab times for consultations with students who require help with the major assignment, while we continue with lectures and exercises based on Chapters 7–12. As indicated by the sample case study report, which was prepared by one of our students and is included in Chapter 13, a high standard of work can be expected.

1   Introduction to cost-benefit analysis

1.1 Introduction

Cost-benefit analysis is a process of identifying, measuring and comparing the benefits and costs of an investment project or program. A program is a series of projects undertaken over a period of time with a particular objective in view. A project is a proposed course of action that involves reallocating productive resources from their current use in order to undertake the project. The project or projects in question may be public projects – projects undertaken by the public sector – or private projects. Both types of projects need to be appraised to determine whether they represent an efficient use of resources. Projects that represent an efficient use of resources from a private viewpoint may involve costs and benefits to a wider range of individuals than their private owners. For example, a private project may pay taxes, provide employment for some who would otherwise be unemployed, and generate pollution. The complete set of project effects are often termed social benefits and costs to distinguish them from the purely private costs and returns of the project. Social cost-benefit analysis is used to appraise the efficiency of private projects from a public interest viewpoint as well as to appraise public projects.

Public projects are often thought of in terms of the provision of physical capital in the form of infrastructure such as bridges, highways and dams, so-called "bricks and mortar" projects. However, there are other less obvious types of physical projects which augment environmental capital stocks and involve activities such as land reclamation, pollution control, fish stock enhancement and provision of parks, to name but a few. Other types of projects are those that involve investment in forms of human capital, such as health, education, and skill development, and social capital through drug use prevention and crime prevention, and the reduction of unemployment. While outside the notion of the traditional kind of project, changes in public policy, such as the tax/subsidy or regulatory regime, can also be assessed by cost-benefit analysis. There are few, if any, activities of government that are not amenable to appraisal and evaluation by means of this technique of analysis.[1]

Gathering and analysing the price and quantity data required to conduct a cost-benefit analysis involves a cost. How detailed, and hence costly, an analysis is justified depends upon the magnitude of the project being analysed and its potential for net gain or loss. This matching of cost of analysis against expected benefit should inform the design of the analysis – in effect, a cost-benefit analysis of cost-benefit analysis! In Chapter 9 we discuss how the expected value of the information to be obtained from the cost-benefit study can be assessed for comparison with the cost of the study, and we discuss the option of a low-cost "rough and ready" appraisal.

DOI: 10.4324/9781003312758-1

Investment involves diverting scarce resources – land, labour, capital and materials – from the production of goods for current consumption to the production of capital goods which will contribute to increasing the flow of consumption goods available in the future. An investment project is a particular allocation of scarce resources in the present which will result in a flow of output in the future: for example, land, labour, capital and materials could be allocated to the construction of a dam which will result in increased hydro-electricity output in the future (in reality, there are likely to be additional outputs such as irrigation water, recreational opportunities and flood control but we will assume these away for the purposes of this example). The cost of the project is measured as an opportunity cost – the value of the goods and services which would have been produced by the land, labour, capital and materials inputs had they not been used to construct the dam. The benefit of the project is measured as the value of the extra electricity produced by the dam. As another example, consider a job training program: scarce resources in the form of classroom space, materials and student and instructor time are diverted from other uses to enhance job skills, thereby contributing to increased output of goods and services in the future. Or, third, consider a proposal to reduce the highway speed limit: this measure involves a cost, in the form of increased travel time, but benefits in the form of reduced fuel consumption and lower accident rates. Chapters 2 and 3 discuss the concept of investment projects and project appraisal in more detail.

The role of the cost-benefit analyst is to provide information to the decision-maker – the official who will appraise or evaluate the project. We use the word "appraise" in a prospective sense, referring to the process of actually deciding whether resources are to be allocated to the project or not. We use the word "evaluate" in a retrospective sense, referring to the process of reviewing the performance of a project or program. Since social cost-benefit analysis is mainly concerned with projects undertaken by the public sector, or with private sector projects which significantly affect the general community and consequently require government approval, the *decision-maker* will usually be a senior public servant acting under the general direction of a politician. It is important to understand that cost-benefit analysis is intended to inform the decision-making process, not supplant it. The role of the analyst is to supply the decision-maker with relevant information about the level and distribution of benefits and costs, and potentially to contribute to informed public opinion and debate. Ideally, the decision-maker will take the results of the analysis, together with other information, into account in coming to a decision about the project. Parties involved in the project may use the results of the analysis to inform their decision as to whether or not to participate under the terms laid down by the decision-maker. The role of the analyst is to provide an objective appraisal, and not to adopt an advocacy position either for or against the project.

An investment project makes a difference and the role of cost-benefit analysis is to measure that difference. Two as yet hypothetical states of the world are to be compared – the world *with* the project and the world *without* the project. The decision-maker can be thought of as standing at a node in a decision tree as illustrated in Figure 1.1. There are two alternatives: *undertake the project* or *don't undertake the project* (in reality, there are many options, including a number of variants of the project in question, but for the purposes of the example we will assume that there are only two).

The world *without* the project is not the same as the world *before* the project; for example, in the absence of a road-building project, traffic flows may continue to grow and delays to lengthen, so that the total cost of travel time in the future *without* the project exceeds the cost *before* the project. The time saving attributable to the project is the difference between

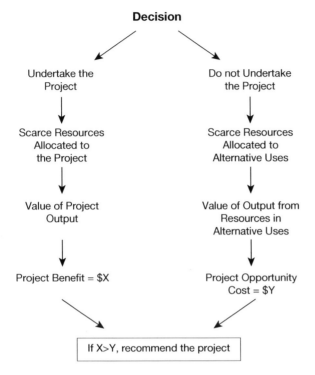

Decision

Undertake the
Project

Do not Undertake
the Project

Scarce Resources
Allocated to
the Project

Scarce Resources
Allocated to
Alternative Uses

Value of Project
Output

Value of Output from
Resources in
Alternative Uses

Project Benefit = $X

Project Opportunity
Cost = $Y

If X>Y, recommend the project

Figure 1.1 The "with and without" approach to cost-benefit analysis. The decision tree has two paths: following the left-hand path by allocating scarce resources to the project will result in output valued at $X being produced. The right-hand path considers alternative uses for these scarce resources which would result in production of output valued at $Y.

travel time with and without the road-building project, which, in this example, is greater than the difference between travel time before and after the project.

Which is the better path in Figure 1.1 to choose? The *with-and-without* approach is at the heart of the cost-benefit process and also underlies the important concept of opportunity cost. *Without* the project – for example, the dam referred to above – the scarce land, labour, capital and materials would have had alternative uses. For example, they could have been combined to increase the output of food for current consumption. The value of that food, assuming that food production is the best (highest valued) alternative use of the scarce resources, is the opportunity cost of the dam. This concept of opportunity cost is what we mean by "cost" in cost-benefit analysis. *With* the dam project we give up the opportunity to produce additional food in the present, but when the dam is complete, it will result in an increase in the amount of electricity which can be produced in the future. The benefit of the project is the value of this increase in the future supply of electricity over and above what it would have been in the absence of the project. The role of the cost-benefit analyst is to inform the decision-maker: if the *with* path is chosen, additional electricity valued by consumers at $X will be available; if the *without* path is chosen, extra food valued by consumers at $Y will be available. If X > Y, the benefits exceed the costs, or, equivalently, the benefit/cost ratio exceeds unity. This creates a presumption in favour of the project, although the decision-maker might also wish to take distributional effects into account – who would receive the benefits and who would bear the costs – and other considerations as well.

The example of Figure 1.1 has been presented as if the cost-benefit analysis directly compares the value of extra electricity with the value of the forgone food. In fact, the comparison is made indirectly. Suppose that the cost of the land, labour, capital and materials to be used to build the dam is $Y. We assume that these factors of production could have produced output (not necessarily food) valued at $Y in some alternative and unspecified uses. We will consider the basis of this assumption in detail in Chapter 5, but for the moment it is sufficient to say that in competitive and undistorted markets the value of additional inputs will be bid up to the level of the value of the additional output they can produce. The net benefit of the dam is given by $(X − Y) and this represents the extent to which building a dam constitutes a better (X − Y > 0) or worse (X − Y < 0) use of the land, labour, capital and materials than the alternative use.

When we say that $(X − Y) > 0 indicates that the proposed project is a better use of the inputs than the best alternative use, we are applying a measure of economic welfare change known as the *Kaldor-Hicks Criterion*. The K-H criterion says that, even if some members of society are made worse off as a result of undertaking a project, the project is considered to confer a net benefit if the gainers from the project could, in principle, compensate the losers. In other words, a project does not have to constitute what is termed a *Pareto Improvement* (a situation in which at least some people are better off and no-one is worse off as a result of undertaking the project) to add to economic welfare, but merely a *Potential Pareto Improvement*. The logic behind this view is that if government believed that the distributional consequences of undertaking the project were undesirable, the costs and benefits could be redistributed by means of transfer payments of some kind. The problem with this view is that transfers are normally accomplished by means of taxes or charges which distort economic behaviour and impose costs on the economy. The decisionmaker may conclude that these costs are too high to warrant an attempt to redistribute benefits and costs. We return to the appraisal of the distributional effects of projects in Chapter 11.

Since building a dam involves costs in the present and benefits in the future, the net benefit stream will be negative for a period of time and then positive, as illustrated in Figure 1.2. To produce a summary measure of the net benefits of the project, all values have to be converted to values at a common point in time, usually the present. The net present value (NPV) is the measure of the extent to which the dam is a better (NPV > 0) or worse (NPV < 0) use of scarce resources than the next-best alternative. Converting net benefit streams, measured as net cash flows, to present values is the subject of Chapters 2 and 3.

When we compute present values for use in a cost-benefit analysis we need to make a decision about the appropriate rate of discount. The discount rate tells us the rate at which we are willing to give up consumption in the present in exchange for additional consumption in the future. It is often argued that a relatively riskless market rate of interest, such as the government bond rate, provides a measure of the marginal rate of time preference of those individuals participating in the market. However, it can be argued that future generations, who will potentially be affected by the project, are not represented in today's markets. In other words, when using a market rate of interest as the discount rate, the current generation is making decisions about the distribution of consumption flows over time without adequately consulting the interests of future generations. This raises the question of whether a (lower) *social discount rate*, as opposed to a market rate, should be used to calculate the net present values used in public decision-making. This issue is considered further below and in Chapters 5 and 11.

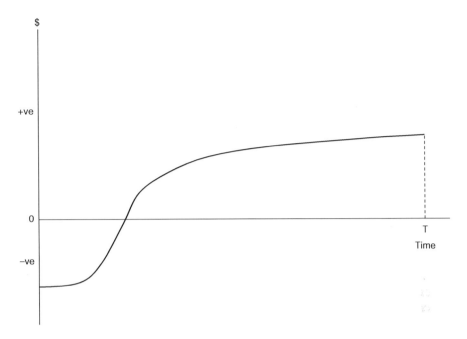

Figure 1.2 Typical time-stream of project net benefits. Investment costs in the initial years of the project's life lead to net benefits being negative (costs exceed benefits). In the later stages net benefits are positive (benefits exceed operating costs and capital replacement costs).

Much of what has been said up to this point about public projects also applies to projects being considered by a private firm: funds that are allocated for one purpose cannot also be used for another purpose, and hence have an opportunity cost. Firms routinely undertake investment analyses using the same techniques as those applied in social cost-benefit analysis. Indeed, the evaluation of a proposed project from a private viewpoint is often an integral part of a social cost-benefit analysis, and for this reason the whole of Chapter 4 is devoted to this topic. A "private" investment appraisal takes account only of the benefits and costs of the project to the project's proponent – its effect on revenues and costs and hence on profit. The project may have wider implications – environmental and employment effects, for example – but if these do not affect the private or public sector proponent's profits – its "bottom line" – they are omitted from the private appraisal. In contrast, a social cost-benefit analysis takes a wider perspective – it measures and compares the costs and benefits experienced by all groups affected by the project. How to obtain these measures is the subject of Chapter 5, and how to identify the costs and benefits to those groups whose economic welfare is to be taken into account in making the decision about the project is the subject of Chapter 6.

The traditional form of social cost-benefit analysis calculated the aggregate net benefits of a project irrespective of which groups were affected. In this book we refer to this approach as an *Efficiency Analysis* as it tells us whether the project is an efficient reallocation of resources according to the Kaldor-Hicks criterion; in other words, does undertaking the project increase the size of the cake, irrespective of who are the gainers and losers? However, as suggested above, decision-makers also require information about the distributional effects of the project: how does undertaking the project affect the sharing out of the cake? It is a relatively easy matter to partition a social cost-benefit analysis to take account of different viewpoints. For example, the analysis can identify the subset of overall benefits and costs experienced by the project's proponent, whether a government department or a private firm; in the case of a firm, it can summarise the benefits and costs to the owners of the equity (the shareholders) in the firm; an analysis from this perspective is referred to in this book as a *Private Analysis*. Alternatively, the analysis can be used to identify and summarise all benefits and costs to members of a subset of those affected by the project, termed the *Referent Group*. The Referent Group, to be discussed below, consists of those entities whose benefits and costs *matter* in coming to a decision about the project.

In what we term a *Market Analysis* (and sometimes refer to as a *project analysis*), estimates of all project benefits and costs are calculated at market prices. The project NPV calculated in this way is generally neither the Private NPV (the value of the project to the equity holders) nor the Referent Group NPV, nor the NPV measured by the Efficiency Analysis. The project NPV generally exceeds the private NPV because the equity holders do not stand to receive all the benefits of the project or incur all of the costs: for example, taxes may be due on project income and loans may be obtained to finance part of the project, with consequent outflows in the form of interest payments. Equity and debt holders may or may not be part of the Referent Group and their net benefits will be treated accordingly. By pricing inputs and outputs at market prices, the Market Analysis ignores the effects of market distortions or missing markets: for example, the wage rate established by a distorted labour market fails to measure the employment benefits of a project, and the market is obviously unable to value all the benefits or costs of non-marketed goods and services such as vaccinations or pollution. Project effects such as these, omitted in the Market Analysis, are included in the Efficiency Analysis which corresponds to the traditional social cost-benefit analysis. We discuss the measurement of benefits and costs in the presence of market distortions in Chapter 5, and in the absence of markets in Chapter 8.

1.2 The Referent Group

Before conducting the cost-benefit study the analyst needs to know whose benefits and costs are to be measured and compared. Which groups of individuals, firms, private organizations and public institutions have "standing" in the decision-making process? The Efficiency Analysis, in effect, gives standing to all groups affected by the project and, as economists, we may prefer that the decision be based on the results of that analysis on the grounds that the chosen project will offer the best prospect for an overall gain in economic welfare. However we noted above that the decision-maker would likely be interested in the distribution of project benefits and costs and he may decide that not all the benefits and costs identified by the social cost-benefit analysis (the Efficiency Analysis) are relevant to the decision about the project. The simplest way to determine the appropriate focus of the study is to ask the decision-maker (the client) who commissioned the study to specify who is to be included and who is to be left out. To conclude the discussion of standing at this point would be to ignore the proactive role envisaged for the analyst and to be described in Section 1.6. The decision-maker may well seek some options and advice as to who should have standing. Because of space limitations our discussion of this complex issue will be brief and readers are referred to the classic discussion by Whittington and MacRae Jr (1986).

We can think of the question of standing as having horizontal and vertical dimensions. The horizontal dimension is the question of how wide to cast the net. Are we to consider anthropocentric values only – the values humans place on the effects of the project – or should we extend the scope of the Referent Group to include fauna and flora? To be clear about this, people may place a value on the conservation of whales, and we would most likely include that value in the cost-benefit analysis, but presumably whales themselves value protection and should that also be taken into account in the analysis? This kind of question has also been raised in respect to the conservation of forests. While we might wish to give such questions sympathetic consideration, we should remind the reader that cost-benefit analysis may be but one of a number of inputs to the decision-making process and we suggest that wider considerations of this kind are best dealt with separately in the decision-making process.

Confining ourselves to anthropocentric values, we still need to establish the boundary of the jurisdiction within which those affected by the project can be considered to have standing. Should the analyst confine the Referent Group Analysis to reporting the project net benefits accruing to inhabitants of the local municipality, or of the State, or of the country as a whole? Should a wider perspective be adopted, considering, for example, net benefits to North America or the European Community, or, in the case of climate change mitigation measures, to the inhabitants of the planet as a whole? In a cost-benefit analysis of treating US uranium mill tailings to reduce emissions of cancer-inducing radon gas, it was estimated that 10% of the adverse North American health effects of gas emissions were experienced by Canadians and Mexicans. In deciding whether the benefits of treatment justified the costs to be incurred by the United States, should Canada and Mexico be included in the Referent Group? (They were!) What about the value of health benefits in other continents, estimated to be around 25% of those in North America? (Europe, Asia and the rest of the world were excluded from the analysis.) A similar issue arises in appraising projects aimed at reducing sulphur dioxide emissions by US firms: should the benefits of a reduction in acid rain precipitation in Canada and northern Europe be included or not?

The vertical dimension is the choice of which inhabitants of the chosen level of jurisdiction are to have their benefits and costs taken into account. Normally the answer to this question would be all the residents of the jurisdiction – the individuals, firms and other private entities located in the jurisdiction, plus the changes in expenditures and revenues of the public institutions which serve those residents. But the relevant subset might be defined more narrowly by a social grouping, such as the poor, the unemployed, the elderly, or people of aboriginal descent. Some might argue that only net benefits to those with citizenship should be considered, and that recent or illegal immigrants should be excluded from consideration. What about the yet to be born? Should foetuses have standing, and what about future generations? For example, the reduction in radon gas emissions canvassed above can be expected to have health benefits for thousands of years to come: do we include these benefits to the unborn in the analysis and what discount rate should we use? In the case of radon gas, the US Environmental Protection Agency opted for a zero discount rate and included undiscounted benefits over a 100-year period.

What about the proceeds of crime? One benefit of the US Job Corps program was held to be a reduction in the rate of crimes against property as participants in the program found employment. A study, quoted in Whittington and MacRae Jr (1986), estimated that burglary earned the perpetrator $1,247 on average and that the cost to the rest of the community was $9,996. Should the $1,247 be netted out of the $9,996 benefit of an averted burglary in a cost-benefit analysis, or do criminals, and the proceeds of their crimes, have no standing in the matter? The above questions raise many thorny legal and ethical issues which we will not pursue further here. Chapter 6 is concerned with differentiating those costs and benefits identified by the social cost-benefit analysis which accrue to members of the Referent Group from those that do not.

1.3 The structure of the cost-benefit model

The important concept of the Referent Group is shown in Figure 1.3, which illustrates an example that will be developed in Chapters 4–6 of this book. Suppose that a wholly foreign-owned company proposes to set up a factory in a developing country. The government wishes to appraise the proposal from the point of view of residents of the host country – the Referent Group in this case. The firm (not a member of the Referent Group) has two questions it may wish consider. First, is the overall project efficient from a market view-point? This is determined by the Market Analysis which compares the benefits and costs associated with undertaking the project, where benefits and costs are calculated at market prices; the present value of the net benefits (which could be either positive or negative) is represented by Area A + B in Figure 1.3 (the interpretation of the breakdown of the project net present value into the components A and B will be explained shortly). Second, is the project profitable from the perspective of the firm's owners, or equity holders? The answer to this question is determined by the Private Analysis. If the project is to be wholly intern-ally financed, the answer to this second question is obtained by deducting tax payments from the NPV identified by the Market Analysis. However, we will assume that there is to be some debt participation in the project in the form of a loan from a financial institution in the host country (a member of the Referent Group). The amount of the loan must be deducted from the project cost and the loan repayments and interest charges deducted from the project's after-tax benefits to give the benefits and costs of the project to the equity holders as measured by the Private Analysis.

In this example, as noted above, we assume that the firm's foreign equity holders are not considered part of the Referent Group. This being so, in Figure 1.3, Area A represents the net present value of those project net benefits, measured by the Market Analysis, and

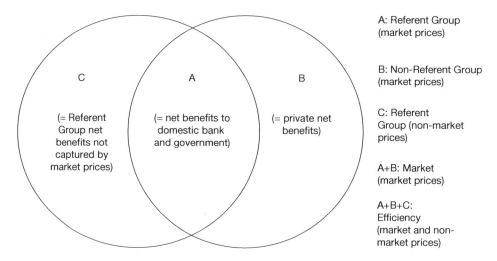

Figure 1.3 Relationship between the Market, Private, Efficiency and Referent Group net benefits.

accruing to members of the Referent Group: the providers of the firm's loan (the domestic bank) and the recipients of the firm's tax payments (the government). The net benefit of the project to the non-Referent Group members (the firm's equity holders) expressed as a net present value, is represented by Area B. Only if the net benefit to equity holders is positive is the project worthwhile from the firm's viewpoint. The sum of the net benefits represented by areas A and B amounts to the project NPV as determined by the Market Analysis.

As noted above, the project may have a wider impact than that summarised by the Market Analysis. For example, some residents who would otherwise have been unemployed may obtain jobs: the pay that they receive from the firm may be higher than the value of their time in some non-market activity, thereby resulting in a net benefit to them. The firm may purchase various goods and services, such as water and electricity, from government agencies, paying prices in excess of the production costs of these inputs, again generating net benefits for this section of the Referent Group. The project may generate pollution which imposes health and other costs on residents of the host country. In Figure 1.3, Area C represents the set of net benefits (present value of benefits net of costs) experienced by the Referent Group as a result of divergences of market prices from Referent Group valuations of benefits or costs, or as a result of non-marketed effects. We shall refer to these as *non-marketed net benefits* accruing to members of the Referent Group. The total Referent Group net benefit is represented by Area A + C. A further category of net benefit will be introduced in Chapter 6 in the form of non-Referent Group net benefits not measured by market prices – pollution of the high seas, for example – but we will ignore this in the meantime.

What then does the whole area, A + B + C represent? This can be thought of as representing the *efficiency net benefits* of the project – the present value of net benefits valued in terms of the value of project outputs and the opportunity cost of project inputs, irrespective of whether or not these benefits or costs are accurately measured by market prices or accrue to members of the Referent Group. Area B represents the net benefits to the non-Referent Group equity holders, which will determine the firm's decision whether or not to undertake the project. Area A + C represents the net benefits to the Referent Group, which will determine the government's decision as to whether or not to allow the project to proceed. This is the main issue which the cost-benefit analyst is called upon to address, although in negotiating with the firm, the decision-maker may also be interested to know how attractive the project is from a private viewpoint.

Apart from measuring the aggregate Referent Group net benefit, the analyst will also need to know how this is distributed among the different sub-groups as the decision-makers will, most probably, want to take into consideration the distribution of gains and losses among the Referent Group members: this breakdown is provided by the detailed *Referent Group Analysis* which is the subject of Chapter 6.

In summary, the hypothetical project discussed above (or any other project) can be appraised from four different points of view:

i   *the Market Analysis*: this is represented by Area A + B and is obtained by valuing all project inputs and outputs at private market prices;

ii   *the Private Analysis*: in the case of a private firm this is accomplished by netting out tax and interest and debt flows from the market appraisal, and, if the firm's equity holders are not part of the Referent Group as in the example illustrated in Figure 1.3, it will be given by area B which is the non-Referent Group project net benefit;

iii   *the Efficiency Analysis*: this is represented by Area A + B + C and is obtained in a similar way to the market appraisal, except that some of the prices used to value inputs or outputs are shadow- or accounting-prices, which are discussed in Chapter 5, or are derived from the application of non-market valuation techniques as discussed in Chapter 8;

iv   *the Referent Group Analysis*: this is represented by Area A + C and can be obtained in two ways as noted below – directly, by enumerating the costs and benefits experienced by all members of the Referent Group; or indirectly, by subtracting non-Referent Group net benefits from the net benefits calculated by the Efficiency Analysis. In our example, the non-Referent Group net benefits are summarised by the private NPV (Area B), though in other cases the private project owners may be part of the Referent Group.

In the course of undertaking a complete social cost-benefit analysis of the project in our example the analyst will need therefore to follow a sequence of steps:

1   Calculate the project net benefits at *market* prices (Area A + B in Figure 1.3).
2   Calculate the *private* net cash flow at market prices (Area B in Figure 1.3).
3   Re-calculate the project net benefits at *efficiency* prices (Area A + B + C).
4   Disaggregate the efficiency net benefits among the *Referent Group* (and non-Referent Group) members.

It is clear that there are two ways of going about the task of estimating Area A + C – the net benefits to the Referent Group: directly, by listing all the benefits and costs to all members of the group – in this example, labour, government organizations, and the general public – and measuring and aggregating them; or indirectly by measuring the efficiency net benefits of the project and subtracting from them the net benefits which *do not* accrue to the Referent Group. Under the first approach, Area A + C is measured directly; under the second approach, Area A + B + C is measured (by the Efficiency Analysis) and the net benefits to those not in the Referent Group (represented in the example by Area B identified by the Private Analysis) are subtracted to give Area A + C.

At first sight it might seem strange to consider using the indirect approach. However, as we will see in Chapters 4 and 5, it is relatively straightforward to measure the net benefits represented by Areas B and A + B + C respectively. The net efficiency benefits of the project are obtained by valuing all project inputs and outputs at their marginal values to the economy: these marginal values are represented by market prices, as discussed in Chapter 4, or by accounting- or shadow-prices, which are artificial rather than observed market prices and which can be calculated, as discussed in Chapter 5, or by prices obtained from the application of non-market valuation techniques, as discussed in Chapter 8. The net private benefits are obtained by using market prices, which can be directly observed, and by deducting tax and debt flows: this calculation simply mimics the process which the firm undertakes internally to decide whether or not to proceed with the project. Measuring Area A + C directly is more difficult because each subset of the Referent Group which is affected by the project has to be identified and their costs and benefits measured. In summary, the indirect approach produces an *aggregate* measure, whereas under the direct approach the net benefits are measured in *disaggregated* form and assigned to various groups. While the disaggregation provides important information which relates to the income distributional concerns of the decision-maker, it is more difficult to obtain than the summary figure.

In this book we advocate the use of both approaches: in terms of the current example, measure Area A + C as A + B + C less B, and then measure its component parts directly and sum them to get Area A + C. If the same answer is not obtained in both cases, an error has been made – some benefits or costs to members of the Referent Group have been omitted or incorrectly measured. A check of this nature on the internal consistency of the analysis is invaluable.

An analogy which may assist in determining what is to be measured and where it belongs in the analysis is to think of the project as a bucket. Costs go into the bucket and benefits come out: however, the range of benefits or costs which are to go in or come out depends on the perspective that is taken. In the *Efficiency Analysis* we count all the costs and benefits measured at the appropriate market and shadow-prices, and the benefits minus the costs – the net benefits of the project – are equivalent to Area A + B + C in Figure 1.3. The *Market Analysis* is similar to the Efficiency Analysis except that all the costs and benefits are measured at market prices, where these exist, to obtain an estimate of Area A + B, and non-marketed benefits and costs are ignored. In the *Private Analysis*, which, in our example measures non-Referent Group net benefits accruing to the foreign firm, we count all sums contributed or received by the firm's equity holders to calculate the non-Referent Group net benefits (Area B). In the example, this consists of the project cost less the loan obtained from the domestic bank, and the project revenues less the interest and principal repayments, and the tax payments to the host country. In the *Referent Group* analysis we count all contributions and receipts by Referent Group members to estimate Area A + C: in our example this consists of Area A – the capital contribution of the domestic financial institution, together with the loan repayments and interest payments, and the government direct tax revenues from profits tax and any indirect tax flows levied on project output – and Area C – indirect taxes levied on project inputs, the employment benefits received by domestic labour, and any other rents generated by the project. (If there are other, non-marketed net benefits or costs accruing to non-Referent Group members, as discussed above, we would need to include these in an additional category, say, D, which would be included in the Efficiency Analysis, but, like Area B, deducted from the aggregate efficiency net benefit to arrive at the total Referent Group net benefit. This possibility is considered in more detail in Chapter 6.)

At this point it should be stressed that many projects will not correspond exactly to the above example, and what is to be included in Areas A, B and C will vary from case to case. Furthermore, additional categories of project effects may be required. For example, suppose, as in the case study developed in the appendices to Chapters 4–6, that part of the cost of the project was met by a loan from a foreign bank which is not part of the Referent Group. To incorporate this possibility we would add the foreign bank's net benefits to Area B in Figure 1.3. Area B still forms part of the Efficiency Analysis, as the loan measures part of the cost of capital goods, but this area has to be subtracted from the total efficiency NPV to calculate the Referent Group NPV. Another possibility is that instead of paying taxes the foreign firm receives an annual subsidy. Then Areas A and B would need to take account of the subsidy: as a credit item in Area B and a debit item in Area A. A comprehensive framework which takes account of all possible categories of benefits and costs is presented in Chapter 6.

1.4 The use of spreadsheets in cost-benefit analysis

An important theme of this book is the use of spreadsheets in cost-benefit analysis. This theme is developed in detail in Chapters 3–6. However, since the structure of the spreadsheet directly reflects the various points of view accommodated in the social cost-benefit analysis, it is instructive to consider the layout of the spreadsheet at this point.

The spreadsheet is developed in five parts in the following order:

i   a *data* or *variables* section containing all relevant information about the project – inputs, outputs, prices, tax rates, etc. It is important to bear in mind that much of these data consists of *forecasts* of future values which need to be obtained by those affected by the project. This is the only part of the spreadsheet which contains the raw data pertaining to the project. All the entries in the remaining four parts of the spreadsheet consist of cell references to the data in this first section.
ii   a section containing the *Market* Analysis;
iii a section containing the *Private* Analysis;
iv   a section containing the *Efficiency* Analysis;
v   a section containing the *Referent Group* Analysis.

The relationships between these sections can be illustrated by means of the following simple example, similar to that discussed in Section 1.3. Suppose that a foreign company proposed to invest $100 in a project which is forecast to produce 10 gadgets per year for a period of 5 years. The gadgets are predicted to sell for $10 each. To produce the gadgets, the firm will have to hire 20 units of labour per year at a wage of $3 per unit. The project is located in an area of high unemployment and the opportunity cost of labour is estimated to be $2 per unit. The firm will pay tax at a rate of 25% on its operating profit (defined here as its total revenue less its labour costs). The project will also generate water pollution which has been estimated to result in increased costs to the local community of $12 per annum from year 1 onwards. There are no other costs or allowances, such as interest deductions or depreciation allowances, and the project has no effect on the market price of any input or output.

Figure 1.4 illustrates the structure of the spreadsheet. The project data, consisting of the capital cost, the output and input flows, the market prices and shadow-price of labour, the pollution cost, the tax rate and two discount rates (5% and 10%) are entered in Section 1.

In Section 2 the Market Analysis of the project is conducted: the flows of costs and benefits, valued at market prices, are calculated for each of the five years of the project's life, using Section 1 as the source of the data. A net benefit stream, represented by a net cash flow, summarises the effects of the project, and net present values at a range of discount rates, and an internal rate of return are calculated. In Section 3, the Private Analysis is conducted, again with reference to the data in Section 1. In this simple example, the benefits of the project to the private firm consist of the after-tax returns. Again the performance of the project is summarised in the form of a net cash flow, and net present values and the internal rate of return are calculated. In Section 4 the Efficiency Analysis is conducted which involves using shadow-prices where appropriate. In the simple example, the only shadow-prices required are that of labour, estimated at $2 per unit, and pollution, at $12 per annum. As in Sections 2 and 3 of the spreadsheet, the net benefit stream is calculated from the data in Section 1 and expressed in the form of a net cash flow. Net present value is calculated for the chosen range of discount rates and an internal rate of return is calculated.

Section 5 contains the Referent Group Analysis. In this case it is assumed that the Referent Group consists of the residents and government of the host country, and does not include the foreign equity holders of the private firm. The first line of the Referent Group analysis, in this example, is simply the difference between the efficiency net benefit stream and the private net benefit stream. While this calculation gives us the aggregate net benefits

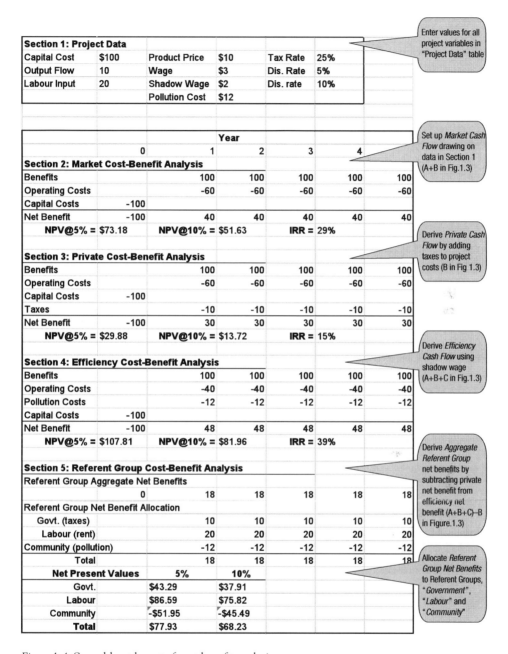

Figure 1.4 Spreadsheet layout of cost-benefit analysis.

to the Referent Group, we want some information about the distribution of these benefits. The two groups of beneficiaries are the government and labour and we enter the net benefit streams accruing to each: tax revenues to government, and rent to labour – wages in excess of the value of the workers' time in the alternative activity. The costs to members of the Referent Group (entered as negative values) derive from the pollution experienced by the general community. When these benefit and cost streams are added, we get an alternative measure of the Referent Group net benefit stream.

If the Referent Group net benefit measures obtained by the two methods of calculation are inconsistent in any year, an error has occurred, perhaps in shadow-pricing or in identifying or measuring benefits or costs to Referent Group members. Once any discrepancy has been accounted for and corrected, summary measures of the performance of the project from the viewpoint of the Referent Group can be prepared. In this case net present values only are presented because the Referent Group net benefit stream does not have an internal rate of return, for reasons to be discussed in Chapter 2.

At this point we might ask why we bother with the Market, Private and Efficiency Analyses when all the decision-maker may be concerned with is the outcome for the Referent Group. There are two reasons. The first is computational in nature: the Market Analysis uses and organises the data required for the Private Analysis and much of the data required for the Efficiency Analysis. The Referent Group Analysis is then constructed by selecting the relevant items from the Private and Efficiency Analyses. The second reason is that the Private or Efficiency outcome of the project may be relevant to the decision. The private investor may be part of the Referent Group, or, alternatively, the Private Analysis may reveal substantial profits accruing to a foreign proponent of the project suggesting, perhaps, that the host country's tax regime is too favourable and should be amended to provide a greater net benefit to the Referent Group. Perhaps an international aid agency, such as the World Bank or the Asian Development Bank, regards economic efficiency as an important criterion in its lending policy and will participate only if this can be demonstrated.

In summary, we have developed a template for conducting a social cost-benefit analysis, using a spreadsheet, which contains a check for internal consistency. Because project data are entered in Section 1 only, we can use our model to perform sensitivity analyses simply by changing one cell entry in the data section. For example, suppose the tax rate were increased to 40%, would there still be sufficient inducement to the foreign firm to proceed with the investment, while providing a reasonable net benefit stream to the Referent Group? To answer this question, all we need do is change the tax rate cell in Section 1 to 40% and review the new set of results.

1.5 The rationale for public projects

It was suggested above that a set of accounting-prices (usually termed shadow-prices), or a set of prices obtained by use of non-market valuation techniques was required to calculate the efficiency net benefit stream of many proposed projects. This raises the question of what is wrong with market prices, or why markets do not exist for some commodities which affect economic welfare, and indirectly raises the issue of the rationale for public projects – and social cost-benefit analysis.

It would be appropriate to use market prices to calculate the efficiency net benefit stream of a project if these prices measured the benefits (opportunity costs) generated by all project outputs (inputs). This condition would be satisfied if all markets were competitive (in the sense that no market participant can have any individual influence on price), undistorted (by taxes and regulation, for example), and complete (in the sense that everything that contributes to economic welfare is traded in a market). If we lived in such a world, market prices would accurately measure efficiency benefits and costs, and since participants in markets are assumed to be utility or profit maximisers, every scarce input would be allocated to its highest value use. The economy would be working efficiently in the sense that no reallocation of scarce resources could make anyone better off without making someone else worse off – a Pareto Optimum would exist.

What then would be the role of the public sector? While governments might be concerned with the fairness of the income distribution generated by the market economy, the argument that public intervention, either in the form of undertaking projects or requiring modifications to proposed private projects, could lead to a more efficient allocation of resources would be unsustainable. In other words, if markets are perfectly competitive and the existing distribution of income is deemed optimal, there is no need for public projects, social cost-benefit analysis, and social cost-benefit analysts!

In fact, as discussed in detail in Chapters 5 and 8, in most economies markets are non-competitive, distorted and incomplete to a greater or lesser degree. Even the so-called "free market economies" are rife with market imperfections – budding social cost-benefit analysts can breathe a sigh of relief! Proposed private sector projects are not necessarily in the public interest, and projects which are in the public interest will not necessarily be undertaken by the private sector and will require government involvement. However, market imperfections constitute a double-edged sword: because some markets are imperfect to a significant degree, we cannot trust the prices they generate to accurately measure the efficiency benefits and costs of a proposed project. This means that for the purpose of undertaking a project appraisal, the analyst needs to modify observed market prices in various ways to ensure that they reflect marginal values to households or firms, or, in cases in which the market does not exist, to generate them in other ways. It should be stressed that these modifications are for the purpose of the social cost-benefit analysis only: the accounting- or shadow-prices are not actually used to pay for project costs or to charge for project benefits.

It was suggested above that even if the market economy succeeded in achieving a completely efficient allocation of resources, there might still be a case for government intervention on income distributional grounds. Recalling that the decision-maker is a public servant acting under the instructions of an elected politician, it should come as no surprise that the income distributional effects of a public project – who benefits and who loses – may be an important consideration in determining whether or not the project should go ahead. In Chapter 11 we outline some approaches to appraising the income distributional effects of projects. Furthermore, the decision-maker may be interested in the wider economic impact of the project in terms of its contribution to GDP or economic growth, and we will discuss this issue in Chapter 12. However, we consider the social cost-benefit analyst's primary roles to be: identifying, measuring and appraising the efficiency effects of the project; using this information to determine whether the aggregate net benefits to the Referent Group are positive; and identifying and measuring, but generally not evaluating, the distributional effects of the project. The first role is fulfilled by means of the aggregated approach to measuring the social net benefits of a project, whereas the latter roles require the disaggregated approach to Referent Group net benefit measurement described earlier.

The cost-benefit analysis should identify sub-groups among the Referent Group who are significantly affected by the project, for example, a proposed new airport will benefit air travellers but will inflict costs on residents in the vicinity of the proposed site. The analyst needs to identify the nature of these effects – reduced travel time for passengers, and increased noise for residents – to measure them in physical units – hours saved, additional decibels endured – and to quantify them in dollar terms as far as possible. The role of the decision-maker is then to determine whether X dollars' worth of gain for one group outweighs Y dollars' worth of pain for another.

1.6 The role of the analyst

To this point we have described the analyst's role as a passive actor in the cost-benefit process – called in to provide information about a particular proposal and then dismissed once the report has been submitted. This view neglects a very important function of the analyst – that of contributing to project design – a function that emerges during the process of *scoping* the analysis. Suppose that the analyst is contacted by the Department of Agriculture: "We are considering building a 30-metre dam at the end of Green Valley to provide irrigation water to farmers. We would like you to do a cost-benefit analysis." At the first meeting with department officials the analyst needs to ask: "Why a *30-metre* dam? Why Green Valley and not somewhere else? Why is the dam not also producing electricity? What provisions have been made for recreational access? Whose costs and benefits are relevant to the decision?" and so on. In the process of this meeting it may emerge that all these questions have been carefully considered and that in fact the proposed project design is the best option. However, often it turns out that in the process of determining exactly what is to be appraised the analyst is able to assist the department officials in clarifying their objectives and refining their proposal. In other words, cost-benefit analysis is not limited to a linear process of project design, from appraisal to implementation, but can be a feedback process as illustrated in Figure 1.5.

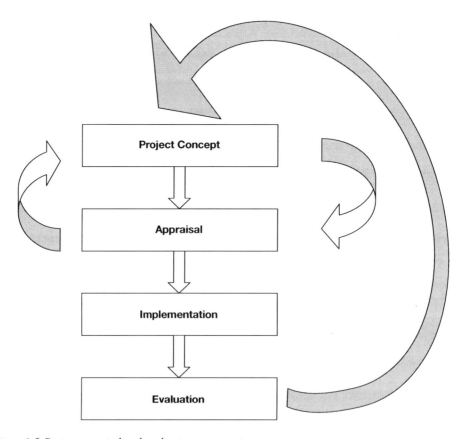

Figure 1.5 Project appraisal and evaluation as a continuous process.

Project *evaluation*– the retrospective assessment of project benefits and costs – may also play a part in this process by providing valuable information about the appropriate design of future projects. However, it is worth noting that government agencies seldom commission retrospective project evaluations, suggesting that governments are more inclined to use cost-benefit analysis as a tool to inform current decisions than for assessing their performance in making past decisions.

We have suggested that the analyst may be able to help the client formulate a better proposal – one which has a higher net present value to the Referent Group than the original proposal – though this does not guarantee that the revised project can be recommended. However, the analyst may also be able to assist with assessing the income distribution aspects of the project: these are sometimes difficult to identify and may emerge only during the drafting of the report. At this stage the analyst may be in a position to suggest measures which may offset the losses which some sub-groups would experience if the proposal went ahead. For example, Green Valley may be a prime area for wildlife viewing and bird-watching. Is there an alternative area which could be developed as a substitute and could the costs of this alternative be justified? This might reduce the net loss suffered by the nature-viewing group and make the project more "fair" in terms of its distributional effects. It might incidentally, and of interest to the politicians, reduce the level of political opposition to the proposed project.

1.7 Further reading

Most introductory- or intermediate-level microeconomics texts contain sections devoted to the three main economic issues arising in this introductory chapter: (1) the notion of scarcity and opportunity cost; (2) how to identify a change in the level of economic welfare; and (3) the failure of markets to allocate resources efficiently. The latter two issues are the main topic of welfare economics, about which many books have been written, some of which are quite technical. An advanced text with a focus on practical issues is R.E. Just, D.L. Hueth and A. Schmitz, *Applied Welfare Economics and Public Policy* (Englewood Cliffs, NJ: Prentice Hall, 1982). The classic article on the question of standing is D. Whittington and D. MacRae Jr (1986), "The Issue of Standing in Cost-Benefit Analysis", *Journal of Policy Analysis and Management*, 5(4): 665–682.

Exercises

1   The casino has given you $100 worth of complementary chips which must be wagered this evening. There are two tables: roulette and blackjack. The expected value of $100 bet on roulette is $83, and the expected value of $100 bet on blackjack is $86. What is the expected value of:

   i   the opportunity cost of betting the chips on roulette?
   ii   the opportunity cost of betting the chips on blackjack?

2   A recent high school graduate has been offered a job at the local supermarket paying $15,000 per year. She believes that if she were to graduate from college her annual earnings would be $6000 more than what she could make from the supermarket job. A three-year degree course would cost her $3000 per year in fees, textbooks and other

expenses. Assuming a zero rate of discount, and making explicit any other assumptions that you consider necessary, answer the following questions:

   i   What dollar amount measures her annual opportunity cost of attending college?
   ii   Construct the net benefit stream expected from her investment in education.
   iii   Explain how you would estimate the net present value of her proposed investment in education.
   iv   What dollar amount is your best estimate of the net present value?

3   A foreign firm is considering a project which has present values of benefits and costs, at market prices, of $100 and $60 respectively. If the project goes ahead, tax of $20 will be paid to the host country, and pollution costing the host country $10 will occur. Assuming that the owners of the foreign firm are not part of the Referent Group, state what are the net present values generated by the following:

   i   the Market Analysis;
   ii   the Private Analysis;
   iii   the Efficiency Analysis;
   iv   the Referent Group Analysis.

4   Suppose that the economy consists of three individuals, A, B, and C, and that there are three projects being considered. The net benefits, measured in dollars, received by the three individuals from each of the three projects are reported in the following table:

Individual	Project		
	1	2	3
A	−10	+20	+5
B	+5	+1	−5
C	+10	+1	−1

Which of the projects, if any, can be described as follows?:

   i   a Pareto improvement;
   ii   a potential (though not an actual) Pareto improvement;
   iii   neither a potential nor an actual Pareto improvement.

Note

1 For convenience, we refer mostly to "project" throughout this book, bearing in mind that the principles and techniques discusses apply equally to the appraisal or evaluation of policies.

2 Project appraisal

Principles

2.1 Introduction

This chapter provides a simple introduction to the principles of investment project appraisal. It starts with an outline of the logic of the appraisal process from the viewpoint of an individual considering a very simple type of project. During the course of the discussion of this appraisal process we develop such important concepts as the discount rate, the discount factor, the net present value, the benefit/cost ratio and the internal rate of return. The discussion then shifts to the economy as a whole and the role of investment appraisal in allocating resources between investment and the production of goods for current consumption. We then present the simple algebra of various investment decision-rules which we apply in the latter part of the chapter, together with a numerical example. Following this, we discuss some special concepts, such as annuities, economic depreciation, and inflation and risk, in the context of investment appraisal.

In Chapter 3, the discussion shifts to applications of the various investment decision-rules. Some of the applications rely on the simple algebraic concepts already discussed in Chapter 2, while some are developed using the basic tool of the cost-benefit analyst – the spreadsheet. Some issues already discussed in the present chapter, such as the time value of money and the calculation of present value and internal rate of return, are explored further in Chapter 3, while new concepts, such as choice among projects and capital rationing are introduced and discussed.

2.2 Project appraisal from an individual viewpoint

Economists start from the proposition that an individual's economic welfare in a given time period is determined by the quantity of goods and services she consumes in that time period; the consumption of goods and services is taken to be the ultimate goal of economic activity. However, sometimes it pays to use scarce resources – land, labour, capital and materials – to produce capital goods in the present so that the flow of consumption goods and services (hereafter referred to as "consumption goods") can be augmented in the future. Two important processes are at work here: (1) saving, which is refraining from consumption so that scarce resources are freed up for a use other than producing goods for current consumption; and (2) investing, which is the process of using scarce resources to produce goods for future consumption. Sometimes an individual undertakes both activities, for example, instead of spending all of her income on current consumption goods, she spends part of it in refurbishing the basement of her house so that it can be rented. However, often the

DOI: 10.4324/9781003312758-2

connection between saving and investment is made through a financial intermediary such as a bank: the individual saves, lends to the bank, the bank lends to an investor who uses the money to construct the capital good.

In the above example of the person considering investing in refurbishing part of their house for letting there are two options under consideration: (1) do not undertake the investment and the required saving, in which case all income received in each period can be used to purchase consumption goods; or (2) undertake the investment, in which case the value of consumption will be lower in the current period by the amount of the saving required to finance the investment project, but higher in future periods by the amount of the rent net of any letting costs. To make it simple, suppose that there are only two periods of time – this year (Year 0) and next year (Year 1) – and assume that all payments and receipts occur at the end of the year in question (this is an assumption about the "accrual date" to which we will return later), and that the reference point (the "present" in a present value calculation) is the end of this year ("now" or Year 0). To anticipate the discussion below, this means that benefits or costs assumed to occur at the end of this year are not discounted, and those occurring at the end of next year (Year 1) are discounted by one period back to the present. The two options available to the individual are illustrated in Figure 2.1, in which present values are represented as vertical distances, and future values as horizontal distances on the axes scales.

How is the investor to decide whether it is worthwhile undertaking the project? To be worthwhile, clearly the project must be at least as good an investment as putting the money to be expended on the project in the bank instead. How does the return on the investment compare with the alternative of placing the savings in the bank? There are two main ways of answering this question: the rate of return on the proposed investment can be compared with the rate of return on money deposited in the bank; or the cost of the project, $Y_0 - C_0$ in Figure 2.1, can be compared with the present value of the project return – the sum of money which would have to be deposited in the bank now to yield a future return equal to the project return, $C_1 - Y_1$ in Figure 2.1. Both these approaches recognise the concept of "time value of money", which means that dollars at different points of time cannot be directly compared.

If the individual puts \$AC in the bank this year, she will have the principal plus interest available to spend on consumption goods next year: this amounts to \$AC$(1 + r)$ where r is the annual *rate of interest* the money earns in the bank. If \$AC$(1 + r) < BC$, then the return from the bank is less than that from the project and the individual will be made better off by choosing the project as compared with the bank. The condition for the project yielding a higher rate of return than that offered by the bank can be rewritten as $BC/(1 + r) > AC$, where $BC/(1 + r)$ is the *present value* of the benefit of the project, AC is its opportunity cost, and r is the rate of interest paid by the bank. The present value is obtained by multiplying the project return, BC, by the *discount factor* $1/(1 + r)$, where r is the *discount rate*. In terms of Figure 2.1, this is accomplished by projecting the length BC on to the vertical line through AC to get a point such as D, where DC is the present value; since $DC = BC/(1 + r)$, it is apparent that the slope DC/BC equals the discount factor $1/(1 + r)$. The *net present value* (NPV) of the project is given by the present value of the project benefit, $BC/(1 + r)$, less the project cost, AC. In mathematical terms both DC/BC and $1/(1 + r)$ in Figure 2.1 are negative values, but in this and subsequent discussion we talk in terms of absolute values; the purist can suppose that we have multiplied both sides of the slope equation by –1. The net present value is given by: $BC/(1 + r) - AC$, and is measured by the distance DA in Figure 2.1.

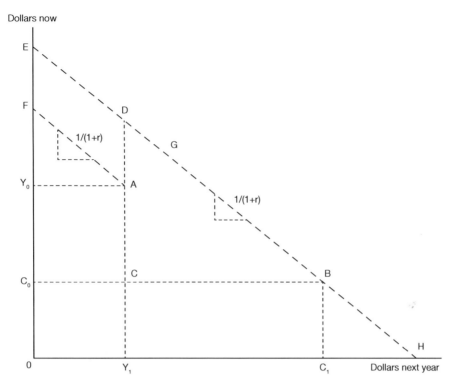

Figure 2.1 Investment appraisal: an individual perspective. The individual's income levels this year and next year, from sources other than the investment project, are given by Y_0 and Y_1, represented by point A. The consumption level in the present year is Y_0 if the project is not undertaken. The cost of the project is $Y_0 - C_0$, where C_0 represents the level of expenditure on consumption goods in year 0 (now) if the investment is undertaken (income of Y_0 less savings of $Y_0 - C_0$ devoted to undertaking the investment); the cost of the investment is represented by the length AC. Undertaking the investment project provides a return which allows the individual to spend more on consumption goods next year than her income, Y_1, from other sources. Consumption next year in that event would be C_1 and this means that the return on the investment is $C_1 - Y_1$, represented by length BC.

The ratio $(BC - AC)/AC$ will be familiar to students of finance as the *return on investment* (ROI) – the ratio of the gain to the initial cost. This concept is not generally used in project appraisal because it fails to take account of the timing of the project returns. In the two-period case, however, the ROI is identical to the *internal rate of return* (IRR) on the project. Also referring to the two-period case, the ratio BC/AC measures the *marginal productivity of capital* – the return per additional unit of investment. More generally the marginal productivity of capital is measured as: 1 + the internal rate of return.

The internal rate of return, which we will denote by r_p, is the discount rate which must be imposed to reduce the project NPV to zero: $BC/(1 + r_p) - AC = 0$. It can readily be ascertained from the above relationships that the investment is worth undertaking if the internal rate of return (IRR) on the investment, r_p, is greater than the rate of interest, r. In the case of a multi-period project, the calculation of the IRR is not so straightforward and we will return to this matter in Chapter 3. While the internal rate of return is sometimes calculated as part of a social cost-benefit analysis, it is not as frequently relied upon as the net present value of the project, for reasons discussed later in this and Chapter 3.

As already noted, if the NPV is positive, the present value of the project's return is greater than the present value of its opportunity cost and the project will make the individual better off. Another way of expressing the NPV > 0 condition is in the form of a *benefit/cost ratio*: $[BC/(1 + r)]/AC > 1$. Yet another way of assessing the project is to compare the project cost, AC, *compounded forward* at the rate of interest, with the project return, BC: if the future net value of the project, $BC - AC(1 + r) > 0$, the project is worthwhile undertaking. The project net benefit compounded forward (known as the terminal value) is seldom used, though it can have a role in project selection under a budget constraint.

Instead of making a direct comparison of the cost, AC, and the present value of the return, DC, of the investment project, the project could have been evaluated indirectly by comparing the consumption streams with and without the project, represented by points B and A respectively in Figure 2.1. The present value of stream B is given by point E and the present value of A by point F. Since OE > OF, the individual is better off with the project than without it. This gives rise to a general rule (which applies when the lending and borrowing rates of interest are the same[1]): the individual can maximise economic welfare by maximizing the present value of consumption; whatever the individual's preference for present relative to future consumption is, she should choose to produce at point B rather than at A because she can trade along the line BE (by lending or borrowing in the initial period) to reach a point superior to A, such as G, for example, where she can consume more in both periods than the income levels Y_0 and Y_1 permit. This rule is very important in investment analysis: it means that the analyst does not need to know anything about the individual's preferences for present versus future consumption in order to recommend that a particular investment project be undertaken or not; if the net present value is positive, the individual can be made better off by undertaking the project.

2.3 Investment opportunities in the economy as a whole

We illustrated some important concepts used in investment appraisal by means of a small project being considered by an individual – the discount rate, discount factor, net present value, internal rate of return, benefit/cost ratio and so on – but how well does the analysis translate to the economy as a whole? From the individual's point of view the rate of interest offered by the bank is the opportunity cost of the funds to be invested – if she decided not to refurbish the house, she could earn interest on her savings in the bank. In the wider economy the market rate of interest also measures the opportunity cost of funds since some investor is willing to borrow at that rate (otherwise it would not be the market rate of interest); this means that if the NPV of the proposed project being appraised by the investor is positive using the market rate of interest as the discount rate, the present value of the benefits of the investor's project is greater than its opportunity cost.

We can assume that among all the investors' projects which are competing for funds there is one project which is marginal in the sense that its NPV is zero. The present value of the benefits of this project is the present value of using the required set of scarce resources in the particular way proposed by this investor. If the scarce resources are used in an alternative way – such as investing in refurbishing the basement of an individual's house – the opportunity cost of that alternative project is given by the present value of the benefits of the marginal project. In other words, discounting the alternative project by the market rate of interest and following the NPV > 0 rule ensures that the benefits of the alternative project are at least as high as those of the marginal project it displaces.

We suggested that, whatever the levels of income the individual earned in the two years illustrated in Figure 2.1, she could lend or borrow at the market rate of interest to obtain the desired consumption profile: this was described as trading along the net present value line passing through the point representing the income profile – point B if she undertakes the project. At one extreme she could consume the entire NPV of her income stream this year (given by OE in Figure 2.1 if she undertakes the refurbishing investment), and at the other she could consume the first year's income plus interest plus the second year's income next year (given by OH).

Suppose the economy as a whole was considering an option equivalent to the latter: this would involve undertaking all the investment projects that could be funded by the scarce resources available this year. As many of us know to our cost, there are good investments and bad investments: projects with high IRRs and projects with low IRRs. In choosing projects

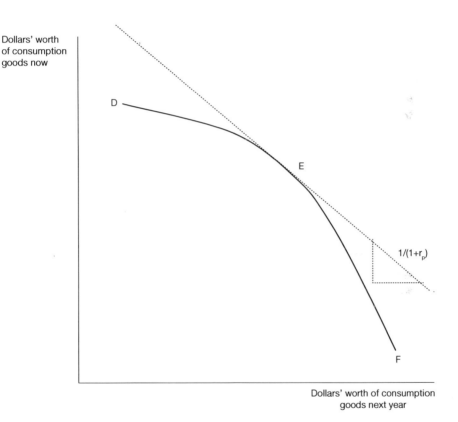

Figure 2.2 A country's inter-temporal production possibilities curve. The inter-temporal production possibilities curve starts at the consumption combination which would occur if virtually no investment were undertaken in the current year, as represented by point D. The slope of the curve DF is the ratio of the cost to the benefit of the marginal project, which in the light of our earlier discussion (consider the ratio AC/BC in Figure 2.1) can be identified as a discount factor based on the project's IRR. As we move along the curve in the direction of increased output of goods next year, say to point E, the internal rate of return on the next project in line falls (the absolute value of the slope of the curve, which is $1/(1 + r_p)$, rises and hence r_p must fall) – this is termed diminishing marginal productivity of capital. As more and more investment projects are undertaken, production of consumption goods falls in the present and rises in the future until a point such as F is reached where present consumption is at subsistence level.

to be undertaken this year it would pay to undertake the better projects first, but as more and more projects were to be undertaken, lower and lower quality projects would have to be accepted. This means that the economy as a whole is unable to trade along an NPV line; rather it can undertake *inter-temporal* trades of the consumption goods it produces along a *curve* such as that illustrated in Figure 2.2 which shows an *inter-temporal* production possibilities curve (IPPC) – a curve showing all possible combinations of values of consumption goods which can be produced in the present and the future (assuming a two-year world).

In reality, the economy will not choose to be at either of the two extremes discussed in Figure 2.2 but somewhere in the middle, such as point E, where the slope of the curve represents a discount factor based on the IRR of the marginal project – the project for which NPV = 0. However, we know that when NPV = 0 the IRR equals the discount rate. Hence the slope of the IPPC at the chosen point is equal to the discount factor based on the market rate of interest. This is the interest rate which should be used in evaluating all potential

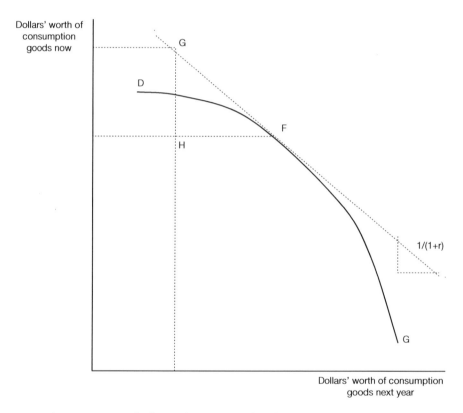

Dollars' worth of consumption goods now

Dollars' worth of consumption goods next year

$1/(1+r)$

Figure 2.3 The inter-temporal effects of international trade. A country produces at point F but consumes at point G; it achieves this by borrowing GH in the present (a capital account surplus for the borrowing country) and spending the funds on GH worth of imported goods and services (a current account deficit). In this two-period analysis the funds have to be repaid with interest the following year, consisting of an amount FH (a capital account deficit) which foreigners spend on exports from the debt repaying country valued at FH (a current account surplus). The ratio GH/FH represents a discount factor based on the interest rate which is established in international financial markets. Of course, in a multi-period world, it is possible to maintain a current account deficit for a period of time.

projects – those with a positive NPV lie to the left of E and those with a negative NPV lie to the right. In summary, the role of the interest rate in investment appraisal is to ensure that no project is undertaken if its benefits are less than its opportunity cost, where opportunity cost is the value of an alternative use of the resources involved.

The above discussion neglects the possibility of international trade: countries can consume combinations of present and future goods which lie outside of the IPPC as illustrated in Figure 2.3.

2.4 The algebra of NPV and IRR calculations

So far the analysis has been in terms of two time periods; we now extend it to three. It will be assumed that there is a project capital cost in the current year (Year 0) and a net benefit in each of Years 1 and 2; a net benefit means a benefit less any project operating costs incurred in the year in question. As noted earlier, the assumption about the accrual date of benefits and costs is important.

Project benefits and costs generally accrue more or less continuously throughout the year. It would be possible to discount benefits and costs on a monthly, weekly or even daily basis using the appropriate rate of interest. However, in practice, such level of detail is unnecessary. Instead it is generally assumed that all costs or benefits experienced during the year occur on some arbitrarily chosen date: it could be the first day of the year, the middle day or the last day. This assumption converts the more or less continuous stream of benefits and costs to a set of discrete observations at one-year intervals. The annual rate of interest is then used to discount this discrete stream back to a present value.

There is no general rule as to which day of the year should be chosen, or whether the calendar or financial year is the relevant period. However, we suggest that costs and benefits should be attributed to the last day of the calendar or financial year in which they occur. For example, following this convention, the net cash flow of a project with a capital cost of $100 during the current calendar year, and net benefits of $60 in each of two subsequent calendar years would be treated as a cost of $100 on 31 December this year, and net benefits of $60 on 31 December in each of the two subsequent years. It should be noted that while this procedure tends to introduce a slight downward bias to a positive NPV calculation (assuming the firm has access to short-term capital markets), any other assumption about accrual dates also introduces bias.

In denoting the project years we generally call the current year "Year 0" and subsequent years "Year 1", "Year 2", etc. Since we do not discount benefits or costs occurring in the current year, we are effectively choosing 31 December of the current year as the reference date – "the present" in the net present value calculation. Benefits or costs occurring in Year 1 (by assumption on 31 December in Year 1) are discounted back one period, those occurring in Year 2 are discounted back two periods, and so on. Thus the number used to denote the year tells us how the appropriate discount factor is to be calculated. This discussion might seem a bit laboured and trivial but consistency in the choice of accrual date is an important practical issue.

Using our assumption about the accrual date, our illustrative project has a capital cost now (K at time zero), and a net benefit one year from now (at the end of the first year), and two years from now (at the end of the second year). The resulting sequence – K, B_1, B_2 – can be termed a net benefit stream, as illustrated in Figure 2.4. To calculate the project NPV, we need to bring B_1 and B_2 to present values, sum these, and then subtract from the sum the project cost, K, which is already a present value (since it is incurred in Year 0).

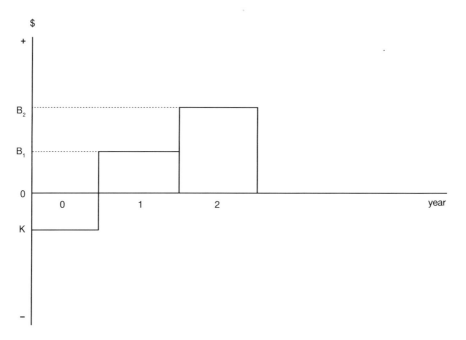

Figure 2.4 Net benefit stream of an investment project. Net benefit is negative in Year 0, representing the cost of the investment, K, and then positive in Years 1 and 2, representing the revenues received less the operating costs incurred.

The net benefit B_2 can be brought to a present value by using the interest rate in Year 2 to bring it to a discounted value at the end of Year 1, and then using the interest rate in Year 1 to bring that discounted value to a present value. B_1 can be discounted to a present value using the interest rate in Year 1:

$$NPV = -K + \frac{B_1}{(1+r_1)} + \frac{B_2}{(1+r_1)(1+r_2)}$$

In practice, it is usually assumed that the rate of interest is constant over time, so that $r1 = r_2 = r$ in the example, though some analysts have argued for the use of time-declining discount rates ($r_2 < r1$) for the net benefits of very long-lived projects – projects which have effects beyond, say, 50 years. This approach is consistent with observations of the behaviour of individuals when confronting choices with effects over a long time interval; it takes account of uncertainty about the level of interest rates far into the future; and it gives relatively more weight to the net benefits of future generations. With a time-invariant rate of discount, however, the NPV formula reduces to:

$$NPV = -K + \frac{B_1}{(1+r)} + \frac{B_2}{(1+r)^2}$$

To take a simple example, suppose the project costs $1.6 and yields net benefits of $10 after 1 and 2 years respectively, and that the rate of interest is 10%:

NPV = −1.6 + 10 * 0.909 + 10 * 0.826 = 15.75

The values of 1/1.1 and $1/(1.1^2)$ can be obtained from a Table of Discount Factors (see Appendix 2), or by using a calculator. The benefit/cost ratio is given by 17.35/1.6 = 10.84 and the net benefit/cost ratio by 15.75 /1.6 = 9.84. (The mechanics of deriving discount factors and using discount tables are discussed in more detail in Chapter 3.) We can see from the NPV formula that the NPV falls as the discount factors fall, and hence as the discount rate rises; this relationship is illustrated in Figure 2.5.

It was explained earlier that the internal rate of return, r_p, is the discount rate which reduces the project's NPV to zero, as denoted by IRR in Figure 2.5. It can be calculated by solving the following equation for r_p:

$$-K + \frac{B_1}{(1+r_p)} + \frac{B_2}{(1+r_p)^2} = 0$$

Multiplying both sides of this equation by $(1 + r_p)^2$ gives an equation of the following form:

$$-Kx^2 + B_1x + B_2 = 0$$

where $x = (1 + r_p)$. This is the familiar quadratic form, as illustrated in Figure 2.6, which can be solved for x and then for r_p.

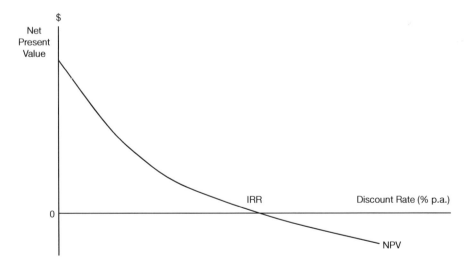

Figure 2.5 Net present value in relation to the discount rate. The present value of a net benefit stream (NPV) falls as the discount rate is increased. The discount rate which makes the NPV zero is termed the internal rate of return (IRR).

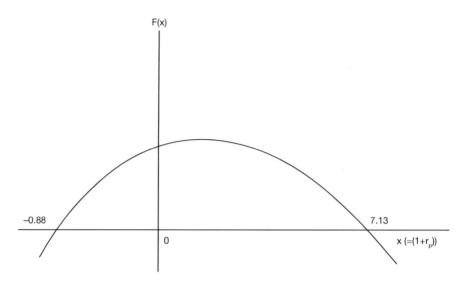

Figure 2.6 Calculating internal rates of return: one positive value. The equation: $F(x) = -Kx^2 + B_1x + B_2 = 0$ has two solutions for x, where x represents $(1 + r_p)$. The negative solution can be ignored. The positive solution indicates that $1 + r_p - 7.13$. In other words, the internal rate of return, r_p, is 613%.

More generally, it can be noted that if we extended the analysis to include a benefit at time 3, we would need to solve a cubic equation, and if we further extended it to time 4, it would be a quartic equation, and so on; fortunately we will be solving only a couple of IRR problems by hand – the rest will be done by computer using a spreadsheet function.

To return to our example, we need to solve:

$$F(x) = -1.6x^2 + 10x + 10 = 0$$

which is a quadratic equation of the general form:

$$F(x) = ax^2 + bx + c = 0$$

where a = −1.6, b = 10, and c = 10, and which has two solutions given by the formula:

$$x = -b \pm \sqrt{\frac{(b^2 - 4ac)}{2a}}$$

A bit of arithmetic gives us the solution values 7.13 and −0.88, which is where the function crosses the x-axis as illustrated in Figure 2.6. This means that r_p is either 6.13 (613%) or − 1.88 (−188%). The negative solution has no meaning in the context of project appraisal and is discarded. Hence the IRR is 613%, which is very healthy compared with an interest rate of 10%.

Let us now change the example by making $B_2 = -10$. For example, the project could be a mine which involves an initial establishment cost and a land rehabilitation cost at the end of the project's life. When we calculate the IRR using the above method, we find that there are two solutions as before, but that they are both positive: $r_p = 0.25$ (25%) or 4 (400%). Which is the correct value for use in the project appraisal process? The answer is possibly neither. As discussed above, when we use the IRR in project appraisal, we accept projects whose IRR exceeds the rate of interest. In the current example the IRR clearly exceeds an interest rate in the 0–24% range, and it is clearly less than an interest rate in excess of 400%. But what about interest rates in the 25–400% range? We cannot apply the rule in this case, and it is for this reason that it is not recommended that the IRR be used in project appraisal when there is more than one positive IRR.

What would have happened if we had asked the computer to calculate the IRR in this case? The computer's approach is illustrated in Figure 2.7.

Referring to Figure 2.7, if you had given the computer x_a, for example, as the initial guess, it would have converged on the solution value x = 1.25 ($r_p = 0.25$). How would you know if there were two positive solutions? Some programs will report no solution or an error if there is more than one positive IRR. However, a simple way of checking

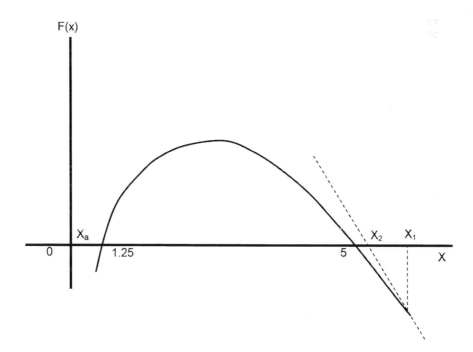

Figure 2.7 Calculating internal rates of return: two positive values. F(x) is the value of the quadratic function, and setting F(x) = 0 gives the solution values – these occur where the function crosses the x-axis. In the previous example this occurred once in the negative range of x and once in the positive range, and the negative value was discarded. However, in the present case there are two positive values of x for which F(x) = 0, 1.25 and 5. The computer solves the quadratic equation by asking for an initial guess, say x = x_1. It then calculates the first derivative of the function at that value of x (the equation of the slope of the function at that point) and works out the value of x, x_2, that would set the first derivative equal to zero. It then uses x_2 as its guess and repeats the process until it converges on the solution value x = 5 (IRR = 4).

for the problem in advance is to count the number of sign changes in the net benefit stream. In the initial example we had a well-behaved cash flow with signs "– + +" i.e. *one* change of sign, hence one positive IRR. In the second example we had "– + –" i.e. *two* changes of sign and two positive IRRs. This is a general rule which applies to higher order equations – the number of positive solutions can be the same as the number of sign changes.[2] If you are analysing a proposed project which has a net benefit stream (including costs as negative net benefits) with more than one change of sign, it is better not to attempt to calculate the IRR. The NPV rule will always give the correct result irrespective of the sequence of signs.

As illustrated in Figure 2.5, the IRR is the discount rate which reduces the project NPV to zero. The case in which there are two positive IRRs is illustrated in Figure 2.8, where the project NPV is positive at first, and falling as the discount rate rises, and then, after reaching some negative minimum value, it starts to rise and eventually becomes positive again. There are two points at which NPV = 0 and these correspond to the two positive IRRs. The discount rate at which NPV is a minimum, 100% in the example, can be solved for from the NPV equation.

Finally, we consider a case in which no IRR can be calculated. Suppose that in the original example we change the initial cost of $1.6 to a benefit of $1.6. The net benefit stream now consists of three positive values and there is no value of the discount rate which can reduce the NPV to zero and hence no IRR. This case is illustrated in Figure 2.9 where the NPV continually falls as the discount rate increases, but never actually becomes zero.

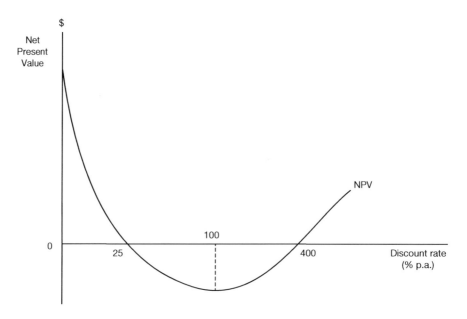

Figure 2.8 Net present value in relation to the discount rate: the two positive internal rates of return case. Net Present Value falls as the discount rate increases, reaches a minimum at a discount rate of 100%, and then starts to rise. There are two internal rates of return: 25% and 400%.

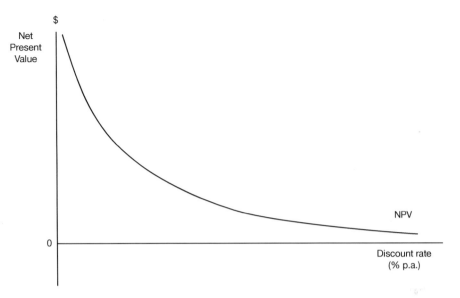

Figure 2.9 Net present value in relation to the discount rate: the no internal rate of return case. Net Present Value falls as the discount rate is increased but the NPV function is asymptotic to the horizontal axis – NPV never actually reaches zero – so there is no Internal Rate of Return.

2.5 Annuities and perpetuities

To this point we have been considering net benefit streams consisting of three values: an initial cost (a negative net benefit) and then net benefits after one and two years respectively. Clearly it would be very tedious to do higher-order examples by hand and we leave these until Chapter 3 where we introduce the use of spreadsheets. However, there are some additional steps we can readily take without needing the help of a computer. Suppose a project has a cost in Year zero and then the same level of benefit at the end of each and every subsequent year for a given time interval: the benefit stream $B_1, B_2, B_3 \ldots B_n$, where $B_i = B$ for all i, is termed an *annuity*. The present value of the annuity (on 31 December of Year 0) is given by:

$$PV(A) = \frac{B}{(1+r)} + \frac{B}{(1+r)^2} + \frac{B}{(1+r)^3} + \cdots + \frac{B}{(1+r)^n}$$

where B is the equal annual net benefit which occurs at the end of each year (the accrual date in this case). This expression is a geometric progression – each term is formed by multiplying the previous term by a constant value: $1/(1 + r)$ in this case. It is relatively easy to work out a formula for the value of the sum, S. Having done this, we can write the present value expression as:

$$PV(A) = -\frac{B\left[(1+r)^n - 1\right]}{\left[(1+r)^n \cdot r\right]}$$

which can readily be computed using a calculator. The expression $[(1 + r)^n - 1]/[(1 + r)^n \cdot r]$ is known as an *annuity factor* and it can be obtained from a set of Annuity Tables, such as those provided in Appendix 2. We sometimes refer to this expression as $AF_{r,n}$, where r denotes the interest rate and n denotes the length of the annuity. (The use of Annuity Tables is discussed in more detail in Chapter 3.)

In the previous example the accrual date was set at the end of each year in accordance with the normal practice. An *annuity due* is a similar stream of payments, but with the accrual date set at the beginning of each year. Its present value is given by:

$$PV(D) = B + \frac{B}{(1+r)} + \frac{B}{(1+r)^2} + \frac{B}{(1+r)^3} + L + \frac{B}{(1+r)^{n-1}}$$

It can be seen from the above expression that $PV(D) = PV(A) + B - B/(1 + r)^n$. We can use this relation to modify the formula for the present value of an annuity to make it apply to an annuity due, or we can take the annuity factor from the Annuity Tables and add B and subtract $B/(1 + r)^n$, obtained from Discount Tables. Finally, as shown in Technical note 2.1, we can use the formula for the present value of an annuity to work out the present value of a *perpetuity* – an annuity that goes on for ever. If we take the limit of $PV(A)$ as n goes to infinity, we get the following simple expression for the present value of a perpetuity:

$$PV(P) = \frac{B}{r}$$

Where a project net benefit stream takes the form of an annuity we can use the Annuity Tables to calculate the project IRR. We know that $NPV = -K + B.AF_{r,n}$, where K is project cost, B is the annual benefit and $AF_{r,n}$ is the annuity factor for interest rate r and life n. To calculate the IRR, we set $NPV = 0$, giving $AF_{r,n} = K/B$. In other words, we divide the cost of the project by the annual benefit to obtain a value for the annuity factor, and then look up the row in the Annuity Tables which corresponds to the life, n, of the project under consideration to see which discount rate, r, gives an annuity factor closest to the one we obtained. That discount rate is the project IRR.

At this point it might be asked why the various formulae discussed above are worth knowing about in the computer age. There are two main advantages: first, it is helpful to know what the computer is doing so that you can recognise an absurd result if it appears because of some programming error you may have made, such as asking for an IRR when there will be at least two positive values which may be difficult to interpret in the project appraisal. Second, you may be parted from your computer – it may have gone to New Caledonia while you went to Papua New Guinea (take it as hand luggage next time), or you may both be in New Caledonia but you forgot to take the adaptor plug and the batteries are low, or you were called unexpectedly into a meeting where you are supposed to provide a "back–of-the-envelope assessment" of the project, but you left your computer at the hotel! In all these cases you will find that you can make quite a lot of progress using a pocket calculator and some elementary understanding of discounting annuities (if you forgot your calculator, stick to perpetuities!).

TECHNICAL NOTE 2.1 The present value of an annuity

An annuity consisting of $1 to be paid one year from now and in every subsequent year until year n has a present value:

$$S = \frac{1}{(1+r)} + \frac{1}{(1+r)^2} + \frac{1}{(1+r)^3} + \frac{1}{(1+r)^4} + \cdots + \frac{1}{(1+r)^n}$$

If we multiply both sides of this equation by $(1 + r)$ we get:

$$S(1+r) = 1 + \frac{1}{(1+r)} + \frac{1}{(1+r)^2} + \frac{1}{(1+r)^3} + \frac{1}{(1+r)^4} + L + \frac{1}{(1+r)^{n-1}}$$

Subtracting the former equation from the latter gives:

$$Sr = 1 - \frac{1}{(1+r)^n}$$

which can be solved for S:

$$S = \frac{\left[(1+r)^n - 1\right]}{\left[r(1+r)^n\right]},$$

which is the formula used to derive the Annuity Factors reported in Appendix 2. This formula can also be expressed as:

$$S = \frac{1 - (1+r)^{-n}}{r}$$

The present value of a perpetuity – an annuity which continues forever – is obtained by taking the limit of S as n tends to infinity:

$$S(n \to \infty) = \frac{1}{r}$$

2.6 The Rule of 72

To estimate the number of years required to double the value of an initial investment at a given annual percentage rate of interest (r), maintained over time, simply divide r into 72.

For example, if r = 6%, divide 6 into 72 and get 12 years which is (approximately) how long it would take for the value of the initial investment to double. Two applications of the rule tell us that at 6% the value of the initial investment would quadruple in 24 years. Similarly, if a given investment doubles its value in, say, 12 years, simply divide 12 into 72 to get an estimate of the annual compound percentage rate of return.

The Rule of 72 can also be used to estimate the number of periods (years) it would take for the purchasing power of money to halve. Simply divide 72 by the expected inflation rate to estimate the number of years. For example, if a country's inflation rate is 3% per annum, it will take approximately 24 years for the purchasing power of money to halve. Similarly the rule tells us that if the discount rate is 3%, a dollar to be received 24 years from now has a present value of 50 cents.

Why it works

The Rule of 72 applies to interest compounded annually. If P is the initial investment and r is the rate of return, we can solve for the value of T (number of years) which will double P. Solve:

$$2P = P(1 + r)^T$$

to get:

$$T = \frac{\ln(2)}{\ln(1+r)} = \frac{K}{r}$$

where $K = [\ln(2) / \ln(1 + r)]r$. For example, if r = 0.075, K = 0.7188 and the doubling time is 9.58 years. While this calculation is exact for the values chosen here, a reasonable approximation for T can be obtained for rates of return in the commonly experienced range by setting K = 72 (a convenient number because of its many factors) and dividing by the rate of return expressed as a percentage.

2.7 Economic depreciation and the annual cost of capital

An illustration of the use of the concept of an annuity in investment analysis is the calculation of annual "economic depreciation" – not to be confused with accounting depreciation used in financial statements, as will be discussed in Chapter 4. Sometimes we want to express the capital cost of a project as an annualised cost over the life, n years, of the project, rather than as an initial lump-sum cost, K. The annual capital cost consists of interest and depreciation, and as always in cost-benefit analysis, the cost is an opportunity cost. For simplicity, we can think of the bank as the investment which is the alternative to the project being appraised. If K dollars are deposited in the bank, the depositor will receive annual interest of rK at the end of each of n years, and at the end of the nth year will be repaid the principal, K. The amount rK is the annual interest cost of the capital investment. While there are various ways of computing annual economic depreciation, we need to distinguish between those methods which are mandated by the Taxation Office, and which may bear no close relation to opportunity cost, and economic depreciation. Expressed as a constant amount per year, economic depreciation is the sum of money, D, which needs to be deposited in the bank each year over the life of the project in order to recoup the initial capital investment at the end of *n* years.

The annual opportunity cost of the capital sum, K, is the sum of interest plus economic depreciation, and the present value of the interest plus depreciation equals the capital sum invested in the project:

$$A_{r,n}\left\{rK+D\right\}=K$$

where Ar,n is the annuity factor which brings an equal annual flow of payments over *n* years to a present value using the interest rate r. By rearranging this equation, we can solve for the annual depreciation, D:

$$D=K\left\{1/A_{r,n}-r\right\}$$

Using the formula developed in Technical note 2.1 for the annuity factor, D can be expressed as:

$$D=\frac{rK}{\left\{(1+r)^{n}-1\right\}}$$

If we want to express the capital cost of a project as an annual cost over the project life, rather than as a capital sum at the start of the project, we can do this by summing the annual interest cost, rK, and the annual depreciation cost, D, calculated according to the above formula.

It should be emphasised that the cost of capital should not be accounted for twice in a net present value calculation. When we include the initial capital cost of the project in the net benefit stream and use a discount rate reflecting the opportunity cost of capital, we are taking full account of the cost of capital. If, in addition to this, we were to deduct interest and economic depreciation cost from the net benefit stream, we would be double-counting the cost of capital. This can readily be seen by recognizing that if interest and depreciation costs have been correctly measured, their present value equals the initial capital cost. Thus to include them in the net present value calculation, as well as in the initial capital cost, is to deduct the capital cost twice from the present value of the net benefits of the project.

2.8 Treatment of inflation in project appraisal

We now turn to the problem of calculating present values when the values of future project benefits and costs are subject to *inflation*. Inflation is a process which results in the nominal prices of goods and services rising over time. The existence of inflation raises the question of whether project inputs and outputs should be measured at the prices in force at the time of the appraisal – today's prices, termed *constant prices* – or at the prices in force in the future when the project input or output occurs – termed *current* or *nominal prices*.

It was argued earlier that the purpose of an NPV calculation is to compare the performance of a proposed project with the alternative use of the scarce resources involved. The
alternative use is taken to be a project yielding an IRR equal to the market rate of interest,
in other words, a project with an NPV = 0 at the market rate of interest. The market rate
of interest is generally taken to be the rate of return on a riskless asset such as a government
bond. If you look up the government bond rate in the financial press, it will be quoted at
some *money (or nominal) rate of interest*, m, which includes two components: the *real rate of
interest*, r, and the *anticipated rate of inflation*, i. The relationship can be expressed approximately as:

$$m = r + i$$

In other words, the anticipated rate of inflation is built into the money rate of interest. For
example, if the rate of interest on government bonds is reported as 5% (the financial press
always quotes the rate of interest on money), this will consist of perhaps 3% real rate of
interest and 2% anticipated inflation. The anticipated inflation rate cannot be observed,
though it can be inferred from an analysis of a time series of money interest rates and rates
of inflation. Often it is assumed that the anticipated inflation rate is the present rate, and
that the real rate of interest is the money rate less the present rate of inflation.

When you calculate the present value of a commodity, such as a ton of coal at time t, you
calculate the value of the ton of coal at time t, B_t, and then calculate the present value of
that value, $B_t /(1 + x)^t$, where x represents the appropriate rate of discount. A simple rule,
which it is vitally important to remember, is that *if you include inflation in the numerator of
the present value calculation, you must also include it in the denominator*. What does this mean?
There are two ways of calculating the value of a ton of coal at time t: we could use a constant price, which does not include the inflation which will occur between now and time t,
or we could use the current price at time t, incorporating inflation. Similarly, there are two
ways of discounting the value of the coal back to a present value: we could use the real rate
of interest – the money rate less the expected rate of inflation – or we could use the money
rate. If inflation is included in the value of coal at time *t*, the money rate of interest (which
incorporates inflation) must be used as the discount rate; if inflation is not included, the real
rate of interest must be used.

Which is the better approach? It can be shown that when the prices of all commodities
inflate at the same rate, the two procedures give more or less the same result. At today's
price, P_0, the value of a quantity of coal at time t, Q_t, is given by $P0Q_t$. The present value
of the coal, using the real rate of interest, r, is given by $P_0Q_t/(1 + r)^t$. At the inflated price
of coal its value at time t is given by $P0(1 + i)^tQt$ where i is the annual inflation rate. The
present value, at the money rate of interest, m, is given by $P_0Q_t(1 + i)^t/(1 + m)^t$, or $P_0Q_t[(1
+ i)/(1 + m)]^t$. Since $(1 + i)/(1 + m)$ is approximately equal to $1/(1 + m - i)$, where $m - i = r$
is the real rate of interest, the two procedures normally give essentially the same result. In
the case of a high rate of inflation the approximation is not so close and it is better to obtain
a more accurate measure of the real rate of interest, as demonstrated by Technical note 2.2.

Are there circumstances in which one approach to accounting for inflation is preferred
to the other? Using constant prices and a real rate of discount is obviously computationally
simpler. However, there may be cases in which all commodities are *not* expected to rise in
price at the same rate over time. For example, the income elasticity of demand for outdoor
recreation services is thought to exceed unity; this means that as a consumer's income rises,
she spends an increasing proportion of it on outdoor recreation. In consequence, rising per

capita incomes will result in a significant increase in demand for these services, relative to that for goods and services with low income elasticity of demand, and hence, given supply limitations, we can expect a rise in the price of outdoor recreation services relative to the prices of other goods and services. Suppose you are appraising an outdoor recreation project and want to take account of the effect of the rising relative unit value of outdoor recreation services. If the unit value is expected to rise at a rate i_c, the present value of a unit of recreation services at time t is given by: $P0[(1 + i_c)/(1 + m)]^t$, where P_0 is the unit value at time zero (expressed as a shadow-price, as discussed in Chapter 8), or equivalently by: $P0/(1 + m - i_c)^t$, where $m - i_c$ is a real rate of discount specific to recreation services; this present value exceeds $P_0/(1 + r)^t$ as long as $i_c > i$. In this example, in which the rate of price increase of the output (or of a significant input) is expected to be different from the general rate of price inflation, omitting inflation completely from the appraisal, in the interests of computational simplicity, will give an incorrect estimate of present value. Such cases are not unusual: as noted above, any good which has a high income elasticity of demand and faces significant supply limitations is likely to experience a relative price increase as per capita income rises; and inputs, such as coal, which have deleterious environmental effects can expect a fall in their real price in the long run.

TECHNICAL NOTE 2.2 Real and nominal rates of interest

The discount factor using the real rate of interest, r, is $1/(1+r)^t$. Using the nominal rate of interest, m, the discount factor is: $(1+i)^t/(1+m)^t$, where i is the expected rate of inflation. Since the two discount factors must generate the same present value, it must be the case that:

$$\frac{1}{(1+r)} = \frac{(1+i)}{(1+m)}.$$

Solving for r yields:

$$r = \frac{(m-i)}{(1+i)}.$$

For small values of i, the real rate of interest can be approximated by the nominal rate less the expected rate of inflation.

2.9 Incorporating a risk factor in the discount rate

It is quite common for private investment projects to be evaluated using a discount rate which includes a risk factor: if the money rate of interest is 5%, the nominal discount rate may be set at 8%, allowing 3% for project risk. We favour a more formal approach to risk analysis, which is described in Chapter 9, but here we examine the basis of this kind of adjustment for risk as the analysis is similar to the discussion of inflation which we have just completed.

Suppose that a project will yield a net benefit of B_t in Year t, provided that some catastrophe has not occurred in the intervening years, for example, a dam will produce electricity if it hasn't burst prior to Year t, or a forest will produce timber if it hasn't been consumed by fire, or an oil well will generate revenues for its private sector owners if it hasn't been nationalised. Suppose that the probability of catastrophe is p per year, for example, a 2% chance (p = 0.02) each year that the catastrophe will occur during that year. The expected value of the project net benefit in Year t is given by:

$$E(B_t) = B_t(1-p)^t$$

where $(1-p)^t$ is the probability of the catastrophe *not* occurring during the time interval to t. The present value of the expected net benefit is given by:

$$PV = \frac{B_t(1-p)^t}{(1+r)^t} = B_t\left(\frac{(1-p)}{(1+r)}\right)^t$$

Since $(1-p)/(1+r)$ is approximately equal to $1/(1+r+p)$ (for small values of p), the expected present value of the net benefit in year t can be obtained by discounting by the interest rate plus the risk factor. There are two reasons why we do not advocate this approach to costing risk: first, the "risk" involved has a very special time profile which projects in general may not exhibit; and, second, under the normal economic interpretation of "risk" as variance around an expected outcome (as discussed in Chapter 9), the procedure does not deal with risk at all, but rather only with calculating the expected value of the outcome.

2.10 Further reading

A classic work on the economic theory underlying investment analysis is J. Hirschliefer, *Investment, Interest and Capital* (Englewood Cliffs, NJ: Prentice Hall, 1970). A useful book which deals with practical problems involving NPV and IRR calculations, and issues such as depreciation is L.E. Bussey, *The Economic Analysis of Industrial Projects* (Englewood Cliffs, NJ: Prentice Hall, 1978).

Exercises

1   A firm is considering a project which involves investing $100 now for a return of $112 a year from now. What are the values of the following variables?

    i   the marginal productivity of the capital investment;
    ii   the internal rate of return on the project;
    iii the net present value of the project, using a 5% discount rate;
    iv the project benefit/cost ratio.

2   A project requires an initial investment of $100,000 and an annual operating cost of $10,000. It will generate annual revenue of $30,000. If the life of the project is 10 years and the discount rate is 6%, decide whether to accept or reject this project using: (i) NPV criterion, and (ii) IRR criterion.

3   Given the following information, use two different methods to calculate the present
value of a ton of coal to be received one year from now:

money rate of interest (m):	8% *per annum*
expected rate of general price inflation (i):	6% per annum
expected rate of increase in the price of coal (i_c):	2% per annum
current price of coal (P_0):	$25 per ton

Notes

1   Individuals often face lower interest rates if they lend than the rates they pay if they borrow. In this
case, the NPV rule still holds if a lender remains a lender after the project is undertaken (use the
lending rate as the interest rate), or a borrower remains a borrower (use the borrowing rate), but
it no longer applies if a lender becomes a borrower, or a borrower becomes a lender, as a result of
undertaking the project.

2   In fact, more than one change in sign is a necessary but not sufficient condition for the existence
of more than one positive IRR. A sufficient condition is that the unrecovered investment balance
becomes negative prior to the end of the project's life. The reader is referred to advanced texts for
a discussion of this concept.

3 Project appraisal

Decision-rules

3.1 Introduction

In this chapter the discussion shifts to applications of the various investment decision-rules described in Chapter 2. Some of the applications rely on the simple algebraic concepts already discussed in Chapter 2, while some are developed using the basic tool of the cost-benefit analyst – the spreadsheet. Some issues already discussed in Chapter 2, such as the time value of money and the calculation of present value and internal rate of return, are explored further, while new concepts, such as comparison of projects under capital rationing are introduced and discussed. Finally, the use of spreadsheets in project appraisal is discussed in detail.

3.2 Discounted cash flow analysis in practice

We now turn to evaluating multi-period investment projects – projects that have a net benefit stream occurring over many years. The remainder of this chapter aims at familiarizing the reader with the practical application of discounted cash flow (DCF) decision-making techniques. By the end of the chapter the reader should know how decisions are made: to accept or reject a particular project; to select a project from among alternatives; and to rank a number of projects in order of priority. Slightly different DCF decision-rules apply in each of these cases. While some examples are given to illustrate the use of these techniques, the exercises at the end of this section are also an aid to understanding.

The widespread availability and relatively low cost of personal computers have transformed the task of the project analyst. Spreadsheet programs, such as Microsoft Excel©, have greatly facilitated the previously laborious, computational side of cost-benefit analysis. Repetitive, mechanical calculations can now be performed at will, which has the enormous advantage of allowing more project options and alternative scenarios to be considered than ever before. However, it should not be assumed that the spreadsheet program can assist in the design or setting-up of the framework for the cost-benefit analysis. This requires both skill and art on the part of the analyst. In this chapter we want to familiarise readers with the necessary techniques, framework, and computer skills.

As noted earlier, cost-benefit analysis (CBA) is a particular method of appraising and evaluating investment projects from a public interest perspective. In later parts of this book, in particular in Chapters 4 and 5, we will examine the main differences between private and social cost-benefit analysis. At this stage these differences are not important because essentially the same principles and techniques of discounted cash flow (DCF) analysis apply to both private and public sector investment analysis.

DOI: 10.4324/9781003312758-3

Common to all DCF analysis is the conceptualization of an investment project as a net benefit stream expressed as a "cash flow". Economists define an investment as a decision to commit resources now in the expectation of realizing a flow of net benefits over a period in the future. The flow of net benefits, measured in terms of a net cash flow, is represented graphically in Figure 2.4 in Chapter 2.

When resources (valued in terms of funds) are allocated as investment outlays, the "cash flow" is negative, indicating that there is a net outflow of funds. Once the project begins operations, and benefits (revenues) are forthcoming, the cash flow becomes positive (hopefully), indicating that there is a net inflow of funds. What is represented here is a *net* cash flow, measuring the annual benefits less costs, so we should not forget that throughout the project's life there are also outflows in the form of operating costs.

We should also note that although we use the term *cash flow*, the monetary values assigned to the costs and benefits in CBA might be different from the actual pecuniary costs and benefits of the project. This point is taken up later in Chapter 5 when we discuss *Efficiency Analysis* and the principles and methods of *shadow-pricing*.

The process of project appraisal and evaluation can be considered in terms of three aspects of cash-flow analysis:

i   *identification* of costs and benefits;
ii   *valuation* of costs and benefits;
iii *comparison* of costs and benefits.

In this chapter we consider item (iii), assuming that the relevant costs and benefits of the project we are appraising have been identified and valued. Chapter 4 deals with items (i) and (ii) from a private perspective, Chapter 5 considers them from an economic efficiency perspective, and Chapter 6 considers them from the perspective of the Referent Group.

3.3 Discounting and the time value of money

By this stage you should be familiar with the concept of the net cash flow of a project (NCF) and aware that, to compare benefits and costs accruing at different points in time, you cannot simply add up all project benefits and take away all project costs, unless of course you are assuming that the discount rate is zero. There is a need for *discounting* when comparing any flow of funds (costs and revenues or net benefits) over time. To explore the process of discounting in practice, consider two investment projects, A and B. The net cash flows of these projects are given as:

	Year			
	0	*1*	*2*	*3*
Project A	−100	+50	+40	+30
Project B	−100	+30	+45	+50

Remember the convention we use to denote the initial year of the project, "Year 0". If project years are also calendar years, then all costs and benefits accruing during that year are assumed to accrue on 31 December of that year. Therefore, any costs or benefits accruing in the course of the next year, "Year 1", are also assumed to accrue on 31 December of that year, i.e. one year from Year 0. Similarly, "Year 2" refers to two years from Year 0, and so on. Of course, there is no reason why the chosen time period need be a year. It could be a quarter, month, week or day – the same principles hold whatever the time period used, but then of course the (annual) discount rate would have to be adjusted accordingly.

Our task is to compare projects A and B. Which would you prefer? A and B both have initial capital costs of $100, but A's net benefits total $120 while B's net benefits total $125. Can we say that B is preferred to A because $125 – $100 is greater than $120 – $100? Obviously not, as this calculation ignores the *timing* of cash inflows and outflows. We need to discount all future values to derive their equivalent *present values*.

How do we accomplish this? From the previous chapter we saw that we need to derive the appropriate *discount factor*. What is the present value (PV) of $100 a year from now assuming a discount rate of 10% per annum?

$$PV = \$100 \times 1 / 1.1$$
$$= \$100(0.909)$$
$$= \$90.9$$

The value 0.909 in this example is the *discount factor*. It tells us the amount by which any value one year from now must be multiplied by to convert it to its present value, assuming a discount rate of 10% per annum.

What about the present values of benefits accruing in years beyond Year 1? These values need to be discounted a number of times, and we will perform these calculations, making the usual assumption that the discount rate is constant from year to year, as illustrated by Example 3.1, dealing with Project A.

EXAMPLE 3.1 Discounting a Net Cash Flow (NCF) stream to calculate Net Present Value

	Year			
	0	*1*	*2*	*3*
NCF	−100	+50	+40	+30

PV of $50 = $50(0.909) = 45.45

PV of $40 = $40(0.909)(0.909) = 33.05

PV of $30 = $30(0.909)(0.909)(0.909) = 22.53

Net Present Value = −100 + 45.45 + 33.05 + 22.53 = 1.03.

By converting all future values to their equivalent present values (discounting) we make them directly comparable. For example, what is the PV of $40, two years from now?

PV in year 1 = $40 × 0.909 = $36.36

PV in year 0 = $36.36 × 0.909 = $33.05.

We have already noted that the discount factors (DFs) do not have to be calculated from our formulae each time we want to perform a calculation, as they are usually available in Discount Tables (see Appendix 2), which give discount factors for different discount rates (interest rates) and years. Table 3.1 shows the DFs for 10% and 15% rates of discount (interest) to three decimal places.

The present value of $100 accruing in 10 years' time assuming a 10% discount rate, is found arithmetically by: $100 × 0.386, where 0.386 is the discount factor for year 10 at a 10% discount rate, and obtained from Table 3.1. This discounting technique can now be used to discount the whole net cash flow of a project to obtain its discounted net cash flow (NCF). In order to obtain this, one simply multiplies the net benefit (or cost) in each year by the respective DF, given a particular discount rate. To illustrate this, the NCF of a hypothetical project is discounted in Example 3.2.

Table 3.1 Some discount factors for 10% and 15% discount rates

	Year									
	1	2	3	4	5	6	7	8	9	10
10%	0.909	0.826	0.751	0.683	0.621	0.565	0.513	0.467	0.424	0.386
15%	0.870	0.756	0.658	0.573	0.497	0.432	0.0376	0.327	0.284	0.247

e.g.

EXAMPLE 3.2 Calculating discounted Net Cash Flow

	Year					
	0	1	2	3	4	5
1. NCF	−1000	200	300	400	500	600
2. DF$_{(10\%)}$	1.000	0.909	0.826	0.751	0.683	0.621
3. Discounted NCF (1×2)	−1000.0	181.8	247.9	300.5	341.5	372.5

To obtain row (3) we simply multiplied row (1) by row (2). Note that for year 0 the DF is always 1: this is so, irrespective of what the discount rate is, because $DF_0 = 1/(1 + r)^0$. We then derive the Net Present Value (NPV) by summing the discounted NCF:

$$NPV_{(10\%)} = -\$1000.0 + \$1444.2 = \$444.2.$$

3.4 Using Annuity Tables

We saw in Chapter 2 that when an investment project produces a cash flow with a regular or constant amount in each year it is possible to calculate the present value of this stream more easily using an *Annuity Table*. For example, consider the following cash flow and discount factors for 10%:

	Year				
	0	1	2	3	4
NCF	−100	35	35	35	35
DF	1.0	0.909	0.826	0.751	0.683

We could calculate the PV of net benefits by multiplying each year's cash flow by its respective DF and then summing these up, as illustrated in Example 3.2. Alternatively, because each year's cash flow is the same, we can simply add up the DFs and then multiply the *cumulative DF* by the constant amount: $(0.909 + 0.826 + 0.751 + 0.683) \times \$35 = 3.169 \times \$35 = \$110.92$. We then subtract the cost in year 0 to get: NPV = $10.92.

In fact, there is an even quicker method. Annuity Tables (see Appendix 2) provide us with the values of the cumulative discount factors for all discount rates and years. We simply look up Year 4 in the 10% column and read off the *annuity factor* (AF) of 3.170.

Note that the Annuity Tables can be used to calculate the PV of a stream of equal annual net benefits occurring over a subset of consecutive years during a project's life. For example, a project may have equal NCF for years, say, 5 to 10 of its life and different NCFs in other years. In this case Annuity Tables can be used for years 5 to 10, and Discount Tables for the other years. The AF for years 5 to 10 is found simply by subtraction of the AF for year 4 from the AF for year 10, i.e.

$$
\begin{aligned}
AF_{5\,to10} &= AF_{10} - AF_4 \\
&= 6.145 - 3.179 \\
&= 2.975 \left(at\,10\%\,rate\,of\,discount \right)
\end{aligned}
$$

e.g.

EXAMPLE 3.3 Using the Annuity Tables

Find the NPV of an investment with an initial cost of $1000 and an annual return of $500 for three years, when the discount rate is 4%.

$$
\begin{aligned}
NPV(A)_{(0,4\%)} &= -1000 + 500(0.962) + 500(0.925) + 500(0.889) \\
&= -1000 + 500(2.776)
\end{aligned}
$$

From the Annuity Table, Year 3 at 4% gives the annuity factor 2.775. The value 2.775 is the Annuity Factor (AF) for a 3-year annuity at a 4% discount rate, and it is obtained from Annuity Tables in the same way as obtaining the DF for a given year and discount rate. (Any slight difference is due to rounding error.)

In general, if you want to find the AF for years X to Y, then use the following rule (where Y > X):

$$AF_{xto} = AF - AF_{x1}$$

i.e. the appropriate AF between any pair of years is the AF for the larger numbered year minus the AF for the year prior to the smaller numbered year.

It is also possible to convert a given present value into an annuity, i.e. an *annual equivalent amount* for a given number of years. For example, if we have a present sum of, say, $3000 and wish to know what constant annual amount for 6 years at 10% discount rate would have the same present value, we simply divide $3000 by the AF. In this case, $3000/4.354 = $689.02. This calculation comes in handy when you want to annualise any given fixed amount, such as working out the annual (or monthly) payments necessary on a loan, or, for calculating economic depreciation as discussed in Chapter 2, Section 2.7.

It is also sometimes necessary to know what the present value of a perpetual annual sum is. The Annuity Tables do not show what the annuity factor is for a payment over an infinite number of years. As noted earlier in Chapter 2, this can be calculated using the formula: $AF = 1/r$, where r is the discount rate. In other words, $100 per annum for an infinite number of years, starting in Year 1, has a present value of $100 × 1/r = $100 × 10 = $1000, if r = 10%; or, $100 × 20 = $2000 if r = 5%, and so on.

3.5 Using investment decision-making criteria

We have already seen that there are a number of variants of DCF decision-rules that are used to appraise or evaluate investment projects. Of these decision-rules the best-known are the *net present value* (NPV) criterion, the *internal rate of return* (IRR), and *benefit/cost ratio* (BCR). In the sections that follow we shall examine each of these criteria and their usefulness in the different project decision-making situations.

3.5.1 The Net Present Value (NPV) criterion

The NPV of a project simply expresses the *difference between* the discounted present value of future benefits and the discounted present value of future costs: NPV = PV (Benefits) − PV (Costs). A positive NPV value for a given project tells us that the project benefits are greater than its costs, and vice versa. When we compare the project's total discounted costs and discounted benefits, we derive the NPV as shown in Example 3.4.

Which project is preferred: A or B? As NPV(B) > NPV(A), B would be preferred to A (at a 10% discount rate).

The effect of changing the discount rate

We have been assuming a 10% discount rate in Example 3.4. What happens to the NPVs of projects A and B if:

i   we increase the discount rate to 20%?
ii   We decrease the discount rate to 5%?

e.g.

EXAMPLE 3.4 Comparison of projects A and B using Net Present Value

Project A

	Year			
	0	1	2	3
(1) Cash flow	−100	+ 50	+ 40	+ 30
(2) Discount factor (at 10%)	1.000	0.909	0.826	0.751
(3) Discounted cash flow (=1×2)	−100.00	45.45	33.04	22.53

$$NPV(A)_{01} = -\$100(1.0) + \$50(0.909) + \$40(0.826) + \$30(0.751)$$
$$= \$101.02 - 100.00$$
$$= \$1.02$$

As NPV(A) > 0, accept the project (when cost of capital = 10%).

Project B

	Year			
	0	1	2	3
(1) Cash flow	−100	+ 30	+ 45	+ 50
(2) Discount factor (at 10%)	1.000	0.909	0.826	0.751
(3) Discounted cash flow (=1x2)	−100.00	45.45	33.04	22.53

$$NPV(B)_{01} = -\$100(1.0) + \$30(0.909) + \$45(0.826) + \$50(0.751)$$
$$= -\$100.00 - \$27.27 + \$37.17 + \$37.55$$
$$= \$101.99 - \$100$$
$$\$1.99$$

As NPV(B) is positive, project B can also be accepted.

Consider Project A at a discount rate of 20%:

$$NPV(A)_{0.2} = -100 + 50(1/1.2) + 40[1/(1.2)^2] + 30[1/(1.2)^3]$$
$$= -100(1.00) + 50(0.83) + 40(0.69) + 30(0.58)$$
$$= -100.0 + 86.5$$
$$= -\$13.5$$

Note that NPV(A) has decreased, and even become negative. Therefore, if the discount rate is 20%, *reject* project A. We leave it as an exercise to show that NPV(A) rises when the discount rate falls to 5%, and to perform similar calculations for Project B.

In Example 3.2 considered earlier, the NPV = $444.2 at a discount rate of 10%. Using the NPV decision-rule, we should accept this project. If we were to increase the discount rate from 10% to 15%, would this investment still be worthwhile? To determine this, we can recalculate the NPV as follows:

$$NPV = -1000(1.0) + 200(0.870) + 300(0.756) + 400(0.658) + 500(0.571) + 600(0.497)$$
$$= -1000 + 174.0 + 226.8 + 263.2 + 285.5 + 298.2$$
$$= \$247.7$$

Clearly the NPV of the project is still positive. Thus, if the appropriate discount rate was 15%, we would still accept it. It should be noted, however, that the NPV at 15% is much lower than the NPV at 10%. This stands to reason, as the project's net benefits accrue in the future whereas the capital costs are all at the beginning. The higher the rate of discount, the lower will be the present value of the future benefits, with the capital cost unchanged, and, therefore, the lower the NPV. We noted earlier that typical investments of this sort have a downward-sloping NPV curve. Can you derive the NPV schedule and plot the NPV curve for this project? (You should derive the NPV for at least three discount rates, including 0%. See Figure 2.5 in Chapter 2 for an illustration of an NPV curve.)

From the NPV schedule you have derived it should be evident that when the discount rate rises to 23% (approximately), the NPV for the project described in Example 3.2 becomes zero. At higher discount rates, the NPV becomes negative; at 25%, the NPV is clearly negative. In other words, if we were to use a discount rate that is 23% or less, the project would be acceptable; if we were to use any rate above that, the viability of the project becomes questionable.

Summary of NPV decision-rules

1 For accept or reject decisions:

if NPV > 0, accept;

if NPV < 0, reject.

2 When choosing or ranking alternatives:

if NPV(A) > NPV(B), choose A;

if NPV(B) > NPV(A), choose B.

In the event that NPV was calculated to be zero, or that NPV(A) = NPV(B), the analyst would seek further information, as discussed in Chapter 9.

At this stage we should point out that no mention has been made of the size of the available investment budget. The *implicit* assumption is that there is no budget constraint, which allows us to look favourably upon ("accept") all projects with a positive NPV. As we shall see later, in situations of a budget constraint, other decision-rules are needed to rank projects as there will not always be sufficient funds to accept all projects with a positive NPV.

3.5.2 The Benefit-Cost Ratio decision-rule

Another form of the NPV decision-rule is the *Benefit-Cost Ratio* (or BCR) decision-rule, which is, in effect, another way of comparing the present value of a project's costs with the present value of its benefits. Instead of calculating the NPV by *subtracting* the PV of Costs from the PV of Benefits, we *divide* the PV of Costs into the PV of Benefits:

$$BCR = \frac{PV(Benefits)}{PV(Costs)}$$

It should be noted that the denominator of the BCR includes the present value of *all* project costs, not just the capital costs. Later on, in Section 3.5.5, we discuss a variant of this rule that includes only the capital costs in the denominator.

If this ratio is equal to or greater than unity, then accept the project. If it is less than unity, then reject the project. It should be clear that when:

NPV > 0, then BCR > 1

and

NPV < 0, then BCR<1.

However, when it comes to *comparing* or *ranking* two or more projects, again assuming no budget constraint, the BCR decision-rule can give incorrect results. For instance, if two projects A and B are being compared, where:

	PV Benefits	PV Costs	NPV	BCR
Project A	$100	$60	$40	1.67
Project B	$ 80	$45	$35	1.78

using the BCR to rank these projects would place B (BCR = 1.78) above A (BCR = 1.67). Using the NPV decision-rule would place A (NPV = $40) above B (NPV = $35). In this situation the NPV decision-rule would be the correct one to use for ranking purposes, unless there is a budget constraint, when another variant of the BCR rule needs to be used, as discussed below.

3.5.3 The Internal Rate of Return (IRR) criterion

We have seen that in the case of "normal" or "well-behaved" investment cash flows, the NPV curve slopes downwards from left to right. At some point the curve intersects the horizontal axis. That is, the NPV becomes zero. The discount rate at which the NPV becomes zero is called the *Internal Rate of Return* (IRR). In the discussion of Example 3.2, the IRR was found to lie between 20% and 25%, approximately 23%. Once we know the IRR of an investment project, we can compare this with the cost of financing the project. Let us say, in this case, that the cost of financing the project is 15%. Now, as the rate of return, the IRR, is *greater than* the cost of financing the project, we should accept the investment. In fact, in the case of this investment, we would accept the investment at any cost of finance that is below 23%. When the IRR is *less than* the cost of finance, the project should be rejected. We can summarise this decision-rule as:

when IRR > r, then accept

and

when IRR < r, then reject

where r = the interest rate (assumed to be the cost of capital).

When considering an individual project, the IRR decision-rule will always give exactly the same result as the NPV decision-rule: from the NPV curve in Figure 2.5 it can be seen that when the discount rate is less than IRR, the NPV will be positive, and vice versa. For instance, provided the discount rate we used to calculate the NPV (i.e. the cost of financing the project) was less than 23%, the NPV would be positive and the project accepted. Similarly, using the IRR decision-rule, we saw that once we knew what the IRR was (i.e. 23%), we would compare it with the given cost of capital, and so long as the latter was lower, we would accept the project. These two decision-rules amount to exactly the same thing in such a situation.

In summary, NPV and IRR give identical results for *accept vs reject* decisions when considering an individual project. As discussed later in this chapter, this may not be the case when a choice has to be made between two or more projects.

The calculation of the IRR assumes, in effect, that the project returns can be reinvested to yield a rate of return equal to the IRR itself. For some projects this may not be the case: the project in question may represent an unusually good opportunity but a similar opportunity may not be likely to arise when funds generated by the project become available for reinvestment. In that case the proceeds from the project may be used simply to pay down loans charged at the market rate of interest. The *Modified Internal Rate of Return* rule (MIRR) calculates the rate of return on the project assuming that its proceeds are reinvested at a specified rate of interest (perhaps the firm's borrowing rate). This is a more advanced topic in financial analysis which we will not pursue further in this book.

A note on calculating the IRR

Today we are generally more fortunate than our predecessors, as financial calculators and spreadsheets have built-in algorithms for calculating the IRR. In the absence of a spreadsheet or programmable calculator with a built-in IRR formula, there are two ways of approaching the IRR calculation. Which of the two is used depends on the type of cash flow the project has:

i   When the cash flow is not regular in the sense that there is not an identical value every year after the initial (Year 0) investment, it must be estimated by trial and error; iteration and interpolation.

ii When the cash flow is regular, the easiest method is to use Annuity Tables, as discussed in Section 2.5 of Chapter 2.

The IRR of an investment can be approximated using a process of *interpolation*. In Example 3.2, the NPV is positive at a 15% discount rate, so we should try a higher rate, say, 20%, which again gives a positive NPV, so we should try a yet higher rate, say, 25%, which gives a negative NPV. Since the NPV is positive at 20% and negative at 25%, the IRR must lie somewhere between these two rates. The actual IRR can be found by interpolation:

$$IRR = 20 + 5\left(\frac{88.73}{88.73 + 41.79}\right) = 23.4$$

i.e., the rule for interpolation is:

$$IRR = \left(lower\ discount\ rate\right) + \left(difference\ between\ the\ two\ discount\ rates\right)$$
$$\times \left(\frac{NPV\ at\ the\ lower\ discount\ rate}{sum\ of\ the\ absolute\ values\ of\ the\ NPVs}\right)$$

Note that the difference between the two discount rates (the range) should be as small as convenient. This is because the larger this range, the more inaccurate the interpolation will be. However, although there is no hard-and-fast rule about this, a range of 5% will usually reduce the amount of work involved and give a reasonable estimate of IRR.

Use of Annuity Tables

It was noted in Chapter 2 that when there is a constant or regular cash flow it becomes very easy to calculate the IRR using the Annuity Tables. Using the Example 3.3, we proceed as follows: noting that the IRR is given by that discount rate which yields an NPV = 0, we write the NPV as:

NPV = $-1000(1.0) + 500(AF_3) = 0$

which implies that

1000 = $500(AF_3)$

or, solving for AF_3, when

$$\frac{1000}{500}, then\ AF_3 = 2$$

In other words, all we need to do now is find out at what discount rate the annuity factor for Year 3 has a value equal to 2.0. To determine this, we simply refer to our Annuity Tables in Appendix 2 and look along the row for Year 3 until we find an AF that is approximately equal to 2.0. This occurs when the discount rate is just above 23%. We can therefore conclude that the IRR is approximately 23%.

e.g.

EXAMPLE 3.5 Using Annuity Tables to calculate the IRR

Find the IRR for an investment (B) of $2362 that generates an annual net return of $1000 for 3 years.

$$\begin{aligned} NPV(B) &= -2362(1.0) + 1000(df_1) + 1000(df_2) + 1000(df_3) \\ &= -2362 + 1000(df_1 + df_2 + df_3) \\ &= -2362 + 1000(AF_3) \end{aligned}$$

Remember, IRR gives NPV = 0.
  Therefore, by setting NPV(B) = 0:

$$AF_3 = \frac{2362}{1000} = 2.362$$

From Annuity Tables, see Year 3 and look for the row in which AF = 2.362. This entry appears in the row corresponding to a discount rate of 13%. Therefore, IRR(B) = 13%.

3.5.4 *Problems with the IRR decision criterion*

Selecting among mutually exclusive projects

In the previous sections we have been examining investment decision-rules in situations in which a decision must be made whether or not to accept a given investment project. In these situations we saw that the NPV and IRR decision-rules gave identical results.

In other situations this need not necessarily hold true. In particular, when faced with the choice between mutually exclusive projects that is, in situations where one has to choose *one* of two or more alternatives (and where the acceptance of one automatically eliminates the other project(s)), the NPV and IRR decision-rules can yield conflicting results. One could think of a road project, for example, where one is considering two or more alternative designs or types of road. Once one of these is accepted, the other potential projects are automatically rejected.

Consider Example 3.6 with two mutually exclusive road projects, A and B, with different initial costs and yielding different net benefit streams as measured by net cash flows. Suppose we are told that the cost of financing the project is 10% per annum. Using the IRR decision-rule, it would appear that Project B is preferable to Project A given that the IRR is 25% for B as opposed to 20% for A. However, if we were instead, to use the NPV decision-rule, we would discount the future net benefits of each investment at 10% and obtain:

NPV (A) = $181.3

NPV (B) = $136.6

e.g.

EXAMPLE 3.6 Choosing between mutually exclusive projects

Net cash flows for two mutually exclusive road projects ($ thousands)

	Year				
	0	*1*	*2*	*3*	IRR (%)
A	−1000	475	475	475	20
B	− 500	256	256	256	25

Using the NPV decision-rule we would naturally prefer A to B, the exact opposite of the ranking using the IRR rule! Which of the two is correct? Why do the two decision-rules conflict? In such situations, *the NPV decision-rule will always give the correct ranking*, so it would make sense to use it in preference to the IRR. The reason for the conflicting result is due to a phenomenon referred to as "switching". Switching occurs when the NPV curves of the two projects intersect one another as illustrated in Figure 3.1. The NPV curves of the two projects cross over at 15%. In other words, at a discount rate of 15%, the NPV of A is equal to the NPV of B. At all discount rates below 15%, the NPV of A is greater than the NPV of B, and at all discount rates above 15%, the NPV of A is less than the NPV of B. Thus, with these two projects, their ranking changes or *switches* as the discount rate changes, hence the term "switching". In conclusion, because of the possibility of switching, it is always safer to use the NPV decision-rule when selecting from among mutually exclusive alternatives.

There is a way, however, in which the IRR rule could still be used to choose between mutually exclusive projects. This is to consider the *incremental project* which is defined as the *difference between* cash flows of projects A and B. We first compute the cash flow of a hypothetical project "A-B". We then calculate the IRR of that cash flow. If the IRR of the incremental project is equal to or greater than the cost of capital, we choose project A. If it is less, we choose project B.

	Year			
	0	*1*	*2*	*3*
Project (A-B)	−500	219	219	219
IRR (A-B) = 15%				

In other words, if the cost of capital is less than 15%, the cross-over point in Figure 3.1, we should accept A rather than B. Note that project A is larger than project B in the sense that it has a larger initial investment cost, of $1000 compared with $500. In effect, we are asking whether it makes sense to invest the *additional* $500 in project A, i.e. project A's cash flow is made up of the equivalent of project B's cash flow plus the incremental cash flow we have labelled "A-B". If the IRR is greater than the cost of capital, then it does make sense to invest the extra amount in project A.

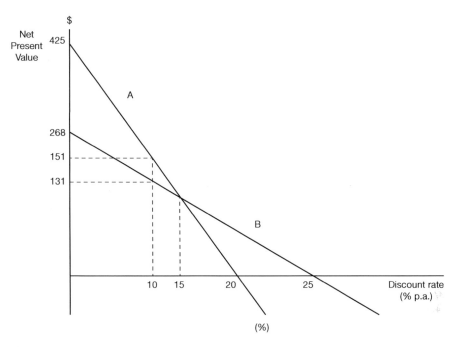

Figure 3.1 Switching when the NPV curves of two projects intersect. The NPVs for Projects A and B decline as the discount rate is increased, but NPV(A) falls more steeply than NPV(B), with the result that the NPV curves intersect at a discount rate of 15%. Project B has a higher IRR than Project A, but A has a higher NPV than B at a 10% discount rate.

As noted above, the IRR of the incremental cash flow (A-B) is also the discount rate at the switching point shown in Figure 3.1. We can see from Figure 3.1 that at the switching point the NPV of project A is equal to the NPV of project B. In other words, NPV(A) *minus* NPV (B) is zero. If the NPV of (A-B) is zero at some rate of discount, then that discount rate must also be the IRR of (A-B). Therefore, when we find the IRR of an incremental cash flow as in the case of (A-B), we are also identifying the discount rate for the switching point of the two NPV curves for projects A and B.

3.5.5 Problems with the NPV decision criterion

While the NPV decision-rule performs better than the IRR in choosing between mutually exclusive projects, it is not problem-free. There are essentially two investment decision-making situations in which the NPV rule, as described above, needs to be modified: namely, (1) under a capital rationing situation in which the objective is to finance the combination of investment projects from a given, constrained budget, in such a way as to obtain the highest overall NPV; and (2) when comparing two or more investment projects that do *not* have equal lives.

e.g.

EXAMPLE 3.7 Interpreting the IRR of an incremental cash flow ($ million)

Year						IRR	NPV
0	1	2	3	4	4%	10%
A −1000	500	500	500	23%	387	243
B −2362	1000	1000	1000	13%	413	125

Which project should be chosen, A or B? (Assume they are mutually exclusive.) Assume the cost of capital is:

i   4%
ii   10%

As can be seen from the example, IRR (A) > IRR (B), which suggests A is preferred to B both at 4% and 10% discount rates, but

$NPV(A)_{004} < NPV(B)_{004}$, therefore accept B, if the discount rate is 4%.

$NPV(A)_{01} > NPV(B)_{01}$, therefore accept A, if the discount rate is 10%.

Which decision-rule gives the correct decision in each case? Try using the IRR of the incremental cash flow (B-A) as a decision-rule in this example.

Under capital rationing

The problem in this situation is that it may be better to accept a combination of smaller projects that are less profitable in terms of the size of their individual NPVs, but which allow for a higher overall NPV from the available budget, than to accept those with the highest individual NPVs. A simple example will serve to illustrate this case. Suppose that we have a budget of $800,000 and we have five potential investment projects described in Example 3.8.

As we can see, all five projects have positive NPVs and all would be acceptable if there were sufficient funds available to finance them all. However, as the budget is limited to $800,000, we must rank them and select only the best combination. Using the NPV rule, we would rank them in the sequence C, D, E, B, A. With the available budget we would be able to fund only C and D, which would use up $600,000. With the remaining $200,000 we would also finance part of E (40% of it) assuming it is a divisible investment project. The total NPV obtained from this package would be $220,000 ($103 + $94 + 0.4*$58) thousand.

If instead, we now calculated the ratio of the present value of each project's net cash inflows (PV(B)) to the initial investment necessary to fund the project (PV(K)), we could then rank the projects according to their *profitability ratios* (NB/K), i.e. the present value of the net benefits generated *per $ of investment*. This is also referred to as the *Net Benefit Investment Ratio* (NBIR). It is very similar to the Benefit-Cost Ratio (BCR) discussed earlier, but should not be confused with this. The important difference is that the NBIR shows the value of the project's discounted *net benefits* (benefits net of operating costs) per dollar of (discounted) investment costs, whereas the BCR shows the value of the project's discounted *gross benefits* per dollar of (discounted) *total costs*, including both investment and operating costs.

e.g.

EXAMPLE 3.8 Ranking of projects where there is capital rationing

All values in thousands of dollars.

Project	PV(K)	PV(NB)	NPV (rank)	NB/K (rank)
A	100	130	30 (5)	1.30 (2)
B	400	433	33 (4)	1.08 (5)
C	200	303	103 (1)	1.52 (1)
D	400	494	94 (2)	1.24 (3)
E	500	558	58 (3)	1.11 (4)

Where:

PV(K) = Present value of the initial investment outlay

PV(NB) = Present value of the net benefits

NPV = Net Present Value (ranked)

$\dfrac{NB}{K}$ = Ratio of PV(NB) over PV(K) (ranked)

Note that using the NBIR changes the project ranking to C, A, D, E, B. With the $800,000 budget we would then finance projects C, A, D, and 20% of E. This would yield a total NPV of $239,000 ($103 + $30 + $94 + 0.2*$58) thousand, which is higher than the NPV obtained from the projects selected according to the NPV rule. Using the NBIRs is clearly a better way of ranking potential investments in situations of capital rationing, in comparison with ranking according to NPVs.

A further complication arises if the projects cannot be partially undertaken because of their so-called "lumpiness". This refers to the fact that many projects are *indivisible*, meaning that it is not possible to undertake a fraction of the project: there is no point in building half a bridge, for example. In this situation it is possible that two smaller projects which have lower profitability ratios but which exhaust the total investment budget will generate a higher overall NPV than, say, one more profitable investment which leaves part of the budget unallocated to any of the projects under consideration. Example 3.9 illustrates this case.

e.g.

EXAMPLE 3.9 Ranking indivisible projects in the presence of capital rationing

All values in thousands of dollars.

Project	PV(K)	PV(NB)	NB/K (rank)
A	125	162.5	1.30 (2)
B	175	189	1.08 (4)
C	200	304	1.52 (1)
D	400	496	1.24 (3)

Where

PV(K) = Present value of the initial investment outlay

PV(NB) = Present value of the net benefits

$$\frac{NB}{K} = \text{Ratio of PV(NB) over PV(K) (ranked)}$$

If we were to follow the NBIR decision-rule, we would rank the projects in Example 3.9 in the sequence C, A, D and B. Assuming an available investment budget of $300,000, and assuming all projects are indivisible or "lumpy", the NBIR decision-rule would tell us to finance project C, which would yield an NPV of $104,000 (plus $100,000 of investible funds which would be the subject of some alternative allocation by the agency concerned). However, if we look at the other projects, we see that it would have been possible to finance A and B instead of C. Even though these have lower NBIRs, we would utilise the full budget ($125,000 + $175,000), and generate a higher *overall* NPV of $51,500 ($162,500 + $189,000 – $300,000) from the projects chosen. Before we came to a final conclusion we would need to consider the net benefits of the alternative allocation of any portion of the budget not used by the agency and add it to the sum of the NPVs of the projects selected.

In conclusion, in situations of *capital rationing* where funds are constrained and potential investments must be ranked, we should use the *profitability ratio* (NBIR) rather than the absolute size of the NPV. However, when investments cannot be divided into smaller parts (lumpy investments), we should give consideration to whether or not our selection utilises the full budget available; and, if not, to whether accepting a combination of lower ranking, smaller projects would generate a higher overall NPV.

COMPARING PROJECTS WITH DIFFERENT LIVES

Another instance in which it would be incorrect to use the NPV decision-rule as presented above, is when we compare two or more projects that have different lives. For example, an infrastructure project such as an irrigation network, could have a number of possible designs, with varying durability, and hence, longevity, with all projects supplying the same annual quantity of water. This case is illustrated in Example 3.10.

e.g.

EXAMPLE 3.10 Comparing projects with different lives

Project	Initial cost	Annual net costs	Life years
A	$40,000	$2,800	4
B	$28,000	$4,400	3

Assuming a discount rate of 10% per annum:

PV of Costs (A) = −40,000(1.0) − 2,800(3.17) = −$48,876

PV of Costs (B) = −28,000(1.0) − 4,400(2.49) = −$38,956

The present values of the costs of the projects reported in Example 3.10 indicate that Project B has a lower aggregate cost in present value terms. However, we should note that project A has a longer life, providing an irrigation network that lasts one year longer than that of project B. How then should we decide between A and B? One solution is to assume the projects are renewed so that they cover the same total number of years. In the above example, a period of 12 years would be common to both projects if A was renewed three times over, and B was renewed four times over. The cash flows would then appear as reported in Table 3.2.

Using a 10% discount rate, and expressing values in thousands of dollars:

PV of Costs (A) = −40(1.0) − 40(0.683) − 40(0.467) − 2.8(6.814)
= −40.00 − 27.32 − 18.68 − 19.08
= −105.08

PV of Costs (B) = −28(1.0) − 28(0.751) − 28(0.564) − 28(0.424) − 4.4(6.814)
= −28.00 − 21.03 − 15.79 − 11.87 − 29.98
= −106.67

This example shows that if we compare the present value of costs of the two alternatives *over identical life spans*, project A is indeed less costly than project B. The rule in this situation, therefore, is that the NPV decision criterion should only be used in project selection once we have constructed cash flows of equal lives for our potential investment projects.

Table 3.2 Cost streams for projects A and B (Example 3.10) under a common life

Project	0	1	2	3	4	5	6	7	8	9	10	11	12
A($000s)	−40.0	−2.8	−2.8	−2.8	−2.8 −40.0	−2.8	−2.8	−2.8	−2.8 −40.0	−2.8	−2.8	−2.8	−2.8
B($000s)	−28.0	−4.4	−4.4	−4.4 −28.0	−4.4	−4.4	−4.4 −28.0	−4.4	−4.4	−4.4 −28.0	−4.4	−4.4	−4.4

The identical life spans comparison might be easy to do in cases like the one above, where the lowest common multiple (the LCM) of the projects' lives is a relatively low figure, such as 12 years in Example 3.10. But, what would we do if one project had a life of, say, 5 years and the other 7 years? We would need to construct a 35-year cash flow for each project. And, what if we wanted to compare these with another alternative that had a 9-year life? We would need to extend each cash flow to 315 years! Aside from the computational problem, we would be dealing with benefit and cost flows in future periods beyond our ability to predict.

For this reason, we use another decision-rule when comparing projects of different life spans. It is called the *Annual Equivalent* (AE) method, which is based on a Present Value (PV) calculation. The rationale of this method is to convert the actual stream of net benefits of a project into an equivalent (in present value terms) stream of *constant* annual net benefits. In other words, in Example 3.10 we take the cash flow of project A and ask: what fixed annual sum incurred over the same 4-year period would give us the same NPV? We then repeat this for project B over a 3-year period and compare the two annual equivalent net benefits as shown in Example 3.11.

To derive the AE, we follow two steps:

i   we compute the NPV for each project *for its own life* at the given discount rate; (ii) we convert the NPV for each project into an annuity by dividing the NPV derived in step (i) by the corresponding annuity factor for that life and discount rate.

ii In conclusion, when projects of unequal lives are being compared, one must *not* simply compare their NPVs. We need to convert the cash flows into an *Annual Equivalent* by dividing the NPV by the annuity factor corresponding to the life of the project.

e.g.

EXAMPLE 3.11 Using the annual equivalent method of project comparison

The annual equivalent method can be illustrated using the data in Example 3.10, and again using a 10% discount rate. We calculate:

NPV of (A) = – $48,876

NPV of (B) = – $38,956

A has a 4-year life and B has a 3-year life. The annuity factor, at 10% (from Appendix Table 2) is: 3.17 for 4 years, and 2.49 for 3 years. The AE is therefore:

$$AE(A) = \frac{-\$48876}{3.17} = -\$15418$$

$$AE(B) = \frac{-\$38956}{2.49} = -\$15645$$

Thus project A is less costly than project B.

Example 3.10 can also be extended to illustrate the *Unit Cost Equivalent* method of comparison. Suppose that each of Projects A and B supplies a constant 100 megalitres of irrigation water per annum. We can define a project's equivalent unit cost of irrigation water as the price, c, that would have to be charged for each unit of water, at the time it is supplied, such that the present value of this hypothetical revenue stream equals the present value of the project's costs. Thus, for Project A:

$$c_A*100*3.17 = 48876, \text{ hence: } c_A = \$154.18$$

and, similarly $c_B = \$156.44$

On the basis of these unit cost equivalents, as before, Project A is slightly preferred over Project B.

3.6 Using spreadsheets

In this section we review some simple operations of spreadsheets, using NPV and IRR calculations as illustrations. The reader who is familiar with the use of spreadsheets in financial analysis may wish to skip this section, but it does contain some important material.

Consider Figure 3.2 which illustrates a simple NPV calculation using an Excel spreadsheet. For the reader who is not already familiar with a spreadsheet, the best way to understand the concept is to think of it as nothing more than a very large calculator that is organised in rows (1, 2, 3 ... etc.) and columns (A, B, C, ... etc.), forming a matrix of *cells*, each with its own address (A1, B3, X75, etc.). One important difference between a calculator and a spreadsheet is that you can use any given cell in a spreadsheet for a variety of operations. For instance, you could use it to enter plain text.

In cell A1 in Figure 3.2 we have written the text "Example 1(a)". We could have entered a whole sentence as you might with a word processor.

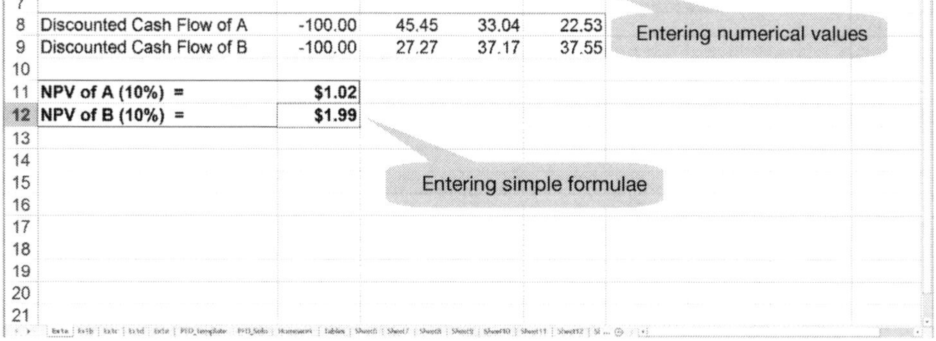

	A	B	C	D	E
1	Example 1(a)				
2					
3	Project/Year	0	1	2	3
4	NCF of A	-100	50	40	30
5	NCF of B	-100	30	45	50
6	Discount factor (10%)	1.000	0.909	0.826	0.751
7					
8	Discounted Cash Flow of A	-100.00	45.45	33.04	22.53
9	Discounted Cash Flow of B	-100.00	27.27	37.17	37.55
10					
11	NPV of A (10%) =	$1.02			
12	NPV of B (10%) =	$1.99			

Figure 3.2 Spreadsheet presentation of DCF calculation.

In this spreadsheet we have also set up the cash flow of two projects used as examples earlier in Section 3.3 on Discounting and the Time Value of Money: Project A and Project B. You will notice that column A contains a set of *labels*. In other cells we have entered *numbers*. In row 3 we have numbered the years: Year 0 in B3, Year 1 in C3, etc. In row 4 we have entered the net cash flow for project A: –100 in B4, 50 in C4, and so on. In row 5 we have entered the NCF for project B. (Note that we have chosen to set up the cash flows in rows running from left to right rather than in columns from top to bottom. Either method will work but we believe, from our own practical experience, that setting up a spreadsheet in rows makes it easier to handle the more complicated sorts of analyses we will be doing later.)

Having set up the two cash flows in this way, we are now ready to compute their NPVs. To do this we use another function of the spreadsheet cell: we enter a formula into a cell. You can enter almost any formula into a cell. For instance, you could simply instruct the spreadsheet to add together, say, "50" and "30" by entering the formula "=50 + 30". (Note that when using Excel we always begin with the "=" sign (or some other operator such as a "+" or "–" sign) when entering a formula. This is to distinguish the entry of a formula from the entry of a label.)

Alternatively, instead of using actual *values* in the formula, you could instruct the spreadsheet to add together the contents of two cells, by giving their reference addresses. For example, if we entered the formula "=C4 + C5" into cell G6, then the amount "80" would appear in that cell; i.e. 50 + 30=80. Although you do not see the formula you have entered in the cell itself, you can see what it is by placing the cursor on the cell and looking in the "window" or *Formula Bar* near the top of the spreadsheet. In the spreadsheet shown in Figure 3.2 the cursor is positioned on cell B12, so the formula in that cell appears in the Formula Bar.

In this simple example we have also entered the discount factors (at a 10% discount rate) for each project year in row 6. To derive the NPV for each project, we first calculate the *discounted* cash flow by multiplying each year's cash flow by the corresponding discount factor. This is shown in rows 8 and 9 for projects A and B respectively. For instance, in cell C8 we have entered the formula "=C4 * C6", i.e. 50 × 0.909. The result displayed in C8 is "45.45". If we do this for each cell of the cash flow, we can then find the NPV by adding up the values in the cells. To do this we place another formula in cells B11 and B12 instructing the spreadsheet to add up the cells B8, C8, D8 and E8 to get the NPV of A, and B9 through to E9 for project B. The solutions are shown in cells B11 and B12: + $1.02 and + $1.99 respectively.

Note that there are two types of formulae we could have used in cells B11 and B12: a "self-made" formula or a "built-in formula". If we had used a self-made formula, we would have written into cell B11, "=B8 + C8 + D8 + E8". (Note that the symbol "*" is used as the multiplication sign, "/" is used as the division sign, and "^" is used as the exponent sign.) When there are a lot of cells to add up, relying on self-made formulae becomes a laborious exercise and one that is prone to careless error. The alternative is to use one of the built-in formulae. In this case we use the "SUM" formula which is displayed in the Formula Bar at the top of the spreadsheet. This tells us that in cell B12 we have entered the formula "= SUM(B9:E9)" where the colon sign indicates that all other cells in between B9 and E9 are to be included in the summation. With built-in formulae the spreadsheet is programmed to know what functions to perform when it reads certain words. In this instance, the word "SUM" instructs it to add up. Similarly, "AVERAGE" would instruct it to find the mean; "STDEV" would instruct it to find the standard deviation, and so on. Information about the range of formulae available and their operation can be found in the "Help" section of the spreadsheet program.

What we have just done demonstrates the use of some of the very basic functions of a spreadsheet. In practice you would not need to enter in the discount factors as we have done in row 6. We could have written our own formulae for the discount factors instead and let the spreadsheet calculate the actual figure, but even this is unnecessary because the spreadsheet has built-in formulae for all the DCF calculations you need to use, including the NPV and IRR formulae.

To demonstrate the use of the built-in NPV formula, look at Figure 3.3 where we show the same NPV calculations as before, but using the built-in formula. Note that we do not have a row of discount factors or a row showing the discounted cash flow of each project. Instead we simply enter one formula into cells B8 and B9 to calculate the NPVs for projects A and B respectively. The first thing you will notice is that the answers are not exactly the same as those shown in Figure 3.2. The reason is due to rounding differences. When we entered the discount factors into the spreadsheet in Figure 3.2, we rounded them off to three decimal places. As the spreadsheet is capable of using very many more decimal places in its computations, the solutions in Figure 3.3 are more precise.

The first part of the formula for the NPV calculation entered into cell B8 is shown in the Formula Bar at the top of the spreadsheet: "=NPV(0.1,C4:E4) + B4". The "0.1" tells the program to use a 10% discount rate, then follows a comma, and then "C4:E4" tells it to discount the contents of cell C4 (Year 1) through to cell E4 (Year 3). Note that we have to include "+ B4" in the formula (but outside the brackets) to instruct it to include the *undiscounted* Year 0 amount of –$100. The reason we have to do this is that *the spreadsheet does not use the same convention that we do when numbering project years*. Our convention is to treat all project costs and benefits accruing in the initial year as "Year 0", which therefore

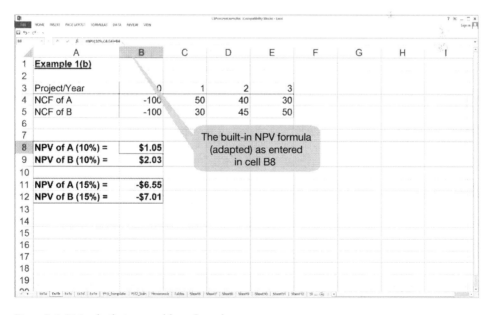

Figure 3.3 Using built-in spreadsheet formulae.

remain *undiscounted* in our NPV calculations. The spreadsheet, on the other hand, is programmed to treat the earliest year of the cash flow as accruing one year later and therefore will always discount the initial year of the cash flow entered. For this reason we have to adapt the formula by adding back the Year zero cash flow on an undiscounted basis, and beginning the NPV computation for our project in Year 1.

If we change the discount rate in the NPV formula, the solution in the corresponding cell will automatically change. In cells B11 and B12 we have entered the identical formulae as in B8 and B9 except that we have set the discount rate at 15%, e.g. "=NPV(0.15,C4:E4) + B4". It should be noted that all entries in column A of the spreadsheet are labels. We have typed into cell A8, for example, "NPV of A (10%) =". This is not a formula! It is only there for the user's convenience. If a label were accidentally used in performing a calculation, the spreadsheet will treat it as a zero. This also applies where numbers have been entered as part of a label as in the case of "10%" in cells A8 and A9. (It is also important to remember that if any cell in a row is left blank, the formula will ignore that cell altogether, in which case the value in each cell after that one will be treated as if it occurred a year earlier. For this reason always enter the numeral zero in any empty cell of a row to be discounted.)

The discount rate is a variable in most cost-benefit analyses as the analyst will generally want to see how the NPV of a project changes with different discount rates. To construct an NPV curve, you will need to recalculate a project's NPV at several discount rates. To simplify undertaking this sort of *sensitivity analysis* there is another "trick" we use when designing a spreadsheet. Instead of entering the actual values in the NPV formula, we can enter a reference to another cell somewhere else in the spreadsheet. This cell might contain the discount rate, for example, so that whenever we change the value of the discount rate in that cell, the value of the NPV will change *wherever in the spreadsheet there is a NPV formula that refers to that cell for the discount rate*. For instance, if we placed the discount rate in cell B10 and then entered the value "0.1" into that cell, we could then change the formulae in cells B8 and B9 to read "=NPV(+ B10,C4:E4) + B4" and "=NPV(+ B10,C5:E5) + B5" respectively. The results will be the same as we have in Figure 3.2 but if we were to change the value in cell B10 to 0.15, the NPVs of projects A and B, in cells B8 and B9 respectively, would immediately change to the values already computed in cells B11 and B12.

Similarly, what we enter into the net cash flows in rows 4 and 5 could also be *references* to other cells in the spreadsheet rather than actual values. For instance, the net cash flow in each year is likely to be derived from other calculations consisting perhaps of a number of different costs and benefits. These calculations can be performed elsewhere in the same spreadsheet, and then, instead of re-entering the values derived from these calculations, we would simply refer to the cells in which the solutions to these "working calculations" appear. An example of this is shown in Figure 3.4 where we have inserted a "Working Table" in the spreadsheet, still using the same example as previously. In the Working Table, rows 21 and 22 contain simple arithmetic formulae that derive the net cash flows for projects A and B from the raw data in the rows above. (These values could themselves be based on calculations elsewhere in the spreadsheet.) In the final table in the upper part of Figure 3.4, that we now label "Net Cash Flow", we no longer have any values in any of the cells. The cells in that table contain only references to cells in the Working Table. Similarly, the NPV formulae in cells B7 and B8 contain a reference to cell B10 where we have entered "15%". (Remember, the rest of the NPV formula is also in the form of references to the appropriate rows in the "Net Cash Flow" table above.) Finally, spreadsheets also contain built-in formula for the IRR. In cells E7 and E8 we have entered the formulae for the IRR of projects A and B respectively.

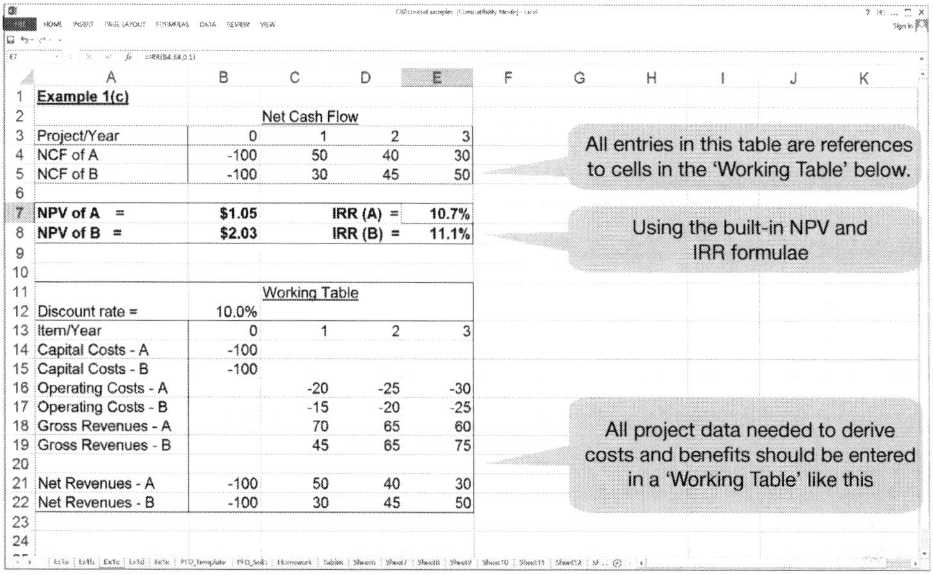

Figure 3.4 Referencing within the spreadsheet.

The IRR formula for project A is displayed in the Formula Bar at the top of the screen in Figure 3.4, as "=IRR(B4:E4,0.1)". Notice, first, that there is no need to leave out of the formula the cash flow in year zero and then add it back in as we had to do with the NPV formula. Second, though not necessary, we have included a "trial" or "prompt" discount rate in the formula. The algorithm uses this rate to begin the iteration process to derive the IRR (this process was illustrated in Figure 2.7 of Chapter 2). In this case it makes no difference what rate we enter. You can try changing it yourself to, say, 5% or 20% and you will always get the same answer, because, as we learned earlier, a cash flow with only one change in sign, as we have here, has only one IRR. There are exceptions, however, where there is more than one solution, as seen in Chapter 2. In that case entering a different "trial" discount rate in the formula could produce a different IRR. (Note that some spreadsheets are programmed to report an error when an attempt is made to use the IRR formula when there is more than one positive IRR.)

Until now, when using the built-in formulae such as "NPV" and "IRR", we have needed to type out the formulae and enter the appropriate parentheses, values, cell references, commas, etc. for the spreadsheet to perform the desired calculation. The spreadsheet contains very many formulae that are extremely useful and time-saving for the project analyst. These include numerous *financial formulae* for performing other commonly used calculations in financial analysis, such as interest and principal repayments on loans (the "IPMT" and "PPMT" formulae)[1] or depreciation allowances, as well as *statistical formulae* and others. It is not necessary to memorise or even look up the appropriate words and format for all these formulae so that each time a function is to be used you may enter the details correctly. In the spreadsheet's main toolbar at the top of the screen there is a button which, when entered, displays all available formula. Once you have selected the one you wish to use, this can be pasted into the appropriate cell in the spreadsheet. These processes are illustrated in Figure 3.5 for the built-in IRR formula.

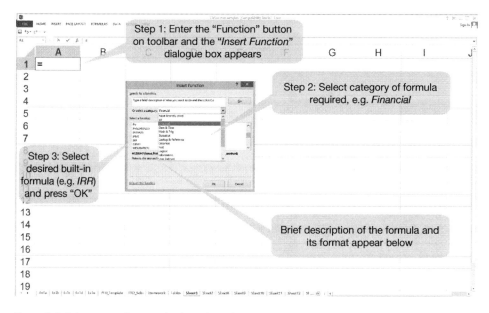

Figure 3.5 Selecting and using a built-in formula.

3.7 Further reading

One example of a book containing discussion of investment decision-rules is H. Bierman and S. Smidt, *The Capital Budgeting Decision*, 9th edition (Basingstoke: Macmillan, 2006).

Further information about the operation of spreadsheets can be obtained from the relevant manual, such as *Microsoft Excel© User's Guide*.

Exercises

1   i   What is meant by the term "mutually exclusive projects"?

     ii Explain why the IRR decision-rule could give the wrong result when comparing mutually exclusive projects.

2   The following net cash flows (in $000s) relate to two projects:

	Year						
	0	1	2	3	4	5	6
Project A	−60	20	20	20	20	20	20
Project B	−72	20	20	20	20	20	20

    i Calculate the NPV for each project, assuming a 10% discount rate.

    ii Assuming that the two projects are independent, would you accept them if the cost of capital is 15%?

iii Calculate the IRR of each project.

iv Which project would you prefer if they are mutually exclusive, given a 15% discount rate?

3 Using a spreadsheet, generate your own set of Discount and Annuity Tables for, say, all discount rates between 1% and 20% (at 1 percentage point intervals) and for time periods 1 to 30 (at one time period intervals), as well as time periods 50 and 100. You should generate these tables by inserting the numbers for the time periods in the first column of each row and the discount rates in the first row of each column, and then inserting the appropriate formula into one cell of the table – year 1 at 1% – and then copying it to all other cells in the matrix. (Hint: Do not forget to anchor the references to periods and discount rates using the "$" symbol.)

4 Consider the six potential public sector projects described below.

Project	Capital investment (year 0) ($)	PV of net benefits (excluding investment cost) ($)
A	125	155
B	55	100
C	150	320
D	45	115
E	165	320
F	130	245

i In a situation of a budget constraint, in what order would you rank these projects, assuming all are perfectly divisible?

ii Assume the government department has a limited budget of $250 to finance the capital costs and all are perfectly divisible, how would you allocate your budget, assuming that the department will "lose" any unspent balance?

iii Assume you have a limited budget of $250 to finance the capital costs and that the projects are indivisible (lumpy), how would you allocate your budget, assuming that the department will "lose" any unspent balance?

5 A firm has a capital budget of $100 which must be spent on one of two projects, each requiring a present outlay of $100. Project A yields a return of $120 after one year, whereas Project B yields $201.14 after 5 years. Calculate:

i the NPV of each project using a discount rate of 10%;

ii the IRR of each project.

What are the project rankings on the basis of these two investment decision-rules? Suppose that you are told that the firm's reinvestment rate is 12%, which project should the firm choose?

6 A firm has a capital budget of $100 which must be spent on one of two projects, with any unspent balance being placed in a bank deposit earning 15%. Project A involves a present outlay of $100 and yields $321.76 after 5 years. Project B involves a present outlay of $40 and yields $92 after one year. Calculate:

i   the IRR of each project;
ii the B/C ratio of each project, using a 15% discount rate.

What are the project rankings on the basis of these investment decision-rules? Suppose that if Project B is undertaken its benefit can be reinvested at 17%; what project should the firm choose? Show your calculations (spreadsheet printout is acceptable as long as entries are clearly labelled).

7   A firm has a capital budget of $30,000 and is considering three possible independent projects. Project A has a present outlay of $12,000 and yields $4,281 per annum for 5 years. Project B has a present outlay of $10,000 and yields $4,184 per annum for 5 years. Project C has a present outlay of $17,000 and yields $5,802 per annum for 10 years. Funds which are not allocated initially to one of the projects can be placed in a bank deposit where they will earn 15%. Project returns can be reinvested at the specified reinvestment rate.

i   Identify six combinations of project investments and a bank deposit which will exhaust the budget.
ii Using a spreadsheet, determine which of the above project combinations the firm should choose:
    a   when the reinvestment rate is 15%?
    b   when the reinvestment rate is 20%?

Explain your answer and show your calculations (spreadsheet printout is acceptable as long as entries are clearly labelled). (Hint: compound forward the returns of each combination at the reinvestment rate to get a Terminal Value in year 10.)

8   A public decision-maker has a budget of $100 which must be spent in the current year. Three projects are proposed, each of which is indivisible (it is not possible to undertake less than the whole project) and non-reproducible (it is not possible to construct two versions of the same project). The discount rate is 10% per annum. The project benefits and costs are summarised in the following:

	Project cost ($)	Benefits ($)	
	Year 0	Year 1	Year 2
A	30	40	0
B	30	0	50
C	70	0	100

i   Work out the Net Present Value (NPV), Internal Rate of Return (IRR) and Benefit/Cost Ratio (B/C) for each project.
ii Rank the projects according to the NPV, IRR and B/C investment criteria.
iii Assuming that project returns can be reinvested, which projects should be undertaken to spend the budget:
    a   if the reinvestment rate is 22% per annum?
    b   if the reinvestment rate is 28% per annum?

9 Three different technologies are being considered for a project to provide a given service. Assume that the quantity and quality of the service are identical for the three options. Project A has a life of 7 years, project B has a life of 5 years and project C has a life of 9 years. The investment and operating costs for the three projects are as shown in the table below:

Project	0	1	2	3	4	5	6	7	8	9
A	−1200	−250	−250	−250	−250	−250	−250	−250		
B	− 800	−400	−400	−400	−400	−400				
C	−1500	−200	−200	−200	−200	−200	−200	−200	−200	−200

Rank the three options using the Annual Equivalent Cost method.

Note

1 The "PMT" formula gives the constant annual amount for an annuity with a given initial loan amount, interest rate and repayment periods. This formula can also be used to convert any present value to an Annual Equivalent value, but remember to reverse the sign as it is designed to convert positive (loan) amounts into negative (repayment) instalments.

4 Private cost-benefit analysis
Financial appraisal

4.1 Introduction

As discussed in Chapter 1, social cost-benefit analysis takes a broad, "social" perspective in the context of project appraisal – it identifies, measures and compares the costs and benefits experienced by all those affected by the project. In contrast, when a project is being evaluated by a private or public enterprise from a purely commercial or "private" perspective, account is taken only of the benefits and costs of the project that accrue to the enterprise itself and that affect its profitability. The project may have wider implications – environmental and employment effects, for example – but if these do not affect the enterprise's financial position, they are omitted from the *Private Analysis*.

Private firms operate by trading goods and services in the relevant markets: they purchase inputs and they sell outputs. It follows that the benefits and costs that determine the profitability of a project to the firm are values calculated at market prices. As a first step to calculating the profitability of a project, therefore, we could simply value all the inputs and outputs at market prices. We refer to this calculation as the *Market Analysis*. However, the equity holders of the private firm do not receive all the benefits nor incur all of the project costs identified by the Market Analysis. The firm's lenders contribute to the initial capital cost and receive a portion of the project's net benefits in the form of interest and capital repayments. The market value of project output may include indirect tax, such as a sales tax or value-added tax, and this portion of project benefit accrues to government rather than the firm. Since the Market Analysis values project inputs at their tax-inclusive price, the net benefits of the project at market prices do not include indirect taxes levied on inputs and received by government. Finally, the project's earnings will be subject to business income tax. The net benefit stream identified by the Market Analysis can be adjusted, in the Private Analysis, to provide an estimate of the profitability of the project to its owners by subtracting debt flows, taxes on output and business income taxes. This analysis of the distribution of the net benefits identified by the Market Analysis, between the firm, lenders and government, is analogous to the approach adopted in Chapter 6 where we analyse the distribution of all project net benefits, whether measured by market prices or not, among all the groups affected by the project.

It is to be expected that the private enterprise would already have undertaken a *private* investment appraisal of the project and will be basing its investment decision on the results of that analysis. However, these results may not be available to the public sector decision-maker, and the Market Analysis, using familiar discounted cash flow techniques and incorporating the private investment analysis, may be important in helping the public decision-maker decide what incentives are required in the form of tax concessions, for

DOI: 10.4324/9781003312758-4

example, to induce the firm to undertake the project. It could also be that the private enterprise is one of the stakeholders included as part of the Referent Group, as we discuss further in Chapter 6. For both these reasons the appraisal of a proposed project from a purely private viewpoint is often an integral part of a social cost-benefit analysis undertaken by the project analyst. This chapter focuses on the methodology of private cost-benefit analysis, using *discounted cash flow* (DCF) analysis.

It can be noted that, in the case of a public project, or a project undertaken by a public–private partnership (PPP), similar considerations of revenue and cost flows to those addressed in the Private Analysis may be relevant in determining whether the government department decides to undertake the project or not. The appraisal of the project would start with a Market Analysis but the subsequent allocation of net benefits to the participating organizations will vary from case to case. Issues relating to government expenditure and tax flows are dealt with in the Referent Group Analysis discussed in Chapter 6, and in the part of Chapter 5 dealing with the opportunity cost of public funds. The remainder of the present chapter focuses on private sector projects.

4.2 Benefits and costs measured as cash flows

Cash flows play a central role in the development and appraisal of almost any investment project: they are a summary presentation of the various costs and benefits expected to accrue over the project's life. As we saw in Chapter 3, it is these cash flows that provide the main basis for deciding whether a project is "worthwhile" (are the benefits greater than the costs?), and/or which of a number of project variants or alternatives is the "best" (shows the greatest net benefits).

In the first part of the present chapter we focus on the *derivation of cash flows*, i.e. how to build up cash flow estimates from the basic technical, economic and social information that is collected and analysed in the course of preparing a project proposal or performing a project appraisal. Here we remind the reader that these estimates are based on *forecasts* of project input and output quantities and future market prices. Since we are mainly concerned in this chapter with private profitability the cash flows we consider here are derived from input or output flows valued at market prices. However, project benefits and project costs do not always result in cash receipts and cash outlays. Obvious examples are the benefits of public roads, which are made available to the general public free of direct charge and which, therefore, do not result in revenues for the entity undertaking the investment (the national or regional government or local council). In such cases, which are considered in subsequent chapters, it might be more correct to use the expressions "flow of benefits and costs" and "net benefit flow" instead of "cash flow" and "net cash flow". However, the technique of analysis remains the same and the discussion of the treatment of cash flows in this chapter will be drawn on subsequently when we consider the full range of a project's costs and benefits.

4.2.1 Identifying project inputs and outputs

The first stage in developing a cash flow is to identify the range of project inputs and outputs and the quantities of each. This is usually accomplished as part of the preliminary assessment of the project proposal, sometimes referred to as *scoping* the project. Inputs can be classified as materials, labour, capital and land. Materials include inputs such as electricity, fuel, chemicals, wood, steel, cement, paint, office supplies and so on. These are sometimes referred to as intermediate inputs. The labour input may include both unskilled and skilled labour, with the latter covering a wide range including managerial, medical, engineering and computational skills. Capital refers to the infrastructure, buildings and equipment associated with the project, and land is usually interpreted to include all types of natural resources used in production. Project outputs can be consumption, capital or intermediate goods. Consumption goods include products such as food and home furnishings, capital goods include tractors and chainsaws, and intermediate goods include fertilisers and sawn timber.

4.2.2 Valuing inputs and outputs at market prices

Once the project's input and output flows have been identified and estimated in physical terms, market prices are used to convert them to cash flows, which are sometimes referred to as time-streams to reflect the fact that outputs and inputs generally occur over a number of years. Since we are concerned in this chapter with only those inputs and outputs that are traded in the market, and can affect the profitability of the firm, the valuation process is relatively straightforward. When subsequently we consider the wider implications of a project, we will usually encounter benefit or cost items which are not traded, such as savings in commuter time, improved health, air or water pollution, for example, and valuing these commodities is a challenge which is explored in Chapter 8.

When we refer to market prices, we mean the prices that buyers actually pay or sellers receive. In the case of project output or inputs, the market price includes any indirect tax such as sales, excise, value-added or payroll tax or tariff on imported goods. Labour input, for example, is priced at the gross wage plus the payroll tax, if any. The gross wage consists of the worker's take-home pay plus any income taxes withheld from the pay packet. When, in subsequent chapters, we consider the implications of a project beyond those affecting the profitability of the private firm, we will consider the incidence and allocation of tax revenues in detail.

4.2.3 Characteristics of cash flows

In Chapter 3 we considered the cash flows of a number of hypothetical projects. We saw how a project's net cash flow shows the difference between benefits and costs for each year of the project's life. The net cash flow of an investment normally has a very distinctive time profile, being negative in the earlier years when the project is under construction, and becoming positive only after construction has been completed and the project has started to produce the goods or services for which it was designed. In previous chapters we deliberately simplified the analysis by using projects with a relatively short life and uncomplicated cash flow profile. In practice, cash flows are not so simple. There are a number of ways in which actual project cash flows might deviate from the profile of the hypothetical examples we have used until now. For instance:

- Cash flows of projects usually cover a relatively large number of years: anything from 10 to 20 years is quite normal, while certain projects have even longer lives (up to 50 years or more), depending on the nature of the project and the useful life of its main assets. The total number of years covered by a project's cash flow we usually call the length of a project's life or "lifetime", though some authors also use the words "planning horizon" or "time horizon".

- Although the net cash flows of investment projects have a distinct time profile, the exact shape and dimensions will differ from one project to another; there can be large variations in project life and in the "gestation period", i.e. the period before a project starts to yield positive net benefits. Moreover, some projects have high initial benefits that gradually taper off towards the end of their life; others (e.g. tree crop projects) require much more time before they reach their maximum net benefits but then maintain these for a relatively long time.
- In Chapter 3, we also saw how a project's costs are usually divided into: (i) investment costs, e.g. the cost of construction and installation of a project's physical assets, the cost of building up stocks of materials and spare parts needed to operate the project and the cost of training the staff; and (ii) operating costs, i.e. the costs of the materials, labour, etc. used to operate and maintain the project. In our simple examples we always assumed that the entire investment cost occurred at the very beginning of the project: in Year 0. In practice, though most investment costs are concentrated in the early years of the project (the initial investment), there is usually also a need for regular replacement investments when certain shorter-lived assets (machinery, transport equipment, etc.) reach the end of their useful life. Some projects even require a large outlay at the end of their life, such as mining projects where the owners are required to rehabilitate the landscape after the mine closes down.
- The relative importance of investment and operating costs can vary from one type of project to another. In the case of infrastructure projects (e.g. roads), a very large part of the total costs consists of the initial investment, while operating cost consists only of maintenance cost. In other projects (e.g. for crop production), the operating cost (fertilisers, pesticides, harvesting, storage, etc.) may be much more important relative to the initial investment.
- Project benefits consist mainly of the value of the output (the goods or services) produced by a project. In addition, there may be incidental benefits such as the scrap value of equipment replaced during the project's life. At the end of the project's life, the scrap or rest value (also called salvage or terminal value) of a project's assets is included as a benefit, though in certain cases this value may be negative, e.g. when a project site has to be brought back to its original condition upon termination of the project. Similarly as working capital is run down towards the end of the project's life, the reductions in value of the stock of working capital appear as project benefits.
- We also noted in the previous chapter (pp. 41–42) that the accounting period used in cash flow accounting is normally a year, but project years do not necessarily (and in reality hardly ever do) correspond to calendar years, and, though costs and benefits normally accrue throughout the year, the standard convention in all cash flow accounting is to assume that all costs and benefits are concentrated (accrue as one single outflow or inflow) at the end of each project year.

In this chapter we also deal with a number of other important issues to do with the derivation of cash flows, including: the treatment of incremental (or relative) cash flows; how to account for inflation; depreciation; how we should deal with flows of funds relating to the financing of a project; and calculating profits taxes and the after-tax net cash flow. We should also note at this stage that the procedures that have to be followed may vary from case to case, depending on, among other factors, the point of view taken by the analyst when appraising a specific investment proposal. In this chapter, however, we will begin by considering the cash flows of projects from the relatively narrow perspective of the firm which plans to undertake the project.

4.3 Inflation and relative prices

Most countries experience *inflation*, i.e. an increase in the general price level. A dollar today will buy less than it did one, five or ten years ago. Rates of inflation – the percentage increases in the general price level per annum – differ widely between countries. At best the rate of inflation will be 1–2% per annum, but in many countries rates of 10–15% per annum have been regularly experienced. At the other end of the spectrum we find that some countries have experienced "hyper-inflation"– up to 100% per annum or sometimes more.

The question that arises is how to deal with inflation in preparing our cash flow estimates. We have seen that the cash flows of development projects might cover anything between 10 and 50 years. We need to make the distinction again between project *appraisal* when we are looking into the future, and project *evaluation* when we are looking back. In appraisal, do we have to forecast inflation that far ahead in order to be able to value our flows of inputs and outputs in a correct way? The answer is luckily "no", since there is no one who would be able to provide us with such estimates. As inflation is an increase in the *general* price level, as opposed to *relative* price changes, we can generally ignore it for purposes of project *appraisal*. In effect, we are assuming that it affects all costs and benefits in the same way, and therefore will not affect the relative returns on different projects in any real sense (recall the discussion of discounting and inflation in Chapter 2). What we do in practice is to use constant or real, Year 0, prices to value all the project's inputs and outputs, i.e. we estimate all costs and benefits throughout the project's life at the price level obtaining at the time of appraisal (or as near to it as is possible).

In project *evaluation* when we are looking at a project's performance in retrospect, we can adjust actual prices by an appropriate deflator such as the CPI (consumer price index), the GDP (Gross Domestic Product) deflator, or the wholesale price index, to convert nominal cash flows to real values. We then compare the project IRR with the real cost of capital. Alternatively, we could leave all cash flows in nominal terms and then calculate a *nominal* IRR which we would compare with a *nominal* cost of capital, or which we could convert to a *real* IRR and compare with the *real* cost of capital, as discussed in Chapter 2. In NPV analysis we have to be careful to use a *nominal* discount rate to discount a *nominal* cash flow. A simple example is shown in Table 4.1.

In this example we have a nominal cash flow (Row 1) where each year's net cash flow is in current prices (i.e. including inflation). When we calculate the IRR on this cash flow, we arrive at rate of 15% per annum; but this is the nominal IRR which includes the effects of inflation on the cash flow. If we know the inflation rate for each year, we can convert this to a real cash flow (Row 2) by converting the nominal values to constant (base year) prices. If we assume that the inflation rate in each year is 5%, and taking Year 0 as the base year, we derive the real cash flow shown in Row 2. If we then calculate the IRR we arrive at a rate of 10% per annum. This is the *real* rate of return. Note that the difference between the real and nominal rate is equal to the inflation rate: 15% minus 10% equals 5%.

If we were interested in knowing the NPV at a given cost of capital, again we have two options. If the real cost of capital is given as 6% per annum, we would need to discount the nominal cash flow (Row 1) using a nominal discount rate of 11% (6% real plus the 5% inflation rate). The NPV is $7.5. Alternatively, we can discount the real cash flow (Row 2) using the real cost of capital (6%). Here the NPV is $6.9. There is a slight difference

Table 4.1 Nominal versus real cash flows ($)

	Year			
	0	1	2	3
1 Net cash flow (current price)	−100	42.0	44.1	46.3
2 Net cash flow (constant Year 0 prices)	−100	40.0	40.0	40.0

Inflation rate = 5% per annum
IRR (nominal) = 15%
IRR (real) = 10%
Cost of capital = 6% (real)
or = 11% (nominal, 6% + 5%)
NPV (row 1 @ 11%) = $7.5
NPV(row 2 @ 6%) = $6.9

between the NPVs which is due to the fact that when we discount in one stage (by 11%), this is not precisely equivalent to discounting in two stages, first by 5% (to convert nominal to real values) and then by 6%. You can check this by looking at the discount tables. Take Year 3 discount rates as an example. The discount factor for 5% is 0.86 and for 6% is 0.84. Multiplying the two we get 0.72. Yet, when we look up the discount rate for 11% we see it is 0.73. We discussed the reason for this discrepancy in more detail in Technical note 2.2 where we demonstrated that, if the discrepancy is likely to be large, it can be eliminated by calculating the real rate of interest, r, as $r = (m - i)/(1 + i)$, where m represents the nominal discount rate and i the expected rate of inflation, instead of as the approximation $r = m - i$.

It can be observed, however, that in periods of inflation not all prices increase at the same rate. There may be goods and, in particular, services, whose prices increase at a much faster rate than the general price level; energy prices, for example, are likely to rise at a faster rate than the rate of inflation. There are also, on the other hand, goods, for example, consumer durables, the prices of which increase much less than the overall rate of inflation, or which even show nominal price decreases (e.g. consumer electronics such as personal computers and pocket calculators). *Relative* price changes, where they are significant, should not be ignored in project appraisal and therefore should be incorporated into our cash flow estimates. The point to remember, however, is that relative price changes would also occur in a world without general price inflation. Relative price changes are not the result of inflation, but occur alongside inflation, because of changes in demand, technological developments (e.g. development of synthetic materials and improved inputs), exhaustion of natural resources, and deterioration of environmental quality. They should, in principle, always – even in the absence of inflation – be taken into account when preparing cash flow estimates. For example, if the rate of general inflation was 3% and energy prices were expected to rise at 5%, we could inflate the Year 0 price of energy at 2% per annum to take account of its rise relative to other prices and discount values back to the present using the real rate of interest.

While it is clear that, in principle, cash flow estimates should be based on *projected* changes in relative prices of inputs and outputs over the project's life, this is easier said than done, and, in most studies, it is *not* done, so that cost and benefit estimates tend to be based on the assumption of a constant price level *and* unchanged relative prices. What can be done, however, and is often done as standard practice, is to apply some form of risk or sensitivity analysis, which is discussed in Chapter 9. At this stage all that we need note is that a sensitivity analysis would identify those prices (and other parameters) on which the overall results of the project are most dependent, and calculate the net benefits of a project making a range of assumptions about possible price movements. This does not yield a clear answer, however, but it shows how project decisions depend on certain assumptions. Risk analysis takes this a step further, in the sense that it attempts to place probabilities on each price scenario and possible project outcome.

4.4 Incremental or relative cash flows

The concept of incremental cash flow is relevant for all types of investment projects. Its meaning and importance can most easily be explained, however, with reference to investments that aim at the improvement of existing schemes. Examples of these are abundant and may range from the simple replacement of an outdated piece of machinery by a more modern model to the complete rehabilitation of a factory or agricultural scheme, such as an irrigation project.

In such instances it is assumed that there is already a cash flow from the existing project and that the main objective of a proposed additional investment is to improve the net cash flow, either by decreasing cost, or by increasing benefits, or by doing both of these at the same time. Such an improvement in the net cash flow we call the *incremental cash flow* (or incremental net benefit flow). Generally defined, it is the difference between the net benefit flow *with* the new investment and the net benefit flow *without* this investment.

If we want to find out whether an improvement or rehabilitation is worthwhile, we should, in principle, look at the incremental cash flow, and the same holds true if we want to find out which of two alternatives is better, e.g. the rehabilitation of existing irrigation facilities or the construction of an entirely new irrigation scheme. The answer to these questions will often turn out to be in favour of rehabilitation. In many case, relatively small investments, if used to improve weak components (or remove bottlenecks) in a much larger system, can have relatively large returns.

At the same time, however, there are pitfalls in appraising or evaluating investments of the above type on the basis of incremental cash flows where we consider only the *with* and *without* rehabilitation scenarios. The main points can best be explained by means of a few simple hypothetical examples (Examples 1, 2 and 3 shown in Table 4.2). In the first two examples we show the net cash flow *with* rehabilitation (line 1), the net cash flow *without* rehabilitation (line 2) and the resulting incremental cash flow (line 3). It will be noted that in both cases rehabilitation yields the same improvement (the same incremental cash flow), but in Example 1 the starting position (the without cash flow) is much better than in Example 2 where it is negative and remains negative even with rehabilitation. In Example 2, rehabilitation does improve things but it merely cuts losses and does not convert these into positive net benefits. The question that arises, of course, is whether we should undertake rehabilitation in the event that the scheme concerned continues to yield negative returns even with the rehabilitation carried out. The answer is "no", assuming of course that the rehabilitation proposed is the best that can be done under the circumstances, and that the existing project can be terminated.

That the answer should be "no" may be further demonstrated with the help of Example 3, which starts from the same assumptions as case 2 (it has the same net cash flow without rehabilitation). In Example 3 we compare the "without" cash flow with the alternative of completely abandoning the scheme (closing down the factory or whatever the asset may be). Line 1 is now the cash flow if the scheme is abandoned (liquidated). We assume it to be

Table 4.2 Incremental cash flows

	0	1	2	3	–	10
Example 1						
1 Net Cash Flow With rehabilitation	−1500	750	750	750	–	750
2 Net Cash Flow Without rehabilitation		500	500	500	–	500
3 Incremental Cash Flow (1–2)	−1500	250	250	250	–	250
Example 2						
1 Net Cash Flow With rehabilitation	−1500	−250	−250	−250	–	−250
2 Net Cash Flow Without rehabilitation		−500	−500	−500	–	−500
3 Incremental Cash Flow (1–2)	−1500	250	250	250	–	250
Example 3						
1 Net Cash Flow With closure	0	0	0	0	–	0
2 Net Cash Flow Without closure		−500	−500	−500		−500
3 Incremental Cash Flow (1–2)		500	500	500	–	500

zero but it might be positive in Year 0 (because of a positive salvage value). The incremental cash flow (line 3) is now the improvement in the cash flow that results from abandoning (closing down) the scheme completely. If we compare this incremental cash flow with the one resulting from rehabilitation (Example 2), it shows, of course, that liquidating the scheme is the better alternative.

These examples illustrate the danger of basing a rehabilitation decision simply on the resulting incremental cash flow – taking as a basis only two alternatives: the scheme *with* and the scheme *without* rehabilitation. The above examples show that it is necessary to consider at least one more alternative – that of liquidating the scheme.

A further look at these examples will also show that there is an alternative to incremental cash flow analysis. Instead of comparing *with* and *without* flows and calculating an incremental (or relative) flow, one could look at the *with* and *without* flows themselves (the absolute flows) and use DCF appraisal techniques to decide which is the better of the two – the *with* or the *without* flow – and, at the same time, see if the *with* flow yields sufficiently large net benefits to make it worthwhile to continue (instead of liquidating the project).

The question that arises is why introduce the incremental cash flow at all or even – as in the case of certain textbooks – lay down the rule that all projects should be appraised on the basis of incremental cash flows? The answer is that in the case of many projects it is extremely difficult to establish the full *with* and *without* situation and that the only feasible alternative may be to estimate directly the improvement (the incremental cash flow) expected to result from the project. Many investments aim at introducing marginal and gradual improvements to existing projects, where it may be more practical to estimate directly the net benefits resulting from the improvements instead of having to compare the complete benefit and cost information in the area of operations affected by the improvements.

4.5 Capital costs and the treatment of depreciation

As a rule, there will usually be two forms of investment or capital cost: *fixed investment* and *working capital*. In this section we discuss issues relating to these two types of capital investment, including replacement investment, salvage or terminal values, and depreciation.

Fixed investment

Fixed investment, as the term suggests, usually refers to the acquisition of all those capital assets such as land, buildings, plant and machinery that effectively remain intact during the production process, apart from the usual wear-and-tear. Investment in fixed capital is usually concentrated in the first year or years of a project's life, but, as different types of fixed investment have different life spans, replacement investment will also occur in various years over a project's life. These costs will, of course, appear as negative items in the cash flow. It is also conceivable that at the end of the project period at least part of the fixed investment will be intact: almost certainly the land (unless this has been destroyed, environmentally, by the project), buildings, equipment, vehicles, etc. As these can be sold off, it is usual for there to be a positive amount (inflow of cash) in the last year of the project representing scrap, salvage or terminal value of the project's fixed assets. In some instances, however, additional investment-related costs could be incurred. For instance, some types of projects are costly to close down, such as nuclear power plants. Mining investments too, will often have a negative cash flow in the final year if environmental regulations require that the mine developers restore the natural environment to its "original" (or at least, pre-mining) state. Of course, not all investments require that the investor purchases physical capital goods. In the case of investment in human capital by an individual, for instance, the "fixed" investment is in the form of an initial outlay on education and training and the benefits consist of the stream of additional earnings flows in the future.

What about depreciation charges? In the financial statements of an enterprise's operations, there will always be a "cost" item labelled "depreciation". Should this also be entered as a capital cost? If it is included in the accountant's profit and loss statement as an item of expenditure, should we leave it in when calculating the project's operating costs? The answer to both questions is "NO!"

A golden rule in discounted cash flow analysis is that all investment costs should be recorded in the cash flow only in the year in which the investment cost was actually incurred. This rule represents a departure from standard bookkeeping practices where an accountant will apportion investment costs over a number of years in the firm's profit and loss statements. This apportionment is an accounting convention and is done for purposes of calculating an operation's *taxable profits*, which, as we are going to see later on, are not necessarily the same thing as net operating revenues in discounted cash flow (DCF) analysis. Although the project accountant will always include depreciation of capital equipment as a cost in preparing a profit and loss (or income and expenditure) statement for the project, depreciation is not like other project expenses or costs in the sense that it does not actually involve a *cash transaction*.

Depreciation is primarily a bookkeeping device – in principle it identifies and records as a cost the part of the project's income or benefit that would notionally be required to be set aside to finance capital expenditure to replace worn-out capital assets or recoup capital costs in the future. In practice, the amount set aside as depreciation in any one year is based on some percentage of the value of the capital equipment already installed. Let us say, for example, that we have a project with an investment in buildings, machinery and transport equipment worth $100,000. Of this, $50,000 is the value of the project building with an estimated life of 50 years; $30,000 is the value of machinery with an estimated life of 10 years, and $20,000 is the value of transport equipment with an estimated life of 5 years. One method of calculating the annual depreciation allowance is the straight-line method,

Table 4.3 Calculating depreciation using the straight-line method

Item	Initial cost ($)	Life	% Annual depreciation	Annual depreciation allowance ($)
Buildings	50,000	50	2	1000
Machinery	30,000	10	10	3000
Transport Equipment	20,000	5	20	4000
Total	100,000	–	–	8000

where we spread the depreciation evenly over the life of the investment. Applying this method to the above example would give us the results shown in Table 4.3.

As can be seen from Table 4.3, the project accountant would enter $8000 as a "cost" or "expenditure" each year when calculating the project's taxable profits. As discussed in Chapter 2, the sum of the depreciation allowance set aside in this manner each year would, in principle, permit the project to finance the replacement of each item of capital equipment at the end of its expected life.

However, from the point of view of the project analyst, who is interested in deriving the net cash flow of the project, depreciation should *not* be treated as a project cost. If we were to include depreciation as a cost, we would be double counting the project's investment cost and therefore undervaluing the project's net benefit. The reason for this is simple. If we look at the examples in Chapters 2 and 3, we will be reminded that the investment costs of the various projects we looked at were recorded as costs *as and when they were actually incurred*, i.e. usually at the beginning of the project's life, and then again when various items of equipment were replaced. If we were now to include depreciation allowances as an annual "cost", we would, in fact, be recording our investment costs twice – once at the beginning of the project's life and then again as a series of annual costs. For this reason, project analysts do not treat depreciation as a cost in calculating a net cash flow. Therefore, if the project accountant provides us with a net income (or net cost) figure that includes depreciation as a cost, we should be careful to *add* this amount back in (i.e. as a "benefit" to our annual cash flow).

Working capital

In addition to fixed investment, it is normally necessary for an operation to have some working capital as part of its operations. Working capital refers essentially to stocks of goods that the business or project needs to hold in order to operate. A manufacturer who is engaged in the production of goods using raw materials will need to hold stocks of raw materials to offset possible interruptions to supply. The producer will also want to hold stocks of finished goods to meet orders without delay. If the producer uses machines, equipment, vehicles, etc. it may also be necessary to maintain a reasonable stock of spare parts, and perhaps fuel, if these are not close at hand from nearby suppliers. At any point in time, a project will be carrying such stocks. As existing stocks are used up, they need to be replaced with new stocks, so the actual components of working capital will be changing during the course of the year, and it is possible that their *total value* will change from one year to the next. The most important point to note about working capital is that it is only a *change in the level* of working capital that should be treated as investment in cash flow analysis.

Table 4.4 Investment in working capital ($)

Item/Year	0	1	2	3	4	5
Total Working Capital	0	50	100	100	100	0
Investment in W/Capital	0	−50	−50	0	0	+ 100

When a project is established and the stocks of working capital are first set up, the full amount of the initial working capital will be recorded as an investment cost (cash outflow, or negative cash flow). If the total value of working capital then remains constant, even though its actual composition may change, there is no additional investment cost. If the stock of working capital were to *decrease*, i.e. a lower value of the total stocks at the end of the year than at the beginning, then investment has fallen and this must be recorded in the cash flow as an *inflow*. This will normally happen in the final year of the project's life when the project will either run down or sell off all its stocks of working capital completely. For example, in Table 4.4, the value of total working capital increases by $50 in Years 1 and 2, remains constant in Years 3 and 4, and falls by $100 in Year 5. These figures imply that $50 was spent on acquiring working capital in each of Years 1 and 2, nothing was spent in years 3 and 4, and the $100 capital stock was run down in Year 5. Since investment expenditures are a cost, they are recorded as negative numbers in Years 1 and 2 of the Investment in Working Capital row, and as a benefit (an avoided cost) in Year 5.

In setting up the cash flow the lower row entitled "Investment in Working Capital" will be added to the cash flow for fixed investment to arrive at a total investment cash flow. As noted above, a positive investment in working capital appears as a *negative* cash flow and vice versa, because investment is always treated as a negative value in cash flow analysis, reflecting its status as a cost.

4.6   Interest charges, financing flows and cash flow on equity

Another item of "capital cost" that the accountant will have included in the project's profit and loss statement is interest paid on money borrowed, perhaps to finance the initial capital investment. Again, we must ask the same questions as in the case of depreciation. Should interest be included as part of investment cost, or, if it has been included as part of the operating costs, should it remain in our cash flow statement when undertaking DCF analysis? Again the answer to both questions is "NO", but we should qualify this by stating that interest charges should not be included in the cash flow *in the initial instance when we are interested in the project's total profitability* – we call this the *Market Analysis*. At subsequent stages, when we are interested in calculating the project's profitability *from the standpoint of the owners' equity capital* – we call this the *Private Analysis* – then *all* cash inflows and outflows relating to the debt financing of the project, including the disbursements and repayments of the loan itself, need to be incorporated into the cash flow. Furthermore, as with depreciation, interest charges are a legitimate tax-deductible cost, so when we come to calculate taxes and the net cash flow *after tax* in the private appraisal, we will need to include interest as a project cost. We return to this point in the next section.

The cash flows of the projects that were discussed in Chapter 3, for example, were all *market* cash flows, i.e. *before* financing. The cash flow used in most cost-benefit analyses is usually the market cash flow which shows the financial flows expected to result from the project itself and contains no information about the way the project might be financed. Any given project can, in theory at least, be financed in many different ways, involving different possible combinations of debt and equity finance, and, within the debt finance category, many different possible loan arrangements and terms. Different loan terms, involving, say, different rates of interest and/or different maturities will generate different profiles of financing inflows and outflows for the project's owners or *equity* holders. It is therefore important that we are able to consider any given project's profitability *independently* of the terms on which its debt finance is arranged –*market appraisal*, using our terminology. If we were to ignore this issue, it is conceivable that a "bad" project could be made to look good, simply by virtue of its sponsors having access to concessional funding on terms much more favourable than what the financial markets offer. Conversely, a potentially "good" project may look "bad" only because its sponsor is unable to secure (or is unaware of) more favourable loan conditions available elsewhere in the market.

Does the foregoing imply that in appraising a project and coming to a decision about its profitability without considering the cost of funding, we are effectively ignoring the "cost of capital"? The answer is "NO!" This is precisely what we are doing when we apply DCF analysis to a project's cash flow, whether we use the NPV or IRR decision rule as discussed in Chapter 3. When we discount a cash flow, we use a discount rate that we believe to be a good indication of the opportunity cost of funds to the party concerned, as determined by the capital market.

As we noted in the discussion of the Market Analysis, the project analyst's focus on a project's overall cash flow (ignoring the manner in which the project is to be financed) should not be interpreted as implying that he is not interested in knowing the financial arrangements or terms of the debt finance for any particular project under consideration. Indeed, it is a very important part of the appraisal process also to derive the cash flows showing the inflows and outflows of debt finance, and then to derive a separate net cash flow for the project *after debt finance*. Whether the project is under consideration by a private firm, a public–private partnership or a state enterprise, this part of the appraisal process is called the *Private Analysis* in our terminology.

The two questions immediately arising in this connection are: (i) what is shown by such a cash flow *after* financing?; and (ii) how can such a cash flow be used in the project appraisal process?

These two questions can best be answered with the help of a simple, hypothetical example, shown in Table 4.5, illustrating the case of a small-scale farmer participating in a large publicly-supported agricultural scheme. Farmers participating in the scheme are expected to make a number of investments from their own funds at the farm level, perhaps for a tractor, some simple machinery and tools, irrigation equipment, and so on, and will also be provided under the scheme with certain inputs (e.g. irrigation water) against payment of a certain fee. All these investments and operating costs, together with the farm-gate value of the crops produced, are reflected in the net cash flow *before* financing (the investment cash flow at the farm level), shown in Table 4.5 (line 1), and valued at constant prices. The *investment cash flow* shows an initial investment of $5000 in Year 0, to be undertaken by the farmer, followed by a stream of net benefits of $1000 per annum for 10 years (we often use the "dash" symbol to indicate continuation of an equal annual flow of funds, as shown in Table 4.5). Farmers participating in the scheme will also qualify for a loan covering part of the required investment and carrying a real interest rate of 7% per annum. The loan has to be repaid over 10 years. The flows connected to this loan are shown in line 2 of Table 4.5. They constitute the *financing flow* (or in the terminology used in corporate

Table 4.5 Deriving the private cash flow on farmers' equity

	Year										
	0	1	2	3	4	5	6	7	8	9	10
1. Market net cash flow *before* financing	−5000	1000	1000	–	–	–	–	–	–	–	1000
2. Debt finance flow	3512	−500	−500	–	–	–	–	–	–	–	−500
3. Private net cash flow *after* financing (1 + 2)	−1488	500	500	–	–	–	–	–	–	–	500

accounting, the *debt finance flow*). The financing flow shows that the farmer receives a loan of $3,512 in Year 0, while the servicing cost of the loan (amortization including interest) will be $500 per annum for 10 years (when you consult your Annuity Tables you will find that an annuity of $500 per year for 10 years is just sufficient to service a loan of $3,512 at an interest rate of 7%).

Note that the profile of a debt finance flow is the mirror image of that of an investment flow. An investment flow starts negative and then turns positive, if everything goes well. A debt finance flow starts positive and then turns negative, from the standpoint of the borrowing party.

The net cash flow *after* financing, shown in line 3, is the sum of the investment cash flow (line 1) and the debt finance flow (line 2). If we examine Table 4.5, we notice that the net cash flow *after* financing shows us two things: (i) which part of the investment has to be financed from the farmer's own funds ($1,488); and (ii) the net return the investor can expect ($500 per annum) after meeting the servicing cost of the loan. While the net cash flow *before* financing (the Market Analysis) shows us the return to the *project* investment at the farm level, the net cash flow *after* financing shows us the return on that part of the investment financed from the farmer's own funds: in the terminology used in corporate accounting this is called the return on *equity*. In our terminology, it is called the *Private* return.

This still leaves us with the question of why this return is calculated and why we do not stick to the general rule of using the cash flows before financing. The answer to the last part of the question is easiest. We are in fact not departing from the rule, as the farm level cash flow after financing is not the basis for public-sector decision-making about the scheme as a whole. The decision to carry out the scheme will be based on the aggregate net benefits of the scheme, of which the net cash flow before financing, accruing to all the farms in the scheme, and identified by the Market Analysis, forms an important component. The Private cash flow after financing is only calculated because it determines whether farmers will be willing to participate in the scheme on the conditions offered to them. What farmers will be primarily interested in are the net returns accruing to them after having met their financial obligations. If these are too low (or perhaps zero in the case in which all net benefits generated at the farm level go towards servicing the loan), farmers will not be interested in participating.

Returns to farmers may be too low for a number of reasons. One reason may be that the potential returns on the scheme *as a whole* as indicated by the Market Analysis are not high enough. That would mean that, on the information available, the scheme should not be carried out or should be revised. If, on the other hand, the aggregate costs and benefits of the

Table 4.6 Market cash flows equal debt plus equity (private) cash flows

	Year										
	0	*1*	*2*	*3*	*4*	*5*	*6*	*7*	*8*	*9*	*10*
1 Market net cash flow	−5000	1000	1000	–	–	–	–	–	–	–	1000
2 Debt finance flow	−3512	500	500	–	–	–	–	–	–	–	500
3 Equity finance flow	−1488	500	500	–	–	–	–	–	–	–	500

scheme in the Market Analysis show that it is potentially attractive, then too low returns to farmers shown by the Private Analysis will mainly be a reflection of the distribution of net benefits between farmers and other parties in the scheme (the public authority running the scheme and general government). It is an indication that the incentives to farmers (the conditions on which they can participate) may have to be reviewed.

We noted above that the profile of the debt financing flow, from the perspective of the borrower – in this case the farmer – was *positive* in the initial, investment period and then *negative* over the repayment period. Of course, from the standpoint of the lender, this debt cash flow has the opposite signs: negative followed by positive. It may be useful to consider the same project example we used previously (Table 4.5), where we now show the two components of the financing flows, debt and equity, from the perspective of the funding sources: *lenders* and *owners* (or, *equity holders*). Their respective financing flows are shown in Table 4.6, rows 2 and 3.

Notice that the two parts of the project's financing flows add up to exactly the same amount as the overall *Market* cash flow. This is true by definition, for that part of the investment that is not funded by the lender (debt) must come from the investor's own funds (equity). Conversely, that part of the positive net cash flow that is not paid out to (received by) the lender (interest plus principal repayments) is left over as profit to the equity holder. This cash flow on equity (the *Private* cash flow in our terminology) is, in effect, derived as the residual cash flow, as we saw previously.

Example: Calculating the return on total investment, the cost of debt finance and the return on equity

We have seen that financing inflows and outflows are of two broad types: *debt* (loans, long- or short-term, trade credit, etc.), and *own funds* (equity, shares, etc.). A project can be funded entirely by debt, entirely by own funds or by a combination of these. The latter is most common, i.e. the investor finances part of the project from borrowed funds and the remainder from "own funds" (whether it is an independent producer, private company or state enterprise).

The point to remember is that the total (or net) inflow and outflow of *finance* must, by definition, correspond exactly with the net inflow and outflow of the project's costs and benefits, as measured by its cash flows, over the project's life. A useful exercise is to calculate the following from Table 4.6:

i   the IRR on the total investment made at the farm level: the *Market* IRR (line 1);
ii   the rate of interest implied in the debt finance flow (line 2);
iii the rate of return on the farmer's own investment: the *Private* IRR (line 3).

If it is then also assumed that, given the alternatives open to farmers, full participation in the scheme will be attractive to them only if they can secure a return on their own (equity) investment of 20% per annum, we can then ask, what flow of annual net benefits *before* financing would be required to give farmers that minimum return? The initial investment would still be $5000 and the amount of debt finance and the conditions on which it can be obtained (line 2) are also unchanged. We can also work out what IRR on total investment (the Market IRR) at the farm level is required to give farmers the minimum return of 20% on their own investment (the Private IRR). Given the nature of the cash and financing flows, the quickest way to compute your answers is by making use of the annuity factors provided in Appendix 2:

i   To calculate the Market IRR on the project's Net Cash Flow before financing, we follow the simple procedure shown in Chapter 3, using the Annuity Tables:

$$NPV = -5000 \, (1.0) + 1000(AF_{10}) = 0$$

$$AF_{10} = 5000/1000 = 5$$

From the Annuity Tables, IRR is approximately 15%.

ii   Assuming we did not already know what the annual rate of interest was on the loan, we can determine this by calculating the IRR on line 2, the "Debt Finance Flow":

$$NPV = 3512(1.0) - 500(AF_{10}) = 0$$

$$AF_{10} = 3512/500 = 7.024$$

From the Annuity Tables, IRR is exactly 7%.

iii   From the residual cash flow (line 3), we derive the Private IRR on the farmer's Own Funds:

$$NPV = -1488(1.0) + 500(AF_{10}) = 0$$

$$AF_{10} = 1488/500 = 2.976$$

From the Annuity Tables, IRR exceeds 30%.

From the results of these calculations we see that although the IRR on the project as a whole (as identified by the Market Analysis) is only 15%, the farmers themselves can expect a return of around 31% on their own investment, due to the availability of loans at a favourable 7% interest rate. We can also ascertain what minimum cash flow on the project in the Market Analysis would be needed to ensure that farmers earned a 20% Private rate of return on their own investment. Again, using the Annuity Tables:

For Private IRR of 20% on farmer's investment

$$NPV = -1488(1.0) + X(AF_{10/0.20}) = 0$$

Where X = Farmers' annual net benefit after financing.

$$(AF_{10/0.20}) = 4.192 = 1488/X$$

$$X = 1488/4.192 = 354.9$$

This example shows us that the annual net benefit required to provide a 20% private rate of return for farmers, after financing, is \$354.9. As the loan is repaid at \$500 per annum, the required annual net benefit before financing will be \$354.9 plus \$500 = \$854.9. We can then calculate what the IRR on this cash flow is:

Market IRR on Net Cash Flow before financing:

$NPV = -5000(1.0) + 854.9(AF_{10}) = 0$

$AF_{10} = 5000/854.9 = 5.849$

$IRR = 11\%$

In other words, if farmers are to earn 20% on their investment, the Market Analysis must show a rate of return of at least 11% per annum.

4.7 Taxation and after-tax net cash flows

Up until now all the examples we have considered ignore company or profits tax. When private investors appraise a project, they will want to know what return to expect on an after-tax basis. From a private investor's standpoint, taxes paid to government are, like any other costs, to be deducted in deriving a net cash flow. (In Chapters 5 and 6 we will see that in social cost-benefit analysis, when a project is appraised from the standpoint of society as a whole, taxes should *not* be treated as a project cost.) Tax laws vary considerably from one country to the next. These laws will determine what can and cannot be treated as a legitimate tax-deductible expense, what rate of taxation will apply to different types of project, and so on. It is not the purpose of this section to consider the implications for cash flow analysis of the tax laws of any country in particular. The main point that needs to be stressed here is that the analyst should be aware that there are certain components of project cost that we, as project analysts, do not consider as "costs" in DCF analysis, but which are considered as allowable costs when calculating taxes liable. In other words, these items of project cost may not enter into our derivation of a project's cash flow *directly* (in the Market Analysis), but they will affect the Private net cash flow *indirectly* through their effect on the project's *taxable profits*.

We refer here to two items dealt with at some length in the preceding sections: *depreciation* and *interest*. We saw that neither of these should be treated as components of project cost when considering the overall project in the Market Analysis. We saw that depreciation is an accounting device, and since the project analyst includes investment costs in the cash flow as and when they are incurred, to also include depreciation would involve double counting. Similarly, interest charges on borrowed capital are not treated as *project costs per se* as these relate to the financing of the project, and the process of discounting in DCF analysis uses the discount rate as the opportunity cost of capital. To include interest costs would again amount to double counting. But as both depreciation and interest charges appear in the accountant's profit and loss statements for the project, and therefore affect the project's taxable profits, we need to perform a "side" calculation of the project's taxable profit, applying the same conventions used by the accountant.

In effect, all that this amounts to is setting up an additional "cash flow" that includes all the same items of operating costs and revenues as in our Market Analysis, to which we must:

i   add back in depreciation and interest charges as deductible "costs" for taxation purposes; and

ii   exclude capital costs (and subsequent inflows of salvage or terminal values, unless it is stipulated that these are to be treated as incomes or losses for taxation purposes).

Having derived a taxable income cash flow we then apply the appropriate tax rate to that stream to obtain a "taxation due" cash flow. *It is only this "taxation due" line that should be added to our main cash flow table in order to derive the after-tax Private net cash flow.*

One issue that is likely to arise is the treatment of *losses* in any year. This will depend on the laws of the country in question, but, by and large it would normally be reasonable to assume that losses incurred in one project could be offset by the enterprise against profits it earns elsewhere, or, that losses in one year may be offset against profits in subsequent years. In the first case, you would then need to calculate the *decrease* in overall tax liability resulting from the loss and treat this as a *positive* value in deriving the *after-tax* cash flow. In the second case you would need to reduce the subsequent year(s) taxable profits by the magnitude of the loss, which again would imply a positive impact on the *after-tax* cash flow in the subsequent year(s).

4.8 The discount rate

In calculating net present values, a range of discount rates, incorporating both the domestic government bond rate and the firm's required rate of return on equity capital, should be used. The Private Analysis will be used to assess whether the project is attractive from an investor's point of view, and this will be determined by comparing the Private IRR with the required rate of return, or by considering the NPV calculated at a rate of discount approximately equal to the required Private rate of return. However, where the private firm is a member of the Referent Group, the Private Analysis may also be used to assess some of the implications of the project for the Referent Group in the social cost-benefit analysis, and, as argued in Chapter 5, the government bond rate is generally appropriate for discounting Referent Group net benefits. The firm's required rate of return on equity may be unknown and the range of discount rates chosen should be wide enough to include any reasonable estimate. If no attempt is made to inflate benefits or costs, a real rate of interest is required as the discount rate, and, as noted in Chapters 2 and 3, the use of a real rate of interest involves the implicit assumption that all prices change at the general rate of inflation. The real domestic bond rate can be approximated by the money rate less the current rate of inflation (which acts as a proxy for the expected rate). The real required rate of return on equity capital is approximated by the money rate of return less the expected rate of inflation. There are also considerations of risk which are deferred until Chapter 9.

In summary, a range of discount rates is used for several reasons: first, in the Private Analysis, the firm's required rate of return on equity capital may not be known; second, as to be discussed in Chapter 5, there may be valid reasons why the market rate of interest should be replaced by a shadow (or social) discount rate in the Efficiency or Referent Group Analysis, and we do not know what that rate should be; and, third, we want to see how sensitive the NPV is to the choice of discount rate (if it is very sensitive, this compounds the difficulty posed by the uncertainty about the appropriate discount rate).

4.9 Summary of the relationship between the Market Analysis and the Private Analysis

Before considering a worked example, a brief review of the approach described above to assessing private profitability will be helpful. We first undertake a Market Analysis in which all project inputs and outputs are valued at market prices in the derivation of the time-stream of net benefits. Inputs or outputs which are not traded in the market (pollution, for example) are effectively priced at zero as they have no effect on the firm's profitability, though they may play an important role in the subsequent Efficiency Analysis. If we subtract from the net benefit stream calculated by the Market Analysis any indirect taxes levied on the output of the project, we obtain the firm's net earnings stream before interest, tax, depreciation and amortization (EBITDA). We now subtract interest and loan repayments (amortization) to get the time-stream of net returns to the firm's equity holders, before business income tax is deducted. While interest and depreciation charges are not relevant in the Market Analysis, they must be considered in calculating business income tax liabilities. A side calculation is undertaken to determine these and the resulting tax liability is deducted from the before-tax earnings stream to calculate the net returns to equity holders, the Private Analysis. The net returns in the Private Analysis are then subjected to IRR and NPV analysis in order to assess private profitability.

4.10 Derivation of project private cash flows using spreadsheets

In this section, using a hypothetical agricultural case study, we illustrate, step-by-step how to derive a project's (Private) net cash flow from basic information, derived from forecasts, about the project's costs, revenues and financing. We have chosen this kind of project for this case study because the range of agricultural inputs and outputs is readily identifiable and familiar to most readers. In undertaking the study we also recommend a method for designing and setting up the spreadsheet in such a way that it is readily amenable to subsequent revisions or refinements, whether these are for purposes of Efficiency Analysis, Referent Group Analysis, or sensitivity and risk analysis. The underlying principle we adhere to in setting up a spreadsheet is that any project datum or value of a variable (or potential variable) should be *entered into the spreadsheet only once*. Any subsequent use of that data should be in the form of a reference to the address of the cell into which the value was first entered.

Worked example: National Fruit Growers (NFG) project

In the year 2015, NFG is considering a potential project to grow apples, peaches and pears in Happy Valley. The current market prices are: apples, $1000 per ton; peaches, $1,250 per ton; and pears, $1,500 per ton. Once the project is at full production, the company expects to grow and sell 100 tons of apples, 90 tons of peaches and 75 tons of pears per annum.

To set up the project NFG plans to rent 100 Ha of land, currently planted with fruit trees, at an annual rent of $30 per Ha. The contract will be for 20 years and the company will begin paying rent in 2016. NFG will purchase the capital equipment immediately (in year 2015) consisting of: (i) 4 units of farm equipment at $100,000 each; (ii) 3 vehicles at $30,000 each; and (iii) 250 m² of storage units at $1000 per m². To calculate depreciation for tax purposes, the equipment has a book life of 10 years; vehicles 5 years; and buildings 20 years. (In practice, the company plans to maintain these assets for the full duration of the investment without replacement and at the end of the project it anticipates a salvage value of 10% of the initial cost of all investment items.)

NFG will also need to invest in various items of working capital in 2016. These are: 2 tons of fertiliser stocks at $500 per ton; 2,500 litres of insecticide at $30 per litre; 10 months'

Table 4.7 NFG's annual operating costs

Item	No. of units	Cost/unit ($)
Rent on land (Ha)	100	30
Fuel (lt)	2500	0.70
Seeds/plants (kg)	250	20
Fertiliser (Tons)	3	500
Insecticides (lt)	3000	30
Water (ML)	900	20
Spares & maintenance	12	1000
Casual labour (days)	100	60
Administration (per month)	12	1000
Insurance	1% investment cost	8263.5
Management salary	12	3000
Miscellaneous (p.a.)		7700

supply of spare parts for equipment and vehicles at $1000 per month; and 500 litres of fuel at $0.70 per litre. The items included in annual operating costs are summarised in Table 4.7.

NFG expects operations to begin in 2016, initially at 25% of capacity, increasing to 50% in 2017, 75% in 2018 and to full capacity in 2019. All operating costs and revenues in the years 2016 to 2018 are assumed to be the same proportion of full-scale costs and revenues as output is of the full-scale level. The project is to be terminated after 20 years of operation.

To finance the project, NFG intends to secure an agricultural loan for $700,000 in 2015, carrying an interest rate of 3.5% per annum (real), repayable over 10 years. NFG will also take out a bank overdraft in 2015 for $40,000, at 5% (real), which it intends to repay after 4 years. The balance of the required funding will be met from NFG's own funds (equity). The current tax rate on agricultural profits is 25%, payable annually after deducting depreciation allowances and interest on debt. NFG's accountant informs us that any losses incurred in this operation may be deducted, for taxation purposes, from profits the company earns in other agricultural projects, in the same year.

You have been asked to assist NFG by calculating the following:

i   the IRR and NPV (at 5%,10% and 15% real rates of interest) for the project, before tax and financing, i.e. a Market Analysis; (ii) the private IRR and NPV (at 5%,10% and 15% real) on NFG's own equity, before *and* after tax, i.e. a Private Analysis.

ii   tax and financing, i.e. a Market Analysis; (ii) the private IRR and NPV (at 5%,10% and 15% real) on NFG's own equity, before *and* after tax, i.e. a Private Analysis.

To qualify for the subsidies on its inputs, NFG may have to satisfy the government that this project is in the broader public interest – is socially worthwhile. It is also conceivable that NFG will want to consider variations of this project in some form of scenario and sensitivity analysis in the future. You should therefore be careful to set up your spreadsheet for the Market and Private Analysis in such a way that it is readily amenable to changes in the values of key variables at a later stage, and to further development to calculate Efficiency and Referent Group benefits and costs. (The cash-flow entries should be in thousands of dollars.)

Step 1: Setting up the "Key Variables" table

As we will be using most of the variables at a number of different places in the spreadsheet, it is important that we enter all the "Key Variables" in one dedicated section of

Table 1: Key Variables		Project			Operating Costs	No.	Price	Cost
Investment Costs	No.	Price	Cost					
(i) Fixed Investment	(units)	($)			Rent on land (Ha)	100	30	3,000
Farm equipment (units)	4	100,000	400,000		Fuel (litres)	2,500	0.7	1,750
Vehicles (units)	3	30,000	90,000		Seeds (Kg)	250	20	5,000
Buildings (m2)	250	1,000	250,000		Fertilizers (tonnes)	3	500	1,500
TOTAL			$740,000		Insecticides (litres)	3,000	30	90,000
(ii) Working Capital					Water (ML)	900	20	18,000
Fertilizer Stocks (tons)	2	500	1,000		Spares	12	1,000	12,000
Insecticide stocks (Litres)	2500	30	75,000		Casual labour (days)	100	60	6,000
Equipment spare parts (units)	10	1,000	10,000		Administration (/month)	12	1000	12,000
Fuel stocks (Litres)	500	0.7	350		Insurance (p.a.)	1%	826,350	8,263
TOTAL			$86,350.0		Management (/month)	12	3,000	36,000
(iii) Salvage Value	10%		$74,000		Miscellaneous	1	7,700	7,700
					TOTAL (Market prices)			$201,213
Depreciation	Life(yrs)	Amount p.a.			Revenues			
Equipmnt.	10	40,000			Apples (tons)	100	1,000	100,000
Vehicles	5	18,000			Peaches (tons)	90	1,250	112,500
Buildings	20	12,500			Pears (tons)	75	1,500	112,500
Financing	Amount	Interest	Life (yrs)		TOTAL			$325,000
Loan	$700,000	3.5%	10			Conversion factor		1000
Overdraft	$40,000	5.0%	4	Capacity Output				
Discount rate =	5.0%	10.0%	15.0%		2016	2017	2018	2019+
Tax rate on profits =	25.0%			%	25%	50%	75%	100%

Figure 4.1 Spreadsheet set-up of "Key Variables" table.

the spreadsheet. We suggest that this section is created at the top of the spreadsheet as a table, in the format shown in Figure 4.1. (In more complex case studies the "Key Variables" section is more likely to be a separate spreadsheet in a workbook.)

Notice that all the information contained in the preceding paragraphs about the project has been arranged systematically under the headings "Investment Costs", "Operating Costs", "Revenues". We have also provided the relevant information about other important variables including: depreciation allowances for each type of fixed investment; the expected salvage value of investment; the details of the loans for financing the project; the rate of taxation; the percentage of full capacity output; and the discount rates. As we want to express all values in thousands of dollars, we have also included a "conversion factor" of 1:1,000 for this purpose. Later on we may wish to add further variables for the purpose of undertaking the Efficiency CBA and doing sensitivity and risk analysis. The values of the variables entered in the table have been used to create some working tables in sub-sections of the main Key Variables table: the data in the table (together with formulae) have been used to derive sub-totals for each of the main categories of the cash flow, i.e. Fixed Investment Costs, Working Capital, Operating Costs, and Revenues. This procedure, which will greatly simplify the subsequent spreadsheet analysis of the project from a private viewpoint, will also be followed when we come to the Efficiency CBA.

Step 2: Setting up the project's cash flow for the Market Analysis

To set up the main cash flow table for the Market Analysis we have listed the categories of cost and benefit in the left-hand column as shown in Figure 4.2, while the project years (from year 2015 through to 2035) are listed across the columns, so that the cash flow runs from left to right rather than top down. We recommend this format for reasons

TABLE 2: MARKET NET CASH FLOW																					
	2015	2016	2017	2018	2019	2020	2021	2022	2023	2024	2025	2026	2027	2028	2029	2030	2031	2032	2033	2034	2035
ITEM/YEAR	0	1	2	3	4	5	6	7	8	9	10	11	12	13	14	15	16	17	18	19	20
Investment costs																					
Fixed Investment	-740.0																				74.0
Working Capital		-86.4																			86.4
Total Investment	-740.0	-86.4	0.0	0.0	0.0	0.0	0.0	0.0	0.0	0.0	0.0	0.0	0.0	0.0	0.0	0.0	0.0	0.0	0.0	0.0	160.3
Operating Costs		-50.3	-100.6	-150.9	-201.2	-201.2	-201.2	-201.2	-201.2	-201.2	-201.2	-201.2	-201.2	-201.2	-201.2	-201.2	-201.2	-201.2	-201.2	-201.2	-201.2
Revenues		81.3	162.5	243.8	325.0	325.0	325.0	325.0	325.0	325.0	325.0	325.0	325.0	325.0	325.0	325.0	325.0	325.0	325.0	325.0	325.0
Net Cash Flow	-740.0	-55.4	61.9	92.8	123.8	123.8	123.8	123.8	123.8	123.8	123.8	123.8	123.8	123.8	123.8	123.8	123.8	123.8	123.8	123.8	284.1
(Before Financing & Tax)																					
	5%	10%	15%																		
Net Present Value =	609.6	100.4	-178.3																		
IRR =	11.5%																				

Figure 4.2 Spreadsheet set-up of market cash flow table.

of convenience when working with a number of additional working tables in the same spreadsheet. It should also be noted that we have chosen, in this instance, to have a single line for major components of the cash flow, such as Operating Costs and Revenues. In some instances, where one may want to have more detail on the composition of these categories, it might make sense to include sub-categories of cash flow components in the main cash flow table: this is essentially a matter of stylistic choice and what you as the analyst consider more user-friendly. For convenience, all entries in Table 2 (see Figure 4.2), and subsequent tables, are in thousands of dollars, converted using the Conversion Factor in Table 1 (Figure 4.1).

The main point to note, however, is that every cell in Table 2 of the spreadsheet is based on a reference back to Table 1 in the spreadsheet (see Figure 4.1) containing the key variables. Once the categories of the cash flow have been entered, we can set up the Net Cash Flow which contains the simple arithmetic formula subtracting all project costs from the revenues, by adding negative cost values to positive revenue values. We then insert the appropriate NPV and IRR built-in formulae. We see from the results that the project has a Market IRR of 11.5%, and that the NPV is therefore negative at a 15% discount rate. If we were to change any of the key variables in Table 1, so too would the cash flows and NPV and IRR results change. The cash flow will then be updated and the results will be recalculated automatically.

Notice also that, at this stage, we have set up the cash flow for the project as a whole – the *Market* CBA – ignoring how it is financed. To ascertain the profitability of the project from NFG's *Private* (or "equity") standpoint, after tax, we need to incorporate the relevant details of debt financing and taxation.

Step 3: Setting up the cash flow for the financing of the project

The *Key Variables* table (see Table 1 in Figure 4.1) also contains details of the loan and bank overdraft NFG intends to use to finance the project. We are informed that NFG will take out a loan of $700,000 in year 2015, that this loan will carry interest at a real rate of 3.5% per annum, and that it will be repaid over 10 years. For the purpose of this exercise, we are operating with real rates of interest and not incorporating inflation in any of the values included in the table. We need to calculate what the annual repayments and interest charges, in real terms, will be on this loan, assuming it takes the form of an annuity: equal annual instalments of interest and principal repayment. However, we must also anticipate that we will need to know what the interest component of the annual annuity is in order to calculate NFG's taxable profits, as interest charges are tax-deductible costs. We use two of the spreadsheet's built-in formulae for this purpose. The annual principal repayments are calculated using the "PPMT" formula, and the interest payments are calculated using the "IPMT" formula illustrated in in Figure 4.3. (These formulae use the "Years" numbers reported at the top of the spreadsheet to calculate "PER" and "NPER" values required by the formulae in each year.)

Notice again that all the entries are references back to the Key Variables table in Figure 4.1. Also notice that we present these cash flows from the standpoint of the lenders, so the signs are the opposite of what they would be from the borrower's perspective. The sum of each year's interest and principal repayment comes to the same constant amount of $84,2000 (which can also be calculated using the "PMT" formula). The cash flow for the bank overdraft is shown in the rows below the main loan: $40,000 is borrowed in 2015 and repaid in 2019. In each year, interest of $2000 in real terms (5% real rate of interest) is paid. Adding these financing flows together gives us the Net Financing Flow. Subtracting this from the Market Net Cash Flow (Figure 4.2) gives the Net Cash Flow (on equity, before tax). Now we need to calculate what tax NFG will need to pay on the profits earned from this project.

Step 4: Calculating business income taxes and after-tax flows

To calculate taxable profits we need to set up a separate working table within Table 3, as shown in the shaded section of Figure 4.3 entitled (ii) Taxes. There is an important reason why we calculate tax separately rather than in the main cash flow table itself: taxable profits are not the same as net cash flow. You will recall from earlier sections of this chapter that for the purpose of calculating taxes, a private investor is allowed to include depreciation and interest charges (but not principal repayments) as project costs. By the same token, investment costs (both fixed and working capital) are not taken account of in the calculation of taxable profits, and salvage value from the sale of assets at the end of the project's life is also excluded on the assumption that it is not subject to tax. Depreciation is calculated from the information provided in the Key Variables table in Figure 4.1, while interest on loans is taken from the first part of Table 3. From these rows we derive taxable profits from which Taxes Liable are derived, using the tax rate of 25% given in the Key Variables table. Note that in the first three years (2016 to 2018), taxable profits and taxes liable are negative, as the project is running at a loss. As NFG is entitled to deduct these losses from profits it earns in other projects, we should still include these "negative taxes" as a benefit in deriving the after-tax net cash flow.

Step 5: Deriving the Private (after-tax) net cash flow

It is now a simple matter to derive the Private after-tax net cash flow by subtracting the Taxes Liable row from the Net Cash Flow (before tax) in Table 3 (see Figure 4.3). The results for the *Private* NPV and IRR after tax are then derived by inserting the appropriate NPV and IRR formulae as shown in Figure 4.3. We see from these results that the after tax real rate of return on NFG's equity is 19.1%, and that the NPV of the investment is $59,600 at a 15% discount rate.

This completes the Private Analysis of NFG's proposed project. In later chapters of this book we return to this case study to demonstrate how we would extend this analysis to include the calculation of *Efficiency* and *Referent Group* net benefits. We also extend the scope of the spreadsheet analysis to consider the treatment of risk.

TABLE 3: PRIVATE NET CASH FLOW

ITEM/YEAR	2015	2016	2017	2018	2019	2020	2021	2022	2023	2024	2025	2026	2027	2028	2029	2030	2031	2032	2033	2034	2035
	0	1	2	3	4	5	6	7	8	9	10	11	12	13	14	15	16	17	18	19	20
(i) Financing																					
Principal																					
Loan	-700.0	59.7	61.8	63.9	66.2	68.5	70.9	73.3	75.9	78.6	81.3										
Overdraft		-40.0				40.0															
Interest																					
Loan		24.5	22.4	20.3	18.0	15.7	13.3	10.8	8.3	5.6	2.8										
Overdraft				2.0	2.0	2.0															
Net Financing Flows	-700.0	44.2	86.2	86.2	86.2	126.2	84.2	84.2	84.2	84.2	84.2										
NCF(equity, pre-tax)	-40.0	-99.6	-24.3	6.7	37.6	-2.4	39.6	39.6	39.6	39.6	39.6	123.8	123.8	123.8	123.8	123.8	123.8	123.8	123.8	123.8	284.1
(ii) Taxes																					
Revenues		81.3	162.5	243.8	325.0	325.0	325.0	325.0	325.0	325.0	325.0	325.0	325.0	325.0	325.0	325.0	325.0	325.0	325.0	325.0	325.0
Operating Costs		-50.3	-100.6	-150.9	-201.2	-201.2	-201.2	-201.2	-201.2	-201.2	-201.2	-201.2	-201.2	-201.2	-201.2	-201.2	-201.2	-201.2	-201.2	-201.2	-201.2
Depreciation		-70.5	-70.5	-70.5	-70.5	-70.5	-52.5	-52.5	-52.5	-52.5	-52.5	-12.5	-12.5	-12.5	-12.5	-12.5	-12.5	-12.5	-12.5	-12.5	-12.5
Interest on loans		-24.5	-24.4	-22.3	-20.0	-17.7	-13.3	-10.8	-8.3	-5.6	-2.8	0.0	0.0	0.0	0.0	0.0	0.0	0.0	0.0	0.0	0.0
Profits (before tax)		-64.1	-33.0	0.1	33.3	35.6	58.0	60.5	63.0	65.7	68.4	111.3	111.3	111.3	111.3	111.3	111.3	111.3	111.3	111.3	111.3
Taxes Liable		-16.0	-8.3	0.0	8.3	8.9	14.5	15.1	15.8	16.4	17.1	27.8	27.8	27.8	27.8	27.8	27.8	27.8	27.8	27.8	27.8
(iii) Equity (after tax)																					
Private Cash Flow	-40.0	-83.6	-16.0	6.6	29.3	-11.3	25.1	24.5	23.9	23.2	22.5	96.0	96.0	96.0	96.0	96.0	96.0	96.0	96.0	96.0	256.3

	5%	10%	15%
NPV =	$483.3	$196.4	$59.6
IRR =	19.1%		

Use the "PPMT" and "IPMT" formulae to calculate debt payments

Depreciation and interest payments are included in the calculation of taxable profits

The private cash flow is derived by subtracting debt financing flows and profits taxes from the market net cash flow in Table 2

Figure 4.3 Spreadsheet set-up for private net benefits.

4.11 Further reading

Most corporate finance texts cover the issues discussed in this chapter in more detail. An example is J.C. Van Horne, *Financial Management and Policy*, 12th edition (published by Kashif Mizra, 2012). This book is also available in more recent editions with a country-specific focus, e.g. J.C. Van Horne, R. Nicol and K. Wright, *Financial Management and Policy in Australia*, 4th edition (Englewood Cliffs, NJ: Prentice Hall, 1995).

Appendix to Chapter 4

Case study of International Cloth Products[1]

The textile industry is very important to developing country economies and gives rise to significant international trade flows. This Appendix discusses and commences the appraisal of the proposed International Cloth Products Ltd project using an Excel© spreadsheet. An appraisal of this project can be undertaken as a series of exercises and the results compared with the solution tables in the appendices to Chapters 5, 6, 7, 9 and 10.

Imagine that you are in the Projects and Policy Division of the Ministry of Industry in Thailand. Using the information which follows and which relates to the activities and plans of a wholly-owned subsidiary of a foreign company, called International Cloth Products Limited (ICP), you are asked to answer the following questions:

i   Should the Government support the proposed yarn-spinning project of ICP in Central Thailand by granting a concession as requested by ICP?

ii   Should the Government induce ICP to locate in the less-developed region of Southern Thailand? If so, should the Government be willing to give the incentives requested to bring this about?

What additional information would you like to have? How would you go about getting such information? What conditions, if any, would you wish to impose on the investment if approval is given?

The project analyst is required to calculate the returns on the investment: (i) to the private investor, ICP; and (ii) to Thailand, under each of seven different scenarios. There are two possible locations, Bangkok and Southern Thailand, and at each of these, there are three possible cases: (i) no concessions at all, or (ii) no import duties to be paid by ICP on cotton imports; or (iii) ICP enjoys a tax holiday on profits taxes. Finally, the analyst should consider the case where ICP enjoys *both* concessions at the alternative location in southern Thailand.

It is also suggested that there should be a written report that:

i   starts with an executive summary;

ii   briefly discusses method of analysis, significant assumptions and the choice of key variables;

iii   indicates the best scenario for ICP and Thailand, and the extent to which each party stands to gain or lose from the different concessions;

iv   calculates IRRs and provide a sensitivity analysis showing how the NPVs will vary
      at different real discount rates between 0% and 20%;
v   is supported by summary tables.

An example of such a report on the ICP Project is provided in Chapter 13.

ICP Project

1 Introduction

International Cloth Products Ltd (ICP) is a foreign-owned textile firm that manufactures
cloth products in the Bangkok area of Central Thailand, including both unfinished
and finished cloth for sale to other textile and garment manufacturers, as well as textile
prints for the domestic and export market. At present, it imports cotton yarn, but it is
thinking of investing at the end of 1999 in central Thailand in a spinning mill with sufficient capacity to supply not only its own demand for yarn (10 million lbs per annum
for the foreseeable future), but also to supply about 5 million lbs for sale to other textile firms in Thailand. Values of inputs or output are expressed either in terms of local
currency units or in foreign exchange. The exchange rate of the Thai Baht (Bt.) is
assumed to be Bt.44 = US$1.00. We will find it useful as a first stage to analyse the prof-
itability of the project to ICP, but we should keep the analysis as simple as possible and
round off all figures to one decimal place of a million Baht. Assume, for simplicity, that all
payments and receipts are made and received at the end of the year in which they occur.

2 The market

A study of the textile industry shows that Thailand has the capacity to weave 125 million
linear yards of cloth. Existing and planned spinning mills, except for ICP's proposed mill,
are expected to be capable of producing enough yarn for 75 million linear yards of cloth,
leaving a deficit of yarn for 50 million yards to be imported. One yard of cloth requires 0.33
lbs of yarn so that the domestic market for ICP's output seems to be reasonably assured. The
Thai government has no plans to change the duties on imported yarn. At present the landed
c.i.f. price of yarn is $0.58/lb and this is expected to remain constant in real terms in the
future (c.i.f. stands for cost, insurance and freight and the concept is discussed in detail in
Chapter 10). There is an import duty of Bt.2.1/lb, and customs handling charges of Bt.0.3/
lb. This determines the Bangkok factory gate price that ICP is paying for its own imported
yarn and also the price at which it expects to be able to sell yarn to other firms in Thailand.

3 Investment costs

ICP plans to purchase 40,000 spindles and related equipment (blowing machine, air condi-
tioning and so on) capable of processing 16.2 million lbs of cotton into 15 million lbs of yarn
each year. The equipment is expected to last for ten years (including the start-up year), and
technical consultants have given a satisfactory report on the company's investment plans.
The estimated investment costs and their expected date of payment are listed in Table A4.1.
   During the year 2000, ICP hopes to produce at about one-half of full capacity, reaching
full capacity in the following year. In 2000, it expects water, power and raw material costs
to be 50%, all other costs to be 100%, and output to be 50% of full-time operational levels.

Table A4.1 Investment costs for yarn-spinning project: ICP

Item	Cost		Date of payment
	$'000	Bt.'000	
Spindles	4000[a]		1999
Ancillary equipment	850[a]		"
Installation	100[b]		"
Construction of buildings		36,000[c]	"
Start-up costs		18,000[b]	"
Working capital			
(i) raw materials (3 months' requirements at c.i.f. value)	1200[d]		2000
(ii) spare parts (1 year's use)	243[a]		"

(a) Plus import duty at 5% of c.i.f. value.
(b) Consisting of a mixture of traded goods at c.i.f. value and skilled labour.
(c) All unskilled labour costs.
(d) Plus import duty at 10% of c.i.f. value.

4 Raw materials

A total of 16.2 million lbs (3,240 bales) of imported cotton will be required to produce 15 million lbs of yarn (1.08 lbs of cotton/lb of yarn). The raw material requirement is expected to be strictly proportional to the output of yarn. The current price of cotton, expected to remain unchanged, is $0.30/lb c.i.f. Bangkok, and the import duty is 10% *ad valorem* (i.e. 10% of the price).

5 Direct labour force

In addition to the labour employed on the investment phases of the project, the mill will employ about 1,600 workers in spinning and associated operations (see Table A4.2). All workers will be recruited from the Bangkok area and paid the going market rate for their labour. But there is considerable unemployment in the area and the shadow wage of unskilled labour recruited in the Bangkok area is assumed by the government to be 50% of the market wage. For the less-developed areas of the country, the government uses a shadow wage for unskilled labour of zero. The concept of the shadow wage will be discussed in Chapter 5. However, because there is very little unemployment among skilled workers, all skilled labour is valued, in efficiency terms, at its market price regardless of its place of employment.

Table A4.2 Employment in the yarn-spinning project: ICP

	No. employed and date of first payment	Average wage salary (Bt. p.a.)	Total annual wage/ salary bill (Bt.'000)
Supervisors and technicians	100 (during 2000)	150,000	15,000
Other skilled workers	1000 (during–2000)	22,500	22,500
Unskilled workers	500 (during–2000)	12,000	6000
Total	1600		43,500

6 Fuel, water, spare parts

Water and electricity each cost Bt.480,000 per million lbs of yarn produced. Power will be supplied from a hydro-power plant at a nearby dam using transmission lines already in place. The variable cost to the Electricity Corporation of Thailand of supplying power to ICP for the spinning mill is about 50% of the standard rate charged in Bangkok. ICP expects to import and use spare parts (in addition to those reported as investment costs) costing 10% of the c.i.f. cost of equipment in each year after the first (that is from Year 2000 onwards). The duty on spare parts is 5%. The labour costs of maintenance are already included in Table A4.2.

7 Insurance and rent

The cost of insurance and rent are each estimated as Bt.3 million p.a. from Year 2000 onwards (it can be assumed that the insurance risk is widely spread so that the social cost of risk is zero – see Chapter 9, Section 9.6.2). The spinning mill's share of overheads (already being incurred) is estimated to be Bt.7.5 million p.a.

8 Financing

ICP intends to finance imported spindles and equipment under the foreign supplier's credit for 75% of the c.i.f. value in 1999. The credit carries a real interest rate of 8% p.a. and is repayable from year 2000 in five equal annual instalments. In addition, once the plant is in operation, ICP expects to finance Bt.60 million of the working capital reported in Table A4.1 from a bank overdraft borrowed from local banks in 2000 and carrying a real 10% p.a. interest rate payable from 2001 onwards. The bank overdraft will be repaid at the end of the project (year 2009). Interest payable on bank overdraft and supplier's credit are allowable against profits tax. The remainder of the investment will be financed from ICP's own sources, but the company expects to be able to (and intends to) remit all profits to its overseas headquarters as and when they are made.

9 Taxes and incentives

ICP claims that this investment is not sufficiently profitable to induce them to invest, at the present profits tax of 50%. (The profits tax is levied on the profits calculated after allowing annual depreciation to be charged at 10% of the initial cost of spindles, equipment, installation and construction.)

The depreciation charge can be levied from 2000 onwards. Losses made by the yarn-spinning project are assumed to be offset against ICP's profits from the existing cloth-producing operations. ICP say that they are unlikely to invest unless they are given some concession; *either* in the form of duty-free entry of cotton, *or* exemption for ten years from profit taxes.

10 Location in "deprived areas"

ICP wants to locate its spinning mill (if it goes ahead with the investment at all) in the Bangkok area, adjacent to its existing plant. (It already owns the vacant land next to its

Table A4.3 Additional cost of locating in Southern Thailand

Item	*Additional costs (Bt.'000)*
1. Additional capital costs to be incurred at end–1999:	
• construction (more factory buildings, workers' housing, site preparation)	30,000[(a)]
• transportation and handling equipment	6000[(b)]
• total additional capital outlay	36,000
2. Changes in recurrent costs (at full capacity):	
• transport of raw materials from Bangkok (about 100 miles at Bt.0.27/ton mile)	1,950
• transport costs for yarn to TFL's cloth factory and warehouse at Bangkok (100 miles at Bt.0.27/ton mile)	1,800
• 20% reduction in wages of unskilled workers	–1,200
• cost of additional supervisors	450
• total additional recurrent costs	3000

(a) all unskilled labour costs;
(b) Plus import duty at 5% of c.i.f. value.

existing factory.) But the government has a policy of dispersing industries to "deprived areas". The government would like to see the spinning mill located in Southern Thailand some distance to the south of Bangkok, but ICP has said that it will locate the spinning mill there only if it is given *both* duty-free entry of cotton *and* a ten-year tax holiday. ICP has provided a table showing the estimated additional costs of locating in southern Thailand (see Table A4.3).

Some solution spreadsheet tables for the ICP case study

This section shows the set-up of the spreadsheet tables for the first part of the case study. This consists of three tables linked together in one spreadsheet, but presented here in three separate Figures, showing: the Basic Data table (Figure A4.1); the Market Cash Flow (Figure A4.2) and the Private Cash Flow (Figure A4.3). (Also linked to these tables are the Efficiency Cash Flow and the Referent Group Cash Flow, presented as appendixes at the end of Chapters 5 and 6 respectively.)

ICP PROJECT: 1999-2009
TABLE 1: KEY INPUT VARIABLES

VARIABLES		INVESTMENT COSTS	000 $'	000 BT'	RUNNING COSTS		MATERIAL FLOWS		FINANCE	
EXCHANGE RATE	44	SPINDLES	4000	176000	WATER BT PER MN LBS	480000	COTTON MN LBS	16.2	SUPPLIERS CREDIT (%)	75%
TARIFF A	5%	EQUIPMENT	850	37400	POWER BT PER MN LBS	480000	YARN MN LBS	15	SUPPLIERS CREDIT (BT)	160,050
TARIFF B	10%	INSTALLATION	100	4400	SPARE PARTS (%)	10%	LABOUR 1	100	OVERDRAFT	60
PROFITS TAX	50%	CONSTRUCTION		36000	INSURANCE&RENTBTMN	6	LABOUR 2	1000	CREDIT REPAY YEARS	5
DEPRECIATION	10%	START-UP COSTS		18000	OVERHEADS	0	LABOUR 3	500		
PRICE YARN CIF US$	0.58								DISCOUNT RATES	4%
YARN TARIFF BT	2.1	WORKING CAPITAL								8%
HANDLING BT	0.3	RAW MATERIALS	1200	52800						12%
TARIFF C	10%	SPARE PARTS	243	10692						16%
PRICE COTTON CIF US$	0.3								CONVERSIONS	
WAGE 1 BT PA	150000	SHADOW PRICES	%						CAPACITY OUTPUT	50%
WAGE 2 BT PA	22500	CONSTRUCTION	50%						THOUSANDS	1000
WAGE 3 BT PA	12000	POWER	50%							
INTEREST RATE 1	8%	INS & RENT	0%							
INTEREST RATE 2	10%	WAGE 3	50%							

All basic data for the project and private cash flow calculations are entered in one table

Figure A4.1 ICP project solution: key input variables.

TABLE 2:   MARKET ANALYSIS

YEAR	1999 MN BT	2000 MN BT	2001 MN BT	2002 MN BT	2003 MN BT	2004 MN BT	2005 MN BT	2006 MN BT	2007 MN BT	2008 MN BT	2009 MN BT
INVESTMENT COSTS											
SPINDLES	-184.800										
EQUIPMENT	-39.270										
INSTALLATION	-4.400										
CONSTRUCTION	-36.000										
START-UP COSTS	-18.000										
RAW MATERIALS		-58.080									58.080
SPARE PARTS		-11.227									11.227
TOTAL	-282.470	-69.307	0.000	0.000	0.000	0.000	0.000	0.000	0.000	0.000	69.307
RUNNING COSTS											
WATER		-3.600	-7.200	-7.200	-7.200	-7.200	-7.200	-7.200	-7.200	-7.200	-7.200
POWER		-3.600	-7.200	-7.200	-7.200	-7.200	-7.200	-7.200	-7.200	-7.200	-7.200
SPARE PARTS		-22.407	-22.407	-22.407	-22.407	-22.407	-22.407	-22.407	-22.407	-22.407	-22.407
RAW MATERIALS		-117.612	-235.224	-235.224	-235.224	-235.224	-235.224	-235.224	-235.224	-235.224	-235.224
INSURANCE		-6.000	-6.000	-6.000	-6.000	-6.000	-6.000	-6.000	-6.000	-6.000	-6.000
LABOUR		-43.500	-43.500	-43.500	-43.500	-43.500	-43.500	-43.500	-43.500	-43.500	-43.500
TOTAL		-196.719	-321.531	-321.531	-321.531	-321.531	-321.531	-321.531	-321.531	-321.531	-321.531
REVENUES											
SALE OF YARN		209.400	418.800	418.800	418.800	418.800	418.800	418.800	418.800	418.800	418.800
NET BENEFIT (Project)	-282.470	-56.626	97.269	97.269	97.269	97.269	97.269	97.269	97.269	97.269	166.576

DISCOUNT RATE	NPV BT(Mn)	IRR (%)
4%	405.3	20.6%
8%	259.8	
12%	152.0	
16%	70.7	
0%	605.6	

Working capital appears at start (–ve) and end of project (+ve)

All entries in cash flow tables are references to cells in 'Key Input Variables' Table A4.1, or are formula drawing on these data and other cells in this table.

NPV and IRRs shown in box.

Figure A4.2 ICP project solution : the market cash flow.

TABLE 3: PRIVATE ANALYSIS

YEAR	1999 MN BT	2000 MN BT	2001 MN BT	2002 MN BT	2003 MN BT	2004 MN BT	2005 MN BT	2006 MN BT	2007 MN BT	2008 MN BT	2009 MN BT
DEBT FLOWS											
SUPPLIERS CREDIT	160.050										
INTEREST		-12.804	-10.621	-8.264	-5.719	-2.969					
REPAYMENT		-27.282	-29.464	-31.821	-34.367	-37.116					
BANK OVERDRAFT	60.000										-60.000
INTEREST		-6.000	-6.000	-6.000	-6.000	-6.000	-6.000	-6.000	-6.000	-6.000	-6.000
NET FINANCING FLOWS	160.050	19.914	-46.086	-46.086	-46.086	-46.086	-6.000	-6.000	-6.000	-6.000	-66.000
DEPRECIATION		-26.447	-26.447	-26.447	-26.447	-26.447	-26.447	-26.447	-26.447	-26.447	-26.447
NET OPERATING REVENUE		12.681	97.269	97.269	97.269	97.269	97.269	97.269	97.269	97.269	97.269
TAXABLE PROFIT		-26.570	56.201	56.558	59.103	61.853	64.822	64.822	64.822	64.822	64.822
PROFIT TAX		13.285	-28.100	-28.279	-29.552	-30.926	-32.411	-32.411	-32.411	-32.411	-32.411
NET FLOW ON EQUITY	-122.420	-23.426	24.069	22.905	21.632	20.257	58.858	58.858	58.858	58.858	68.165

DISCOUNT RATE	NPV BT(Mn)	IRR (%)
4%	154.5	16.7%
8%	88.7	
12%	40.8	
16%	5.4	
0%	246.6	

The IRR and NPVs show the profitability from the firm's private viewpoint.

Inflows and outflows on loans and profits taxes are then deducted from the "Market" cash flow to derive cash flow on the firm's own funds or "equity" the Private cash flow.

This table shows how the firm calculates its taxable profits. Depreciation and interest payments are included as "costs" only for the purpose of calculating taxable income.

Figure A4.3 ICP project solution : the private cash flow.

Exercises

1   A project's net cash flow, valued at current prices, is given by:

	Year					
	0	1	2	3	4	5
Net Cash Flow ($):	−1000	250	250	250	250	250

i   The nominal or money rate of interest is 8% and the anticipated rate of general price inflation over the next five years is 3%. Calculate the NPV of the net cash flow:

    a   using the real rate of interest as the discount rate;
    b   using the nominal rate of interest as the discount rate.

ii   Suppose that you are told that all prices used to calculate the project net cash flow in years 1–5 are expected to rise at 5% (2% above the general rate of inflation). Recalculate the NPV of the net cash flow:

    a   using the real rate of interest as the discount rate;
    b   using the nominal rate of interest as the discount rate.

2   The dam currently providing irrigation water to farmers in Happy Valley cost $100 million to build, and the valley currently produces 15 million kilos of sugar per annum which sells at the world price of $1 per kilo. The height of the dam can be raised at a cost of $8 million in order to provide more irrigation water to farmers. It is estimated that, as a result of the extra water supplied in this way, sugar production would rise to 15.5 million kilos per annum in perpetuity (i.e. forever), without any increase in the levels of other inputs such as fertiliser, labour, equipment, etc. Using a 10% rate of discount, what is the net present value, as indicated by the Market Analysis, of raising the height of the dam?

3   If a project involving an initial investment of $20 million is funded partly by debt (75%) and partly by equity capital (25%), and the internal rate of return on the overall cash flow of the project is 15%, while the rate of interest on the debt capital is 10%, what is the rate of return on equity capital? (Make any additional assumptions explicit.)

4   Suppose that a firm has borrowed $1000 in the current year at a 10% interest rate, with a commitment to repay the loan (principal and interest) in equal annual instalments over the following five years. Calculate:

    i   the amount of the annual repayment;
    ii   the stream of interest payments which can be entered in the tax calculation of the private cost-benefit analysis.

5   A firm's stock of working capital at the end of each year in the life of a project was as follows:

	Year					
	0	*1*	*2*	*3*	*4*	*5*
Working Capital ($)	0	50	75	70	30	20

Calculate the working capital cost stream which should be entered in the Market Analysis.

6   Explain why depreciation and interest should not be included as costs in a discounted cash flow (DCF) analysis of a project.

Note

1   This is a fictitious company. Any similarity to any actual company is unintended and accidental.

5 Cost-benefit analysis and economic efficiency

5.1 Introduction

In Chapter 4 we valued the project's outputs and inputs at market prices to calculate a net benefit stream which we then split into components measuring the net benefits to various stakeholder groups – the private firm and/or public corporation, the banks and the tax authorities in our example. In the present chapter we argue that the net benefit stream valued at market prices may be an incomplete measure of the effect of the project for two reasons: in some cases, the market price of an output or input may not accurately measure its value or cost to the economy as a whole; and in some cases an output or input, which confers a benefit or a cost, is not traded in a market, has no market price, and consequently is valued at zero in the Market Analysis. The effects of the production of goods and services in the latter category are termed external effects, or externalities, and they are classified as either positive, if they contribute to economic welfare – "goods", or negative if they detract from it – "bads". We need a complete measure of the net benefit stream if we are to calculate the net benefits to all groups affected by the project – the stakeholder groups identified in Chapter 4 plus the general public – and the present chapter considers the set of prices which will produce that measure.

Just as Chapter 4 was concerned with the distribution of net benefits identified by the Market Analysis, Chapter 6 will be devoted to analysis of the distribution of the overall net benefit identified by the analysis of economic efficiency. In the present chapter we focus on the circumstances in which market prices are not appropriate measures of benefit or cost and suggest procedures for determining *shadow-prices* for use in the cost-benefit analysis. In addition to substituting for some market prices in the Efficiency Analysis, shadow-prices will also be required for non-marketed goods and services, but the details of calculating those are left to Chapter 8.

When we refer to the overall net benefit of the project, we mean the aggregate of the benefits and costs experienced by all the various groups affected by the project. As noted in Chapter 1, a project involves a reallocation of scarce resources resulting in increases in output of some goods and services and reductions in output of others. These changes in output are calculated by comparing the *world with the project* with the *world without the project*. While some groups will benefit from the project and others will lose, the project is regarded as *efficient* if a redistribution of the benefits and costs could, in principle, make someone better off, with no one worse off. A project which could be rendered a *Pareto Improvement* through redistribution is judged to be a *Potential Pareto Improvement*, as discussed in Chapter 1. Each person who benefits from the project could, potentially, give up a sum of money so that they remain at the same level of economic welfare as they would

DOI: 10.4324/9781003312758-5

have been without the project; these sums are money measures of the project benefits. Similarly, each person who bears a net cost as a result of the project could, potentially, be paid a sufficient sum to keep them at the same level of well-being as they would have been without the project; these sums are money measures of the project costs. If the sum of the monies which could notionally be collected from beneficiaries exceeds that required to compensate those who are affected adversely by the project, then, according to the *Kaldor-Hicks Criterion* discussed in Chapter 1, the project is a Potential Pareto Improvement, which is our criterion of economic efficiency. We will return to the concept of measuring gains and losses in Chapter 7, but for the moment we simply note that a project which is revealed to have a positive NPV in an Efficiency Analysis, which includes all project benefits and costs appropriately measured, is an efficient use of resources according to this welfare criterion.

Before we consider how to appraise a project from the perspective of economic efficiency, we need to remind ourselves that the sums received from gainers and paid to losers in applying the Kaldor-Hicks criterion are notional and that these side payments are usually *not* part of the project design. Where some transfers are included in the project proposal, such as road tolls, for example, they usually impose costs and may detract from the overall net benefit of the project. In the absence of such side payments, the project will usually make some groups better off and some worse off, but the Efficiency Analysis ignores these distributional consequences, which may be important in the decision-making process for several reasons. First, the fact that the decision-maker has specified a particular subset of the economy as the Referent Group tells us that some groups "matter" more than others in the comparison of benefits and costs. Second, even if a project is inefficient from an overall perspective, it may still be undertaken because of its positive aggregate net benefits to the Referent Group. Third, even if the social cost-benefit analysis indicates that the aggregate net benefits to the Referent Group are negative, the decision-maker may still favour the project because it benefits a particular subset of the Referent Group. For example, an irrigation project in a low-income area may be undertaken to alleviate rural poverty even though the state as a whole incurs a negative NPV. The analyst needs to accept that some "efficient" projects will be rejected and some "inefficient" projects accepted on distributional grounds, and needs to feel comfortable with this outcome as long as the distributional consequences have been drawn to the attention of the decision-maker as clearly as is feasible in the analysis. The appraisal of the distribution of project net benefits identified in Chapter 6 is discussed in Chapter 11.

5.2 The competitive market

When trying to gauge the efficiency of resource allocation, economists commonly compare actual outcomes with the outcome that would be produced by a perfectly competitive market economy. In such an economy each market consists of a large number of well-informed buyers and sellers of the good or service which is being traded; there are no distortions, such as taxes or regulations; and there are no externalities, meaning that all goods and services which affect economic welfare are traded. All real-world economies depart from this ideal to a greater or lesser extent: some markets have a small number of buyers or sellers; taxes are levied on outputs and inputs; regulations such as rent control or minimum wages are in place; and some goods and services which affect economic welfare, such as clean air or water, may not be marketed. Economists use the competitive market not as a representation of reality but rather as a standard of efficiency against which actual outcomes may be judged. In a competitive market, trading would continue as long as economic agents could benefit from exchange of goods and services. Once all opportunities for mutual gain had been exhausted, it would not be possible to make anyone better off without making someone else worse off. In economic jargon, a state of *Pareto Optimality* would exist and there would be no scope for *Pareto Improvement*. Of course, the outcome might not be considered as "fair", but for the moment we are not concerned with equity but rather with economic efficiency.

If we lived in the perfectly competitive economy of the textbooks, with no external effects, and were concerned solely with economic efficiency, we could have terminated the discussion of cost-benefit analysis at the end of Chapter 4. The Market Analysis would have accurately measured all the gains and losses to all parties and a positive NPV would have indicated a Potential Pareto Improvement. However, all real-world markets are non-competitive to some degree and some depart significantly from the competitive ideal. Furthermore, markets in some critical goods and services do not exist at all. This means that the Market Analysis will be an imperfect judge of whether a project contributes to an improvement in the efficiency of resource allocation: some market prices fail to measure the value of an extra unit of output or the opportunity cost of an extra unit of input; and values of changes in output or input of some critical goods and services will not be represented in the analysis. We tackle these problems by developing *shadow-prices* for valuing certain goods and services in the analysis. These are not prices that are actually paid to, or received by any economic agent; they are purely to account for effects that are not measured by market prices. In this chapter we discuss how market prices can be adjusted, where appropriate, to provide a measure of benefit or cost to the economy; we will proceed by considering a series of examples of project outputs or inputs that are traded in imperfectly competitive markets and we will develop a rule which will enable us to shadow-price these goods and services. In Chapter 8 we consider how prices can be developed to measure the values of non-marketed goods and services.

If we are to use the competitive market process, however theoretical a concept this may be, to develop a standard of economic efficiency we need to understand its virtues. Figure 5.1 illustrates a competitive market in equilibrium where the demand (D) and supply (S) curves for a product intersect (the analysis applies equally to the market for an input). The demand curve tells us, at each level of output of the product, how much someone is willing to pay for one extra unit of it; the supply curve tells us, at each level of output, what an extra unit of the product will cost to produce. The cost of producing an extra unit of the product is the market value of the extra factors of production which will be needed. However, in a competitive market system, profit maximization ensures that factor prices are bid to a level where they are equal to the value of their contribution to additional output in their highest value use. In other words, the cost of an extra quantity of factors used to produce an extra unit of product, as measured by the supply curve for a particular product, is equal to the value of what that extra quantity of factors could add to production in their best alternative use. This means that the supply curve measures, at each possible output level, the opportunity cost of producing an additional unit of the product.

Since the demand curve slopes downward, it must be the case that the value of an extra unit of a product, its marginal benefit, declines as the level of output increases: this follows from the principle that the more of something you have, the less you value an additional unit of it. Since the supply curve slopes upward, it must be the case that the opportunity cost of producing an extra unit, the marginal cost of the product, increases as the level of output rises; this follows from the fact that producing more and more units of the product involves diverting scarce resources from progressively higher and higher value uses elsewhere in the

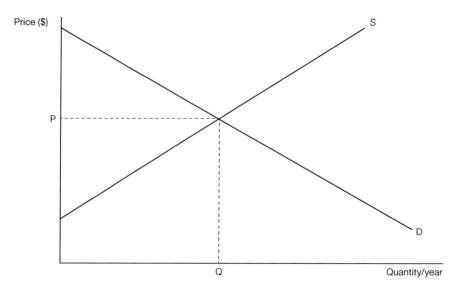

Figure 5.1 Competitive market equilibrium. Market equilibrium occurs at the intersection of the demand curve, D, and the supply curve S. At the equilibrium price, P, the quantity buyers wish to purchase equals the quantity sellers wish to sell. This equilibrium quantity traded is denoted by Q.

economy. The equilibrium point (P,Q) is efficient because at that point the last unit of the product supplied is worth exactly its opportunity cost of production; to the left of the equilibrium point an additional unit is valued higher than its opportunity cost – the gain from producing it outweighs the loss, and hence producing more of it contributes to efficiency; to the right of the equilibrium point, the reverse holds.

As illustrated in Figure 5.1, a feature of equilibrium in a competitive and undistorted market is that the demand price (the amount buyers are willing to pay for an additional unit of output) equals the supply price (the opportunity cost of an additional unit of output). In this situation no question arises as to whether the demand or supply price should be used to value an extra quantity of an output because the two prices coincide. In a non-competitive or distorted market, as we shall see, the demand and supply prices for a product are different from one another in equilibrium and the cost-benefit analyst must apply a pricing rule to determine which price provides the appropriate valuation of the output. The same argument applies to the market for an input, such as labour.

Returning to the competitive and undistorted market, suppose that a large-scale public project is proposed which will cause a change in the price of an input or an output: for example, a large hydro-electric project shifts the supply curve of electricity to the right, as illustrated in Figure 5.2, from S_0 to S_1 and this results in a fall in the market price from P_0 to P_1. We know that at output level Q_0 an extra unit of electricity is worth P_0, and at Q_1 it is worth P_1. Which price should be used to value the project output in an Efficiency Analysis? This question has a simple answer (i.e. use the average of the two prices) but involves some complicated issues which are postponed for discussion in Chapter 7. For the moment we will assume that we are dealing with a relatively small project which will have no effect on market prices of outputs or inputs – many projects are of this nature.

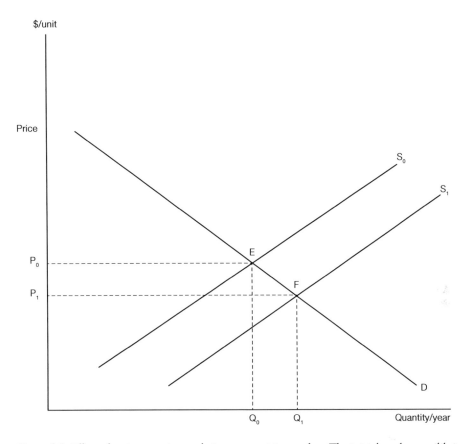

Figure 5.2 Effect of an increase in supply in a competitive market. The initial market equilibrium is at price P_0 and quantity traded is at Q_0. An increase in supply from S_0 to S_1 causes price to fall to P_1 and quantity to rise to Q_1. In each of the two equilibria, represented by points E and F, in this competitive market the demand price is the same as the supply price.

5.3 Shadow-pricing project inputs and outputs

We have argued that actual economies are not perfectly competitive in the sense of having a complete set of undistorted markets, without any individual buyer or seller having the power to set or influence price. In imperfectly competitive markets, which lack at least one of these characteristics, the observed demand price may not represent the marginal value of a good or service, and the observed supply price may not represent marginal cost. Indeed, in some cases, such as, for example, the market for wilderness recreation, the market (and the price) may not exist at all. As noted earlier, we will defer consideration of what to do in the case of missing markets until Chapter 8, and in the meantime focus on how to use the information provided by prices generated in imperfect markets in an analysis of economic efficiency.

The approach we adopt is known as *shadow-pricing*: where appropriate, we adjust the observed imperfect market price so that it measures marginal benefit or marginal cost, and we use the adjusted price – the *shadow-price* – in the Efficiency Analysis. Shadow-prices are sometimes also referred to as "efficiency prices" or "accounting-prices". It must be stressed again that these prices are used for the purposes of project appraisal and evaluation only, and are not prices at which project inputs or outputs are actually traded. It should not be forgotten that the purpose of shadow-pricing is to bring investment decision-making and resource allocation into line with what would be the situation in the absence of market distortions.

5.4 Shadow-pricing marketed inputs

In general, additional units of an input, produced to augment the current quantity supplied, are costed at the input's marginal cost of production which we will refer to from now on as its *supply price*, which measures its marginal cost of supply. In conditions of full employment, however, an input can be diverted from its existing use to be employed in the project. In that case, its opportunity cost is the value of its marginal product in that existing use, represented by its *demand price*. In a competitive undistorted market, the demand price of an input measures the value of its marginal product to its buyer, but indirect taxes, market regulation and lack of competition can result in a gap between the supply price of the input and the value of its marginal product.

Indirect taxes provide a useful example of how this gap is created and its implications for shadow-pricing. Consider an industry which is competitive apart from the existence of indirect taxes: an input is subject to an *ad valorem* indirect tax, and the output is also subject to such a tax. The profit-maximizing firm will employ the input up to the point at which:

$$P_{QS} * MPP_X = P_X(1 + t_X)$$

where:

P_{QS} is the price suppliers of the output receive
MPP_X is the marginal physical product of the input (the amount of extra output resulting from adding one extra unit of the input)
P_X is the price suppliers of the input receive
t_X is the rate of indirect tax on the input.

Now suppose that there is also a tax on the firm's output so that:

$$P_{QB} = P_{QS} (1 + t_Q)$$

where:

P_{QB} is the price buyers pay for the product produced by the firm
t_Q is the rate of indirect tax on that product.

By simple substitution we can work out the opportunity cost of diverting a unit of the input from use by the firm:

$$\text{Opportunity Cost} = P_{QB} * MPP_X = P_X(1 + t_X)(1 + t_Q)$$

where opportunity cost is the value buyers of the firm's product would have placed on the output forgone as a result of diverting a unit of the input to another use. As we can see from the above expression, the indirect taxes create a gap between the market price and the opportunity cost of the input. The expression for opportunity cost illustrates the concept of *tax cascading* – the taxing of taxes – where the output tax, t_Q, is applied to the supplier's price, P_{QS}, which already incorporates the input tax, t_X.

There are many jurisdictions in which indirect taxes on *both* inputs and outputs do not need to be taken into account in calculating the opportunity cost of an input. All OECD countries, except the United States, and many developing economies have adopted a Value Added Tax (VAT) system which avoids the so-called cascading effect by having taxes on inputs refunded as deductions against tax on output. In these circumstances it is effectively only the VAT rate which needs to be applied to the input price, implying that, in terms of the above expression for opportunity cost, $t_X = 0$. An exception is the United States where a traditional sales tax on retailed goods is levied in most states. However, an attempt is made to avoid tax cascading by exempting from such sales taxes the purchase of goods for further manufacture (inputs) or for resale, though goods that are retailed more than once, such as automobiles, can have tax levied on tax. As an approximation, sales taxes in the USA can be treated similarly to VAT by setting $t_X = 0$ as above. In many countries excise duties levied on a small range of goods and services are paid by manufacturers and then VAT is paid by the consumer. In situations in which sales taxes or excise duties apply to both inputs and outputs, we suggest approximating the tax premium on an input as: $T = t_X + t_Q$. For simplicity in the following discussion, we will maintain the assumption that only one tax rate is responsible for the divergence between input price and its opportunity cost. We also assume that the tax is distorting, leaving the discussion of corrective taxes until later.

5.4.1 Materials

A material subject to an indirect tax

In a market distorted by an indirect tax, there will be, in effect, two prices generated by the market at the equilibrium quantity of output. In Figure 5.3 these two prices are the demand price, P_B (the price buyers pay), and the supply price, P_S (the price suppliers receive). Which price should the cost-benefit analyst use to cost the quantity of the input to be used in the project under consideration?

If a small number of additional units of the input good or service are produced to meet the demand of the project, their marginal cost is measured by P_S. If, on the other hand, additional units cannot be produced in the time available, they will have to be diverted to the project from existing users who value them at P_B per unit, which is equal to the value of their marginal product in their existing use. Hence, in the latter case, P_B measures their opportunity cost – the value of output forgone by diverting them from their existing use.

At this point it will be helpful to be clear about the terminology used in describing the effect of indirect taxation. When we talk of the *gross of tax* price, or the *after-tax* price, or the *tax-inclusive* price of a good or service, we mean its price including the tax – P_B in Figure 5.3. The *net of tax*, or *before-tax*, or *tax-exclusive* price is its price *not* including the tax – P_S in Figure 5.3. Similarly when we talk of a *gross of subsidy*, or *before-subsidy*, or *unsubsidised* price, we mean the price before the subsidy is applied – P_S in Figure 5.5. The *net of subsidy*

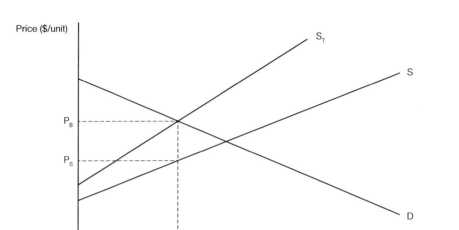

Figure 5.3 Market for a good or service subject to an indirect tax. The supply curve, S, shows the cost of an extra unit of output (marginal cost), and hence the price to be paid to producers (the supply price) for that extra unit, at each level of output of the good or service. The supply curve S_T shows the effect of an *ad valorem* tax (a tax levied as a proportion of supply price) on the price paid by buyers. Equilibrium is where the supply curve, S_T, and the demand curve, D, intersect. At the equilibrium level of output, Q, there are two prices for the good or service – the price paid by buyers, P_B, and the price received by sellers. P_S, and the amount of tax per unit of the good or service supplied is $(P_B - P_S)$.

or *after-subsidy* or *subsidised* price is the price once the subsidy has been applied and is given by P_B in Figure 5.5.

An imported material subject to a tariff

If domestic production of an input is protected by a tariff, there will be, in effect, two supply prices at the equilibrium quantity established by the market. In Figure 5.4 these two prices are denoted by P_B and P_W respectively: P_B is the cost of an extra unit of domestic production and P_W the cost of an extra unit of imports. Assuming the absence of a quota on the volume of imports, the small increase in input demand contributed by the project will be met by imports. While the firm will pay the market price, P_B, for the input, the cost to the economy of an additional unit of imports of the good is P_W. A portion, $(P_B - P_W)$, of the market price is not a real cost but rather a transfer of tariff revenue from the firm to the government. If, on the other hand, further imports were prevented by an import quota, the quantity of the good or service used by the firm would be supplied by additional domestic production, or, if this was not possible, would be diverted from other uses. In either case the opportunity cost of the good or service would then be measured by P_B.

A material subject to a subsidy

The effect of a subsidy is to lower the market price of a good or service, thereby encouraging its use. There can be reasons of economic efficiency for subsidizing the use of certain

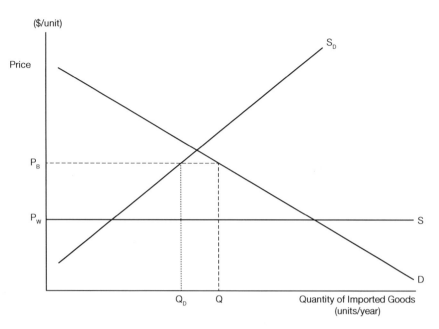

Figure 5.4 An imported good or service subject to a tariff. The supply curve, S, of the imported good is drawn as perfectly elastic, reflecting the fact that the level of the country's imports will not affect the world price of the good, P_W. Production from domestic sources is represented by the supply curve, S_D, which slopes upward because as quantity supplied increases the cost of supplying an extra unit, the *marginal cost of production*, increases as ever higher prices have to be offered to attract additional resources into the industry. The effect of the import tariff is to increase the price paid by domestic consumers for imports to P_B, and quantity demanded at this price is Q. At the price P_B domestic producers find it profitable to supply Q_D units of the good, with the remaining demand, represented by the distance QQ_D, being satisfied by imports.

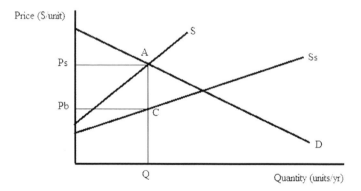

Figure 5.5 The supply curve S measures the marginal cost of producing the good or service. The supply to customers is represented by S_S, which is the marginal cost of production less the *ad valorem* subsidy (a percentage of the cost of production) paid by government to producers. The market demand, D, for the good or service is downward sloping, and market equilibrium is at Q. At equilibrium the marginal cost of production is $P_S = QA$, the subsidy is $(P_S - P_B) = AC$, and the price paid by consumers is $P_B = QC$.

products as described below in the discussion of corrective taxation, but here we will assume that the purpose of the subsidy is simply to support the incomes of the producers. Figure 5.5 illustrates equilibrium in the market for a product, such as sugar, which is subject to an *ad valorem* subsidy. While a project using sugar as an input would pay the price P_B per unit, there are in effect two prices at the equilibrium quantity, Q, and the analyst needs to decide which price to use in the Efficiency Analysis. If extra sugar were produced to meet the project's requirements, the opportunity cost of production would be the supply price, P_S, and this would be the shadow-price used in the Efficiency Analysis. Alternatively, if there was a quota on total production, the sugar required for the project would have to be diverted from an alternative use valued at the market price, P_B, and this would be the price used in the analysis.

A material produced under conditions of decreasing unit cost

One of the features of electricity production is the economies of scale over a wide range of plant output; this means that the average cost of production of power continues to fall as the chosen level of output increases until that level becomes quite large. Power plants come in discrete units: when the generating capacity has to be increased, it usually involves construction of a large plant which has the capacity to provide more power than is currently demanded of it. This means that the plant may be operating on the declining portion of its average (AC) and marginal cost (MC) curves, as illustrated in Figure 5.6. Since the power utility is a monopoly, the government regulates the price of electricity at a level sufficient to cover the utility's costs – at price P. Needless to say, there is a substantial literature and

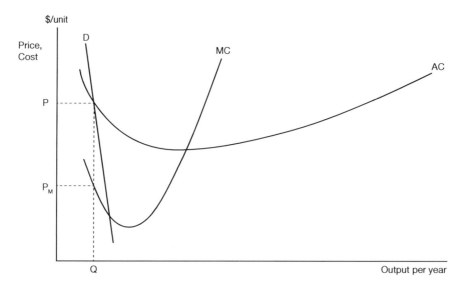

Figure 5.6 A good produced under conditions of decreasing unit cost. D represents the market demand curve for the good, which cuts the average cost curve, AC, in its declining region. P is the price, set by the market regulator, which allows the firm to recover its costs. MC is the marginal cost schedule which tells us the cost of producing an extra unit of output at each output level. Since equilibrium output, Q, occurs in the region of declining average cost, marginal cost, P_M, is less than average cost, P.

debate in economics about the complicated issue of regulating power utility pricing but we will not discuss it here.

When a good is produced under conditions of decreasing cost, the marginal cost of production, P_M in Figure 5.6, will be lower than the average cost. At the equilibrium quantity of output, Q, there are, in effect two prices of the good or service – the regulated price, P, paid by customers, and the marginal cost or supply price P_M. From the viewpoint of the analyst, what is the appropriate price of electricity, used as a project input, in an Efficiency Analysis? If the small quantity of power used in the project will be in addition to the total supply, it is valued at its supply price, or marginal cost – the value of the extra quantity of scarce resources required to produce the extra power. Hence power would be shadow-priced at the level of marginal cost – P_M in Figure 5.6. If, on the other hand, the utility is operating at capacity, the power used in the project will be diverted from some other use, and the electricity input to the project should be priced at its market price, P, which measures its value in its alternative use.

5.4.2 Labour

Labour subject to a minimum wage

In Figure 5.7, the labour market is regulated by a minimum wage. By the *wage* in this context we mean the unit cost of labour to the firm consisting, perhaps, of take-home pay, income tax withheld, superannuation contributions by the employer and payroll tax. In principle, as discussed earlier, if the market wage is to measure the value of the marginal product of labour, it may need to be augmented to take account of indirect tax on the output produced by labour, but we ignore this complication in the following discussion. It can be seen from Figure 5.7 that there are two prices of labour at the market equilibrium, Q_D – the supply price W_S (the wage which a marginal unit of labour would be willing to accept) and the market price, W_M (the cost to the employer of a marginal unit of labour). Suppose that a public project involves hiring labour in such a market (public projects are often designed to provide work for the unemployed), at what wage should the labour be costed in the Efficiency Analysis? The relevant cost is the opportunity cost – the value of what the labour would produce in its alternative use. If the worker hired for the public project would otherwise be unemployed, it might be tempting to say that the opportunity cost is zero. However, in the case illustrated in Figure 5.7, where the labour supply elasticity exceeds zero, that conclusion is incorrect: an extra unit of labour has an opportunity cost measured by W_S, the supply price which represents the value to the economy of the marginal hour of unemployed labour in its alternative occupation – work in the informal economy, such as gardening, or just plain leisure. Hence W_S is the appropriate wage at which to price the otherwise unemployed labour used in the project. (Remember that we are assuming that the project is not large enough to cause a change in the supply price of labour.) In other words, W_S is the shadow-price of labour.

Now suppose that unemployed labour receives an unemployment benefit. This payment is effectively a subsidy (a negative tax) on leisure which has the effect of reducing the supply of labour at all wage levels, and reducing the level of unemployment. Referring to Figure 5.7, it will be left to the reader to draw the new supply curve of labour, S_s, to the left of the original supply curve, S, and to locate the gross of subsidy return to the unemployed, W_{SS}. There are now two opportunity costs of otherwise unemployed labour in market equilibrium: the opportunity cost to the economy is the value, W_S, measured

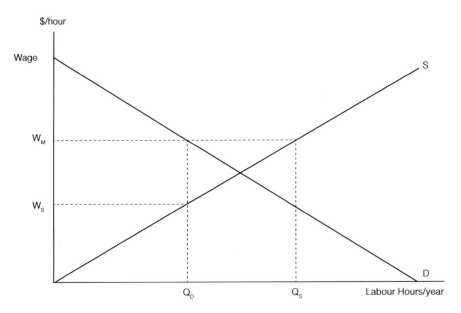

Figure 5.7 A minimum wage for labour. If the labour market were unregulated, the market wage
would be set by the intersection of the Demand (D) and Supply (S) curves. At this wage
level the market would *clear*– meaning that the quantity of labour demanded would equal
the quantity supplied. When a minimum wage, W_M, is set, by government or by a trade
union, at a level above the market-clearing price – in a bid to make workers better-off –
the quantity of labour supplied at that wage, Q_S, exceeds the quantity demanded, Q_D.
The difference between quantity supplied and quantity demanded, measured by the length
Q_DQ_S, represents involuntary unemployment – hours that people would like to work at the
minimum wage but that are not taken up by employers. From the viewpoint of employers,
hiring labour beyond the level Q_D would reduce profits because the value of the marginal
product of labour (VMP), represented by the demand curve, D, is lower than the wage
which has to be paid, W_M.

by the supply curve S, as before; and the opportunity cost to the unemployed person is
the gross of subsidy value, W_{SS}, measured on the new supply curve, S_s. The private opportunity cost per unit of labour is the value of leisure time, W_S, plus the unemployment
benefit, $(W_{SS}- W_S)$, which the person will forgo if they accept a job. However, in determining the social opportunity cost of employing the individual, to be used as the shadow-
price of labour in the Efficiency Analysis, we must set the gain to government in the form
of the reduction in unemployment benefit to be paid against the corresponding loss of
unemployment benefit to the individual. In other words, the unemployment benefit is a
transfer which nets out in the Efficiency Analysis. The opportunity cost of labour is the
value of output forgone as a result of employing the person on the project, which remains
W_S as before.

If the labour employed on the project is sourced from an alternative market use, its
opportunity cost is the value of its marginal product in its alternative job, which is the
gross of tax wage, W_M. Because of the taxation of labour income, the marginal unit of
labour employed in the alternative occupation does not receive the full benefit of its

contribution to the value of output in that sector of the economy; instead this contribution is shared between labour (the after-tax wage) and the government (the income tax). If we were to price labour at the after-tax wage, we would be understating its opportunity cost to the economy by the amount of the tax. The sharing of the benefits of employment between workers and the government, through the workings of the tax and unemployment insurance system is discussed further in Chapter 6, dealing with the distribution of project net benefits.

The examples of unemployment benefits and income tax provide an instructive lesson on the treatment of transfers. In the case of the subsidy on leisure, the unemployment benefit netted out of the Efficiency Analysis. However, the income tax is also a transfer, so why does it not net out of the efficiency calculation as well? Since the before-tax wage, W_M, is the unit cost of labour to the employer, it measures the contribution of the marginal unit of labour to the value of output, labour's value of the marginal product (VMP). The effect of the income tax is to share out the VMP between labour and the government; in other words, the income tax transfers a real benefit from labour to government. Contrast this with the unemployment benefit which is not matched by any output in the real economy. To put it in a colloquial way: the unemployment benefit represents a redistribution of the national cake, with no corresponding increase in its size; whereas a tax levied on earned income, while it also redistributes the cake, is part of the increase in the size of the cake resulting from work.

In developing economies the alternative to paid employment is sometimes subsistence agriculture, hunting or fishing. The value of such activity to the economy is often less than the value of the goods which the worker acquires as a result. The reason for this is that where resources, such as land or fishing grounds, are subject to open access, the total value of the output produced is effectively divided among the workers who exploit them. This means that each worker, on average, takes home goods equal in value to the value of the average product of labour in the occupation in question. Since labour applied to land or fisheries has a diminishing marginal productivity, the value of the *average* product of labour, which is the worker's "take-home pay", exceeds the value of the *marginal* product of labour, which is the value of the contribution of the worker's labour to the economy. In other words, even if W_S is what the worker could earn in a subsistence activity, it will exceed the opportunity cost of labour. In such circumstances the opportunity cost could be zero, or close to zero.

The fact that unemployment exists in the economy does not necessarily mean that the labour hired to construct the project would have otherwise been unemployed – a worker could be hired from some other job which pays the market wage. In this case the relevant question is, who would replace that worker in that other job? If the job is relatively unskilled, it is reasonable to assume that it would be filled by someone from the ranks of the unemployed, in which case the opportunity cost of labour is still W_S, or less if the job is filled by a person who has been sharing the output from exploiting open-access resources. Alternatively, if the worker hired is skilled, it may be that there will be no-one available to fill the resulting vacancy; this situation is known as structural unemployment – where the skill profile of the labour force does not match the skill profile of the job vacancies. In this case, the opportunity cost of labour would be W_M because the previous employer was willing to pay W_M to secure the worker's services, implying that the worker would have added the amount W_M to the value of output in that use.

In summary, the cost of an input, such as labour, is generally measured either by a point on the supply curve of the input (the marginal cost of supplying an extra unit), or by a point on the demand curve (the marginal value of the output resulting from an extra unit of input), at the current market equilibrium: which of these two points indicates the relevant price depends on the alternative use of the labour – the "*without*" part of the "*with and without*" analysis. If a project hires several workers – some of whom are replaceable in their current occupation or who are currently unemployed, and some who are not replaceable because of skill shortages – then the quantity of labour supplied by the former group needs to be shadow-priced at the supply price, W_S, while the quantity supplied by the latter group can be valued at the demand price, W_M. Generally this procedure is accomplished by differentiating between skilled and non-skilled workers and shadow-pricing the labour of the latter at the supply price. As noted above, the supply price tends to overstate the opportunity cost of subsistence activity, and labour diverted from that source may be shadow-priced at close to zero.

While regional pockets of unemployment are a feature of many jurisdictions in the developed world, we usually associate significant and persistent levels of unemployment or under-employment with developing economics where the employment benefits of large-scale projects can be significant. These benefits are often held out as an important advantage of a foreign investment project and as an inducement to the host government to approve it. For example, a tuna cannery will often require over 2000 employees, many of whom are disadvantaged in the labour market because of gender or lack of skills. Taking account of the alternative opportunities available to people such as these, a study of the opportunity cost of the labour employed in a tuna cannery in Papua New Guinea (Campbell, 2008) estimated that the value of these alternatives could be measured at the level of around half of the market wage. In other words, the shadow-price of labour in this case was W/2, where W is the market wage, and, consequently, roughly half of the cannery's wage bill consisted of employment benefits.

Labour diverted from a monopoly

It was argued above that the market wage measures the opportunity cost of labour diverted to a project from an alternative market use. However, the next two examples reveal that, in unusual circumstances, the opportunity cost of labour may be higher than the market wage.

Suppose that a public project will hire labour, with perhaps special skills, away from a monopoly. A monopoly producer can influence the price of its product by varying the quantity supplied, and maximises profit by restricting output to force up price. Like any other profit-maximizing producer, the monopoly will hire labour as long as the extra revenue from doing so exceeds the extra costs. The extra cost of a unit of labour is the market wage, and the extra revenue is derived from the additional sales generated by employing the labour. However, in computing the extra revenue, the monopolist has to take into account the fact that hiring more labour and supplying more output will result in a fall in product price. The lower price will apply not only to the additional units of output, but also to all the current units as well. This means that the addition to revenue as a result of selling an extra unit is less than the price received for that unit.

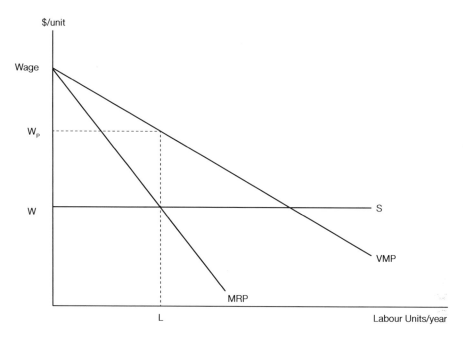

Figure 5.8 The value of marginal product of labour employed by an unregulated monopoly. The competitive firm's demand curve for labour is the value of the marginal product of labour curve (VMP). Value of the marginal product is defined as the additional contribution to the value of output of an extra unit of labour input. VMP is measured by the market price of output multiplied by the marginal physical product of labour (MPP), which is the quantity of additional output produced by an additional unit of labour input. Additional output supplied by a monopoly, on the other hand, has the effect of driving down the product price because of the downward sloping demand curve for output: not only does an extra unit of output receive a price lower than the current market price, but the price of all existing units of output supplied is lowered as well. This means that the extra revenue to a monopoly from supplying an additional unit of output, termed the Marginal Revenue (MR), is less than the current output price. Employing an extra unit of labour adds an amount of monopoly revenue equal to the MPP of labour multiplied by MR; this value is termed the Marginal Revenue Product (MRP). The profit-maximizing monopolist will hire an additional unit of labour as long as its contribution to revenue (MRP) is at least as large as its contribution to cost, measured by the wage (W). The monopoly profit-maximizing point is at labour input L, where MRP = W. At this level of input the opportunity cost of labour, VMP = Wp, is higher than the market wage, W.

Figure 5.8 illustrates the monopolist's demand curve for labour, the marginal revenue product curve, MRP. At the market wage, W, the monopolist will hire L units of labour. Suppose that a public project hired labour away from the monopoly. In reality this would be accomplished by bidding up the market wage, thereby reducing the profit-maximizing amount of labour for the monopolist to use and releasing labour to be hired for the public project. However, in this chapter we are assuming that the public project is not large enough to significantly affect the market price of any input or output, and so we have to assume that any increase in the market wage is so small as to be able to be ignored. When the monopoly releases a unit of labour, its revenue falls by W, which equals the marginal revenue product of labour, but the value of output falls by W_p, the value of the marginal product of labour. The opportunity cost of the labour released from the monopoly is the value of the output forgone – the VMP measured by W_p in Figure 5.8 – and this would be the appropriate shadow-price to be used in an Efficiency Analysis.

It can be seen from Figure 5.8 that the Value of the Marginal Product (VMP) of labour diverted from use by a monopolist exceeds the market wage. By restricting the supply of its product, the monopoly drives up its market price and creates a gap $(W_p - W)$ between VMP and the wage, which is a measure of the monopoly profit on the marginal unit of labour employed. When a unit of labour with, say, special skills is hired away from the monopoly, the value of output falls by the amount W_p. Marginal revenue can be expressed as:

$$MR = P*\left(1 - \frac{1}{e_d}\right)$$

where e_d is the elasticity of the product demand curve, expressed as a positive number. At the price, P, set by the monopolist the marginal revenue product, MRP, can be expressed as:

$$MRP = P*\left(1 - \frac{1}{e_d}\right)*MPP$$

And since W = MRP, it can be established by simple substitution that:

$$VMP = P*MPP = \left(\frac{W}{(1 - 1/e_d)}\right)$$

which is the shadow-price of the labour diverted from the monopoly.

Labour diverted from a monopsony

A monopsonist is the sole buyer of a product or a factor of production; for example, if a firm is the only employer of labour in the market in which it operates – such as the company in a one-company town – it maximises profit by restricting the amount of labour it hires in order to drive down the market wage. Even though the firm sells its product in a competitive market, it is able to make a monopsony profit which can be represented in Figure 5.9 as $(W_p - W)$ on the marginal unit of labour. Analogously with the case of monopoly, at the equilibrium quantity of labour hired, L, the value of the marginal product of labour, W_p, exceeds the wage, W.

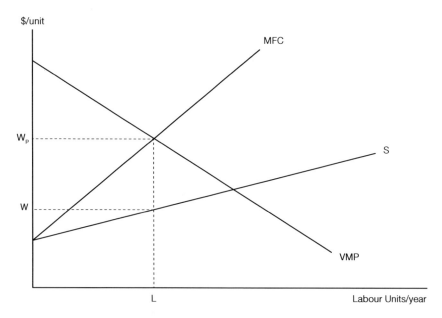

Figure 5.9 The value of marginal product of labour employed by a monopsony. The supply curve, S, shows that if the monopsony decided to hire more labour it would have to offer a higher wage, and that higher wage would apply to all units of labour currently supplied as well as to the additional quantity hired. The result is that the marginal factor cost (MFC) – the increase in the wage bill as a result of hiring an extra unit of labour – exceeds the wage. Assuming that the monopsony sells in a competitive product market, it maximises profit by hiring labour up to the point, L, at which the value of the marginal product of labour (VMP) – the amount an extra unit of labour adds to the value of output, and hence revenue – equals the MFC – the amount hiring the extra unit of labour adds to cost.

Suppose that, in a similar way as that discussed in Figure 5.8, a public project was going to hire a small quantity of labour away from the monopsonist. The public project would need to pay the wage, W, for a marginal unit of labour, but the opportunity cost would be W_p, the value of the marginal product of labour employed by the monopsonist, which exceeds the market wage. The shadow-price of labour, measuring VMP, can be expressed as:

$$VMP = P*MPP = W\left(\left(1 + 1/e_s\right)\right)$$

where e_s is the elasticity of the labour supply curve at the equilibrium employment level.

$$f(x)$$

TECHNICAL NOTE 5.1 Expressions for marginal revenue and marginal factor cost

Marginal revenue is the increase in revenue resulting from an additional unit of sale. For a competitive firm, facing a perfectly elastic demand curve, marginal revenue equals market price, but a monopoly faces a downward sloping demand curve and must lower price slightly to secure an extra sale. The change in total revenue resulting from an extra sale can be expressed as:

$$\Delta(Q*P) = \Delta Q*P + Q*\Delta P$$
$$= \Delta Q*P*\left(1 + \frac{Q*\Delta P}{P*\Delta Q}\right)$$

Hence:

$$MR = \frac{\Delta(Q*P)}{\Delta Q} = P*\left(1 - \frac{1}{e_d}\right)$$

where e represents elasticity of product demand, expressed as a positive value:

$$e_d = \frac{\Delta Q/Q}{\Delta P/P}$$

which, for monopoly profit-maximization, will always be less than unity.

Analogously the marginal factor cost facing a monopsonist can be derived as:

$$MFC = \frac{\Delta(W*L)}{\Delta L} = W\left(1 + \frac{1}{e_s}\right)$$

where e_s represents elasticity of labour supply:

$$e_s = \frac{\Delta L/L}{\Delta W/W}$$

5.4.3 Capital

The term "capital" is often used in two ways in cost-benefit analysis: one is to refer to financial capital – the dollars used to meet the costs of establishing the project; the other is to refer to the capital goods – the infrastructure, buildings and equipment – that constitute the project and are funded by the financial capital. By providing financial capital for the project, investors relinquish their claims on resources which could have been used to produce other goods, thereby freeing them up to produce the capital goods to be used in the project. The opportunity cost of capital is the present value of the consumption goods forgone as a result of financing the project. Consumption goods are forgone directly when funds are diverted from consumption to investment, or indirectly when funds are diverted from one capital expenditure to another.

Suppose that each dollar of financial capital raised, by, say, an issue of bonds displaces values of production of capital and consumption goods in the proportions b and $(1 - b)$. Where the funds raised by the bond issue displace consumption expenditures, the project capital is costed at its supply price, measured by the market cost of borrowing, while in the case in which the raising of funds displaces other capital expenditures, the project capital is, in principle, costed at the before-tax rate of return it would have earned elsewhere in the economy. The logic underlying choice of the before-tax of return is that if the Efficiency Analysis includes as a project benefit the share of the project's return that accrues to government in the form of business income tax, it must also include as an opportunity cost business income taxes that would have been paid on an alternative displaced investment project. This means that it is the tax-inclusive rate of return that is appropriate for measuring the opportunity cost of capital when it is diverted from another investment.

It should be stressed that the issue is not the forgone business income taxes *per se* from a displaced project, which, after all, are simply a transfer from business to government, but rather the fact that these taxes represent government's share in the real returns to investment. Treating private investment projects as perpetuities, to reflect the reinvestment of project returns, and ignoring debt financing, the annual after-tax rate of return on equity capital, net of the cost of risk, is given by $r = r * (1 - t)$, where r is the borrowing rate, $r *$ is the before-tax rate of return on the project and t is the business income tax rate. This means that \$1 of equity capital diverted from elsewhere in the economy would have produced an annual stream of output valued at \$r *; and the annual value of this output would have been shared between the firm and government in the amounts \$r and \$r * t respectively. The annual opportunity cost of the diverted capital is the full value of the forgone output, r * per annum.

In summary, when the capital contributed to the project is in addition to current supply (diverted from consumption), it is costed at its nominal value in the Efficiency Analysis: \$1 of capital would have yielded an annual return of r to its suppliers, with a present value of \$1. When the capital is diverted from an alternative project, there is an additional opportunity cost to the economy measured by the forgone business income taxes, and the opportunity cost is measured by r *, with a present value of r * /r. In this example, the shadow-price which adjusts the nominal one dollar capital cost to reflect the social opportunity cost of project capital, taking account of displaced consumption and investment activity, is: SOC = $(1 - b)$ + br * / r. However, in a global economy with relatively free flows of capital, it is reasonable to assume that, in general, the capital involved in a project is in addition to current supply (i.e. b = 0) and that the nominal capital cost measures the opportunity cost. Only in special circumstances, where lending is constrained by sovereign risk or capital goods are in inelastic supply, for example, would it be necessary to shadow-price the capital cost of a private project. As we shall see, however, project funds contributed by the public sector are routinely shadow-priced.

5.4.4 Land

The opportunity cost of land is the net value of the output which the land would have produced in its alternative use. Land use may be subject to taxes/subsidies and regulations: in the former case, the value of forgone output is calculated according to the procedures discussed above; if, in the latter case, the regulations would continue to apply in the absence of the proposed project, the value of output in its constrained use measures the opportunity cost of the land. Often the alternative use of land is in agriculture, in which case its opportunity cost is the time-stream of net returns – value of output less cost of inputs other than land – that would have been generated. Sometimes the net returns from marginal areas of land decline over time as soil degrades as a result of cropping and erosion. In that event the shadow-price representing the annual opportunity cost of the land also declines over time. Sometimes it is argued that the activities which would have occurred on the area of land selected for the project will be undertaken on some other, currently vacant, area of land, in which case the opportunity cost is the net value that would have been generated by that currently vacant area of land.

It might seem, at first sight, strange to attribute net value to an otherwise vacant area of land, but when land is described as vacant, it simply means that it is currently not traded in the market because it is unable to generate goods or services with a net market value. However, goods or services which are valued by the community, but not traded in markets, are often generated on vacant land. Ecosystems in so-called wilderness areas help to prevent soil erosion, stream sedimentation and salinity, act as a carbon sink, provide a refuge for wildlife, supply fodder, firewood and food for hunter-gatherer societies, and recreational opportunities for members of the community, to name but a few types of ecosystem goods and services produced. On the other hand, vacant areas of land sometimes impose costs in the form of a refuge for agricultural pests, a source of weeds, risk of fire, and so on, which offset to some extent their beneficial effects. While such externalities, whether goods or bads, are not priced in the market, they have a value or cost which must be reflected in the Efficiency Analysis through the non-market valuation techniques to be discussed in Chapter 8. Instead of it seeming strange that a positive opportunity cost is attributed to otherwise vacant land, the analyst should question the use of a zero shadow-price in the case of land.

The opportunity cost of land is best entered as a series of annual values over the life of the project. In cases in which the capital value of land is used to measure cost, the residual value of the land must be entered as a benefit at the end of the project's life.

5.4.5 Rules for shadow-pricing marketed inputs

In summary, the rules for shadow-pricing marketed inputs in an Efficiency Analysis of a proposed project are as follows:

- If the project input is sourced from an increase in supply, its marginal cost is the supply price (the marginal cost of supplying an extra unit). In the case of an indirect tax on the input, the supply price is the before-tax price; in the case of a subsidy, the supply price is the unsubsidised price;
- If the project input is diverted from employment elsewhere in the economy, its marginal cost is the forgone benefit in its alternative use, which is measured by the demand price (the price of buying an extra unit). In the case of an indirect tax, the demand price is the after-tax price; in the case of a subsidy, the demand price is the subsidised price which buyers actually pay.

5.5 Shadow-pricing marketed outputs

Output subject to a tax

If the output of a project satisfies additional demand for a good or service, it is valued at the price consumers are willing to pay for an additional unit – the buyer's price measured by the relevant point on the demand curve. Thus, in Figure 5.2 at equilibrium output level Q_0, consumers are willing to pay P_0 for an additional unit, and P_1 at equilibrium level Q_1. In Figure 5.3, at the equilibrium level of output Q, consumers are willing to pay the after-tax price, P_B, for an extra unit of the good or service and hence that is the value placed on additional output. The fact that, in this example, consumers are willing to pay more for an extra unit than its cost of production illustrates why we generally refer to indirect taxes as *distorting* – they distort the outcome of the competitive market, producing an equilibrium in which both buyers and sellers could be better off (by increasing output) in the absence of the tax. However, it should be recognised that this analysis does not take into account the value of the uses to which the tax revenues are put.

Project output does not always satisfy additional demand for a good or service. Consider Figure 5.4 which could refer to the output of an import-replacing project intended to satisfy an additional portion of existing demand from domestic production. Since consumers remain at the same price-quantity equilibrium (P_B,Q), and assuming that they are indifferent between domestically produced and imported product, they are unaffected by the project. The value of additional output consists of the avoided cost of the imports displaced, and hence the world price, P_W, is used to value project output.

Output subject to a subsidy

Since output that satisfies additional demand is valued at the price consumers are willing to pay for an additional unit, the price, P_B, in Figure 5.5 would be used to value output of a subsidised good or service, such as an agricultural product covered by a farm support system. If project output replaced a portion of current demand from an alternative unsubsidised source, such as imports, on the other hand, it would be valued at the unsubsidised price, P_S, which measures the avoided cost of supplying the product from the subsidised source.

Output supplied by a monopoly

In each of the above examples there is only a single imperfection in the market. What happens if a market suffers two or more distortions? We are now venturing into an area known as *the theory of second best*. What it says, simply, is that where there are several imperfections in the market economy, such as market power, tax and regulatory distortions, and open-access to resources, removing one of these distortions will not necessarily result in a more efficient allocation of resources. The implication for Efficiency Analysis is that we need to take account of *all* the *significant* market distortions in order to construct the appropriate shadow-prices for all inputs and outputs for which there is a distorted price. This point can be illustrated by the following example of a monopoly which is also receiving an output subsidy.

Figure 5.10 illustrates a monopoly producer, operating under conditions of constant unit cost of production, who is receiving a subsidy on each unit of output produced. The marginal cost of production is represented by MC, and the marginal cost net of the subsidy is represented by MC_S. There are two imperfections in this monopoly market: the monopoly can influence market price – it faces a downward sloping demand curve for its product – and therefore has an incentive to limit production to drive up price; and the monopolist receives a subsidy on output which creates an incentive to increase production. In the absence of the subsidy, the monopolist maximises profit by producing output level Q_M, where marginal revenue (MR) – the increase in revenue as a result of selling an extra unit of output – equals marginal cost (MC). This involves charging price P_M. However, the effect of the subsidy is to lower the effective marginal cost so that the profit-maximizing level of output is Q_S which fetches a price of P per unit. The example has been contrived to make Q_S– the appropriate level of output from an efficiency viewpoint – the profit-maximizing output level for the monopolist.

Suppose that a project is to use some of the monopoly output as an input. Consider first the no-subsidy case. If additional output can be produced to meet the requirements of the

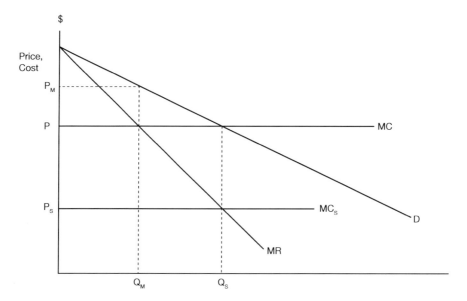

Figure 5.10 Monopoly output with and without a subsidy. The monopolist's cost of production is P
per unit, as measured by the marginal cost curve, MC, which, in this case, also represents
average cost. If monopoly output is subsidised by an amount $(P - P_S)$ per unit the effect of
the subsidy received by the monopolist is to reduce the unit cost of supply to P_S. Without
the subsidy the monopolist maximises profit where marginal revenue, MR, equals marginal cost, MC; this occurs at output quantity Q_M and market price P_M. With the subsidy
the monopolist's marginal cost falls to MC_S and the new equilibrium is at quantity Q_S and
price P. The example has been constructed so that the subsidised monopoly outcome (P,
Q_S) is the outcome that would have been achieved by a competitive industry under the
same demand and cost conditions.

project, its opportunity cost is measured by a point on the supply curve or, in the case of an
imperfectly competitive industry which does not have a supply curve, at a point on the marginal cost curve. In this case, the project input would be shadow-priced at the true marginal
cost, P. If, on the other hand, the project input had to be diverted from other uses, it would
be priced at the market price, P_M. In contrast, in the case of the subsidy, the quantity of the
input would be priced at the market price, P, whether it was in addition to current supply
or diverted from an alternative use. Since the combined effect of monopoly power and the
subsidy is to mimic the competitive outcome (P = MC) the market price measures opportunity cost regardless of the source of the input.

Output of rental units

In Figure 5.7 we considered the effect of minimum wage regulation and its implications for
pricing labour in the Efficiency Analysis. Now we consider the effect of maximum price
regulation and the implications for pricing the output of a project. Suppose that a maximum
value is set on the price which can be charged for an input or output. A common example
is a maximum value set on an exchange rate – an official exchange rate, in terms of the
number of units of the local currency per US dollar, which is lower than required by the
equilibrium price. This issue is considered further in Chapter 10. Another example is rent

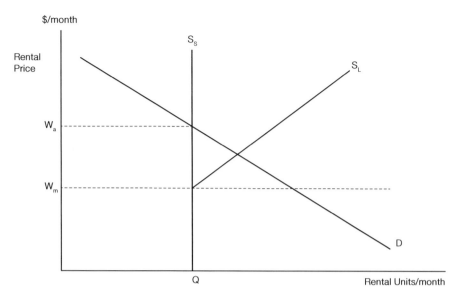

Figure 5.11 The market for rental units with rent control. The short-run supply curve of rental units, S_S, is vertical at the current quantity supplied, Q, reflecting the fact that no additional units can be supplied in the short-run. The long-run supply curve, S_L, offers the possibility of adding to the current stock but a regulated maximum price, W_m, is less than the marginal cost of supplying additional units, thereby preventing adjustment of quantity supplied to meet demand. According to the demand curve would-be renters value an extra unit at W_a. Since quantity demanded exceeds quantity supplied at the market price, W_a, apartments are rationed according to criteria adopted by landlords.

control – the setting of a maximum price that can be charged for the services of a rental unit. The distorted rental market is illustrated in Figure 5.11, which shows the demand and supply curves for rental units; the relevant supply curve illustrated is a short-run supply curve (S_S) – it is perfectly inelastic (vertical) to reflect the fact that in the short run (the space of a few months), no additional units can be completed and offered for rent.

W_m is the maximum price – the level of the controlled rent. Suppose that a public project involves bringing some specialised workers into the area and accommodating them for a few months while they complete the project. What is the opportunity cost of the accommodation? An apartment occupied by one of the project workers is not available for rent to a local resident. We can see from Figure 5.11 that while a local resident could expect to rent an apartment for W_m, they actually place a value of W_a on it – at the current level of supply W_a is the marginal value of a month's accommodation in a rental unit, if one were available. Hence, in this case, W_a is the relevant shadow-price of the input in an Efficiency Analysis. Since the input is diverted from alternative market demand, it is priced at an equilibrium point on the demand curve.

One of the features of a controlled housing market is that it is difficult to find a rental unit at the controlled price; since the marginal unit is worth more to the tenant than the rent which is actually paid, the unit tends to get handed on informally when the tenant leaves. Hence the above example begs the question of how the manager of the public project managed to secure some of these scarce units for their staff. While the example deals only with the short run, it can be seen that a long-run analysis is irrelevant because of rent control. Suppose the long-run supply curve is S_L as illustrated in Figure 5.11. This curve tells us that if the market rent rose, additional units would be supplied in the long run because the higher rent would cover the extra costs involved in supplying the additional units. If an additional rental unit were to be produced for use in the project, then the pricing rule would tell us to cost it at W_m: the project input is an addition to supply and is priced at the equilibrium point on the supply curve. Under effective rent control, however, rents cannot rise and hence quantity supplied will not increase.

Time saved

Many transportation investments, such as urban rail systems, for example, have reductions in travel times as a principal output, and some projects, such as relocating an airport, for example, may involve extra travel time on the part of travellers. Figure 5.12 shows an individual's demand curve for leisure time, with the supply of leisure regulated by the 8-hour work day. Figure 5.12 illustrates a kind of market imperfection which can be termed the "all or nothing" situation: a person might like to work 6 or, alternatively, 10 hours a day but they have a job which requires 8 hours – "take it or leave it". In Figure 5.12 the price of leisure (its opportunity cost) is the hourly wage that could be earned, W. As can be seen from the intersection of the wage with the demand curve, this individual would ideally like to work 10 hours a day, but the employer has specified 8 hours or nothing – no job. At this level of leisure (16 hours), an extra unit of leisure is worth W_L to the individual.

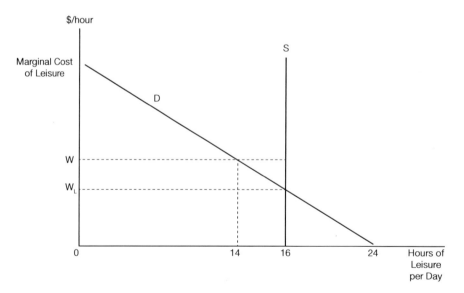

Figure 5.12 An individual's leisure supply and demand. The individual's demand curve for leisure is denoted by D, and the market wage is W. At the wage W the individual would prefer 14 hours of leisure, with the remaining 10 hours in the day spent at work. However he is constrained to accept 16 hours of leisure because his employment contract limits him to an 8-hour day. When he consumes 16 hours of leisure he values an extra hour of leisure at W_L which is less than his hourly wage.

Now suppose that a public project is proposed which will reduce the time taken to travel to work. There will be many kinds of benefits – reduced vehicle wear and tear, reduced risk of an accident, etc. – but one important benefit is a reduction in travel time resulting in an increase in time the individual can devote to other purposes. Since the individual is still required to work 8 hours, and no more, the travel time saved will be devoted to additional leisure – getting up 15 minutes later, for example. How is this time to be valued in an Efficiency Analysis? At the risk of oversimplifying a complicated issue, we can refer to Figure 5.12 and argue that W_L is the appropriate shadow-price. This is because the worker would prefer to be working more hours at the current wage, but the work is not available and hence the marginal value of leisure time is less than the market wage. The project output satisfies additional demand for leisure and hence should be priced at an equilibrium point on the demand curve for leisure (W_L), which is a proportion of the market wage.

A landmark study in the 1960s of the costs and benefits of an additional airport for London costed urban travel time at one-third of the market wage. A more recent survey of studies involving savings in commuting time by Waters (1996) found that savings in travel time to work are generally valued at between 40% and 50% of the after-tax wage. Time savings in the course of work, by truck drivers, for example, are generally valued at the wage paid by the employer. The value of time saved is an important feature of the Highway Project case study presented in Appendix A1.9.

5.5.1 *Rules for shadow-pricing marketed outputs*

In summary, the rules for shadow-pricing marketed outputs in an Efficiency Analysis of a proposed project are as follows:

- If the project output satisfies additional demand, it is valued at the buyer's price, sometimes referred to as the demand price. In the case of an indirect tax on the output, the demand price is the after-tax (tax-inclusive) price; in the case of a subsidy, it is the subsidised price.
- If the project output satisfies existing demand from an alternative source, it is valued at the current supply price of the product. In the case of a product subject to an indirect tax, the supply price is the before-tax price; in the case of a subsidy, it is the unsubsidised price.

5.6 The efficiency pricing rules: summary

We now draw together, in Figure 5.13, the two sets of rules we have established for shadow-pricing the marketed inputs and outputs of a project subject to an Efficiency Analysis, and suggest a procedure for determining which price should be used in each set of circumstances. In order to apply the rules to the valuation of a good or service, the first step is to identify the good or service as either a project output or an input. If the good or service is an output, the output row in Figure 5.13 then requires the project output to be classified as either satisfying additional demand, or satisfying existing demand from an alternative source. In the former case, the appropriate price is given by a point on the demand curve for the good (the buyer's price), and in the latter, by a point on the supply curve (the supplier's price). In the case of an input, the good or service is classified as either being sourced from additional supply, or being diverted from use elsewhere in the economy. In the former case, the appropriate price

ITEM TO BE VALUED	VALUED AT EQUILIBRIUM POINT ON A:	
	DEMAND CURVE (BUYER'S PRICE)	SUPPLY CURVE (SELLER'S PRICE)
OUTPUT	SATISFIES ADDITIONAL DEMAND	SATISFIES EXISTING DEMAND FROM ALTERNATIVE SOURCE
INPUT	SOURCED FROM AN ALTERNATIVE MARKET USE	SOURCED FROM ADDITIONAL SUPPLY

Figure 5.13 The efficiency cost-benefit analysis pricing rules.

is given by a point on the supply curve of the good (the supplier's marginal cost), and in the latter by a point on the demand curve (the buyer's price).

5.7 Corrective taxation: the modified efficiency pricing rules

Up to this point in our discussion of the treatment of taxes in an Efficiency Analysis we have been implicitly assuming that the purpose of indirect taxes is to raise government revenue, and that market distortion is an unintended by-product of revenue-raising through taxation or of economic assistance to producers through subsidies. However, another possible role for indirect taxes or subsidies is to discourage or encourage consumption of certain goods and services, or use of certain inputs in production. For example, is the tax on tobacco aimed at raising revenue or at discouraging smoking? Because the demand for tobacco is very inelastic, many commentators have concluded that the primary aim of taxing tobacco is raising revenue. Suppose, however, that the aim of the tax was to discourage smoking because of the cost of its adverse health effects; some of these costs are borne by the smoker, but some are borne by the general community through passive smoking and the associated costs to the health care system. The latter category of cost is termed an *external cost* – in this case a cost of consuming the product which is not borne by the consumer of the good.

Economic efficiency requires that each good or service should be consumed to the point where its marginal benefit equals its marginal cost. We supposedly can rely on smokers to balance their own health costs against the satisfaction obtained from smoking in determining their marginal benefit. However, we cannot rely on them taking the external cost of smoking into account unless we incorporate it into the price of a packet of cigarettes. If the tobacco tax is set at a level equal to the marginal external health cost of smoking, the price of a packet of cigarettes to the consumer will equal the marginal cost of producing the cigarettes plus the marginal external health cost. An indirect tax set at this rate, in effect, forces the smoker to take the marginal external health cost into account in choosing their level of consumption – it *internalises* the externality. Setting the rates of indirect taxes in this way is termed *corrective taxation*.

If we assume that the tax on tobacco is a corrective tax, set at the efficient level, what are the implications for valuing the output of tobacco in an Efficiency Analysis? Suppose that the market illustrated by Figure 5.3 is that for tobacco. It was argued that the output of a project producing a small additional quantity of the product to satisfy additional demand should be valued at P_B, the gross of tax price. However, the gross of tax price, P_B, measures only the marginal benefit to the smoker. The additional consumption of tobacco will impose an additional health care cost on the community, measured by the indirect tax ($P_B - P_S$). This means that the net benefit to the community of the additional output of tobacco is the benefit to the smoker less the external cost to the community, giving a net benefit of P_S per unit of output. In other words, where the indirect tax is a corrective tax, additional output supplied by the project should be valued at the net of tax price.

Let us now consider an example dealing with a subsidy. Consider Figure 5.5 which illustrates the market for a subsidised good. Suppose that the good in question is an influenza vaccination. The reason for the subsidy is that persons who pay for and take the vaccination not only protect themselves but also protect others who otherwise might have caught the infection from them. The subsidised price measures the benefit of the vaccination to the private individual, and the subsidy measures the marginal external benefit to the community as a whole. If a project were to produce additional flu vaccinations, which price should be used to value them? Marginal benefit to society of an additional vaccination is measured by the unsubsidised price, which is the sum of the marginal private benefit plus the marginal external benefit, P_S.

The above two examples deal with shadow-pricing outputs which satisfy additional demand, and are subject to corrective taxes or subsidies. It was concluded that the appropriate shadow-price is a net of tax price, or a gross of subsidy price. In the case of inputs which are in addition to existing supply, and subject to corrective taxes or subsidies, this conclusion is reversed: the appropriate shadow-price is gross of tax and net of subsidy. For example, if coal were subject to a carbon tax, the shadow-price that measures the opportunity cost of an additional unit of coal is the marginal cost of production plus the marginal external cost i.e. the supply price plus the tax, represented by P_B in Figure 5.3. If electricity produced from alternative energy sources, such as biofuel, is subsidised, the marginal cost of use is the supply price less the subsidy which measures the marginal external benefit of the cleaner air which results from the use of the alternative energy source as compared with conventional sources. The net opportunity cost is measured by P_B in Figure 5.5.

In contrast to the above examples, which deal with outputs that satisfy additional demand or inputs which are in addition to current supply, when a project output satisfies existing demand from an alternative source, and the output is subject to a corrective tax (subsidy), the output should be shadow-priced at the net of tax (gross of subsidy) price since that value measures the savings in production cost as a result of the alternative source of supply of the good, and *there is no change in the total level of the externality* associated with consumption of the good. When a project input is diverted from an alternative use, and the input is subject to a corrective tax (subsidy), the input should be shadow-priced at the gross of tax (net of subsidy) price, since that price measures the value of its marginal product in the alternative use, and, again, there is no change in the overall level of the negative (positive) externality as a result of reallocating the input.

In summary, when indirect taxes or subsidies play a corrective role, the efficiency pricing rule described in Figure 5.13 needs to be modified: when a project output satisfies additional demand, the sellers' price – the net of tax price (gross of subsidy price) – is used to shadow-price the output; and when an input is in addition to existing supply, the buyers' price gross of tax (net of subsidy) is used to shadow-price the input. Since cost-benefit analysis is concerned with incremental effects, there is no change in the rule in the case of an output which replaces existing supply or an input which is sourced from an alternative use because this involves no change in the level of the externality and hence no change in the external benefit or cost.

How is the analyst to determine whether indirect taxes or subsidies play a corrective role? Unlike income taxes or general sales taxes, corrective taxes or subsidies are targeted at particular consumption goods, such as alcohol, tobacco or flu jabs, or particular factors of production, such as fuel, pesticides or clean energy. It will be shown in Chapter 6 that if a tax or subsidy is treated as corrective in the analysis of efficiency, a corresponding external cost or benefit must be included in the analysis of distribution of net benefits and perhaps included in the Referent Group Analysis. This suggests that it is only where an externality is judged to be significant in impact that the analyst would treat indirect taxes or subsidies as corrective in the Efficiency Analysis.

5.8 How to determine which pricing rule to follow

In Chapter 4 it was suggested that the first step in undertaking a cost-benefit analysis of a proposed project is to identify the relevant inputs and outputs in physical terms. The analyst then classifies these commodities into two groups – those which are traded and have observable market prices – and those which are non-marketed. Consideration of the latter was deferred while the information provided by the market was utilised, initially, to perform the Market and Private Analysis as described in Chapter 4. Some of the prices generated in markets may be amended in the Efficiency Analysis because of market imperfections – principally taxes, regulations and lack of competition. The process of adjusting these prices, where appropriate, has been described in the initial sections of the present chapter.

A set of pricing rules, summarised in Figure 5.13 was developed, with amendments to deal with corrective taxes or subsidies, to help the analyst choose the appropriate price or shadow-price to value the good or service in question. The application of those rules depended on whether a project input was in addition to current supply or diverted from existing use, and whether an output satisfied additional demand, or met existing demand from an alternative source. In general, we would expect project inputs to be in addition to current supply and outputs to satisfy additional demand, but there may be exceptions. In the case of inputs, supply constraints, such as skill shortages or capacity constraints, mean that project inputs have to be diverted from other uses. In the case of outputs, some projects are proposed to displace existing sources of supply, as in the case of an import-competing project for example, so that some existing demand is met from project output.

In Chapter 6 it will be argued that shadow-pricing marketed inputs or outputs, where appropriate, is necessary if the net benefits of the project are to be properly allocated in the analysis of the project's distributional effects. For example, if otherwise unemployed labour is valued at a shadow-wage lower than the market wage, the efficiency net benefit of the project is increased and this increase is registered as an employment benefit in the analysis of distribution. The same argument applies to land rents. When we elect *not* to shadow-price private capital in the Efficiency Analysis, we are creating space in the analysis of distribution for a higher level of business income tax receipts.

While similar arguments apply to the shadow-pricing of materials we must be mindful of the value of the contribution of more precise information to the eventual decision about the project. The value of information is discussed further in Chapter 9, but it is obvious that in some cases more precise detail about the opportunity cost of inputs or the marginal value of outputs will not change the decision about the project. In some developing economies with rudimentary tax systems, the indirect tax consequences of a project, such as the import duties levied on project inputs, may be an important feature of the project from the viewpoint of the Referent Group. In developed economies, however, these issues are less important and it is worth asking what is lost in the cost-benefit analysis by pricing all marketed inputs and outputs, with the exception of labour and land, at market prices. Would overstating the costs of inputs which are in addition to current supply, or the benefits of a project which replaces an alternative source of supply, make a difference to the outcome of the analysis? Would understating the costs of inputs diverted from alternative uses make a difference? A sensitivity analysis might reveal that varying such costs or benefits within the range typical of indirect tax rates, say, 5–15%, would make no difference to the outcome of the analysis. This issue is explored further in Chapter 9 dealing with the value of information.

Before we proceed to discuss the treatment of price changes induced by the project in Chapter 7, and the valuation of non-marketed goods and services in Chapter 8, it remains to consider some more markets which generate prices or values which may have to be adjusted by means of shadow-prices – the markets in public funds and foreign exchange for example – and there may be doubts about the suitability of using a market-based rate of interest as the discount rate. Adjustments to the spreadsheet model to deal with public funds and foreign exchange draw on the analysis of distribution of net benefits, to be considered in Chapter 6, and sensitivity analysis, as described in Chapter 9, can be used to deal with uncertainty about the appropriate rate of discount. Since these matters are not dealt with in the worked examples described in the concluding sections of the current chapter, the reader can defer study of them if desired and can proceed directly to the examples at this point.

5.9 Shadow-pricing public funds

Almost all projects involve public funds to some degree. A public sector project aimed at improving health outcomes, for example, may have its capital and operating costs fully funded by government. A road project may be partially funded out of public funds, with government meeting the construction and maintenance costs but with road users contributing to costs through a road toll. A private project will generally be assessed for business income tax and may also contribute through indirect taxes on its inputs and outputs. The flows of public funds can be identified in the analysis of the distribution of project benefits and costs, the subject of Chapter 6, and, by aggregating these flows, the net inflow or outflow of public funds can be calculated.

As argued below, each dollar of public funds raised though borrowing, taxation or other means inflicts more than a one dollar cost on the economy. This means that in the analysis of economic efficiency the flow of public funds associated with a project should be shadow-priced: a net outflow is costed at more than its nominal value, and a net inflow is valued at more than its nominal value because it reduces the amount of funds to be raised through the traditional methods. The mechanism for carrying out this revaluation in the spreadsheet model is to enter a *public funds premium* as a cost (in the case of a net outflow) or as a benefit (in the case of a net inflow) in the Efficiency Analysis of the project. The premium is calculated by applying the factor (MCF – 1) to the net public funds flow identified in the Referent Group Analysis described in Chapter 6, where MCF is the marginal cost of public funds.

Each method of financing public expenditure – borrowing, taxation and "printing money" – imposes costs on the economy which are not included in the measure of the opportunity cost of project inputs at market or shadow-prices. As we have seen, most taxes distort markets, resulting in prices not reflecting opportunity costs; higher tax rates, imposed to fund additional public expenditure, increase the degree of distortion and move the allocation of resources further from the efficient pattern. Selling bonds diverts funds from private investment which, because of the tax system, offers a higher return to the economy (the before-tax return) than to the private investor (the after-tax return). Financing government expenditure through monetary expansion can lead to inflation causing distortion of the operation of market pricing mechanisms and resulting in an inefficient pattern of resource allocation in the economy.

If government were rational and informed it would use each of these three sources of funds up to the point at which its marginal cost is equal to the marginal cost of each of the other two. In this way the total cost of collecting any given quantity of public funds would be minimised. This implies that if we work out the marginal cost of funds obtained through one source, taxation or selling bonds, for example, we can assume that this is the marginal cost of public funds from any source. There is some evidence that governments *are* rational and informed in the way that we are assuming: there has recently been much less reliance on inflation to fund public expenditures than previously. The reason is not so much that the cost of this source has increased but rather that governments are better informed about the costs in terms of economic instability and resource misallocation. Since bond finance eventually leads to higher taxes to pay for interest and principal repayments, we may be on reasonably solid ground if we use the marginal cost of tax revenues to approximate the marginal cost of public funds. However, in this discussion we will consider both bond and tax finance.

There are three main costs of raising tax revenues: collection costs, compliance costs and deadweight loss: collection costs are costs incurred by the private and public sectors in the battle over the amount of tax due; compliance costs are costs of tax-form-filling incurred by the private sector; and deadweight loss is the cost of changes in economic behaviour induced by the structure of the tax system, with the consequent loss of consumer welfare arising from the less efficient allocation of resources. It can be argued that, while collection and compliance costs may be substantial, they do not increase significantly with an increase in tax rates. In other words, in calculating the shadow-price to apply to *changes* to tax flows, only the deadweight loss needs to be taken into account.

The calculation of the deadweight loss depends on the assumptions made about the incidence of taxation: does it mainly fall on labour, on goods and services, or on capital? Many studies have assumed that since eventually all taxes are borne by households, the main deadweight loss stems from distortions to the incentive to work. Some studies have argued that indirect taxes distort consumers' choices among goods and services, and some have emphasised the distorting effect of direct taxes on the level and pattern of investment. Some have taken all these effects into account in a general equilibrium analysis. Notwithstanding the various assumptions made, most studies of the marginal cost of public funds in developed economies such as Australia, Canada, New Zealand, the United Kingdom, and the United States find that the cost premium is around 20–25%. It might be thought that the cost of public funds in developing economies would be higher than indicated by this range, because of the additional difficulties associated with designing and running an efficient tax system in such economies, but a study of African countries concluded with a similar range of results.

The opportunity cost of bond finance

As an illustration of the cost to the economy of an additional distortion to the allocation of capital, consider a situation in which a public project is to be financed through the sale of government bonds to the public. Ignoring risk, we can expect that private investors will allocate funds between government bonds and private projects until the yields are equalised at the margin. Since the private rate of return on a private project is net of business income tax, the after-business income tax return on private projects will equal the government bond rate. If investment in government bonds displaces investment in private projects, through the "crowding out" effect discussed in Chapter 12, the opportunity cost to the economy of financing the public project is the before-tax return on the displaced private projects. This argument mirrors the previous discussion of the opportunity cost of the private capital allocated to a project. If investors require a private project to yield an after-tax rate of return equal to the government bond rate of r, and the business income tax is levied at rate t, then the before-tax rate of return on private projects is given by $r^* = r/(1 - t)$. The funds obtained through a government bond issue which displaced private consumption and investment in the proportions $(1 - b)$ and b respectively would have a social opportunity cost measured by:

$$SOC = (1-b) + b\left(\frac{r^*}{r}\right)$$

per dollar raised, implying that MCF = $(1 - b) + b (r^* /r)$. For example, if b = 0.5 and the business income tax rate is 33%, the marginal cost of public funds raised in this way is $1.25 per dollar.

The opportunity cost of funds obtained from a tax increase on labour income

A significant form of deadweight loss associated with a tax increase is generally thought to result from the effect of the tax increase on work incentives. An increase in the tax rate lowers the after-tax wage (the price of leisure) and makes additional leisure relatively more attractive. However, the lower after-tax wage also lowers disposable income and makes additional leisure less affordable. While these two effects work in opposite directions, it is likely that a reduction in the after-tax wage will reduce the aggregate amount of labour supplied.

The effect of an increase in a proportional tax rate on labour income can be illustrated by the labour supply curve in Figure 5.14. The opportunity cost of each extra dollar of public funds obtained through a tax increase is the opportunity cost of the funds raised by the tax increase divided by the additional amount of tax revenues obtained. The cost of the tax increase to the worker is the change in after-tax labour income less the change in the cost of supplying the labour. The cost of supplying an additional unit of labour is the value of that unit of the worker's time in an alternative use, which is measured by the supply curve. In terms of Figure 5.14, the loss of after-tax labour income as a result of the tax increase is given by ACDE + DCL0L1 but this is partially offset by increased leisure valued at DCL0L1. The net cost to the worker of the tax increase is given by area ACDE, which represents a loss of producer surplus, a concept that will be discussed in Chapter 7.

The additional tax revenue obtained from the tax increase is measured by area ABDE − FGCB in Figure 5.14. This is the additional tax levied on the new level of labour supply, less the loss of tax levied at the original rate as a result of the reduction in labour supply. The opportunity cost per dollar of public funds is given by ACDE/(ABDE − FGCB). This value can also be expressed as 1 + FGCD/(ABDE − FGCB), where FGCD is the deadweight loss

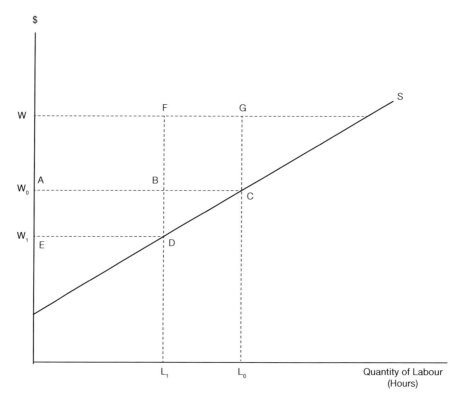

Figure 5.14 Taxation and labour supply. S represents the supply curve of labour, W is the before-tax wage, which is assumed to be unaffected by the tax increase, W_0 is the initial after-tax wage, and W_1 is the after-tax wage following a tax increase, measured by $W_0 - W_1$ per unit of labour. The effect of the tax increase is to reduce the quantity of labour supplied by an amount measured by the distance $L_1 L_0$.

caused by the tax increase. The term FGCD/(ABDE – FGCB) represents the cost premium to be applied to public funds in the Efficiency Analysis.

Why is area FGCD called a deadweight loss? Looking at the question from the point of view of the economy as a whole, there has been a reduction in labour supply and, since the value of the marginal product of labour is measured by the before-tax wage, W, a reduction in the value of output (GDP) of FGL0L1. This is partially offset by the value, DCL0L1, of the additional quantity of leisure, L_0L1. The net loss to the economy is FGCD.

The cost of public funds can be estimated from information about labour supply and tax rates. However, the relevant tax rates are effective marginal tax rates (EMTRs) which incorporate both direct and indirect taxes and include the effects of benefit programs as well as taxation: the EMTR is the increase in tax paid less social security benefit received when household income rises by a dollar. In effect, the household may be penalised twice when income rises as a result of additional work – it pays more tax and it experiences a reduction in those government benefits which are related to income level. When the marginal tax rate on labour income is calculated on this basis, EMTRs can be significant in most income brackets; for example, an Australian study calculated an average EMTR of 47% across income deciles.

As noted above, a rise in the tax rate in order to fund an additional public project causes a decline in the quantity of labour supplied, thereby reducing before-tax incomes and making some households eligible to receive increased benefit levels. How should these increased benefits be financed? It can be argued that the additional revenues resulting from the tax increase should be sufficient to fund the additional public project *and* pay any additional benefits required under current programs because of the reduction in earned income resulting from the reduction in employment caused by the tax increase. When this requirement is imposed on the funding of an additional public project, a larger tax increase is required, with a consequent increase in the deadweight loss and the cost premium.

The size of the reduction in labour supply depends upon how responsive labour supply is to a change in the after-tax wage. An Australian study assumed that, while households in different income deciles would respond differently to a tax increase, the average household would reduce labour supply by 6% in response to a 10% fall in its after-tax wage. Using a simple model of labour supply incorporating this assumption, values corresponding to areas FGCD and (ABDE – FGCB) in the labour supply diagram of Figure 5.14 can be calculated for the average household in each income decile. Aggregate deadweight loss and aggregate additional tax revenues can be calculated by summing across households and their ratio used to calculate the deadweight loss per dollar of additional revenue available to fund public expenditure. A value of $0.19 was obtained in the Australian study, implying that the cost of public funds in Australia is $1.19 per dollar of tax revenue.

The effect of the tax increase on labour supply was considered above. However, undertaking the public project may independently affect labour supply through project-specific effects. For example, a new road could decrease labour supply by making it easier to get to the beach, or could increase labour supply by making it easier to get to work: in the first case, the public project complements leisure activities and, in the second, it is a substitute for leisure. Furthermore, having access to an additional quantity of a public good as a result of a public project increases households' economic well-being, thereby making leisure seem more affordable; this sort of reaction is termed an "income effect". The net effect of the output of the public project will vary from case to case, but may generally tend to be in the direction of reduced labour supply. In terms of Figure 5.14, this means that the supply curve of labour shifts to the left when the project is undertaken, thereby causing an additional reduction in tax revenues.

If undertaking a project causes a reduction in labour supply, for the reasons discussed above, the consequent loss in tax revenues will need to be made up by a further increase in taxation with its associated deadweight loss. Allowing for the income effect of increased public good provision added a further 5% to the Australian estimate of deadweight loss, bringing the estimate of the marginal cost of public funds up to $1.24 per dollar of additional tax revenue. The results of the Australian study are consistent with the estimates of marginal cost of public funds obtained for a range of developed economies as reported in Table 5.1.

The conclusion of this analysis of the taxation of labour income is that changes in levels of public revenues or expenditures should be shadow-priced in the Efficiency Analysis in order to take account of the marginal cost of public funds (MCF). In a developed economy the size of the appropriate shadow-price will probably be in the 1.2–1.25 range, with similar values for developing countries. In principle, the shadow-price should be project-specific, depending on the nature of the project's output as a substitute for (complement to) leisure, but in practice this kind of information is seldom available and never taken into account. Where project-specific labour supply effects are localised, they may be insignificant in

Table 5.1 Estimates of the marginal cost of public funds

Range	Country
>2	Denmark, Sweden
2–1.5	Belgium, the Netherlands, Luxembourg, Germany, Japan, Austria
1.5–1	France, Finland, Czech Republic, Canada, Switzerland, Spain, NZ, Portugal, UK, Australia, Poland, USA

(Source: B. Dahlby, *The Marginal Cost of Public Funds: Theory and Applications*, Cambridge, MA: MIT Press, 2008)

comparison with the overall cost of the project and, in any event, are usually ignored. Shadow-pricing is implemented by applying a cost premium to public funds, where the cost premium is calculated as $(MCF - 1)$.

In practice, guidelines for cost-benefit analysis tend to recommend using sensitivity analysis to take account of the marginal cost of public funds. For example, the US Office of Management and Budget guidelines suggest that where a project involves additional public expenditures, these should be shadow-priced at 1.25 and the project NPV recalculated and reported. The Australian guidelines suggest that where a project requiring public funds has a relatively small NPV, the presumption should be against the project, and for other projects a sensitivity analysis, incorporating the 1.25 value, should be used. The European Commission guidelines recommend a default value of 1 for the MCF unless the relevant national government has specified some other value.

5.10 Shadow-pricing foreign exchange

Projects sometimes involve internationally traded goods or services – imports of project inputs or exports of project outputs. Exports of goods or services earn foreign exchange while imports must be paid for in foreign currency. If the analyst is confident that the exchange rate accurately measures the cost or benefit of foreign exchange to the economy, these foreign exchange flows can be converted to domestic currency by means of the market rate of exchange and treated like any other benefit or cost stream in the analysis. Just as the markets for project inputs or outputs may be imperfect, however, so there may be imperfections in the foreign exchange market, rendering the exchange rate an inaccurate measure of the cost or benefit associated with traded commodities.

Historically the principal foreign exchange market imperfections have been import and export duties, which distort the demand and supply curves for foreign exchange, and exchange rate regulation which cause a divergence between the demand and supply prices. As a result of floating currency regimes and trade liberalization, these imperfections have become relatively insignificant for developed economies, such as those of the OECD, but they may still be relevant for some developing countries. Some observers have argued that China's Yuan has been undervalued in recent history. In circumstances of currency undervaluation, if the official exchange rate is used to measure costs of imports or benefits of exports in the Efficiency Analysis, the calculated NPV will tend to overstate the net value of a project which is a net earner of foreign exchange and understate the value of a project with a net foreign exchange requirement.

The theory of shadow-pricing foreign exchange is quite technical, and the procedures for determining the shadow-price are quite complicated. Since these issues may not be relevant to the majority of the projects the analyst is likely to encounter, they are postponed for discussion to Chapter 10.

5.11 The discount rate

Several roles have been proposed for the discount rate in cost-benefit analysis. We have already seen, in Chapter 2, that it can incorporate inflation in discounting net benefits which are expressed in future prices, and some analysts conceive a role for it in accounting for the opportunity cost of public funds. As we saw in Chapter 2, some propose using a discount rate which incorporates a premium to represent the cost of risk. The main function of the discount rate, however, is to represent time preference and in our approach to cost-benefit analysis we prefer to limit it to that role.

In applying the cost-benefit model we will use a range of discount rates, including the market rate of interest, defined as the rate of interest on a widely traded relatively riskless asset, such as a government bond, to calculate the net present value in the Efficiency and Referent Group Analyses. In selecting an interest rate to represent the market rate it is sensible to choose the rate on a government security with approximately the same time to maturity as the life of the typical public project.

It has been argued, however, that the market rate of interest may not represent society's preference for present, as compared with future, consumption goods, particularly in the case of long-lived investment projects, such as dams or timber plantations, which may span generations. This argument is based on the view that the capital market, in which borrowing and lending takes place, is subject to a form of market failure. Specifically, the problem is that, while future generations are affected by the investment decisions of the present, they are unable to influence the market outcome through participation in the capital market. In consequence, the market fails to take account of the economic welfare of future generations.

Another way of looking at the issue is to recognise that, in theory, the market rate of interest – the rate of discount we use to trade present for future consumption goods – incorporates two components, as discussed in detail in Technical note 5.2. One component – the utility discount rate – measures the extent to which individuals prefer utility from consumption now over utility from consumption in the future. This form of utility time preference is thought to be innate in human nature. The other component – the growth discount factor – reflects the fact that we expect to be better off in the future and that an additional quantity of consumption goods in the future will be worth less to us than the same additional quantity now and should be discounted accordingly. The size of the growth discount factor is the product of how much richer we expect to be in the future (the per capita economic growth rate) and the extent to which the marginal utility of consumption falls (as measured by the elasticity of the marginal utility of income) as we get richer. In a survey of 50 countries. Layard et al. (2008) found values of the elasticity of the marginal utility of income in the range 1.19 to 1.34, with an average of 1.26. If we combine an economic growth rate of 2% with an elasticity of marginal utility of consumption of, say, 1.5 we obtain a growth discount rate of 3%. If the market rate of interest is 4%, the utility discount rate, which is not directly observable, must on these figures, be 1%. It is then argued that, in making investment decisions which affect future generations we should not be discounting their utility, and that the appropriate discount rate, in the case of the above example is 3%, rather than the market rate of 4%. An alternative approach is to allow the rate of discount to decline over time, as discussed in Chapter 4: we might use a 4% rate for discounting the first 50 years of net benefits, and then lower rates thereafter.

Table 5.2 International real rates of discount for cost-benefit analysis

Country	Agency discount rate (%)
The Philippines	15[a]
India	12[a]
Pakistan	12[a]
New Zealand	Treasury and Finance Ministry 8[g]. (From 1982 to 2008 used 10[abf])
Canada	Treasury Board 8[c]. (From 1976 to 2007 used 10 (and test 8–12)[ab])
China (PRC)	8[a]
South Africa	8 (and test 3 and 12)[d]
United States	Office of Management and Budget 7 (and test 3). (Used 10 until 1992.[a]) Environmental Protection Agency 2–3 (and test 7)[a]
European Union	European Commission 5 (From 2001 to 2006 used 6[a])
Italy	Central Guidance to Regional Authorities 5[a]
The Netherlands	Ministry of Finance 4 (risk free rate)[e]
France	Commissariat Général du Plan 4 (From 1985 to 2005 used 8[ab])
United Kingdom	HM Treasury 3.5 from 2003[a] (1 for values occurring > 300 years in the future) (1969–78 used 10[a])
Norway	3.5 (From 1978 to 1998 used 7[ab])
Germany	Federal Finance Ministry 3 (From 1999 to 2004 used 4[ab])
International multilateral development banks	World Bank 10–12[a]; Asia Development Bank 10–12[a]; Inter-American Development Bank 12[a]; European Bank for Reconstruction and Development 10[a]; African Development Bank 10–12[a]

Notes:
a Zhuang et al. (2007, table 4, pp. 17–18, 20).
b Spackman (2006, table A.1, p. 31).
c Treasury Board of Canada (2007, p. 37, 1998, p. 45).
d South African Department of Environmental Affairs and Tourism (2004, p. 8).
e van Ewijk and Tang (2003, p. 1).
f Use of the 10 rate by New Zealand government departments is confirmed by Young (2002, p. 12); Abusah and de Bruyn (2007, p. 4).
g New Zealand Treasury (2008) recommends a default rate of 8 (after adjusting the market risk premium of 7 for gearing).

(Source: Australian Government Productivity Commission, *Valuing the Future: The Social Discount Rate in Cost-Benefit Analysis*, Mark Harrison, 22/4/2010, Table 2–1, for all these notes)

There is a voluminous literature concerning the social time preference rate of discount and, as illustrated here, it can be quite technical. Fortunately for the cost-benefit analyst, many governments and international organizations specify the interest rate they want used in calculating the present value of efficiency and Referent Group net benefits. Table 5.2 reports some of these interest rates and the analyst can take these into account in choosing the band of discount rates used in the analysis of the sensitivity of net present value to the discount rate, as discussed in Chapter 9.

The range of discount rates reported in Table 5.2 offers some support for our analysis of the factors which determine time preference. The discount rates used by the developing countries are generally higher than those for the developed countries. We can attribute this pattern at least partially to the effect of the growth discount factor which is expected to be higher for developing economies because of their higher expected rates of economic growth. Of course this analysis of time preference is based on demand-side considerations only and the level of interest rates will also be affected by supply-side considerations such as productivity.

TECHNICAL NOTE 5.2 The consumption rate of interest

The consumption rate of interest, r, is used to construct the discount factor to be used to bring dollars' worth of goods to be consumed in the future to a present value. In discrete time spreadsheet applications the discount factor is $1/(1 + r)^t$, but in continuous time it is expressed as e^{-rt}. It can be seen from the latter expression that the discount *rate*, r is the negative of the growth rate of the discount *factor*.

The consumption discount factor tells us the rate at which we are willing to give up consumption today in exchange for additional consumption in the future. Utility can be expressed as:

$$U = F(U_t), t = 0...t...T.$$

where t represents time periods, T is the time horizon, and utility at time t depends on dollars' worth of commodities consumed at time t, $U_t = U(C_t)$. A change in present consumption, ΔC_0, coupled with a change in consumption at time t, ΔC_t, which would leave the level of utility, U, unchanged is defined by:

$$\Delta U = \frac{\Delta U}{\Delta C_0} \Delta C_0 + \frac{\Delta U}{\Delta U_t} \frac{\Delta U_t}{\Delta U_t} \Delta C_t = 0$$

Solving this equation gives the discount factor expressed as a positive number:

$$-\frac{\Delta C_0}{\Delta C_t} = \frac{\Delta U}{\Delta U_t} \frac{\lambda_t}{\lambda_0}, \text{ where } \lambda_t = \frac{\Delta U_t}{\Delta C_t} \text{ and } \lambda_0 = \frac{\Delta U_0}{\Delta C_0}$$

The growth rate of the negative of the discount factor in response to a change in C_t is:

$$-G\left[\frac{\Delta U}{\Delta U_t}\right] - G\left[\frac{\lambda_t}{\lambda_0}\right]$$

The first term is the negative of the growth rate of the utility discount factor (the utility discount rate, ρ). Since λ_0 is a constant with respect to changes in C_t, the second term can be expressed as:

$$-\left[\frac{\Delta l_t}{\Delta C_t} \frac{C_t}{l_t}\right]\left[\frac{\Delta C_t}{\Delta t} \frac{1}{C_t}\right]$$

which is the growth discount rate – the elasticity of the marginal utility of consumption, E_{MU}, expressed as a positive number, multiplied by the expected consumption growth rate, g. In summary, the consumption rate of interest can be expressed as:

$$r = \rho + \varepsilon_{MU} g$$

5.12 Worked examples

5.12.1 *Efficiency analysis of the National Fruit Growers (NFG) project*

In Chapter 4 we introduced the NFG project and started the cost-benefit analysis. In the Market Analysis we calculated the net present value of the NFG project at market prices, and then in the Private Analysis we identified the shares of NPV accruing to the project proponent, the financial institutions and the government in the form of business income taxes. In undertaking an analysis of economic efficiency, the market prices of some inputs and outputs may have to be adjusted, as described in the present chapter, and values will have to be placed on the non-marketed effects of the project. In particular, the market prices of inputs which are in addition to current supply will have to be expressed on a before-tax or subsidy basis. If the market price of an input and the rate of tax or subsidy that has been applied to it are known it is a simple matter to calculate the before-tax or before-subsidy price for use in the analysis of efficiency. *Once the efficiency price has been calculated, it is fixed as a value in the spreadsheet and the market price is amended to a formula based on the efficiency price and the rate of tax or subsidy.* The reason for this procedure is that we may be asked to undertake an analysis of the sensitivity of the private NPV of the project to changes in tax rates. For example, suppose that the project proponent asked for a concessional tariff rate on imports of equipment: a reduction in tariff rate would have no effect on the world price (the efficiency price) but it would reduce the domestic market price.

Further details about the efficiency costs of the NFG project presented in Chapter 4 are provided in Table 5.3. Apart from the need to shadow-price some of NFG's *internal* costs, we learn that this project also generates some *external* costs for downstream users. Nutrient, chemical and sediment run-off from NFG's orchard will pollute the adjacent King River and will adversely affect the local commercial and recreational fishery. From this additional information it is possible to re-estimate the project's net benefits, by means of techniques described in Chapter 8, using efficiency prices instead of the actual market prices paid by NFG where appropriate, and including non-marketed effects. In our analysis of the NFG project we assume that all inputs to the project represent additional quantities

Table 5.3 Data for calculating shadow-prices of NFG Project inputs

Input item	Efficiency price information
Opportunity cost	
– land	$0
– labour	20%
Tax on fuel	10%
Subsidies	
– seeds	25%
– fertiliser	30%
– insecticides	20%
– water	40%
Import duties	
– equipment	10%
– vehicles	20%
– spares	15%
External costs (from yr.2)	$10,000 pa

supplied rather than reallocations from other uses, and that project output satisfies additional demand.

On the right-hand side of the Key Variables Table (Table 1 of Figure 5.15) the details of the adjustments to the project's operating costs are entered as shown in columns P to T. In row 4 the opportunity cost of land is entered as zero; cells R5 to R9 provide details of *ad valorem* taxes and subsidies on various inputs; cell R10 provides details of the duties on spare parts; and, cell R11 shows that labour has an opportunity cost equal to 20% of its market wage. Where there is a tax or import duty on the input, the price in column N is divided by $(1 + t)$ to derive the efficiency price (in column S), where t is the tax rate or rate of import duty. For subsidies, the efficiency price is derived by dividing the market price by $(1 - s)$ where s is the rate of subsidy. These formulae should first be entered in the respective cells in column S and then saved as values, as noted above. The market prices of taxed or subsidised inputs reported in column N should then be replaced by formulae based on the values in column S so that if, as suggested above, the analyst at some later stage wishes to vary the rate of tax, subsidy or import duty (in column R), the market prices (in column N) will change accordingly. To value these inputs, the efficiency prices (column S) are multiplied by the respective quantities (column M) to arrive at the efficiency cost (column T). It is assumed that the efficiency prices of the remaining components of operating cost – insurance, management fees and miscellaneous costs – are the same as their respective market prices, so the same values are carried across (from column N to column S).

Investment costs also need to be shadow-priced as shown in columns G and H in Table 1 of Figure 5.15. Equipment and vehicle prices are recalculated using the information on import duties in cells S21 and S22. We price buildings at their market price and arrive at an efficiency cost for total fixed investment of $688,636 (cell H8). Note that the salvage value of fixed investment is also re-calculated at 10% of its efficiency cost (cell H15). Finally, components of Working Capital need to be shadow-priced taking import duties and subsidies into account. The same duty and subsidy rates shown in Column R are used to derive the efficiency costs for Working Capital ($104,192) as shown in Cell H14. As in the case of operating costs, the efficiency prices in column G are fixed as values and the market prices in column E are amended to formulae.

Economists from the local university have estimated the reduction in value of the fish catch resulting from pollution of the river at $10,000 per annum from Year 2 onwards (see Cell T18), and believe that there will be no corresponding reduction in fishing costs. To take this external cost into account when undertaking the Efficiency Analysis, an additional row (External costs row, Figure 5.16) is added to the cost section of the efficiency cash flow.

NFG's fruit output is valued at its market price as we assume that it satisfies additional demand, rather than being an alternative way of meeting existing demand.

We are now ready to undertake the Efficiency Analysis as shown in Table 4 of Figure 5.16. This Table is identical to the Market Analysis cash flow that we derived in Chapter 4 (Figure 4.2), except that we now use efficiency prices instead of actual market prices to value inputs and outputs, and we have added an extra row showing the external costs imposed on the downstream fishers (External costs row). Note that when we calculate the return on this project at efficiency prices, it is much lower than the value obtained from the Market Analysis. The IRR is now only 7.8%, which also means that the NPV is negative at 10%. At a cost of capital above 7.8%, the project would not be worthwhile from an overall efficiency perspective.

By comparing the NPVs generated by the Efficiency Analysis with those of the Market Analysis, we could see that the proposed NFG project was less attractive from an efficiency

Table 1: Key Variables

Investment Costs	No. (units)	Market Price ($)	Cost	Efficiency Price ($)	Cost
(i) Fixed Investment					
Farm equipment (units)	4	100,000	400,000	90,909	363,636
Vehicles (units)	3	30,000	90,000	25,000	75,000
Buildings (m2)	250	1,000	250,000	1,000	250,000
TOTAL			740,000		688,636
(ii) Working Capital					
Fertilizer Stocks (tons)	2	500	1,000	714.29	1,429
Insecticide stocks (Litres)	2500	30	75,000	37.50	93,750
Equipment spare parts (units)	10	1,000	10,000	870	8,696
Fuel stocks (Litres)	500	0.7	350	0.64	318
TOTAL			86,350.0		104,192
(iii) Salvage Value	10%		74,000		68,864

Depreciation	Life(yrs)	Amount p.a.
Equipmnt	10	40,000
Vehicles	5	18,000
Buildings	20	12,500

Financing	Amount	Interest	Life (yrs)
Loan	$700,000	3.5%	10
Overdraft	$40,000	5.0%	4
Discount rate =	5.0%	10.0%	15.0%
Tax rate on profits =	25.0%		

Operating Costs	No.	Market Price	Cost		Efficiency Pricing %	Price	Cost
Rent on land (Ha)	100	30	3,000	Rent op. cost	0%	0.00	0
Fuel (litres)	2,500	0.7	1,750	Fuel tax	10%	0.64	1,591
Seeds (Kg)	250	20	5,000	Seed subsidy	25%	26.67	6,667
Fertilizers (tonnes)	3	500	1,500	Fertilizer Subsidy	30%	714.29	2,143
Insecticides (litres)	3,000	30	90,000	Insecticide subsidy	20%	37.50	112,500
Water (ML)	900	20	18,000	Water subsidy	40%	33.33	30,000
Spares	12	1,000	12,000	Spares -duty	15%	869.57	10,435
Casual labor (days)	100	60	6,000	Labour op. cost	20%	12.00	1,200
Administration (/month)	12	1000	12,000	Administration (/month)			12,000
Insurance (p.a.)	1%	826,350	8,263	Insurance (p.a.)		826,349.6	8,263
Management (/month)	12	3,000	36,000	Management		3,000	36,000
Miscellaneous	1	7,700	7,700	Miscellaneous		7,700	7,700
TOTAL (Market prices)			$201,213	TOTAL (Efficiency Prices)			$228,499

Revenues	No.	Market Price	Cost
Apples (tons)	100	1,000	100,000
Peaches (tons)	90	1,250	112,500
Pears (tons)	75	1,500	112,500
TOTAL			$325,000

External Costs	$10,000

Import duties	
Equipment	10%
Vehicles	20%

Conversion factor	1000

Capacity Output				
Year	2016	2017	2018	2019+
%	25%	50%	75%	100%

Figure 5.15 Spreadsheet showing key variables table with efficiency prices.

TABLE 4: EFFICIENCY CASH FLOW																					
	2015	2016	2017	2018	2019	2020	2021	2022	2023	2024	2025	2026	2027	2028	2029	2030	2031	2032	2033	2034	2035
ITEM/YEAR	0	1	2	3	4	5	6	7	8	9	10	11	12	13	14	15	16	17	18	19	20
Investment costs																					
Fixed Investment	-688.6																				68.9
Working Capital		-104.2																			104.2
Total Investment	-688.6	-104.2	0.0	0.0	0.0	0.0	0.0	0.0	0.0	0.0	0.0	0.0	0.0	0.0	0.0	0.0	0.0	0.0	0.0	0.0	173.1
Operating Costs		-57.1	-114.2	-171.4	-228.5	-228.5	-228.5	-228.5	-228.5	-228.5	-228	-228.5	-228.5	-228.5	-228.5	-228.5	-228.5	-228.5	-228.5	-228.5	-228.5
Revenues		81.3	162.5	243.8	325.0	325.0	325.0	325.0	325.0	325.0	32	325.0	325.0	325.0	325.0	325.0	325.0	325.0	325	325.0	325.0
External costs			-10.0	-10.0	-10.0	-10.0	-10.0	-10.0	-10.0	-10.0		-10.0	-10.0	-10.0	-10.0	-10.0	-10.0	-10.0		-10.0	-10.0
Net Cash Flow	-688.6	-80.1	38.3	62.4	86.5	86.5	86.5	86.5	86.5	86.5	.5	86.5	86.5	86.5	86.5	86.5	86.5	86.5	.5	86.5	259.6

	5%	10%	15%
NPV=	$231.3	5.9	-$333.8
IRR =	7		

External costs to downstream users included · Market cash flow is re-calculated using efficiency prices · Salvage value at efficiency price assuming it is internationally tradeable

Figure 5.16 Spreadsheet showing efficiency net benefit flow.

viewpoint than from the perspective of the market. Why is this so? By comparing the information in Table 4 of Figure 5.16 with that reported in Figure 4.2, Chapter 4, we can see that the net effect of shadow-pricing various inputs to the project is to increase the estimate of annual operating cost and investment costs, while annual revenues are unchanged. The main reason for the increase in the cost estimate is the existence of subsidies on inputs of seed, fertiliser, insecticides and water. An additional cost, in the form of an environmental cost, is also included in the Efficiency Analysis of Figure 5.16. Hence, with the same benefits, but higher costs, except for labour and fuel costs, the NFG project is less attractive from an efficiency perspective.

The Efficiency Analysis includes the benefits and costs to all agents affected by the project, irrespective of whether they form part of the specified Referent Group or not. The decision about whether or not the project should be supported will depend on the size of the net benefits to the Referent Group. The calculation of these for the NFG project is discussed in Chapter 6.

5.12.2 Cost-benefit analysis of the 55 mph speed limit

This example illustrates the use of cost-benefit analysis to appraise a proposed *policy* rather than an investment project. During the 1973 OPEC oil export embargo, the US Congress imposed a temporary 55 mph speed limit on federal highways in order to reduce gasoline consumption and the nation's dependence on foreign oil. The new speed limit was made permanent in 1974, was relaxed for selected highways in 1987, and was repealed in 1995.

As noted, the 55 mph speed limit is not a "project" of the traditional kind we have been discussing but it is amenable to cost-benefit analysis. It has many of the features we have been discussing: it involves outputs or inputs with prices generated in distorted markets, or with no market prices at all. Because these features are the rule in this case, rather than the exception, it makes sense to forgo the Market Analysis and go straight to the Efficiency Analysis. Unlike a project with an uneven flow of net benefits over time, the net benefits of the speed limit will be a relatively even flow, but perhaps increasing gradually in real terms over time because of the forecasted increase in traffic volumes. For this reason it is not necessary to enter benefits and costs specific to individual years, but an approximate NPV could be calculated by dividing the initial annual net benefit by the real rate of interest less the predicted growth rate of traffic volume, a technique discussed in Chapter 2. Since by far the majority of the benefits and costs of the regulation are felt by the general US public, the Efficiency Analysis also doubles as the Referent Group Analysis. In summary, the CBA model tailored for the evaluation of this proposed change in regulation consists of: a Table of Variables, including a Working Table, and the Efficiency Analysis expressed in per annum terms. Clearly there are conflicting views about the efficacy of the 55 mph speed limit, and the aim of this example is not to decide in favour of one side or the other, but to illustrate the application of CBA techniques to this public policy question.

In addition to reduced gasoline consumption, the 55 mph limit was predicted to lower the number of traffic fatalities and injuries. Other benefits include lower exhaust emissions and perhaps reduced road and vehicle maintenance costs. The main cost of the policy is that of increased travel times for private and commercial vehicles, but there may also have been additional enforcement costs.

The variables used in the CBA are obtained from Forester et al. (1984), are based on 1981 data, and are reported in the spreadsheet shown in Table 5.4. In addition to estimates of traffic fatalities, injuries and gasoline consumption avoided annually, Table 5.4 reports an estimate of extra travel time in person years; this estimate is obtained by calculating the extra vehicle time required to cover the total annual distance travelled on federal highways at the predicted post-regulation average speed of 55.5 mph as compared to the observed pre-regulation speed of 60.3 mph, and then multiplying by average vehicle occupancy. Table 5.4 reports three different estimates of the value of a life saved: one estimate is the present value of income which would have been earned from the average age of the accident victim (33.5 years) until retirement; another is the latter estimate less 30% to account for the value of the goods and services that would have been consumed by the victim; and a third is based on markets involving risk. The benefit of lives saved will be considered further in Chapter 8 on non-market valuation.

Table 5.4 also reports estimates of the cost of additional travel time. These measures are based on the kind of analysis described in Figure 5.12 and vary from 100% to 30% of the hourly average wage. While no distinction is made between private and commercial vehicle time, it would be expected that the cost of the time of the latter would be based on 100% of the wage, and might also include costs of operating the larger commercial fleet required to supply the existing volume of goods and services. Of course, as discussed in Chapter 7 on the effects of price changes, the increase in freight costs might reduce the volume to be transported.

The price paid by motorists for gasoline is reported as $1.20 per gallon. Since the original purpose of the regulation was to reduce the quantity of gasoline supplied to the US market, the benefit of fuel savings should be measured at the supply price. While gasoline consumption in the US is taxed both federally and by states (in 2013 the average composite tax rate was around 14%), refining is also subsidised. The appropriate price for use in the CBA is the before-tax and subsidy price, which we will take to be $1.20 although it may well be lower.

The data entered at the top of the Variables Table are used in a Working Table to generate estimates of the annual benefits and costs of the speed limit, based on alternative values of life saved and travel time. To start with, we select the values most favourable to the policy for transfer to the Efficiency Analysis and we find that it imposes an annual net cost of close to $2 billion. Any combination of values reported in Table 5.4 would produce a bigger net cost so it seems that there is little point in spending time trying to obtain more precise information about the values of these variables; the value of more precise information is a concept discussed in Chapter 9. In terms of omitted variables, the most significant

Table 5.4 Benefit-cost analysis of the 55mph speed limit

Variables Table	
Reduction in Number of Traffic Deaths p.a.	7466
Reduction in Number of Injuries p.a.	198,000
Reduction in Gasoline Use (million gals p.a.)	659.7
Price of Gasoline ($/gal)	1.2
Present Value of Income Saved per Avoided Fatality ($)	527,200
Present Value of Income Saved Less Consumption ($)	369,040
Consumer Valuation of Cost of Death ($)	561,300
Cost of Injury per person	15,504
Total Highway Vehicle Miles p.a.	1,548,213
Average Speed without 55mph limit	60.3
Average Speed with 55mph limit	55.5
Average Vehicle Occupancy	1.8
Average Wage ($/hour)	7.45
Working Table	
Million	1,000,000
Hours per year	8760
Total Value of Life Saved ($ million p.a.)	
Based on Value of Income	3936.08
Based on Value of Income less Consumption	2755.25
Based on Consumer Valuation	4190.67
Value of Injuries Prevented ($million p.a.)	3069.79
Benefit of Reduced Gas Consumption ($ million p.a.)	683.64
Cost of Extra Travel Time ($ million p.a.)	
Extra Travel Time (million person hours)	3997.00
Cost at Market Wage	29,777.65
Cost at 50% of Market Wage	14,888.83
Cost at 1/3 Market Wage	9925.88
Efficiency Analysis (most favourable)	$ millions p.a.
Value of Lives Saved	4190.67
Value of Injuries Prevented	3069.79
Savings in Gasoline	683.64
Cost of Extra Travel Time (1/3 Wage)	−9925.88
Annual Net Benefit	**−1981.79**

is the reduction in pollution. It was estimated that health and non-climate related damages from vehicle pollution in the year 2005 would fall by $1–1.6 billion annually if the speed limit were reintroduced. This figure can be calibrated to 1981 levels, for comparison with the results discussed above, by discounting for inflation and traffic volume growth – say, 3% combined – to give an estimate of benefit from this source valued at $995 million, still not sufficient to provide a positive annual net benefit of the policy to the Referent Group. The reduction in greenhouse gas emissions would provide an additional annual benefit which would be distributed globally.

5.13 Further reading

The analysis of market equilibrium, indirect taxation, market regulation and market failure is included in most microeconomic theory texts. These issues, and their relevance to cost-benefit analysis, are also discussed in most works on public finance, such as, for example,

R.W. Boadway and D.E. Wildasin, *Public Sector Economics* (New York: Little, Brown and Co., 1984).

A dated but useful survey of shadow-pricing techniques is R.N. McKean, "The Use of Shadow Prices", in S.B. Chase (ed.), *Problems in Public Expenditure Analysis* (New York: Brookings Institution, 1968), pp. 33–77.

For an analysis of the shadow-wage in a developing economy, see H.F. Campbell, "The Shadow-Price of Labour and Employment Benefits in a Developing Economy", *Australian Economic Papers*, 47(4) (2008): 311–19. W.G. Waters II, "Value of Travel Time Savings in Road Transportation Project Evaluation", in D.A. Hensher, J. King and T.H. Oum (eds), *World Transport Research: Proceedings of the 7th World Conference on Transport Research, Vol. 3* (New York: Elsevier, 1996) reports estimates of the value of travel time saved.

Classic articles on the social discount rate and the opportunity cost of public funds are W.J. Baumol, "On the Social Rate of Discount", *American Economic Review*, 58 (1968): 788–802, and two papers by S.A. Marglin, "The Social Rate of Discount and the Optimal Rate of Investment", *Quarterly Journal of Economics*, 77 (1963): 95–111, and "The Opportunity Costs of Public Investment", *Quarterly Journal of Economics*, 77 (1963): 274–287.

For various approaches to estimating the opportunity cost of public funds in a range of OECD countries, see, for example, E.K. Browning, "The Marginal Cost of Public Funds", *Journal of Political Economy*, 84 (1976): 283–298; H.F. Campbell, "Deadweight Loss and Commodity Taxation in Canada", *Canadian Journal of Economics*, 8(3) (1975): 441–447; H.F. Campbell and K.A. Bond, "The Cost of Public Funds in Australia", *The Economic Record*, 73(220) (1997): 22–34, and W.E. Diewert and D.A. Lawrence, "The Excess Burden of Taxation in New Zealand", *Agenda*, 2(1) (1995): 27–34. For African countries, see E. Auriol and M. Warlters, "The Marginal Cost of Public Funds in Developing Countries: An Application to 38 African Countries", CEPR Discussion Paper No. 6007 (London: Centre for Economic Policy Research, 2006). Guidelines relating to the treatment of the cost of public funds include: United States Office of Management and Budget, "Circular No. A-94 Revised" (Washington, DC, 1992); Commonwealth of Australia, *Handbook of Cost-Benefit Analysis* (Canberra, 2006), and European Commission, *Guide to Cost-Benefit Analysis of Investment Projects* (Luxemburg: European Commission, 2008).

R. Layard, G. Mayraz and S. Nickell, "The Marginal Utility of Income", *Journal of Public Economics*, 92 (2008): 1846–1857, report a range of values of the elasticity of the marginal utility of income.

For an analysis of the 55 mph speed limit, see T.H. Forester, R.F. McNown and L.D. Singell, "A Cost-Benefit Analysis of the 55 MPH Speed Limit", *Southern Economic Journal*, 5(3) (1984): 631–641.

Appendix to Chapter 5

Economic Efficiency Analysis of the ICP case study

We are now in a position to undertake an Efficiency Analysis of the proposed ICP textile project which we analysed from a Market and Private viewpoint in the Appendix to Chapter 4. Since our focus is on efficiency, we will ignore all distributional aspects of the project: this means ignoring its tax implications for the government and the firm, ignoring the project's financial structure and its implications for equity holders, ignoring the distributional effects on various other groups such as public utilities, domestic financial and

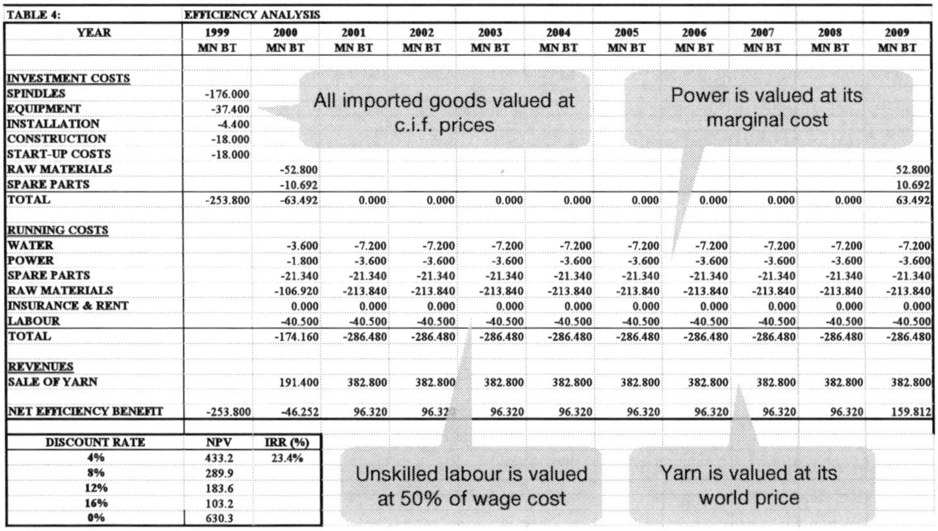

TABLE 4:	EFFICIENCY ANALYSIS										
YEAR	1999	2000	2001	2002	2003	2004	2005	2006	2007	2008	2009
	MN BT	MN BT	MN BT	MN BT	MN BT	MN BT	MN BT	MN BT	MN BT	MN BT	MN BT
INVESTMENT COSTS											
SPINDLES	-176.000										
EQUIPMENT	-37.400										
INSTALLATION	-4.400										
CONSTRUCTION	-18.000										
START-UP COSTS	-18.000										
RAW MATERIALS		-52.800									52.800
SPARE PARTS		-10.692									10.692
TOTAL	-253.800	-63.492	0.000	0.000	0.000	0.000	0.000	0.000	0.000	0.000	63.492
RUNNING COSTS											
WATER		-3.600	-7.200	-7.200	-7.200	-7.200	-7.200	-7.200	-7.200	-7.200	-7.200
POWER		-1.800	-3.600	-3.600	-3.600	-3.600	-3.600	-3.600	-3.600	-3.600	-3.600
SPARE PARTS		-21.340	-21.340	-21.340	-21.340	-21.340	-21.340	-21.340	-21.340	-21.340	-21.340
RAW MATERIALS		-106.920	-213.840	-213.840	-213.840	-213.840	-213.840	-213.840	-213.840	-213.840	-213.840
INSURANCE & RENT		0.000	0.000	0.000	0.000	0.000	0.000	0.000	0.000	0.000	0.000
LABOUR		-40.500	-40.500	-40.500	-40.500	-40.500	-40.500	-40.500	-40.500	-40.500	-40.500
TOTAL		-174.160	-286.480	-286.480	-286.480	-286.480	-286.480	-286.480	-286.480	-286.480	-286.480
REVENUES											
SALE OF YARN		191.400	382.800	382.800	382.800	382.800	382.800	382.800	382.800	382.800	382.800
NET EFFICIENCY BENEFIT	-253.800	-46.252	96.320	96.32	96.320	96.320	96.320	96.320	96.320	96.320	159.812

DISCOUNT RATE	NPV	IRR (%)
4%	433.2	23.4%
8%	289.9	
12%	183.6	
16%	103.2	
0%	630.3	

Callout annotations on the table:
- *All imported goods valued at c.i.f. prices*
- *Power is valued at its marginal cost*
- *Unskilled labour is valued at 50% of wage cost*
- *Yarn is valued at its world price*

Figure A5.1 ICP project solution: the efficiency net benefit flow.

insurance agencies, and suppliers of labour, and ignoring the distributional effects among countries.

The efficiency cash flow for the ICP project is based on the Market cash flow which we set up in Chapter 4, Figure A4.2, and is shown in Figure A5.1. We start by enumerating the "real" aspects of the project. It involves employing a range of inputs – land, capital goods, labour, water, electricity, raw materials – and using them to produce an output – yarn. The costs of the project are the opportunity costs of the inputs, and the benefits are the value of the output. Since the project is an extension of an existing factory involving the use of vacant land which will otherwise remain vacant, it will be assumed that the land used in the project has no opportunity cost, even though the firm pays a nominal rent on the land. (Similarly, we should ignore the overhead cost that the firm chooses to allocate to the project for its own internal accounting purposes.) The remaining inputs need to be shadow-priced because of various market imperfections, and we consider each input in turn. It should be noted that all input data relating to the calculation of efficiency prices are drawn from the Key Input Variables section (Table 1) of the spreadsheet as shown in Figure A4.1 in Chapter 4. Adhering to this convention ensures that all cell entries in Table 4, Figure A5.1, are formulae or references to the preceding tables, and therefore also, that any subsequent changes to the values of the input variables will result in automatic changes to the respective cash flows.

Cost of capital goods

Since capital goods are imported at various times to provide equipment and inventories for the project we can assume that they are in addition to current supply and, accordingly, will be valued at their supply prices. These imports are purchased using foreign exchange obtained by the firm at the official exchange rate. We will consider shadow-pricing the foreign exchange in Chapter 10, but for the present we will assume that the official rate represents the opportunity cost of foreign exchange, and, accordingly, we can convert the US dollar price of capital goods to domestic currency using the official exchange rate. The US dollar prices of imported goods landed in the country are what are termed c.i.f. prices – prices including cost, insurance and freight; this means that the US dollar price is the price of the good at the country's border. Once that price is converted to domestic currency a tariff is applied – the price to domestic buyers is raised by some percentage above the border price in domestic currency. A tariff is an indirect tax as illustrated in Figure 5.3. The opportunity cost of the imported capital good is its supply price which is net of the tariff. Hence, while the price the firm pays for the inputs is the gross of tariff price (the c.i.f. price plus tariff), we use the net of tariff price (the c.i.f. price) in local currency as the shadow-price in the Efficiency Analysis.

Cost of labour

The project uses skilled, semi-skilled and unskilled workers in both its construction and operations phases. The unskilled labour market exhibits significant unemployment and the Ministry of Industry suggests that a shadow-price of 50% of the unskilled wage be used. What is the basis of this figure? It could be that the Ministry is assuming, or has information to the effect that half of the unskilled workers engaged in the project would otherwise be unemployed (with a zero value placed on their non-market activities) and that the other half are hired away from jobs in which they cannot be replaced. The pricing rule would tell us to value the former at their supply price (zero) and the latter at the demand price (the market wage). Alternatively, and more likely, the assumption could be that all unskilled workers are effectively drawn from the ranks of the unemployed (an addition to the current supply) and that the unit value of the non-market activities of those workers (the supply price) is assumed to be 50% of the wage. In either case, and perhaps after some discussion with officials, the analyst adopts a shadow-price for unskilled labour of 50% of the unskilled wage. This means that only half of the unskilled labour costs incurred by the firm are counted as an opportunity cost in the Efficiency Analysis of the project.

Cost of utilities

Water and electricity are supplied by state-owned public utilities. We are informed by Ministry officials that the additional electricity required by the project can be produced at a lower unit cost than the price charged to customers. This is the kind of situation discussed in Figure 5.6. The opportunity cost of an input, the supply of which is to be increased, is measured by an equilibrium point on its supply or marginal cost curve. The marginal cost of electricity is reckoned to be 50% of the price paid by the firm, and hence only 50% of the firm's power bill will be included as an opportunity cost in the Efficiency Analysis. A similar situation may exist with water supply, but it appears that after discussion with officials it has been concluded that the firm's additional water bill does in fact represent the cost of the additional supply, or, alternatively, that the water will be diverted from another use. Hence the water input is valued at the market price, and the full extra water bill is included as an opportunity cost.

Cotton and yarn inputs and outputs

The final input to be considered is raw materials. As in the case of capital goods, and for the same reasons, the appropriate shadow-price is the c.i.f. dollar price converted to domestic currency by means of the official exchange rate. We now turn to the output of the project – spinning yarn. When valuing output we look for an equilibrium point on a demand curve or a supply curve. In this case the project simply replaces an alternative source of supply – the world market which determines the equilibrium world price of yarn. The latter price is the US dollar c.i.f. price, converted to domestic currency by means of the official exchange rate. The efficiency pricing rule tells us that the output of an import-replacing project should be valued at this net of tariff world price.

The Efficiency Analysis

Now that we have decided upon the appropriate shadow-prices for the Efficiency Analysis, we need to enter the values in the spreadsheet and perform the NPV calculation which will tell us if the project is an efficient use of resources. It will be recalled from Chapter 4 that three sections of the spreadsheet have already been completed: Part 1 contains the data available to perform the analysis (which must now be augmented by the shadow-prices we have just computed); Part 2 contains the Market Analysis; and Part 3 – the Private Analysis – contains the analysis from the point of view of the equity holders in the firm. We now open Part 4 of the spreadsheet – the Efficiency Analysis – by entering a set of rows containing the value of output and the opportunity costs of inputs as shown in Figure A5.1. These rows correspond to the categories of "real" output and inputs enumerated and discussed above. The efficiency benefit or cost represented by each category of output or input in each year is calculated by multiplying the quantity by the shadow-price. Each entry in this part of the analysis is a formula incorporating values contained in the first part of the spreadsheet as developed in Figures A4.1 to A4.3 in the previous chapter.

Once the various categories of efficiency benefits and costs have been valued in each year, the net benefit in each year is calculated by subtracting costs from benefits. This single row of net benefits will be negative in the early years of the project – the construction and development phase – and then generally positive in later years. The net benefit stream is discounted using a range of discount rates to yield a set of NPV estimates, and the IRR is calculated where appropriate. The reason for using a range of discount rates is that there may be some dispute about the appropriate rate as discussed earlier in this chapter. For the moment, since we have not inflated the benefit and cost streams, we can use a real market rate of interest – the observed money rate less the observed current rate of inflation – as a reference point. Discount rates below and above this rate tell us how sensitive the NPV result is to the choice of discount rate.

The next task, to be taken up in Chapter 6, is to examine the distribution of the net benefits of the project – the Referent Group Analysis. In particular we will want to determine an aggregate measure of the net benefits, in each year, to members of the Referent Group, taken to be residents (households, firms and government and non-government organizations) of Thailand, and then to determine, by a different process, a set of net benefits to each sub-group which, when aggregated, correspond to our aggregate measure.

Exercises

1   Which of the following conditions must hold in the equilibrium of a competitive market in which a specific tax is levied on the commodity? (A *specific tax* is a fixed dollar amount per unit to be paid on sale of the product, as opposed to an *ad valorem* tax which is calculated as a percentage of the sale price.)

i   the point representing the quantity sold and the price paid by the buyer must lie on the demand curve;

ii the point representing the quantity sold and the price received by the seller must lie on the supply curve;

iii the quantity demanded must equal the quantity supplied;

iv the difference between the price the buyer pays and the price the seller receives must equal the specific tax;

v   all of the above.

2   Draw an upward sloping supply curve and a downward sloping demand curve for labour. Label the axes. Assuming that the labour market is distorted by a minimum wage (which is a binding constraint), draw in the level of the minimum wage, labelled w_m, and identify the quantity of labour demanded, labelled Q. On the supply curve of labour, identify the level of the wage corresponding to Q units of labour supplied, and label it w_a.

i   State what the Efficiency Analysis rule tells us about valuing labour input in a cost-benefit analysis.

ii Suppose that a project will use a small amount of labour drawn from the above market. Under what circumstances should the labour be costed in the Efficiency Analysis at: (a) w_m per unit? or (b) w_a per unit?

3   Suppose that the market wage is $100 per day and the value placed on the leisure time of unemployed labour is $60 per day. A project is expected to take 50 worker-days to complete. It is expected that 40% of the workers hired will be diverted from employment elsewhere in the economy, and that 60% would otherwise be unemployed.

i   What dollar measure of the cost of labour used in the project should be included in the Market Analysis?

ii What dollar measure of the cost of the total amount of labour used in the project should be included in the Efficiency Analysis:

a   assuming that the jobs vacated by workers being hired for the project cannot be filled by labour which would otherwise remain unemployed?

b   assuming that the jobs vacated by workers being hired for the project are filled by labour which would otherwise remain unemployed?

4   A worker values her leisure time at $7 per hour. She is offered work on a public project at a wage of $22 per hour, but if she accepts she will forfeit unemployment benefits equivalent to $13 per hour. Identify:

i   the private opportunity cost of her time;

ii the social opportunity cost of her time;

iii the employment benefit to the economy per hour of her time on the project;

iv the portion of the employment benefit accruing to the government.

5   Suppose that the elasticity of demand for a monopolist's product is 1.5. Calculate the ratio of the value of the marginal product of labour to the marginal revenue product of labour. What is the opportunity cost of labour expressed as a multiple of the wage? What is the appropriate shadow-price of labour diverted from the monopoly's production process?

6   For each of the 16 cases indicated in the table below state whether you would use the demand curve or the supply curve to determine the appropriate price for valuation purposes in an Efficiency Analysis, and state whether the appropriate price is gross or net of the tax or subsidy. In each case briefly justify your answer.

Item to be valued	Indirect tax		Subsidy	
	Distorting	Corrective	Distorting	Corrective
Project Output				
– Satisfies additional demand	(i)	(ii)	(iii)	(iv)
– Satisfies existing demand from alternative source	(v)	(vi)	(vii)	(viii)
Project Input				
– From additional supply	(ix)	(x)	(xi)	(xii)
– Diverted from alternative use	(xiii)	(xiv)	(xv)	(xvi)

Hint: Use Figure 5.13 to work out the answers in the distorting tax and subsidy cases. Then amend Figure 5.13 as discussed in the Section 5.7 on Corrective Taxation and apply the amended version to the corrective tax and subsidy cases.

7   Suppose that a $100 public project is to be financed by the sale of bonds to the public. Purchasers of the bonds increase savings by $80 and reduce private investment by $20. The only tax in the economy is a 50% business income tax.

   i   Work out the opportunity cost to the economy of raising the $100 in public funds.
   ii The cost premium on public funds represents a deadweight loss to the economy. Which group in the economy bears this cost? Explain.

8   In Figure 5.14 showing taxation and labour supply: S represents the aggregate supply curve of labour, W is the market wage, W_0 is the after-tax wage before the tax increase, and W_1 is the after-tax wage after the tax increase. L_0 and L_1 represent quantity of labour supplied before and after the tax-increase respectively.

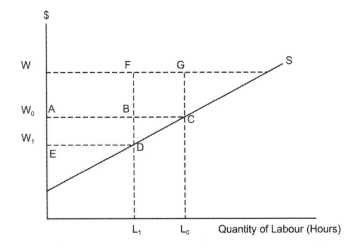

Answer the following questions by identifying the appropriate areas in the diagram:

i   What will be the deadweight loss as a result of the proposed tax increase?

ii   How much additional revenue will the earnings tax produce as a result of the tax increase?

iii What is the cost to the economy of each extra dollar of tax revenue?

9   Suppose that a competitive economy's labour supply consists of 100 units at the after tax wage of $3 per unit. The elasticity of labour supply is 0.25, and there is a flat-rate 40% tax on labour earnings. It is proposed to raise the tax on labour earnings to 50%, and the before-tax wage (which measures the value of the marginal product of labour) is not expected to change.

i   How much extra revenue would the tax change yield?

ii   Work out the cost of the additional deadweight loss imposed on the economy by the tax increase.

iii Work out the marginal cost of public funds.

6 The distribution of project net benefits

6.1 Introduction

In Chapter 4 we calculated project net benefits at market prices (the Market Analysis) and then disaggregated these values to work out the net benefits accruing to selected groups – the private and/or public owners, the banks and the government (a process referred to as the Private Analysis). In Chapter 5 we considered the project from the viewpoint of economic efficiency, broadening the analysis to include the benefits and costs not adequately measured by market prices (the Efficiency Analysis). We recognised that a complete analysis of economic efficiency would have to include values of non-marketed goods and services associated with the project. The Efficiency Analysis can be thought of as identifying the net benefits of the project on a world-wide basis. In the present chapter we discuss the allocation of the net benefits identified by the Efficiency Analysis.

In Chapter 1 we introduced the concept of the Referent Group – the group of individuals, firms and institutions who have "standing" in the analysis – those whose costs and benefits the decision-maker considers relevant to the decision about the project. The Referent Group in a social cost-benefit analysis is unlikely to consist only of the equity holders of the private firm, at one extreme, or of everyone in the whole world affected by the project, at the other extreme. In this chapter we concentrate on identifying the subset of project net benefits which accrue to the Referent Group. In discussion with the client who commissioned the cost-benefit analysis it will be established who is to be included in the Referent Group: as noted in Chapter 1, the client will normally nominate all groups who are resident in her State or country and who are affected by the project (sometimes referred to as "stakeholders"), including effects on government receipts or payments. In the NFG example, which we have been developing in previous chapters, the Referent Group consists of the citisens of a particular region (Happy Valley), while in the ICP case study discussed in Appendix 2 in this chapter, residents of Thailand make up the Referent Group.

As noted in Chapter 1, it is sometimes easier to measure the net benefits of those who are *not* members of the Referent Group than to measure the net benefits to all the relevant sub-groups; in the NFG case study, for example, of those groups who benefit or incur costs, only the equity holders in the private investment project (NFG) and the interstate financial institution which lends to the project are *not* members of the Referent Group. This means that the aggregate net benefits to the Referent Group can be calculated by subtracting the net cash flows experienced by these two groups from the total efficiency net benefits of the project calculated in Figure 5.16 in Chapter 5. Thus we can open Part 5 of our spreadsheet – the Referent Group Analysis – by entering the efficiency net benefits row less the equity holders' and inter-state financial institution's net benefit rows. (See Section 6.7, first row of

DOI: 10.4324/9781003312758-6

Table 5 in Figure 6.2.) We have now completed the aggregate Referent Group cost-benefit analysis.

We want a disaggregated analysis of the net benefits to the Referent Group for two reasons. The main reason is that our client will want information about the distribution of net benefits or costs among sub-groups because this will influence the project's attractiveness to the political decision-maker. The other reason is that if we enumerate all the sub-groups constituting the Referent Group, measure the net benefits to each, sum them and get the same answer as our aggregate Referent Group measure, we can be fairly sure the analysis is correct, or, at least, is internally consistent. It is very common to omit some benefits or costs in the first run of the analysis, and/or to double count some of them, and having a check of this kind is extremely useful.

6.2 How to identify Referent Group net benefits in practice

It is sometimes difficult to identify all the sub-groups within the Referent Group who are affected by the project, and it is not unusual for some group or category of net benefit to be omitted from the first draft of the Referent Group Analysis. Fortunately, as noted above, this kind of error can readily be detected within our project appraisal framework by the existence of a discrepancy between the measure of aggregate Referent Group net benefits, computed by subtracting non-Referent Group benefits from the efficiency net benefits, and the measure computed by aggregating the net benefits to various sub-groups within the Referent Group.

In principle, there is a four-way classification of net benefits, illustrated by Table 6.1, distinguishing net benefits which accrue to the Referent and non-Referent Groups respectively, and net benefits which either are, or are not accurately measured by market prices. Since the net benefit flows associated with various project effects can be either positive or negative the aggregates reported in the cells of Table 6.1 can be of either sign depending on the nature of the project.

Areas A, B and C correspond to the specific example illustrated by Figure 1.3 in Chapter 1. However, area D is a further potential category of net benefit which may be encountered in general – net benefits which are not measured by market prices and do not accrue to the Referent Group. Figure 1.3 can be up-dated to include this additional category, and the revised diagram is presented here as Figure 6.1.

The difference between Figures 6.1 and 1.3 is that *Area D* has been added to allow for a situation in which there are net benefits (costs) to non-Referent Group members arising because of divergences between market prices (including the zero prices implied by missing markets) and efficiency prices. For example, if a negative externality, such as carbon dioxide emissions or pollution of the High Seas, arising from a project is borne predominantly by stakeholders outside the Referent Group, across a state or international boundary,

Table 6.1 Classification of net benefits

	Net Benefits accruing to:	
	Referent Group	*Non-Referent Group*
Net Benefits Measured by Market Prices	A	B
Net Benefits *not* Measured by Market Prices	C	D

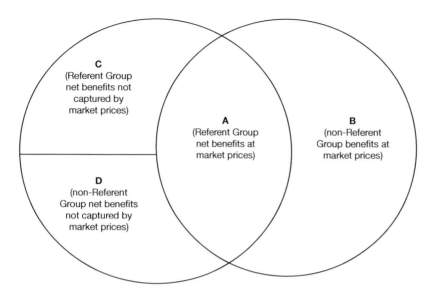

Figure 6.1 The relationship between Referent Group and non-Referent Group net benefits at market prices and efficiency prices. Area A + B includes all the net benefits identified and measured by the Market Analysis. Area A + B + C + D includes all the net benefits identified and measured by the Efficiency Analysis. Area A + C represents the share of efficiency net benefits accruing to the Referent Group.

for example, that cost should be included in the Efficiency Analysis (now defined as area A + B + C + D), but subtracted along with area B from the efficiency net benefit flow to derive the aggregate Referent Group net benefit flow (area A + C).

As illustrated in Figure 6.1 and detailed in Figure 6.3 (see Section 6.7), this framework of analysis can be used to consider the categories of Referent Group and non-Referent Group net benefits in the NFG example:

- *Area A* contains the net benefits to local government in the form of direct taxes identified by the Market Analysis discussed in Chapter 4.
- *Area B* contains the returns to NFG's equity holders and the interstate lenders, which constitute that part of the net benefits, as measured by market prices as discussed in Chapter 4, which accrues to non-Referent Group members.
- *Area C* contains Referent Group net benefits in the form of rents to the landowners in excess of the opportunity cost of land, wages to labour in excess of its opportunity cost, and the costs to downstream fishers due to the pollution generated by the project, all of which are not measured by market prices. (Area C also contains all indirect taxes, subsidies and tariffs paid to or by government. The reason for their inclusion here is that they are not calculated separately in the Market Analysis, as we discuss in more detail in the next sub-section, and in the more detailed discussion of the NFG example in Section 6.7.)
- *Area D* is empty because, in this case, there are no non-Referent Group net benefits not measured by market prices. If, however, the NFG project generated, for example, external costs or benefits for residents of other regions or countries, such as the cost of water-borne pollution for example, these would be allocated to area D.

TABLE 5: REFERENT GROUP CASH FLOW

ITEM/YEAR	2015	2016	2017	2018	2019	2020	2021	2022	2023	2024	2025	2026	2027	2028	2029	2030	2031	2032	2033	2034	2035
	0	1	2	3	4	5	6	7	8	9	10	11	12	13	14	15	16	17	18	19	20
Efficiency-Non ReferentCroup Cash Flow (NFG+Banks)																					
Total Ref. Gp.	51.4	−40.7	−31.9	−30.4	−29.0	−28.4	−22.8	−22.2	−21.5	−20.9	−20.2	−9.5	−9.5	−9.5	−9.5	−9.5	−9.5	−9.5	−9.5	−9.5	3.2
Distribution by Group																					
Total Govt.	51.4	−42.6	−25.8	−26.3	−26.8	−26.2	−20.6	−20.0	−19.3	−18.7	−18.0	−7.3	−7.3	−7.3	−7.3	−7.3	−7.3	−7.3	−7.3	−7.3	5.4
− imp. duties	51.4	1.7	0.8	1.2	1.6	1.6	1.6	1.6	1.6	1.6	1.6	1.6	1.6	1.6	1.6	1.6	1.6	1.6	1.6	1.6	−4.9
− indirect taxes		0.1	0.1	0.1	0.2	0.2	0.2	0.2	0.2	0.2	0.2	0.2	0.2	0.2	0.2	0.2	0.2	0.2	0.2	0.2	0.1
− susidies		−28.4	−18.4	−27.6	−36.8	−36.8	−36.8	−36.8	−36.8	−36.8	−36.8	−36.8	−36.8	−36.8	−36.8	−36.8	−36.8	−36.8	−36.8	−36.8	−17.6
− company taxes		−16.0	−8.3	0.0	8.3	8.9	14.5	15.1	15.8	16.4	17.1	27.8	27.8	27.8	27.8	27.8	27.8	27.8	27.8	27.8	27.8
Landowners' rent		0.8	1.5	2.3	3.0	3.0	3.0	3.0	3.0	3.0	3.0	3.0	3.0	3.0	3.0	3.0	3.0	3.0	3.0	3.0	3.0
Labour rent		1.2	2.4	3.6	4.8	4.8	4.8	4.8	4.8	4.8	4.8	4.8	4.8	4.8	4.8	4.8	4.8	4.8	4.8	4.8	4.8
Downstream users		0.0	−10.0	−10.0	−10.0	−10.0	−10.0	−10.0	−10.0	−10.0	−10.0	−10.0	−10.0	−10.0	−10.0	−10.0	−10.0	−10.0	−10.0	−10.0	−10.0
Referent Group	51.4	−40.7	−31.9	−30.4	−29.0	−28.4	−22.8	−22.2	−21.5	−20.9	−20.2	−9.5	−9.5	−9.5	−9.5	−9.5	−9.5	−9.5	−9.5	−9.5	3.2

	5%	10%	15%		
Salvage Value					
Net Present Value =	−$201.9	−$143.7	−$105.9	Individual Referent Group net benefits sum-up to aggregate Referent Group net benefits; i.e. row 15 = row 5.	Aggregate RG net benefits calculated as a residual (A+B+C+D) − (B+D).
IRR =	66.0%				
Disaggregated Net Benefits					
	5%	10%	15%		
Government	−$173.2	−$124.1	−$91.5	Note IRR of 66% but negative NPVs. Note also that referent Group cash flow is +ve in year 0 then −ve thereafter: effectively a high interest 'loan'.	
Landowners' rent	$33.2	$21.7	$15.2		
Labour rent	$53.2	$34.7	$24.3		
Downstream users	−$115.1	−$76.0	−$53.9		
Total Referent Gp	−$201.9	−$143.7	−$105.9	PV of net benefits to each Referent Group stakeholder.	
Non-Referent Gp	$433.3	$7.8	−$227.9		

Figure 6.2 Spreadsheet showing Referent Group Analysis for the NFG Project.

A (+$92.6) Government (Direct taxes)	**B (+$7.8)** +$196.4 NFG −$188.6 Bank
C (−$236.3) −$216.7 Govt. (Ind. taxes) −$76 Downstream fishers +$21.7 Landowners +$34.7 Labour	**D ($0)** None

Figure 6.3 Distribution of efficiency net benefits for the NFG Project.

Note: In $ thousands, @10% discount rate.

The scope for error in identifying Referent Group net benefits can be narrowed by following some simple guidelines as to where to expect to find them. There are two main ways of identifying Referent Group benefits: one way is to follow the tax and financing flows generated by the project[1], and the other is to examine the shadow-prices used in the analysis. Financial flows distribute the net benefits of a project identified by the Market Analysis described in Chapter 4 among private sector stakeholders, and between the private and public sector. The public sector is normally part of the Referent Group, but some private sector agents, such as foreign firms or banks, may be excluded. Shadow-prices identify differences between market and efficiency valuations of inputs or outputs, and the differences may represent benefits or costs to members of the Referent Group.

Information provided by financial flows

Consider first changes in tax or subsidy flows as a result of the project. The project may result in changes in direct tax revenues, such as income or company tax, and changes in indirect tax revenues, such as tariffs, sales taxes and excise duties. When these tax revenue changes are solely transfers among members of the Referent Group, they net out of the aggregate Referent Group net benefit calculation, although their distribution may be of interest to the decision-maker: referring to Figure 5.5 in Chapter 5, for example, a project resulting in an increase in sugar production would involve a loss to government in the form of increased subsidies paid, and a gain to farmers in the form of increased subsidies received. When changes in tax flows involve transfers between the Referent Group and the rest of the world, on the other hand, the Referent Group experiences a net gain or loss, depending on the direction of the flow. The net benefit resulting from these changes in tax flows is recorded as a gain or loss to the domestic government.

Now consider the private financing flows associated with the project. We have seen in Chapter 5 that these flows are not relevant to the Efficiency Analysis as they simply shift benefits and costs from one group to another. However, they are relevant to the construction of the Referent Group Analysis if they transfer net benefits between members and non-members of the Referent Group, or between members of the Referent Group, but are not relevant if they transfer net benefits among non-Referent Group members. An example of a financial flow which transfers net benefits between members and non-members of the Referent Group is provided by a domestic bank which lends money to a project to be undertaken by a foreign company. The bank advances a loan and then receives a series of interest payments and principal repayments from the foreign entity. The initial loan is a cost to the domestic financial institutions section of the Referent Group Analysis, while the interest and principal repayment flows are benefits. The present value of the net benefit to this sub-group will vary depending on the interest rate charged on the loan and the discount rate used in the social cost-benefit analysis.

Information provided by shadow-prices

Another clue to the existence of Referent Group net benefits lies in the rationale for shadow-pricing. Suppose that an input, such as labour, is assigned a shadow-price lower than its market price (refer to Figure 5.7). This tells us that the wage exceeds the opportunity cost of labour, and, hence, that the labour employed on the project is receiving a net benefit. Since domestic labour is one of the sub-groups within the Referent Group, that net benefit should be recorded among the Referent Group net benefits. Now suppose that an input was assigned a shadow-price in excess of its market price. For example, labour to be diverted from a monopoly or monopsony to work on the project would be shadow-priced at the value of the marginal product of labour (refer to Figures 5.8 and 5.9). The fact that the shadow-price of the input exceeds the market price tells us that the project is imposing a loss somewhere in the economy and this loss will generally be experienced by a member of the Referent Group. In the present example, there is a loss of profit to the domestic monopoly or monopsony which is assumed to be part of the Referent Group.

We have considered cases in which the shadow-pricing of inputs may help us identify categories of Referent Group net benefits. The same applies to the shadow-pricing of outputs. For example, suppose that a project produces an import replacing good: instead of valuing output at its market price, which is the border (c.i.f.) price plus the tariff,

Table 6.2 Using shadow-prices to identify Referent Group benefits and costs

	Input	Output
Market price greater than shadow-price	Benefit to owner of the input, e.g. otherwise unemployed labour	Cost to government or public, e.g. loss of tariff revenue, cost of pollution generated by product use
Market price less than shadow-price	Cost to previous user of the input, e.g. monopoly or monopsony firm	Benefit to public, e.g. value of vaccination to those other than the vaccinated

we shadow-price it at the border price (refer to Figure 5.4). When a project output is shadow-priced at a lower price than the market price this generally indicates a loss to some members of the Referent Group. In the example of an import-replacing project, the loss is incurred by the government in the form of a reduction in tariff revenue. Now suppose that a project output is shadow-priced at a price above its market price (refer to Figure 5.5). This will generally indicate a benefit to members of the Referent Group. An example is influenza vaccinations which may command a price in the market, but will have a social value in excess of the individual willingness to pay for them (perhaps matched by a subsidy as in Figure 5.5) because, in addition to the benefit they confer on the buyer, they reduce the chance of others catching the disease. While it may be difficult to place a dollar value on a non-marketed service of this nature, it is nevertheless clearly a net benefit to the Referent Group.

In summarizing the relationship between shadow-prices and Referent Group net benefits, we can conclude that where the market price of an input exceeds (is less than) its shadow-price, there are likely to be Referent Group benefits (costs); and where the market price of an output exceeds (is less than) its shadow-price, there are likely to be Referent Group costs (benefits). Table 6.2 summarises this simple rule and provides an example of each of the four cases.

6.3 Some examples of the classification of net benefits

It is relatively straightforward, within the spreadsheet framework, to transfer net benefits from the Market or Private Analysis to the Referent Group Analysis where appropriate: tax flows such as revenues from direct taxes or indirect taxes on outputs can be credited to government, loan interest and repayments to domestic (foreign) banks can be included (excluded), and the return to equity holders can similarly be treated in an appropriate manner. We now turn to some shadow-pricing examples similar to those considered in Chapter 5 to illustrate the identification, measurement and classification of project benefits or costs, using the framework described in Table 6.1. We consider shadow-prices on project inputs first followed by outputs. Initially we assume that indirect taxes or subsidies are market distortions and then subsequently we consider the case of corrective taxes and subsidies. The examples are hypothetical, and the numbers cited are purely illustrative. It is to be assumed that all the hypothetical values in these examples are in the form of present values and, hence, comparable. Category totals in the summary tables are reported in bold type. The emphasis of the discussion is on *social accounting* – how are the project's net benefits distributed among the various groups affected by the project?

6.3.1 Shadow-prices on project inputs

> **EXAMPLE 6.1 A minimum wage (refer to Figure 5.7)**
>
> A domestic firm hires $100 dollars' worth of labour at the minimum wage and constructs a walking track for which it is paid $110 by the government. No fee is charged to access the track, and it has a present value of $120 to consumers. The labour would otherwise be unemployed and would be engaged in non-market activities valued at $50.

The Efficiency Analysis tells us that the efficiency NPV of the project is $70: the present value of the output ($120) of the project less the present value of the opportunity cost ($100 worth of labour shadow-priced at 50% of the wage). We now enumerate the sub-groups comprising the Referent Group: the domestic firm; government; labour; and consumers. The net benefits to each group are, respectively, $10 (profit), −$110 (expenditure), $50 (labour rent) and $120 (use value). The aggregate net benefits sum to $70 which corresponds to the result of the Efficiency Analysis.

The example illustrates the simple premise on which the disaggregated analysis is based: if a project has an aggregate net benefit of $70 some groups, whether Referent or not, must be receiving the benefits and incurring the costs of that project; the sum of the net benefits to all groups must equal the aggregate net benefit. In the example above, all groups affected by the project are members of the Referent Group, and hence the sum of Referent Group net benefits equals the aggregate net benefit.

Using the categories and symbols presented in Table 6.1, this example can be summarised as follows:

Market net benefit (A + B) = −$100

Efficiency net benefit (A + B + C + D) = $70

Referent Group net benefit (A + C or (A + B + C + D) − (B + D)) = $70

Non-Referent Group net benefit (B + D) = $0

Distribution of Net Benefits in Example 6.1

A = −$100	**B = $0**
Profit to firm	None
($110 – $100 = $10)	
Cost to government (–$110)	
C = $170	**D = $0**
Rent to labour ($100 – 50 = $50)	None
Benefits to users ($120)	

e.g.

EXAMPLE 6.2 A maximum price (refer to Figure 5.11)

The state government hires a computer programmer from another state for a month, at a cost of $200, which is $50 more than she would normally expect to earn, plus $100 for accommodation in a rent-controlled apartment with a market value of $125. The programmer revises some of the government's programs leading to a cost saving of $400. The Referent Group is defined as the residents of the state.

The efficiency benefit is $400 and the opportunity cost is $275 – the wage forgone by the programmer plus the market rental value of the apartment – so the efficiency net benefit is $125. The net benefits to the Referent Group are the efficiency net benefits less the net benefits to the non-Referent Group; in this case the programmer is not part of the Referent Group (not a resident of the state) and their net benefit of $50 is subtracted from efficiency net benefits to give $75. Two sub-groups within the Referent Group are affected by the project: the government has a net benefit of $100 and potential renters of the rent-controlled apartment lose a service with a net value of $25. The sum of these disaggregated Referent Group benefits ($75) equals the aggregate measure as required.

Using the categories and symbols presented in Table 6.1, this example can be summarised as follows:

Market net benefit (A + B) = $100
Efficiency net benefit (A + B + C + D) = $125
Referent Group net benefit (A + C or (A + B + C + D) – (B + D)) = $75
Non-Referent Group net benefit (B + D) = $50

Distribution of Net Benefits in Example 6.2

A = $100	**B = $0**
Net Benefit to government department [$400 – (200 + 100) = $100]	None
C = –$25	**D = $50**
Loss to potential renter ($25)	Rent to labour ($200 – 150 = $50)

e.g.

EXAMPLE 6.3 An indirect tax: a tariff on imports (refer to Figure 5.4)

A domestic firm proposes to import $100 worth of gadgets for use in a manufacturing project. The $100 includes the c.i.f. (landed) value of the imported goods of $75 plus $25 in tariffs. Other costs are $200 for skilled labour, and the resulting output will sell on the domestic market for $400.

The efficiency net benefit is given by the value of the output less the opportunity costs of the inputs – $75 worth of imports and $200 worth of skilled labour – giving a net value of $125. In this example, there are no non-Referent Group net benefits. The members of the Referent Group are the government, gaining $25 in tariff revenue, and the firm, gaining a profit of $100.

Using the categories and symbols presented in Table 6.1, this example can be summarised as follows:

Market net benefit (A + B) = $100

Efficiency net benefit (A + B + C + D) = $125

Referent Group net benefit (A + C or (A + B + C + D) – (B + D)) = $125

Non-Referent Group net benefit (B + D) = $0

Distribution of Net Benefits in Example 6.3

A = $100	B = $0
Profit to firm	None
($400 – 200 – 100 = $100)	
C = $25	D = $0
Tariff revenue to government	None
($25)	

e.g.

EXAMPLE 6.4 A subsidy (refer to Figure 5.5)

The government spends $100 on improving access to wheat silos, the services of which are supplied free to users. This results in farmers saving $75 worth of diesel fuel at the subsidised price. The subsidy paid on that quantity of fuel amounts to $25.

In the efficiency net benefit analysis the saving in fuel is valued at its unsubsidised cost, reflecting the opportunity cost of the inputs required to produce it, so the efficiency net benefit is zero. In terms of Referent Group benefits, the farmers benefit by $75 and the government incurs costs of $100, but benefits by $25 in reduced fuel subsidies. The sum of the Referent Group net benefits ($0) equals the total efficiency net benefit.

Using the categories and symbols presented in Table 6.1, this example can be summarised as follows:

Market net benefit (A + B) = –$25

Efficiency net benefit (A + B + C + D) = $0

Referent Group net benefit (A + C or (A + B + C + D) – (B + D)) = $0

Non-Referent Group net benefit (B + D) = $0

Distribution of Net Benefits in Example 6.4

A = –$25 Net benefits to farmers ($75) Project cost to government (–$100)	**B = $0** None
C = $25 Government's saving on subsidies ($25)	**D = $0** None

EXAMPLE 6.5 A regulated utility (refer to Figure 5.6)

A foreign company proposes to construct a smelter at a cost of $100 which will use $50 worth of power at the current price. The output of the smelter will be worth $400 on the world market and the company will pay $100 in income tax. The cost of producing the extra power is $30.

The efficiency net benefits are given by the value of output, $400, less the opportunity cost of inputs – $100 construction cost plus $30 power cost – giving a net value of $270. The net benefits of the Referent Group, which excludes the foreign firm, are the efficiency net benefits less the firm's after-tax profit of $150, giving a total of $120. The government and the power producer are the only Referent Group members affected, and their net benefits consist of $100 tax revenue and $20 profit respectively.

Using the categories and symbols presented in Table 6.1, this example can be summarised as follows:

> Market net benefit (A + B) = $250
>
> Efficiency net benefit (A + B + C + D) = $270
>
> Referent Group net benefit (A + C or (A + B + C + D) – (B + D)) = $120
>
> Non-Referent Group net benefit (B + D) = $150

Distribution of Net Benefits in Example 6.5

A = $100 Income tax paid to government ($100)	**B = $150** Profit to foreign company ($400 – 150 – 100 = $150)
C = $20 Extra profit on power supply ($20)	**D = $0** None

e.g.

EXAMPLE 6.6 Monopoly power (refer to Figure 5.8)

A domestic company is the country's sole producer of cellular phones, which require skilled labour to produce. The government is considering establishing a factory to produce computers, which require the same type of skilled labour. Domestic skilled labour resources are fully employed, and skilled labour cannot be imported. The proposed computer plant would employ $100 worth of skilled labour and produce $300 worth of computers.

As a monopoly, the phone producer maximises profit by hiring labour to the point at which its marginal revenue product equals the wage. However, the value of the marginal product of labour (its opportunity cost) is higher than this; in fact, as Technical note 5.1 shows, $VMP = w * [1/(1 – 1/e_d)]$, where VMP is the value of the marginal product of labour, w is the wage rate, and e_d is the elasticity of demand for the product, expressed as a positive number, at the current level of output of phones.

Suppose that, under the proposed computer project, the demand elasticity is 1.5 (we know it must exceed 1 since marginal revenue, and hence marginal revenue product, fall to zero as demand elasticity declines to 1); this means that an additional $100 worth of skilled labour used in phone production results in an extra $300 worth of output. In the Efficiency Analysis we cost the labour at $300 because, under the assumption of full employment, the labour is diverted from phone production and $300 is the value of its marginal product in that activity, and we find that the net value of the project is zero. In terms of disaggregated Referent Group net benefits, the proposed computer operation earns a profit of $200, but this is offset by a $200 fall in the profits of the phone company.

Using the categories and symbols presented in Table 6.1, this example can be summarised as follows:

Market net benefit (A + B) = $200

Efficiency net benefit (A + B + C + D) = $0

Referent Group net benefit (A + C or (A + B + C + D) – (B + D)) = $0

Non-Referent Group net benefit (B + D) = $0

Distribution of Net Benefits in Example 6.6

A = $200 Profit to computer firm ($300 – 100 = $200)	**B = $0** None
C = –$200 Loss to telephone company (–$200)	**D = $0** None

e.g.

EXAMPLE 6.7 Monopsony power (refer to Figure 5.9)

A domestic company which produces fabricated metal for the export market is the main employer in a remote town. The government proposes establishing a new factory in the town which will use $100 worth of local labour to produce output of goods valued at $300.

Because the company has monopsony power in the labour market, it maximises profit by restricting its hiring to drive down the market wage: it hires labour to the point at which the marginal factor cost (the addition to cost as a result of hiring an extra unit of labour) equals the value of the marginal product of labour (the addition to revenue as a result of selling the extra output on the world market). As shown in Technical note 5.1, the marginal factor cost, $MFC = w(1 + 1/e_s)$, where w is the market wage and e_s is the elasticity of supply of labour at the current level of employment. Suppose that the labour supply elasticity is 1 (it can generally take any positive value), then $100 worth of extra labour adds $200 to the value of output. In the Efficiency Analysis the labour is costed at $200, on the assumption that it is diverted from the existing monopsony workforce, and the net value of the project is found to be $100. In terms of Referent Group net benefits, the proposed factory earns a profit of $200 which is partially offset by a $100 reduction in profit to the company already operating in the town.

Using the categories and symbols presented in Table 6.1, this example can be summarised as follows:

Market net benefit (A + B) = $200

Efficiency net benefit (A + B + C + D) = $100

Referent Group net benefit (A + C or (A + B + C + D) – (B + D)) = $100

Non-Referent Group net benefit (B + D) = $0

Distribution of Net Benefits in Example 6.7

A = $200	B = $0
Profit to new company	None
($300–100)	
C = –$100	D = $0
Loss to existing company	None
(–$100)	

6.3.2 *Shadow-prices on project outputs*

e.g.

EXAMPLE 6.8 An import-replacing project (refer to Figure 5.4)

A project uses $200 worth of skilled labour to produce output valued on the domestic market at $400. Domestic consumers could obtain that same quantity of output from abroad at a cost of $300 c.i.f. plus $100 tariff.

The efficiency net benefits consist of the value of the output at the c.i.f. price, $300, less the opportunity cost of the skilled labour, giving a net value of $100. Two sub-groups of the Referent Group are affected: the firm makes a profit of $200 and the government loses tariff revenue of $100 because of the import replacement. Since there are no non-Referent Group beneficiaries, the sum of Referent Group net benefits ($100) equals the total efficiency net benefit.

Using the categories and symbols presented in Table 6.1, this example can be summarised as follows:

Market net benefit (A + B) = $200

Efficiency net benefit (A + B + C + D) = $100

Referent Group net benefit (A + C or (A + B + C + D) – (B + D)) = $100

Non-Referent Group net benefit (B + D) = $0

Distribution of Net Benefits in Example 6.8

A = $200	**B = $0**
Net profit to firm	None
($400 – 200 = $200)	
C = –$100	**D = $0**
Tariff loss to government	None
(–$100)	

e.g.

EXAMPLE 6.9 Output of a product subject to a subsidy (refer to Figure 5.5)

A government project, costing $100, will increase the supply of irrigation water to sugar producers. While there is no charge for the extra water, farmers will increase the quantity of other inputs used in the production process by $50 worth at market prices. As a consequence of the increase in water and other inputs the value of the output of sugar, at market prices, will rise by $125, and farmers will receive an additional $25 in subsidies.

The efficiency net benefit of the project is the value of the additional output at the market (subsidised) price of sugar, $125, less the cost of the additional water, $100, and less the opportunity cost of the complementary inputs, $50, giving a net value of –$25. Two subgroups of the Referent Group are affected: farmers make an extra profit of $100, while the cost to government consists of $100 to increase water supply and $25 in additional subsidy payments. Since there are no non-Referent Group beneficiaries, the sum of Referent Group net benefits (–$25) equals the total efficiency net benefit.

Using the categories and symbols presented in Table 6.1, this example can be summarised as follows:

Market net benefit (A + B) = $0
Efficiency net benefit (A + B + C + D) = –$25
Referent Group net benefit (A + C or (A + B + C + D) – (B + D)) = –$25
Non-Referent Group net benefit (B + D) = $0

Distribution of Net Benefits in Example 6.9

A = $0	B = $0
Net profit to farmers	None
($125 + $25 – $50 = $100)	
Cost of extra water	
(–$100)	
C = –$25	D = $0
Extra subsidy payment	None
(–$25)	

6.4 Corrective taxation

To this point in the discussion of Referent Group net benefits we have assumed that indirect taxes or subsidies are distortionary in nature – their sole purpose is to raise revenue for government or support the incomes of producers. However, as discussed in Chapter 5, some indirect taxes (subsidies) are intended to discourage (encourage) use of harmful (beneficial) goods and services. In principle, the indirect tax (subsidy) per unit should be set at the level of the marginal external cost (benefit) resulting from the use of an extra unit of the good or service. If we assume this is the case, then for every dollar of indirect tax (subsidy) included in the Distribution of Net Benefits Table there is a corresponding cost (benefit) to the general community. We now consider some examples.

e.g.

EXAMPLE 6.10 Cost of fuel subject to a corrective tax (refer to Figure 5.3)

The government spends $100 on improving port facilities to provide sawmills with access to a new market. This results in sawmills receiving an extra $200 for their timber but spending an extra $125 on diesel fuel to transport their product to the port. The market value of the extra fuel includes $25 tax which is imposed to discourage consumption of fuel because of its related health effects.

In the Efficiency Analysis of net benefits extra fuel is costed at its tax-inclusive price, reflecting the opportunity cost of the inputs required to produce it, plus the increase in health costs associated with its use, so the efficiency net benefit of the project is –$25. Contrast this result with the case of a distortionary tax where fuel would be costed at its net of tax price and the project would break even on efficiency grounds. In terms of Referent Group benefits, sawmills benefit by $75, and the government incurs a cost of $100, but receives an extra $25 in fuel tax. However, there is a corresponding cost to the general community of $25 in extra health costs. The sum of the Referent Group net benefits (–$25) equals the total efficiency net benefit.

Using the categories and symbols presented in Table 6.1, this example can be summarised as follows:

Market net benefit (A + B) = –$25

Efficiency net benefit (A + B + C + D) = –$25

Referent Group net benefit (A + C or (A + B + C + D) – (B + D)) = –$25

Non-Referent Group net benefit (B + D) = $0

Distribution of Net Benefits in Example 6.10

A = –$25	B = $0
Net benefit to sawmills ($75)	None
Project cost to government (–$100)	
C = $0	D = $0
Government gain in tax revenue (+ $25)	None
Rise in community health cost (–$25)	

e.g.

EXAMPLE 6.11 Cost of fuel subject to a corrective subsidy (refer to Figure 5.5)

Now suppose that the extra fuel used in Example 6.10 was biofuel – such as a blend of diesel and ethanol – which results in lower emissions of particulates and lower associated health costs as compared with regular fuel. We know from the above example that the production cost of the diesel saved is $100 – the market price less the $25 tax. Suppose that the production cost of the biofuel is $110 and that a corrective subsidy of $10 is paid to producers to render the biofuel competitive with diesel at their tax-inclusive prices. If both the tax and the subsidy are corrective, we can infer that, while burning the diesel fuel imposes a health cost of $25, burning the biofuel has a $15 external cost. We can redo the net benefit calculations on the assumption that the fuel saved was biofuel.

Using the categories and symbols presented in Table 6.1, this example can be summarised as follows:

Market net benefit (A + B) = −$25

Efficiency net benefit (A + B + C + D) = −$25

Referent Group net benefit (A + C or (A + B + C + D) − (B + D)) = −$25

Non-Referent Group net benefit (B + D) = $0

It can be seen from these figures that the overall results of Examples 6.10 and 6.11 are the same. However, the Distribution of Net Benefits Table shows that Example 6.11 involves lower health costs offset by a higher fuel production cost as compared with Example 6.10.

Distribution of Net Benefits in Example 6.11

A = −$25	**B = $0**
Net benefit to farmers	None
($75)	
Project cost to government	
(−$100)	
C = $0	**D = $0**
Extra fuel tax revenue	None
(+ $25)	
Extra biofuel subsidy	
(−$10)	
Extra health costs	
(−$15)	

Why does using biofuel, as opposed to diesel, in these examples appear to confer no net benefit on the Referent Group? The reason is that the subsidy rate on biofuel is assumed to be set at the level of marginal external health benefit, so that the extra health benefits of using biofuel, as opposed to diesel, are exactly offset by the extra production costs. This offset follows from the assumption, adhered to so far in the text, that the projects we are considering are relatively small and have no effect on marginal values. However, to anticipate the analysis to be presented in Chapter 7, and referring to Figure 5.5, we could conclude that while there is no net benefit from substituting a small quantity of biofuel for diesel, because the marginal health benefit, AC, plus the marginal benefit to the fuel user, CQ, equals marginal production cost, AQ, net benefits are generated by the *infra-marginal* units of biofuel for which the marginal health benefit is higher and the marginal production cost lower than the marginal values at market equilibrium.

EXAMPLE 6.12 Production of cigarettes subject to a corrective tax (refer to Figure 5.3)

A private firm plans to produce cigarettes to satisfy additional demand. The market value of its output will be $250, which includes indirect tax of $50. The cost of the extra production is $150 and the firm will pay income tax of $25.

Using the categories and symbols presented in Table 6.1, this example can be summarised as follows:

Market net benefit (A + B) = $100

Efficiency net benefit (A + B + C + D) = $50

Referent Group net benefit (A + C or (A + B + C + D) – (B + D)) = $50

Non-Referent Group net benefit (B + D) = $0

If the tax on cigarettes was distortionary, the efficiency net benefit would be $100, but the $50 indirect tax on output, which is part of consumers' willingness to pay for the product, is offset by an equivalent health cost borne by the community. In general, note that the market analysis of the project values outputs and inputs at their tax-inclusive prices. In consequence, indirect taxes on output are included in the market value of project output, and indirect taxes on inputs, where they occur, are included in Area C of the Distribution of Net Benefits Table.

Distribution of Net Benefits in Example 6.12

A = $100	**B = $0**
Net profit to firm	None
($25)	
Output tax revenue to government	
($50)	
Income tax revenue to government	
($25)	
C = –$50	**D = $0**
Extra health costs	None
(–$50)	

EXAMPLE 6.13 A subsidy on production of vaccine (refer to Figure 5.5)

A pharmaceutical company plans to supply additional influenza vaccinations worth $150 at the market price. In addition, it will receive a $50 subsidy from government. The cost of the extra quantity produced is $175.

Using the categories and symbols presented in Table 6.1, this example can be summarised as follows:

> Market net benefit (A + B) = −$25
> Efficiency net benefit (A + B + C + D) = $25
> Group net benefit (A + C or (A + B + C + D) – (B + D)) = $25
> Non-Referent Group net benefit (B + D) = $0

If the subsidy on supply of vaccine was distortionary, the efficiency net benefit would be – $25, but in the case of a corrective subsidy the health benefit to the community offsets the subsidised portion of the cost of producing the vaccine.

Distribution of Net Benefits in Example 6.13

A = −$25	B = $0
Net profit to firm	None
($25)	
Subsidy from government	
(−$50)	
C = $50	D = $0
Reduced health costs	None
($50)	

6.5 Further examples

To complete the discussion of combinations of categories of project net benefits, two further examples follow.

EXAMPLE 6.14 Where the market and efficiency net benefits are identical

A local firm engages a foreign contractor to assist with the upgrade of an existing plant. The foreign contractor is paid a management fee at the going international rate of $300. The firm spends a further $500 on new technology, purchased abroad at the current world price. The result is an increase in the firm's output of exports, sold at the going world price, equal in value to $1200. The firm pays $100 additional taxes to government and shares the remaining profit 50:50 with the foreign contractor, who is not considered as part of the Referent Group.

In this case, the market net benefit is $400, and the private net benefit to the local firm after tax and profit sharing is $150. As, in this example, all market prices relevant to the project reflect opportunity costs, and there are no non-marketed effects, the efficiency net benefit is also $400. The aggregate Referent Group benefits are $250: the government receives $100 in taxes and the local firm's equity holders receive $150.

Using the categories and symbols presented in Table 6.1, this example can be summarised as follows:

Market net benefit (A + B) = $400

Efficiency net benefit (A + B + C + D) = $400

Referent Group net benefit (A + C or (A + B + C + D) – (B + D)) = $250

Non-Referent Group net benefit (B + D) = $150

Distribution of Net Benefits in Example 6.14

A = $250	B = $150
Profit to Local Firm ($1200 – 800 – 150 – 100 = $150) Direct taxes to government ($100)	Profit to foreign contractor [50% × ($1200 – 800 – 100)]
C = $0 None	D = $0 None

e.g.

EXAMPLE 6.15 Where there are market and non-market non-Referent Group net benefits

A foreign-owned agricultural enterprise invests $420 in an irrigation project, including $220 for a new pump from an inter-state supplier which cost $200 to supply. It employs some local casual labour costing $100, and an inter-state irrigation consultant costing $150 and pays $30 in income tax to the state government. The value of its output of fruit at domestic prices increases by $800, but the project causes $10 in lost output in a neighbouring state due to reduced water flow to downstream users. The opportunity cost of the local labour is $50 and the opportunity cost of the inter-state consultant is $100. The value of the enterprise's increased output at world prices is $750, which is the value at domestic prices less the tariff on imports of fruit. Assume that all foreign and inter-state parties are non-Referent Group members, and that the additional output of fruit replaces imports.

In this case, the market net benefit is $130 consisting of $800 worth of fruit less the cost of inputs at market prices: investment ($420), additional labour ($100) and the consultant ($150). Note that since cost at market prices includes the rents obtained by the inter-state pump supplier and the consultant, these project benefits are not measured by the Market Analysis. In calculating the efficiency net benefit ($190), we value the fruit at $750, and cost the investment at $400, the labour at $50, and the consultant at $100, and we include the $10 cost to downstream water users. The Referent Group net benefits consist of the $30 income tax (already included in the net benefit measured at market prices), the employment benefit valued at $50, and the loss of tariff revenue of $50.

Using the categories and symbols presented in Table 6.1, this example can be summarised as follows:

Market net benefit (A + B) = $130

Efficiency net benefit (A + B + C + D) = $190

Referent Group net benefit (A + C or (A + B + C + D) – (B + D)) = $30

Non-Referent Group net benefit (B + D) = $160

Distribution of Net Benefits in Example 6.15

A = $30 Taxes to government ($30)	**B = $100** Profit to agricultural enterprise ($800 – 420 – 100 – 150 – 30 = $100)
C = $0 Rent to local labour ($100 – 50 = $50) Loss of tariffs to government ($750 – 800 = –$50)	**D = $60** Profit to inter-state pump supplier ($220 – 200 = $20) Rent to inter-state consultant ($150 – 100 = $50) Loss to downstream water users –$10

6.6 Lessons from the examples

What these examples show is that a proposed project can have two sorts of effects: (1) efficiency effects resulting from the reallocation of scarce resources; and (2) distributional effects determined by externalities as well as by market imperfections, government interventions and distortions. The Efficiency Analysis measures the overall net benefits of a project, whereas the Referent Group Analysis measures the distribution of benefits among the Referent Group (and non-Referent Group) members. Members of the Referent Group experience benefits and costs mainly through external effects and changes in the levels of tax/transfer payments or private rents (the difference between the private and social perspective on rent is examined later in the discussion of the ICP case study in Appendix 2 in this chapter).

A *rent* in this context is a term used in economics for a payment made to a factor of production in excess of the payment required to keep it in its current use: economic profit and wages in excess of opportunity cost are examples of rents. The net benefits of a project measured by market prices – the value of its output less its opportunity cost – are shared out in the form of changes in rents or taxes: the sum of the changes (whether positive or negative) in the levels of the various kinds of rent and taxes equals the net value of the marketed effects of the project from an efficiency perspective. However, not all those affected by changes in levels of rent or taxes are members of the Referent Group and an important role of the Referent Group Analysis is to detail changes in the levels of these payments accruing to that group.

The term "transfer" is often used to refer to changes in the levels of rents or tax receipts: the reason is that these changes simply transfer (or redistribute) the real net benefits of the project among various groups. While the changes in these receipts or payments are very real to those affected, from the point of view of the economy as a whole they are purely pecuniary: in other words, their sole function is to distribute the project's marketed net benefits. No matter how complicated and involved the set of transfer payments is, the sum total of the net benefits to all the groups affected must equal the net benefits of the project as a whole.

6.7 Worked example: Referent Group Analysis of National Fruit Growers' (NFG) project

In Chapter 5 it was found that NFG's orchard project was unlikely to be considered worthwhile from an overall economic efficiency perspective. However, this does not necessarily mean that the project is not worthwhile from the perspective of the relevant Referent Group – the residents of Happy Valley where NFG's orchard is located. Indeed, you now learn that NFG is not a locally owned company. It is based in another state across the King River, in Goldsville. It is expected that NFG will remit all its after-tax earnings there. Furthermore, the agricultural loan for the project and NFG's overdraft facility are both with the Goldwealth Bank based in Goldsville. As you are required to appraise this project from the perspective of the residents of Happy Valley, you would not include NFG or the Goldwealth Bank as part of the Referent Group.

The Referent Group, in this case, consists of all project beneficiaries and losers other than the company and bank identified above. It will be assumed that the labour employed by NFG, the landowners who rent the land to NFG, and the fishers whose catches will be affected by the project all live in Happy Valley. Thus, in summary, the Referent Group, in this example, consists of:

1   the local government who receives the taxes and duties and pays the subsidies;
2   the landowners who rent out their land to NFG;
3   the labour NFG employs who earn wages;
4   the downstream fishers who lose some net income because of the run-off from the orchard.

The next step in the analysis of the NFG project is to calculate the total Referent Group net benefits. We can do this in two ways. We can simply subtract the net benefits to non-Referent Group members from the efficiency net benefit estimate; in the example this is done by subtracting the cash flows representing NFG's equity after tax (Private Cash Flow) and bank debt (see Figure 4.3 in Chapter 4) from the total efficiency net benefit (Figure 5.16 in Chapter 5). However, this will provide us with only the aggregate Referent Group net benefit, which is insufficient if we want to know how this component of the project's benefits and costs is distributed among the different sub-groups within the Referent Group. The alternative method is to calculate the benefits and costs to each of the four sub-groups and to add these up. As a consistency check, the two methods should provide the same aggregate net benefit stream. This is demonstrated in Figure 6.2.

In Figure 6.2 we derive the aggregate Referent Group net benefits as a residual by subtracting from the efficiency net benefits the net benefits of the two non-Referent Groups: NFG and the Goldwealth Bank. The rest of the spreadsheet derives the respective Referent Group benefits and costs from the key input data in Figure 5.15, where information on taxes, subsidies, duties, opportunity cost and external cost is provided. Note that rent to landowners is calculated by subtracting the opportunity cost (zero) from the rent paid, and similarly, rent to labour is found by subtracting labour's opportunity cost (20% of the wage) from the wage actually paid. Also note that import duties on imported machinery, vehicles and spare parts as part of fixed and working capital are included as revenues to government when the project is undertaken.

What then is the bottom line? It might appear from the IRR that the Referent Group does very well – the IRR is 66%. However, care is required in interpreting this estimate; note also that the total Referent Group NPVs are all *negative*! Why is this? The reason is that the signs of the Referent Group's cash flow measure of net benefits are reversed in this instance as compared with those of a normal investment project; the cash flow in this instance begins with a positive sign in Year 0 and thereafter is negative, with the exception of the last value. This flow corresponds to the situation of a loan from the borrower's perspective. In this case it is as if the Referent Group receives an initial cash inflow (loan) of $51,400, and then makes a series of cash payments (loan repayments) starting off at $40,700 in year 1 and then levelling off at $9,500 over the latter years. In this instance the cost of the "loan" is 66% per annum. The reasons why the project is seen as being profitable from a private investor's standpoint (as was shown in Chapter 4) are because the inputs are heavily subsidised, the agricultural loan is at a very low interest rate (3.5%) as compared with the cost of capital (10%), and the costs imposed on the downstream fishing industry are not borne by the investor, NFG. The people of Happy Valley would be unlikely to support this project given the availability of loans at lower interest rates than 66% per annum!

The left-hand bottom corner of the spreadsheet shows the distribution of the project's net benefits among the Referent Group stakeholders (and the non-Referent Group). At a discount rate of 10% we see that, within the Referent Group, the two major losers are: the government ($124,100); and the downstream fishers ($76,000). The gainers are: the landowners ($21,700); and labour ($34,700). If each dollar's worth of gains or losses to each stakeholder group is valued equally, the losses clearly outweigh the gains and the government will not support the project: NFG will not qualify for the subsidies. If the government does support it, the implication is that it values the interests of the gainers much more highly than the interests of the losers. The important question of appraising the distribution of a project's gains and losses is dealt with in Chapter 11.

To summarise the results of the whole analysis, it is useful to return to the multiple-account framework introduced in Chapter 1, and developed further in the present chapter, showing how any project's efficiency net benefits can be allocated among the four categories as shown in Figure 6.3 (see p. 152). Note from Figure 6.3 that the direct tax flow is captured in the market measure of project net benefit (A + B), but that the indirect tax and subsidy flows relating to inputs are not. The reason for this, as noted earlier, is that the Market Analysis net benefit is calculated on the basis of market prices which are gross of indirect taxes and net of subsidies on inputs i.e. the price to the firm is the market price. This is why input tax and subsidy flows appear under Category C in Figure 6.3.

In this example there are no taxes or subsidies on output. If there were, their effects would be recorded in Category A: in the case of an output tax, the Market Analysis (A + B) values output at the tax inclusive price, but the value of output to the firm (included in B in this example) would be calculated at the before-tax price; in the case of a subsidy on output, the Market Analysis values output at the net of subsidy price, but the value of output to the firm would be calculated at the market price plus subsidy. In summary, an output tax would be treated as a benefit to government (area A) and a cost to the firm (area B), whereas a subsidy is a cost to the government and a benefit to the firm.

To recap:

- Category A shows the Referent Group net benefits that are captured by market prices.
- Category B shows the non-Referent Group net benefits captured by market prices.
- Category C shows the Referent Group net benefits that are *not* captured by market prices.
- Category D shows the non-Referent Group net benefits that are *not* captured by market prices (none in this case).

- Categories A + B + C + D = Aggregate Efficiency Net Benefits (NPV of –$135.9 at 10%) as calculated in Chapter 5, Figure 5.16.
- Categories A + B = Market Net Benefits (NPV of + $100.4 at 10%) as calculated in Chapter 4, Figure 4.2.
- Categories A + C = Referent Group Net Benefits (NPV of –$143.7 at 10%) as calculated in Chapter 6, Figure 6.2.

6.8 Further reading

The further reading suggested for Chapter 5 is also relevant for Chapter 6. The term "Referent Group" does not appear to be used much in contemporary studies of cost-benefit analysis though we consider it to be a useful concept. An early use of this term in discussion of the principles of cost-benefit analysis occurs in V.W. Loose (ed.), *Guidelines for Benefit-Cost Analysis* (Vancouver, BC: British Columbia Environment and Land Use Secretariat, 1977).

Appendix 1 to Chapter 6

Referent Group net benefits in the ICP case study

Referent Group net benefits

Let us now turn to the analysis of the Referent Group net benefits in the ICP case study. The first step is to enumerate those groups which have "standing" and are likely to be affected. Following the procedure outlined above, we first calculate the total Referent Group net benefit by subtracting the flows of funds on ICP's equity and the foreign credit to derive the cash flow for *Total Referent Group Net Benefits* as shown near the top of Table 5 of the spreadsheet in Figure A6.1.

We then need to identify and quantify the various components of the Referent Group net benefit, beginning with tax flows: government will be affected through changes in various

Import duties gained or lost on all imports including fixed and working capital.

Aggregate Referent Group net benefits derived by subtracting flows on ICP's equity and the foreign loan from the efficiency cash flow in Table 4.

TABLE 5:   REFERENT GROUP ANALYSIS

YEAR	1999 MN BT	2000 MN BT	2001 MN BT	2002 MN BT	2003 MN BT	2004 MN BT	2005 MN BT	2006 MN BT	2007 MN BT	2008 MN BT	2009 MN BT
Total referent group – import duties	28.670	–62.911	32.151	33.330	34.603	35.977	37.462	37.462	37.462	37.462	91.647
Yarn		–18.000	–36.000	–36.000	–36.000	–36.000	–36.000	–36.000	–36.000	–36.000	–36.000
Cotton		10.692	21.384	21.384	21.384	21.384	21.384	21.384	21.384	21.384	21.384
Spindles	8.800										
Equipment	1.870										
Materials inventory		5.280									–5.280
Spares inventory		0.535									–0.535
Spares (annual)		1.067	1.067	1.067	1.067	1.067	1.067	1.067	1.067	1.067	1.067
– Profits taxes		–13.285	27.100	28.279	29.552	30.926	32.411	32.411	32.411	32.411	32.411
(i) Total government	10.670	–13.711	13.551	14.730	16.003	17.377	18.862	18.862	18.862	18.862	13.047
(ii) Power supplier		1.800	3.600	3.600	3.600	3.600	3.600	3.600	3.600	3.600	3.600
(iii) Labour	18.000	3.000	3.000	3.000	3.000	3.000	3.000	3.000	3.000	3.000	3.000
(iv) Domestic bank	0.000	–60.000	6.000	6.000	6.000	6.000	6.000	6.000	6.000	6.000	66.000
(v) Insurance		6.000	6.000	6.000	6.000	6.000	6.000	6.000	6.000	6.000	6.000
Total referent group	28.670	–62.911	32.151	33.330	34.603	35.977	37.462	37.462	37.462	37.462	91.647

NPV(BT. Mn)

DISCOUNT RATE	AGGREGATE	GOVERNMENT	POWER CO.	LABOUR	DOM. BANK	INSURANCE CO.
4%	260.4	116.2	27.5	42.3	25.7	48.7
8%	201.3	93.4	22.5	38.1	6.9	40.3
12%	158.3	76.5	18.7	35.0	–5.7	33.9
16%	126.6	63.6	15.8	32.5	–14.3	29.0
0%	343.3	147.1	34.2	48.0	54.0	60.0

Unskilled labour 'gains' half of what it earns.

Referent Group net benefits shown on aggregated and disaggregated basis.

The power company 'gains' half of what it charges.

Figure A6.1 ICP project solution: the Referent Group net benefit stream.

kinds of tax revenues (business income taxes and tariffs in this case). We then consider inputs or outputs which have been shadow-priced: labour will be affected through provision of jobs for people who would otherwise be unemployed; public utilities will be affected through sale of extra power at a price exceeding production cost; and (by assumption) the domestic insurance sector benefits by charging in excess of cost for an insurance policy. Finally, we examine financial flows: domestic financial institutions will be affected through provision of a loan.

Each of these sub-groups should be represented by a sub-section in Part 5 of the spreadsheet analysis as shown in Figure A6.1. For example, in the government sub-section, we enter rows representing each category of input or output which has consequences for government tariff revenue: for example, forgone tariff revenue as a result of import replacement, and additional tariff revenues as a result of imports of various kinds of capital goods and raw materials; and we also enter a row reporting changes in business income tax revenues. We also need a row representing benefits to domestic labour and one representing the electricity utility. In the financial institutions sub-section we need a row representing the domestic bank and another row representing the domestic insurance company. As the various categories of Referent Group net benefits are calculated, it can be seen that what is recorded as a cost in one part of the cost-benefit analysis can appear as a benefit in another. For example, the wage bill is a cost to the firm, but a portion of the wage bill is a net benefit to domestic labour; taxes and tariffs are costs to the firm but benefits to the government; and the power bill is a cost to the firm, but a portion of it is a net benefit to the domestic utility. The loan received from the domestic bank is a benefit to the firm but a cost to the bank, and the interest and loan repayments are costs to the firm but benefits to the bank.

We now proceed to fill in the rows we have provided for the various Referent Group benefits and costs over the life of the project. As in the case of Parts 2 to 4 of the spreadsheet, all entries are cell addresses taken from the data reported in Part 1. Tariff revenues obtained when the project imports an input, or forgone when the project output replaces an import, are entered in their various categories. Business tax revenues forgone when project costs are written off against other company income, or obtained when the project makes a profit are entered for each year. Similarly, half of the firm's power bill is entered as a benefit as it represents extra profit for the electricity utility. The amounts advanced by the domestic bank are entered as costs, and the interest and principal repayments are entered as benefits. Finally, the amount paid to the domestic insurance company is entered as a benefit.

The merits of treating the full amount of the insurance premium as a Referent Group benefit can be debated: we are treating the payment as if it were a straight transfer but of course the firm is also shifting some of its risk to the domestic insurance company and this must represent some cost to the latter. An alternative approach is to assume that the cost of risk exactly matches the insurance premium, in which case there is no net benefit to the insurance company. If this approach is taken, then the insurance premium has to be included as a cost in the Efficiency Analysis. In reality the true nature of the payment is somewhere between these two extremes: a part of it represents the cost of extra risk borne by the insurance company, and a part is simply a contribution to overhead costs or profits.

It was stated that unskilled labour would be paid twice the value of what it could otherwise earn, so half of the unskilled labour wage bill in each year is entered as a net benefit to labour. In the ICP case study we assume that the value of what labour could otherwise earn is also the value of what it could otherwise produce. However, as discussed in Chapter 5, the value of what labour can earn in subsistence activities may be different from the value of the extra production it generates because of the open access, or communal nature of some land and water resources. Under open access, it is usually a safe assumption that extra labour applied to these resources adds nothing to the value of total output. In that case the labour would be shadow-priced at zero in the Efficiency Analysis, thereby increasing the aggregate net benefit of the project.

This example illustrates the difference between the private and social perspectives on rents. When a worker leaves subsistence activity for a more rewarding job in the city, the rent generated from a private perspective is measured by the difference between the wage earned in the city and the value of the worker's share in the output of the subsistence activity (we are ignoring any non-monetary net benefits such as improved living conditions). From a social or economy-wide point of view, if the value of the worker's marginal product is zero in the subsistence activity, the rent generated by the move from subsistence activity to participation in the labour market is measured by the whole of the worker's wage in the city. The reason for this is that there is no cost to the economy, in the form of reduced output, as a result of a worker leaving the subsistence activity. In summary, while the worker's private net benefit is the difference between her remuneration in the two types of activity, the social or economy-wide net benefit, in this case, is the value of the additional output in the market activity.

How would this difference between private and social rents show up in the Referent Group Analysis? We have already assigned the difference between the wage and the value of what the workers would have taken home as the result of subsistence agriculture as a net benefit to unskilled workers on the project. If withdrawing these workers from subsistence activities has no effect on the value of total output of such activities, the remaining subsistence workers experience an increase in the quantity and value of the produce they take home: in effect the "subsistence cake" is divided among fewer participants and each gets a slightly larger share. If subsistence workers are part of the Referent Group, then we would need to add a new row in the spreadsheet to record the value of this net benefit, if it occurred.

Assuming that the total value of subsistence output has not changed as a result of some workers quitting subsistence activity to work on the project, the amount of the transfer to subsistence workers is measured by the share of subsistence output which would have been received by the workers who left to participate in the ICP project. If we add this amount to Referent Group net benefits, it turns out that, while the wage bill is a cost to the firm, it is a net benefit to the Referent Group, some of which goes to workers employed on the project, and the balance to those who remain in subsistence activity. In the Efficiency Analysis the zero opportunity cost of a marginal amount of subsistence labour would be recognised by shadow-pricing the labour at zero, thereby increasing the efficiency net benefits by the amount paid to unskilled labour, as compared with the situation in which project labour is diverted from market activity.

Once the various rows representing referent sub-group benefits and costs have been filled in, the rows can be aggregated to give a summary statement of the net benefits to the Referent Group as a whole. The row representing this aggregate value should be identical in every year to the Total Referent Group row at the top of the table in Figure A6.1 which we obtained by subtracting non-Referent Group net benefits from the efficiency net benefits of the project. If an error has occurred, it can usually be readily identified: if the two net benefit streams are different in each and every year, then a whole category of net benefits may have been omitted or an inappropriate shadow-price adopted. On the other hand if the difference is in one year only, there may be some problem with costing capital or with some other irregular payment or receipt. In their first attempt at the case study the authors noted a discrepancy in the last year of the project: they forgot to allow for the import-replacing effect on tariff revenues of running down working capital stocks, thereby replacing some annual imports of materials and spare parts.

Once the stream of net benefits to the Referent Group has been calculated, a summary figure is normally obtained by calculating a net present value (NPV), or, sometimes but infrequently, an internal rate of return (IRR). The net benefits to sub-groups within the Referent Group can also be summarised in this way as shown at the bottom of Figure A6.1, although it should be noted that for some groups, such as domestic labour, there may be no project costs and hence no IRR to be computed. The NPV is the appropriate measure at this level of disaggregation, and it has the advantage of corresponding to the sums, specified under the Potential Pareto Improvement criterion, notionally to be taken from beneficiaries or paid to those who bear the costs in order to maintain pre-existing levels of economic welfare.

We have now completed the basic cost-benefit analysis of the ICP case study. Figure A6.2 assembles Tables 1–5, which were derived in Chapters 4–6, in the form of a single spreadsheet. As discussed below, and illustrated in the Appendix to Chapter 13, this spreadsheet can be used for various types of sensitivity analysis. It will also serve as a basis for comparison when we consider more advanced topics: in Appendix 2 in this chapter we amend the ICP spreadsheet to incorporate a premium on public funds; in Chapter 7 we consider the implications of a rise, induced by the project, in the market wage of skilled labour; in Chapter 9 we incorporate risk analysis; and in Chapter 10 we shadow-price foreign exchange. In each of these cases the reader will be invited to compare the amended Figure with the base case described by Figure A6.2. We now proceed to discuss sensitivity analysis, focussing on the discount rate and other variables.

The discount rate

In calculating net present values a range of discount rates, incorporating the domestic government bond rate, should be used. Since no attempt was made in the ICP case study to inflate benefits or costs a real rate of interest is required and the use of a single real rate of interest involves the implicit assumption that all prices change at the general rate of inflation. We use a range of discount rates in the analysis for two reasons: first, as discussed in Chapter 5, there may be valid arguments as to why a market rate of interest should be replaced by a shadow (or social) discount rate, and we may not know at this stage what that rate should be; and, second, we want to see how sensitive the NPV is to the choice of discount rate (if it is very sensitive, this compounds the difficulty posed by the uncertainty about the appropriate discount rate). Sensitivity of the NPV to the values of other key variables, such as the exchange rate, can also be ascertained at this stage, and this issue is explored further in Chapter 9. Because of the way we have constructed the spreadsheet, with all entries in Tables 2 to 5 being composed of cell references to Table 1, all we have to do is to change the value of a variable, such as the exchange rate, in Table 1 to have the spreadsheet recalculate the NPV.

Comparing alternative scenarios

In addition to seeing how sensitive the project NPV is to the choice of discount rate and other key variables, we want to see how sensitive it is to the project design. This is part of the cost-benefit analyst's pro-active role in helping to design the best project from the client's

ICP PROJECT: 1999–2009

TABLE 1: KEY INPUT VARIABLES

VARIABLES		INVESTMENT COSTS	000 $'	000 BT'	RUNNING COSTS		MATERIAL FLOWS		FINANCE	
EXCHANGE RATE	44	SPINDLES	4000	176000	WATER BT PER MN LBS	480000	COTTON MN LBS	16.2	SUPPLIERS CREDIT (%)	75%
TARIFF A	5%	EQUIPMENT	850	37400	POWER BT PER MN LBS	480000	YARN MN LBS	15	SUPPLIERS CREDIT (BT)	160,050
TARIFF B	10%	INSTALLATION	100	4400	SPARE PARTS (%)	10%	LABOUR 1	100	OVERDRAFT	60
PROFITS TAX	50%	CONSTRUCTION		36000	INSURANCE&RENT BT MN	6	LABOUR 2	1000	CREDIT REPAY YEARS	5
DEPRECIATION	10%	START-UP COSTS		18000	OVERHEADS	0	LABOUR 3	500		
PRICE YARN CIF US$	0.58								DISCOUNT RATES	4%
YARN TARIFF BT	2.1	WORKING CAPITAL								8%
HANDLING BT	0.3	RAW MATERIALS	1200	52800						12%
TARIFF C	10%	SPARE PARTS	243	10692						16%
PRICE COTTON CIF US$	0.3									
WAGE 1 BT PA	150000	SHADOW PRICES	%						CONVERSIONS	
WAGE 2 BT PA	22500	CONSTRUCTION	50%						CAPACITY OUTPUT	50%
WAGE 3 BT PA	12000	POWER	50%						THOUSANDS	1000
INTEREST RATE 1	8%	INS & RENT	0%							
INTEREST RATE 2	10%	WAGE 3	50%							

TABLE 2: MARKET ANALYSIS

YEAR	1999 MN BT	2000 MN BT	2001 MN BT	2002 MN BT	2003 MN BT	2004 MN BT	2005 MN BT	2006 MN BT	2007 MN BT	2008 MN BT	2009 MN BT
INVESTMENT COSTS											
SPINDLES	−184.800										
EQUIPMENT	−39.270										
INSTALLATION	−4.400										
CONSTRUCTION	−36.000										
START-UP COSTS	−18.000										
RAW MATERIALS		−58.080									58.080
SPARE PARTS		−11.227									11.227
TOTAL	−282.470	−69.307	0.000	0.000	0.000	0.000	0.000	0.000	0.000	0.000	69.307
RUNNING COSTS											
WATER		−3.600	−7.200	−7.200	−7.200	−7.200	−7.200	−7.200	−7.200	−7.200	−7.200
POWER		−3.600	−7.200	−7.200	−7.200	−7.200	−7.200	−7.200	−7.200	−7.200	−7.200
SPARE PARTS		−22.407	−22.407	−22.407	−22.407	−22.407	−22.407	−22.407	−22.407	−22.407	−22.407
RAW MATERIALS		−117.612	−235.224	−235.224	−235.224	−235.224	−235.224	−235.224	−235.224	−235.224	−235.224
INSURANCE		−6.000	−6.000	−6.000	−6.000	−6.000	−6.000	−6.000	−6.000	−6.000	−6.000
LABOUR		−43.500	−43.500	−43.500	−43.500	−43.500	−43.500	−43.500	−43.500	−43.500	−43.500
TOTAL		−196.719	−321.531	−321.531	−321.531	−321.531	−321.531	−321.531	−321.531	−321.531	−321.531
REVENUES											
SALE OF YARN		209.400	418.800	418.800	418.800	418.800	418.800	418.800	418.800	418.800	418.800
NET BENEFIT (Project)	−282.470	−56.626	97.269	97.269	97.269	97.269	97.269	97.269	97.269	97.269	166.576

TABLE 3: PRIVATE ANALYSIS

YEAR	1999 MN BT	2000 MN BT	2001 MN BT	2002 MN BT	2003 MN BT	2004 MN BT	2005 MN BT	2006 MN BT	2007 MN BT	2008 MN BT	2009 MN BT
DEBT FLOWS											
SUPPLIERS CREDIT	160.050										
INTEREST		−12.804	−10.621	−8.264	−5.719	−2.969					
REPAYMENT		−27.282	−29.464	−31.821	−34.367	−37.116					
BANK OVERDRAFT	60.000										−60.000
INTEREST		−6.000	−6.000	−6.000	−6.000	−6.000		−6.000	−6.000	−6.000	−6.000
NET FINANCING FLOWS	160.050	19.914	−46.086	−46.086	−46.086	−46.086	−6.000	−6.000	−6.000	−6.000	−66.000
DEPRECIATION		−26.447	−26.447	−26.447	−26.447	−26.447	−26.447	−26.447	−26.447	−26.447	−26.447
NET OPERATING REVENUE		12.681	97.269	97.269	97.269	97.269	97.269	97.269	97.269	97.269	97.269
TAXABLE PROFIT		−26.570	54.201	56.558	59.103	61.853	64.822	64.822	64.822	64.822	64.822
PROFIT TAX		13.285	−27.100	−28.279	−29.552	−30.926	−32.411	−32.411	−32.411	−32.411	−32.411
NET FLOW ON EQUITY	−122.420	−23.426	24.083	22.905	21.632	20.257	58.858	58.858	58.858	58.858	68.165

TABLE 4: EFFICIENCY ANALYSIS

YEAR	1999 MN BT	2000 MN BT	2001 MN BT	2002 MN BT	2003 MN BT	2004 MN BT	2005 MN BT	2006 MN BT	2007 MN BT	2008 MN BT	2009 MN BT
INVESTMENT COSTS											
SPINDLES	−176.000										
EQUIPMENT	−37.400										
INSTALLATION	−4.400										
CONSTRUCTION	−18.000										
START-UP COSTS	−18.000										
RAW MATERIALS		−52.800									52.800
SPARE PARTS		−10.692									10.692
TOTAL	−253.800	−63.492	0.000	0.000	0.000	0.000	0.000	0.000	0.000	0.000	63.492
RUNNING COSTS											
WATER		−3.600	−7.200	−7.200	−7.200	−7.200	−7.200	−7.200	−7.200	−7.200	−7.200
POWER		−1.800	−3.600	−3.600	−3.600	−3.600	−3.600	−3.600	−3.600	−3.600	−3.600
SPARE PARTS		−21.340	−21.340	−21.340	−21.340	−21.340	−21.340	−21.340	−21.340	−21.340	−21.340
RAW MATERIALS		−106.920	−213.840	−213.840	−213.840	−213.840	−213.840	−213.840	−213.840	−213.840	−213.840
INSURANCE & RENT		0.000	0.000	0.000	0.000	0.000	0.000	0.000	0.000	0.000	0.000
LABOUR		−40.500	−40.500	−40.500	−40.500	−40.500	−40.500	−40.500	−40.500	−40.500	−40.500
TOTAL		−174.160	−286.480	−286.480	−286.480	−286.480	−286.480	−286.480	−286.480	−286.480	−286.480
REVENUES											
SALE OF YARN		191.400	382.800	382.800	382.800	382.800	382.800	382.800	382.800	382.800	382.800
NET EFFICIENCY BENEFIT	−253.800	−46.252	96.320	96.320	96.320	96.320	96.320	96.320	96.320	96.320	159.812

TABLE 5: REFERENT GROUP ANALYSIS

	1999	2000	2001	2002	2003	2004	2005	2006	2007	2008	2009
TOTAL REFERENT GROUP	28.670	−62.911	32.151	33.330	34.603	35.977	37.462	37.462	37.462	37.462	91.647
– IMPORT DUTIES											
YARN		−18.000	−36.000	−36.000	−36.000	−36.000	−36.000	−36.000	−36.000	−36.000	−36.000
COTTON		10.692	21.384	21.384	21.384	21.384	21.384	21.384	21.384	21.384	21.384
SPINDLES	8.800										
EQUIPMENT	1.870										
MATERIALS INVENTORY		5.280									−5.280
SPARES INVENTORY		0.535									−0.535
SPARES (ANNUAL)		1.067	1.067	1.067	1.067	1.067	1.067	1.067	1.067	1.067	1.067
– PROFITS TAXES		−13.285	27.100	28.279	29.552	30.926	32.411	32.411	32.411	32.411	32.411
(i) TOTAL GOVERNMENT	10.670	−13.711	13.551	14.730	16.003	17.377	18.862	18.862	18.862	18.862	13.047
(ii) POWER SUPPLIER		1.800	3.600	3.600	3.600	3.600	3.600	3.600	3.600	3.600	3.600
(iii) LABOUR	18.000	3.000	3.000	3.000	3.000	3.000	3.000	3.000	3.000	3.000	3.000
(iv) DOMESTIC BANK	0.000	−60.000	6.000	6.000	6.000	6.000	6.000	6.000	6.000	6.000	66.000
(v) INSURANCE		6.000	6.000	6.000	6.000	6.000	6.000	6.000	6.000	6.000	6.000
TOTAL REFERENT GROUP	28.670	−62.911	32.151	33.330	34.603	35.977	37.462	37.462	37.462	37.462	91.647

	MARKET		PRIVATE		EFFICIENCY		REFERENT GROUP	
DISCOUNT RATE	NPV BT(Mn)	IRR (%)	NPV BT(Mn)	IRR (%)	NPV BT(Mn)	IRR (%)	NPV BT(Mn)	IRR (%)
4%	405.3	20.6%	246.6	16.7%	433.2	23.4%	260.4	n/a
8%	259.8		246.6		289.9		201.3	
12%	152.0		246.6		183.6		158.3	
16%	70.7		246.6		103.2		126.6	
0%	605.6		246.6		630.3		343.3	

Figure A6.2 ICP project solution : consolidated tables.

point of view. We have just completed the analysis of the basic project, but in discussions with the client we asked a whole series of questions: what about varying the location of the project; what about varying the tax regime – offering various types of concessions to make the project more attractive to the foreign firm. We can take the base case analysis just completed and edit it to analyse the project NPV and the Referent Group net benefits under various alternative scenarios. For example, suppose that the firm were allowed to import raw materials free of import duty: we change the raw materials tariff rate in its cell entry in Table 1 to zero, save the spreadsheet as a new file, and we now have a complete set of results under this scenario. Suppose that the government insisted that the firm locate in a depressed area; this would involve additional costs which can be entered in Table 1 and edited into Tables 2 to 5 to give a set of results under this scenario, and so on. A set of calculations investigating these issues is reported in Chapter 13.

It would also be possible to further develop the basic analysis to determine the effects on the firm and the Referent Group of various ownership structures for the project. For example, the domestic government may acquire equity in the project, either by purchasing it or simply by being given it in return for allowing the project to proceed. The project might be developed as a joint venture with members of the Referent Group contributing both equity capital and management expertise, or the domestic government might acquire all of the equity in the project and obtain the foreign firm's expertise under a management contract.

While it is very easy to generate results for a range of scenarios thought to be of interest to the client, it is not so easy to present the results in a readily digestible way. There is a danger in swamping the client with detail so that the wood cannot be seen for the trees! While Chapter 13 discusses in detail the presentation of the results of a social cost-benefit analysis, we consider some issues relevant to the ICP case study here.

A useful way to proceed is to think in terms of trade-offs. Some scenarios will be more advantageous to the firm – those involving tax and tariff concessions, for example – and some will be more advantageous to the Referent Group – location of the plant in a depressed area, for example. A *Summary Table* can be prepared which shows how the net benefits to the various parties change as we move from one scenario to another. For example, scenarios could be ranked in descending order of Referent Group net benefits; this may correspond loosely to ascending order of foreign firm net benefits. Project feasibility is established by some cut-off value for the net benefits of the private agent (the individual, the firm, the farm or whatever); scenarios ranked lower than this simply will not occur. This establishes the feasible set of scenarios from among which the client can elect to choose the project design offering the highest aggregate Referent Group net benefits, or perhaps the preferred distribution of net benefits among Referent Group members. While it is useful to think in terms of trade-offs – one group getting more at the expense of another group – in other instances there are likely to be "win–win" situations where the additional net benefits of a more efficient project can be shared to the benefit of all groups.

Where there are trade-offs among Referent Group members, the distribution of net benefits among stakeholder groups could be a criterion for comparing and ranking alternatives. This might be the case where the government is committed to raising the economic welfare of disadvantaged groups such as residents of deprived regions, or particular socio-economic classes. The disaggregated Referent Group Analysis provides important information for assessing the distributional implications of the project. The issue of distribution objectives and how these can be incorporated explicitly into social cost-benefit analysis is discussed further in Chapter 11.

Appendix 2 to Chapter 6

Incorporating the public funds cost premium in the ICP case study

While the concept of the cost premium on public funds was developed in Chapter 5, dealing with the Efficiency Analysis, treatment of this cost item in the CBA was deferred until the Referent Group Analysis was completed. In the Efficiency Analysis no explicit account was taken of direct taxes, such as business income taxes, and outputs and inputs were priced at their marginal values or opportunity costs, which as discussed in Chapter 5, could be either net (excluding) or gross (including) of tax. In either case, no specific account is taken of indirect tax flows in the Efficiency Analysis. The Referent Group Analysis, on the other hand, takes explicit account of direct and indirect tax flows because the government is usually a member of the Referent Group. The tax flows identified in the Referent Group Analysis can be aggregated to provide an estimate of the net outflow or inflow of public funds resulting from the project. A net outflow of public funds means that the project increases the public sector's demand for funds, thereby increasing the deadweight loss imposed on the economy, as discussed in Section 5.9 of Chapter 5. A net inflow, on the other hand, reduces the demand for funds from other sources, thereby reducing the overall deadweight loss. The net effect of the project on the flow of public funds can be multiplied by the public funds cost premium to provide an estimate of the cost (benefit) of the project in terms of the increase (reduction) in deadweight loss in the case of a net outflow (net inflow).

It will now be assumed that the marginal cost of public funds in Thailand is 1.2: this means that a real cost of 0.2 Baht is imposed on the economy for every Baht transferred from the private to the public sector in the form of borrowing or taxes. To take account of this in the ICP case study, we start by entering the public funds premium in the variables section of Table A6.2. Since shadow-pricing public funds to reflect a cost premium will affect the Efficiency and Referent Group Analyses, but not the Market and Private Analyses, we confine the discussion and Figure A6.3 to the former two accounts.

To see how the cost premium on public funds can be integrated into the cost-benefit analysis framework, consider first the Referent Group Analysis. In Figure A6.2, Table 5 of the consolidated tables for the ICP project identified a set of indirect and direct tax flows accruing to government as a member of the Referent Group. These flows were summed to give a Total Government flow. It is a simple matter to create new rows to incorporate the 20% cost premium on public funds. In Table 5 of Figure A6.3 the costs or benefits of public funds, including the 20% premium, are reported in new rows labelled Indirect or Direct Taxes plus Premium. It can be seen that the net effect of the indirect tax flows is an outflow of funds, caused by the import-replacing effect of the project's yarn production. However this outflow is more than offset by the inflow of direct tax revenues, so that the net effect of the project, summarised in the revised "Total Government" row, is an inflow of public funds which attracts the 20% premium. The effect on Referent Group net benefit is to increase it, as can be seen by comparing the Total Referent Group net benefit row in Figure A6.2 with the corresponding row in Figure A6.3. The net inflow of public funds reduces the government's tax or borrowing requirement from other sources thereby reducing the level of deadweight loss in the economy.

It is now clear that some adjustments are required to the Efficiency Analysis to reflect the project's net contribution to public revenues. In the absence of a cost premium on public funds, direct and indirect taxes net out in the Efficiency Analysis because they are a cost to one group (the firm) and a benefit to another (the government). However. if public funds

TABLE 4: EFFICIENCY ANALYSIS

YEAR	1999 MN BT	2000 MN BT	2001 MN BT	2002 MN BT	2003 MN BT	2004 MN BT	2005 MN BT	2006 MN BT	2007 MN BT	2008 MN BT	2009 MN BT
INVESTMENT COSTS											
SPINDLES	−176.000										
EQUIPMENT	−37.400										
INSTALLATION	−4.400										
CONSTRUCTION	−18.000										
START-UP COSTS	−18.000										
RAW MATERIALS		−52.800									52.800
SPARE PARTS		−10.692									10.692
TOTAL	−253.800	−63.492	0.000	0.000	0.000	0.000	0.000	0.000	0.000	0.000	63.492
RUNNING COSTS											
WATER		−3.600	−7.200	−7.200	−7.200	−7.200	−7.200	−7.200	−7.200	−7.200	−7.200
POWER		−1.800	−3.600	−3.600	−3.600	−3.600	−3.600	−3.600	−3.600	−3.600	−3.600
SPARE PARTS		−21.340	−21.340	−21.340	−21.340	−21.340	−21.340	−21.340	−21.340	−21.340	−21.340
RAW MATERIALS		−106.920	−213.840	−213.840	−213.840	−213.840	−213.840	−213.840	−213.840	−213.840	−213.840
INSURANCE & RENT		0.000	0.000	0.000	0.000	0.000	0.000	0.000	0.000	0.000	0.000
LABOUR		−40.500	−40.500	−40.500	−40.500	−40.500	−40.500	−40.500	−40.500	−40.500	−40.500
TOTAL		−174.160	−286.480	−286.480	−286.480	−286.480	−286.480	−286.480	−286.480	−286.480	−286.480
REVENUES											
SALE OF YARN		191.400	382.800	382.800	382.800	382.800	382.800	382.800	382.800	382.800	382.800
INDIRECT TAX PREMIUM	2.134	−0.085	−2.710	−2.710	−2.710	−2.710	−2.710	−2.710	−2.710	−2.710	−3.873
DIRECT TAX PREMIUM		−2.657	5.420	5.656	5.910	6.185	6.482	6.482	6.482	6.482	6.482
NET EFFICIENCY BENEFIT	−251.666	−48.994	99.030	99.266	99.521	99.795	100.092	100.092	100.092	100.092	162.421

TABLE 5: REFERENT GROUP ANALYSIS

	1999	2000	2001	2002	2003	2004	2005	2006	2007	2008	2009
TOTAL REFERENT GROUP	30.804	−65.654	34.862	36.276	37.803	39.453	41.234	41.234	41.234	41.234	94.257
IMPORT DUTIES											
YARN		−18.000	−36.000	−36.000	−36.000	−36.000	−36.000	−36.000	−36.000	−36.000	−36.000
COTTON		10.692	21.384	21.384	21.384	21.384	21.384	21.384	21.384	21.384	21.384
EQUIPMENT	1.870										
SPINDLES	8.800										
MATERIALS INVENTORY		5.280									−5.280
SPARES INVENTORY		0.535									−0.535
SPARES		1.067	1.067	1.067	1.067	1.067	1.067	1.067	1.067	1.067	1.067
INDIRECT TAXES+PREMIUM	12.804	−0.512	−16.259	−16.259	−16.259	−16.259	−16.259	−16.259	−16.259	−16.259	−23.236
DIRECT TAXES+PREMIUM		−15.942	32.520	33.935	35.462	37.112	38.893	38.893	38.893	38.893	38.893
(i) TOTAL GOVERNMENT	12.804	−16.454	16.262	17.676	19.203	20.853	22.634	22.634	22.634	22.634	15.657
(ii) POWER SUPPLIER		1.800	3.600	3.600	3.600	3.600	3.600	3.600	3.600	3.600	3.600
(iii) LABOUR	18.000	3.000	3.000	3.000	3.000	3.000	3.000	3.000	3.000	3.000	3.000
(iv) DOMESTIC BANK		−60.000	6.000	6.000	6.000	6.000	6.000	6.000	6.000	6.000	6.000
											60.000
(v) INSURANCE		6.000	6.000	6.000	6.000	6.000	6.000	6.000	6.000	6.000	6.000
TOTAL REFERENT GROUP	30.804	−65.654	34.862	36.276	37.803	39.453	41.234	41.234	41.234	41.234	94.257

PROJECT			PRIVATE			EFFICIENCY			REFERENT GROUP		
NPV	MN BT	IRR (%)	NPV	MN BT	IRR (%)	NPV	MN BT	IRR (%)	NPV	MN BT	IRR (%)
4%	405.3	20.6%	4%	154.5	16.7%	4%	456.5	24.2%	4%	283.6	n/a
8%	259.8		8%	88.7		8%	308.6		8%	219.9	
12%	152.0		12%	40.8		12%	198.9		12%	173.6	
16%	70.7		16%	5.4		16%	115.9		16%	139.3	
0%	605.6		0%	246.6		0%	659.7		0%	372.7	

Figure A6.3 ICP project solution with premium on public funds.

are to be shadow-priced, the benefit (cost) of a given tax inflow (outflow) will exceed the cost (benefit) of the same tax flow to the firm. The difference between the two values is measured by the premium attached to the tax flow. If the cost premium is 20%, then 20% of the net tax inflow (outflow) identified in the Referent Group Analysis of the ICP project should be credited as a benefit (cost) in the Efficiency Analysis, as shown in the Indirect and Direct Tax Premium rows in Table 4 of Figure A6.3. The intuition behind this procedure is that increased tax revenues resulting from the project will allow the government to cut, or not to raise, taxes elsewhere, thereby generating a benefit, or avoiding a cost, to the economy. And, conversely, reduced tax revenues will require government to raise, or not to cut, taxes elsewhere, thereby imposing a cost, or forgoing a benefit.

An important feature of the spreadsheet framework is its adding up property – the whole is the sum of its parts. If we increase the Referent Group net benefit, in this case, by applying the public funds cost premium to a net inflow of funds, without there being any adjustment to the non-Referent Group net benefits, we must also increase the efficiency net benefit by the same amount. Thus the inclusion of the Public Funds Premium on indirect and direct taxes as a net benefit (in this case) in the Efficiency Analysis in Table 4, Figure A6.3, is essential to the integrity of the cost-benefit model.

The effect of these adjustments, as reported in the summary table at the bottom of Figure A6.3, is a net increase in the project's efficiency net benefits, resulting in an increase in the IRR from 23.4% to 24.2%. This increase reflects the fact that the project is a net contributor to public funds. The Referent Group net benefits increase from $260.4 million to $283.6 million at a 4% discount rate. As a consistency check, to verify that the analyst has not made an error in making these adjustments, the aggregate Referent Group benefits at the top of Table 5 (derived by subtracting ICP's cash flow and the overseas lender's cash flow from the revised efficiency cash flow) can be confirmed to be the same as those on the bottom line (derived by aggregating the sub-Referent Group members' net benefit flows).

Exercises

1   A foreign firm, which is not part of the Referent Group, proposes to establish a food processing plant in Australia (the Referent Group). With one exception, market prices measure all the efficiency benefits and costs of the project. The exception is the input of fuel which is subject to a 20% indirect tax. The plant is to be fully financed by the owners of the firm. The NPV to the firm is $220 million; NPV of business income taxes amounts to $30 million and the indirect fuel tax revenues have an NPV of $10 million. Use the above information to work out the values which belong in areas A, B, C and D of the following table.

Classification of Net Benefits

	Net Benefits accruing to:	
	Referent Group	Non-Referent Group
Net Benefits Measured by Market Prices	A	B
Net Benefits *not* Measured by Market Prices	C	D

2   A local firm hires $150 of unskilled labour at the minimum wage and constructs a walking track, for which it is paid $180 by the state government. The firm incurs no other cost, apart from the wage bill, in constructing the track. The government is to charge a fee to users of the track, and the present value of the fees paid is estimated to be $100. Consumers place a net present value (net of the fees and any other costs incurred) of $50 on the track's services. The labour employed to construct the track would otherwise be unemployed, and the value of its non-market activity is estimated to be $50.

   i   Calculate the net benefit of the project to each member of the Referent Group:
      a   the firm
      b   consumers
      c   labour
      d   the government.

ii   Using the appropriate shadow-price of labour, calculate the efficiency net benefit of the project.

3   The state government hires a computer programmer from another state for a month at a cost of $30,000, which is $10,000 more than she could earn during that period in her home state. The state government also pays $1000 for accommodation in a rent-controlled apartment with a market value of $1200. The programmer edits some programs and saves the state government $40,000. Assuming the Referent Group consists of residents of the state:

i   Calculate the net benefit to the Referent Group.
ii   Using the appropriate shadow-prices, calculate the net benefit of the project according to an Efficiency Analysis.

4   A foreign firm is considering a project which has, at market prices, a present value of benefits of $200 and a present value of input costs of $130. If the project goes ahead, tax with a present value of $50 will be paid to the host country, domestic labour which would otherwise be unemployed will be paid wages with a present value of $50 for working on the project (the wage bill is included in the input costs referred to above), and pollution caused by the project will reduce the value of output elsewhere in the host country's economy by $10. Assuming that the owners of the foreign firm are not part of the Referent Group, and that the opportunity cost of unemployed labour is 50% of the market wage, what are the net present values generated by the following?:

i   the Market Analysis;
ii   the Private Analysis;
iii   the Efficiency Analysis;
iv   the Referent Group Analysis.

5   A local contractor hires $5000 of unskilled labour at the minimum wage and constructs a children's adventure playground, on otherwise vacant land, for which it is paid $7500 by the local council. The cost of materials, in addition to current supply, required to construct the facility was $1500, including $100 in federal sales tax (a distortionary tax), and the firm paid an extra $100 in federal company tax as a result of undertaking the project. (All values are present values.) The council is to charge a fee to users of the playground, and the present value of the fees to be paid by consumers is estimated to be $3000. A survey indicates that consumers' total willingness to pay for the use of the facility (including the fee) has a present value of $5500. The labour employed to con-struct the facility would otherwise be unemployed, receives no unemployment benefit, and the present value of its non-market activity is estimated to be $2000.

i   Calculate the net benefit of the project to each of the following members of the Referent Group:
   a   the contractor
   b   consumers
   c   labour
   d   the council
   e   the federal government.

ii   ii. Calculate the efficiency net benefit of the project.

6   For the ICP case study, prepare a summary table of net benefits (discounted at 8%) in the same format as Table 6.1. Hints: note that the NPV of direct tax receipts is 164.2 million Baht, the NPV of indirect tax receipts forgone is 70.8 million Baht, and the NPV (discounted at 8%) of the domestic bank's financial flow is 6.9 million Baht. Also note that the interest rate charged for the supplier's credit is the same as the discount rate for this problem. All the other information required can be read from Figure A6.2.

Note

1   A reviewer comments that the financial approach described here is similar to *la methode des effets* developed by Marc Chervel in the 1960s for use by the French Ministry of Cooperation in project appraisal.

7 Consumer and producer surplus in cost-benefit analysis

7.1 Introduction

So far in this study we have assumed that undertaking a proposed project would have no effect on the market prices of goods and services. However, since the market economy consists of a complex network of inter-related output and input markets, it is possible, in principle, that undertaking the project will have wide-ranging effects on market prices. If the project's output or input quantities are small relative to the amounts traded in the markets for these goods and services, the effects on market prices will be small enough to be ignored, on the grounds that including them would have no bearing on the outcome of the analysis. While this "small project assumption" is a reasonable one for many of the projects which the analyst will encounter, it is important to be able to identify circumstances in which price changes are relevant, and to know how to deal with them in the cost-benefit analysis.

A project increases the aggregate supply of the output produced by the project, and increases the aggregate demand for the inputs used in the project. Significant changes in quantities supplied or demanded can result in changes in the prices of goods and services traded in the markets affected. A significant change in quantity is one that is large relative to the quantity that would be bought and sold in the market in the absence of the project being undertaken. It is evident that there is little chance of such a change occurring where the output or input in question is traded on international markets, as will be discussed in Chapter 10. Few countries, let alone individual projects, have the capacity to alter world prices by changing their levels of supply of, or demand for, goods and services.

Where non-traded goods are concerned, however, a project can cause a change in the market price of its output or its inputs where the level of project output or input is large relative to the total quantity currently exchanged in the domestic market. This is most likely to occur where the relevant market is local or regional, rather than national in extent. Examples are local markets in transport services or irrigation water, and regional labour markets. Such goods and services have, at a national level, some of the characteristics which make goods and services non-tradeable at the international level – principally relatively high transportation costs.

If a project is large enough at the local or regional level to cause changes in the market prices of the goods and services it produces as outputs, or consumes as inputs, these price changes may affect the prices of complementary or substitute goods and services traded in other markets. For example, a project which provides a new route into the city through provision of a bridge may result in increased demand for apartments in the area served by the bridge. Given a relatively inelastic supply of apartments, the increase in demand for rental

DOI: 10.4324/9781003312758-7

property, which is complementary to the service provided by the bridge, will result in a price increase. Apartments which are not favourably located relative to the bridge offer substitute services, and demand for these may fall once the bridge is constructed, leading to a fall in market price. The analyst needs to be able to determine the relevance, if any, of such price changes to the cost-benefit analysis.

This chapter illustrates the implications of price changes for the cost-benefit analysis by considering a number of examples, with further examples to be discussed in Chapter 8. In the first set of examples, undertaking the project in order to produce a good or service results in a significant increase in the total quantity of the good or service produced and sold. Three cases are considered: (1) where the market supply curve of the good or service is horizontal (perfectly elastic); (2) where it is upward sloping (positive elasticity of supply); and (3) where it is vertical (perfectly inelastic). An urban transportation project is used to illustrate the first case; an investment in job training the second; and, an increase in the supply of irrigation water the third. In each case the emphasis will be on measuring project benefits, with little attention paid to costs. Following these examples, in which the price of the project output changes as a result of the project, a case in which undertaking the project causes a change in an input price is considered. This case is illustrated by an extension to the irrigation water example to include a rural labour market which serves farms which benefit from the increased supply of irrigation water.

In considering the above examples we will assume that there are no market imperfections as described in Chapter 5. This means that there will be no need to shadow-price the market values discussed in the examples in order to calculate benefits from an efficiency viewpoint. In general, however, the procedures described in Chapter 5 can be applied to obtain the appropriate valuations in the presence of price changes.

Two concepts which will be important in the discussion are the difference between real and pecuniary effects, and the notion of surplus accruing to producers or consumers. We have already encountered the notion of pecuniary effects in our discussion of transfers in Chapter 6 and, together with the notions of consumer and producer surplus, they will feature prominently in the examples to be discussed below. We now review these concepts briefly before considering the examples.

7.2 Real versus pecuniary effects

We can think of Efficiency Analysis as being concerned with the "size of the cake": does the project provide a net addition to the value of goods and services available to the world economy? Changes to the quantities of goods and services available are termed "real" effects, in contrast to "pecuniary" effects which represent changes in entitlements to shares in a "cake" of given size. For example, suppose that the price of a good or service falls, as in the case of the rental price of unfavourably located apartments in the bridge example discussed above. Buyers of the service (renters) are better off, but suppliers (landlords) are worse off to the same extent. The gain to buyers and the loss to sellers are termed pecuniary effects and they have the characteristic that they net out in an Efficiency Analysis. Because pecuniary effects measure changes in entitlement to an existing flow of goods and services, they are referred to as transfers – they transfer a portion of existing benefits or costs from one group to another. While they are not relevant in the aggregate Efficiency Analysis, they may, as we saw in Chapter 6, need to be accounted for in the Referent Group Analysis which is concerned with the benefits and costs of the project to a subset of agents in the world economy.

7.3 Consumer surplus

A surplus is generated when a consumer is able to buy a unit of a good at a price lower than her willingness to pay for that unit, or when a producer is able to sell a unit of a good or factor of production at a price higher than that at which he would willingly part with that unit. The concept of *consumer surplus* can be explained by means of Figure 7.1, which illustrates an individual consumer's demand for a recreational service measured by number of trips per annum: according to the diagram, the consumer is willing to pay OE dollars for the first visit she makes, and lower amounts, as measured by the demand curve, for subsequent visits until she is just willing to pay the current price, P_0, and no more, for the last visit she makes. The reason the consumer is not willing to purchase more than Q_0 units (visits) at that price is that additional units are worth less to her, as measured by the height of her demand curve, than the current price, P_0. The consumer's total willingness to pay for OQ_0 units of the service is measured by the area EAQ_0O, while the actual amount she pays is measured by P_0AQ_0O. The difference between these two amounts is termed the *consumer surplus*– the value to the consumer of the quantity OQ_0 over and above what she actually has to pay – and it is measured, in dollars per period of time, by area EAP_0.

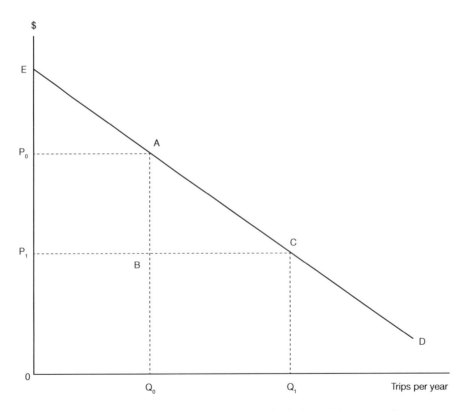

Figure 7.1 Consumer surplus. ED represents the individual's demand for trips each year to a recreation site. At price P_0 the individual chooses to consume Q_0 visits, and at price P_1 she chooses Q_1. A fall in price from P_0 to P_1 would increase consumer surplus by an amount measured by area P_0ACP_1.

Now suppose that the market price of the service falls from P_0 to P_1, as illustrated in Figure 7.1. The consumer will increase the quantity of trips she undertakes from Q_0 to Q_1 because, as indicated by her demand curve, some additional trips are worth more to her than the new price, P_1. The area of consumer surplus at the new price and quantity demanded becomes ECP_1, representing an increase in consumer surplus of P_0ACP_1. The increase in the amount of consumer surplus can be regarded as a measure of the annual benefit the consumer has received as a result of the fall in the price of the service.

An intuitive explanation of why the area P_0ACP_1 measures the benefit of the price fall to the individual consumer runs as follows. When the price falls from P_0 to P_1 the consumer benefits by the amount $(P_0- P_1)$ for each unit of the service she originally purchased, thereby saving her an amount measured by area P_0ABP_1. This amount is effectively extra "cash in hand" which can be used to purchase additional quantities of any goods or services the consumer fancies. As an aside, it can be noted that, since we are concerned, at present, only with the benefit to this individual consumer it is not necessary for us to inquire whether the benefit is real or pecuniary as discussed in Section 7.2 above. When the price of the service falls from P_0 to P_1 the consumer purchases the additional quantity $(Q_1- Q_0)$. The reason she increases the quantity purchased is that, as indicated by her demand curve, each of the extra trips is worth at least as much to her as the new price. For example, the first additional trip she purchases is worth P_0 and the last additional trip is worth P_1. The extra consumer surplus obtained from the first additional trip purchased is measured by $(P_0- P_1)$, and that from the last additional trip by $(P_1- P_1)$, or zero. The amount of consumer surplus contributed by the additional quantity purchased is measured by area ACB. When this amount is added to the additional surplus obtained from the original quantity purchased, the total annual benefit to the consumer of the fall in price is measured by area P_0ACP_1.

We can interpret the extra consumer surplus generated by a price fall/quantity rise as a measure of the sum of money we could take from the individual and still leave her as well off *with* the change induced by the project as she was *without* it. Similarly, the loss in consumer surplus generated by a price rise/quantity fall measures the sum of money notionally to be paid to her to keep her as well off *with* the price rise induced by the project as she would be *without* it. In terms of the Kaldor-Hicks (K-H) criterion of economic welfare change, discussed in Chapter 1, these consumer surplus sums measure the compensating variations associated with the price/quantity changes resulting from undertaking a project. The concept of *compensating variation* is discussed in more detail in Appendix 2 to this Chapter, together with the related concept of *equivalent variation*. Exploring these concepts requires a rather technical discussion of various types of consumer demand curves which is left to the Appendix.

7.3.1 Aggregating consumer surplus measures

To this point we have been considering the measurement of the value of the gain or loss to an individual as a result of a change in market price. However, a change in price will affect all consumers in the market and in Efficiency Analysis we want a measure of aggregate benefit or cost. The market demand curve for a private good is the lateral (horizontal) summation of the demand curves of all the individuals participating in the market. When price falls, existing consumers purchase more of the good or service, as illustrated in Figure 7.1, and some individuals, who did not previously consume the good at its initial price, become purchasers of the good at its new lower price. Both existing and new customers benefit from the additional consumer surplus generated by the lower price of the good. The area of consumer surplus change measured under the market demand curve is the aggregate of the changes which could be measured under the individual curves. This means that the area of consumer surplus identified by the market demand curve is the sum of the amounts hypothetically to be paid to (paid by) losers (gainers) to maintain each person's utility at the level it would be in the world without the project, as specified by the K-H criterion.

7.3.2 The significance of income distribution

We should note that the size of the compensation hypothetically to be paid to or taken from an individual as a result of a price change (the compensating variation) depends on the position of her demand curve. Since an increase in income shifts the individual's demand curve for a normal good to the right, it is obvious that individuals with high incomes will have larger compensating variations than individuals with low incomes, in response to the same price change. A change in the distribution of income would change the positions of the individual demand curves, the individual compensation measures, and possibly the aggregate measure. This means that the measures used in a cost-benefit analysis are not value-neutral: they are based on, and tend to protect, the *status quo* in terms of income distribution. Furthermore potential and actual compensation are very different concepts, and since compensation is rarely paid or collected in practice, generally some individuals will be better off and some worse off as a result of the project. The issue of income distribution can be addressed in a cost-benefit analysis, but consideration of this question is deferred to Chapter 11.

7.4 Producer surplus

We have already encountered the concept of producer surplus in the discussion of the marginal cost of public funds (MCF) in Chapter 5. In Figure 5.14 an increase in the tax on the wage of labour resulted in workers losing producer surplus, measured by area ACDE. The cost to the economy per dollar of funds raised (the MCF) was measured as the ratio of the loss in producer surplus to the value of additional tax revenue.

Another illustration of the concept of producer surplus is provided by Figure 7.9, in Section 7.6, which describes the supply of labour in a market serving the irrigation project discussed later in this chapter: the first unit of labour can be drawn into the market at a wage of OZ dollars, but subsequent units will present themselves only at higher wage rates. A wage rate of W_0 is required to induce the quantity supplied OL_0. While the total amount of wages paid to that quantity of labour is measured by area W_0UL_0O, that amount of labour could be hired for ZUL_0O if each unit was paid the minimum amount required to induce its supply. The difference between the wages actually paid and the minimum sum required to induce the number of units supplied is measured by area W_0UZ in Figure 7.9, and this amount is termed a *producer surplus*.

7.5 Accounting for output price changes

We now consider situations in which undertaking the project results in a significant increase in the total supply of the good or service produced by the project, and in consequence a change in its market price. As noted earlier, three cases are considered: where the market supply curve of the good or service is horizontal (perfectly elastic), upward sloping (positive elasticity of supply), and vertical (perfectly inelastic). Urban transportation projects are used to illustrate the first case; investment in job training the second; and, an increase in the supply of irrigation water the third.

7.5.1 Benefits of urban transport projects

Building a bridge

Suppose that a municipality is considering building a bridge which will reduce trip length and travel time from the suburbs into the centre of the city. While the size of the reductions in distance and time will vary from one suburb to another, the analysis can focus on the effects on the residents of an "average" suburb. If the bridge is built, the cost of a trip into the city falls: a shorter travel distance results in savings in car running costs, and a shorter trip length confers a benefit by reducing the cost of travel time (in other words, releasing time for other purposes). Suppose that the value of these savings can be measured by means of the appropriate market and shadow-prices and expressed as a dollar amount per trip. Figure 7.2, which applies the Consumer Surplus Analysis presented in Figure 7.1, can be interpreted as illustrating the aggregate demand for trips into the city by residents of the

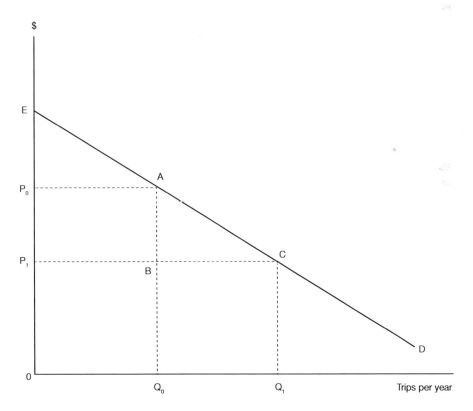

Figure 7.2 Benefits of a bridge. ED represents the market demand curve for trips. Building the bridge reduces the price of a trip from P_0 to P_1, resulting in an increase in the number of trips from Q_0 to Q_1. The increase in consumer surplus generated by the fall in price is measured by area P_0ACP_1.

average suburb, and the fall in price per trip from P_0 to P_1 measures the reduction in average trip cost as a result of building the bridge. If we assume no traffic congestion the reduction in per trip cost will be independent of the number of trips undertaken. Because the demand curve for trips, labelled D in Figure 7.2, is downward sloping, the fall in price will result in an increase in the quantity of trips demanded from Q_0 to Q_1 per year. It is important to note that the costs of building, maintaining and operating the bridge are ignored in the following discussion, which deals with measuring consumer benefits only, and these costs would have to be taken account of in a full cost-benefit analysis.

Consider first the Q_0 annual trips which would be undertaken in the absence of the bridge. The consumers of those trips will experience a per-trip cost reduction of $(P_0 - P_1)$ if the bridge is built. The annual benefit to commuters undertaking these existing trips is measured by the cost reduction $(P_0 - P_1)Q_0$, which is represented by area P_0ABP_1 in Figure 7.2. Now consider the additional trips generated as a result of the reduced cost of travel. These trips were not considered worth undertaking at price P_0 because their gross value, measured by the relevant point on the demand curve, was lower than their price, making their net value to the consumer negative. However, once the price falls to P_1 the extra trips from Q_0 to Q_1 begin to have a positive net value. The net value of the first additional trip above Q_0 is close to $(P_0 - P_1)$, and the net value of the last additional trip (the Q_1th in Figure 7.2) is close to zero. Assuming that the demand curve is a straight line, the average net value of the additional trips is $(P_0 - P_1)/2$, and hence the total annual value of these generated trips is given by $(P_0 - P_1)(Q_1 - Q_0)/2$, which is represented by area ACB in Figure 7.2. We now have an estimate of the approximate annual value derived by consumers as a result of building the bridge: it is area P_0ACP_1, or $(P_0 - P_1)Q_0 + (P_0 - P_1)(Q_1 - Q_0)/2$. In other words, the annual benefit of the bridge consists of the sum of the cost savings on existing trips plus half of the cost savings on generated trips.

As we have seen, the area P_0ACP_1 represents the increase in annual consumer surplus accruing to consumers as a result of the fall in the price of a trip. It was calculated by considering the cost savings experienced by commuters using the bridge. An alternative way of calculating it is to compare the net value to consumers of the amounts of the good consumed with and without the bridge. Net value is the total amount consumers are willing to pay for the good less the amount that they actually pay. Total willingness to pay for a quantity of the good is measured by the area under the demand curve up to the point representing the current level of consumption; in Figure 7.2, total willingness to pay for Q_0 trips is measured by area EAQ_0O. The logic underlying this measure is that someone is willing to pay OE for the first trip, and that person and others are willing to pay amounts measured by the height of the demand curve for subsequent additional trips until the Q_0th trip which is worth AQ_0. Net willingness to pay for these trips is given by total willingness to pay less the amount actually paid for Q_0 trips, which in the initial equilibrium is measured by area P_0AQ_0O. The initial amount of consumer surplus accruing to consumers is therefore measured by EAP_0. Applying a similar line of argument to the equilibrium after the fall in price from P_0 to P_1, consumer surplus in the new equilibrium is measured as ECP_1. Hence the increase in consumer surplus, measured by the new level less the initial level, is measured by area P_0ACP_1 which represents the annual benefit to consumers of the reduction in the cost per trip resulting from access to the bridge. If the demand curve for trips shifts outwards over time, as a result of economic growth for example, the measure of annual benefit may increase, depending on whether congestion costs develop, thereby generating a stream of rising annual benefits over time.

Let us now suppose that the municipality decides to levy a toll from motorists using the bridge in order to recoup some of the construction and operating costs. The price of a trip, once the bridge is in operation, will be the travel cost, P_1, plus the toll, as illustrated by P_2 in Figure 7.3, where $(P_2 - P_1)$ is the amount of the toll. The rise in the price of a trip, as compared with the no-toll case, will reduce the number of trips demanded from Q_1 to Q_2. The annual benefit to consumers, measured by the increase in consumer surplus as compared with the no-bridge equilibrium, now becomes P_0AFP_2, which is less than if the bridge were provided toll-free. However, introducing a toll also provides a benefit to the municipality in the form of toll revenues, measured as P_2FGP_1 in Figure 7.3. The total annual benefit of providing a toll bridge (ignoring the cost) is measured by area P_0AFGP_1, consisting of P_0AFP_2 to users and P_2FGP_1 to the municipality.

Introducing a toll has had two effects: it has reduced the total annual benefit derived from the bridge by the amount GFC; and it has redistributed some of the remaining benefit from consumers to the municipality. Despite the apparent reduction in aggregate benefit as a result of the toll, tolls may be justified on both efficiency and distributional grounds. From an efficiency point of view, in the absence of a toll additional public funds would have

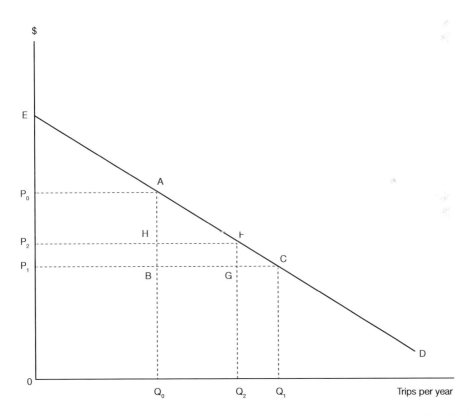

Figure 7.3 Effect of a bridge toll. ED represents the market demand curve for trips. Building the bridge reduces the cost of a trip from P_0 to P_1. However a toll in the form of a charge $(P_2 - P_1)$ for each trip means that the cost of each trip to motorists falls only to P_2, resulting in an increase in the number of trips from Q_0 to Q_2. The increase in consumer surplus generated by the fall in price is measured by area P_0AFP_2, and the toll operators receive P_2FGP_1 in annual revenues.

to be provided through increased taxes or government borrowing and, as we have seen in Chapter 5, there may be a premium on the opportunity cost of such funds which may exceed the inefficiency cost of tolls. And, from an equity point of view, why should the general taxpayer meet all the costs of improving the access of suburban residents to the city centre?

An important issue, referred to earlier, and relevant to the discussion in this and the next section, is the distinction between real and pecuniary effects in an Efficiency Analysis. A pecuniary effect is a gain/loss to one group which is exactly offset by a loss/gain to another. When total gains and losses are summed to obtain an estimate of the net benefit of the project, pecuniary effects cancel out. Pecuniary effects are sometimes referred to as transfers, although this terminology can be misleading. In the case of the toll bridge, for example, it could be argued that the toll transfers some of the benefit of the bridge from motorists to the government. However, since, in the example above, benefits to motorists have been measured net of tolls, both the toll revenues and the cost savings to motorists represent real benefits of the project. When the gains and losses to all groups, measured in this way, are summed, the tolls are not offset by a corresponding loss elsewhere in the calculation of net benefit.

It should be stressed that the above discussion has not dealt with many issues which would arise in the cost-benefit analysis of a proposal to build a bridge. Costs have not been considered, and some of the benefits are not incorporated in the measure of the reduced cost of a trip. For example, when traffic is diverted from existing routes to the bridge, congestion on these routes falls and people who continue to use them benefit as a result. On the other hand, there may be increased congestion costs in the city centre, and congestion may develop on the bridge, as a result of the increase in the number of trips. Since the new route is shorter and quicker, one of the savings resulting from the provision of the bridge is lower fuel consumption. While the reduced fuel bill experienced by users of the bridge is included in the fall in trip price from P_0 to P_1, the benefits of improved air quality experienced by all citizens are not.

Cutting bus fares

We now consider the example of a proposal to cut bus fares, which places a different interpretation on the relationships illustrated in Figure 7.3. Suppose that the municipality supplies bus services as illustrated in Figure 7.4: there is a downward sloping demand curve for bus services, D, and the supply curve, S, is perfectly elastic, reflecting the assumption that any number of additional buses can be assigned to the routes at the current unit cost. Suppose that the municipality proposes to lower the bus fare in order to encourage more trips: it may be argued that this will reduce traffic congestion and pollution. Obviously a cost-benefit analysis of the proposal would need to value the latter effects, and valuing this sort of non-marketed output is the topic of Chapter 8. However for the present we will ignore these third-party effects and concentrate on the benefits and costs of the proposal to bus users and the municipality. Lowering the bus fare below unit cost is, in effect, a subsidy to bus users. As illustrated in Figure 7.4, the effect of the lower fare is to encourage additional trips ($Q_1 - Q_0$) as intended.

What are the efficiency net benefits of the proposal, ignoring the effect on congestion and pollution costs? The cost of the extra output of trips is given by area ACQ_1Q_0 in Figure 7.4: this area consists of Q_1Q_0 extra trips costing P_0 each. At the initial equilibrium, denoted by point A in Figure 7.4, an extra bus trip was valued by consumers at P_0. Since the bus fare was also P_0, consumers were not willing to undertake additional bus travel at

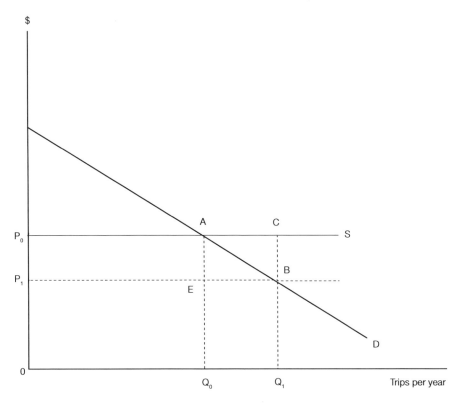

Figure 7.4 Subsidizing bus fares. D represents the market demand for trips and P_0 is the fare which generates revenue equal to the cost of providing the service. A subsidy allows the fare to be set at P_1 which generates $(Q_1 - Q_0)$ additional trips. The increase in consumer surplus is measured by area P_0ABP_1, and the amount of the subsidy payment is P_0CBP_1.

that price: the value of an extra trip, as given by a point on the demand curve, was less than the cost to them of that trip, given by P_0. Once the bus fare is reduced to P_1, consumers are willing to take additional trips which, while they are worth less than P_0, are worth at least the new fare, P_1. As the number of trips rises from Q_0 to Q_1, the gross value of an additional trip falls from P_0 to P_1. On average, the additional trips have a gross value of roughly $(P_0 + P_1)/2$ each. (This is an overestimate if the demand curve is convex to the origin, and an underestimate if it is concave, but we will ignore this complication.)

It follows from this analysis that the gross value to bus users of the extra trips is an amount represented by area ABQ_1Q_0 in Figure 7.4. Ignoring the effects of the proposal on groups other than bus users and the municipality, the net efficiency benefit of the proposal is given by the benefits, area ABQ_1Q_0, less the costs, area ACQ_1Q_0, which amounts to a negative net benefit measured by area ACB. This is the annual amount which would have to be set against the benefits to other groups in terms of reduced congestion and pollution to determine whether the proposal involved an efficient use of scarce resources.

We now turn to the Referent Group benefits. Again we will assume that we are concerned only with the effects on two groups: bus users and the municipality. We have already seen that bus users value the extra trips at approximately area ABQ_1Q_0 in Figure 7.4. At the new fare they pay a total amount measured by area EBQ_1Q_0 for the extra trips. Hence the net benefit consumers derive from the extra trips is measured by area ABE. However, this is not the total net benefit to consumers because they also experience a fall in the price of all trips they take, including the trips originally undertaken at the fare P_0. They now save $(P_0 - P_1)$ on each existing trip, with a total saving of P_0AEP_1. Thus, the net benefit to bus users is a saving of P_0AEP_1 on existing trips plus a net benefit of ABE on the generated trips, giving a total net benefit of P_0ABP_1, which, as we saw earlier, is the increase in consumer surplus generated by the price reduction. The effect on the municipality is a change in its profit position: in the absence of the fare reduction it is breaking even on its bus service operation, with total revenue equal to total cost measured by area P_0AQ_0O in Figure 7.4. With the fare reduction total cost rises by ACQ_1Q_0 and total revenue changes by $EBQ_1Q_0 - P_0AEP_1$. The net effect on the municipality is to generate a loss of $ACQ_1Q_0 - (EBQ_1Q_0 - P_0AEP_1)$, equivalent to area P_0CBP_1.

Since, by assumption in this example, the Referent Group – bus users and the municipality – consists of all those affected by the proposal (remember that we are ignoring pollution and congestion benefits in this discussion), the sum of Referent Group net benefits must equal the efficiency net benefit, a loss of ACB, and it can be verified that this is the case by reference to Figure 7.4. An important feature of the example is that the area P_0AEP_1 featured twice in the Referent Group Analysis, but played no part in the Efficiency Analysis. The reason for this is that the fare reduction on existing trips, unlike the travel cost reduction in the bridge example, is a pecuniary effect only: it is a transfer from the municipality to bus users in the sense that every dollar saved by consumers on the cost of the existing trips is lost revenue to the municipality. When these gains and losses are summed, together with the other effects of the proposal, over the members of the Referent Group to give the efficiency net benefit position, they cancel out. In other words, as noted above, pecuniary effects net out in an Efficiency Analysis – one group's gain is another group's loss.

If pecuniary effects net out, why do we bother recording them? The answer is that we are interested in the distributional effects of the project – who benefits and who gains – as well as the efficiency effects, and we will return to the issue of distribution of benefits and costs in Chapter 11. For example, suppose that in the above illustration the bus company was not part of the Referent Group, and was obliged for some reason to cut the bus fare below cost: subtracting its (negative) net benefit resulting from the fare cut from the efficiency net benefit would give an estimate of the net benefit of the fare reduction to consumers. Continuing to ignore the congestion and pollution costs, the net benefit to the Referent Group is then given by area P_0ABP_1 in Figure 7.4 since the Referent Group now consists only of bus users.

It can be concluded from the above examples that price changes can sometimes be ignored in the Efficiency Analysis, and sometimes not! They can be ignored if they represent pecuniary effects, also termed transfers, although pecuniary effects will resurface in the Referent Group Analysis. However, if price changes represent *real effects*, as in the discussion of Figure 7.2, they must also be taken account of in the Efficiency Analysis. A price change which represents a real effect is matched by an equivalent cost change: this is a case where one person's gain (loss) is not another person's loss (gain).

In summary, when undertaking an Efficiency Analysis, we must classify each price change predicted to result from the project as either a real or a pecuniary effect. If it is a pecuniary effect the predicted fall (rise) in price is matched by an equivalent rise (fall) in price somewhere else in the economy: one group's gain is another group's loss, and the effect of the price change is simply to transfer a benefit or cost from one group to another; when benefits and costs to all groups are summed in an Efficiency Analysis the pecuniary effects, or transfers, net out. If a price change is not identified as a pecuniary effect, then it is a real effect and it will have consequences for the aggregate net benefit outcome measured by an Efficiency Analysis.

7.5.2 Benefits of worker training

Governments sometimes offer training programs which enhance the skills of people on low incomes and help them to get better paying jobs. The effect of the training program on the local labour market is to shift out the supply curve of skilled labour by the number of graduates from the training program. If we assume that the skilled labour market clears so that there is no involuntary unemployment, the effect of the increased supply of skilled labour is to lower its market wage. This situation is illustrated in Figure 7.5: the initial equilibrium involved L_0 units of skilled labour employed at wage w_0; when the supply of labour shifts out from S_0 to S_1, as a result of the training program, employment rises from L_0 to L_1 and the market wage falls from w_0 to w_1. We can use Figure 7.5 to measure the benefits of the training program, which must be compared with the costs in assessing its economic

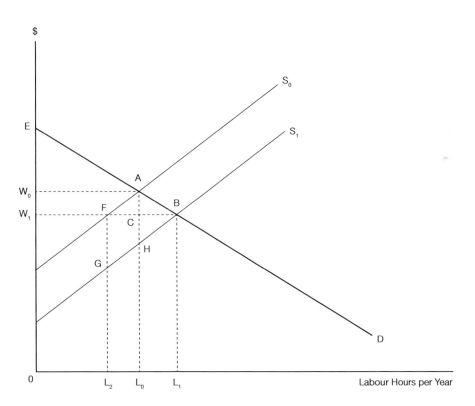

Figure 7.5 Effects of worker training. ED represents the market demand for skilled labour and S_0 represents the initial supply curve. A training program shifts the supply curve from S_0 to S_1 by graduating trainees able to supply FB hours per year. However, the resulting fall in wage from W_0 to W_1 results in some original suppliers in the market shifting to alternative occupations, so that the net increase in annual skilled employment is $(L_1\text{-}L_0)$ hours.

efficiency. In the following discussion we will ignore the costs of the training program and concentrate on the benefits.

One group of beneficiaries of the program consists of firms which are able to employ more labour at a lower wage. Firms save w_0ACw_1 on the wage bill for the L_0 workers they initially employed. They are also able to hire L_0L_1 extra workers at a wage w_1 which is lower than the value of the marginal product of these workers as measured by segment AB of the demand curve for labour. Area ABC measures the net benefit to firms of employing these additional workers: it is calculated by subtracting the cost of the extra labour, given by area CBL_1L_0, from the value of the resulting additional output, ABL_1L_0. The total benefit to firms is measured by the area of surplus w_0ABw_1 (to avoid confusion we have not termed this a consumer surplus even though the firm could be regarded as a consumer of labour services). The surplus will accrue to the firm as an increase in profit in the short run, though in the long run competition between firms might drive the market price of the product down, thereby transferring some of this benefit from the owners of firms to consumers.

We now consider why area ABL_1L_0 in Figure 7.5 measures the increase in value of output as a result of the increased labour input L_0L_1. The competitive firm's demand curve for labour, illustrated by the demand curve ED in Figure 7.5, is the value of the marginal product (VMP) of labour schedule. Given the firm's capital stocks – their buildings and machines, etc. – and the levels of their variable inputs – energy, materials, etc. – the VMP schedule shows what each successive unit of labour input will add to the value of output. This means that the total value of output is given by the area under the VMP schedule from the origin to the level of labour input demanded. Thus area $OEAL_0$ is the total value of the firm's output at labour input level L_0, and area $OEBL_1$ is the total value of output at input level L_1. The difference between these two values, ABL_1L_0 is the increase in the value of output as a result of the increase in labour input from L_0 to L_1.

When the market wage falls from w_0 to w_1, some of the workers originally employed in the skilled occupation leave to take other jobs; these are the workers with the skills and opportunities to earn at least a wage of w_1 in other occupations. In Figure 7.5 the number of workers leaving the occupation is given by L_2L_0. However, these departing workers are more than compensated for by the graduates of the training program who shift out the local supply curve of the skill by the amount L_2L_1, so that, as we saw above, there is a net increase in this type of employment of L_0L_1. Each worker who remains in the industry suffers a wage reduction from w_0 to w_1, while the workers who leave the industry find employment elsewhere at wages ranging from w_0 to w_1. The average loss to workers leaving the industry is $(w_0 - w_1)/2$. Thus the total net loss to the workers originally employed in the industry is measured by area w_0AFw_1, which represents a loss of *producer surplus*. Producer surplus is defined as an amount paid to suppliers of a quantity of a good or service in excess of the minimum amount required to induce them to supply that quantity.

The graduates of the training program find employment at wage w_1 which is higher than the value of their time in their previous situation, whether that was an unskilled job or unemployment. The value of their time, measuring the opportunity cost of the labour attracted to the training program, is given by the segment GB of the supply curve S_1: some workers have relatively low opportunity cost, measured by L_2G, whereas some have an opportunity cost close or equal to w_1. The net benefit to the graduates of the training program is given by area FBG: this area measures the income earned by the graduates in the new occupation, FBL_1L_2, less the value of their time in the absence of the training program, measured by GBL_1L_2. Area FBG represents a gain in producer surplus accruing to the trainees.

In summary, employers of labour benefit by an amount measured by the area of surplus w_0ABw_1 in Figure 7.5: of this amount, w_0AFw_1 represents a transfer from previously employed labour to employers, and consequently nets out of the efficiency cost-benefit calculation. The graduates of the program benefit by an amount measured by the area of producer surplus FBG. The net benefits of the program, which must be compared with the cost of the training program, are measured by area FABG: this value is calculated as the employer benefit (w_0ABw_1) less the loss suffered by previously employed skilled labour (w_0AFw_1) plus the net benefit to the trainees (FBG). While our main concern at present is with measuring the benefit of the program, it should be noted that the program costs would also include the opportunity cost of workers' time while training proceeds, as well as the costs of instructors, materials and facilities.

As we noted in Chapters 5 and 6, the sum of the net benefits of a project to each and all of the groups affected by it must equal the net efficiency benefit, which, in this case, is the net value of the additional output resulting from increased employment of skilled labour. In the above example, the value of the extra output produced in the industry employing the trainees is measured by area ABL_1L_0. A quantity of labour previously employed in the industry, measured by L_2L_0, leaves to seek higher wages elsewhere and raises the value of output elsewhere by an amount measured by FAL_0L_2. Since these workers are replaced by the trainees there is no corresponding fall in the value of output in the industry in question. Thus, the total increase in the value of output in the economy is measured by area $FABL_1L_2$. Against this increase must be set the opportunity cost of the activities which would have been undertaken by the trainees in the absence of training, measured by area GBL_1L_2. The net efficiency benefit (ignoring the cost of the project) is then measured by area FABG, which, as we determined above, is the sum of the benefits to each group in the economy affected by the project.

The foregoing discussion was based on the assumption that the labour market cleared. However, in practice, there may be some sort of wage regulation which prevents wages falling to market clearing levels, thereby resulting in involuntary unemployment in both the market for the skill provided by the training program, and in the alternative occupation for unskilled workers. In this situation there will be no change in the wage of labour, and the effect of the job training program may simply be an increase in the level of involuntary unemployment of this type of labour as would be illustrated by an outward shift of the labour supply curve in Figure 5.7 of Chapter 5.

7.5.3 *Producer benefits from an irrigation project*

We now turn to the example of a project which supplies additional irrigation water to farmers by raising the height of a dam. The opportunity cost of the resources used to raise the dam will not be considered in the discussion, which focuses on measuring the benefits of the project. Initially it will be assumed that the water authority sells irrigation water to farmers at whatever price the market will bear, and then subsequently it will be assumed that a regulated quantity of water is supplied to each farm at a regulated price. It will also be assumed, initially, that only two factors of production are involved in food production – land and water. Subsequently, labour will be introduced as a third factor. As a result of the increased availability of water more food is produced by the region's farmers. We will assume that the increase in the supply of food is small relative to the existing quantity supplied and that the market price of food is unaffected. Given this assumption, no additional consumer surplus is generated by the project, so that the example is concerned solely with producer surplus.

Figure 7.6 illustrates the supply and demand for water in the region of the irrigation project. As in the case of a consumer demand curve, the demand curve for an input shows, at

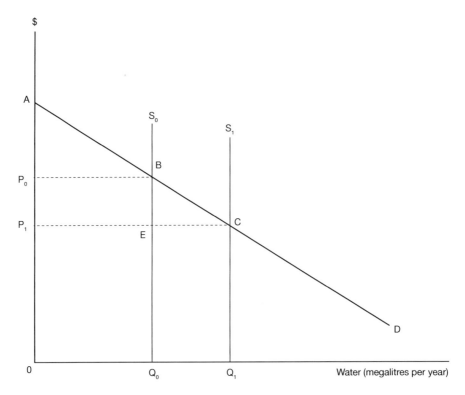

Figure 7.6 Effect of an irrigation project. AD represents the market demand curve for irrigation water, which is the horizontal sum of the value of the marginal product of water schedules of the individual farmers. The supply of water shifts from Q_0 to Q_1 and the market price falls from P_0 to P_1. The area P_0BCP_1 is a measure of the increase in annual producer surplus resulting from the fall in price.

each level of quantity demanded, how much buyers are willing to pay for an extra unit of the input. Assuming that the agricultural industry in the region is competitive in the product market, buyers of the input will be willing to pay the value of the marginal product of water for an extra unit of water; in other words, they are willing to pay a price equal to the value of the extra output that will result from using that extra unit of the input. The supply curve is drawn as perfectly inelastic reflecting the assumption that, given the variability of rainfall, a given infrastructure can only reliably supply a limited quantity of irrigation water. In addition to the current supply curve, S_0, the curve S_1 is also drawn to reflect the effect of raising the level of the dam on the quantity of water supplied. Figure 7.6 is concerned only with analysing the producer benefits of the proposed project, and, as noted above, the opportunity cost associated with the project will be ignored in the discussion.

As shown in Figure 7.6, the effect of increasing the quantity of water supplied is to lower the market price. The reason for this is summarised by the *law of diminishing marginal productivity*: when successive units of an input, such as water, are combined with a fixed quantity of the other input, such as land, then the increments in output resulting from the successive additional units of input (the marginal product of the input) become smaller and smaller (diminish). Since the value of the marginal product of water (the demand curve for water) consists of the price of output multiplied by the marginal physical product of water, the demand curve is downward sloping. In this analysis it is assumed that the price of output remains constant – the area affected by the proposed irrigation project is small relative to the total area involved in producing the product, and hence changes in the quantity of output it produces can have no influence on the market price of the product.

Since the market price of water is predicted to fall as a result of the irrigation project, the cost-benefit analyst has a choice of two prices, P_0 and P_1, with which to value the extra water produced by the project. As in the case of the transport examples, the initial price P_0 measures the value of the first extra unit of water to farmers, and the price P_1 measures the value of the last additional unit of water supplied by the irrigation project; an average price, $(P_0 + P_1)/2$, multiplied by the extra quantity of water, Q_1Q_0 in Figure 7.6, will generally closely approximate the total value of the extra water, and will provide an exact measure when the demand curve for water is linear as shown in the figure.

Recalling that the demand curve for water shows at each level of input the value of the extra output that can be obtained by adding one extra unit of input of water, holding constant the input of the other factor (land in this example), it is evident that the area under the demand curve for the input equals the value of total output of food from the relevant area of land. In other words, area ABQ_0O measures the value of food produced by the region when the quantity Q_0 of water is used, and area ACQ_1O measures the value when the quantity Q_1 is used. This means that the value of the extra food produced as a result of using the extra quantity of water Q_1Q_0 is measured by area BCQ_1Q_0 in Figure 7.6. This is the area obtained by multiplying the additional quantity of water by the average of the prices before and after the project. In other words, the gross benefit, in an Efficiency Analysis, of raising the dam is the value of the extra food produced.

Let us now consider the Referent Group benefits in this example. There are three groups that might be affected: consumers of food, the water authority and landowners in the region. Since the price of food is assumed not to change as a result of the project, consumers receive no net benefit: they pay exactly what the extra food is worth to them. The water authority benefits by an amount measured by area $ECQ_1Q_0 – P_0BEP_1$, which is the change in water revenues; this amount could be positive or negative depending on the elasticity of the demand curve over the range BC (remember that in this discussion of project benefits we are ignoring the opportunity cost of raising the dam which would presumably be borne by the water authority). To analyse the benefit to landowners we need the help of Figure 7.7 which illustrates the value of the marginal product schedule of land – the demand curve for land – together with the supply of land in the region, which is fixed at quantity Q. The demand curve for land shifts out from D_0 to D_1 as a result of the increased availability of water: potential farmers are willing to pay more rent per hectare the lower is the price of irrigation water. The market rental value of the land rises from R_0 per hectare to R_1 as a result of the demand shift, and the annual return to landowners rises by R_1FGR_0.

As we found in the case of the training program, discussed in Section 7.5.2 above, the sum of the net benefits to all parties affected by the project must equal the aggregate net benefit calculated by the Efficiency Analysis. We have seen in the case of the irrigation project that this net benefit is measured by BCQ_1Q_0 in Figure 7.6. Assuming that the Referent Group consists of consumers (who receive a zero net benefit because the price of food is assumed not to change), the water authority and landowners, the aggregate net benefit of the project, ignoring the cost of raising the dam, is given by area $ECQ_1Q_0 – P_0BEP_1$ (in Figure 7.6), representing the change in water revenues, plus R_1FGR_0 (in Figure 7.7) representing the increase in land rent, and together these two values should equal the benefit, in terms of the

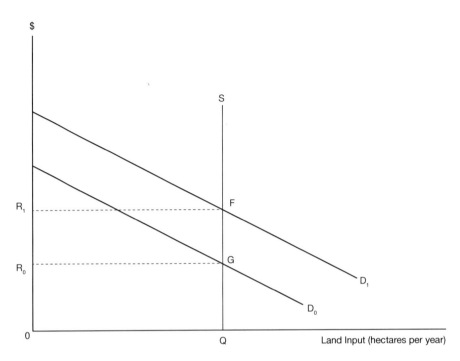

Figure 7.7 Change in the rental value of land. The effect of increased availability of irrigation water is to shift out the value of the marginal product schedule of land from D_0 to D_1. The supply of land in the region, SQ, is perfectly inelastic, and the effect of the demand shift is to raise the annual rental value from R_0 to R_1 per hectare.

value of extra food production, already calculated by the Efficiency Analysis, and we now proceed to demonstrate that this is the case.

Recall that area ACQ_1O in Figure 7.6 measures the total value of food production in the region once the additional quantity of water is available. Note also that area P_1CQ_1O measures water authority revenues once the dam is raised. A simple rule in economics is that the value of output equals the value of the incomes generated in producing the output: in aggregate this rule is reflected in the equality of the Gross National Expenditure (GNE) and Gross National Income (GNI) measures of Gross National Product (GNP); at a micro level the rule reflects the fact that the revenues earned by each firm are paid out as income to suppliers of inputs and to the owners of the firm. This means that area ACP_1 in Figure 7.6 must represent the income earned by landowners after the supply of water has increased; it represents the part of the revenue from food production that does not accrue to the water authority, and must therefore accrue to the owner of the other factor of production, land. Since area ABP_0 (in Figure 7.6) measures landowner income before the increase in availability of water, and area ACP_1 landowner income after the increase in the supply of water, the area P_0BCP_1, in Figure 7.6, must represent the increase in income to landowners, which is also measured as R_1FGR_0 in Figure 7.7. If we now substitute P_0BCP_1 for R_1FGR_0 in the expression for the aggregate net benefits to the Referent Group derived in the previous paragraph, we find that it reduces to area BCQ_1Q_0 in Figure 7.6, which was the net benefit estimate obtained from the Efficiency Analysis.

In the above example a competitive market price was charged for water. Suppose that water is allocated to farms on the basis of some kind of formula, and that a nominal charge, lower than the competitive market price, is levied. Since the observed price of water no longer equals the value of its marginal product, the water price cannot be used to calculate the value of the extra food produced as a result of the extra water. Indeed, when water is allocated by some formula, this usually means that the water levy or price is lower than the price of water in a competitive market. As an aside, it can be noted that if the allocation formula produces a different distribution of water among farms than would be the case under the competitive market outcome, the increase in the value of food production will be less than in the competitive outcome. In Figure 7.8 the supply of water is shown as increasing from Q_0 to Q_1, as a result of raising the level of the dam, but the extra water is sold at price P, yielding revenue of KLQ_1Q_0 to the water authority. Recalling that the area under the demand curve, or value of the marginal product schedule of water, measures the value of the output of food, it can be seen that the value of the extra food is measured by MNQ_1Q_0, which is the project benefit in an Efficiency Analysis.

How is the analyst to measure project benefit in the absence of a competitive market price for water? Since area MNQ_1Q_0 in Figure 7.8 represents the value of additional output, there must be increases in income summing to an equivalent amount. The water authority's

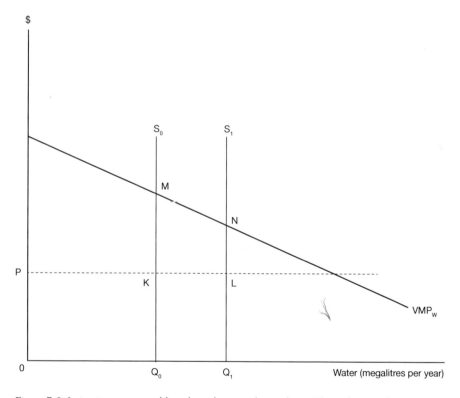

Figure 7.8 Irrigation water sold at less than market value. When the supply or irrigation water increases from Q_0 to Q_1, the price remains at P because water is allocated by a rationing system. The value of extra output produced is measured by MNQ_1Q_0, the increase in area under the value of the marginal product of water schedule resulting from the increase in water supply.

income rises by KLQ_1Q_0 so the income of the other factor of production, land, must rise by MNLK. In other words, as compared with the example in which the market price was charged for water, charging a lower price simply further increases land rent. This makes intuitive sense: the cheaper the water can be obtained, the more the land is worth. This suggests that there are two *alternative* ways of measuring the project benefit: first, through studies of agricultural production, relating physical inputs to output, we can predict the effect of additional inputs of water on the annual output and value of the crop; second, through studies of land values, relating land value to factors such as climate, soil and the availability of irrigation water, we can predict the effect of additional water on the capital value of land, and hence on the annual rental value; the sum of the increase in the annual rental value of land plus any change in revenues of the water authority is an alternative measure of the annual value of the increased output of food. It should be stressed that these two benefit measures are *alternatives*, and to add them together, as some analysts have mistakenly done in the past, would result in *double-counting* of project benefits.

Now suppose that other factors of production besides land and water are involved in food production. We can add as many factors as we like, but for simplicity assume that food is produced by combining land, labour and irrigation water. Just as an increase in the availability of water increases the demand for land, so it also increases the demand for labour. For the moment, assume that the extra workers are drawn from the labour force employed by farms in nearby regions, but that the impact of the project is not sufficiently large as to cause an increase in the wage. As before, the value of the extra output of food is equal to the sum of the values of the extra incomes to land, labour and the water authority. While there is no offsetting fall in production elsewhere as a result of the increased availability of water (ignoring the opportunity cost of the inputs used to raise the height of the dam) or the more intensive use of land, diverting labour from employment in other regions does have an opportunity cost in terms of forgone output. Assuming a competitive and undistorted labour market, the wage measures the opportunity cost of labour, and the extra wage bill must be subtracted from the increase in incomes generated by the project (measuring the value of the additional output of food in the region) in order to calculate the project benefit in an Efficiency Analysis.

The Referent Group is now expanded to include labour as well as food consumers, landowners and the water authority, but, similar to food consumers, labour receives no net benefit since it continues to supply labour at the same market wage as before. Hence the aggregate benefits of the Referent Group, consisting of the additional incomes to landowners and the water authority, sum to equal the project benefit in the Efficiency Analysis.

Other assumptions could be made about the labour market. For example, suppose that the extra labour would otherwise be unemployed. In the Efficiency Analysis we would then cost additional labour at its shadow-price as discussed in Chapter 5. Since the shadow-price is lower than the wage this will increase the estimate of the project benefit. This increase will be matched by an increase in Referent Group benefits to reflect the fact that the extra labour is receiving an employment benefit in the form of a wage higher than that required to induce it to work in the region.

7.6 Accounting for input price changes

We now consider a case in which undertaking the project causes a change in an input price. This case is illustrated by a rural labour market which serves farms which benefit from an increased supply of irrigation water, as discussed above.

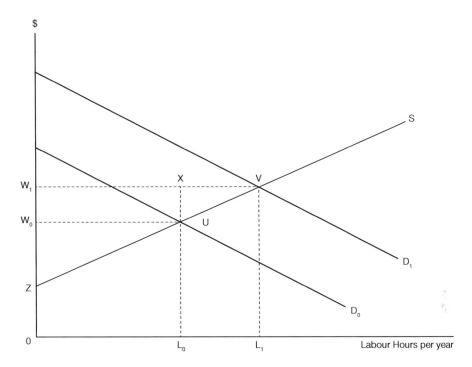

Figure 7.9 Effect of an increase in demand for labour. The demand curve for labour is the Value of the Marginal Product Schedule. Provision of additional irrigation water shifts the demand from D_0 to D_1. The supply curve of labour is denoted by S. The increase in market demand results in a wage increase from W_0 to W_1.

Suppose that the labour market is competitive and undistorted, with no unemployment, and that the extra demand for labour pushes up the market wage as illustrated in Figure 7.9. The first additional unit of labour hired has an opportunity cost of W_0, as shown in Figure 7.9, and the last extra unit hired has an opportunity cost of W_1. Multiplying the additional labour by an average of the two wages, $(W_0 + W_1)/2$, gives a rough measure of the opportunity cost of the total amount of extra labour (an exact measure if the supply curve of labour is linear as shown in Figure 7.9). This amount, given by area UVL_1L_0 in Figure 7.9, must be subtracted from the estimate of the value of the extra food to calculate project benefit in an Efficiency Analysis. Recall that the value of the extra food produced as a result of the irrigation project can be measured by the sum of the increases in income to the owners of the three factors of production involved – land, labour and water. To measure the efficiency gross benefit (i.e. the value of the extra food produced by the economy as a whole), we subtract from that sum the area UVL_1L_0 which represents the opportunity cost of the extra labour employed in the irrigation district. Labour still benefits by an amount measured by the increase in producer surplus W_1VUW_0 in Figure 7.9, and since this is a transfer from landowners the increase in income of the latter as a result of the project will be correspondingly lower.

When we turn to the Referent Group Analysis, we find that total wages paid to labour in the region have risen by the amount $W_1VUW_0 + UVL_1L_0$ as illustrated in Figure 7.9. However, the net benefit to labour is measured by area W_1VUW_0 since area UVL_1L_0

represents the value of time in alternative employment or some other activity. The net benefit to labour measured by area W_1VUW_0 is a pecuniary benefit since the benefit of the higher wage received by labour is offset by the cost of the higher wage paid by farms. In other words, as noted above, the area W_1VUW_0 is a transfer from employers in the region affected by the project to the regional labour force. Thus, the amount represented by area W_1VUW_0 would appear as a benefit to labour and a cost to landowners in the Referent Group Analysis and, in this example, would net out in both the aggregate Referent Group and Efficiency Analysis. If, on the other hand, the employers were not members of the Referent Group, the benefit to labour would not be offset by a cost to employers in the Referent Group Analysis. However, in the Efficiency Analysis pecuniary effects always net out. We will return to this issue in Section 7.8, and Appendix A7.1 illustrates the effect on the cost-benefit analysis of the ICP project of allowing the wage of skilled labour to rise as a result of the increased demand for labour caused by the project.

The same argument as above would apply to the measurement of project cost if raising the dam involved a rise in the market wage as a result of the increased demand for labour. The opportunity cost of the labour, assuming it would be employed elsewhere, is approximated by the wage bill computed at an average of the initial and new equilibrium wage rates.

7.7 Price changes in other markets

Suppose that undertaking a project results in a lower price for a good or service. It could be expected that the demand for a substitute (complementary) product would fall (rise). If this substitute or complementary product is in perfectly elastic supply, there will be no change in its price. However, since the demand curve for the product has shifted in (out), the area of consumer surplus measured under the demand curve *appears* to have fallen (risen). Should the apparent change in consumer surplus be measured and included in the analysis as a cost (benefit)? The answer is *no* because the shift in demand in the market for the substitute or complementary good is simply a reflection of consumers rearranging their expenditures to take advantage of the lower price of the good supplied by the project. The benefit of that lower price is fully measured by the consumer surplus change measured in the market for the project's output. A simple way to remember this point is to note that the benefit (cost) of a price change is measured by an area under a demand curve delineated by the original and the new price. If there is no price change, as in the case of a substitute or complementary good in elastic supply as considered above, then there is no benefit or cost to be measured in that market. The same argument applies in the case in which the project results in a *higher* market price of the good or service it produces.

Suppose now that when the demand for a substitute (complementary) good falls (rises) there is a change in its market price because of an upward sloping supply curve for the substitute (complementary) good. Now we have a benefit (cost) to consumers of that good measured as the area under the new demand curve delineated by the difference between the original and new prices. While this area is a benefit (cost) to consumers, it is also a cost (benefit) to producers and nets out in the Efficiency Analysis. In other words, where price changes occur in other markets as a result of a project they may simply reflect transfers to (from) consumers from (to) producers.

To illustrate the latter point, suppose that a bridge is to be built across a river that is already served by a ferry. Figure 7.10(a) shows the demand for bridge crossings, and Figure 7.10(b) shows the reduction in the demand for ferry crossings, from D_{F0} to D_{F1}, resulting from the provision of the bridge. Because the supply curve for ferry trips is upward sloping the price

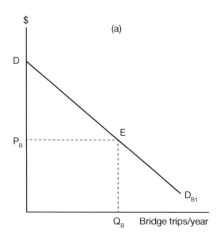
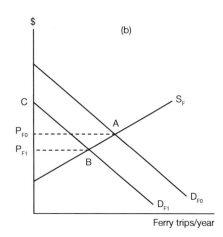

Figure 7.10 Effects of building a bridge on the benefits from a ferry. In Figure 7.10(a) DD_{B1} represents the demand for bridge trips, conditional upon the ferry remaining available. P_B is the price of a trip and Q_B is the annual number of crossings. In Figure 7.10(b) D_{F0} is the demand for ferry trips when the bridge is not available, and D_{F1} is the demand once the bridge is operating. The fall in demand from D_{F0} to D_{F1} results in a fall in the price of ferry trips from P_{F0} to P_{F1}.

of a ferry trip falls as a result of the reduction in demand for the services of the ferry. The area P_{F0} AB P_{F1} in Figure 7.10(b) represents an annual benefit to ferry users in the form of a lower price, but a cost to the ferry operator in the form of a loss in producer surplus. Since it represents a transfer, it can be ignored in the Efficiency Analysis. Area DEP_B in Figure 7.10(a) measures the annual benefit to bridge users, net of the unit cost, P_B, of a bridge trip, given that the alternative of the ferry is available at price P_{F1}.

Now suppose that, once the bridge becomes available, repair work is to be carried out on the ferry, putting it out of service for a year. In a cost-benefit analysis this is equivalent to raising the price of a ferry trip to a level which chokes off all demand. The annual cost to ferry users, as a result of its withdrawal from service, is measured by the area under the new demand curve for the ferry, D_{F1}, between the original price, P_{F1}, and the new infinitely high "price"; in other words, the cost is the entire amount of consumer surplus, CBP_{F1}, generated annually by the ferry. Clearly this cost will be smaller than it would have been if the bridge had not been constructed, because the demand for ferry crossings has decreased. When the ferry closes, demand for the bridge will increase, but in the absence of any cost change for bridge crossings (such as might result from increased congestion, for example), there is no change in net benefit to be measured in that market.

Consider one final extension to the analysis. The price of a bridge trip is a composite variable incorporating a variety of costs such as travel time and vehicle operating costs. Suppose that as the volume of traffic using the bridge increases, the cost of travel time rises because of increased congestion. This effect could be represented by an upward sloping supply curve of bridge trips. An increase in demand for bridge trips (not shown in Figure 7.10(a)), as a result of the closure of the ferry, would result in increased congestion and a higher price. This would result in a loss to bridge users which would be measured by an area similar to area W_1VUW_0 in Figure 7.9. In that example the loss in consumer surplus suffered by farms was offset by an equivalent gain in producer surplus accruing to suppliers of labour, and the area netted out in the Efficiency Analysis. However in the present example the loss in consumer surplus due to increased congestion is a real, as opposed to a pecuniary effect. There is no offsetting gain to suppliers and hence the increased congestion cost would be deducted from the measure of annual consumer benefit from the bridge in the Efficiency Analysis. Thus, the cost of withdrawing the ferry from service, in this extension to the analysis, would consist of the loss of consumer surplus in the ferry market plus the cost of increased congestion on the bridge.

7.8 Classification of consumer and producer surplus changes

A four-way classification of project net benefits was presented in Figure 6.1: net benefits were either measured by market prices (Areas A + B) or not (Areas C + D); and they accrued either to the Referent Group (Areas A + C) or to the non-Referent Group (Areas B + D). In this section we consider where changes in the amount of consumer or producer surplus resulting from undertaking a project fit into this classification. First, we consider changes in consumer and producer surplus resulting from changes in market prices, as illustrated, for example, in Figures 5.2 and 7.9, and then subsequently we consider changes in surplus associated with non-marketed goods or services, as illustrated by Figure 7.1.

Suppose that a project supplies a sufficient amount of output so as to cause a fall in the market price of the output good or service, or, as considered in Appendix 1 to this chapter, hires a sufficient amount of an input to cause a rise in its market price. The changes in price affect all units of the good or service traded, including the quantity traded *without* the project (such as Q_0 in Figure 5.2) and the additional quantity traded *with* the project (such as Q_1Q_0), and surplus is associated with each of these quantities. We start by considering the surplus generated by the additional quantity traded as a result of the project, and will subsequently consider the surplus associated with the original quantity traded.

Since the Market Analysis of the project, the results of which are summarised by Area A + B in Figure 6.1, is conducted at the market prices which will exist in the world *with* the project, additional output resulting from the project is priced at the new lower price, and additional input supplied to meet the demands of the project is priced at the new higher price. This means that the consumer or producer surplus accruing to consumers of the project's output or suppliers of project inputs is not included in the project net benefit as measured by market prices.

In the Efficiency Analysis, in contrast, output is valued at, and input costed at, an average of the *without* and *with* prices, which will be higher than the *with* price of output and lower than the *with* price of an input in the case of the price changes considered here. As compared with the Market Analysis, the higher price of output (lower price of input) used in the Efficiency Analysis reflects the benefit in the form of consumer (producer) surplus generated by the project's output or input; we can think of the pricing procedure in the Efficiency Analysis as "making room" in the Referent Group Analysis for the benefit of the additional surplus. This benefit can be entered under the appropriate category, such as consumer or labour benefits, in Area C + D of Figure 6.1. Since, as noted in Section 7.1, changes in output or input prices induced by a project are likely to be localised, it could be expected that benefits in the form of increased consumer or producer surplus will usually accrue to members of the Referent Group, and consequently belong in Area C.

The Efficiency Analysis tells us whether the project is a more productive use of the inputs involved than their alternative uses in the absence of the project. It should not come as a surprise, therefore, to learn that there is a close relationship between the methods of the

Efficiency Analysis and those of productivity analysis in general. Productivity analysis is often used to determine whether and to what extent resources are being used more productively as time passes. There is an expectation that technical change should, over time, lead to higher levels of outputs from given levels of inputs. Unlike the cost-benefit analysis this is a *before-and-after* comparison rather than the *with-and without* approach of project appraisal. In the Efficiency Analysis project outputs and inputs are valued at an average of their estimated prices with and without the project. The net value of the project under this approach corresponds to the Bennet Indicator of technical change, as described by Diewert and Fox (2017), which compares outputs and inputs valued at an average of the *before-and-after* prices in a similar way to the *with-and-without* comparison conducted in the Efficiency Analysis. In both cases a positive value indicates a more productive use of inputs.

We now turn to consumer and producer surplus changes associated with the quantity of the output or input traded in the world *without* the project. As noted above, a fall in output price will benefit the consumers of the original quantity of the good traded, but it will also be to the detriment of the firms supplying the good. Similarly, a rise in input price, such as an increase in the wage, will benefit the suppliers of the original quantity of the input, but will similarly be to the detriment of the firms employing the input and may result in consumers paying higher prices for output These effects are pecuniary effects, as discussed in Section 7.2, and they net out of the Efficiency Analysis. If they are included in the Referent Group Analysis, it is likely to be in the category described by Area C in Table 6.1 (net Referent Group benefits not measured by market prices), given the localised nature of the project's effects as discussed above. Whether they are relevant to the decision about the project is a matter for the decision-maker, who may not be interested in changes in the distribution of surplus among consumers, labour and firms, especially given that most households represent more than one of these categories of economic agents, but if they *are* included, and these agents are all members of the Referent Group, they are entered twice in the Referent Group Analysis, once as a benefit and once as a cost, so that they cancel out in the measure of overall project effects, thereby preserving the adding-up property of the cost-benefit model.

Suppose now that the output of a project, such as an improvement to a public road leading to a national park, is not marketed. As we saw from Figure 7.1, the project benefit takes the form of an increase in consumer surplus accruing to users of the park. As in the case of changes in consumer and producer surplus induced by changes in market prices and associated with project output or inputs, this form of net benefit is not measured by the Market Analysis. In the case of improved access to the national park, the change in surplus is included in area C or D in Table 6.1, depending on whether or not the consumers are members of the Referent Group. In contrast to the case of a marketed output, surplus accruing to consumers of the *without* quantity of the service (Area P_0ABP_1 in Figure 7.1) is included as a net benefit in the Efficiency Analysis, along with the surplus associated with the generated traffic (area ABC). In this case there is no corresponding cost to offset the benefit to existing users (the cost of the road improvement is treated separately in the analysis) as in the case of a pecuniary externality.

7.9 Further reading

The discussion of compensation in this chapter has been simplified to bring out the essential points. A more technical summary is available in H. Mohring, "Alternative Welfare Gain and Loss Measures", *Western Economic Journal*, 9 (1971): 349–368.

A useful book on welfare economics which includes a diagrammatic exposition of the concepts of compensating and equivalent variation is D.M. Winch, *Analytical Welfare Economics* (Harmondsworth: Penguin Books, 1971).

A paper by R.D. Willig, "Consumer Surplus without Apology", *American Economic Review*, 66 (*1976*): 589–597, discusses the significance of income effects in measuring compensating variation in cost-benefit analysis.

D. Pearce, G. Atkinson and S. Mourato, *Cost-Benefit Analysis and the Environment* (Paris: OECD, 2006) discuss divergences between WTP and WTA measures, as well as many of the issues to be covered in Chapter 8.

D. Kahneman, *Thinking Fast and Slow* (New York: Farrar, Straus and Giroux, 2013).

E. Diewert and K. Fox, *The Difference Approach to Productivity Measurement and Exact Indicators* (University of British Columbia, Vancouver School of Economics, Discussion Paper 17-05, 2017).

Appendix 1 to Chapter 7

Allowing for an increase in the skilled wage in the ICP case study

We now consider how to deal with price changes induced by the project in the spreadsheet framework. As noted earlier, the summary analysis of the ICP project reported in the Appendix to Chapter 6 serves as a basis for comparison with a series of spreadsheet analyses in subsequent chapters, each amended to include an additional feature of cost-benefit analysis. The following discussion concerns how to deal with an anticipated increase in an input price as a result of a project's implementation.

As discussed at the beginning of Chapter 7, undertaking a project is unlikely to cause changes in the prices of tradeable inputs or outputs. However, an example of a non-tradeable input whose price might be affected is skilled labour. Economies often experience skill shortages in particular areas of expertise which cannot be filled quickly because of the time required for training and the absence of alternative sources of supply resulting from migration restrictions.

Suppose that the International Cloth Products (ICP) project is predicted to increase the demand for skilled labour by an amount illustrated by L_0L_1 and to drive up the wage of skilled labour from W_0 to W_1 as shown in Figure 7.9. How would this predicted effect be accounted for in the cost-benefit analysis? In the Market and Private Analyses, labour would be priced at the wage paid by ICP, which is now W_1. As compared with the situation in which the wage remains at W_0, the net benefits measured by the Market and Private Analyses of the project decline by $(W_1 - W_0)$ per unit of skilled labour employed (converted to present value terms) as a consequence of the increase in the wage.

In the Efficiency Analysis, labour should be priced at its opportunity cost; the first unit of labour drawn into the project has an opportunity cost of W_0, while the last unit has an opportunity cost of W_1. As suggested in Section 2 of Chapter 5, an average value for the opportunity cost of labour of $(W_0 + W_1)/2$ can be used in the Efficiency Analysis. As compared with the situation in which the wage remains at W_0, the annual efficiency net benefit falls by $(W_1 - W_0)/2$ per unit of skilled labour employed.

In the particular circumstances of the ICP case study, the aggregate Referent Group net benefit is obtained by subtracting the private net benefit to ICP, together with the net cash flow of the foreign financial institution, from the efficiency net benefit stream. Since the private net benefit has fallen by $(W_1 - W_0)$ per unit of skilled labour employed in the project, and the efficiency net benefit has fallen by $(W_1 - W_0)/2$ per unit, the aggregate Referent Group net benefit has risen by $(W_1 - W_0)/2$ per unit. In terms of the disaggregated Referent Group net benefits, the higher skilled wage generates producer surplus for suppliers of skilled labour used in the project. This producer surplus can be measured by an area such as UVX in Figure 7.9. This area can be approximated by $(W_1 - W_0)/2$ multiplied by the amount of labour employed in the project. In other words the additional benefit to the skilled labour employed resulting from the wage increase caused by the project is measured by $(W_1 - W_0)/2$ per unit. When this value is entered in the disaggregated Referent Group benefit section, the adding-up requirement for consistency of the cost-benefit analysis is satisfied.

If undertaking the project drives up the skilled wage from W_0 to W_1, skilled labour employed elsewhere in the economy will also benefit by $(W_1 - W_0)$ per unit. However, employers of that labour will experience a cost of $(W_1 - W_0)$ per unit. In other words the increase in the skilled wage results in a transfer of benefit, measured by W_1XUW_0 in Figure 7.9, from employers to workers. As explained in Chapter 5, transfers net out and can be ignored in the Efficiency Analysis. However, the transfer could be relevant in the Referent Group Analysis if skilled labour was mainly employed by foreign firms, in which case it would be entered as a benefit in Area C and a cost in Area D of Figure 6.1.

Figure A7.1 shows the ICP case study spreadsheet of Chapter 6 revised to take account of a project-induced rise in the skilled wage rate. Referring to Table 1 in Figure A7.1, the wage rates of supervisors and technicians and other skilled workers are assumed to increase by 20% because of the increased labour demand resulting from the project: the annual salary of individuals working as supervisors and technicians is assumed to increase from 150,000 Bt. p.a. to 180,000 Bt. p.a., and the annual salary of other skilled workers from 22,500 Bt. p.a. to 27,000 Bt. p.a. These higher values are entered as the New Wages in Table 1 of Figure A7.1, and result in an increased flow of labour benefits, reported in Table 5 of Figure A7.1, as compared with those reported in Figure A6.2.

By comparison with Figure A6.2 it can be seen that introducing this assumption has a number of effects besides the increase in benefits to labour. The higher labour costs imply that both the Market and Private NPVs and IRRs are lower; the Market IRR falls from 20.6% to 18.3% and the private IRR falls from 16.7% to 14.3%. Similarly, the efficiency IRR falls from 23.4% to 20.8% due to the higher opportunity cost of skilled labour. Total Referent Group benefits are also lower. The reason for this is that the government's tax revenues from ICP decline as a result of ICP's profits being lower. In other words, the higher opportunity cost of skilled labour reduces the net benefit to both the non-Referent Group private investor and to the government. As a consistency check to verify that the analyst has not introduced an error in making these adjustments, the aggregate Referent Group benefits at the top of Table 5 in Figure A7.1 (derived by subtracting ICP's private cash flow and the overseas lenders' cash flow from the efficiency cash flow) should be the same as those on the bottom line (derived by aggregating the sub-Referent Group members' net cash flows).

ICP PROJECT: 1999–2009

TABLE 1 KEY INPUT VARIABLES

VARIABLES		INVESTMENT COSTS	000 $'	000 BT'	RUNNING COSTS		MATERIAL FLOWS		FINANCE	
EXCHANGE RATE	44	SPINDLES	4000	176000	WATER BT PER MN LBS	480000	COTTON MN LBS	16.2	SUPPLIERS CREDIT	75%
TARIFF A	5%	EQUIPMENT	850	37400	POWER BT PER MN LBS	480000	YARN MN LBS	15	SUPPLIERS CREDIT	160050
TARIFF B	10%	INSTALLATION	100	4400	SPARE PARTS (%)	10%	LABOUR 1	100	OVERDRAFT	60
PROFITS TAX	50%	CONSTRUCTION		36000	INSURANCE & RENT BT MN	6	LABOUR 2	1000	CREDIT REPAY YEARS	5
DEPRECIATION	10%	START-UP COSTS		18000	OVERHEADS	0	LABOUR 3	500		
PRICE YARN CIF US$	0.58								DISCOUNT RATES	
YARN TARIFF BT	2.1	WORKING CAPITAL								4%
HANDLING BT	0.3	RAW MATERIALS	1200	52800						8%
TARIFF C	10%	SPARE PARTS	243	10692						12%
PRICE COTTON CIF US$	0.3									16%
WAGE 1 BT PA	150000	SHADOW PRICES			NEW WAGE 1 BT PA	180000			CONVERSIONS	
WAGE 2 BT PA	22500	CONSTRUCTION	50%		NEW WAGE 2 BT PA	27000			% CAPACITY OUTPUT	50%
WAGE 3 BT PA	12000	POWER	50%						THOUSANDS	1000
INTEREST RATE 1	8%	INSURANCE & RENT	0%							
INTEREST RATE 2	10%	WAGE 3	50%							

TABLE 2 MARKET CALCULATION

YEAR	1999 MN BT	2000 MN BT	2001 MN BT	2002 MN BT	2003 MN BT	2004 MN BT	2005 MN BT	2006 MN BT	2007 MN BT	2008 MN BT	2009 MN BT
INVESTMENT COSTS											
SPINDLES	−184.800										
EQUIPMENT	−39.270										
INSTALLATION	−4.400										
CONSTRUCTION	−36.000										
START-UP COSTS	−18.000										
RAW MATERIALS		−58.080									58.080
SPARE PARTS		−11.227									11.227
TOTAL	−282.470	−69.307	0.000	0.000	0.000	0.000	0.000	0.000	0.000	0.000	69.307
RUNNING COSTS											
WATER		−3.600	−7.200	−7.200	−7.200	−7.200	−7.200	−7.200	−7.200	−7.200	−7.200
POWER		−3.600	−7.200	−7.200	−7.200	−7.200	−7.200	−7.200	−7.200	−7.200	−7.200
SPARE PARTS		−22.407	−22.407	−22.407	−22.407	−22.407	−22.407	−22.407	−22.407	−22.407	−22.407
RAW MATERIALS		−117.612	−235.224	−235.224	−235.224	−235.224	−235.224	−235.224	−235.224	−235.224	−235.224
INSURANCE		−6.000	−6.000	−6.000	−6.000	−6.000	−6.000	−6.000	−6.000	−6.000	−6.000
LABOUR		−51.000	−51.000	−51.000	−51.000	−51.000	−51.000	−51.000	−51.000	−51.000	−51
TOTAL		−204.219	−329.031	−329.031	−329.031	−329.031	−329.031	−329.031	−329.031	−329.031	−329.031
REVENUES											
SALE OF YARN		209.400	418.800	418.800	418.800	418.800	418.800	418.800	418.800	418.800	418.800
NCF (Project)	−282.470	−64.126	89.769	89.769	89.769	89.769	89.769	89.769	89.769	89.769	159.076

TABLE 3 PRIVATE CALCULATION

YEAR	1999 MN BT	2000 MN BT	2001 MN BT	2002 MN BT	2003 MN BT	2004 MN BT	2005 MN BT	2006 MN BT	2007 MN BT	2008 MN BT	2009 MN BT
FINANCIAL FLOWS											
DEPRECIATION		−26.447	−26.447	−26.447	−26.447	−26.447	−26.447	−26.447	−26.447	−26.447	−26.447
INTEREST											
SUPPLIERS CREDIT	160.050	−12.804	−10.621	−8.264	−5.719	−2.969					
REPAYMENT		−27.282	−29.464	−31.821	−34.367	−37.116					
BANK OVERDRAFT	60.000	−6.000	−6.000	−6.000	−6.000	−6.000	−6.000	−6.000	−6.000	−6.000	−6.000
REPAYMENT											−60.000
NET OPERATING REVENUE		5.181	89.769	89.769	89.769	89.769	89.769	89.769	89.769	89.769	89.769
TAXABLE PROFIT		−34.070	46.701	49.058	51.603	54.353	57.322	57.322	57.322	57.322	57.322
PROFIT TAX		17.035	−23.350	−24.529	−25.802	−27.176	−28.661	−28.661	−28.661	−28.661	−28.661
FLOW OF OWN FUNDS	−122.420	−27.176	20.333	19.155	17.882	16.507	55.108	55.108	55.108	55.108	64.415

TABLE 4 EFFICIENCY CALCULATION

YEAR	1999 MN BT	2000 MN BT	2001 MN BT	2002 MN BT	2003 MN BT	2004 MN BT	2005 MN BT	2006 MN BT	2007 MN BT	2008 MN BT	2009 MN BT
INVESTMENT COSTS											
SPINDLES	−176.000										
EQUIPMENT	−37.400										
INSTALLATION	−4.400										
CONSTRUCTION	−18.000										
START-UP COSTS	−18.000										
RAW MATERIALS		−52.800									52.800
SPARE PARTS		−10.692									10.692
TOTAL	−253.800	−63.492		0.000	0.000	0.000	0.000	0.000	0.000	0.000	63.492
RUNNING COSTS											
WATER		−3.600		−7.200	−7.200	−7.200	−7.200	−7.200	−7.200	−7.200	−7.200
POWER		−1.800		−3.600	−3.600	−3.600	−3.600	−3.600	−3.600	−3.600	−3.600
SPARE PARTS		−21.340		−21.340	−21.340	−21.340	−21.340	−21.340	−21.340	−21.340	−21.340
RAW MATERIALS		−106.920		−213.840	−213.840	−213.840	−213.840	−213.840	−213.840	−213.840	−213.840
INSURANCE & RENT			0.000	0.000	0.000	0.000	0.000	0.000	0.000	0.000	0.000
LABOUR		−44.250		−44.250	−44.250	−44.250	−44.250	−44.250	−44.250	−44.250	−44.250
TOTAL		−177.910		−290.230	−290.230	−290.230	−290.230	−290.230	−290.230	−290.230	−290.230
REVENUES											
SALE OF YARN		191.400		382.800	382.800	382.800	382.800	382.800	382.800	382.800	382.800
NET EFFICIENCY BENEFIT	−253.800	−50.002		92.570	92.570	92.570	92.570	92.570	92.570	92.570	156.062

TABLE 5 REFERENT GROUP CALCULATION

	1999 MN BT	2000 MN BT	2001 MN BT	2002 MN BT	2003 MN BT	2004 MN BT	2005 MN BT	2006 MN BT	2007 MN BT	2008 MN BT	2009 MN BT
TOTAL REFERENT GROUP	28.670	−62.911	32.151	33.330	34.603	35.977	37.462	37.462	37.462	37.462	91.647
– IMPORT DUTIES											
YARN		−18.000	−36.000	−36.000	−36.000	−36.000	−36.000	−36.000	−36.000	−36.000	−36.000
COTTON		10.692	21.384	21.384	21.384	21.384	21.384	21.384	21.384	21.384	21.384
EQUIPMENT	1.870										
SPINDLES	8.800										
MATERIALS INVENTORY		5.280									−5.280
SPARES INVENTORY		0.535									−0.535
SPARES		1.067	1.067	1.067	1.067	1.067	1.067	1.067	1.067	1.067	1.067
– TAXES		−17.035	23.350	24.529	25.802	27.176	28.661	28.661	28.661	28.661	28.661
(i) TOTAL GOVERNMENT	10.670	−17.461	9.801	10.980	12.253	13.627	15.112	15.112	15.112	15.112	9.297
(ii) POWER SUPPLIER		1.800	3.600	3.600	3.600	3.600	3.600	3.600	3.600	3.600	3.600
(iii) LABOUR	18.000	6.750	6.750	6.750	6.750	6.750	6.750	6.750	6.750	6.750	6.750
(iv) DOMESTIC BANK		−60.000	6.000	6.000	6.000	6.000	6.000	6.000	6.000	6.000	60.000
											6.000
(v) INSURANCE		6.000	6.000	6.000	6.000	6.000	6.000	6.000	6.000	6.000	6.000
TOTAL REFERENT GROUP	28.670	−62.911	32.151	33.330	34.603	35.977	37.462	37.462	37.462	37.462	91.647

PROJECT			PRIVATE			EFFICIENCY			REFERENT GROUP		
NPV	MN BT	IRR (%)	NPV	MN BT	IRR (%)	NPV	MN BT	IRR (%)	NPV	MN BT	IRR (%)
4%	344.5	18.3%	4%	124.1	14.3%	4%	402.8	22.1%	4%	260.4	n/a
8%	209.5		8%	63.5		8%	264.8		8%	201.3	
12%	109.7		12%	19.6		12%	162.4		12%	158.3	
16%	34.4		16%	−12.7		16%	85.1		16%	126.6	
0%	530.6		0%	209.1		0%	592.8		0%	343.3	

Figure A7.1 ICP project solution with an increase in the skilled wage.

Appendix 2 to Chapter 7

Compensating and equivalent variation

In this appendix we discuss the notions of *compensating* and *equivalent variation* and relate these measures to areas under the Marshallian and Hicksian demand curves (named after the famous economists who originated them). The Marshallian demand curve is the familiar "ordinary" demand curve which traces the effects of price changes on quantities demanded holding money income constant, while the Hicksian demand curve deals with the effects of price changes holding the individual's level of utility (or satisfaction) constant. We will use these individual demand curves to consider the relationship between consumer surplus measures and measures of compensating variation used in cost-benefit analysis

As we noted in Chapter 1, a cost-benefit analysis is an attempt at implementing the Kaldor-Hicks (K-H) criterion for an improvement in economic welfare: the project contributes to an increase in welfare if the gainers from the project could, in principle, compensate the losers – in other words, if the project represents a Potential Pareto Improvement. The gain to an individual is measured by the sum of money which, if the project were undertaken, could be taken from her while still leaving her as well off as she would have been in the absence of the project. Conversely, a loss is measured as a sum of money which must be paid to an individual in compensation for the effects of the project. We now consider what the relationship is between the sum of money, hypothetically to be paid to or received by a person affected by a project, and termed the *compensating variation*, and the concept of consumer surplus. It is important to stress that these are hypothetical sums, and that cost-benefit analysis does not require that compensation is actually paid or received.

To investigate this issue we must digress to one of the more arcane areas of economic science in order to pursue the issue of compensation. Economics deals with "rational" consumers. One of the things a rational consumer exhibits is a *preference ordering*: offered a choice between two bundles of goods and services, she can tell us either she prefers one bundle to the other, or that she is indifferent between the two bundles because they offer her the same amount of utility or satisfaction. From the individual consumer's point of view, we can consider the world *with*, and the world *without* the proposed project as alternative bundles of goods. In principle, we could ask the consumer which bundle she prefers. If she chooses the "with" bundle, it means that she considers that she will be a gainer from the project, and if she chooses the "without" bundle, she believes she will be a loser as a result of the project. In the former case (a gainer) we can obtain a money measure of the amount the consumer gains by attaching a money *penalty* to the "with" bundle, and successively increasing the amount of that penalty until she tells us she is indifferent between the "with" bundle, together with its attendant money penalty, and the "without" bundle. At that point the amount of the money penalty measures the compensating variation. In the latter case (a loser) a money *premium* must be attached to the "with" bundle to obtain, in a similar manner, the compensating variation (CV) measure of the loss.

A critical feature of the above conceptual experiment is the attainment of indifference – the amount of the money penalty or premium is altered until the individual tells us that she is indifferent between the "with" and "without" bundles. Indifference between these two hypothetical states of the world implies that the individual gets the same amount of utility or satisfaction from each. Relying on the notion of indifference suggests that if we want to measure the compensating variation by an area under a demand curve, the relevant demand curve is one which shows all price and quantity combinations which provide the same amount of utility as that yielded by the initial combination (P_0, Q_0). As noted above,

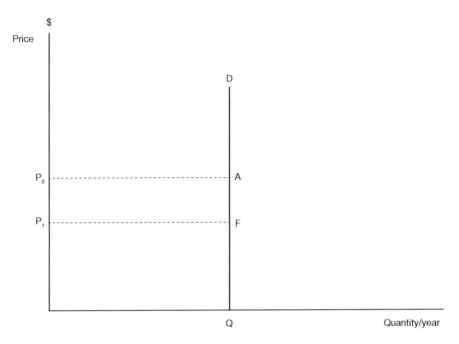

Figure A7.2a Consumer surplus with inelastic demand. DQ represents the consumer's perfectly inelastic demand for a good. A fall in price from P_0 to P_1 has no effect on the quantity demanded, Q, but benefits the consumer by the amount P_0AFP_1.

ordinary (Marshallian) demand curves do not have the property of holding utility constant as price falls and quantity demanded increases; as price falls, and the individual moves down the ordinary demand curve, utility increases. For this reason, we will be relying on the Hicksian demand curves to measure compensating variation.

Consider first an individual's ordinary demand curve that is perfectly inelastic, as illustrated in Figure A7.2(a). When price falls from P_0 to P_1, the consumer receives a windfall gain amounting to area P_0AFP_1. This latter amount could be taken away from the consumer, following the fall in price, and she *could* still buy the same quantities of goods as she did before the fall in price, and *would*, in fact, still choose those quantities. In other words, if the price falls from P_0 to P_1 *and* the consumer was made to give up P_0AFP_1 dollars, she would be exactly as well off after the fall in price as she was before. Hence, in terms of the K-H criterion, P_0AFP_1 dollars measures the benefit of the price fall, and, conversely, measures the cost to the individual (the amount she would have to be paid in compensation) if price rose from P_1 to P_0.

Now consider an individual's elastic demand curve for a good as illustrated in Figure A7.2(b). If price fell from P_0 to P_1, and no sum of money was taken in compensation, the individual would increase quantity demanded as a result of two effects. First, when the price of a good falls, the individual experiences an increase in purchasing power similar to that which would result from an increase in income. When income increases, the individual will increase the quantity consumed of all normal goods, including, we assume, the good whose price has fallen. This type of increase in the quantity of the good consumed as a result of the fall in its price is termed the *income effect* of the price change. The second type of effect, termed the *substitution effect* of the price change, refers to the fact that when the price of

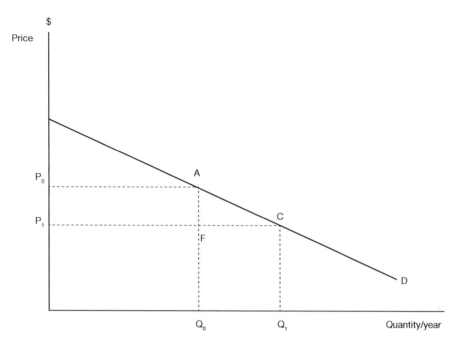

Figure A7.2b Consumer surplus with elastic demand. The line D represents the individual's demand
curve for a good. At price P_0 she consumes quantity Q_0, and at price P_1 she consumes
Q_1. The consumer surplus gain as a result of a fall in price from P_0 to P_1 is measured by
area P_0ACP_1.

a good falls, but purchasing power is held constant, an individual will be able to increase
her economic welfare by purchasing more of the now relatively cheaper good, and less of
other goods.

If the price of the good fell from P_0 to P_1 and the sum P_0AFP_1 dollars was taken from the
individual in compensation, the income effect of the price change would be nullified, but
the substitution effect would still occur. The individual *could* still buy the same quantities
of goods as she did before the fall in price, but, because of the substitution effect, *would not*
choose that bundle of goods. Instead she would elect to improve her economic welfare by
consuming more of the now relatively cheaper good, and less of other goods. The net result
is that, even if she had to surrender the sum P_0AFP_1 dollars, the individual would still be
better off after the price fall because of the substitution effect. This means that, if we are
to implement the K-H compensation criterion, we need, in principle, to take a little more
money than the sum represented by P_0AFP_1 away from her, represented by the area ACF
in Figure A7.2(b). This area measures the gain resulting from the substitution effect – the
difference between her willingness to pay for the extra quantity of the good and the amount
she actually pays.

To recap, holding utility constant in the face of a fall (rise) in price requires that we
take some income away from (give some additional income to) the individual. One effect
of taking or giving income in compensation for the price change is to nullify the income
effect of the price change. In the case of a *normal good* (defined as a good which the individual will purchase more of as income rises), the income effect is positive: the increase in
income received as compensation for a price rise causes an increase in quantity of the good

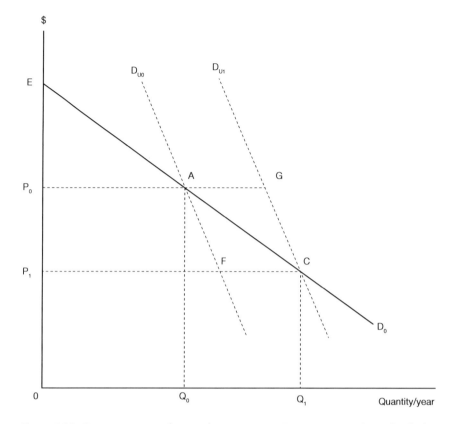

Figure A7.3 Compensating and equivalent variation. ED_0 represents the individual's Marshallian demand curve for a good, and D_{U0} and D_{U1} are Hicksian demand curves. As price falls, a move down the Marshallian demand curve involves an increase in utility, whereas utility remains constant for moves down a Hicksian demand curve. Compensating and Equivalent Variations of a price fall or rise are represented by areas under the demand curves as discussed in the text.

demanded, and the reduction in income as a result of having had to pay compensation for a fall in price causes a reduction in quantity demanded. This means that the utility-constant demand curve passing though the point A in Figure A7.3 is to the right of the observed demand curve for prices higher than P_0 (because compensation for the higher price has notionally been paid to the individual, thereby increasing quantity demanded, as compared with the ordinary demand curve) and to the left of it for prices lower than P_0 (because compensation for the lower price has notionally been taken from the individual, thereby reducing quantity demanded, as compared with the ordinary demand curve). In other words, the utility-constant demand curve passing through point A is steeper than the observed demand curve, and this is true for all price quantity combinations on the ordinary demand curve. This is illustrated in Figure A7.3 where two utility-constant demand curves, denoted by D_{U0} and D_{U1} are drawn through points A and C on the ordinary demand curve, D_0.

If we start at price P_0 in Figure A7.3 and reduce price to P_1, the area P_0AFP_1, under the utility-constant demand curve D_{U0}, measures the amount we can take from the individual in compensation for the fall in price, while maintaining her original level of utility. This is the compensating variation measure of the benefit of a price fall/quantity rise required by the K-H criterion. If we start at price P_1 and raise price to P_0, the area P_0GCP_1, under the utility-constant demand curve D_{U1}, measures the amount we must pay the individual in compensation for the rise in price, to maintain her original level of utility. This is the compensating variation measure of the cost of a price rise/quantity fall required by the K-H criterion. It can be seen from Figure A7.3 that the area of surplus measured under the ordinary (Marshallian) demand curve (P_0ACP_1), which we have been referring to as the Consumer Surplus (CS), overstates the measure of compensating variation for a price fall, and understates the measure for a price rise. However, since the extent of the shift in the individual's demand for a commodity in response to a relatively small increase in income is likely to be slight, the utility constant demand curves are likely to be very close to one another and similar in position to the ordinary demand curve (contrary to Figure A7.3 which is drawn for expositional purposes), and the size of the error is, in theory, likely to be small.

The notion of compensation requires that money is paid to those who will lose as a result of a project, and taken from those who will gain (negative compensation). A related notion is that of equivalence: the *equivalent variation* is the sum of money to be paid to an individual to make her as well off in the absence of the project as she would have been if the project had gone ahead. In other words, the equivalent variation is a sum of money to be paid to (taken from) a potential gainer (loser) from a project *in lieu* of the project. In the example illustrated in Figure A7.3 the fall in price from P_0 to P_1 enables the consumer to increase her consumption of the commodity from Q_0 to Q_1. Suppose instead that the price had remained at P_0. Money would have had to have been paid to the individual to make her as well off as she would have been if the project had gone ahead with its effect of lowering price. As she receives additional income in the form of compensation, her utility-constant demand curve shifts to the right of the original curve, D_{U0}, until it passes through the point (P_1,Q_1) where she is as well off as she would have been if the price of the commodity had fallen from P_0 to P_1. The sum of money required to achieve this increase in utility, termed the equivalent variation, is measured by the increase in surplus measured under the demand curve, D_{U1}, at the new level of utility, between the prices P_0 and P_1, and is given by P_0GCP_1.

Now suppose that the original equilibrium is at (P_1,Q_1) in Figure A7.3 and that price rises from P_1 to P_0. We can measure the compensating variation associated with this change in price by the area of reduced surplus between P_1 and P_0 under the utility constant demand curve passing through (P_1,Q_1), measured by area P_0GCP_1. This is a sum of money to be paid to the individual in compensation for the price increase. However, recalling the discussion in the previous paragraph, it is also the sum of money which is equivalent to the fall in price from P_0 to P_1 when the consumer is at the equilibrium point (P_0,Q_0) – the sum of money to be paid to the individual *instead* of the fall in price from P_0 to P_1. In other words, the compensating variation for a rise (fall) in price is the same, in absolute value, as the equivalent variation for a fall (rise) in price over the same range.

It follows from the above discussion that, perhaps contrary to intuition, the compensating and equivalent variations for a given change in price are not equal in absolute value. For example, starting at the equilibrium (P0,Q0), the sum of money which must be taken from the individual to make her as well off when price falls to P1 as she was in the initial equilibrium (the compensating variation for the price fall from P_0 to P1), is smaller than the sum of money which must be paid to the individual to make her as well off as she would have been in the equilibrium (P1,Q1) had the price risen from P1 to P_0 (the equivalent variation for the price rise from P1 to P_0). Question 6 in the Exercises section of this chapter illustrates this point.

The set of relationships between the measures of Compensating Variation (CV), Consumer (or Marshallian) Surplus (CS) and Equivalent Variation (EV) which have been established in the above discussion, and illustrated in Figure A7.3, can be summarised as follows:

>For a price fall/quantity rise: CV < CS < EV
>For a price rise/quantity fall: CV > CS > EV

We now relate this rather technical discussion to the important concepts of willingness-to-pay (WTP) and willingness to accept (WTA) which are widely used in cost-benefit analysis. We have argued that the compensating variation is the measure of WTP that is required to implement the K-H criterion for measuring a change in economic welfare. Despite its name, the WTP measure is, as we have seen, actually *negative* in the case of a price rise/quantity fall: in this case it is sometimes referred to as willingness-to-accept (WTA). For a price fall/quantity rise, the CV measures the *maximum* amount the individual is willing to pay, and for a price rise/quantity fall, it measures the *minimum* amount the individual is willing to accept as compensation. Thus the CV fulfils the role required of it by the K-H criterion.

What role then remains, in the analysis of changes in economic welfare, for the notion of Equivalent Variation? While the CV tells us the sum to be paid or received to make the individual's level of economic welfare in the world *with* the project the same as that in the world *without* the project, the EV, on the other hand, tells us the sum to be paid or received to make the level of economic welfare in the world *without* the project the same as it would be in the world *with* the project. In other words, the payment identified by the EV is an *alternative* to the project. Some analysts have suggested that the proper role of the EV is in dealing with the effects of proposed changes to the structure of property rights. For example, the English countryside is criss-crossed by a network of rights of way which allow walkers to traverse private land. Understandably, but regrettably, landlords sometimes find these tracks to be a nuisance and seek ways to eliminate the public right of way. If such a right were to be taken from an individual, the EV would measure the minimum sum of money she would accept to relinquish it.

We argued earlier that, contrary to the picture painted by Figure A7.3, the Hicksian demand curves are, in theory, likely to be very close to the ordinary or Marshallian demand curve. This proximity would imply that the WTP for a price fall/quantity rise, measured by P_0AFP_1 in Figure A7.3, should be very similar (although smaller) in size to the WTA in compensation for a price rise/quantity fall, measured by P_0GCP_1. In fact, empirical studies have found ratios of WTA/WTP for public or non-marketed goods averaging around 10. These ratios are typically lower for private goods but are still significant. While Pearce, Atkinson and Mourato (2006, p. 160) suggest some reasons for these divergences, and Kahneman (2013) presents experimental evidence indicating that individuals' preferences display asymmetry in the valuation of gains and losses, they remain a conundrum yet to be fully resolved.

Exercises

1   A new publicly financed bridge is expected to reduce the cost of car travel between two areas by $1 per trip. This cost reduction to motorists consists of a reduction in travel time, car depreciation and petrol expenses totalling $1.50 per trip, less the $0.50 bridge toll which the government will collect. Before the bridge was built, there were

1 million trips per year between the two areas. Once the bridge is in operation, it is estimated that there will be 1.5 million trips per year between the two areas.

   i   In terms of areas under the demand curve, what is the annual benefit of the bridge:
       a   to motorists?
       b   to the government?
   ii   What other Referent Group benefits would need to be considered in a social benefit/cost analysis?

2   i   Briefly outline the model and principles you would use to conduct a cost-benefit analysis of a highway project which will reduce the travel time between two cities (include the appropriate diagram showing the demand for trips in your explanation).

   ii   Using the above framework, together with the following information and any reasonable assumptions you need to make about the values of variables and the timing of benefit flows, work out a rough estimate of the present value of the benefit to users of an extension to the Pacific Motorway:

       Current usage rate: 85,000 trips per day
       Estimated future usage: 170,000 trips per day by 2020
       Estimated time saved per trip at current usage rate: 10 minutes.

   iii   The capital cost of the project is reported to have a present value of $850 million. Using this value, work out a rough estimate of the net present value of the project.

   iv   Briefly discuss the accuracy of your net present value estimate as a measure of the net efficiency benefit of the project, taking account of any significant omissions from your analysis.

3   The Happy Valley Water Control Board (HVWCB) will undertake an irrigation project which will provide 10 million gallons of water annually to 100 wheat farms of 100 acres each. The HVWCB will charge $0.02 per gallon for the water, which will be distributed equally among the 100 farms. It is estimated that:

   i   land rent in Happy Valley will rise by $70 per acre per annum;
   ii   50 farm labourers will be attracted to work on Happy Valley farms, at the market wage of $2000 per annum, from the vineyards of nearby Sunshine Valley;
   iii   production of wheat on Happy Valley farms is subject to constant returns to scale.

Assuming the Referent Group is the economy as a whole, work out the annual Referent Group benefit of the irrigation project in terms of areas defined by demand and supply curves for output or inputs. Explain your answer.

4   Comment critically on the following approaches to measuring project benefits:

   i   An analyst measures the annual benefit of a bridge as the gain in consumer surplus to bridge users plus the increase in annual rental value of properties served by the bridge.
   ii   An analyst measures the annual benefit of an irrigation project by the increase in the value of food production, plus the increase in rental value of the land served

by the project. He takes the present value of the annual benefit, calculated in this way, and adds the increase in capital value of the land to obtain an estimate of the present value of the benefit.

5   When a public project results in lowering the price of a particular commodity (Commodity A), it impacts on the markets for complementary or substitute commodities. Give some examples of this type of relationship. Choose one of the examples (call it Commodity B) and use it to explain in detail the relevance of the changes in the market for Commodity B for the cost-benefit analysis of the project which results in a lower price for Commodity A.

6   Ellsworth spends all of his income on two goods, x and y. He always consumes the same quantity of x as he does of y, $Q_x = Q_y$. Ellsworth's current weekly income is $150 and the price of x and the price of y are both $1 per unit. Ellsworth's boss is thinking of sending him to New York where the price of x is $1 and the price of y is $2. The boss offers no raise in pay, although he will pay the moving costs. Ellsworth says that although he doesn't mind moving for its own sake, and New York is just as pleasant as Boston, where he currently lives, he is concerned about the higher price of good y in New York: he says that having to move is as bad as a cut in weekly pay of $A. He also says he wouldn't mind moving if when he moved he got a raise of $B per week. What are the values of A and B?

a   A = 50 B = 50.
b   A = 75 B = 75.
c   A = 75 B = 100.
d   A = 50 B = 75.
e   None of the above.

Which value (A or B) measures the compensating, and which the equivalent variation? Explain. (Hint: see Appendix 2 to Chapter 7).

8   Non-market valuation

8.1 Introduction

This chapter deals with the methods and techniques used by economists to put monetary values on non-marketed goods and services, or, more specifically, those project inputs or outputs which affect the level of economic (material) welfare but which do not have market prices. While the discussion will focus mainly on environmental goods in order to illustrate the issues involved, the chapter also includes a brief discussion of various methods of valuing human life. The methods of cost-benefit analysis are then applied to public policy towards the COVID-19 pandemic and to responding to climate change.

In previous chapters we have discussed how cost-benefit analysis involves the identification and valuation of a project's costs and benefits. Our analysis has so far been confined mainly to those inputs and outputs that are exchanged through markets and where, in consequence, market prices exist. We recognised in Chapter 5 that the market does not always provide the appropriate measure of value and that shadow prices sometimes have to be generated in order to better reflect project benefits or opportunity costs. This discussion was essentially about adjusting existing market prices where these were believed to be distorted, due perhaps to government intervention or to imperfections in the structure and functioning of markets. The present chapter deals with another type of market failure – where there is no market for the input or output in question, and, therefore, no market price at which to value the cost or benefit.

8.2 Causes of market failure

As a starting point, it is reasonable to ask why there is no private market for the good or service in question. Markets exist in order to trade ownership of goods and services – hamburgers and machine tools, haircuts and labour services. For economic agents to exchange ownership of a good or service, there must exist a reasonably well-defined set of property rights that specify what is being traded. In particular, the good or service must be *excludable*, meaning that it is feasible for the owner to prevent others from enjoying the good or service unless they pay for it. If a commodity is not excludable, a private agent would be unlikely to be able to recover the cost of supplying it by charging a price for it – even if some customers purchased the good, others could act as *free riders*, enjoying the good without paying for it. Of course excludability is partly a matter of cost – while free-to-air TV is non-excludable, excludability can be achieved by installing a cable system. However if access to a good or service is not excludable at a reasonable cost it will not be traded in private markets. Examples of non-excludable goods are national parks, national defence, and some forms of fire and police protection.

DOI: 10.4324/9781003312758-8

Air and water pollution are further examples of non-excludable goods, which could more accurately be described as "bads", since they detract from rather than contribute to economic welfare. While there is no incentive for economic agents to produce bads – no one would pay for them – they are sometimes generated as a by-product of economic activity and are termed *externalities* to reflect the fact that they are external to the market system. For example, firms and consumers who generate air pollution as a by-product of their production and consumption activities impose a *negative externality* on other consumers and firms. Since access to clean (or polluted) air is non-excludable, there is no market mechanism that enables members of the general public to purchase cleaner air, or requires firms or individuals to pay for the air pollution they cause. The former mechanism would provide the "carrot" and the latter the "stick" that would persuade the profit-maximizing firm or utility-maximizing individual to reduce their level of emissions. In the absence of a market, there is no price that can be used directly to measure the cost of the air pollution generated by a project.

While air pollution is an example of a negative externality, there also exist *positive externalities* – beneficial side-effects of production or consumption activity that are not captured by the market. Examples are pollination of fruit trees as a by-product of honey production, or enjoyment of the output of the neighbour's stereo system. While economic agents might, in principle, be willing to purchase commodities such as these, in practice, there is no need to pay for them because of their non-excludable nature, and hence there is no market for them.

Non-excludability is one of the two main characteristics of a *public good*, with the other main characteristic being non-rivalness. A good or service is *non-rival* in nature if each individual or firm's consumption of a unit of the good or service does not affect the level of benefit derived by others from consumption of the same unit. Examples of non-rival services are weather forecasts, views and film shows. The latter example provides an instructive illustration of the market's failure to provide the efficient quantities of non-rival goods and services, even if these goods or services are excludable. Since the costs of showing the film are fixed, regardless of the number of viewers, the theatre owner's incentive is to charge an admission price that maximises revenue. If that price does not result in filling the theatre, additional customers could benefit from viewing the film without imposing any additional cost. In other words, the service has not been provided up to the efficient level at which marginal benefit equals marginal cost. In principle, this problem could be solved by price discrimination – charging different prices to different customers – but, in practice, theatre owners may be unable to obtain detailed information about customers' willingness to pay.

While most of the above examples deal with non-excludable negative and positive public externalities, it can be noted that externalities can also be excludable and private in nature. The time-honoured example of an excludable (or private) externality concerns a smoker and a non-smoker sharing a railway compartment. In principle, this problem can be resolved by negotiation: if it is a "no-smoking" ("smoking") compartment, the smoker (non-smoker) pays the non-smoker (smoker) to be allowed to smoke (to limit their smoking). While there is no market in smoking, the efficient level can be achieved through negotiation between the two parties. To appreciate the importance of excludability in this example, suppose that a second non-smoker were present: a three-way negotiation would be more difficult (costly) because of the temptation for free-riding, and if we add more non-smokers to the example, the bargaining costs would eventually outweigh any benefit that could be achieved by negotiation. In the case of private externalities, such as those involving disputes between neighbours, an efficient solution can often be found though bargaining, provided that transactions costs are not too high.

In summary, the private market does not supply efficient levels of goods or bads which are non-excludable and/or non-rival. Provision of public goods is one of the important activities of governments, and is one of the reasons for social CBA: proposed government programmes or projects which supply public goods need to be appraised. Furthermore, because many private projects produce non-excludable external effects, it cannot be assumed that because they are in the private interest of their proponents they are also in the public interest. Social CBA is used to appraise such projects from an efficiency and public interest viewpoint before they are allowed to proceed. To be of use, the social CBA needs to be able to assess the project's external effects in terms commensurate with its private net benefits. Thus, the market failure which provides the major rationale for social CBA at the same time poses one of its major challenges – that of valuation in the absence of market prices.

8.3 Valuing environmental costs and benefits

Environmental economics has made substantial progress in recent years in devising and refining non-market valuation methods. The subject area has become extremely wide and increasingly technical in nature. It would not be possible, in a single chapter, to equip the reader with a working knowledge and technical capacity to apply the full range of non-market valuation techniques. This has become a highly specialised aspect of economics, an in-depth study of which could constitute a separate course in itself. In this chapter the reader will be familiarised with:

- the issues and arguments giving rise to the need for the incorporation of non-market environmental values in the Efficiency Analysis;
- the range of approaches and techniques available, together with illustrations of how non-market values can be incorporated into the CBA framework developed in the first part of this book;
- some of the more important theoretical concepts underlying the valuation methods; and
- some of the strengths and weaknesses of the various methods.

Environmental resources as public goods

The distinguishing feature of environmental goods is that they possess *public good* characteristics. One of the most commonly cited examples of an environmental good is the air we breathe. This is classified as a *pure public good* as it possesses all three characteristics of a public good. These are:

- *non-rivalry in consumption*, implying zero opportunity cost of consumption in the sense that one person's consumption does not affect the availability of the good to others;
- *non-excludability by producers*, implying that the suppliers cannot exclude any consumers or other producers who want access to the good; and
- *non-excludability by consumers*, implying that the consumer cannot choose whether or not to access or consume the good.

Many environmental commodities such as air quality, flood protection, noise and views are pure public goods. At the other end of spectrum are *private goods*, which meet none of the above three criteria. The goods we buy at the local supermarket are examples of private goods in that each item that any one of us purchases and consumes leaves one less for others to consume. A producer has the right to decide whether or not to sell the good, and we as consumers have the right to decide whether or not we wish to purchase it. In such cases, there is rivalry in consumption and excludability on the part of both producers and consumers.

In between the two extremes lie various forms of "impure" public goods where only one or two of the three criteria are satisfied. A *semi-public good* is defined as one where criteria (1) and (2) are satisfied. There is zero opportunity cost of consumption and producers cannot exclude anyone from using the good, but consumers have the right and ability not to use the good. An example is free-to-air broadcasting: the reception of a station's broadcast by one viewer does not weaken the signal for others; the producer cannot exclude some would-be viewers and include others, but consumers can decide whether or not they want to tune in to the broadcast. As noted earlier, technological change involving the invention and marketing of the decoder by pay-TV broadcasters has allowed some producers to effectively remove the non-excludability characteristic of their TV channels.

An example of a semi-public environmental good is the *common*. In this instance, exploitation of the resource is rival but non-excludable by producers. The resulting form of market failure, described in the economics literature as "the tragedy of the commons", accounts for the tendency for unregulated competitive market forces to result in the overexploitation of natural resources, such as fish stocks. As each producer catches fish, he reduces the fish stock, and the catch rates of all fishers fall as a result; in other words, each fisher imposes a negative public externality in production on his fellow fishers. Each fisher could benefit his colleagues by reducing his impact on the fish stock, but has no way of benefiting himself by so doing. Put another way, each fisher could invest in the fish stock (by reducing his catch), but, while the net benefit to the economy of such investment is positive, there is no net benefit to an individual investor. While all fishers may recognise the possibility of increasing the aggregate net benefit derived from the fishery by reducing total fishing effort, the non-excludability of the fishery makes it impossible for any individual to benefit in this way. The result is that the fish stock is over-exploited from an economic viewpoint: the marginal benefit of effort is lower than its marginal cost.

A further case is that of the *non-congestion* public good, where there is no opportunity cost of consumption, but where both producer and consumer excludability apply; uncongested roads are an example of a non-congestion public good. The owners of the road can exclude potential users by introducing a charge, and consumers can elect to use the road or not. As in the earlier example of the movie theatre, charging for access to an uncongested road leads to a lower level of use than the efficient level.

Even where a competitive market economy is free of distortions in the form of regulations or taxes in its markets for *private goods*, there will still be market failure associated with the existence of the various types of *public goods* described above. In these instances there is either no market price at all, or, where there is one, it will not reflect the true value or opportunity cost of the good or service in question. It is this situation that typically characterises the markets pertaining to environmental goods. To take proper account of environmental costs and benefits in project appraisal it is therefore necessary for the analyst to use some alternative means to the market for eliciting the values society attaches to such goods and services. In recent years a wide range of non-market valuation methods have been developed through economics research and are available in the literature, particularly in the field of environmental economics. These methods are examined in later sections of this chapter.

Externalities and the environment

As noted above, externalities arise where there is no market connection between those taking an action, which has consequences for material welfare, and those affected by that action. The action could be, say, the run-off of nutrients and chemicals from irrigated farmlands into a river resulting in downstream pollution damage. The costs are borne by others, such as the fishers or tourists whose benefits are determined by the quality of the water downstream and perhaps at the adjacent coast where there might be a coral reef. In other words, the costs are *external* to the person who causes them and who has no direct financial incentive to avoid imposing them.

Externalities can also be *positive*. In this case the person taking the action results in another (or others) external to her benefiting from the action. For example, if the nutrients in agricultural run-off stimulate the growth of fish stocks, the fishers may benefit from larger catches but the farmers who provided the extra nutrients do not. As the unregulated market either altogether ignores or inaccurately reflects the external costs and benefits associated with use of a resource, there is a need for some form of non-market valuation method to allow the decision-maker to measure the dollar value of the external effect and to allow for the external cost or benefit (to *internalise* it) in appraising alternative project or policy options. Failure to take account of Referent Group external costs generally results in the over-utilization of an environmental resource, while failure to account for external benefits leads to under-utilization.

Internalizing externalities, or accounting for them in CBA, is therefore necessary for economic evaluation in the same way as shadow-pricing was required to account for distortions in private markets as discussed in Chapter 5. In this regard, environmental valuation and the inclusion of environmental costs and benefits in CBA should, in principle, be treated as a necessary part of the "bottom line" in Efficiency Analysis. Until recently economics lacked the methods and techniques for non-market valuation which meant that externalities were largely ignored in formal CBA, or, at best, treated qualitatively in the form of an *Environmental Impact Assessment* (EIA), and therefore remained below the "bottom line". With recent advances in non-market valuation techniques there can be no justification today for failing to attempt to quantify and value externalities as part of the shadow-pricing process of Efficiency Analysis.

The development of the CBA of the National Fruit Growers project in Chapter 5 showed how environmental costs could be integrated into the Efficiency Analysis section of the spreadsheet framework. In Figure 5.16 the annual cost of water pollution was entered in Table 4 of the spreadsheet in the form of a cash flow, along with other project costs estimated at efficiency prices. These costs appeared subsequently in the Referent Group Analysis described in Chapter 6 as an annual cost to downstream users of the river. In Figure 6.2 this cost was entered in the form of a cash flow in Table 5 of the project spreadsheet. It can be seen from this example that, in principle, environmental costs can be incorporated as an integral part of a social CBA. For example, environmental costs are included in the Olives and Walnuts case studies in Appendices A1.1 and A1.2, and these costs feature in more detail in the Tasmanian Pulp Mill Study in Appendix A1.7.

Total economic value

It is now generally acknowledged that environmental resources contribute value not only to those who use the resource, but also to non-users, who may value the conservation of the resource. The value to non-users could arise for reasons of altruism – the value to one individual from knowing that the asset can be used and enjoyed by others – or for reasons of self-interest – the value to an individual from knowing that the asset will continue to be accessible in the future. For instance, you may be unsure whether you will ever be able to visit Australia's Great Barrier Reef, yet you might feel a significant loss of value if it, and the ecosystem it supports, were degraded.

Total economic value (TEV) is the term used by economists to describe the range of use and non-use values of a resource, where:

> TEV = Direct use value + Indirect use value + Option value + Existence value +
> Bequest value

Use values can be *consumptive*, meaning that the user benefits by removing part of the asset, or *non-consumptive*, meaning that the benefit is derived from contact alone.

It should be noted that the TEV concept is limited to *anthropocentric* values only. In other words, in CBA, the resource is valued exclusively in terms of the values it yields to humans; no intrinsic value is attributed to it.

As an illustrative example consider the values generated by a coral reef. The sorts of values it provides can be categorised under the components of TEV as shown in Figure 8.1. As we move from the left to the right of Figure 8.1, the values become more difficult to quantify:

- *Direct uses* usually include the most obvious and important market-based uses such as fisheries (a consumptive use that can include subsistence, artisanal inshore fishing, recreational fishing and large-scale commercial fishing) and tourism (mainly a non-consumptive use in the form of viewing, though it sometimes involves commercially-organised recreational fishing trips). Other consumptive uses can include coral mining for building materials, as well as shell and coral collecting.

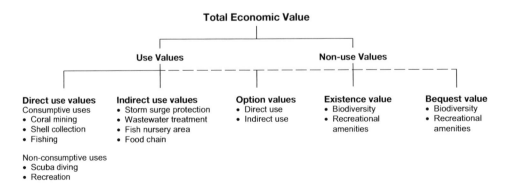

Figure 8.1 Total economic value of coral reef ecosystems.

- *Indirect uses* include regulatory functions such as storm surge protection, fish nursery and food chain regulation, and, where mangroves form part of the reef's ecosystem, wastewater treatment. Only a few studies have made attempts to value these uses, most notably storm protection values.

- *Option value* is increasingly recognised as a significant form of benefit to be derived from natural resource stocks, such as coral reefs and forests, partly because of the increasing number of pharmaceutical and medicinal values that are being derived from them. Option value is the value of preserving an environmental asset for possible future use. Since it is the value attached to potential use, its current "non-use" value, is attributable to its potential use value in the future, which is why we draw it as straddling the "use" and non-use" categories in Figure 8.1. Option value exists independently of any *uncertainty* about the benefits to be derived from the resource or *irreversibility* of decisions we take about its exploitation.

  There are uncertainties about the values of possible future discoveries and bio-technological advances to be gained from ecosystems, values which may be diminished if we allow damage to occur. Some adverse environmental effects may be, for all practical purposes, irreversible: examples are heavy metals pollution and carbon dioxide emissions. In appraising projects with irreversible and uncertain consequences, analysts should take account of the value of delaying the project while more information is obtained. The value placed on delay of irreversible projects which may have adverse environmental consequences is termed a "*quasi-option value*" or "*real option value*". It is similar to the value of a call option in the financial asset markets: the value of the right, without obligation, to purchase an asset at a specified price within a specified time in the future. Quasi-option values will be considered in detail in Chapter 9 in the context of determining the value of delaying an irreversible decision while additional information about the effects of the project accumulates. Unlike option value, quasi-option value is *not* an additional component of TEV but rather the value of making irreversible decisions rationally in the face of uncertainty, taking account of their effect on TEV.

- *Existence and bequest values:* Economists have come to recognise that individuals may derive value from something even though it is accepted that these individuals will probably never get to "use" or "consume" it – consider the whale, the panda, or South American rainforests, for instance. Individuals are prepared to commit funds to the conservation of such natural assets for no reason other than their belief that these natural resources should continue to remain in existence. Where the individual's satisfaction arises purely from the knowledge that the environmental resource will continue to exist it is labelled *existence value*. Where the individual's satisfaction is attributable to the continued existence of the resource for the future possible benefit of others, either known or unknown to her, we label it a *bequest value*.

8.4 Incorporating non-market values in cost-benefit analysis

It is important to remember that in a CBA framework we are analysing the effects of changes to an existing regime (the world *without* the project), and that it is the "trade-off" or marginal value of the resource that is relevant, not the total economic value. Again, let us consider the example of a coral reef. If the purpose of the valuation is to provide a dollar value for use in a CBA of alternative management options, what is required is the trade-off of values under competing management scenarios. Resource management is usually about introducing changes at the margin. Most decisions mean that some users (or non-users) gain at the expense of others. Declaring a section of a coral reef that was previously used for, say, commercial fishing and tourism, as a "no-go" marine park implies that the fishing and tourism industries may forgo net benefit, while others may gain from the increase in, say, scientific, biodiversity, existence and/or option values.

What is needed, therefore, is not the total value of each component of TEV but rather their *marginal* values. In the context of our coral reef example, this could mean the amount of TEV gained or lost per unit of reef usage or non-usage. If a re-zoning management plan proposes to introduce a no-go marine park of 200 km² which means that the area that may be accessed for purposes of tourism is reduced from 2000 km² to 1800 km², we need to know by how much this marginal reduction in access for tourism affects the value to tourists who are prevented from using it. Knowing only the total value of tourist expenditure or income from tourism is of little help in addressing this issue.

In a CBA of alternative management options, both the incremental benefits (in terms of changes in TEV) and the incremental costs need to be quantified and compared. The net benefits of a given management option (relative to some base-case scenario) can be written:

$$NB(M) = (B_m + B_{nm}) - (C_m + C_{nm})$$

where

$NB(M)$ = net benefits *with* management option vs *without* the option

B_m = incremental benefits for which there is a market

B_{nm} = incremental benefits for which there is no market

C_m = incremental costs for which there is a market

C_{nm} = incremental costs for which there is no market.

In the absence of a budget constraint, the decision-rule would be to adopt the option provided that $NB(M) > 0$.

Having established the appropriate decision-support framework within which the estimate of non-market value is to be used, the question of which approach and method of non-market valuation is appropriate can then be addressed. In the context of this book, the decision-support framework is CBA. If another type of framework such as multi-criteria analysis is being used, it may not be appropriate to rely on the same kind of trade-off values or non-market valuation methods.

8.5 Methods of non-market valuation

Various types of non-market values suggest the need for a variety of valuation methods, which we discuss here in rather broad terms without going into too much detail of the techniques and methodologies involved. It is useful to think in terms of two broad approaches to economic valuation: the *production approach* and *the utility approach*. The production approach can be thought of as a "supply side" approach, while the utility approach is a "demand side" approach. We now consider briefly each of these approaches.

8.5.1 The production approach

In a CBA the net benefits of a development project, D, involving an environmental cost can be written:

$$NB(D) = (B\text{-}C) - EB$$

where

B = the benefits of the development project

C = the cost of the project (excluding the environmental cost)

EB = the environmental cost of the project, measured as the value of the forgone benefits resulting from the decline in quantity or quality of the environmental resource.

Following the production approach, there are three types of methods for taking account of the environmental cost, EB:

- *the dose/response method* measures the forgone environmental benefits in terms of the resulting decline in market value of output in the economy;
- *the opportunity cost method* involves working out the cost of preventing the environmental damage by either modifying or forbidding the proposed development project and making a judgement as to whether that cost is worth incurring;
- *the preventative cost method* involves working out the cost of avoiding or mitigating the environmental damage if the project does go ahead, and assuming that that cost is a measure of the environmental benefits at stake.

We briefly consider each of these approaches.

The dose/response method

The dose/response method estimates value by measuring the contribution of the environmental resource to the output of those who rely on it for the production of goods or services for sale in markets. For instance, a firm combines environmental resources of a given quality with man-made capital goods and labour to produce an output such as tourism services or a harvest of fish. The production function combines natural capital in the form of, say, water quality (Q_1) and reef quality (Q_2), with man-made capital (K), labour (L) and other purchased inputs (M) to produce the output (X):

$$X = f(K, L, M, Q_1, Q_2)$$

The effect of a change in Q_1 or Q_2 is then gauged by the size of the impact it has on X, and possibly K, L and M, and the benefit or cost of that impact measured in the markets for these commodities.

For example, to value the impact of agricultural development on land adjacent to a coral reef and a fishery, the analyst would need to model the physical relationship between a pollutant dose, such as run-off of nitrates, and the change in water quality and its subsequent impact on fish stocks and reef quality. They would then need to estimate the production response of users of the fishery and calculate the reduction in net value of the catch. Similarly the value lost or gained from reef-based tourism as a result of changes in reef quality, as opposed to changes in other input levels such as man-made capital, material inputs, and labour, would need to be estimated. This approach is most readily applicable to environmental resources which have direct market-based uses, such as tourism and fishing. It cannot be used to estimate non-use values.

Another example of the dose-response method is the human capital approach, which measures the value of a change in environmental quality in terms of its impact on human health and the earnings capacity of the individuals affected. This could be used to calculate the costs of air or water pollution (or benefits of reduced pollution) to the extent that these can be accounted for in terms of their impact on human health and productivity.

The opportunity cost method

The opportunity cost method is often used when it is not feasible to value the environmental benefits or costs of a project or policy. *Opportunity cost* refers to the cost of limiting or preventing the loss of amenity resulting from the environmental damage associated with a proposed project. Loss of amenity can be avoided in several ways, including not proceeding with the project, modifying the project, or providing a replacement for the environmental asset affected. The opportunity costs associated with each of these ways can be compared to determine the least-cost alternative, which is then defined as the opportunity cost of avoiding the environmental damage. It is left to the judgement of the decision-maker whether the benefit of avoiding the loss of environmental amenity is large enough to justify the opportunity cost. If so, the least-cost remedial measure is adopted, whether that is not undertaking, or modifying the proposed project, or providing a substitute for the asset in question. Example 8.1 illustrates this approach applied to the question of water quality improvement.

EXAMPLE 8.1 The opportunity cost method

Irrigators in a water catchment area are currently growing 10,000 tons of sugar cane per annum. They receive $150 per ton and their variable costs are $50 per ton. As a result of cane cultivation, the water quality in the catchment has fallen to a level of 40 points measured on a 0–100 point indicator. Three mutually exclusive proposals are being considered to improve the catchment's water quality as summarised in Table 8.1. Each proposal involves a combination of direct costs, such as those of improvements in irrigation channels, and indirect costs, such as reduction in the yield of cane resulting from the creation of riparian non-cultivation zones. Option 1 involves annual direct costs of $50,000 which raises the index of water quality to 50; option 2 costs $30,000 per annum which raises the water quality index to 65; and option 3 costs $20,000 per annum and raises the water quality index to 85. In addition, irrigators are required to reduce their output of sugar by: 10% under option 1, 20% under option 2 and 40% under option 3. Three questions are posed:

i   Which option has the lowest total opportunity cost?
ii Which option has lowest cost per unit of water quality improvement?
iii Why might neither of these represent the most efficient option?

Table 8.1 Water quality improvement project options

	Input Values		Direct Cost p.a ($)	Output reduction p.a.(%)	Quality Index	Total Cost p.a. ($)	Unit Cost/ Quality Change p.a. ($)
Output (tons)	10,000	Option 1	50,000	10	50	150,000	15,000
Price ($/ton)	150	Option 2	30,000	20	65	230,000	9200
Costs ($/ton)	50	Option 3	20,000	40	85	420,000	9333

From Table 8.1 the answers to the three questions posed by Example 8.1 can be provided as follows:

1   Option 1 has lowest total opportunity cost equal to $150,000. The value of forgone output is $100,000, being 10% of 10,000 at $100 net revenue per annum ($150–50), to which direct costs of $50,000 are added.
2   Option 2 has the lowest cost per unit of water quality improvement equal to $9,200 per annum. Total cost is $230,000 per annum, which is divided by the increase in quality index of 25 (65–40).
3   The opportunity cost method does not provide an estimate of the environmental benefit from the improvement in water quality, but leaves it to the judgement of the decision-maker whether the value to society of a particular amount of improvement in environmental quality is worth the additional opportunity cost.

The decision-maker can employ the incremental project method discussed in Chapter 3, Section 3.5.4, to choose from among the three options. Each option can be regarded as providing an incremental benefit at an incremental cost as compared with the previous option. Thus, Option 1 provides 10 additional units of water quality at a cost of $150,000, as compared with the option of doing nothing; Option 2 provides a further 15 units of quality at an additional cost of $80,000 ($230,000–$150,000); and Option 3 provides a further 20 units of quality at an additional cost of $190,000 ($420,000–$230,000). However, without a dollar measure of environmental benefit, we cannot say whether additional units of water quality are worth the additional cost.

In a threshold analysis, discussed in more detail in Example 8.2, the analyst turns that question around by calculating what *minimum* value the associated (unknown) environmental benefits would need to have to justify the choice of a particular option. If decision-makers consider the environmental benefits due to the water quality improvement to be worth at least this amount, the option can be selected.

The answer to the questions posed by Example 8.2 involves choosing from among three mutually exclusive projects. As discussed in Chapter 3, the appropriate decision-rule is to choose the project with the highest NPV. If the project does not proceed (Option (i)), the man-made capital invested elsewhere will yield an NPV of $15 million and the wetland will yield an NPV of $EB million. If the project proceeds without replacement (Option (ii)), it yields an NPV of $40 million. If it proceeds with replacement (Option (iii)), the NPV is $(40 + EB – K) million. Hence we would choose:

e.g.

EXAMPLE 8.2 Threshold analysis

Suppose the government is considering an application for a development project that will involve degradation of a wetland. The wetland plays an important role in the provision of ecosystem services, such as breeding habitat for some rare species of birds. To replace the wetland elsewhere is estimated to cost $K million. The development project is expected to generate a net income to the Referent Group (in economic efficiency prices) with a NPV of $40 million (not taking account of the cost of the loss of the wetland). If the same capital was invested elsewhere it would generate net benefits with a NPV of $15 million, implying that the contribution of the wetland's *natural capital* to the project has a NPV of $25 million. What would the present value to society of the wetland's ecosystem services, denoted by $EB million, need to be to justify a decision:

i   to preserve the wetland intact, not allowing the project to proceed?;
ii to allow the project to proceed without replacing the wetland elsewhere?;
iii to allow the project to proceed, provided the wetland is replaced elsewhere?

---

i   Option (i) if $15 + EB > 40 (i.e. EB > 25) *and* $15 + EB > 40 + EB − K (i.e. K >
ii   25); (ii) Option (ii) if $ 40 > 15 + EB (i.e. EB < 25) *and* $40 > $40 + EB − K (i.e.
     EB < K);
iii Option (iii) if $40 + EB − K > 15 + EB (i.e. K < 25) *and* $40 + EB − K > 40 (i.e. EB > K).

Suppose we have an estimate that K < $25 million: that immediately rules out Option (i); the choice between Options (ii) and (iii) then turns on whether EB > K, and it may be possible for the decision-maker to make that decision without further research.

Both of the above applications of the opportunity cost method provide dollar measures of the cost of environmental preservation, and invite the decision-maker to choose the appropriate level of environmental quality by trading these costs off against benefits measured in physical units. An alternative way of applying the opportunity cost method is to select an objective or target, such as a given level of water quality, and then choose the management option that achieves the target at the least cost. This approach is usually referred to as *Cost-Effectiveness Analysis* (CEA). For example, if the target level of water quality is set at, say, 50 points and its current level is 40 points, a CEA would compare the cost of alternative means of raising water quality by 10 points. The target level of 50 points is treated as a given (by policy-makers), implying no need to measure the value of the benefits associated with the improvement in water quality.

The preventative cost method

Preventative cost refers to the cost of preventing loss of environmental amenity by restoration or replacement of the environmental asset (replacement cost), or by not undertaking the project in question. In this approach the value of the services of the environmental asset is inferred from the cost of restoring or replacing the degraded environment to its level before the damage occurred; for example, the cost of restoring lost vegetation, forests, wetlands, or coral reefs is sometimes used to estimate damages in legal proceedings. In terms of Example 8.2, only Options (i) and (iii) would be considered: Option (i) would be selected if K > 25, i.e. if the replacement cost is too high to be borne by the proposed project; and Option (iii) would be chosen if K < 25, i.e. the project can incur the replacement cost and still record a positive NPV.

Since the environmental cost inflicted by a project is, in general, the sum of the mitigation cost and the residual damage cost, the least cost option will rarely involve 100% mitigation (i.e. prevention). Consider a household subjected to an increase in aircraft noise; it will generally not be optimal to invest in a level of soundproofing sufficient to block out all the additional noise, but rather less costly overall (in terms of the mitigation plus residual damage costs) to block out some of the additional noise and to suffer the cost of the remaining noise. This means that the observed expenditure on soundproofing is generally a minimum estimate of the cost of the environmental damage. Conversely, when 100% mitigation is mandated under the preventative or replacement cost approach, the observed cost is generally a maximum estimate of the cost of the environmental damage inflicted by the project. This point is illustrated by Figure 8.2 which records dollar values of restoration benefits and costs against the degree of restoration undertaken.

In Figure 8.2, restoring beyond level Q_L would imply incurring total costs greater than the benefits. At Q_L there is no net benefit to this level of restoration as total costs and benefits are equal. At Q_e, net benefits of restoration are maximised. From an

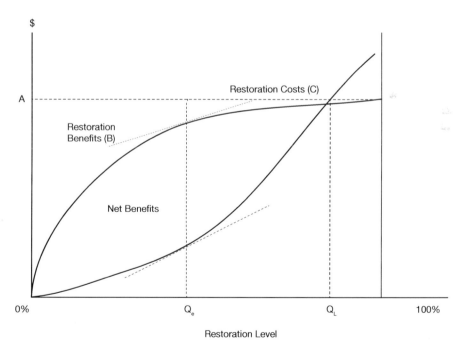

Figure 8.2 Measures of value using the replacement cost method. B represents the total benefit, and C the total cost, of restoration. Net benefit is maximised at restoration level Q_e where the slopes of the schedules are equal (marginal benefit equals marginal cost). At restoration level Q_L net benefit is zero.

economic efficiency perspective, the appropriate level of restoration costs would be Q_e, where marginal benefit equals marginal cost. This analysis assumes that the level of total economic value at each level of restoration, represented by B in Figure 8.2, is known.

It is important to note that all cost of production methods assume that incurring a cost to prevent, avoid or replace produces a benefit at least as great as the value of the environmental cost, or loss of benefit, to society. This could be considered a reasonable assumption when the cost is incurred by private individuals or conservation groups, and when the objective is to assess the value to those individuals or groups. These methods become a lot less acceptable when used to estimate the value of benefits to society as a whole, and to calculate how much governments should spend on the environment.

While the production approaches usually use actual market prices as the basis for calculating costs for valuation purposes, there is no reason why the analyst could not use measures of opportunity costs to derive shadow prices, following the methodology developed in Chapter 5 of this book. For instance, if the cost of avoiding pollution damage from agricultural run-off is a loss of agricultural output, the loss of net income to the farmers will not reflect the opportunity cost to the economy if agricultural production is subsidised. In this case, appropriate shadow-prices should be used.

8.5.2 *The utility approach*

Neoclassical economic theory is based on the working assumption that consumers are rational and make consumption decisions in accordance with the objective of maximizing their individual utility, subject to their income or budget constraint, and given the market prices of all goods and services. As discussed in Chapter 7, the net benefit consumers obtain from consumption activity is measured by *consumer surplus* which is the difference between the amount actually paid for the quantity of the good consumed and the amount that the consumer would be *willing to pay*. The concepts of *consumer surplus* and *willingness to pay* can be illustrated by means of a market demand curve for a product, as shown in Figure 8.3.

As discussed in Chapter 7, we can measure the total annual WTP for a given consumption level, Q_1, by the area $OABQ_1$. If the actual charge to users for access to, say, a park, is P_1 the area OP_1BQ_1 measures the amount actually paid, or total cost. The total benefit, measured by the amount of WTP, exceeds the cost by the area of the triangle P_1AB. This area represents the *consumer surplus* (CS), and is the measure of annual net benefit to the consumers who pay price P_1 for use of the park.

What happens when the good in question is a public good for which there is no market price? First, if the public good has the non-rival characteristic then, unlike in the case of a private good, a given quantity can simultaneously be consumed by any number of consumers; a free-to-air radio or TV signal is an example. In that case, the market demand for the good is obtained by summing the individual consumer demand curves *vertically*, instead of horizontally (or laterally) as described in Chapter 7 in the case of a rival private good. Second, if there is no market price, we cannot actually observe a relationship between price and quantity demanded from market information. Nonetheless it can be assumed that a demand curve for the use of the public good exists, and it can be estimated by means of non-market valuation techniques, even though the amount actually paid to use it is zero. In the absence of a price, annual Consumer Surplus (CS) is given by the entire amount of the Willingness to Pay (WTP) – the area of the whole triangle ODA under the demand curve

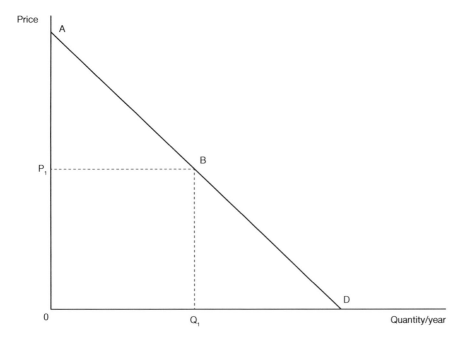

Figure 8.3 Willingness to pay and consumer surplus. AD is the market demand curve for the product, P_1 is the market price, and Q_1 the annual quantity consumed. The net benefit of consuming quantity Q_1 is measured by the area of consumer surplus, ABP_1.

in Figure 8.3. It is for this reason that the net benefit of an environmental asset is often equated with WTP, because, in the absence of a price, CS = WTP.

As noted previously, when valuing the goods and services generated by environmental assets we are more likely to be interested in comparing *changes* in net benefits resulting from changes in the management of these assets at the margin, than in the total asset value. The change in consumer surplus to users of an environmental resource, associated with a proposed project to improve the quality of the resource is measured by the maximum amount of money that they are willing to pay to obtain the improvement rather than do without it. A management option that improves the quality of a resource is likely to increase consumers' demand for that resource, resulting in a change in consumer surplus. This is illustrated in Figure 8.4, where tourists have access to a coral reef at a cost or price of P_0, and an improvement in reef quality (measured, say, in terms of percentage of reef cover) causes the demand for the reef's services to shift from D_0 to D_1. The resulting change in WTP is given by the entire area between the two demand curves, and the change in CS by the area ABCE.

Note that this interpretation of the significance of a demand shift is different from that discussed in Chapter 7 where the demand shift was the result of a change in the price of a substitute or complementary good and where there was no consumer surplus change to measure. In the present case the demand shift is a result of a real improvement in the quality of the resource, and reflects an increase in the value of the service offered.

If the management option is a change in price, such as the entrance fee to a park, without any change in the quality of the service offered, the demand curve and WTP for the services

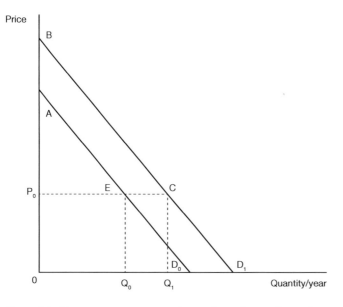

Figure 8.4 Change in consumer surplus resulting from a demand curve shift. Market demand shifts from AD_0 to BD_1 as a result of a quality improvement. The annual net benefit, measured by the consumer surplus, increases from AEP_0 to BCP_0.

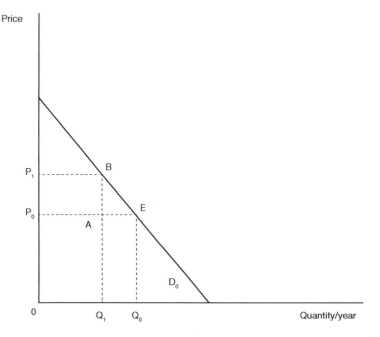

Figure 8.5 Change in consumer surplus resulting from a price change. D_0 represents the market demand curve. When price rises from P_0 to P_1 quantity consumed (i.e.. number of park visits) falls from Q_0 to Q_1 and the amount of annual consumer surplus falls by P_1BEP_0.

of the park do not change. However, the change in net benefit to users is measured by the change in CS as illustrated in Figure 8.5. If the access fee rises from P_0 to P_1 and quantity demanded falls from Q_0 to Q_1, CS falls by the amount given by area P_0P_1BE.

It should not be forgotten that, while the change in CS measures the change in net benefit to consumers as a result of the change in price, in a CBA framework we are interested in the net benefit from the perspective of all members of the Referent Group, including government. In the above example, access fee revenues change by $P_0P_1BA - Q_0Q_1AE$ and this change must be included along with the change in consumers' net benefit to arrive at the net benefit or cost to the economy as a whole in the Efficiency and Referent Group Analyses. Furthermore, any reduction in management costs resulting from the lower number of visitors would also be included as a benefit.

Up to this point the discussion has been limited to instances in which changes in WTP and CS are attributable only to changes in the value to the users of an environmental resource. It was noted previously that TEV can include both use and non-use values. However, as will be noted in the following discussion about different utility-based valuation methods, not all are capable of capturing and measuring both use and non-use values.

8.6 Revealed and stated preference methods of applying the utility approach

There are essentially two categories of valuation methods based on the utility derived by consumers: the "*Revealed Preference Method*" and the "*Stated Preference Method*". The former uses observations from consumer behaviour as revealed in actual or surrogate markets, while the latter uses a survey instrument to construct hypothetical markets in which the respondents are asked (usually in an indirect kind of way) to express their preferences. Coincidentally, both these two main approaches to estimating the demand for an environmental service, such as outdoor recreation were proposed in the late 1940s: Hotelling (1949) suggested the Travel Cost Method (TCM) and Ciriacy-Wantrup (1947) suggested the Contingent Valuation Method (CVM).[1] The TCM is one of a range of methods that rely on revealed preference. Similar methods include the Random Utility Model (RUM) and the Hedonic Price Model (HPM). The CVM is one of a range of methods that rely on stated preferences. A similar technique is Discrete Choice Modelling (DCM), sometimes referred to as Choice Experiment Modelling.

8.6.1 *Revealed preference methods*

The Travel Cost Method

An example of a surrogate market which provides information for valuing the services of an environmental asset is expenditure on travel. If we observe the amount individuals spend on travel and other associated costs in getting to and from the non-market activity they participate in, such as a day's recreational fishing around the lagoons of the reef, we can infer the value to those individuals of the activity; hence the term Travel Cost Method (TCM). By collecting survey-based information from different groups of recreational fishers about the distances travelled, time spent and associated costs, such as the hire of equipment, as well as the frequency of their visits, it is possible to construct a demand curve for recreational fishing. The TCM uses information on total visitation and associated travel costs to sites to estimate the relationship between the prices or costs of trips experienced by different individuals and the numbers of trips they choose to undertake. This method assumes that an individual's demand for access to a site is related to travel (and associated) costs in the same way as demand would be related to the level of an entrance fee. As individuals must be using the site to reveal a preference for it, it follows that this method is not capable of measuring non-use values. For example, if you value an area of rainforest for both the recreational value it offers you and its continued existence in its current state for your (and future generations') possible use, any valuation based exclusively on the costs of your current visit would capture only its use value to you.

e.g.

EXAMPLE 8.3 Travel Cost Method

Suppose that individuals A and B have the same money income, have the same tastes, and face the same set of prices of all goods and services except that of access to a National Park. Individual A lives further away from the park than Individual B and hence incurs a higher travel cost per visit. There is no admission charge to enter the park. The data in Table 8.2 summarise their annual use of the park.

You are asked to calculate:

i   approximately how much consumer surplus does Individual A receive per annum from her use of the park?

ii   approximately how much consumer surplus does Individual B receive per annum from his use of the park?

iii what is the approximate measure of the annual benefits to A *and* B from their use of the park?

iv   why are the measures in (i), (ii) and (iii) "approximate"?

Table 8.2 Hypothetical travel cost example

Individual	Travel cost per visit	Visits per year
A	$15	10
B	$5	20

To answer the questions posed in Example 8.3, note that in this simple example there are only two visitors, and they are assumed to have identical characteristics, implying identical demand curves. Also note that each individual's demand for visits to the park is assumed to be independent of the number of visits chosen by the other individual. This suggests that the park has the non-rival characteristic of a public good, implying that there are no congestion costs. If the two demand curves are identical, the pair of (price, quantity) observations given in Table 8.2 must lie on the same individual demand curve as shown in Figure 8.6.

Assuming that no park entrance fee is charged:

i   consumer surplus for A = 0.5(10×10) = $50

ii   consumer surplus for B = 0.5(20x20) = $200

iii total consumer surplus = $50 + 200 = $250

iv   the estimate of total consumer surplus is approximate as the demand curve will not generally be a straight line as assumed here.

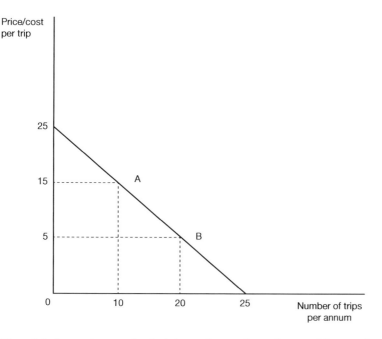

Figure 8.6 Approximate individual demand curve for park visits. The price/quantity combinations (15,10) and (5,20) are observed for a pair of individuals A and B. Assuming they have the same linear demand curve for the service the individual facing the higher price receives consumer surplus of $50 per annum, and the individual facing the lower price receives $200.

Since the individuals using the park will not have identical demand curves, because of different tastes, income levels and prices of the other goods which they consume, statistical methods must be used to accommodate these differences. For example, the number of visits can be regressed against travel cost, education level, household income, and prices of other goods (such as alternative sites) to get a conditional relationship between the number of visits and price.

The main strengths of the TCM are:

- it is relatively easy to use;
- it is well suited to cases where current use values are required;
- it is based on actual revealed preference observations.

The main weaknesses of the TCM are:

- it has limited ability to measure the values of individual site attributes;
- there are complications when valuing a trip with multiple destinations or purposes (though the multiple site TCM has also been developed);
- biases arise if possible substitute sites are not properly incorporated into the analysis.

The Random Utility Method

The Random Utility Method (RUM) deals with the situation in which the question facing the consumer is not whether or not to visit an individual site, but rather which of a group

TECHNICAL NOTE 8.1 The random utility model

Suppose that the utility an individual gets from visiting beach i is given by:

$$U_i = bP_i + a_1 x_{1i} + a_2 x_{2i} + \cdots + a_n x_{ni} + e_i$$

where

P_i is the cost of visiting the site (the individual's travel cost to beach i);

X represents various characteristics of site i, e.g. length and breadth of the beach, sandy or rocky, wave pattern, presence of lifeguards; availability of facilities, etc.;

e_i is a random variable that picks up the effects of characteristics known to individual i but which cannot be observed by the researcher;

b and a_k are parameters.

The individual chooses to visit site i if: Ui > Uj, where j represents all the other possible sites to visit. The probability that the individual chooses site i is given by:

$$\Pr\left(bP_i + a_1 x_{1i} + a_2 x_{2i} + \cdots + a_n x_{ni} + e_i > bP_i + a_1 x_{1j} + a_2 x_{2j} + \cdots + a_n x_{nj} + e_j\right)$$

where j represents all the possible choices other than i.

The probability distribution chosen for ei determines the explicit form of the probability of choosing site i. The logit model gives the following expression for the probability of the individual choosing site i:

$$\Pr(i) = \frac{\exp\left(bP_i + a_1 x_{1i} + a_2 x_{2i} + \cdots + a_n x_{ni} + e_i\right)}{\sum \exp\left(bP_i + a_1 x_{1j} + a_2 x_{2j} + \cdots + a_n x_{nj} + e_j\right)}$$

Values of the parameters, b and a_1, a_2, etc. are chosen so as to maximise the log-likelihood of observing the actual pattern of visits observed:

$$\ln L(b,a) = \sum_{k=1}^{m} \sum_{i=1}^{n} r_{ki}\left(\frac{\exp\left(bP_i + a_1 x_{1i} + a_2 x_{2i} + \cdots + a_n x_{ni} + e_i\right)}{\sum \exp\left(bP_i + a_1 x_{1j} + a_2 x_{2j} + \cdots + a_n x_{nj} + e_j\right)}\right)$$

where there are m observations on individual choice and n sites; r_{ki} = 1 if individual k chooses site i, and zero otherwise.

How do we use the model? Suppose that we want to know what value individuals place on having lifeguards on the beaches sampled. We can evaluate U for each individual when they make their best choice of site with (x_L=1) and without (x_L=0) lifeguards. Their utility will be lower if the lifeguards are removed (coefficient a_L> 0). The fall in utility can be converted to dollars by dividing by the negative of the coefficient on P (travel cost), representing the marginal utility of income. We do this calculation for all individuals in the sample, and then extrapolate to the population as a whole.

of sites to visit: for example, which beach to go to? The individual's choice will depend on the cost of travel to the site, and the characteristics of the site. We are often interested in the costs or benefits of changes to the characteristics of recreational sites: for example, what is the cost of beach erosion? (and what is it worth spending to prevent such erosion?); what value do users place on the services of lifeguards? The RUM can help answer such questions.

As with the TCM, the RUM uses trip data and travel costs, together with consumer characteristics such as income level and tastes (level of education or experience) to estimate a demand function for a recreational site. Unlike the TCM, which assumes the number of visits is a continuous function of price (travel cost), the RUM is a model of discrete choice among substitute sites which analyses consumer choice among alternative recreational sites. It is often used to value fishing or hunting trips, where the probability of visiting one site, given information about all other sites and the individuals' characteristics, can be modelled. Since the RUM takes explicit account of site characteristics in modelling choice among sites, it can be used to value the individual attributes of a site, and changes in site value per trip as a result of changes in these attributes. The application of the RUM is illustrated by Technical note 8.1 where it is used to estimate the value beach-goers place on the presence of lifeguards.

The main advantage of the RUM is that, unlike the TCM, it takes explicit account of the existence of substitute sites in situations where individuals change site choice in response to changes in opportunities available. It also allows for valuation of each attribute of a recreational site visit. The main problem with the RUM is that it does not account for the number of trips made, which means that when the quality of the site changes only the estimate of the per trip benefit to each user changes; the number of trips made is assumed to remain constant.

The Hedonic Price Method

As an alternative to the TCM or the RUM we can use the behaviour of consumers as revealed in related markets to infer their preferences for the non-marketed good we are interested in. The underlying proposition is that an individual's utility for a good or service is derived from the attributes of the good or service in question, and that it is possible to distinguish the value of each attribute. For example, if the quality of an environmental resource, such as air or water, is considered an important attribute entering our choice of house, variations in air or water quality should directly affect relative house prices. For instance, the value to the resident of property frontage on a waterway could be affected by the quality of the water. If we compare house prices in polluted vs non-polluted situations, with controls for price differences attributable to other factors, we should be able to measure the dollar value of differences (and changes) in water quality. When all other effects have been accounted for, any difference in property price is attributed to the differential water quality. This is called the Hedonic Price Method (HPM).

To apply the HPM in this instance, data on house prices are gathered to estimate a model that explains variations in house prices in terms of a whole set of attributes, one of which is the environmental attribute in question. For example, observed house price can be modelled as a function of house and site characteristics, neighbourhood characteristics, and environmental quality characteristics, such as exposure to traffic noise or proximity to a park. This is the *hedonic price function* which can be expressed in general terms as:

$$P_i = F(H_i, S_i, N_i, Q_i)$$

where

P_i = price of house i
H_i = a vector of building characteristics
S_i = a vector of site characteristics
N_i = a vector of neighbourhood characteristics
Q_i = a vector of environmental quality characteristics.

If a linear form of the hedonic price function is fitted to the house price data the coefficient on environmental quality characteristic Q_j tells us the effect of a change in the quality measure on the price of an average house. For example, if the quality measure was aircraft noise the (negative) coefficient would tell us the effect of an extra decibel (resulting, perhaps, from construction of an additional runway) on the price buyers are willing to pay for a house with otherwise average characteristics. This reduction in WTP is an implicit measure of the unit cost of the additional noise inflicted on the average house. When multiplied by the number of houses affected, it provides an estimate of the present value of the cost of the noise inflicted by the additional runway. If a logarithmic form of the function is fitted, the coefficient on characteristic Q_j tells us the percentage change in WTP for the average house in response to a 1% increase in the amount of the characteristic; for example, a 1% increase in noise level might cause a 2% fall in price. In this case the estimate of the present value of the cost of the additional noise is 2% of the current market value of the housing stock affected by the new runway.

---

e.g.

EXAMPLE 8.4 Hedonic pricing

It is proposed to build a new city ring road which will increase noise levels in two suburbs, A and B, by 6 and 8 units respectively, and reduce noise levels in two others, C and D, by 3 and 10 units respectively. You are required to calculate the environmental cost (or benefit) of the change in traffic noise due to the project. A hedonic pricing model based on recent sales prices and house characteristics has estimated the percentage fall in the price of the average house in each area which can be attributed to a one unit increase in noise level: these percentage changes are Area A = −0.45; Area B = −0.18; Area C = −0.92; Area D = −0.08. Table 8.3 shows the number of houses in each area and the mean house price prior to the construction of the new road, together with the estimates of the changes in noise levels and the changes in house prices attributable to a one unit change in noise level.

Table 8.3 Impact of road noise changes on property values

Area	Mean house price ($)	Number of houses	Change in noise level	% change in house price per unit change in noise level	Total change in cost of noise ($ thousand)
A	250,000	76	6	−0.45	513.0
B	125,000	124	8	−0.18	223.2
C	400,000	64	−3	+0.92	−706.6
D	100,000	60	−10	+0.08	−48.0
		Net cost of change in noise levels			−$18.4

The changes in noise levels will result in a net increase in the value of properties of $18.4 thousand (a negative cost) as calculated in Table 8.3. This sum is an estimate of the present value of the net environmental benefit of the proposed road.

The main strengths of the HPM are:

- it is conceptually intuitive;
- it is based on actual revealed preferences.

The main weaknesses of the HPM are:

- it requires a relatively high degree of statistical knowledge and skill to use;
- it generally relies on the assumption that the price of the house is given by the sum of the values of its individual attributes, implying a linear relationship among attributes (though, as noted above, a log-linear form can be used);
- it assumes that there is a continuous range of product choices containing all possible combinations of attributes available to each house buyer.

8.6.2 Stated preference methods

Stated preference methods use surveys or other instruments such as focus groups in an attempt to get consumers to express their view of use and non-use values of an environmental asset on the same basis as they would reveal their preferences in actual markets through the price mechanism. A hypothetical situation for the use of an environmental resource is described and the interviewees are asked, contingent on the existence of the situation described to them, how much they would be willing to pay for the use and/or non-use services of the resource, such as those derived from recreation or existence.

The Contingent Valuation Method

The most commonly used stated preference method is the *Contingent Valuation Method* (CVM). CVM uses surveys to ask people directly how much they would be willing to pay for a change in the quality or quantity of an environmental resource. The replies, together with information on income level and personal characteristics, are subjected to statistical analysis to determine the WTP of a typical respondent. The resulting sample mean (or median) WTP is then multiplied by the relevant population to estimate total WTP. CVM is, in principle, a relatively simple method, although state-of-the-art applications have become quite complex.

CVM is susceptible to a number of response biases. These include:

- *hypothetical market bias*: where responses are affected by the fact that it is a hypothetical and not a real market choice, and where individuals may enjoy, for example, "warm glow" effects from overstating their true preferences for an environmental good, or where they simply want to please the interviewer – "yeah-saying";
- *strategic bias*: where respondents believe that their survey response bids could be used to determine actual charges or expenditures they may understate or overstate their true WTP;
- *design bias*: the way in which the information is presented to the respondents can influence the individuals' responses, especially concerning the specification of the payment vehicle (a "tax" or a "contribution"), raising the question of how far preferences can be considered exogenous to the elicitation process; and
- *part-whole bias*: individuals have been found to offer the same WTP for one component of an environmental asset, say, recreational fishing in one river, as they would for fishing in the entire river system.

With refinements in the design of surveys, most of these biases can be avoided or at least minimised, though the debate still continues as to whether individuals are indeed capable of expressing preferences for environmental and other public goods and services on the same basis as they would for private goods in, say, the context of a supermarket or shopping mall.

One of the main advantages of CVM, as with other stated preference methods, is that it is capable of estimating both use and non-use values and it can be applied to many types of situation, one of which is illustrated by Example 8.5.

We can see that Option 3 generates the largest annual net benefit of $750,000; the ranking of options on this measure is Option 3, then 1, then 2. However, Option 1 yields the highest net benefit per dollar of cost, of $0.25, followed by Option 3, then Option 2.

e.g.

EXAMPLE 8.5 Contingent Valuation Method

A local authority is considering alternative mutually exclusive proposals to improve the quality of the water in the river on which the city is located. The three options are: Option 1, improve river quality to a level suitable for recreational boating; Option 2, improve river quality to a level suitable for recreational fishing; and, Option 3, improve river quality to a level suitable for swimming. The annual costs of the options are estimated at $2 million, $3.25 million, and $4.25 million respectively. A contingent valuation study based on a representative sample of households, including both users and non-users of the river, estimates average annual WTP for the three options at $12.5, $17.50 and $25 per household respectively. The city has a total of 200,000 households. Suppose you are required to rank the options in terms of:

i   maximum aggregate net benefit
ii   maximum net benefit per $ invested.

Which of the three options would you recommend and why? The calculation of project net benefit (in $ thousands) and net benefit per dollar cost is shown in Table 8.4.

Table 8.4 Estimating net benefits of improved water quality using CVM

	No. of households in population (000's)	Sample WTP/ HH ($)	Estimate of total WTP ($000's)	Project cost ($000's)	Net Benefit ($000's)	NB/$cost
Option 1	200	12.5	2500	2000	500	0.25
Option 2	200	17.5	3500	3250	250	0.08
Option 3	200	25	5000	4250	750	0.18

If funds are not rationed, then Option 3 would be preferred. If public funds are rationed, it would be necessary to compare the net benefit per dollar of cost with other proposed public projects to determine the net benefit per dollar cost of the best alternative project which could be funded, which we refer to as the *cut-off level*. For instance, if the cut-off level is more than $0.18 but less than $0.25, Option 1 would be chosen. If it was less than $0.18, then Option 3 would be chosen. Option 2 would never be preferred as Option 1 costs less and yields a higher value of net benefit.

Discrete Choice Modelling

One of the main limitations of CVM is that it is usually restricted to comparison of only one or two options with the status quo. The *Discrete Choice Modelling* (DCM) method, which is another stated preference approach, provides for the comparison of a much broader range of options. DCM is a form of conjoint analysis using choice experiments. The environmental good is described in terms of a set of attributes, where each attribute is allowed to take on a number of possible values over a defined range. All possible combinations of attributes are assembled to derive a large matrix of hypothetical scenarios for dealing with the environmental resource. Surveyed individuals are presented with approximately 8–10 choice sets, each containing (usually) two hypothetical options together with an option, which describes the status quo, and are asked to select their most preferred. The survey data are used to estimate trade-off values for each attribute, using a multinomial logit model. Provided that one attribute is expressed in monetary terms (e.g. an entrance fee, a tax, or an addition to some existing charge), then the net benefit, measured in money terms, of each management option can be calculated from the regression results.

e.g.

EXAMPLE 8.6 Discrete Choice Modelling

In a study by Morrison and Bennett (2000), the DCM method was used to value a wetland that provides a number of important ecosystem services including habitats for water birds and water filtration for downstream users. With the construction of a dam on the adjacent river, the area of wetland fell from $5000km^2$ to $1000km^2$ and the number of endangered and protected bird species using the wetland fell from 34 to 12. A DCM questionnaire was developed where respondents were informed about the problem and management options were specified. Through a series of focus group meetings with stakeholders, the key attributes characterizing the wetland were identified and, according to an experimental design, possible combinations of levels for each attribute were combined into a number of choice sets, an example of which is shown in Table 8.5.

Table 8.5 Hypothetical example of a choice set

Attribute	Status quo	Option 1	Option 2
Water rates (one-off increase)	No change	$20 increase	$50 increase
Irrigation related employment	4400 jobs	4350 jobs	4350 jobs
Wetlands area	1000km^2	1250km^2	1650km^2
Water birds breeding	Every 4 years	Every 3 years	Every year
Endangered and protected species present	12 species	25 species	15 species

(Source: Morrison, M. and Bennett, J. (2000) "Choice Modelling, Non-Use Values and Benefit Transfer", *Economic Analysis and Policy*, 30(1): 13–32)

Respondents were asked to select their most preferred of the three options in each choice set. This enabled estimation of the relative importance of each attribute. The study found, for instance, that respondents were willing to pay 13 cents for an extra irrigation-related job preserved and about $4 for an additional endangered species to be present. Once the marginal values for each attribute were estimated, the actual levels of each attribute corresponding to each policy option under consideration can be used to compute the relative value of each option.

Variants of the DCM method include *Contingent Ranking* and *Contingent Rating* where, instead of making a single choice from each set of options, the respondent is required to rank or rate (on a given scale) each hypothetical option in the set.

DCM may be less prone than CVM to bias because of its emphasis on choice among specific alternatives. One advantage of the DCM is that it enables the analyst to derive the trade-off values for, and compare the net incremental benefits of, a number of alternative management interventions in one survey questionnaire. The main disadvantage is that there is no rule as to how many attributes should be chosen and over what range their possible values should be allowed to vary. Increasing these features complicates the choice experiment enormously. Another restriction is that DCM assumes that the value of each option is given simply by the linear sum of the values of its attributes. This does not allow for the possibility that, in some situations, the whole is greater or less than the sum of its parts.

8.7 Benefit Transfer and Threshold Analysis

It will be obvious from the above discussion that estimating non-market values can be a very costly affair. Collecting data from cross-sectional surveys is time-consuming and expensive and significant expertise is required to process and interpret the data. Recalling the discussion in Chapter 1 of the appropriate level of resources to be devoted to the cost-benefit study, an issue to be revisited in Chapter 9 where we discuss the value of information, it is prudent to weigh the cost of additional and more precise information against the benefit of the resulting improvement in the quality of the decision which will be made.

If elaborate non-market analysis of the effects of a project cannot be justified on cost-benefit grounds an alternative valuation method, known as *Benefit Transfer*, can be used. Benefit Transfer simply means adopting a non-market value from a study, based on one of the methods discussed above, which has already been conducted elsewhere. To use this method, the analyst reviews the literature to look for estimates of the non-market value of the effect in question: air or water pollution, soil salinity reduction, carbon sequestration, traffic accidents and so on. Care must be taken that the context of the study selected for comparison is similar to that of the project in question: a country at a similar level of development and with a similar natural resource endowment, and a project with similar characteristics. Given that the study from which the non-market value was derived would have been undertaken some time ago, it will also be necessary to inflate the value to its present-day equivalent using an appropriate index such as a GDP deflator (or a real exchange rate index in the case of a value from another country).

While a search engine could be used to identify suitable studies, some on-line sites are available that are specifically intended to facilitate implementation of the Benefit Transfer Method. These sites list references to non-market valuation studies by country and by type and can be used to search for an appropriate comparison. In Australia, the government of New South Wales operates *Envalue*:

www.environment.nsw.gov.au/envalueapp/

which offers studies of air and water quality, noise, radiation, land quality, natural areas, non-urban amenity and risk of fatality. For example, the user can choose air quality (101 studies) and then search sub-categories, such as odour (one study).

Canada operates the Environmental Valuation Reference Inventory (*EVRI*):

www.evri.ca/Global/HomeAnonymous.aspx

which contains around 1700 studies grouped by geographic area, environmental focus asset, and valuation technique. Focus asset has four sub-categories: air, land, infrastructure and water; and valuation technique has three: revealed preference, stated preference and simulated market price.[2]

A simple technique which can be used on its own or in partnership with Benefit Transfer is *Threshold Analysis* (sometimes termed *Break-Even Analysis*) illustrated by Example 8.2 and further discussed in Chapter 9. Here the analyst uses the spreadsheet model to determine what level the benefit (cost) of the non-marketed effect would have to attain to result in the project being approved (rejected); the level can be established by trial and error or by using the *Goal Seek* function in *Excel*©. If this level is sufficiently low (high), then it may be reasonable to recommend approval (rejection). Alternatively, the analyst might prefer to indicate to the decision-maker(s) what the threshold value is without passing judgement on whether the project should be accepted or not. If values obtained through Benefit Transfer are available for comparison, confidence in interpreting the result of the Threshold Analysis is increased.

8.8 Alternative approaches to environmental valuation

Criticism of CBA and orthodox methods of environmental valuation has led some economists to propose totally different approaches to appraisal and evaluation of environmental resource management options.

Deliberative Value Assessment

Deliberative Value Assessment (DVA) methods treat the *process* of value elicitation as an important factor in determining the validity of the information acquired from surveys. Unlike the traditional methods described above, individuals' valuations are treated as endogenously determined and adaptive, being determined as part of an ongoing activity within the decision-making process itself. In a DVA framework the elicitation of values does not occur independently of the decision-making process itself; it forms part of it. For instance, Citizens Juries (CJ) can be formed to deliberate on matters of public policy such as the use and management of environmental resources. These juries will include expert witnesses with scientific knowledge, including economists, along with, perhaps, the results of their non-market valuations and CBA studies. The jury is required to deliberate and to come to an informed, considered decision on the most acceptable course of action or management intervention to be adopted by the respective authority responsible for the utilization of the resource. A CJ can be used to pronounce upon a value or to make a specific decision.

Multi-Criteria Analysis

A similar approach to a CJ is a multi-criteria analysis (MCA) where each of the relevant stakeholder groups is required to rate or rank alternative performance criteria, with the ranking depending on their preferences and constraints, and then to rate the extent to which they believe alternative management interventions are likely to succeed in achieving the criteria identified. In an issue of *Agenda* devoted to questions of good governance, the MCA approach to project selection was criticised by Dobes and Bennett (2009) as being arbitrary and susceptible to manipulation by vested interests. In the same issue, Ergas (2009) argues for the use of traditional cost-benefit methods in public decision-making.

It could be argued that there is not necessarily any inconsistency between a DVA approach to appraisal and evaluation, and what has been prescribed in this chapter and elsewhere in this book. It is recognised that CBA is essentially an approach to inform the decision-making process. There is no reason why CBA could not also be used in a deliberative, democratic decision-making process and, where that process gives rise to changes in stakeholders' preferences and choices, there is no reason why the CBA could not be conducted on an iterative and interactive basis, as stakeholders' preferences and valuations evolve. In other words, rejection of utility theory's assumption of fixed preferences, which underlies utility-based non-market valuation methods, does not in itself imply rejection of CBA as a decision-support tool. CBA could and perhaps should play an important part in a Citizens Jury or other forms of DVA. Similarly, valuation methods such as CVM and DCM could equally be adapted to decision-making contexts in which it is recognised that values can be endogenously determined and adapted in response to new information.

8.9 Non-market valuation: the value of life

We chose environmental assets to illustrate the use of the production and utility approaches to determining values of non-marketed goods and services because environmental valuation is currently a rapidly expanding area of research and application. However, in traditional CBA, the aspect of non-market valuation which received much of the attention was the valuation of life. A significant benefit of investment in health services or transport infrastructure is the saving of lives and prevention of injuries, and CBA of these kinds of investments requires that values be placed on lives saved.

Modern developed economies spend as much as 20% of GDP on investments aimed at promoting the health and safety of their citizens. This total includes the national health budget together with expenditures on such goods and services as police and fire protection, design of road and building construction, establishing and enforcing road and air traffic regulations, and many other examples. If one dollar in five is allocated to policies dealing with health and safety, it is important that these expenditures are subject to the scrutiny of program and project appraisal. For example, the value of prevention of injury and death was a critical input into the CBA of the 55 mph speed limit discussed in Chapter 5.

Two main questions confront the policy-maker: is the current level of expenditure on health and safety efficiently allocated among projects and programs? and how much in total should be allocated to such projects? The former question can be answered by application of Cost-Effectiveness Analysis (CEA) which compares lives saved and injuries avoided with costs incurred for the various programs and projects involved: the chosen budget is allocated efficiently when the marginal dollar spent in each area yields the same additional health and safety output. The question of how much should be expended in total, and which projects should be funded, is a wider one as it involves a trade-off between health and safety, on the one hand, and other goods and services on the other – the realm of cost-benefit analysis.

As we saw in Chapter 4, the first step in undertaking a project appraisal is to identify the outputs and inputs involved, and the second step is to put dollar values on these quantities. Some of the inputs involved in promoting health and safety have been referred to above and resources such as these can be valued at market prices, or shadow-prices where appropriate. Defining, measuring and valuing the output of health and safety projects are much more difficult tasks and our discussion focuses on these aspects of a CEA or CBA of such projects. We have space here for a cursory introduction only and the reader is referred to the health economics literature for a fuller account.

The output measure generally used in health economics is the Quality Adjusted Life Year (QALY) which refers to a year of life for a person in full health. For a person in less than full health the unit of output measure can be adjusted downwards by a weight reflecting the lower quality of life: for example, if a person has a quality of life assessed at 75% of the quality they would enjoy in full health, the weight to be attached to a year of life saved is set at 0.75. And if a project saves a person in full health from a debilitating injury which would reduce their quality of life by 25%, that project output could be weighted at 25% of a QALY. Such weights are determined by medical experts in consultation with a wide range of community representatives and, needless to say, are the subject of much controversy.

If we accept the QALY as a measure of output, and are able to value the inputs involved, we can proceed with a CEA of a health and safety project. In order to conduct a CBA of such a project, however, we need to establish the value of the output of the project. If we can determine the value of a year of life in perfect health (VLY denoting the value of a QALY) then the value of any particular quality of life year is given by w_i VLY where w_i is the weight to be attached to that quality of life, $1 \geq w_i > 0$ (some analysts argue that some health states are worse than death and that, consequently, w_i could be negative).

Recalling the discussion of Chapters 2 and 3, the value of a life saved can then expressed as the present value of the expected number of additional life years:

$$VSL = w_i \; VLY \; A(r,n)$$

where $A(r,n)$ is the annuity factor incorporating the chosen rate of discount, r, and the expected number of years of life remaining, n, for the person concerned, usually considered to be a young adult with a life expectancy of 40 years. VSL stands for the *Value of a Statistical Life* to emphasise the point that we are not valuing the life of any particular individual, but rather our willingness to pay to save the life of a category of person – young, old, sick or well.

It can be seen that, for a given rate of discount, VSL assigns a higher value to life saved, the younger the person, and the better their quality of life.

We are all familiar with situations in which the community is apparently willing to spend seemingly limitless sums to save the life of a lone yachtsman shipwrecked in the Southern Ocean, or of a miner trapped underground. The concept of a statistical life is intended to exclude such situations in which we know the identity and circumstances of the individual concerned. It refers to the saving of the life of a person selected at random. For example, we cannot say in advance whose life will be saved by a project which straightens out a corner in a road – it could be mine, yours, someone you know, or more likely someone you have never heard of – but we can predict the *number* of lives that will be saved or injuries prevented, and the likely characteristics, in terms of age and quality of life, of the persons involved, and it is those lives that we seek to value. Of course the prediction does not carry certainty with it but, like any statistic, is subject to some confidence interval. What we are paying for when we decide to undertake the road project is a reduction in the *risk* of death or injury.

Suppose it is established that the community is willing to pay $V to reduce the probability of the loss of a statistical life by a small amount, for example, from 2% to 1%. It has been argued that the value of life, VSL, can then be expressed as $V/\Delta p$ where, in this example, $\Delta p = 0.01$. A problem with this argument is that using this measure presupposes a linear relationship between probability and willingness to pay. It seems to imply that if we were willing to pay, say, $10,000 for a 1% reduction in risk the willingness to pay for a 99% reduction in risk (avoiding the near certainty of death) would be capped at $990,000. However, we are rarely faced with choices at the extreme end of the scale, and it may not be unreasonable to assume a linear relationship over the range in which the analysis is likely to operate.

The spending decisions of health and transport departments implicitly place values on lives saved, and estimates of these values can be ascertained by comparing expenditures and outcomes at the margin. However, there are at least two problems with this approach to placing a value on life. First, the project outcome is rarely the single product "a life saved". We have already seen that transport departments are also interested in time savings and, in considering the trade-off between these two objectives, information about the value of time saved (another non-marketed good which was discussed in Chapter 5) would be required to estimate the value placed on life. In the case of health expenditures, there is a research component to medical procedures that may justify the discrepancy between the costs of lives saved by transplanting various kinds of organs, such as, for example, hearts and kidneys. The second problem is similar to the one we will encounter in Chapter 11 when we attempt to infer income distribution weights: the decision-maker is seeking independent advice about the value that "should" be attached to life as opposed to information about the values that are implicitly used. These implicit values are a product of expenditure allocation decisions which ideally should be informed by prior information about the value of life.

We can estimate the value of life using non-market valuation techniques and, as discussed earlier in this chapter, there are two general approaches: the production- and utility-based approaches. Early attempts to estimate the cost of illness or death were based on the production approach and focused on medical expenses and forgone earnings as a result of lost work time. This type of measure has some shortcomings: a significant proportion of the population is not in the work force; and it takes no account of the cost of pain and suffering. It is principally a measure of the value of livelihood rather than life. In order to conduct cost-benefit analysis we need to establish the community's willingness to pay (WTP) to save or improve lives.

Two utility-based approaches to establishing the WTP for a non-marketed good or service were discussed earlier in this chapter: Revealed Preference methods, which use observations of market behaviour to infer the value that individuals place on a good or service; and Stated Preference methods, which ask people how much they are willing to pay for a life saved or a better health outcome.

The Revealed Preference approach can be applied to the behaviour of consumers either as sellers of an input, such as labour, or buyers of an output, such as airbags in a motor vehicle. Significant wage differentials are observed in the labour market and some portion of these may reflect the different levels of risk attached to various jobs. If we regress wage rates against the personal characteristics of workers, the characteristics of jobs, and the occupational risks of injury or death we can identify the WTP for lower risk in the form of the willingness to accept a lower wage. Such studies assume that all individuals in the sample have the same attitude to risk. Alternatively we could examine consumer demand for such products as seat belts, airbags, cigarettes, smoke alarms, bicycle helmets, swimming pool fencing, etc. and use the hedonic pricing method (HPM) to get a relationship between product price and risk reduction, by, for example, comparing the prices of motor vehicles with and without airbags. Stated preference approaches can be applied to virtually any situation involving risk of injury or death but are subject to the problems associated with hypothetical questions as discussed earlier in this chapter.

Abelson (2008, Table 1) presents the results of a survey of numerous studies of the value of a statistical life (VSL). The studies are for developed economies in the period 1991–2005 employing both revealed and stated preference approaches. Values obtained from these studies, in US dollars at the time of the study, range from around $0.5 million to $20 million, though most results are in the $2 million–$5 million range. The wide range of results points to the serious limitations associated with empirical studies of the value of life. Nevertheless for consistency of appraisal across projects in a variety of health and safety related areas, an estimate is required. For Australia, Abelson recommends a VSL of $3.5 million (2007 Australian dollars) which, based on 40 years of life and using a 3% discount rate, implies a VLY of $151,000. If these VSL and VLY estimates were to be used in current studies they would have to be adjusted for inflation using an appropriate index. As discussed above, VSL can be adjusted downwards with increasing age, so that at 20 years remaining life expectancy, for example, VSL would fall to $2.25 million.

8.10 The pandemic

We can expect a plethora of research studies, including cost-benefit analyses, on various aspects of the Covid-19 pandemic. Since the full range of data required for analysis is not available at the current stage of the pandemic, but is gradually accumulating over time, these studies will be *ex post* and, in the case of cost-benefit analyses, can be classed as *evaluations* rather than *appraisals*. The results of these studies will inform the policy response during the next pandemic, which assuredly lies ahead. Bearing in mind this lack of information, we draw on our experience to date to suggest a brief outline of the issues cost-benefit studies might address.

A cost-benefit analysis always involves a comparison of the effects of one course of action with those of one or more alternative courses. At the start of the pandemic two opposing extreme policies towards the pandemic were mooted. Under the "let-it-rip" approach nothing was to be done to prevent the spread of the virus; the number of infections would follow a Bell Curve until "herd immunity" was attained. Under the "flattening-the-curve" approach interactions amongst the population would be restricted, thereby placing a limit on the number of infections – in effect truncating the height of the Bell Curve and spreading the pandemic further out over time.

It was argued that the benefit of the let-it-rip approach lay in the avoided cost of limiting interactions amongst people, and hence curtailing economic and social activity, and that the cost lay in numbers of infections beyond the capacity of the health system to treat. Suppressing the number of infections under the flattening-the-curve approach might involve the cost of significant reduction in economic activity, due to "stay home" measures, referred to as "lockdowns", but, as a benefit, would keep infections at a level that could be treated in the health care system. It should be noted that whatever course a government chose to follow could be only partially implemented: under the let-it-rip approach segments of the population might voluntarily self-isolate, and under the suppression strategy segments might ignore public health directives.

In the absence of the development of a vaccine the costs of maintaining the suppression strategy until herd immunity was attained would likely be intolerably high. Early in the pandemic, however, it became apparent that there was a reasonable likelihood that an effective vaccine could be developed. The suppression strategy could then be considered as a holding action designed to limit cases until the level of vaccination in the population was sufficient to limit the spread of the disease. It would have the further advantage of providing time for the development of improved treatment strategies. The cost of pursuing the suppression strategy would depend on how long it took to implement an adequate vaccination program in the population, and on the degree of transmissibility of emerging strains of the virus. In the event, vaccination programmes in developed economies started within twelve months of the start of the pandemic, but the emergence of more readily transmissible strains of the virus raised the extent and duration, and hence the cost, of the isolation policies required to suppress the spread.

Since the presence of the pandemic involves costs, but no benefits, the cost-benefit studies will be in the spirit of case studies A1.4 and A1.5 in that they will try to determine which of various policies involves the least cost to society. In comparing costs we are essentially balancing the costs of additional suffering and loss of life, due to the inability of the health system to cope under the let-it-rip approach, against the costs of additional loss of output in the economy, and the various social costs resulting from enforced isolation, under the suppression policy. While adopting a money measure of the cost of suffering and death may be repugnant to many, we have argued in Section 8.9, where we discuss methods for estimating the value of life, that, whether or not such a measure is made explicit, it is implicit in the decisions we make as a society.

It will be noted that, in this rough comparison, we have not included the costs of operating the health system itself. It can be assumed to be operating at capacity under both approaches, and that, while the pandemic rages, the level of additional non-Covid-19 related suffering and death, owing to resources being diverted from regular health treatments, is similar under the two approaches. The timing of the costs of adverse health effects, however, is different under the two approaches with delay favouring the suppression approach because of the effect of discounting. The costs of running a test-trace-isolate regime, under a suppression approach consist of diverting public health and police resources, at the cost of neglect of some of their usual duties, to supervising isolation in quarantine facilities, and enforcing stay-home orders.

Many governments have adopted fiscal and monetary stimulus measures to mitigate the effect of the pandemic on aggregate demand, and hence on the level of employment in the economy. As discussed in Chapter 12 such measures can be effective to some extent in a crisis, but they can involve an increasing debt burden, depending on how they are financed.

Since the costs fall, to some extent, on different sections of the population the comparison raises interesting distribution issues: the burden of suffering and earlier death tends to fall disproportionately on the elderly; unemployment and loss of economic output is felt most keenly in parts of the service sector, such as tourism, restaurants and bars, and sports and entertainment venues, which are generally staffed by young and relatively lowly paid workers. The young can also expect to bear much of the burden of repaying any borrowing required to finance economic stimulus measures. The social costs of isolation are felt most keenly amongst the disadvantaged in society.

We have framed this rough comparison as a choice for government between health and the economy, although we recognise limits to governments' ability to maintain economic activity under the let-it-rip approach, or to maintain suppression measures under the flattening-the- curve approach. However an early snapshot of the relationship between the intensity of the pandemic, as measured by Covid-19 cases per capita, and losses to the economy, as measured by the percentage decline in employment, tends to cast doubt on the significance of the implied trade-off. Table 8.6 reports the values of these variables for a range of advanced economies. While the outcomes reported in Table 8.6 are the result of a broad range of policy measures, including fiscal and monetary stimulus, they tend to suggest that countries which managed to suppress case numbers suffered smaller economic losses, at least initially. Detailed modelling of the transmission of the virus and the performance of the Australian economy tends to support the view that a suppression strategy was likely the least cost policy in relation to the pandemic (Kompas et al., 2021), at least in its initial stages. This view came to be recognised by various jurisdictions which significantly modified their initial relaxed approach to the spread of the virus.

As new strains of the virus emerged and the pandemic entered its third year it became apparent that the experience of Covid-19 would be an extended one, and the continued

Table 8.6 Covid cases and employment changes in advanced economies

COUNTRY	COVID CASES (Total per million)	CHANGE IN EMPLOYMENT (since February 2020)
Sweden	108,317	-1.4%
USA	102,639	-4.4%
Israel	98,119	-3.4%
France	90,319	-1.0%
Spain	87,032	-2.4%
UK	78,086	-1.3%
Italy	70,761	-3.2%
Germany	44,739	-1.6%
Canada	37,881	-1.8%
Japan	6575	-1.4%
South Korea	3384	+0.9%
Australia	1236	+1.2%
New Zealand	580	+0.6%

(Source: Data taken from *The Weekend Australian*, July 17–18, 2021)

viability of suppression strategies started to come into question. On the one hand, concerns were raised about the effectiveness of lockdowns in limiting transmission of the virus, and, on the other hand, it became evident that the costs of lockdowns were greater and wider in extent than initially appreciated. These costs include an increase in poverty and inequality together with adverse effects in the form of disruptions to the provision of general health care and education, increasing loneliness and family violence, deterioration of mental health, and civic unrest. A review of a wide range of studies, employing standard techniques of cost-benefit analysis, such as QALY and WELLBY indices of wellbeing, conducted by Joffe and Redman (2021) concluded that the costs of lockdowns were many times greater than the benefits. These authors suggest that the normal suite of public health measures is best suited to provide a continuing response to the virus.

8.11 Climate change

We know that carbon dioxide (CO_2) levels in the atmosphere have risen considerably over pre-industrial levels and that the increase is primarily a result of human behaviour. We also know that the planet is warming as a result and that sophisticated climate models predict more frequent, and more severe weather events such as droughts and floods, together with their attendant costs. A wider range of effects, such as sea level rise and ocean acidification, is also predicted. Reduction of CO_2 emissions is a public good, and production of such goods generally requires coordinated action by government and society. Consideration of the costs foreshadowed by the results of these climate models has led many countries to agree on a target of zero net CO_2 emissions by the year 2050.

Climate models also raise the possibility of so-called "tipping points" being reached which may result in runaway, irreversible and catastrophic warming of the planet. This alarming prediction is subject to significant uncertainty because of the limitations of such models, but it lends support to those who invoke the precautionary principle in their approach to climate policy. The prospect is sufficiently grave and uncertain to be classed as *radical uncertainty*, as discussed in Chapter 9. There we argue that cost-benefit analysis is not well suited to dealing with broad uncertainty attached to significant events. However, the methods of cost-benefit analysis, outlined in Chapters 2–5, are well suited to comparing the costs of attaining a stated objective such as the 2050 goal.

The three broad categories of CO_2 emissions which need to be reduced result from electricity generation, transport and machinery operation, and land use practices. While the latter category deals mainly with agricultural methods and forest cover, it could be broadened to include the inter-tidal and coastal zones in which mangrove and kelp forests are important CO_2 sinks. We will not try here to deal with the highly complex issues raised by this category. In the case of transport and machinery operation we are interested in the costs of energy sources which are alternatives to fossil fuels. Electricity and "green" hydrogen, generated by electricity, are the main contenders for these purposes at present. This brief discussion suggests that methods of electricity generation are critical to achieving the 2050 target, and should, therefore, be the focus of our cost comparisons. A recent study (Graham et al., 2020) by Australia's CSIRO compared the costs of a range of options, including fossil fuels, solar, wind and nuclear power.

The CSIRO study was conducted in collaboration with the Australian Energy Market Operator (AEMO). It draws on a very wide range of information and research and we do not attempt to describe it in detail. It suggests that solar photovoltaic and wind power generation will be the least cost methods of power generation consistent with achieving the 2050 goal. The analysis includes some interesting applications of principles already discussed in the text:

- *in making cost comparisons we must compare like with like.* The CSIRO study calculates a Levelised Cost of Electricity (LCOE) for each method of electricity generation, which is the estimated unit cost of electricity, at point of delivery, with a specified degree of reliability. The costs to be included are those of generation, storage and transmission. Unit cost reflects both relevant capital and operating costs and can be calculated using the unit equivalent cost method described in the extension to Example 3.10. A levelised cost of electricity comparison is the subject of Case study A1.11;

- *in cost-benefit analysis we are interested only in incremental costs and benefits.* Hence, as discussed in Chapter 4, sunk costs are not included in the comparison of costs. This procedure gives power generation by fossil fuels a cost advantage because of their existing plant and distribution networks as compared with alternative energy resources requiring investment in new assets. This advantage holds for the period of the remaining economic life of the existing assets, assumed to extend out to 2050. As noted in Chapter 4, as well as sunk costs, depreciation allowances and borrowing costs are excluded from the discounted cash flow model which is used to calculate unit equivalent cost;

- *all relevant costs must be included in making the cost comparisons.* Many categories of capital and operating costs are measured by market prices. However, as discussed in Chapter 5, some projects involve commodities which are not traded in traditional markets, but nonetheless affect the level of economic welfare. CO_2 emissions are one such commodity, albeit with a negative value or cost, and to exclude them from the costs of power generation would miss the main point of the cost comparison exercise – somewhat like staging *Hamlet* without the Prince of Denmark! The process of pricing emissions in the CSIRO analysis is complex and we offer only a brief description here of the method adopted. In order to achieve zero net emissions by the year 2050 emissions must decline on some path from their present level towards zero. This path specifies the emissions allowable in each year. We know from the market models referred to throughout the text that each market price implies a corresponding quantity, and vice versa: we can set a carbon price and that price will determine the level of emissions, or set a quantity of emissions in a trading system and that will determine the price. If we know the allowable quantity of emissions in each year, we can solve for the implied carbon price which, ideally, measures the marginal cost of emissions. When this exercise is conducted, and the resulting price path of emissions is included in the cost comparisons, fossil fuels lose the cost advantage described above and solar and wind become the least cost methods of electricity generation according to the CSIRO study.

The social cost of carbon dioxide emissions in cost-benefit analysis

As discussed above, a quantity constraint is associated with an implicit price, but the cost-benefit analyst requires an explicit price of CO_2 if she is to include the social cost/benefit of emissions/abatement in the Efficiency Analysis. Many governmental agencies in OECD

Table 8.7 Social cost of CO_2, 2020–2050 (in 2020 US dollars per metric ton of CO_2)

Emissions Year	Discount Rate and Statistic			
	5% Average	3% Average	2.5% Average	95th Percentile
2020	14	51	76	152
2025	17	56	83	169
2030	19	62	89	187
2035	22	67	96	206
2040	25	73	103	225
2045	28	79	110	242
2050	32	85	116	260

(Source: Technical Support Document: Social Cost of Carbon, Methane, and Nitrous Oxide Interim Estimates under Executive Order 13990; Interagency Working Group on Social Cost of Greenhouse Gases, United States Government, February 2021)

countries, as well as international organisations, use estimates of the Social Cost of Carbon (SCC) in cost-benefit analyses to value the long-term cost generated/avoided by a ton of CO_2 emissions/abatement in a particular year (see Table 8.7).

Some models estimate annual costs of emissions as far out as the year 2300, and these costs, up to the terminal year, are then discounted back to the present giving an estimate of the SCC of today's emissions. For example, the SCC of one ton of CO_2 released in the year 2022 is an estimate of the present value in 2022 of the cost of climate change damages attributable to that ton of CO_2 over the period 2022–2300. The cost estimate will depend critically on the assumptions and parameters of the models used to estimate damages, leading to a wide range of estimates, and on the interest rate used to discount future costs to a present value. Since there is no consensus on the appropriate forecasting model or discount rate, it may be advisable to use a range of possible SCC values in a sensitivity analysis, as discussed in Chapter 9.

In 2009–2010 the US government considered three integrated assessment models each of which generated SCC estimates for each year of emissions. From these values a range of SCC estimates was calculated for use in cost-benefit analysis: three sets of estimates were based on the average SCC from each of the models, at discount rates of 5%, 3% and 2.5%. A fourth estimate, described as High Impact, was based on the 95th percentile of the range of estimates, using a 3% discount rate, thereby providing an 'upper bound' estimate of SCC. These estimates have been revised periodically since then. Table 8.7 summarises the 2021 range of estimates for four sets of SCC values based on four assumptions about the cost of CO_2 emissions, using the three discount rates.

From Table 8.7 it can be seen, for example, that an extra ton of CO_2 emitted in the year 2025 was assigned a SCC in the range \$17 to \$169. In 2021 President Biden's administration adopted a set of estimates of the SCC in the middle of the range, implying, for example, a cost of \$56 per metric ton for 2025 emissions.

8.12 Further reading

Most public finance texts include a detailed discussion of public goods and externalities, for example, R. Boadway and D. Wildason, *Public Sector Economics* (New York: Little, Brown and Co, 1984). A classic paper on externalities is R. Coase, "The Problem of

Social Cost", in R.D and N.S. Dorfman (eds), *Economics of the Environment* (New York: W. W. Norton, 1972) and a classic paper on the commons is G. Hardin, "The Tragedy of the Commons", *Science*, 162 (1968): 1243–1248. A useful introduction to non-market valuation techniques is J.L. Knetsch and R.K. Davis, "Comparisons of Methods for Recreation Evaluation", in R.D and N.S. Dorfman (eds), *Economics of the Environment* (New York: W.W. Norton, 1972). More recent studies are: R. Cummings, D. Brookshire and W.D. Schulze (eds), *Valuing Environmental Goods* (Lanham, MD: Rowman and Allanheld, 1986); R.C. Mitchell and R.T. Carson, *Using Surveys to Value Public Goods: the Contingent Valuation Method* (Resources for the Future, 1989); N. Hanley and C.L. Spash, *Cost-Benefit Analysis and the Environment* (Cheltenham: Edward Elgar, 1993); G. Garrod and G. Willis, *Economic Valuation and the Environment* (Cheltenham: Edward Elgar, 1999); D. Pearce, G. Atkinson and S. Mourato, *Cost-Benefit Analysis and the Environment* (Paris: OECD, 2006). P. Abelson (2008), "Establishing a Monetary Value for Lives Saved: Issues and Controversies", *Office of Best Practice Regulation*, Department of Finance and Deregulation, New South Wales, Working Paper 2008–02, discusses valuing life and provides some estimates. A critique of alternatives to CBA is provided by L. Dobes and J. Bennett, "Multi-Criteria Analysis: 'Good Enough' for Government Work?", *Agenda*, 16(3) (2009): 7–30; and support for traditional CBA techniques is included in the same issue: H. Ergas, "In Defence of Cost-Benefit Analysis", *Agenda*, 16(3) (2009): 31–40. J. Hassel, "Which countries have protected both health and the economy in the pandemic?" (Our World in Data, 1 September 2020). Kompas T, Grafton RQ, Che TN, Chu L, Camac J (2021), "Health and economic costs of early and delayed suppression and the unmitigated spread of COVID-19: The case of Australia". (PLoS ONE 16(6): e0252400. https://doi.org/10.1371/journal.pone.0252400.) A.R. Joffe and D. Redman (2021), "The SARS-Cov-2 Pandemic in High Income Countries Such as Canada: A Better Way Forward Without Lockdowns", *Front. Public Health* 9:715904, doi: 103389/fpubh.2021.715904. P. Graham, J. Hayward, J. Foster and L. Havas, *GenCost 2020–21: Consultation Draft* (CSIRO: December 2020, pp. viii + 79).

Appendix to Chapter 8

The annual benefits of the Virginia Creeper Trail as measured by the Travel Cost Method

The Virginia Creeper Trail (VCT) is a 34-mile walking, biking and horse riding trail based on the abandoned rail track between the towns of White Top and Abingdon in Grayson and Washington counties in south-west Virginia. It is named after the well-known climbing ivy which originated in the eastern United States and which is a popular horticultural plant because of its flowers, foliage and beautiful autumn colours. Responsibility for maintenance and management of the trail is shared roughly evenly between federal and local governments. It is reasonable for these public authorities to ask whether the expenditures and any other opportunity costs associated with providing this facility are justified by the benefits it generates. A study by Bowker, Bergstrom and Gill[3] applied the Travel Cost Method (TCM) to estimate the annual benefits accruing to users of the trail.

Bowker and his colleagues conducted an exit survey which yielded a sample of users, stratified by seasons, exit points and times of day, which was used to estimate the total

Table A8.1 Annual number of VCT person-trips

Type of trip	PPDU	NPDU	PPON	NPON
Tourist Person-Trips	33642	7578	5725	3918
Local Person-Trips	61503	N/A	N/A	N/A
Total Person-Trips	95145	7578	5725	3918

annual use of the trail. Detailed survey questionnaires were completed by a sub-sample of users which recorded the number of trips made annually by each user, the distance travelled to access the trail, the type of recreation activity engaged in, the availability to the user of a substitute facility, the user's income and other socioeconomic characteristics. Each user was classified as either local (living within 25 miles of the trail) or non-local (a tourist), and each trip was classified as primary purpose day (PPDU) or overnight use (PPON), or non-primary purpose day (NPDU) or overnight use (NPON). The reasons for making these distinctions are that only primary purpose trips will be included in the demand model used to estimate annual benefits, and that while both primary and non-primary purpose trips are relevant in the subsequent analysis of economic impact, overnight, as opposed to day, users have different local economic effects because of their different spending patterns. Table A8.1 summarises the estimates of the annual number of person-trips (one person making one trip).

The aim of the TCM is to estimate the annual consumer surplus generated by the VCT in the form of an area under a representative individual's demand curve, such as that illustrated in Figure 7.1: if, for example, P_0 is the price of a trip, the annual consumer surplus is measured by the area of triangle EAP_0. Another way of measuring the area of consumer surplus is to multiply the average consumer surplus associated with a trip, given by the length $(E - P_0)/2$ in Figure 7.1, by the number of trips $0Q_0$ ($= P_0A$). The VCT study, which is based on a non-linear demand curve, uses the latter approach.

Application of the TCM proceeds by estimating an individual demand curve for trips, using travel cost as a proxy for the price of accessing the facility. As noted above, only primary purpose trips are included in the demand estimation to avoid attributing to the VCT consumer surplus generated by a trip which has an alternative site as its primary purpose. The travel cost incurred by each person-trip was estimated as a per-mile cost of transport ($0.131) multiplied by the distance from the person's home to the trail. A variant of the travel cost variable was also calculated which included an estimate of the cost of travel time as well as transport costs: travel time was costed at 25% of the user's wage rate (as discussed in Chapter 5, Figure 5.12). Other variables included in the specification of the demand curve are: availability of a substitute recreation site, income and other socioeconomic variables such as age, sex, household size, tastes and the type of recreation activity chosen.

A sample of 800 respondents obtained from the detailed questionnaire provided information on the price and number of trips annually undertaken by each primary purpose respondent, as well as the values of the socioeconomic and other variables noted above. The sample was used to estimate a demand curve of the following form:

$$\ln Q = \beta_1 + \beta_2 X_2 + \beta_3 X_3 + \cdots + \beta_n X_n$$

where ln Q represents the logarithm of the number of trips taken annually by each member of the sample, X_2 represents the price of each trip, and $X_3 \ldots X_n$ represent the other variables in the demand equation as discussed above. Since the demand curve relating quantity demanded to price slopes downward, the coefficient, β_2, of the price variable will be negative in sign.

As demonstrated in Technical note 8.2, the average amount of consumer surplus per person-trip is measured by the term $-1/\beta_2$, which takes the value \$22.78 when the opportunity cost of time is excluded from the travel cost estimate, and \$38.90 otherwise. When these values are combined with the primary purpose person trip data reported in Table A8.1 total annual consumer surplus is estimated (in 2003 US dollars) at \$2.3 million when travel cost is valued at a zero opportunity cost of time, and \$3.9 million otherwise.

TECHNICAL NOTE 8.2 Measuring consumer surplus in the VCT study

Consumer surplus consists of an area such as EAP_0 under the demand curve illustrated in Figure 7.1. This area is calculated as the integral of the curve between the limits P_0 and E, at which price quantity demanded falls to zero. Since the demand curve estimated in the VCT study is asymptotic to the price axis, quantity demanded falls to zero as price approaches infinity. Thus, in this case, the limits of integration are, P_0, the current price of a trip, and infinity.

The demand curve can be written as:

$$Q = \exp\left(\beta_1 + \beta_2 X_2 + \beta_3 X_3 + \cdots + + \beta_n X_n\right)$$

And the indefinite integral as:

$$\int_{X_2 = P_0}^{\infty} \left(e^{\beta_1 + \beta_2 X_2 + \beta_3 X_3 \ldots \ldots \ldots \beta_n X_n}\right) dX_2 = \left(\frac{1}{\beta_2}\right)\left[e^{\beta_1 + \beta_2 X_2 + \beta_3 X_3 \ldots \ldots \ldots \beta_n X_n}\right]$$

Since the coefficient, β_2, on the price variable has a negative sign the amount of consumer surplus represented by the value of the definite integral reduces to:

$$CS = -\left(\frac{1}{\beta_2}\right) Q_0$$

where Q_0 is the observed number of trips, as calculated from the demand curve at the price P_0, and $-(1/\beta_2)$ is the average consumer surplus per trip.

As noted earlier, the estimates cited above under-estimate the annual benefit derived from the VCT: they include only the use values derived by primary purpose users, and exclude the benefits derived by non-primary purpose users and any non-use values. On the other hand, they include both the user benefits derived by tourists as well as by local residents. If the Referent Group were defined as residents of Grayson and Washington counties, local primary purpose users only would be relevant to the consumer surplus calculation and the annual consumer surplus estimates would fall to $1.4 million and $2.4 million depending on the treatment of the opportunity cost of travel time. The latter values are those that local government might consider pertinent in an appraisal of its contribution to the provision of the VCT, whereas federal authorities might consider the consumer surplus estimates as a whole.

A further contribution of the VCT which may be of interest to the Referent Group is the stimulus provided to the local economy and this issue is considered in the Appendix to Chapter 12.

Exercises

1   Manufacturers in an urban environment are currently producing 25,000 widgets per annum. Their gross revenue is $300 per widget and their variable costs are $125 per widget. Air quality in the city has fallen to a level of 20 points measured on a 0 to 100 point indicator. Three proposals are being considered to improve the city's air quality. Option 1 involves annual direct costs of $100,000 which raises the index of air quality to 32; Option 2 costs $130,000 per annum which raises the air quality index to 42; and Option 3 costs $150,000 per annum and raises the water quality index to 50. In addition, producers are required to reduce their output of widgets by: 5% under Option 1, 10% under Option 2, and 15% under Option 3. Three questions are posed:

   i   Which option has the lowest total opportunity cost?
   ii Which option has lowest cost per unit of air quality improvement?
   iii Why might neither of these represent the most efficient option?

2   Individuals A and B have the same money income, have the same tastes and face the same set of prices of all goods and services except that of access to a National Park. Individual A lives further away from the park than Individual B and hence incurs a higher travel cost per visit. There is no admission charge to enter the park. The following data summarise their annual use of the park:

Individual	Travel cost per visit	Number of visits p.a.
A	$20	5
B	$10	10

   i   Approximately how much consumer surplus does Individual A receive per annum from her use of the park?
   ii Approximately how much consumer surplus does Individual B receive per annum from his use of the park?

iii   What is the approximate measure of the annual benefits to A *and* B from their use of the park?

iv   Why are the measures in (i), (ii) and (iii) "approximate"?

3   A contingent valuation study was used to value improvements to a coral reef, where, at present, the reef quality, measured in terms of percentage of reef cover, is so poor that the local residents can no longer rely on the reef for their supply of fish. The study sample, which included users (one-third) and non-users, found the following estimates of average WTP.

	Mean WTP ($ per annum)		
Scenario	Whole sample	Local users	Others
1. Increase reef cover to 50%	25	45	15
2. Increase reef cover to 75%	40	70	25
3. Increase reef cover to 100%	53.5	95	33

i   Construct a demand curve for reef quality for: users; others; and the whole sample and discuss the forms of these curves (remember that a demand curve describes willingness to pay for an *additional* unit of a good or service).

ii   Explain how this information could be used in a BCA to compare management options to improve the quality of the reef.

4   Using the data provided in Example 8.4, calculate the net benefits of an alternative road proposal that results in increased noise levels in Areas A and B of 5 and 10 decibels respectively, and decreased noise levels in Areas C and D of 2 and 8 decibels respectively.

Notes

1   Ciriacy-Wantrup, S.V. (1947) "Capital returns from soil-conservation practices", *Journal of Farm Economics*, Vol. 29, pp. 1181–96; Hotelling, H. (1949), Letter in "An Economic Study of the Monetary Evaluation of Recreation in the National Parks", Washington, DC: National Park Service.

2   Unfortunately as these websites are not updated regularly, they will not always include recent studies.

3   See J.M. Bowker, J.C. Bergstrom and J. Gill (2007) "Estimating the Economic Value and Impacts of Recreational Trails: A Case Study of the Virginia Creeper Trail", *Tourism Economics*, 13(2): 241–260.

9 Uncertainty, information and risk

9.1 Introduction

If cost-benefit analysis is to assist in the decision-making process, the analysis must be conducted in advance of the project being undertaken. This means that the value of none of the variables involved in the analysis can be observed, but rather has to be predicted. In the preceding chapters, exercises and case studies we have implicitly assumed that all costs and benefits to be included in a cash flow are known with *certainty*. While this assumption might be acceptable in the context of project evaluation where the analyst is undertaking an *ex post* assessment of what has already occurred, it is clearly an unrealistic assumption to make where the purpose of the analysis is to undertake an appraisal of a proposed project where one has to forecast future cost and benefit flows. The future is *uncertain*: we do not know what the future values of a project's costs and benefits will be, although we can usually make an informed guess.

Uncertainty arises either because of factors internal to the project – we do not know precisely what the future response will be to, say, some management decision or action taken today – or because of factors external to the project – for instance, we do not know precisely what the prices of the project's inputs and outputs will be. In brief, uncertainty implies that there is more than one possible value for any project's annual costs or benefits. The range of possible values a variable can take may vary considerably from one situation to another, with the two extremes being complete certainty, where there is a single known value, to complete uncertainty where the variable could take on any value. Most situations lie somewhere between the two extremes.

An economic model, such as a cost-benefit analysis, is a simplified version of reality designed to assist in decision-making. The "real world" is too complex a place for analysis to comprehend all its multiple facets, and the purpose of a model is to single out the essential elements of a problem and their relationships. The choice of which variables are essential to the model is a matter of judgement on the part of the modeler. As noted above, the values of some, if not most, included variables are not known with certainty while the model is being applied – these are the "known unknowns". There is another category of possible variables, however, which can be described as the "unknown unknowns" – variables which are not included in the model but, perhaps, we may find in retrospect, should have been. This is the problem of *radical uncertainty* described by Kay and King (2020). Which of us, in the year 2019, contemplating opening a business offering personal services, such as a restaurant or a beauty salon, could have taken the looming pandemic into account? The pandemic can be categorised as a "black swan event" – significant and unanticipated. In fact, pandemics are anticipated by the World Health Organisation (2015), but, like predictions of a California

DOI: 10.4324/9781003312758-9

earthquake or an asteroid strike, the predictions are not sufficiently precise to be taken account of in project analysis. In consequence, Kay and King argue that models, such as CBA, offer a false sense of precision and should be backed up by judgement based on experience and widespread consultation. While we support their timely reminder that formal models should not have the last word, we believe that CBA is a useful tool in decision-making and, in this chapter, we outline how risk and uncertainty can be accommodated within its framework.

Where we believe that the range of possible values could have a significant impact on the project's profitability, our decision about the project will involve taking a *risk*. In most situations there will be some information on which to base an assessment of the probabilities of possible outcomes (or values) within the feasible range. If we can estimate the probability, or likelihood, of the values that an outcome could take, we can quantify the degree of risk. In some situations the degree of risk can be *objectively* determined, for instance, when flipping a fair coin: there is a 50% chance it will be heads and a 50% chance it will be tails, about which there can be no disagreement.

On the other hand, estimating the probability of an *el Niño* effect next year involves some judgement on the part of the analyst on the basis of the current information available. There is not enough information for there to be an unequivocal answer, as with flipping a coin, but there is enough to place some *estimate* on the probability. In these situations estimating the probability of an event in order to quantify the element of risk involves *subjectivity*. In other words, it is something about which there can be disagreement; some may argue that there is a 20% chance of rain while others may argue, on the basis of the same information, that there is a 60% chance of rain. Subjective risk characterises most situations we will encounter in project appraisal. There will be some information which can be used to assign numerical probabilities, which can in turn be used to undertake a more rigorous analysis of possible project outcomes. As new information comes to hand we may revise our probability estimates accordingly.

An analyst may be confronted with a range of values that a variable critical to the CBA may take. The possible values may be described by a probability distribution, such as the familiar Normal distribution, but unless the analyst understands the underlying process which generates the possible values of the variable the values of the mean and variance of the distribution will not be known. Nevertheless the analyst wishes to choose the "most likely" value of the variable for use in the CBA. In the case of a symmetric distribution, such as the Normal distribution, the most likely value is the mean value, but, as we have already suggested, the analyst will be uncertain about what that value is. If it is possible to gather additional information about the values the variable may take some of that *uncertainty* about the mean may be resolved. However, even with partial resolution of the uncertainty there will still remain *risk* in the form of a spread of possible values around whatever mean value emerges from the information gathering process. While this nice distinction between uncertainty and risk may not be helpful to the analyst confronting a wide range of possible values of a variable, it will be useful in organizing the discussion of measures we can take to cope with this problem.

In general, we cope with uncertainty by gathering additional information, and we cope with risk by diversifying our portfolio of assets. Confronted with a single project, however, the analyst is usually not in a position to provide advice about the effect of undertaking the project on the variance of the overall return of the decision-maker's portfolio. Instead the cost-benefit analysis is confined to a descriptive role through risk modelling – describing the range and probability of project outcomes. Additional information may contribute to

Table 9.1 The value of information

Prior decision	Accept the project	Reject the project
CBA Result = Good (NPV > 0)	Value of information = 0	Value of information > 0
CBA Result = Bad (NPV < 0)	Value of information > 0	Value of Information = 0

reducing both uncertainty and risk but here we confine our discussion to its former role. We need to recognise that gathering additional information has a cost and, according to the principles of cost-benefit analysis, the process of information gathering should be continued only as long as the marginal benefit of information is at least equal to the marginal cost.

Since cost-benefit analysis (CBA) is a process of information gathering and analysis, it is itself subject to the cost-benefit calculus: what is more information worth? What does it cost? We can argue that the value of CBA lies in averting a bad decision: if the project is found to have a positive NPV and would have gone ahead in the absence of the analysis the CBA has contributed nothing. Similarly, if the project would not have gone ahead and is found to have a negative CBA, again nothing has been gained. Conversely, as illustrated in Table 9.1, information has value when it contributes to overturning a potentially wrong decision: avoids rejecting a good project (Type I Error) or avoids undertaking a bad project (Type II Error). While the analyst cannot know in advance the outcome of the CBA, a view needs to be formed about its likely result if a judgement is to be made about whether to proceed with it. In forming this judgement it may be recognised that there are many more potentially wrong than correct decisions and that prudence requires an analysis to be undertaken. The same argument about the value of additional information applies to assessing the likely contribution of individual pieces of information to the decision-making process. In other words, we must compare the expected value of information with the cost of collecting it in deciding what information to collect.

9.2 The value of information

In this section we discuss how to employ sensitivity analysis to identify variables whose values are critical to the outcome of a cost-benefit analysis. Further information about the values of these variables might be worth obtaining, either through delaying the decision about the project to allow for more precise estimates to present themselves, or by actively seeking such estimates through further research. Of course this approach presumes that a CBA is to be undertaken – on the basis that the expected value of the contribution of the analysis exceeds its cost – and we also discuss a framework for predicting the net benefit of undertaking the CBA.

Sensitivity analysis

The analyst usually has the option of gathering additional information about many values to be used in the study. However, as noted above, additional information comes at a cost and should be obtained only when it has a corresponding benefit in terms of the likely quality of the decision to be made, based on the likely outcome of the CBA. If the estimate of project NPV is likely to remain positive (or negative) over the feasible range of values of a variable there may be no benefit to be gained from a more precise estimate. Where the value a variable takes within its feasible range can affect the outcome of the analysis, it may be worth acquiring additional information. Sensitivity analysis is a useful technique for identifying such variables.

   The term "sensitivity analysis" refers to the process of establishing how sensitive the outcome of the CBA is in response to changes in assumptions made about the values of selected variables involved in the analysis. For instance, in a number of the previous examples and exercises we calculated the NPV of a project over a range of discount rates. In performing these calculations we were, in effect, conducting a sensitivity analysis. In this case we were testing how sensitive the NPV of the project is to the choice of discount rate. Our spreadsheet framework is ideally suited to this kind of analysis as the values in the Variables Table can readily be altered to obtain a revised set of results. More formal techniques, such as the *Goal Seek* routine in *Excel©*, can be used to identify a threshold or break-even value – the value of a variable at which NPV = 0, or at which the IRR is equal to some target value. At higher values of the variable in question, project NPV may be positive and at lower values, negative, or vice versa. If the threshold value is within the range of feasible estimates, there may be a case for gathering additional information, but otherwise perhaps not.

   The analyst can allow the values of particular key variables (variables likely to influence the outcome of the CBA) to vary over the full range of their possible values. Let us assume that in relation to the construction of a road the engineers have indicated that the actual cost could vary by up to 25% above or below their 'best guess' estimate. In this case the analyst may wish to calculate the present value of the road's costs or net benefits at 75%, 100% and 125% of the best guess value of cost. The decision-maker is then provided with a range of possible project outcomes corresponding to the range of possible cost values.

   It is likely that the analyst will want to ascertain the effect of changes in the value of more than one of the variables involved in the CBA. Let us assume that there is uncertainty about both the construction costs and the future usage of the road. Again, there could be, say, three estimates of future usage: low; medium; and high, each corresponding to a different level of annual benefit. The analyst can again test the significance of changes in this variable independently of others – by calculating the net present value for each level of road usage – or in conjunction with others, for example, by allowing both the capital costs and the road usage levels to vary simultaneously over their respective ranges. This case is illustrated in Table 9.2.

   Table 9.2 provides some information useful to the decision-maker. We can see by how much the NPV varies (at a 10% discount rate) when we allow one or both inputs to the CBA to vary across their respective possible ranges. For instance, if we hold the level of road usage at its "best guess" or "Medium" level, we see that the NPV varies from $25 to $47 million, while, if road construction costs are held at their "best guess" level (100%), the NPV varies from $32 to $40 million. Clearly, the outcome is more sensitive to changes in assumed construction cost values than to road usage values, across their possible ranges as specified in Table 9.2. When we allow both input values to vary simultaneously, we see that NPV can vary from a minimum value of $20 million – the most pessimistic scenario when road usage is assumed to

Table 9.2 Sensitivity analysis results: NPVs for hypothetical road project

Road usage benefits	Construction costs ($)		
	75%	*100%*	*125%*
High	50	40	30
Medium	47	36	25
Low	43	32	20

Note: $ millions at 10% discount rate.

be at its lowest and construction costs are at their highest level – to $50 million – where road usage is at its highest and construction costs at their lowest level. This calculation may provide the decision-maker with sufficient information to incorporate uncertainty in the decision-making process, provided that she is satisfied that these are the only two variables that can be considered "uncertain" and that all she is concerned about is knowing that even in the worst case scenario, the project's NPV is still positive at a 10% discount rate.

However, what if the range of NPVs produced by the sensitivity analysis varies instead from, say, *negative* $20 million to positive $50 million? In other words, there are some scenarios under which the project would not be undertaken on the basis of the simple NPV decision criterion. Or what if the decision-maker had to decide between a number of project alternatives where the range of possible NPV values was very large yet still positive for all projects? Unfortunately the main problem with sensitivity analysis is that the matrix of values as shown in Table 9.2 does not contain any information about the *likelihood* of the project NPV being positive or negative, or, in the case of choice among alternatives, of being higher than that of some other project.

Furthermore, it is conceivable that there will be other inputs to the analysis for which the forecast values are also uncertain and which the analyst will want to include as variables in the sensitivity analysis. Adding one more input to the above sensitivity analysis with, say, 3 possible levels will increase the number of scenarios to 27; adding 3 more inputs increases the number of scenarios to 81, and so on. Very soon we reach a situation of information overload where there is just too much raw information and far too many possible combinations for the decision-maker to use sensibly in reaching a decision.

Another issue concerns possible correlations between variables representing levels of uncertain outputs and inputs. The sensitivity analysis described above assumes that the variables in question are all independent of one another. As we discuss in more detail below, it is likely that variations in some variables will be correlated with variations in others, such as, say, road usage and road maintenance costs. Allowing these to move in opposite directions could produce nonsensical results in the sensitivity analysis.

For these reasons sensitivity analysis should be used with care and discretion. It can be a useful first stage in determining:

- how sensitive project NPV is to the values of the variables used in the analysis; and
- to which variables the project outcome appears most sensitive.

This preliminary stage can be useful for the analyst in deciding whether or not to try to obtain additional information which will narrow the range of estimates for some variables and in identifying which forecast variables are to be investigated further with a view to including them in a more formal risk modelling exercise. In that process values of several variables can be specified in the form of probability distributions, and correlations between values can be included.

9.3 An abbreviated cost-benefit analysis

Practitioners often point to the lack of time and other resources available to undertake a detailed CBA and ask whether there is a "rough and ready" approach. An informal sensitivity analysis, incorporating the analyst's judgement about the range and values of the key variables, can be used as the basis of a "back-of-the-envelope" study. The brief CBA of the 55 mph speed limit in the USA described in Chapter 5 is an example of such a study. Some methods of simplifying the computational tasks, such as treating net benefit streams as perpetuities where appropriate, have already been suggested in Chapters 2 and 3. It will be apparent that such a "rough and ready" analysis should normally be undertaken to determine whether a detailed CBA is required; recall Table 9.1 which outlined the circumstances in which such a detailed analysis could be useful, and, conversely, worthless.

9.4 The option of delay

CBA tends to focus more on *whether* a project should be undertaken rather than *when* it should start if it is approved. However, it is easy to envision circumstances in which the NPV of a project (its NPV now, at time zero) might rise if its start was delayed by a year or so. Furthermore, if the sensitivity analysis has identified a key variable, the value of which merits further research, the process of information gathering will inevitably force delay; an environmental impact statement (EIS) associated with a major project may take several years to complete. However, delay in deciding about the project may also arise as a result of the decision-maker adopting a "wait-and-see" approach: the predicted project NPV may depend critically on the value of a variable, such as an exchange rate or a price, which is likely to resolve itself in the near future. In this case, even though no resources need be devoted to acquiring additional information, learning occurs by waiting. It can be argued that the simple passage of time may impose a cost by deferring the expected project net benefits, but it can also be argued that undertaking the project now involves a cost in the form of a loss of option value – the value of the option of *not* undertaking the project, especially when the decision is effectively irreversible, such as the construction of a large dam.

We have argued that, in deciding whether to perform a CBA, the decision-maker must weigh the benefits of the analysis against its costs – in other words, at least implicitly perform a CBA of the CBA. The question is whether the value of the information obtained through the CBA is high enough to justify incurring the cost of performing the analysis. The value of information can be defined as the excess of the value of the decision which is made with the information over the value of the decision made without the information. Since these values cannot be known with certainty, the value of information is expressed as an expected value. We now consider a simple example of the question of whether or not to delay implementing a project in order to obtain better information before making a decision; delay has a cost, but it also has an expected benefit and the difference between these two values is the expected value of the information obtained through delay.

Suppose that a project involving an irreversible investment of $K in Year 0 will yield an expected operating profit of E(R) in perpetuity starting in Year 1. The expected annual value in perpetuity of operating profit is given by:

$$E(R) = qR_H + (1 - q)R_L,$$

where q is the probability of a high price environment, and R_H and R_L are the annual operating profits in the high and low price environments respectively. The net present value rule developed in earlier chapters would suggest a risk-neutral decision-maker would undertake the project if:

$$\frac{E(R)}{r} > K$$

where y is the rate of discount.

Now suppose that in Year 1 information will become available about the project's future profitability; for example, it will become known whether prices will be high or low and hence whether annual operating profit will be R_H or R_L. Assume that the firm has the option of delaying the project for one year. If it does so and finds that prices turn out to be low it need not suffer a loss by undertaking the project in Year 1; alternatively if prices turn out to be high it can invest \$K in Year 1 and receive R_H in perpetuity, starting in Year 2. The benefit of waiting derives from the possibility of avoiding a mistake; the cost of waiting is the cost of delaying the start of the future benefit stream by one period net of the benefit of postponing the capital cost. If the expected benefit of waiting exceeds the cost it is better to wait.

Figure 9.1 illustrates the decision-maker's choice of investing now or deferring the decision for one period. The expected net present value of investing now is:

$$NPV(0,0) = \left(\frac{qR_H}{r} + \frac{(1-q)R_L}{r} \right) - K$$

where NPV(0,0) is the NPV at time 0 of undertaking the project at time 0.

If the decision-maker defers the decision for one period, he will have the option of not investing if the low price environment eventuates. Assuming that $R_L/r < K$, he will opt not to undertake the project in that event. This means that there is probability $(1 - q)$ of a zero payoff if he waits. If the high price environment eventuates the project is undertaken in period 1 with a net present value in period 1 of $R_H/r - K$, or $(R_H/r - K)/(1 + r)$ at time 0. The expected net present value at time 0 of the investment decision deferred to Year 1 is given by:

$$NPV(0,1) = \left(\frac{q}{1+r} + \frac{R_H}{r} - K \right)$$

where NPV(0,1) is the expected NPV at time 0 of undertaking the project in Year 1. The value to the decision-maker of waiting —"keeping the option open"— is given by the difference between NPV(0,1) and NPV(0,0):

$$V_W = K\left(1 - \frac{q}{1+r}\right) - \frac{qR_H}{1+r} - \left(\frac{R_L}{r}\right)(1-q)$$

which is sometimes referred to as a *quasi-option value*.

By means of partial differentiation it can be ascertained that the value of waiting increases as K rises, and falls as q, R_L, or R_H increase. In other words, the larger the capital investment, the more an option to delay is worth; however, the value of the option falls as the probability of high operating profit increases and as the level of operating profit in each of the two possible price scenarios increases.

For the decision-maker to decide to undertake the project now, the option of waiting must be worthless: Vw < 0. From the expression for Vw, and using the expression for E(R) derived earlier, it can be shown that the option is worthless if:

$$\left(\frac{E(R)}{r}\right) - K > q\left(\frac{R_H}{r} - K\right)\left(\frac{1}{1+r}\right)$$

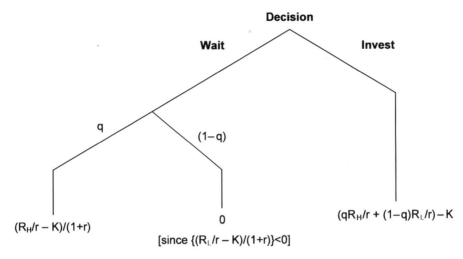

Figure 9.1 A decision tree with two options for timing an investment decision. If the decision-maker decides to undertake the project now – "Invest"– he incurs the capital cost, K, in Year 0 and receives a net benefit in perpetuity starting in Year 1 with the expected net present value shown. If he chooses to "Wait" there are two possibilities: additional information reveals that the price, and hence the return, will be high, R_H, or that it will be low, R_L; at time zero he assigns the subjective probabilities q and (1 – q) respectively to these possibilities. If the high return scenario eventuates he will undertake the project in Year 1, with the return R_H starting in Year 2, and with NPV at time zero of $(R_H/r – K)/(1 + r)$. We suppose that in the low return scenario he will abandon the project (otherwise why wait?) because the NPV is negative.

This inequality expresses the not unexpected result that there is no point in a risk-neutral decision-maker delaying the project if the NPV of undertaking the project now is greater than the expected NPV of waiting. This approach to investment appraisal is sometimes referred to as *real options analysis* to reflect its focus on the value of the option to delay a physical, as opposed to a financial, investment.

9.5 Calculating the value of information

In 2009, the Australian Government decided on a massive communications infrastructure project called the National Broadband Network (NBN) involving fibre-to-the-premises internet connection to the majority of homes, businesses and other organizations in the country. Since the cost, spread over several years, was estimated to be many billions of dollars, surprise was expressed at the government's decision not to undertake a CBA before proceeding with the project.[1] As we have seen, it would be rational for the government not to undertake a CBA if it had already decided to proceed with the project no matter what. In this section we consider a simple numerical example of the value of information, such as that provided by a CBA.

In the analysis of the value of delay we assumed that the decision-maker was risk-neutral, meaning that he makes his choice among alternatives on the basis of expected value (mean), with no consideration for the risk (variance) associated with the outcome. We also assumed that perfect information could be obtained – that after a delay, the size of the return would be known with certainty. While analysis of the value of information can accommodate risk-averse decision-makers acquiring imperfect information, the aim here is clarity of exposition and we will maintain the assumptions of risk-neutrality and perfect information in this discussion of the value of information.

Figure 9.2 summarises the consequences, from the perspective of a risk-neutral decision-maker, of alternative courses of action: proceed immediately with a project which has an expected NPV of + $1, or perform a CBA before deciding whether or not to undertake the project. Assume that the CBA will reveal the value of the NPV with certainty, and that in the event that NPV turns out to be negative the project can be dropped from consideration.

The value of the CBA lies in avoiding a bad decision – deciding to undertake a project that will turn out to have a negative NPV (Type II Error in Table 9.1). In dollar terms, the expected value of information is the difference between the expected values of the alternative courses of action, $2 in the example. If the cost of performing the CBA is less than $2 then it is worth undertaking according to the cost-benefit criterion.

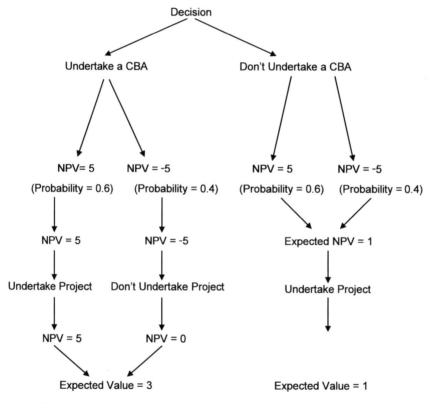

Figure 9.2 The value of perfect information to a risk-neutral decision-maker. The decision-maker judges that there is a 60% chance that the project NPV will be + $5 and a 40% chance that it will be –$5. Since the expected value of the project is + $1 the *prior optimal act* is to proceed. However a CBA would reveal whether the NPV will be + $5 or –$5, and in the latter event the project could be dropped from consideration. The expected value of the project once the extra information provided by the CBA is available is + $3.

For a further example, consider the range of forecasts confronting Australia's Murray Darling Basin Authority in its consideration of proposed major water-related investment projects aimed at ameliorating the effects of climate change. The level of annual rainfall in the Basin in the future is predicted to lie somewhere between 70% higher and 70% lower than the current level. This degree of uncertainty creates a case for delaying the decision about an irreversible project, such as a major dam, until estimates of future rainfall become more precise in the light of further information.

e.g.

EXAMPLE 9.1 Maximin, Maximax and Minimum Regret strategies for dealing with uncertainty

In considering undertaking construction of a dam, how might a decision-maker cope with the level of uncertainty presented by the Murray Basin climate forecast described above? The task is to compare the present values of agricultural output, net of input costs and less the cost of the dam, if built, generated in the two hypothetical worlds *with* and *without* the dam project, and under different climate outcomes. In this example there are four possible outcomes for Action/Climate (although in a real application there could be many more dam designs and climate conditions considered): Build/ Wet, Build/Dry, Don't Build/Wet and Don't Build/Dry. Hypothetical values associated with each are reported in the following pay-off table:

Net Benefits Associated with Different Actions under Different Conditions ($Millions)

ACTION \ CLIMATE	WET CONDITIONS	DRY CONDITIONS
BUILD THE DAM	13(3)	10
DON'T BUILD THE DAM	16	8(2)

Corresponding to each Action there are two possible values (ignore values in brackets for the moment), depending on the climate outcome. These are referred to as the maximum and minimum pay-offs associated with that Action: for example, if the Action is "Build the Dam" the maximum pay-off is 13 and the minimum is 10. Three strategies are considered:

Maximin: choose the Action with the higher minimum pay-off: Build and get at least 10;

Maximax: choose the Action with the higher maximum pay-off: Don't Build and get 16, if Dry;

Minimum Regret: clearly Build is the best decision if Dry conditions turn out to prevail, and Don't Build is best under Wet. The Regret value (in brackets) is the relative cost of having made the wrong, as compared with the correct, decision: a loss of 3 if we Build but it turns out Wet, and a loss of 2 if we Don't Build and it turns out Dry. Choose Don't Build.

9.6 The cost of risk

In the discussion of the value of information we considered the case of perfect information, but in reality, irrespective of the quantity and quality of information that has been gathered, there will always be residual risk associated with a proposed project. If the decision-maker were risk neutral, as assumed in our discussion, the remaining risk associated with the project would be irrelevant to the decision to be made. However, in practice, decision-makers are generally risk-averse – they perceive risk as a cost to be set against the expected net benefits of the project. We now examine why individuals tend to be risk-averse and now this characteristic affects project selection. We then consider various ways of dealing with risk in the decision-making process.

9.6.1 The theory of risk aversion

In the private sector risk is taken into account in investment analysis by means of the *Capital Asset Pricing Model* (CAPM). The amount of risk associated with an investment is measured by its *Beta Factor*, which is an industry-specific measure of the extent to which the asset contributes to the overall risk of a widely diversified portfolio, taking account of the variance of its return and its covariance with the returns on other assets. The unit cost of risk is measured by the *market risk premium*, which is the additional rate of return required to compensate an investor for each additional unit of risk. The risk premium assigned to the project is expressed as a rate of return measured by the quantity of risk (the Beta Factor) times the unit cost of risk (the market risk premium). The required rate of return on a project is the risk-free rate plus the project's risk premium rate.

The market puts a price on risk because, in general, individual investors are risk-averse where decisions are to be made about projects involving significant sums of money. This is not necessarily inconsistent with the observed risk-loving behaviour of gamblers at the race track where typically small sums of money are wagered in order to expose the bettor to risk.[2] Risk aversion follows from the proposition that the utility derived from wealth rises as wealth rises, but at a decreasing rate; at a low level of wealth an extra dollar can provide a significant amount of additional utility, whereas at a high level it may make little difference. The concept of *diminishing marginal utility of wealth* implies that taking a fair gamble will reduce the individual's level of utility. For example, if the individual bet $1000 on the toss of a fair coin, the loss in utility if the call was incorrect would be greater in absolute terms than the gain in utility if the call was correct. This observation provides a useful characterization of risk-averse behaviour – unwillingness to accept a fair gamble because the expected change in utility is negative.

Figure 9.3 illustrates the utility of wealth function of a risk-averse individual, with wealth $W, considering a fair gamble of $h. The levels of utility associated with the wealth levels $W + h$, W, and $W - h$ are identified on the vertical axis. The expected utility of wealth if the bet is accepted, $E[U(\tilde{W})]$, is also identified on the vertical axis: since there is a 50% chance of winning and a 50% chance of losing, this value occurs exactly half-way between $U(W + h)$ and $U(W - h)$. Figure 9.3 shows that, because of diminishing marginal utility of wealth, $E[U(\tilde{W})] < U(W)$. From this we can conclude that the individual will choose not to accept the bet.

We can use Figure 9.3 to work out the cost of the risk associated with taking the bet. It can be seen from Figure 9.3 that the level of wealth with certainty that yields the same utility as the expected utility of wealth with the gamble is W_1: $U(W_1) = E[U(\tilde{W})]$. It

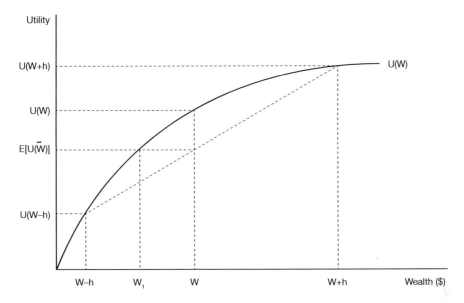

Figure 9.3 The relationship between utility and wealth for a risk-averse individual. The current level
of wealth is denoted by W, and the individual has been offered a fair bet consisting of a
50% chance of winning $h and a 50% chance of losing $h. Since h is a random variable it
follows that, if he accepts the bet, his level of wealth will also be a random variable, which
we denote by $\tilde{W}$. Since the bet is "fair" in the sense that the expected value of h, E(h),
is zero, it follows that the expected value of the individual's wealth if he accepted the bet,
E($\tilde{W}$), equals W. If he were risk-neutral he would base his decision on the expected value
of wealth and would be indifferent between accepting or not accepting the bet, since E($\tilde{W}$)
= W. However, under the *expected utility hypothesis*, we assume that he will choose the
alternative that has the higher *expected utility*. The expected utility of the random variable
$\tilde{W}$ is less than that of W with certainty: E[U($\tilde{W}$)] < U(W), and so he will refuse the bet.

follows that the individual would be indifferent between giving up an amount of wealth
(W – W_1) and accepting the bet. This means that the cost of risk associated with having to
take the bet is (W – W_1), and that the individual would be willing to pay up to this amount
to avoid the risk. It will be left to the reader to demonstrate that if the size of the bet is
increased to h_1> h, the cost of risk rises. When the bet is increased from h to h1, with the
odds unchanged, the amount of risk, measured by the variance of the random variable $\tilde{W}$,
increases, and hence the total cost of the risk rises.

We can infer from Figure 9.3 that the cost of risk is higher with more curvature of the
utility function. The curvature of the utility function tells us how quickly the marginal
utility of wealth is falling as wealth increases. We have already encountered this concept
in Chapter 5 where Technical note 5.2 explored the relationship between elasticity of marginal utility of income or wealth, ε_{MU}, and the consumption rate of interest. Now we find
that ε_{MU} indicates the degree of the individual's aversion to risk. In fact readers can satisfy themselves that the Pratt measure of risk aversion, discussed in Technical note 9.1, is
r(W) = ε_{MU}/W.

It makes intuitive sense that, for a risk-averse individual at a given level of wealth, the
cost of risk rises as the amount of risk rises. But what happens to the cost of risk if the

individual's wealth is increased? In general we cannot tell *a priori* because at a higher level of wealth both the increase in utility as a result of a win, and the fall in utility as a result of a loss, are reduced. However, it seems to make sense that the higher the level of wealth, the lower the cost of risk to the individual: a rich person worries less about hazarding a given sum of money than does a poor person. This is a characteristic of the logarithmic utility function discussed in Technical note 9.1. The implication is that the amount of additional expected wealth that would be required to compensate the individual for a given increase in the risk associated with his wealth portfolio falls as the expected value of the portfolio rises.

To illustrate a decision-maker's attitude to risk we construct what is known as an individual's *indifference map*, which treats the level of wealth as a random variable and shows the trade-offs the individual is willing to make between the expected or mean value of wealth, and the variance of wealth, measuring the degree of risk. Figure 9.4 provides an example. The expected value of the individual's wealth, $E(W)$, is measured on the horizontal axis, and the degree of risk, as measured by the variance of wealth, $VAR(W)$, is shown on the vertical axis. The indifference curves show the individual's attitude to risk. Each curve traces the locus of possible combinations of outcomes, as measured by $E(W)$ and $VAR(W)$, that provide the individual with a given level of utility; since these combinations provide the same level of utility the individual is said to be *indifferent* between them.

TECHNICAL NOTE 9.1 Utility and the cost of risk

The cost of risk is calculated by equating $E[U(W + \tilde{h})]$ with $U(W - p)$ and solving for p, where W is the current level of wealth, p is the cost of risk, and h is the random variable representing the risky prospect. Taking a first order Taylor Series approximation to $U(W - p)$, and a second order approximation to $U(W + \tilde{h})$, to account for the randomness of h, then taking the expected values and equating the resulting expressions, it can be shown that the cost of the risk associated with the risky prospect is approximately $p = E(h^2)r(W)/2$. In this expression $E(h^2)$ is the variance of h, measuring the amount of risk, and $r(W)$ is the Pratt measure of risk aversion: $r(W) = \varepsilon_{MU}/W$, where $\varepsilon_{MU} = -(\Delta\lambda/\Delta W) * (W/\lambda)$, the elasticity of the marginal utility of wealth, λ. For the special case in which utility is measured by the natural logarithm of wealth, $U = \ln W$, the Pratt measure is $1/W$. In other words, according to this measure, the cost of risk is the amount of risk, measured by the variance of the risky prospect, times the cost per unit of risk, measured by $1/2W$. As can be seen from the formula, the cost of the risk falls, in the logarithmic utility case, as W increases.

The shape of the indifference map is clearly a reflection of how the decision-maker perceives risk. The slope of an indifference curve at any point indicates the trade-off the individual is willing to make between risk and return at this point: it measures the additional variance of wealth the individual will tolerate in return for a higher mean value. The reciprocal of the slope value can be interpreted as the cost of risk: the extra mean return required to compensate for extra variance. It can be seen from the curvature of the indifference curves in Figure 9.4 that this amount rises as variance increases along an indifference curve. In other words, at a given level of utility the marginal cost of risk rises as the level of risk increases.

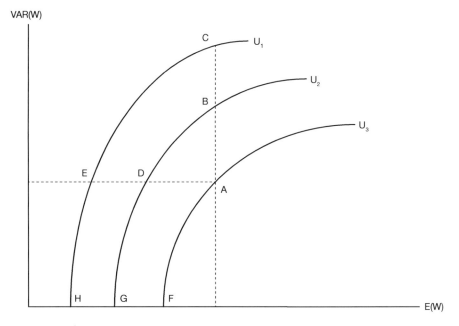

Figure 9.4 A risk-averse individual's indifference map between mean and variance of wealth. The individual's wealth, W, is described by a random variable with mean E(W) and variance Var(W). Each indifference curve plots mean/variance combinations that provide a given level of utility. The three levels of utility illustrated are ranked in order of preference $U_3>$ $U_2>U1$. In terms of the combinations identified: A, B and C have the same mean but A > B > C on the basis of variance; similarly, A, D and E have the same variance, but A > D > E on the basis of mean. Combinations that provide the same level of utility are C= E = H and B = D =G, and A = F. The points associated with zero variance, F, G and H are certainty equivalent values corresponding to the utility levels U_3, U_2 and U_1 respectively.

9.6.2 Dealing with project risk

While indifference curves are useful for conceptualizing the cost of risk and the subjective attitude to risk, constructing such a map would require that the analyst engage in a fairly complex process of choice experimentation with the relevant decision-maker with a view to eliciting from the choices she makes the information necessary to construct the indifference map – a lot easier said than done!

A further problem is that the indifference curves are defined over total wealth, not just the value of a particular project. The risk associated with a particular project is what it adds to the total risk of the portfolio of wealth. Thus adding to the portfolio a project whose NPV has a high variance, but is negatively correlated with the NPVs of other projects in the portfolio, might actually *reduce* the variance of the total NPV of the portfolio. Indeed, in a large, well-diversified investment portfolio it is possible, in principle, for portfolio risk to be eliminated completely. This is the basis of the argument that governments need not take risk into account in project appraisal because the portfolio of public projects is large and well diversified and any additional project adds nothing to the risk of the overall portfolio. To appreciate this argument, suppose that the fair gamble of $h illustrated by Figure 9.3 was to be shared equally among two individuals with identical utility of wealth functions. It can be seen from Figure 9.3 that, because of the curvature of the utility of wealth function, the individual cost of risk associated with the fair gamble h/2 is less than half the original cost. This means that sharing the risk between two individuals has reduced its total cost. If the risk were shared among all members of the population, as in the case of a public project, the total cost of risk could, in principle, fall to close to zero. Needless to say, few individual decision-makers in the public sector subscribe to this view!

Portfolio risk can be reduced by spreading risk through judicious selection of projects and the use of financial derivatives traded in the market. As noted earlier, the amount of risk associated with a given asset, reflecting the covariance of its rate of return with that of the portfolio, can be multiplied by the market price of risk to reveal the premium in expected return that is required to make the asset attractive to investors; just as individuals trade-off risk against return along an indifference curve, the market undertakes a similar trade-off along a security market line which graphs required expected rate of return on assets against their risk, measured by the Beta Coefficient as discussed earlier. When the risk associated with the project in question (its Beta value) is multiplied by the price of risk, the risk premium over the risk-free rate of return required to render the project viable from the investor's point of view is obtained.

Adding a risk premium to the discount rate as a method of accounting for risk is a common approach in the analysis of private sector projects: the required IRR consists of the risk-free rate plus the risk premium. As discussed in Chapter 2, Section 9, under this approach NPV is calculated using a higher discount rate than that which would be used in the absence of risk, with the risk premium representing the analyst's perception of the degree of riskiness of the investment. As a higher discount rate implies a lower NPV, *ceteris paribus*, it will be more difficult for the project to pass the NPV decision criterion the higher is the risk premium. Providing the decision-maker with NPVs for a given project over a range of discount rates provides the information necessary to assess the significance of applying a risk premium, and deciding whether the project is marginal or not.

In the public sector, some governments attempt to mimic the private sector investment appraisal process by specifying a required rate of return for public projects; for example, the State of Victoria, Australia, sets the market risk premium at 6% and uses Beta Factors of 0.3, 0.5 and 0.9 for very low, low and medium risk projects respectively. The risk-free rate of interest can be approximated by the real rate of return on top quality government bonds – say, around 3% – so that, on the basis of this information, Victoria would require a 6% real rate of return from a project judged to be low risk, i.e. $(0.5 \times 6\%) + 3\%$.

$$f(x)$$

TECHNICAL NOTE 9.2 Effect of adding a risk premium to the discount rate

Using the methodology discussed in Chapter 2, it can be seen that the discount factors $(1/(1 + r + p))$ and $(1 - p)/(1 + r)$ are approximately the same. The former corresponds to adding the risk premium, p, to the discount rate, r, whereas the latter corresponds to incorporating the chance of project failure into the expected net present value calculation. To see this, suppose that there is a chance, p, of the project benefit stream terminating in any given year. The expected project benefit in year t is then $B(1-p)t$ and the expected present value of this benefit is $B[(1-p)/(1+r)]t$ which is approximately equal to $B/(1 + r + p)t$.

Using a risk premium on the discount rate, though common practice in private sector investment analysis, suffers a few important drawbacks:

- Often the analyst will be relying on a totally subjective estimate of what value the risk premium should take, though we have seen that inter-sectoral comparisons of rates of return are sometimes used to estimate sector-specific risk premiums.
- By attaching a constant premium to the discount rate over the life of the investment we are assuming that the further into the future the forecasted value, the more risky the outcome becomes, whereas it may be the case that it is the earliest years of an investment project that are the riskiest.
- Applying the premium to the net cash flow effectively assumes that the forecast cost and benefit streams are affected equally by the risk, whereas it is likely that each is subject to a different degree of riskiness.

In the following discussion we consider how the decision-maker can, in practice, obtain basic information about the variance of a project's NPV for use in the decision process. The costing of the risk so identified depends on the decision-maker's attitude to risk about which it is impossible to generalise.

9.7 Risk modelling

As discussed in the previous sections of this chapter, the preliminary steps in risk analysis allow the analyst to determine the extent to which the outcome of the project under consideration can be considered risky, the particular variables that have the greatest influence on project risk, and whether a more detailed and rigorous form of risk analysis can be justified. Assuming the decision has been made to proceed further, the next stage is to identify and describe the nature of the uncertainty surrounding the relevant project variables. To do this we use *probability distributions*. A probability distribution takes the description of uncertainty one level beyond that which we used in the sensitivity analysis. There we described a variable's uncertainty purely in terms of the range of possible values, e.g. high, medium, low; or, maximum, mean, minimum; or optimistic, best-guess, pessimistic.

A probability distribution does this, but it also describes the likelihood of occurrence of values within the given range. When only a finite number of values can occur, the probability distribution is described as *discrete*, and when any value within the range can occur, it is termed *continuous*. An example of the former type of distribution is the probabilities of heads or tails in the toss of a coin, and an example of the latter is the familiar Normal distribution.

9.7.1 Use of discrete probability distributions

An example of a discrete probability distribution for the NPV of road construction costs is shown in Table 9.3.

Once a probability distribution for a project input variable has been chosen, the analyst is able to calculate the *expected value* of the variable, and to use this in the cash flow for the project rather than the *point estimate* that we would have used if uncertainty had been ignored. Table 9.4 provides a simple numerical example showing how the expected value of net road usage benefits is derived. We assume for simplicity that construction cost is the only uncertain variable, and that the net benefits exclusive of road construction costs are estimated at $136 million.

Table 9.3 A discrete probability distribution of road construction costs

	Road construction cost (C) ($ millions)	Probability (P) (%)
Low	50	20
Best Guess	100	60
High	125	20

Table 9.4 Calculating the expected value from a discrete probability distribution

	Road Construction Cost (C) ($ millions)	Probability (P) (%)	E(C)= P × C ($ millions)	NPV ($ millions)	E(NPV) ($ millions)
Low	50	20	10	86	17.2
Best Guess	100	60	60	36	21.6
High	125	20	25	11	2.2

Note: Present Value of net benefits, excluding road construction cost, is assumed to be $136 million.

The expected cost of road construction can be derived as E(C) = $10 + $60 + $25 = $95 million. Rather than using the point (best guess) estimate of $100 million, the analyst would use $95 million as the cost in calculating the net cash flow for the project. Moreover, it will also be possible to derive a probability distribution for the NPV of the project: a 20% probability of NPV = $86 million; 60% probability of $36 million; and 20% probability of $11 million. If we use these values to calculate the expected NPV, we get:

E(NPV) = 17.2 + 21.6 + 2.2 = $41 million.

9.7.2 Joint probability distributions

In the previous example it was assumed that there was uncertainty about only one of the project's inputs, which made the task of deriving a probability distribution for the outcome – the NPV, in this instance – a straightforward matter. More realistically however, the analyst will find that there is uncertainty about the level of more than one project input or output. The probability distribution for the outcome (NPV) will then depend on the aggregation of probability distributions for the individual variables into a joint probability distribution. In aggregating probability distributions we must first distinguish between *correlated* and *uncorrelated* variables. It will often be the case that we have two uncertain variables relevant to the project outcome that are closely correlated; for instance, if road usage increases so too do road maintenance costs. Modelling risk in this situation requires the analyst to take account of the dependence of possible variations in the level of one variable on the possible variations in the level of the other. The two variables cannot be assumed to vary independently of each other.

In this situation deriving the joint probability distribution is quite easy as shown in Table 9.5 in which it is assumed that the net benefit of the services of a road is given by the value of user benefits less the cost of maintenance. In this example there is a 20% chance of road maintenance costs being $50 *and* road user net benefits being $70, a 60% chance of road maintenance costs being $100 *and* road user net benefits being $125, and so on. The expected values are calculated in the usual way, and are reported in brackets in Table 9.5.

Table 9.5 Joint probability distribution: correlated variables

	Probability (P) (%)	Cost ($ millions)	Benefits ($ millions)	Net Benefits ($ millions)
Low	20	50(10)	70(14)	20(4)
Best Guess	60	100(60)	125(75)	25(15)
High	20	125(25)	205(41)	80(16)
(Expected Value)		(95)	(130)	(35)

Table 9.6 Joint probability distribution: uncorrelated variables

Level	Probability(P) (%)	Cost ($ millions)	Benefits ($ millions)
Low (L)	20	50	70
Best Guess (M)	60	100	125
High (H)	20	125	205

Combination	Joint Probability	Net Benefit ($)
LC-HB	0.2 × 0.2 = 0.04	155(6.2)
LC-MB	0.2 × 0.6 = 0.12	75(9.0)
LC-LB	0.2 × 0.2 = 0.04	20(0.8)
MC-HB	0.6 × 0.2 = 0.12	105(12.6)
MC-MB	0.6 × 0.6 = 0.36	25(9.0)
MC-LB	0.6 × 0.2 = 0.12	30(3.6)
HC-HB	0.2 × 0.2 = 0.04	80(3.2)
HC-MB	0.2 × 0.6 = 0.12	0(0.0)
HC-LB	0.2 × 0.2 = 0.04	−55(−2.2)

Where the variables are uncorrelated (can vary independently of each other), the situation becomes a lot more complicated. If we assume, for example, that the benefits of the project are determined exclusively by movements in international prices while the costs depend on domestic factors, there may no longer be any correlation between the variables. In this situation we need to consider all possible combinations of values for the costs and benefits on a pair-wise basis, and to calculate the value of net benefits for each combination. This procedure will produce a whole range of possible values for the project's net benefits which then need to be ordered and arranged as a probability distribution: an example is shown in Table 9.6, in which all possible combinations such as Low Cost, High Benefit (LC-HB) are assigned joint probabilities.

The upper panel of Table 9.6 shows the independent or uncorrelated probabilities for the costs and benefits respectively. The lower panel shows how the joint probability of each possible combination of cost and benefit is calculated; since *some* combination *must* occur, the joint probabilities sum to 1. Table 9.6 also reports the estimated net benefit associated with each probability. From these net benefit estimates and joint probability values the expected values of the respective outcomes (net benefits) are then derived and reported in brackets in Table 9.6. The expected or mean value is $42.2 million, which is the sum of the individual expected values associated with each event. The two extreme values are *negative* $55 million and *positive* $155 million, each with a probability of 4%.

9.7.3 *Continuous probability distributions*

A continuous probability distribution assigns some probability to the event that the outcome, the project NPV, for example, lies in any particular segment within the range of the distribution; for example, the probability that NPV will be greater than zero, or less than $100. The continuous probability distribution familiar to most readers is the Normal distribution which is represented as a bell-shaped curve. This distribution is completely described by two parameters – the *mean* and the *standard deviation*. While its *range* is the full extent of the real line, i.e. from minus infinity to plus infinity, implying that, in principle, *anything* is possible, the bell shape indicates that only events in the vicinity of the mean are likely. The degree of *dispersion* of the possible values around the mean is measured by the *variance* (S^2) or, the square root of the variance – the *standard deviation* (S). When the standard deviation is divided by the mean, we get the *coefficient of variation* which is a useful measure for comparing the degree of dispersion for different variables when their means differ. The variance or standard deviation is a useful measure of the amount of risk: a high standard deviation implies a reasonable probability of the outcome being significantly higher or lower than the mean value, whereas a low standard deviation implies a relatively small range of likely outcomes in the vicinity of the mean.

The characteristics of the variable's probability distribution are important inputs into formal risk modelling using spreadsheet 'add-ins' such as *ExcelSim*© which we use in this text.[3] From analysis of relevant data, perhaps time series or cross–sectional data, the analyst must decide what type of probability distribution best describes the uncertain variable in question. Some public sector jurisdictions offer the analyst advice about the appropriate form of distribution to be used: in the State of South Australia, for example, analysts are advised to use the Poisson or Binomial Distribution for a discrete variable, a Normal Distribution for a symmetric and continuous variable, a Triangular Distribution for an asymmetric and bounded variable, and a Uniform Distribution if no information about the value of the variable, apart from its likely bounds, is available.

Risk modelling programs, such as *ExcelSim*©, offer the analyst a whole range of distributions to choose from, including: the linear distribution in which each value within the specified range is equally likely; the triangular distribution which assigns higher probabilities to values near some chosen point within the range, and lower probabilities to values near the boundaries of the range; and, of course, the familiar Normal distribution. If the analyst has a reasonable set of past observations on the uncertain variable in question, but is unsure as to what type of probability distribution best describes that data, software add-ins such as *BestFit*© can be used to identify the most appropriate probability distribution.

Quite often, however, the analyst finds that there may be no reliable historical information about the variable in question and that he has no information beyond the range of values the variable could reasonably be expected to take, as described, for instance, in the sensitivity analysis discussed previously. Here the analyst may still undertake a more formal risk modelling exercise than a sensitivity analysis by adopting what is called a Triangular (or 'three-point') Distribution, where the distribution is described by a high (H), low (L) and best-guess (B) estimate, which determine the maximum, minimum and modal values of the distribution respectively. Each event in the range between L and H is assigned some probability, with values in the vicinity of B being most likely. The precise specification and statistics for the three-point distribution can vary, depending on how much weight the analyst wishes to give to the mode in relation to the extreme point values. The Triangular Distribution is particularly useful because often information about the distribution of the variable has to be obtained from experts in areas relevant to the project: industry experts can supply information about likely prices and exchange rates, foresters can supply information about likely tree-growth rates, doctors can supply information about the likely effects of medical intervention, and so forth. The parameters of the Triangular Distribution can often be elicited by a series of simple questions: What do you regard as the most likely value this variable will take? What is the lowest value the variable could take? What is the highest value this variable could take? Clearly, adopting a Triangular Distribution of this type is a 'rough-and-ready' form of risk modelling and should be used only when limited resources prevent obtaining sufficient information to identify the characteristics of the uncertain variable's probability distribution more rigorously.

The triangular probability distribution can be represented graphically as shown in Figure 9.5. Using the information provided by the triangular distribution, the decision-maker now knows not only the range of possible values the variable could take, but can also see what the probability is of its value lying within any particular range of possible values.

Selected variables involved in the CBA can be described by probability distributions, and, using a program such as *ExcelSim*© as described below, a probability distribution can be derived for the project's NPV. A useful format for presenting this information is a *cumulative probability distribution*. Here, the vertical axis is scaled from 0 to 1 showing the cumulative probability corresponding to the NPV value on the horizontal axis, as shown in Figure 9.6. The cumulative distribution indicates the probability of the NPV lying below (or above) a certain value.

The advantage of presenting the results of a risk analysis in this way is that it provides the relevant information to the decision-maker in a user-friendly, and easy to interpret

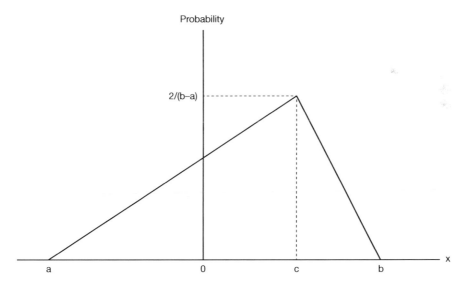

Figure 9.5 The triangular probability distribution. The *range* of the values of the random variable, x, shown by the distribution is from a (a negative value in this example) to b. The distribution is *skewed* to the left. The most likely value (the *Mode*) is c. The probability of c is determined by the property that the area under the distribution has to be 1 (*something* has to happen). The expected value of x (the *Mean*) is given by (a + b + c)/2. The *Median* value (half the possible values of x are lower, and half are higher, than the Median) lies between the Mean and the Mode.

graphical format. It is relatively simple for a decision-maker, unfamiliar with basic statistical concepts, to read from this distribution the probability of the project achieving at least a given target NPV value, or the probability of failure, defined as a negative NPV. To undertake these sorts of probability calculations manually, even with the aid of a spreadsheet, can become complicated and very time-consuming, especially when the values of several variables are uncertain. When one is working with a cash flow of, say, 20 to 25 years, and where the value of a variable is allowed to vary randomly in each and every year of the project, the task becomes unwieldy. But with add-ins such as *ExcelSim*© the process is relatively straightforward. To represent the risk modelling output from *ExcelSim*© in graphical form, it is necessary to use the add-in *Analysis Toolpak* in *Excel*©.[4]

The use of *ExcelSim*© is illustrated with a simple example later in this chapter. At this stage it needs to be noted that what such programs do is perform a simulation, known as a *Monte Carlo* analysis, wherein the NPV or IRR of the project is recalculated over and over again, using each time a different, randomly chosen, set of values for the variables in the project's cash flow calculation. The random selection of values is based on the characteristics of each input variable's probability distribution, where this information, rather than a single point estimate is entered into the relevant cell of the spreadsheet by the analyst. The program effectively instructs the spreadsheet to randomly sample values of the uncertain variables from their specified probability distributions, to calculate the project NPV using the sample values, to save the calculated NPV value and to repeat this process many times over. The saved NPV calculations for all combinations of sampled values for the input variables can be used to develop a probability distribution of NPV. It is as if the analyst had run and saved thousands of NPV calculations by changing the input values across the full range of

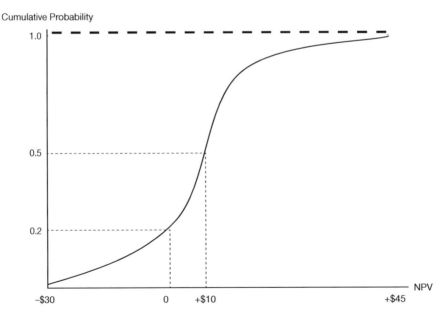

Figure 9.6 A cumulative probability distribution. The distribution shows the probability of NPV being less than the value shown on the horizontal axis. For example, there is a 20% chance of NPV being negative. The chance of NPV lying between 0 and + $10 is calculated by subtracting the probability of a negative value from the probability that it will be less than $10, to give 30%. There is virtually no chance of NPV exceeding $45.

likely combinations implied by the probability distributions. The program then assembles all the saved results, presenting them as a probability distribution which can be displayed in numerical or in graphical format using the *Analysis Toolpak* in *Excel©*.

9.8 Using risk analysis in decision-making

Producing a probability distribution for a project's range of possible outcomes is only part of the risk analysis process. As discussed at the beginning of this chapter, the decision-maker needs to interpret this information and use it in the decision-making process. It must be emphasised that this process necessarily involves the decision-maker determining her subjective attitude towards risk. The same risk analysis results provided to a number of different individuals is likely to be interpreted differently and to result in different decisions being made. Figure 9.7 illustrates this point.

Which of the two projects should the decision-maker choose? There is no definitive answer to this question. It will depend on the decision-maker's attitude towards risk, as discussed earlier and illustrated in Figure 9.4 which showed the trade-off between risk and return. The choice may depend on whether the decision-maker is willing to accept the higher variance of Project B's NPV in exchange for its higher expected value. Alternatively the risk-averse decision-maker might prefer the project with the lower probability of failure. The probability of failure is measured by the area of the probability distribution to the left of the NPV = 0 point on the horizontal axis. On this basis the decision-maker might prefer Project A.

9.9 Modelling risk in spreadsheet applications using *ExcelSim©*

The aim of this section is to provide the reader with a quick introduction and overview of how the *ExcelSim©* add-in can be used to undertake a risk analysis for a net present value calculation.

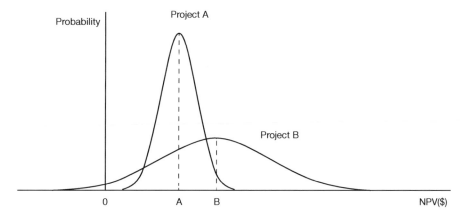

Figure 9.7 Two projects with different degrees of risk. This figure illustrates symmetrical probability distributions for the NPVs of two projects, A and B. Project B has a higher expected NPV than project A. However, the dispersion of values around the mean is much wider in the case of project B than of A. Indeed, it can be seen that there is a (small) chance that the NPV of project B could be negative. Its NPV could also potentially be very much higher than that of project A. In other words, project B has a higher expected NPV (a good thing) but has a higher variance or risk (a bad thing) from the viewpoint of the risk-averse decision-maker.

A simple numerical example is developed, with the reader taken through a number of steps as shown in the screen dumps presented in the following series of figures with text boxes. The example concerns a project with an initial cost of $100, which is known with certainty, and an annual net benefit over a five-year period consisting of a random variable with an expected value of $30. Three different probability distributions are used to characterise this

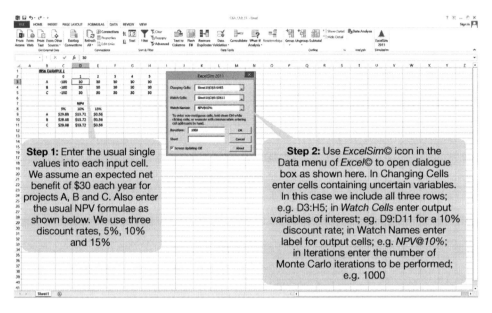

Figure 9.8a Entering the data and simulation settings.

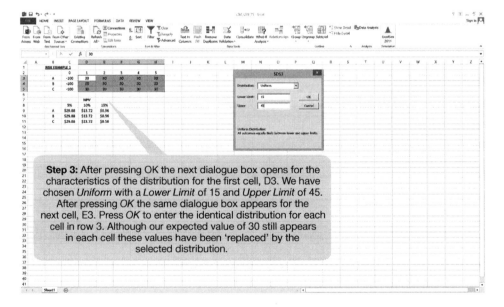

Figure 9.8b Entering the characteristics of the first distribution.

random variable and a probability distribution of NPV is calculated for each and expressed both as a histogram (the discrete version of the probability distributions illustrated in Figure 9.7) and a cumulative distribution (similar to Figure 9.6). The spreadsheets reported in this and following sections of this chapter can be accessed through the Support Material website. Dialogue boxes can be opened by following the steps indicated in Figure 9.8.

Panel A

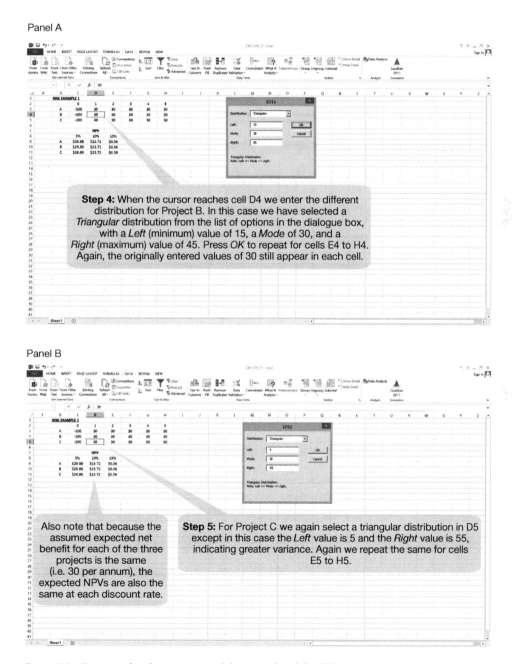

Step 4: When the cursor reaches cell D4 we enter the different distribution for Project B. In this case we have selected a *Triangular* distribution from the list of options in the dialogue box, with a *Left* (minimum) value of 15, a *Mode* of 30, and a *Right* (maximum) value of 45. Press *OK* to repeat for cells E4 to H4. Again, the originally entered values of 30 still appear in each cell.

Panel B

Also note that because the assumed expected net benefit for each of the three projects is the same (i.e. 30 per annum), the expected NPVs are also the same at each discount rate.

Step 5: For Project C we again select a triangular distribution in D5 except in this case the *Left* value is 5 and the *Right* value is 55, indicating greater variance. Again we repeat the same for cells E5 to H5.

Figure 9.8c Entering the characteristics of the second and third distributions.

Figure 9.8d Generating the simulation report.

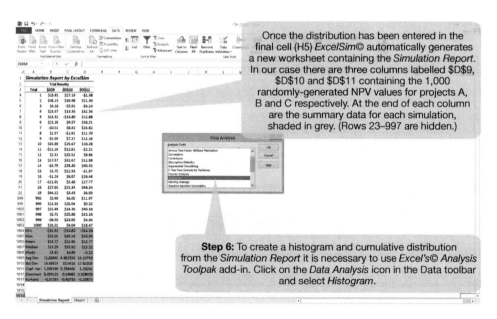

Figure 9.8e Generating the histogram and cumulative distribution.

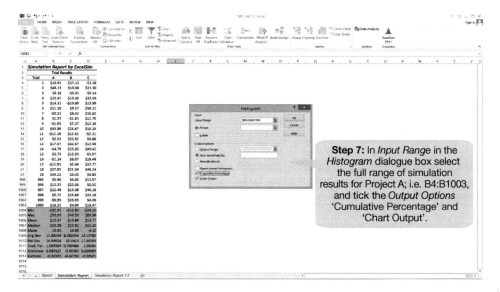

Figure 9.8e Continued

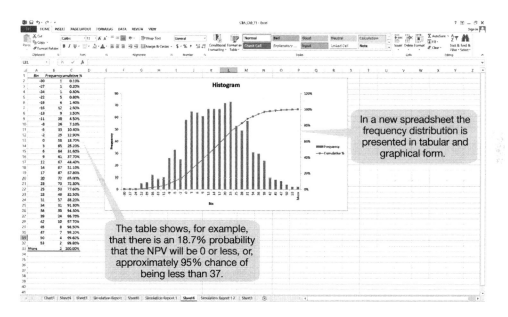

Figure 9.8e Continued

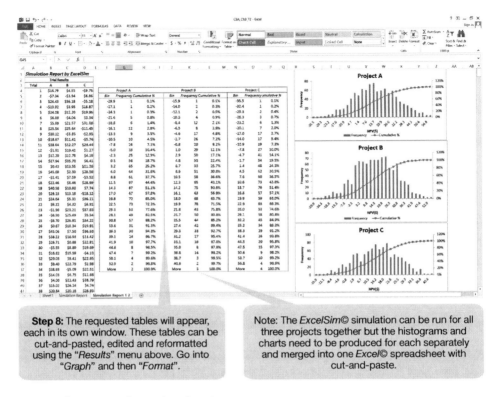

Step 8: The requested tables will appear, each in its own window. These tables can be cut-and-pasted, edited and reformatted using the *"Results"* menu above. Go into *"Graph"* and then *"Format"*.

Note: The *ExcelSim©* simulation can be run for all three projects together but the histograms and charts need to be produced for each separately and merged into one *Excel©* spreadsheet with cut-and-paste.

Figure 9.8f Presenting the combined results.

9.9.1 Modelling a "random walk"

In many instances it is unrealistic to assume that an uncertain variable varies over time around some given "best-guess" value. For example, variables such as interest rates or exchange rates, or house or share prices, are often more appropriately modelled to vary within some limits around their value in the preceding time period. This type of model is usually referred to as a "random walk". The purpose of this section is to demonstrate how to perform a random walk-type simulation using *ExcelSim©*.

We use Project B in our previous example. In the original example we assumed that the uncertainty of the annual net benefit of 30 could be characterised by a triangular distribution with minimum, mode, and maximum values of 15, 30 and 45 respectively, i.e. 50% variation around the mode.

Now let us assume that that type of uncertainty characterises the distribution in year 1 only. Thereafter, the value of net benefits can be expected to vary 50% each side of the preceding year's value. The easiest way to set up this scenario is to create a new row in the spreadsheet into which the *ExcelSim©* simulation data are entered, row 6 in the example in Figure 9.9a. However, for practical reasons it is best to leave the entry of the *ExcelSim©* code and data in row 6 until the final stage before running the simulation as the *ExcelSim©* entries cannot be saved.

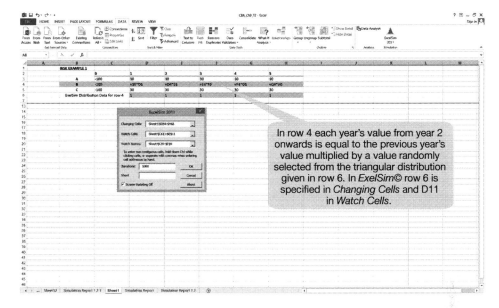

Figure 9.9a Modelling a random walk: simulation setting.

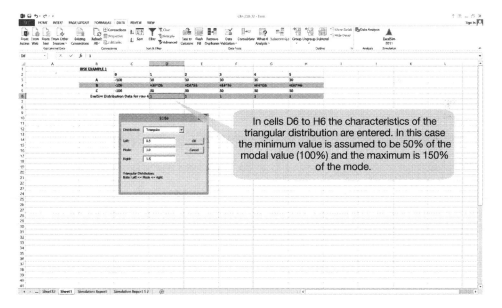

Figure 9.9b Modelling a random walk: probability characteristic.

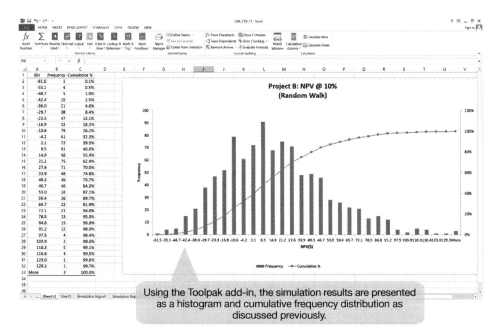

Figure 9.9c Modelling a random walk: presenting results.

In the cells in row 4 for Project B we replace the originally entered values of 30 with a new formula, multiplying the value in the preceding year's cell by the yet-to-be-entered *ExcelSim©* distribution formula in row 6, for the respective year. We have entered a value of '1' provisionally in each cell in row 6 which will be replaced later with the *ExcelSim©* data. In this case we enter 0.5, 1.0 and 1.5 for the *left*, *mode* and *right* values respectively in the *ExcelSim©* dialogue box in each cell of row 6 as shown in Figure 9.9a.

The *ExcelSim©* icon in the *DATA* menu is then activated and the initial dialogue box for the simulation parameters appears as shown in the first panel of Figure 9.9a. In *Changing Cells* row D6 to H6 is entered. The *Watch Cells* are specified as the NPV for Project B at the three discount rates, 5%, 10% and 15%, and the *Watch Names* as the labels for the discount rates. (Neither of these can be seen in Figures 9.9a and 9.9b as the NPV outputs have been hidden in the spreadsheet for purposes of exposition.) The number of *Iterations* has been set at 1000.

Once the first dialogue box has been completed the next dialogue box will appear for entering the simulation data in row 6 as indicated previously, and now shown in Figure 9.9b.

The results for Project B at a 10% discount rate are presented in Figure 9.9c in tabular and graphical format. It is shown that the project has about a one-third probability of not achieving a positive NPV.

9.10 Further reading

J. Hirschleifer, *Investment, Interest and Capital* (Englewood Cliffs, NJ: Prentice Hall, 1970) includes several chapters on investment decision under uncertainty. A useful article on the value of delaying an investment decision in order to gather more information is R.S. Pindyck, "Irreversibility, Uncertainty and Investment", *Journal of Economic Literature*, 29 (1991): 1100–1148, especially pages 1100–1118. A paper by H.F. Campbell and R.K. Lindner, "Does Taxation Alter Exploration? The Effects of Uncertainty and Risk", *Resources Policy*, 13(4) (December, 1987): 265–278, provides a numerical example of the role of imperfect information in reducing both uncertainty and cost of risk borne by a risk-averse decision-maker. Two books with a practical focus on risk and project selection are S. Reutlinger, *Techniques for Project Appraisal under Uncertainty* (Baltimore, MD: Johns Hopkins Press, 1970) and J.K. Johnson, *Risk Analysis and Project Selection* (Manila: Asian Development Bank, 1985). A classic paper on dealing with the risk associated with public sector projects is K.J. Arrow and R.C. Lind (1970), "Uncertainty and Evaluation of Public Investment Decisions", *American Economic Review*, 60: 364–378. John Kay and Mervyn King, *Radical Uncertainty: Decision-Making Beyond the Numbers* (Norton, 2020, pp. 384)

World Health Organisation, *Anticipating Emerging Infectious Disease Epidemics* (Ref. WHO/OHE/PED/2016.2, 2015, pp. 68).

The *@RISK* manual accompanying the software package *@RISK: Advanced Risk Analysis for Spreadsheets* (Newfield, NY: Palisade Corporation, 1997) provides a useful discussion of risk modelling.

Appendix 1 to Chapter 9

Incorporating risk analysis in the ICP case study

Additional information

In the ICP case study we were given an exchange rate of Bht.44 to the US$, a world price of cotton of US$0.3 per bale, and a world price of yarn of US$0.58 per pound (lb). Suppose you are concerned about the uncertainty surrounding future values for each of these variables and decide to undertake a risk analysis using *ExcelSim*© with the following additional information.

- The exchange rate could appreciate or depreciate by 3% in any year, in relation to the previous year's value.
- The US$ world price of cotton and yarn could rise or fall by up to 5% in any year in relation to the previous year's price.
- World cotton and yarn prices can be assumed to be closely correlated with each other, and unrelated to movements in the exchange rate.

In this case we model all three variables using a random walk as discussed in the previous section. For example, if the exchange rate depreciates to, say, Bht.47 to the US$ in one year, then the simulation will base the following year's value on a random variation of 3% around this new value, and so on.

For the purpose of this analysis we also assume that in all cases a triangular distribution represents a reasonable description of the variable's uncertainty.

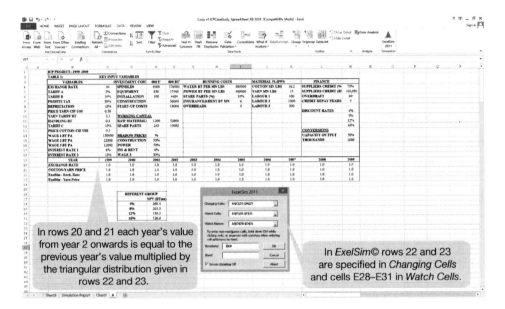

Figure A9.1a Entering the risk analysis data in the ICP case study.

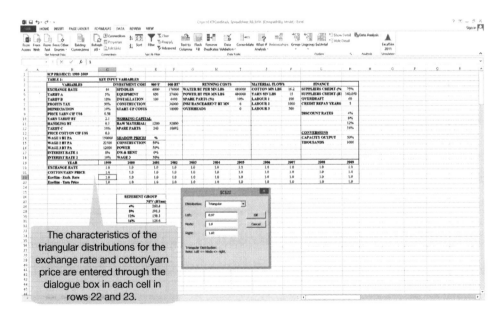

Figure A9.1b Programming a random walk in the ICP case study.

Entering the risk analysis data

Four additional rows need to be entered into the Variables Table of the original spread-sheet to accommodate entries of the simulation data for each project year. These entries are shown in rows 20 to 23 of the spreadsheet in Figure A9.1a. The exchange rate is modelled in rows 20 and 22, and the cotton and yarn prices in rows 21 and 23. In row 20 we enter the exchange rate factor such that after year 1999, each year's exchange rate is a function of the previous year's exchange rate, and in row 22 we enter the *ExcelSim*© simulation data. In the first year of the project (cell C20) the formula '1 * C22' is entered, and in cell D20 we enter 'C20 * D22' which is then copied across all cells in the row to the final year. In cell C22 the *ExcelSim*© data for the *Triangular* distribution are entered for the first project year, 1999, as: *left*=0.97, *mode*=1, and *right*=1.03, as shown in the Dialogue Box in Figure A9.1b. The same *ExcelSim*© data are then copied across all cells in the row to the final year.

For the world prices of cotton and yarn, both of which are assumed to rise or fall each year by 5% of the previous year's price, a similar process is followed. In the first year of row 21 (cell C21) the formula '1 * C23' is entered and then in the following year (cell D21) we enter 'C21 * D23' which is copied across to the last project year. In row 23 the *ExcelSim*© data are entered following the same process as in row 22, except the three points of the triangular distribution are given as: *left*=0.95, *mode*=1, and *right*=1.05. However, before entering the *ExcelSim*© data in rows 22 and 23, and running the simulation, it is necessary to complete the programming of the random walk by linking the contents of rows 20 and 21 to the respective risk variables used elsewhere in the spreadsheet. (We return to the instructions for entering the *ExcelSim*© data in rows 22 and 23 when discussing Figure A9.1b. It is suggested that the value '1' is entered provisionally in all cells in rows 22 and 23. These will subsequently be replaced by the *ExcelSim*© code for the triangular distributions.) For row 20, the exchange rate, whenever there is a cell reference anywhere in the spread-sheet to the value of the exchange rate (in cell C4) this should be multiplied by the cell in row 20 corresponding to the particular year. Similarly, wherever there is a cell reference to the world price of cotton (C13) or yarn (C9), this value is multiplied by the corresponding cell for that year in row 21.

The *ExcelSim*© icon in the *DATA* menu is activated and the initial dialogue box for the simulation parameters appears as shown in the first panel of Figure A9.1a. In *Changing Cells* rows 20 and 21 are entered. The *Watch Cells* are specified as the NPV for the aggregate Referent Group at each of the four discount rates, and the *Watch Names* as the four cells containing the labels, 4% to 16%. The number of *Iterations* has been set at 3000.

The next dialogue box will appear for entering the simulation data in rows 22 and 23, as indicated previously, and now shown in Figure A9.1b. In row 22 the formula for the tri-angular distribution for the exchange rate (*left*=0.97, *mode*=1, and *right*=1.03) is entered in each cell, and in row 23 the different triangular distribution for prices of cotton/yarn (*left*= 0.95, *mode*=1, and *right*=1.05) is entered in each cell.

The results of the risk analysis

The results of the simulation are reported in Figure A9.1c. For convenience, only the results for the aggregate Referent Group net benefits at an 8% discount rate are shown here. The table in the left-hand panel shows the data for the cumulative frequency

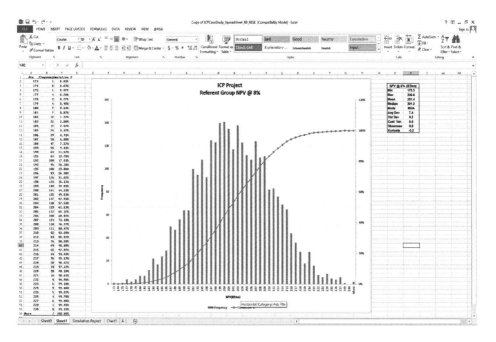

Figure A9.1c Risk analysis: summary statistics for Referent Group net benefits.

distribution and the chart shows both the histogram and the cumulative distribution. To the right of the chart we have copied the summary statistics for the project's NPV at an 8% discount rate: a maximum of $336.1 million, a minimum of $89.6 million and a mean of $201.1 million. The cumulative probability values indicate that there is only a 5% probability that the NPV will be less than $190 million, approximately (or, a 95% chance that the NPV will be greater than $190 million, approximately). There is a 95% chance that NPV will be less than $334 million, or a 5% chance that it will be greater than this amount.

This example shows how it is possible to present the decision-maker with the full range of possible project outcomes and their associated probabilities, thereby providing her with a lot more useful information on which to base her final decision than a single point estimate (i.e. mean or mode NPV), or a sensitivity analysis which would have indicated the possible range of values, but not the probability of the NPV achieving a given level.

Appendix 2 to Chapter 9

Using the @RISK© (Palisade) risk modelling program

The purpose of this section is to provide the reader with a quick introduction and overview of how the @RISK© software add-in can be used to undertake a risk analysis for a net present value calculation.

A simple numerical example is developed, with the reader taken through a number of steps as shown in the screen downloads presented in the following series of figures with text boxes (Figure A9.2–Figure A9.5). The example concerns a project with an initial cost of $100,

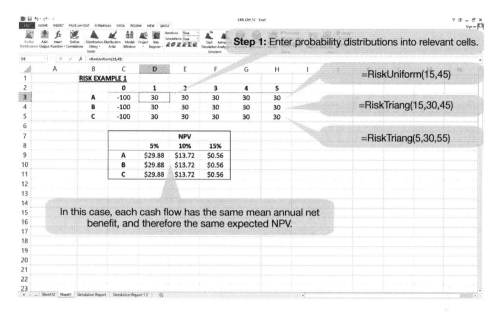

Figure A9.2 Entering the data.

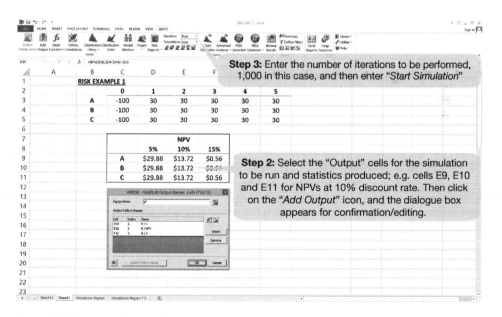

Figure A9.3 Entering the simulation settings.

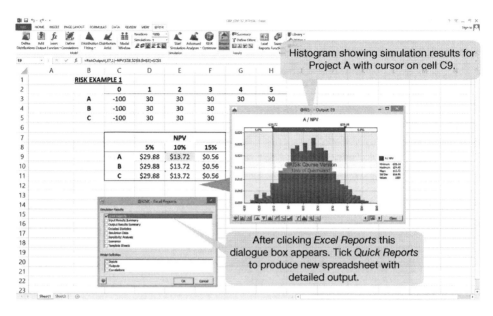

Figure A9.4 Generating simulation output results.

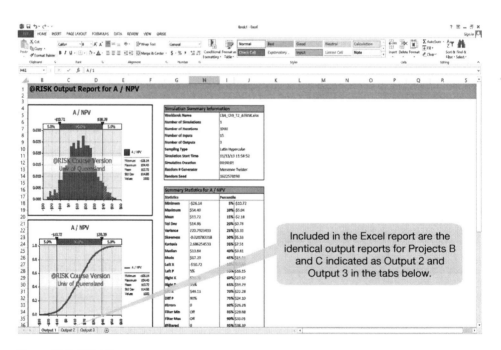

Figure A9.5 @Risk© output report.

which is known with certainty, and an annual net benefit over a five-year period consisting of a random variable with an expected value of $30. Three different probability distributions are used to characterise this random variable and a probability distribution of NPV calculated for each and expressed as both a histogram and a cumulative distribution. A further graph showing how the mean and variance of NPV change as the discount rate is raised is also generated. Some comments on the procedures used follow the presentation of the tables.

Once the simulation has run, a histogram appears showing the results for Project A (cell E9, Figure A9.4). The histograms for Projects B and C will also appear when the cursor is moved to cells E10 and E11. Note that these are NOT the final results to be saved.

To access the downloadable spreadsheet with the complete set of statistics and charts click on the "*Excel Reports*" icon in the toolbar. A dialogue box appears as shown on the screen. Tick the box for "*Quick Reports*", click OK and a new spreadsheet as shown in Figure A9.5 appears. This shows the results for Project A. The same outputs for Projects B and C can be accessed through the other two tabs labelled *Output 2* and *Output 3*.

Using @Risk© to model a random walk: the ICP case study

A detailed explanation of modelling a random walk for the exchange rate and cotton and yarn prices using *ExcelSim©* was given earlier in this chapter. For the reader using @Risk© the process is exactly the same, except for entering the characteristics of the probability distributions in rows 22 and 23 of Figure A9.1b. The corresponding @Risk© formulae for the triangular distributions should be entered into each cell; in row 22, "=*RiskTriang* (0.97,1.0, 1.03)" and in row 23, "=*RiskTriang*(0.95,1.0,1.05)". The "*Output*" cells are again the Referent Group NPVs at the four discount rates in cells D28 to D31 of Figure A9.1b. In the *Iterations* box "3000" should be entered. The results for each of the four outputs can be accessed and downloaded as discussed above.

Some additional points to note when using @Risk©

In the above example we have assumed in most cases a triangular distribution (*TRIANG*) using estimates of the "minimum", "best guess" and "maximum" values for each input. One problem with this probability distribution is that the two extreme values are assigned a probability of zero and will therefore never be sampled in the simulation. An alternative to this is the *TRIGEN* distribution which allows the analyst to specify the bottom and top per-centile values giving the percentage of the total area under the triangle that falls to the left of the entered point. Another alternative to the triangular distribution is the *BETASUBJ* function which requires a minimum, most likely, mean, and maximum value.

In some situations there may be a range of *discrete* values where each value has the same probability of occurrence. Here we would choose the *DUNIFORM* function. On the other hand, if the uniform distribution was *continuous* – one in which any value within the range defined by the minimum and maximum values has equal probability of occurrence – we would use the *UNIFORM* function. (Be careful not to confuse the two.)

With more information about a variable, we could use one of the many other distributions available in @Risk©, such as "*Normal*" for a Normal distribution, where we enter the mean and standard deviation for the variable in question. Use of triangular distributions is a rough-and-ready form of risk analysis.

In the preceding example we set our number of iterations at 3000 for the simulation. During the simulation a screen appears indicating whether *convergence* has occurred. If the simulation has converged, an icon showing a smiling face appears; if not, it frowns! Make sure that there is convergence for each output cell. If not, the number of iterations should be increased until convergence is achieved.

We have undertaken one simulation in the preceding examples. It is possible to program @Risk© to run the same set of iterations under a number of scenarios. For instance, the analyst may wish to allow one of the variables to take on three possible values. This is not a variable that varies randomly, as described by a given probability distribution, but rather is a 'controllable' or 'exogenous' variable, such as a policy variable like a tax rate, a tariff or a regulated price. It could even be the discount rate. In this instance we use the function "=Risksimtable". This allows us to enter a number of alternative values in a particular cell (let us say three alternative tax rates) and then the simulation is automatically re-run using the same sampled input data for each tax rate scenario. Having entered the three tax rates (or values of other parameters) it is necessary to set the number of simulations required (in this case, three) in the @Risk "Settings" menu, shown earlier when we set the number of iterations. For each simulation a separate and consistent set of probability statistics is produced.

When selecting variables to which an @Risk© probability distribution is to be attached be careful not to introduce an inconsistency by attaching separate distributions to variables that are dependent on each other, or correlated. As noted earlier in this chapter, if, in analysing a road transport project, you attached a separate distribution to maintenance cost and road usage when the former depends (positively) on the latter, you could end up with a nonsensical simulation; another example of correlated variables is where rainfall rises and agricultural yield falls. @Risk© is able to incorporate joint probability distributions by establishing probabilities for each variable conditional on the values selected for others.

Be careful to set up your simulation to allow the program to treat each cell separately in each simulation. If you attach the @Risk© distribution to a variable located in the "Key Variables" or "Inputs" table of your spreadsheet, and then this variable is referred to in each year of the cash flow, what will happen is that during the simulation, the variable will take on the identical value *in each year of the project*. Normally what is required in a simulation is for each year's value to be independent of previous years' values, unless the analyst is programming a scenario where a variable should be related to its value in the previous year. Interest rates or share prices, for instance, are unlikely to take on a random value in each period, as in the case of a Random Walk discussed earlier.

Exercises

1   An analyst has calculated the expected net present value in millions of dollars (Mean NPV) and the variance of net present value around the mean (Variance) for a number of projects, each involving a capital cost of $5 million:

Project	Mean NPV	Variance
A	2.1	1.5
B	1.9	1.2
C	2.0	1.1
D	2.2	0.9
E	3.0	1.4

She knows that the decision-maker is risk averse, but has no detailed information about the degree of risk aversion. Which project(s) should she recommend for further consideration?

2   The following table describes the projected costs and net benefits for two projects (all values in $ millions.)

	Project A			Project B		
	Pessimistic	Best Guess	Optimistic	Pessimistic	Best Guess	Optimistic
Year 0	−80	−75	−70	−100	−75	−50
Year 1	−110	−100	−90	−150	−100	−50
Year 2	18	20	22	12	20	28
Year 3	28	30	32	20	30	40
Year 4	28	30	32	25	45	65
Year 5	40	45	50	25	45	65
Year 6	40	45	50	25	45	65
Year 7	40	45	50	25	45	65
Year 8	40	45	50	25	45	65
Year 9	40	45	50	25	45	65
Year 10	55	65	75	35	65	95

i   Using the NPV decision rule and assuming a 10% discount rate, which of the two projects would you prefer using the "best guess" estimate of cash flow?

ii   Using the available range of estimates apply an *ExcelSim*© simulation with a triangular distribution (1000 iterations) to derive the following:

   a   the mean, minimum, and maximum NPVs for each project;

   b   a graph of the probability distribution of NPVs for each project;

   c   a graph of the cumulative (ascending) distribution of NPVs for each project.

iii Which project would a risk-averse decision-maker favour, and why? (Assume a 90% confidence level.)

iv   Which project would a risk-taking decision-maker favour, and why? (Assume a 20% confidence level.)

v   Would your answers to questions (i), (iii) and (iv) be any different if the discount rate was set at: (a) 5%; and (b) 15%?

3   Nicole leads an exciting life. Her wealth is a risky prospect which will take the value $100 with probability 0.5, or $36 with probability 0.5. Her utility of wealth function is: $U = W^{0.5}$.

i   What is the expected value of her wealth?

ii   What is the expected utility of her wealth?

iii What level of wealth with certainty will give her the same utility as the risky prospect?

iv   What is the cost of the risk associated with the risky prospect?

4   A risk-neutral decision-maker is considering making an irreversible investment costing $1200. The proposed project has no operating costs, and will produce 100 units of output per annum in perpetuity, starting one year after the investment cost is incurred. The price at which the output can be sold will either be $1 per unit, with probability 0.5, or $3 per unit, with probability 0.5. The rate of interest is 10% per annum.

   i   What is the expected NPV of the project if it is undertaken now?

   ii   One year from now the price at which the output can be sold will be known with certainty. The project can be delayed for one year, in which case the investment cost is incurred a year from now, and the flow of output starts two years from now.

   iii What is the expected NPV of the project *now* if it is to be delayed for one year pending the receipt of the price information?

   iv Should the project be delayed for a year? Explain why or why not.

   v   What is the value *now* of the option of delaying the project? Explain.

Notes

1   This omission was rectified in August 2014 with the publication of the "Independent Cost-Benefit Analysis of Broadband": www.communications.gov.au/broadband/national_broadband_network/cost-benefit_analysis_and_review_of_regulation/independent_cba_of_broadband

2   In *Investment, Interest and Capital* J. Hirschliefer discusses the Friedman-Savage utility of wealth function which implies risk aversion where large gains or losses are possible, but incorporates a relatively small concave segment in the vicinity of the current wealth level which indicates a willingness to gamble small amounts on a fair, or even unfair, bet.

3   We use the *ExcelSim©* 2011 add-in program developed and copyrighted by Timothy R. Mayes, Ph.D., Department of Finance, Metropolitan State University of Denver, Denver, Colorado. This add-in is downloadable free from the companion website to his textbook *Financial Analysis with Microsoft© Excel©*, *6th Edition*, T.R. Mayes and T.M. Shank, Cengage Learning, Connecticut, 2012. www.cengage.com/cgi-wadsworth/course_products_wp.pl?fid=M20bI&product_isbn_issn=978111 1826246.

4   We discuss the installation of the *ExcelSim©* and *Analysis Toolpak* add-ins on the companion website. The reader preferring the use of an integrated risk analysis package with a more extensive range of programming options and professional graphics outputs is advised to consider purchase of one of the commercial programs such as @RISK (Palisade, Corp.) or *Crystal Ball* (Decisionengineering, Inc.)

10 Valuing traded and non-traded goods in cost-benefit analysis

10.1 Introduction

A project undertaken in an open economy may result in changes in the flows of goods and services which are exported or imported. It is not important whether the actual output of the project is exported, or the actual inputs imported. If the output is a traded commodity (a commodity which is exchanged in international trade), it may be exported or it may replace imports; similarly, if the inputs are traded goods and services, they may be imported or they may come from domestic sources which are replaced by increased imports or reduced exports. The changes in international trade flows resulting from the project need to be valued, and the prices which measure the benefit or cost to the economy of changes in exports or imports are international prices.

Projects which involve outputs or inputs of traded goods and services are likely also to involve outputs and inputs of non-traded goods and services (goods and services which are not traded internationally), and these non-traded goods and services are valued at domestic prices. The cost-benefit analysis of such projects requires comparisons of values of traded and non-traded goods. Since traded good prices are denominated in foreign currency (often US$) and non-traded good prices are denominated in domestic currency, an exchange rate is required to convert from one to the other. In the absence of significant market distortions, such as exchange rate regulation, tariffs, import quotas, and taxes and subsidies the market exchange rate can be used. In the presence of such distortions, which are more often encountered in developing economies, the domestic price structure may differ from the international price structure, and care must be taken to value benefits and costs under the same price structure if they are to be compared. In this case, either world prices of traded goods are converted to domestic prices using a shadow-exchange rate, or domestic prices of non-traded goods are shadow-priced to convert them to world prices.

10.2 Traded and non-traded goods

We start this discussion of the distinction between traded and non-traded goods by considering a broader distinction – that between tradeable and non-tradeable goods. A tradeable good or service is one that is *capable* of being traded in international markets. Many goods and services are tradeable – steel, cement, wheat, consulting services, etc. The main reason some goods and services are non-tradeable is that they have high international transport costs relative to their production cost. For example, services such as haircuts are traditionally regarded as non-tradeable because no one is willing to pay the cost of travelling abroad to get a haircut; perishables such as fresh bread and milk tend to be non-tradeable

DOI: 10.4324/9781003312758-10

because of the high cost of keeping them fresh while they are being transported; and bulky goods such as sand and gravel are non-tradeable because of high transport costs relative to their market prices. The fact that cost is the key factor in determining tradability is illustrated by the example of a wealthy New Yorker paying for Vidal Sassoon to fly over from Paris to do a hair style, thereby possibly removing haircuts as the time-honoured example of a non-tradeable service!

Where goods and services do not currently enter into international trade for reasons of cost, either:

- the domestic price of the good is lower than the c.i.f. import price and hence no domestic customer wants to import the good; or
- the domestic price of the good is higher than the f.o.b. export price and hence no foreign customer wants to import the good (i.e. export it from the producing country).

The c.i.f. (cost insurance and freight) price is the price of an imported good at the border, and the f.o.b. (free on board) price is the price of an exported good at the border. The basic condition for a commodity to be a non-tradeable commodity under current cost conditions is:

f.o.b. price < domestic price < c.i.f. price.

Of course there are other trade-related costs, such as domestic handling, transport and storage costs and agents' fees, but we will ignore these in this discussion.

In considering the costs of international trade we have to decide how to treat tariffs. Tariffs are not a social cost in themselves, as they involve no real resources beyond those required to collect them (leaving aside their contribution to resource misallocation). However, they are a private cost and do inhibit the imports of the country which levies the tariff, and the exports of potential supplying countries. Suppose that a good is tradeable in the absence of a tariff, but is made prohibitively expensive in the domestic market as a result of the tariff. Should we treat it as a traded or non-traded good? This is a fundamental question as it forces us to choose between dealing with the world as it "ought to be" from the viewpoint of economic efficiency, and the world as it is. The former is the world of the economic planner, whereas, we argue, the latter is the world inhabited by the cost-benefit analyst. The cost-benefit analyst takes the world as it is, warts and all. The relevant distinction is not between tradability and non-tradability, but rather between goods and services that are actually traded and goods and services that are not. The basic condition for a good or service to be non-traded is:

f.o.b. price less export tax < domestic price < c.i.f. price plus import duty.

If the good in question is not traded because of tariffs or other taxes, and if it will remain not traded once the project is undertaken, then we treat it as a non-traded good in the analysis.

10.3 Valuing traded and non-traded goods and services

Once we introduce international trade into the cost-benefit framework, we must take account of both traded and non-traded outputs and inputs. The value or cost to the economy of project output or inputs of traded goods is established in international markets. This is true irrespective of whether the particular units of output or inputs enter into international trade: for example, if a project produces a quantity of an importable good, it replaces that quantity of imports, thereby avoiding the cost of imports measured at the c.i.f. price; if it uses a quantity of an importable good as an input, the economy incurs a cost measured by the value of that quantity at the c.i.f. price; if it produces a quantity of an exportable good, exports valued at the f.o.b. price per unit can be increased; and if it uses a quantity of an exportable good as an input, exports valued at the f.o.b. price per unit have to be curtailed.

While traded goods have to be valued at international prices and non-traded goods at domestic prices, values of traded and non-traded goods have to be compared in the cost-benefit analysis. In order to compare like with like, it is necessary either to convert domestic prices to equivalent international prices, or to convert international prices to equivalent domestic prices. Furthermore, international prices are denominated in foreign currency (often US$) whereas domestic prices are denominated in domestic currency. An exchange rate is required to convert from one to another, and while there is normally only one official exchange rate, there is also a notional exchange rate, devised by economists to take account of the effects of tariffs and other taxes or subsidies on domestic relative to international prices, and known in the cost-benefit literature as *the shadow-exchange rate*. The cost-benefit analyst has to decide which set of prices to use to value project inputs and outputs – world prices or domestic prices – and which rate of exchange – the official rate or the shadow-rate – to use in converting foreign currency prices to domestic currency or vice versa.

10.4 Worked example: domestic and international price structures

We consider a simple example of how tariff protection can result in two sets of prices – domestic and international – and two exchange rates – the official rate and the shadow rate – and the implications for project evaluation. The example is first presented in real terms, without using domestic and foreign currencies to value outputs and inputs, in order to emphasise that the important issue is that of *opportunity cost*. Currencies and prices are then introduced to illustrate how CBA addresses this issue in practice. To keep the example simple and focused on the domestic and international price structures, some important issues are omitted, such as the determinants of the balance of trade, the question of comparative advantage and the role of government in taxation and expenditure.

10.4.1 *Evaluation of an import-replacing project in real terms*

Suppose that a country has one factor of production, labour, which is non-traded and fully employed. Two traded goods are produced – food and clothing – and there is a 100% tariff on imported clothing. We can choose units of food and clothing such that the international price ratio is 1; for example, if a unit of clothing – 30 suits – costs the same as a unit of food – 1 ton of wheat – then the ratio of the price of a unit of clothing to a unit of food is 1. On the domestic market, on the other hand, the price of a unit of clothing will be twice the price of a unit of food because of the tariff on clothing. This establishes the fact that there are two sets of relative prices – domestic and international. The international price is usually referred to as the *border price* – the price of exports f.o.b. or the price of imports c.i.f.

Now suppose that the domestic economy is competitive, with no tax distortions apart from the tariff on clothing. This means that a unit of labour transferred from food to clothing production would reduce the value of output of food by the same amount as it would increase the value of output of clothing, where both values are measured at current domestic prices. Since these values are equal at domestic prices, and since the domestic price of clothing is twice that of food, this means that if we choose the quantity of labour to be transferred such that food production is reduced by one unit by the reallocation of labour, clothing production will increase by half a unit. What would be the efficiency net benefit of a project involving such a reallocation of labour?

The extra half unit of clothing produced domestically could be used to replace imports of half a unit of clothing, so that clothing consumption remained unchanged. Since units of food and clothing trade on a 1:1 basis on world markets, exports of food could be reduced by half a unit. The net effect of the reallocation of labour in this case is a loss of half a unit of food – a reduction of one unit of domestic production partially offset by a reduction in exports by half a unit. Alternatively, exports of food could be reduced by one full unit, to keep the domestic supply of food constant, in which case imports of clothing would have to fall by one unit. The net loss to the economy would then be half a unit of clothing – an increase of half a unit of domestic production which is more than offset by the fall in imports of clothing by one unit.

We can see from the example that the country is worse off as a result of the proposed import-replacing project: the loss is some weighted average of half a unit of clothing and half a unit of food, with the weights depending on how the economy wishes to absorb the loss. If food and clothing are both normal goods (goods whose consumption rises/falls when income rises/falls) then consumers will opt for a reduction in consumption of both goods. In that case, if the proportion of the loss absorbed by reduced food consumption is denoted by the fraction "a", such that $0 < a < 1$, the reduction in food consumption is $a/2$ units, and the reduction in clothing consumption $(1 - a)/2$ units.

Figure 10.1 illustrates the options available to consumers with and without the project: *without* the project the initial consumption bundle is denoted by point E; *with* the project a consumption bundle along E_1E_2 must be chosen. Clearly such a point will represent less of at least one of the goods, implying that undertaking the project makes the country worse off. However, as we shall see later, the project breaks even when appraised at domestic prices without taking the tariff distortion into account.

In project evaluation we argue that the efficiency net benefit is given by the value of the extra clothing less the cost of the labour transferred to clothing production. At first sight, as noted above, the project appears to have a net benefit of zero: at domestic market prices the value of the extra clothing equals the cost of the extra labour in clothing production. However, we are not comparing like with like: the value to the economy of traded goods such as food and clothing is expressed at international prices, whereas the cost of a non-traded good such as labour is expressed at domestic prices. We have to either convert the value of traded goods to equivalent domestic prices, or convert the value of the non-traded good to world prices to make an appropriate comparison.

10.4.2 Evaluation of an import-replacing project in money terms

Now let us attach some money prices to the quantities in the previous example. Suppose the tariff on clothing is 100% as before and that the project involves transferring 1000 Rupees worth of labour from food to clothing production. Suppose that the international prices of food and clothing are both \$1000 per unit, and that the official rate of exchange (OER) is 1 Dollar (\$) = 1 Rupee (R). The exchange rate tells us the price of the domestic currency in terms of the foreign currency (usually US dollars), or, equivalently, the price of dollars in terms of the domestic currency.

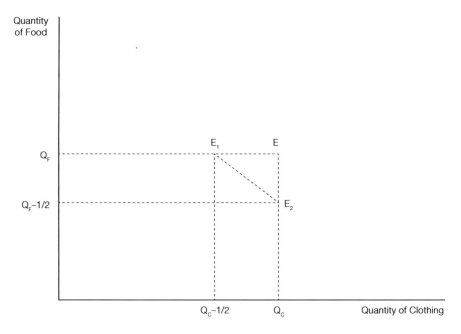

Figure 10.1 Consumption opportunities with and without an import-replacing project. Without the proposed project the economy consumes the combination of food and clothing represented by point E. With the import-replacing project consumers will have to choose a food and clothing combination on the line segment E_1E_2.

In our simple example the exchange rate is unity whichever way the official rate is specified. In general, however, the price of domestic currency in terms of dollars is the reciprocal of the price of dollars in terms of domestic currency, and if we wish to convert dollars to domestic currency, we divide by the former and multiply by the latter. Because this can cause a lot of confusion we will adopt the convention that the exchange rate is always expressed as the price of domestic currency in dollars, which is the form in which it is usually quoted in the financial press. In working with exchange rates you may find it helpful to be explicit about the units involved, for example, OER ($/R) equals (1/OER) (R/$).

At domestic prices 1000 Rupees worth of labour is used to produce half a unit of clothing worth 1000 Rupees in the domestic market. Since the benefit appears to equal the cost, the project seems to be a marginal one from the viewpoint of the economy. However, we have already seen that undertaking the project makes consumers worse off. Clearly we need a way of evaluating the project that takes account of the distorting effect of the tariff and identifies the project as one that should be rejected on economic efficiency grounds.

We have seen that the project in the example results in a loss to the economy in the form of a weighted average of half-a-unit of food and half-a-unit of clothing: a/2 units of food plus $(1 - a)/2$ units of clothing, where $0 < a < 1$, and a is the weight attached to food in the reduction of domestic consumption of food and clothing as a result of the project. The value, at world prices, of the reduced consumption of food is 1000a/2 Rupees and the value of the reduced consumption of clothing is $1000(1 - a)/2$ Rupees; these values are obtained by taking the border prices of the goods in dollars and converting to Rupees using the official exchange rate. At domestic prices, however, the values are 1000a/2 and 2000(1 – a)/ 2 Rupees respectively. This means that the ratio of domestic to international value of the consumption goods forgone is [a + 2(1 – a)] > 1. This ratio can be used, together with the official exchange rate, OER expressed in $/Rupee, to construct a shadow-exchange rate, also expressed in $/Rupee: SER = OER/[a + 2(1 – a)], which can be used to convert values at international prices to values at domestic prices. In the example, the official exchange rate is $1/Rupee, but the shadow-exchange rate is $1/ [a + 2(1 – a)] per Rupee, which is lower than the official rate.

Now let us consider how the shadow-exchange rate would be used in an Efficiency Analysis to convert the border prices of food and clothing to domestic equivalent prices. The proposed project produces half a unit of clothing which is worth $500 at border prices. The $500 worth of clothing is converted to domestic prices using the shadow-exchange rate to give a value of 500/SER Rupees. Since we know that 0.5 < SER < 1 in this case, because 0 < a < 1, we know that the value of the extra clothing at domestic prices is less than 1000 Rupees. Since the cost of the labour used to produce the extra clothing is 1000 Rupees, the project has a negative efficiency net benefit. For example, if a = 0.4, the SER is $0.625/Rupee, the value of the project output is 800 Rupees, and the net efficiency benefit at domestic prices is –200 Rupees, or a loss of 200 Rupees.

Suppose that we decided to express the project's net benefits in terms of world rather than domestic prices. The output of half a unit of clothing could be valued at the border price of $500 and converted to 500 Rupees at the official exchange rate. However, this value could not be compared with the 1000 Rupees labour cost of the project since the latter is denominated in domestic prices. We need to work out the opportunity cost of labour at border prices: if the extra labour had been left in food production, it would have produced an extra unit of food which could have been exported to earn foreign exchange. That quantity of foreign exchange could have been used to fund imports of quantities of food and clothing summing to one unit. The value of these imports at border prices is 1000a/2 + 1000(1 – a)/2 Dollars, where 'a' is defined as above. The border value at dollar prices can be converted to Rupees by dividing by the OER (OER = 1 in this example) to give 1000[a/2 + (1 – a)/2] Rupees. At domestic prices, however, the imports are worth 1000[a/2 + 2(1 – a)/2] Rupees. Since the ratio of the opportunity cost of labour at domestic prices to its cost at border prices is [a + 2(1 – a)], we need to multiply the value of labour at domestic prices by the reciprocal of this amount to convert to a value in border prices. However, we have already seen that 1/[a + 2(1 – a)] is the ratio SER/OER. Thus, to convert from values at domestic prices to values at border prices, we multiply by the ratio SER/OER. For example, if OER is 1 and SER is 0.625 ($/Rupee), the opportunity cost, at border prices, of the labour used in the project is 1000 * (0.625/1) Rupees. This cost of 625 Rupees exceeds the value of the extra clothing output at border prices by 125 Rupees. Hence the net efficiency benefit is –125 Rupees, a negative net benefit of the project, or, in other words, a net loss.

It can be seen that, while both approaches to measuring efficiency net benefit assign a negative net benefit to the project, they give different numerical results: using domestic prices as the *numeraire*, the loss is 200 Rupees, and in border prices it is 125 Rupees. However, this difference is not surprising – different relative prices give different valuations. We can convert a value expressed in domestic prices to a value at border prices by multiplying by SER/OER, or a value at border prices to a value at domestic prices by multiplying by OER/ SER. Thus, 200 Rupees multiplied by 0.625/1 equals 125, and 125 Rupees multiplied by 1/ 0.625 equals 200 Rupees.

10.5 Summary of the two approaches to valuation: border versus domestic prices

In summary, once we recognise that markets may be distorted by regulation or tariffs, we can no longer rely on the official exchange rate to generate the appropriate relative valuation of traded and non-traded goods. There are two solutions to this problem: either convert world valuations of traded goods to domestic price equivalents, or convert domestic valuations of non-traded goods to world price equivalents. These alternatives amount to nothing more than the analyst's choice of the numeraire in which to express the net benefits calculated by the cost-benefit analysis – the underlying principles and methodology are identical. Under the former approach we convert the world prices of traded goods to equivalent domestic prices, using the shadow-exchange rate (SER); under the latter approach we use the OER to generate domestic currency prices of traded goods, and use SER/OER to shadow-price domestic non-traded goods prices to account for trade distortions. In both approaches we shadow-price non-traded goods and services to take account of other distortions and imperfections in domestic markets where appropriate. As the above example demonstrated, the two approaches give the same result in terms of project appraisal, and we need to discuss why this is in general the case. However, it should be emphasised that it is only a difference between the OER and the SER, arising from distortions in the traded goods markets or in the foreign exchange markets, which underlies the two approaches. Distortions in the foreign exchange market and the calculation of the SER are discussed later in this chapter. First, we discuss the two approaches in general and demonstrate their equivalence.

10.6 The equivalence of the two approaches

As noted above, the analyst has a choice of whether to use domestic prices or border prices to value benefits and costs. It is important to understand that what is at issue is not the currency in which values will be denominated – this will usually be the domestic currency in either case – but the set of prices which will be used. The two conventions described above are known as:

- the *UNIDO approach*, named after the United Nations International Development Organization guidelines, which advocate the use of non-traded goods (domestic) price equivalents (see United Nations, *Guidelines for Project Evaluation*, New York: UN, 1970);
- the *LM approach*, named after its authors Little and Mirrlees and adopted by the Organization for Economic Cooperation and Development (OECD), which advocates using traded goods price equivalents (border prices) (see OECD, *Manual of Industrial Project Analysis in Developing Countries*, Paris: OECD, 1968). This approach was also recommended by Squire and van der Tak in their (1975) World Bank research publication cited at the end of this chapter.

It should be emphasised that while these approaches generally yield different estimates of the absolute magnitude of NPV in an analysis of economic efficiency, a project which is accepted (NPV > 0) by one method is always accepted by the other, provided that the same data are used in the calculation of the relevant shadow-prices and the shadow-exchange rate.

To recap, under the UNIDO approach, non-traded goods and services are valued at domestic prices (shadow-priced where appropriate to reflect domestic market imperfections), and traded goods and services are valued at international prices (US$) which are converted to domestic prices using the shadow-exchange rate (SER). Under the LM approach, traded goods and services are valued at international prices converted to domestic currency by means of the official exchange rate (OER), and non-traded goods and services are shadow-priced to account for imperfections in their domestic markets, and also adjusted for foreign exchange market distortions (caused by factors such as regulation, tariffs or export taxes) using the SER. The OER is either the fixed rate set by the government or the market rate established by the foreign exchange (FOREX) market. The best way to illustrate the way the two approaches operate is by the kind of example discussed above, but first we need to convince ourselves that they will always give the same result in terms of whether the project has a positive or negative NPV.

Suppose that a project uses imports, M, and a non-traded commodity such as unskilled domestic labour, N, to produce an exported good, X. The traded goods X and M are denominated in US$ and the non-traded good, N, is denominated in domestic currency. Suppose that there are two types of distortions in the domestic economy: tariffs, resulting in a divergence between the OER and the SER; and a minimum wage resulting in unemployment of unskilled labour. The two approaches to calculate the project NPV in an Efficiency Analysis can be summarised as follows:

(i) $\text{UNIDO} : \text{NPV}(1) = \left(\dfrac{1}{\text{SER}}\right) X - \left(\dfrac{1}{\text{SER}}\right) M - bN,$

where SER is the shadow-exchange rate (measured in US$ per unit of local currency, and discussed later in this chapter) and b is the shadow-price of domestic labour which takes account of unemployment as described in Chapter 5;

(ii) $\text{LM} : \text{NPV}(2) = \left(\dfrac{1}{\text{OER}}\right) X - \left(\dfrac{1}{\text{OER}}\right) M - cN,$

where OER is the official exchange rate (measured in US$ per unit of local currency), and c is the shadow-price of domestic labour which takes account of both unemployment and trade distortions, as discussed below.

To see the equivalence of the two approaches, multiply NPV(1) by the ratio of SER/OER:

(iii) $\text{NPV}(1)\left(\dfrac{\text{SER}}{\text{OER}}\right) = \left(\dfrac{1}{\text{OER}}\right) X - \left(\dfrac{1}{\text{OER}}\right) M - b\left(\dfrac{\text{SER}}{\text{OER}}\right) N$

It can be seen that expression (iii) is the same as expression (ii) if b(SER/OER) = c. In the expression for c, the parameter *b* shadow-prices labour to account for unemployment in the domestic labour market, and the ratio SER/OER shadow-prices the opportunity cost of labour at domestic prices to convert to border prices. In consequence, NPV(1)(SER/

OER) = NPV(2): the NPV under the LM approach is the NPV under the UNIDO approach multiplied by the ratio of the shadow to the official exchange rate. Hence when NPV(1) is positive, NPV(2) must be positive. Of course if there are no imperfections (tariffs or regulations) in the traded goods or FOREX markets, OER = SER and the distinction between the two approaches collapses.

In the above example the input supplied locally was unskilled labour, a non-traded input. However, in the absence of the project, some portion of that unskilled labour might have been employed elsewhere in the economy and might have produced traded goods, and, if so, that portion should be treated as a traded good in the CBA. Furthermore, suppose that the input had been concrete blocks produced locally from imported cement and unskilled labour. In principle, we would identify the share of the cost of the blocks represented by imported cement and treat it as a traded good in the CBA, with the labour cost share being treated as a non-traded good. These refinements to the analysis require some very detailed information about the structure of the domestic economy which may not always be readily available, and they are often ignored in practice.

What is the logic, under the LM approach, of shadow-pricing the non-traded good, domestic labour, to account for distortions in international markets as well as those in the domestic labour market? When we shadow-price to account for unemployment, we convert the wage of labour to a measure of its opportunity cost in the domestic economy – the value of what it would produce in its alternative occupation. However, under the LM approach, the goods or services produced by domestic labour in its alternative occupation have to be valued at border (international) prices. The question is: what is the equivalent in domestic currency of the US$ value of what the labour could produce if it were not employed on the project? The domestic currency overstates the opportunity cost of labour if the currency is over-valued relative to the US$ (i.e. SER < OER) because of trade distortions. This problem can be solved by converting the measure of opportunity cost to US$ using the SER, to get an accurate measure of opportunity cost at world prices, and then converting the US$ value back to domestic currency by means of the OER, so that the opportunity cost of labour is measured in the same currency as the other project costs and benefits. For example, suppose that, in Papua New Guinea, unskilled labour is willing to work for 3 Kina per hour (its domestic shadow-price) and that the shadow-exchange rate is 1.55 Kina = 1US$. This suggests that an hour of unskilled labour can produce US$1.94 worth of goods at world prices. In common with other US$ values in the CBA, this figure is converted to domestic currency by means of the OER, say, 1.33 Kina to the US$, to give an opportunity cost, at border prices, of 2.57 Kina.

The UNIDO and LM approaches are summarised in Table 10.1 which illustrates an example of a simple project in Papua New Guinea (PNG). The project uses US$1 of imports and 5 Kina's worth of domestic labour at the market wage to produce US$6 worth of exports. Because of the high rate of unemployment in PNG, the opportunity cost of domestic labour, in terms of the value of forgone output at domestic prices, is assumed to be 60% of the market wage. It can be seen from Table 10.1 that the two approaches give the same result in the sense that the NPV calculated by the Efficiency Analysis is positive in both cases, but different results in terms of the actual value of the NPV. It can be verified from the data in Table10.1 that if the NPV under the LM approach is multiplied by the ratio of (SER/OER), the NPV under the UNIDO approach is obtained: 3.99K multiplied by 1.49/1.33 equals 4.47K.

Table 10.1 The UNIDO and LM approaches to Efficiency Analysis in CBA

UNIDO		LM	
Tradeables	*Non-tradeables*	*Tradeables*	*Non-tradeables*
Use border prices in US dollars converted to domestic currency using the SER	Use domestic prices in domestic currency shadow priced for domestic distortions	Use border prices in US dollars converted to domestic currency using the OER	Use domestic prices in domestic currency shadow priced for domestic distortions and adjusted for FOREX market distortions using SER/OER

Example: OER: $0.75/Kina, implying that 1.3333 Kina = 1 US$
SER: $0.67/Kina, implying that 1.4925 Kina = 1 US$
Shadow-price of labour: 60% of market wage

Exports	*Imports*	*Labour*	*Exports*	*Imports*	*Labour*
$6	$1	K5	$6	$1	K5
K8.96	K1.49	K3	K8.0	K1.33	K2.68
Net Benefit =	K4.47		Net Benefit=	K3.99	

The example can be further developed to identify and measure Referent Group net benefits. The Referent Group generally consists of the residents of the domestic economy, or some subset of it. It is natural for members of the Referent Group to measure benefits and costs at domestic prices, and hence the UNIDO approach is more suitable for this purpose. Suppose that the Referent Group is defined as the residents of Papua New Guinea and that the firm is domestically owned. There are three sub-groups potentially affected by the project: the firm, domestic labour, and the government. If we assume, for simplicity, that there are no taxes or tariffs (the distortion in the FOREX market is a fixed exchange rate), government revenues will not be affected by the project. The firm reports a profit of 1.67 Kina (its revenue less its costs, with foreign exchange converted to Kina at the OER), and labour receives a rent of 2 Kina (the wage less what labour could earn in an alternative occupation). However, because of the distortion in the FOREX market, the sum of these two measures of Referent Group net benefits understates the aggregate net benefits of the project.

The project's net foreign exchange earnings of US$5 ($6 of exports less $1 of imports) need to be adjusted upwards to reflect the fact that local residents are willing to pay 0.16 Kina above the OER for each US$ (a foreign exchange premium of 12%), implying a benefit in the form of a total foreign exchange premium valued at 0.80 Kina. The total benefit to the Referent Group is then 4.47 Kina, consisting of profit plus labour rent plus foreign exchange premium, which is the efficiency NPV estimated under the UNIDO approach. The distribution of the foreign exchange premium depends on the rationing system used to allocate foreign exchange. For example, if the firm were able to sell its net earnings of foreign currency on the parallel market it would appropriate the 0.80 Kina of foreign exchange benefits under the UNIDO approach. On the other hand, if the firm is forced to sell the foreign currency to the Central Bank at the OER, the Bank appropriates the benefit.

10.7 Determinants of the shadow exchange rate

We now analyse the factors which determine the SER. The need for a SER arises from imperfections in international markets which have two main sources: fixed exchange rates and tariffs, taxes and subsidies on imports and exports. We consider fixed exchange rates first.

Suppose that a country decides to regulate the rate at which its currency can (legally) be exchanged for foreign currency. For example, in the 1980s PNG operated a fixed exchange rate system under which the Kina was set at an artificially high rate against the US$ so that foreign companies wishing to do business in PNG had to pay above market value for Kina. Figure 10.2 illustrates a foreign exchange market distorted by a fixed exchange rate (note the resemblance to Figure 5.11 which illustrated the effect of a maximum price). The horizontal axis shows the quantity of foreign exchange (US$) traded, and the vertical axis shows the price in Kina/US$. Suppose the regulated price is 1.33 Kina per US$, which implies an OER of $0.75 per Kina, and which overvalues the Kina since the market equilibrium price in terms of Kina per US$ is higher than 1.33. In the distorted market equilibrium illustrated in Figure 10.2 an additional US$ is actually worth 1.55 Kina, implying an exchange rate of $0.65 per Kina. If there was a small unofficial market in foreign exchange, this is the price

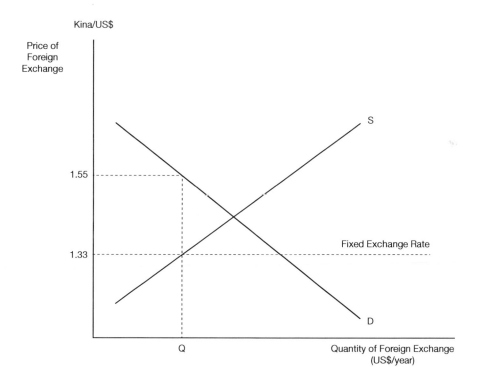

Figure 10.2 The foreign exchange market with a fixed exchange rate. D represents the demand for foreign exchange (traders wishing to buy dollars in exchange for Kina) and S represents the supply of foreign exchange (traders wishing to buy Kina in exchange for dollars). The Official Exchange Rate is regulated at 1.33 Kina to the dollar in this illustration.

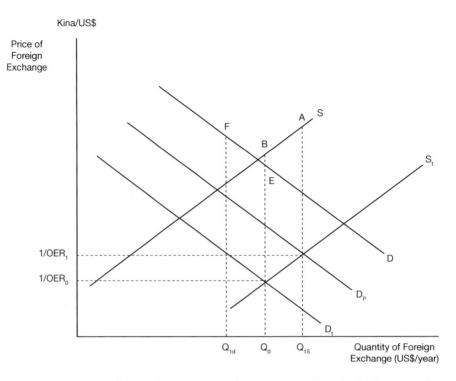

Figure 10.3 Supply and demand for foreign exchange with tariffs and subsidies. D and S are the demand and supply curves for foreign exchange in the absence of import tariffs and export subsidies. OER represents the market exchange rate in $/Kina. A tariff reduces the demand for foreign exchange to D_t, and an export subsidy increases the supply of foreign exchange to S_t. Imports associated with a large project increases the demand from D_t to D_p. The effect of the increase in demand is to increase the price of foreign exchange (in Kina/$). The increase in price reduces quantity demanded by buyers by $Q_{1d}Q_0$, and increases quantity supplied by sellers by $Q_{1s}Q_0$ in order to meet the project's foreign exchange requirement $Q_{1d}Q_{1s}$.

that we would expect to observe: a tourist could get 1.55 Kina for every US$. In terms of the UNIDO and LM approaches, the OER is $0.75/Kina and the SER is $0.65/Kina.

Some foreign exchange markets are distorted by both fixed exchange rates and tariffs and subsidies. However, to keep our discussion of the effects of tariffs simple we will now assume that there is a floating exchange rate, but that exports and imports are subject to taxes and subsidies. The effect of these taxes and subsidies is to shift the demand for imports and the supply of exports, and hence to shift the demand and supply curves for foreign exchange. In Figure 10.3, the demand and supply curves for foreign exchange in the absence of taxes and subsidies are denoted by D and S respectively, and those in the presence of the market distortions by D_t and S_t. It can be seen from Figure 10.3 that, in this example, the distortions have reduced the demand for foreign exchange – tariffs discourage imports – and increased the supply of foreign exchange – export subsidies promote exports.

Suppose that a proposed project involves importing some equipment. What value should be placed on the foreign exchange required to purchase the equipment? The effect of the project is to cause a small increase in the demand for foreign exchange, exaggerated for expositional purposes and illustrated by the demand curve D in Figure 10.3, and a small fall in the market exchange rate (the OER, measured in $/Kina) which we will disregard as being too small to measure. However, this small fall in OER will reduce the quantity of foreign exchange demanded (because, for example, more Kina are required to buy each US$), and increase the quantity of foreign exchange supplied (because at the new OER foreigners get more Kina per US$). In this way the proposed project diverts some foreign exchange from the purchase of other imports, and draws in some foreign exchange earned through additional exports.

To calculate the SER, we need to value the foreign exchange used in the project at its opportunity cost. In Figure 10.3, the opportunity cost of the quantity of foreign exchange $(Q_0 - Q_{1d})$, which is diverted from the purchase of imports, is given by FEQ_0Q_{1d} in Figure 10.3, while the opportunity cost of earning the additional quantity of foreign exchange $(Q_{1s} - Q_0)$ is given by $BAQ_{1s}Q_0$. The sum of these two areas is the opportunity cost, measured in domestic currency, of the foreign exchange to be used in the project. It can be seen that the opportunity cost, in terms of Kina, of the foreign exchange is higher than the nominal cost as measured by the OER. To get a measure of the opportunity cost, we convert the value of foreign exchange to Kina using the SER. Following our convention, the SER is expressed in US$/Kina and, in this example, is lower than the OER. When foreign exchange is converted to domestic currency by dividing by the SER, a higher value is obtained than if the OER were used.

The previous analysis can be developed in more detail to identify the effect of taxes and subsidies on the SER. Recall that the proposed project diverted some foreign exchange from imports ($Q_{1d}Q_0$ in Figure 10.3, which we will denote by Δ_{FEM}) and some from exports (Q_0Q_{1s} which we denote by Δ_{FEX}). To calculate the opportunity cost (denominated in domestic currency) to the economy of the additional foreign exchange, we value Δ_{FEM} at the price given by the demand curve, D, and Δ_{FEX} at the price given by the supply curve, S, in Figure 10.3.

In Figure 10.3, the demand curve D shows where the demand curve for foreign exchange to purchase imports would be if there were no tariffs. The effect of tariffs is to reduce the price importers are willing to pay for foreign exchange by the amount of the tariff. In other words, the opportunity cost, in Kina, of Δ_{FEM} is $\Delta_{FEM}(1+t)/OER$ where t is the tariff expressed as a proportion of the US$ c.i.f. price converted to domestic currency at the OER. The supply curve S shows where the supply curve of foreign exchange to purchase exports would be if there were no taxes or subsidies on exports. In Figure 10.3 S_t is to the right of S, suggesting that the net effect of taxes and subsidies on exports is to reduce their price to foreign purchasers, thereby increasing the quantity demanded, and hence increasing the supply of foreign exchange. The opportunity cost, in Kina, of Δ_{FEX} is given by $\Delta_{FEX}(1+s-d)/OER$ where s is the proportional subsidy and d the proportional tax on exports. Since we are assuming that the net effect of the tax/subsidy regime is to lower the price of exports, $s > d$, and the opportunity cost of the foreign exchange derived from this source is higher than its value at the OER.

The SER is a notional or accounting exchange rate, as opposed to a market rate, measured in US$ per unit of local currency, which, when divided into the amount of foreign exchange used in a project, will give a measure of the opportunity cost, measured in units of local currency, of that amount of foreign exchange. We have worked out an expression for the social opportunity cost of foreign exchange:

$$SOC = \frac{\Delta_{FEM} \cdot (1+t)}{OER} + \frac{\Delta_{FEX} \cdot (1+s-d)}{OER}$$

Suppose that the amount of foreign exchange to be used in the project is represented by QF and that $\Delta_{FEM} = b.Q_F$ and $\Delta_{FEX} = (1-b).Q_F$, i.e. that the foreign exchange is obtained from reduced imports and increased exports in the proportions b and $(1-b)$ respectively. We can now rewrite our expression for SOC as:

$$SOC = \{b \cdot (1+t) + (1-b) \cdot (1+s-d)\} \cdot \frac{Q_F}{OER}.$$

This expression tells us that we can calculate the opportunity cost of a quantity of foreign exchange (US$) by dividing it by OER/{b.(1 + t) + (1 – b).(1 + s – d)}. It follows that OER/{b.(1 + t) + (1 – b).(1 + s – d)} is the shadow-exchange rate, SER, because this is the exchange rate, in US$ per unit of local currency, which, when divided into the quantity of foreign exchange, converts it to a measure of opportunity cost in local currency. Since SER < OER, the effect of shadow-pricing foreign exchange, under the assumptions made here, is to raise the cost of foreign exchange above its value at the OER.

A reasonable assumption may be that the proposed project diverts foreign exchange from the purchase of imports in the same proportion as imports are in total foreign trade: in other words, that b = M/(X + M) where M and X are the values of total imports and exports respectively. On this basis we can rewrite the SER as:

$$SER = \frac{QER}{\left(\dfrac{M}{X+M}\right) \cdot (1+t) + \left(\dfrac{X}{X+M}\right) \cdot (1+s-d)}.$$

where the expression {(M/(X + M)).(1 + t) + (X/(X + M)).(1 + s – d)} is (1 + FEP) where FEP is the foreign exchange premium. In the example of Figure 10.2, OER = $0.75/Kina, SER = $0.65/Kina and (1 + FEP) = (OER/SER) = 1.17, and FEP = 0.17.

10.8 Further reading

The two approaches to project evaluation in developing economies are summarised in I.M.D. Little and J.A. Mirrlees, *Manual of Industrial Project Analysis in Developing Countries* (Paris: OECD, 1968) and P. Dasgupta, A. Sen and S. Marglin, *Guidelines for Project Evaluation* (Geneva: United Nations, 1970). L. Squire and H.G. van der Tak, *Economic Analysis of Projects* (Baltimore, MD: Johns Hopkins University Press, 1975) also provide a comprehensive review.

Appendix to Chapter 10

Shadow-pricing foreign exchange in the ICP case study

We now consider how to shadow-price foreign exchange in the cost-benefit analysis of the proposed International Cloth Products Project. As noted above, we prefer to use the UNIDO approach as the focus is usually on Referent Group net benefits and it is more informative to express these in domestic prices. Using the UNIDO approach implies that in the Efficiency Analysis we will value tradeable goods at the SER rather than the OER. The other shadow-prices in the analysis will not be affected.

While we have no reason to believe that the Baht (Bt) is overvalued on foreign exchange markets, we will assume for the sake of argument that the SER is 50Bt per US$, which means that the currency is around 14% overvalued in the foreign exchange market. The overvaluation has no effect on the Market or Private cost-benefit analyses which are based on market prices. However, in the Efficiency Analysis, the exchange rate of 44Bt per $US will be replaced with 50Bt wherever it occurs. Considering only the tradeable inputs and outputs of the project, benefits are higher relative to costs than is the case for the project as a whole. These changes are shown in the adjusted ICP spreadsheets in Figure A10.1, showing only the Efficiency and Referent Group Analyses (since the Market and Private Analyses are unchanged), which should be compared with Figures A5.1 and A6.2 in Chapters 5 and 6. A new input is entered into Table 1 of the spreadsheet (not shown) indicating that the SER is 50 Baht.

Using the SER to calculate the efficiency cost of imported capital goods, spare parts and raw materials, increases the efficiency costs as shown in Table 4 of Figure A10.1. Conversely, using the SER to calculate the efficiency benefits from the sale of yarn increases the project's gross efficiency benefits as shown in the table. As the increase in gross efficiency benefits is greater than the increase in efficiency costs as a result of this adjustment, the net efficiency benefit is higher when the SER is used. It can be seen by comparing the results reported in Table 4 of Figure A10.1 with those in Table 4 of Figure A6.2, in Chapter 6, that the net effect of introducing the shadow-price of foreign exchange is to increase the size of the net efficiency benefit. The efficiency IRR increases from 23.4% to 25.8% and the NPV, at a 1% discount rate rises from 433.2 to 553.3 million Baht, an increase of 28% (see Figure A10.1). Since the aggregate Referent Group net benefit is calculated by subtracting the (unchanged) private and foreign financial institution net benefits from the efficiency net benefit, it is clear that net Referent Group net benefits will increase.

The question which we now have to address is where the gain in net Referent Group benefit is experienced. There is no change in the measure of any of the Referent Group benefits which were reported in Table 5 of the spreadsheet in Figure A6.2 because under the UNIDO approach these are calculated using the OER of 44 Baht per dollar. However, one additional line needs to be added to the calculation of *Total Government* net benefit. Each time the Central Bank sold a US dollar to the project to pay for imports, such as spindles, for example, it sold the dollar at 44 Baht when its true value was 50 Baht; and each time it bought a US dollar obtained from exports of yarn, it paid 44 Baht when the true value was 50 Baht. While these losses and gains will not show up directly as an item in the Central Bank accounts (since changes in foreign exchange reserves are valued at the OER), they nevertheless represent a gain for the economy as a whole: when the project bought US$ from the Central Bank it got them for less than their true value, and when it sold US$ it

TABLE 4: EFFICIENCY CALCULATION

YEAR	1999 MN BT	2000 MN BT	2001 MN BT	2002 MN BT	2003 MN BT	2004 MN BT	2005 MN BT	2006 MN BT	2007 MN BT	2008 MN BT	2009 MN BT
INVESTMENT COSTS											
SPINDLES	−200.000										
EQUIPMENT	−42.500										
INSTALLATION	−5.000										
CONSTRUCTION	−18.000										
START-UP COSTS	−18.000										
RAW MATERIALS		−60.000									60.000
SPARE PARTS		−12.150									12.150
TOTAL	−283.500	−72.150	0.000	0.000	0.000	0.000	0.000	0.000	0.000	0.000	72.150
RUNNING COSTS											
WATER		−3.600	−7.200	−7.200	−7.200	−7.200	−7.200	−7.200	−7.200	−7.200	−7.200
POWER		−1.800	−3.600	−3.600	−3.600	−3.600	−3.600	−3.600	−3.600	−3.600	−3.600
SPARE PARTS		−24.250	−24.250	−24.250	−24.250	−24.250	−24.250	−24.250	−24.250	−24.250	−24.250
RAW MATERIALS		−121.500	−243.000	−243.000	−243.000	−243.000	−243.000	−243.000	−243.000	−243.000	−243.000
INSURANCE & RENT		0.000	0.000	0.000	0.000	0.000	0.000	0.000	0.000	0.000	0.000
LABOUR		−40.500	−40.500	−40.500	−40.500	−40.500	−40.500	−40.500	−40.500	−40.500	−40.500
TOTAL		−191.650	−318.550	−318.550	−318.550	−318.550	−318.550	−318.550	−318.550	−318.550	−318.550
REVENUES											
SALE OF YARN		217.500	435.000	435.000	435.000	435.000	435.000	435.000	435.000	435.000	435.000
NET EFFICIENCY BENEFIT	−283.500	−46.300	116.450	116.450	116.450	116.450	116.450	116.450	116.450	116.450	188.600

TABLE 5: REFERENT GROUP ANALYSIS

YEAR	1999 MN BT	2000 MN BT	2001 MN BT	2002 MN BT	2003 MN BT	2004 MN BT	2005 MN BT	2006 MN BT	2007 MN BT	2008 MN BT	2009 MN BT
TOTAL REFERENT GROUP	−1.030	−62.959	52.281	53.460	54.733	56.107	57.592	57.592	57.592	57.592	120.435
– IMPORT DUTIES											
YARN		−18.000	−36.000	−36.000	−36.000	−36.000	−36.000	−36.000	−36.000	−36.000	−36.000
COTTON		10.692	21.384	21.384	21.384	21.384	21.384	21.384	21.384	21.384	21.384
SPINDLES	8.800										
EQUIPMENT	1.870										
MATERIALS INVENTORY		5.280									−5.280
SPARES INVENTORY		0.535									−0.535
SPARES (ANNUAL)		1.067	1.067	1.067	1.067	1.067	1.067	1.067	1.067	1.067	1.067
PROFITS TAXES		−13.285	27.100	28.279	29.552	30.926	32.411	32.411	32.411	32.411	32.411
FOREIGN EXCHANGE	−29.700	−0.048	20.130	20.130	20.130	20.130	20.130	20.130	20.130	20.130	28.788
(i) TOTAL GOVERNMENT	−19.030	−13.759	33.681	34.860	36.133	37.507	38.992	38.992	38.992	38.992	41.835
(ii) POWER SUPPLIER		1.800	3.600	3.600	3.600	3.600	3.600	3.600	3.600	3.600	3.600
(iii) LABOUR	18.000	3.000	3.000	3.000	3.000	3.000	3.000	3.000	3.000	3.000	3.000
(iv) DOMESTIC BANK	0.000	−60.000	6.000	6.000	6.000	6.000	6.000	6.000	6.000	6.000	6.000 60.000
(v) INSURANCE		6.000	6.000	6.000	6.000	6.000	6.000	6.000	6.000	6.000	6.000
TOTAL REFERENT GROUP	−1.030	−62.959	52.281	53.460	54.733	56.107	57.592	57.592	57.592	57.592	120.435

PROJECT			PRIVATE			EFFICIENCY			REFERENT GROUP		
Discount Rate	BT(Mn)	IRR (%)		BT(Mn)	IRR (%)	NPV	BT(Mn)	IRR (%)	NPV	BT(Mn)	IRR (%)
4%	405.3	20.6%	4%	154.5	16.7%	4%	553.3	25.8%	4%	380.4	n/a
8%	259.8		8%	88.7		8%	380.6		8%	292.0	
12%	152.0		12%	40.8		12%	252.4		12%	227.1	
16%	70.7		16%	5.4		16%	155.4		16%	178.8	
0%	605.6		0%	246.6		0%	790.4		0%	503.4	

Figure A10.1 ICP Project solution with shadow exchange rate.

received less than their true value. Since the ICP project is a net foreign exchange earner, the net effect in the example is a gain to the Central Bank, and this is included as a benefit under the heading *Foreign Exchange* in an additional row under *Total Government* in the Referent Group benefits, as shown in Table 5 of Figure A10.1. This restores the equality in the spreadsheet between aggregate Referent Group benefits and the sum of net benefits to all members of the Referent Group.

The gain or loss from a foreign exchange transaction is the difference between the value of the transaction, at border prices, calculated at the SER and its value at the OER. If the transaction is a cost, it is entered as a negative value in the spreadsheet and hence appears as a negative value in the Foreign Exchange row. The gain or loss associated with a trans-action is computed by multiplying its value, calculated at the SER, by (1 – OER/SER). For example, installation of imports of spindles and equipment cost the project $4.95 million c.i.f. which translates to a cost of 247.5 million Baht at the SER. However, at the OER, this transaction was valued at 217.8 million Baht, understating its cost to the economy by 29.7 million Baht. The latter value is entered as a cost in the Foreign Exchange row (in Figure A10.1) of the Referent Group Analysis. Similarly sales of yarn in the first full year of the operation of the project are valued at 382.8 million Baht at the OER, but at 435 million Baht using the SER.

In this example, use of a SER makes a big difference to the total Referent Group net benefits because the project is a significant net earner of foreign exchange. Comparing Figures A6.2 and A10.1, it can be seen that the net present value of aggregate Referent Group net benefits, using a 4% discount rate, increases from 260.4 to 380.4 million Baht. As a consistency check, to verify that the analyst has not made an error in making these adjustments, the aggregate Referent Group net benefits at the top of Table 5 in Table A10.1 (derived by subtracting ICP's private cash flow and the overseas lender's cash flow from the efficiency cash flow) should be the same as those on the bottom line (derived by aggregating the sub-Referent Group members' net cash flows).

While import-replacing projects have some adverse consequences for the efficiency of resource use, as discussed earlier in this chapter and in Chapter 5, it was clear from Figure A6.2 that the proposed ICP project was of net benefit to Thailand. Since the project is a significant net earner of foreign exchange, the measure of the net benefit to the country is increased once it is recognised that there is a premium on foreign exchange.

Exercises

1   A project in Bhutan uses $100 of imported goods, and 2000 Rupees worth of domestic labour, paid at the minimum wage, to produce exported goods worth $200 on world markets. The opportunity cost of labour in Bhutan is estimated to be 40% of the minimum wage. The official exchange rate is 15 Rupees = $1, and the shadow-exchange rate is 20 Rupees = $1.

   i   Work out the NPV of the project:
       a   using the UNIDO approach
       b   using the LM approach.
   ii What is the relationship between the two values?

2   A proposed project in Papua New Guinea (PNG) involves importing T-shirts at a cost of $6 each, painting tribal logos on them, and selling them on the world market at $9 each. The cost, at market prices, of painting each T-shirt is 4 Kina's worth of local materials, and 12 Kina's worth of local unskilled labour (PNG's currency is the Kina). It is estimated that, because of the high rate of unemployment in PNG, the opportunity cost of unskilled labour is 50% of the market wage. It is also known that the production of local materials does not involve unskilled labour. The official exchange rate (OER) is $0.3 = 1 Kina, and the shadow-exchange rate (SER) is $0.2 = 1 Kina.

   i   Calculate the net benefit of the project, expressed in Kina per unit of output, at domestic market prices. Explain your answer.

ii   Calculate the net benefit of the project, expressed in Kina per unit of output, at Efficiency prices:

a   using the UNIDO method;

b   using the LM method.

iii In each case briefly explain your answer.

iv   Using the relationship between the UNIDO and LM methods, state how you would check your result and perform and report the result of the check.

3   Suppose that the annual value of a country's imports is $1 million and the annual value of its exports is $750,000. It imposes a tariff of 20% on all imported goods, and exported goods receive a 10% subsidy. The official exchange rate is 1500 Crowns = $1. Work out the appropriate value for the shadow-exchange rate.

11 Appraisal of the distribution of project benefits and costs

11.1 Introduction

We have seen in previous chapters how market prices can be adjusted in order to accurately measure project benefits and opportunity costs to the economy as a whole, and how benefit and cost measures can be further adjusted to account for non-market values and externalities. In this chapter we discuss how the valuation of project costs and benefits, and hence the appraisal, evaluation and choice of projects, can also be made to reflect social objectives with respect to the *distribution* of a project's benefits and costs.

From the decision-maker's point of view the distribution of project net benefits may simply be a by-product of a project whose primary goal is an improvement in economic efficiency, and this is the situation mainly dealt with in this chapter. Some projects, however, have distribution objectives as a primary goal, although we may still want to assess their efficiency as well, For example, the Defarian Early Childhood Intervention Program case study, described in Appendix A1.6, is intended to increase the educational attainment of children deemed "high risk" because of their family situation, and to reduce their propensity to engage in behaviour that could lead to poverty in their later life. Similarly, the Drug Court Program case study in Appendix A1.14 is directed at reducing recidivism amongst drug offenders.

We have already discussed in Chapter 5 how concern for the inter-generational distribution of wealth may be reflected in cost-benefit analysis through the choice of discount rate. In this chapter we are mainly concerned with the distribution of income among members of current generations – the *atemporal distribution* – but we will return to the question of inter-generational equity – the *inter-temporal distribution* in the concluding section.

It is probably true to say that there is no country in which the government does not claim to be committed to narrowing the gap in the distribution of income between rich and poor. If there is a choice where two projects have the same net present value, but one results in a more equitable distribution of net benefits, then the latter will generally be preferred. The first part of this chapter concerns the *atemporal* distribution of income. It examines data on the distribution of income for a hypothetical country, shows how the degree of inequality of income distribution is measured, and then shows how income distribution weights can, in principle, be calculated. In the second part of the chapter the *inter-temporal* distribution of income is also considered.

DOI: 10.4324/9781003312758-11

Table 11.1 Distribution of households by annual income

Annual income (= $189)	% Households
Less than $100	30.7
$100 to 200	42.9
$200 to 300	13.4
$300 to 400	5.8
$400 to 500	2.7
More than $500	4.5

Table 11.2 Income distribution by deciles

Households	% of Income
Top 1%	8.31
Top 2%	22.81
Top 10%	33.73
2nd decile	15.49
3rd decile	11.61
4th decile	9.22
5th decile	8.12
6th decile	7.32
7th decile	6.47
8th decile	2.94
9th decile	2.67
Bottom decile	2.45

Interpersonal distribution

We can look at income distribution from a number of different points of view. For example, there is *interpersonal distribution* which refers to distribution between different individuals or households. Table 11.1 shows how income might be distributed among households in a given country. Here we see, for example, that 30.7% of the population earned less than $100 per annum, while 4.5% earned more than $500 per annum. We can refer to these categories as *income groups*.

Another way of presenting this kind of information is given in Table 11.2 which shows how the country's total income was divided among the different income groups. For example, we see that the richest 10% of the population received more than 33% of total income, while the poorest 10% received only 2.45% of total income.

Inter-sectoral distribution

Another way of looking at income distribution is by sector, or *inter-sectoral income distribution*. Here we see how income is distributed among households in different sectors of the economy. For example, we might be interested in comparing household income distribution between the rural and urban sectors. Such a comparison is shown in Table 11.3, from which we can see, for example, that a much higher percentage of rural households fall into the poorest income group; 34.2% of rural households had incomes less than $100 compared with 3.8% of urban households. On the other hand, only 1.3% of rural households earned more than $500, compared with 22.2% of urban households.

Table 11.3 Distribution of income by sector

Annual income	Percentage of households in			
	Urban areas	Semi-urban areas	Rural areas	All areas
Less than $100	3.8	15.5	34.2	30.7
$100 to 200	24.4	34.0	47.7	42.9
$200 to 300	25.1	22.3	11.1	13.4
$300 to 400	14.8	11.8	4.3	5.8
$400 to 500	9.7	5.8	1.4	2.7
More than $500	22.2	10.6	1.3	4.5
Mean annual income Y($)	411	270	148	189

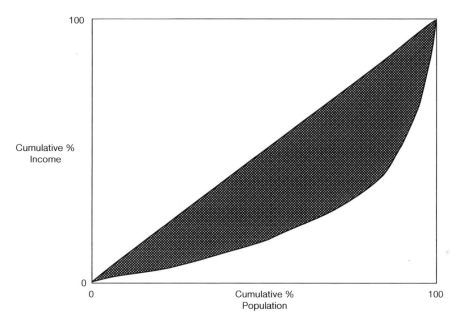

Figure 11.1 The Lorenz Curve. The Lorenz Curve shows the relationship between the cumulative percentage of income and the cumulative percentage of the population (individuals or households). If there were perfect income equality the Lorenz Curve would correspond to the diagonal, indicating that x% of the population earns x% of income for all values of x between 0 and 100.

11.2 Measuring the degree of inequality

To measure the degree of equality or inequality, we use a Lorenz Curve as shown in Figure 11.1. The deviation of the Lorenz Curve from the diagonal indicates the degree of income inequality; the greater the extent of the deviation, as shown by the shaded area in Figure 11.1, the greater the degree of inequality. To compare the degree of inequality among countries, or to measure changes in income distribution within a country over time, a *Gini Coefficient* is used. This is a measure of the area between the Lorenz Curve and the diagonal (the shaded area) expressed as a percentage of the area of the whole triangle containing

Table 11.4 Income distribution by percentile

Households	% of total income		
	Urban	*Rural*	*Whole country*
Top 1%	8.1	6.3	8.3
Top 2%	20.9	17.2	22.8
Top 10%	31.8	27.3	33.7
2nd decile	15.2	15.6	15.5
3rd decile	11.6	12.0	11.6
4th decile	9.7	19.5	9.2
5th decile	8.1	9.8	8.1
6th decile	6.6	8.8	7.3
7th decile	6.1	6.5	6.5
8th decile	4.4	3.6	2.9
9th decile	3.8	2.3	2.7
Bottom decile	2.7	3.0	2.5
Gini Coefficient	0.4	0.3	0.4

it. The value of the Gini Coefficient ranges between zero and one. Where there is perfect equality the Gini Coefficient will have a value of zero; the higher its value, the more unequal the distribution of income.

Table 11.4 presents the data used to construct a Lorenz Curve. It shows the proportion of total income (in each sector) received by each income group. It shows, for example, that the richest 10% of the rural population received 27.3% of total rural income, while almost 32% of total urban income was received by the richest 10% of the urban population. On the other hand, we can also see that in both the rural and the urban sectors, the poorest 10% of the population received around 3% of total income.

What can we conclude from such information about the distribution of income between the rural and urban sectors? First, we can see that the rural sector contains a higher proportion of poor people, while the urban sector contains a relatively higher proportion of rich people. Second, we can see that within the rural sector income seems to be more evenly distributed than it is within the urban sector. We also see that the Gini Coefficient for income distribution among urban households has a value of 0.4 while for rural households it is 0.3. This confirms that the degree of income inequality is higher among urban than rural households.

11.3 Alternative measures of income distribution

Other ways of comparing the distribution of income include "functional" income distribution and "international" income distribution. "Functional" distribution concerns the way in which income is distributed among the different factors of production in the economy. For example, we might need to compare the relative share of total income accruing to wage-earners, profit-earners, rentiers, etc. In a poor country it is likely that wage-earners will save a lower proportion of their income than wealthier profit-earners and businesses. Therefore, the higher the share of total income earned by wage-earners, the lower the overall savings level is likely to be. In these circumstances, changing the distribution of income between wage- and profit-earners will change the overall savings rate in the economy.

Within each of these categories of owners of factors of production there are further sub-categories. For example, within the wage-earning category, we have "professionals", "skilled-labourers", "semi-skilled labourers", "unskilled labourers", etc. We might be interested in knowing how total wage-income is distributed among these sub-categories. Similarly, we might need to distinguish between different categories of profit-earners. In particular, it is important to know how profits and dividends are shared among nationals and non-nationals, especially since their propensities to remit their earnings overseas are likely to differ.

11.4 Policies to change the income distribution

If a government is committed to improving the distribution of income between rich and poor households, regions or sectors, it has a number of policy instruments that could, in principle, be used. Use could be made of fiscal policies to promote a more equitable income distribution. Tax policies could be designed so that the overall incidence of taxation is progressive, in the sense that the higher the individual's level of income, the higher the proportion of income they contribute in tax. For example, a progressive income tax could be introduced, regressive indirect taxes could be avoided to the extent possible, and subsidies could be paid on essential items of consumption. Moreover, policies of expenditure on health, education, co-operatives and social welfare could be designed so as to deliver a range of social services to low-income groups.

In the economics literature there has been much debate on the relationship between economic growth and income distribution and about how a more egalitarian distribution of income can best be achieved. It is often the case that political opposition or resistance to income redistribution on the part of the wealthy makes measures such as tax reforms impractical or infeasible. Some commentators argue that only a more egalitarian distribution of productive resources such as land and capital can bring about a more egalitarian distribution of income.

Another way in which changes in income distribution can be effected is through the choice of investment projects. Clearly, the government can affect the distribution of income between sectors and regions of the economy through the sectoral and regional spread of public investment projects. Similarly, the type of investment undertaken can influence the distribution of income among various categories of income-earners. An investment that uses relatively more labour than capital will imply more employment, and perhaps, a larger share of income for the wage-earner versus the profit-earner, and vice versa. Similarly, different types of investment will have different implications for the employment (and income) of different types of labour.

It follows that, if the government is aware of the implications of different projects for the distribution of income (whether it be among individuals, regions or sectors), it can indirectly affect the pattern of income distribution through its influence over the choice of investments in the public and private sectors. For instance, the government could choose between, say, one large capital-intensive scheme and a number of smaller, more labour-intensive projects, to produce a given level of output. Such a choice will have implications for the distribution of income among individuals, categories of income earners and perhaps regions.

While one investment option could be superior to another from an economic efficiency viewpoint, it might be considered inferior to the other from an income distribution perspective. The final choice among projects will depend on the relative importance that the policy-makers attach to the economic efficiency objective versus the income distribution objective. It is to the explicit incorporation of distributional objectives in CBA that we now turn.

11.5 The use of income distribution weights in project appraisal: some illustrative examples

Suppose that we are required to provide advice on the best choice of project taking into consideration the government's commitment to the twin objectives of economic efficiency and improving income distribution. In Table 11.5 we have before us three possible projects, A, B and C, of which only one can be undertaken.

Project B can be rejected purely on the ground that its aggregate net Referent Group benefits (measured at efficiency prices) are less than those of both A and B, and, in addition, the distribution of benefits among the rich and poor is less egalitarian than that of either A or C. So, the question is whether to choose A or C. As long as there is a commitment to select projects in conformity with the objective of improving income distribution, project C will be preferred. Aggregate net Referent Group benefits are the same, but the distribution of benefits is more favourable towards the poor in the case of project C.

Which project would we choose from the two options D and E in Table 11.6?

Here the choice is much less straightforward. Project D would be preferred on purely economic efficiency grounds, whereas Project E might be preferred on purely income distribution grounds. As long as there is a commitment to the objectives of economic efficiency and income distribution, a conflict arises. Choose D and we sacrifice distribution; choose E and we sacrifice efficiency. This choice is a classic example of what economists call a *trade-off*.

At this stage the analyst needs further information to make the relative importance of these objectives explicit in the project selection process. This information takes the form of an assessment of the distribution of the Referent Group net benefits among income classes and the assignment of "weights" or "factors" which can be applied to the net benefits which accrue to the various groups. It should be stressed that the weighted Referent Group net benefit table does not replace the unweighted table, but, rather, complements it. Let us

Table 11.5 Comparing projects with different atemporal income distributions

Project	Referent Group Net Benefits ($NPV)		
	Rich	Poor	Total
A	60	40	100
B	50	30	80
C	20	80	100

Table 11.6 Comparing projects with different aggregate net benefits and distributions

Project	Referent Group Net Benefits ($NPV)		
	Rich	Poor	Total
D	60	40	100
E	40	50	90

Table 11.7 Applying distribution weights to project net benefits

Project	Referent Group Net Benefits ($NPV)			Weighted (social) benefits ($)		
	Rich	Poor	Total	Rich	Poor	Total
D	60	40	100	(60x1.0)	+ (40x3.0)	= 180
E	40	50	90	(40x1.0)	+ (50x3.0)	= 190

assume for the moment that we weight each additional dollar of net benefit received by the poor by three times as much as each additional dollar of benefit received by the rich. For example, let us assign a weight of 1.0 to the net benefits accruing to the rich, and a weight of 3.0 to the net benefits accruing to the poor. We can now make an explicit decision on the choice between the two projects as shown in Table 11.7.

Clearly, Project E is favoured, as the total value of its net benefits, adjusted according to the weights described above, exceeds that of Project D. (Distribution weighted net benefits are sometimes referred to as 'social net benefits' in the CBA literature.) This need not have been the case, however; it is possible that the weight given to the net benefits accruing to the poor could have been much less than 3.0.

What these simple examples reveal is that:

- in order to make a choice between projects taking into account both the economic efficiency objective *and* the income distribution objective, we could attach explicit weights to the net benefits accruing to the different categories of project beneficiaries;
- the difference between the weights attached to the net benefits to the rich and the poor would vary according to how much importance the policy-makers gave to the equality of income distribution objective. The greater their commitment to this objective, the greater the difference between the weights.

If the government decided not to attach different income distribution weights to the net benefits accruing to different groups, projects would be selected purely on the basis of their aggregate Referent Group net benefits. What would this imply about the government's objective concerning income redistribution? From what we have seen, it could mean one of two things: either

- it does not regard project selection as an important means of redistributing income, but prefers other more direct measures such as a progressive taxation system; or,
- it does not care about income distribution, i.e. it attaches equal weight (1.0) to net benefits accruing to all Referent Group members.

11.6 The derivation of distribution weights

In the previous section we saw how a system of distribution weights could be used to compare projects that have different income distribution effects. We now need to discuss how, in practice, we might go about the task of deriving such weights.

The approach to this problem most commonly proposed is based on the concept of diminishing marginal utility of consumption which we have already encountered in

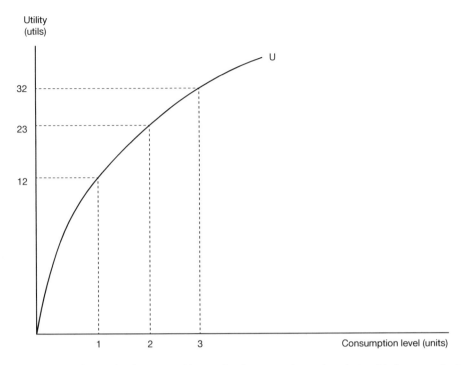

Figure 11.2 The total utility curve. The total utility curve shows the relationship between the level of
utility, measured in units of utility (utils), and the level of consumption, usually measured
in dollars.

Chapter 9 in the analysis of risk aversion. Diminishing marginal utility of consumption
means simply that the more one has of a particular good, or of consumption in general, the
less utility or satisfaction one gets from consuming an additional unit. As one's consumption increases, so the total satisfaction one derives also increases, *but* the amount by which
satisfaction increases, as a result of each extra unit of consumption, gradually declines. This
can be represented graphically as in Figure 11.2.

For consumption of any individual good or service, or of a range of goods, at a level of,
say, 1 unit, total utility might be, say, 12 utils. When consumption increases to 2 units, total
utility increases to, say, 23 utils. If consumption were now further increased to 3 units, utility
rises to 32 utils. In other words, the *extra* utility per *additional* unit gradually decreases – from
12 to 11 to 9 and so on. We call this type of relationship *diminishing marginal utility*, and the
concept is represented graphically as in Figure 11.3.

Figure 11.3 shows that the higher the level of consumption of the good, the lower the
marginal utility of an extra unit of consumption. This concept of diminishing marginal
utility applies to consumption *in general*. In other words, instead of just considering one
good, we can apply this concept to the individual's consumption of all goods together by
stating that, as our *total* consumption level increases, so the extra utility we get from each
additional unit of consumption declines. Total consumption can be measured in terms of
money (dollars) spent on consumption goods and services. In effect, we are then saying that
the higher the level of consumption expenditure of an individual, the less additional utility
they will get from each extra dollar's worth of consumption.

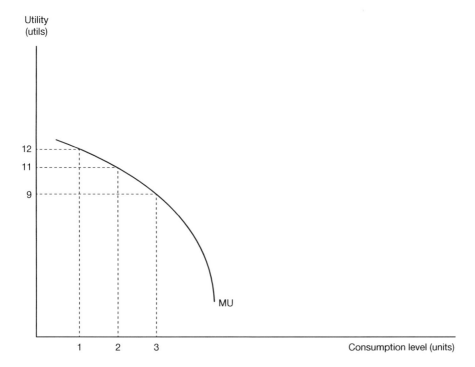

Figure 11.3 The marginal utility curve. Marginal utility (the extra utility derived from an extra unit of consumption) declines as the level of consumption increases.

We can extend the notion of diminishing marginal utility of consumption to that of diminishing marginal utility of income. The difference between consumption and income is savings. We can assume that the individual chooses her savings level such that the marginal utility of a dollar saved is equal to the marginal utility of a dollar consumed. In other words, the marginal utility of an extra dollar of income equals the marginal utility of an extra dollar saved or consumed. An increase in income would lead to an increase in consumption and savings, and a decline in the marginal utility of each.

There is plenty of empirical evidence to support the concept of diminishing marginal utility; for example, as noted above, observed risk-averse behaviour follows from the concept of diminishing marginal utility of wealth, as discussed in Chapter 9. It is tempting to draw on the concept to support the following conclusion: if the income of a rich landowner increases by $100 per annum the extra utility he will get will be much less than in the case of a $100 per annum increase in the income of a poor unskilled worker. However, that conclusion would be unwarranted as utility itself is not measurable and not comparable across individuals. On the other hand, if all individuals were assumed to be identical in their tastes and capacity for enjoyment, we would not need to be able to measure utility to conclude that an extra dollar to the rich generates less additional utility than an extra dollar to the poor; in other words, as income level increases, marginal utility of income declines.

It is from the concept of diminishing marginal utility that distribution weights are rationalised and derived, the idea being that the weight attached to additional consumption by an individual should be based on the marginal utility that they receive at their particular level of income, relative to some base level, and given that marginal utility declines as income and consumption increase, the *higher* the level of their income and consumption, the *lower* is the distribution weight. To illustrate this point, consider an individual with the utility function:

$$U = Y^\alpha,$$

where U is the level of utility, Y is the income level, and α is a parameter. To simplify the discussion, we shall assume, for the time being, that all of income is consumed. The extra utility, ΔU, obtained from a small increase in income, ΔY, is given by:

$$\Delta U = \alpha Y^{(\alpha-1)} \Delta Y,$$

where $\alpha Y^{(\alpha-1)}$ is the marginal utility of income, denoted by λ. Marginal utility is positive, but is assumed to diminish as income rises:

$$\left(\frac{\Delta \lambda}{\Delta Y} \right) = \alpha(\alpha - 1)Y^{(\alpha-2)}$$

from which expression it is evident that we must set α in the range $0 <\alpha< 1$. The elasticity of the marginal utility of income, ϵ_{MU}, expressed as a positive number, can be calculated as:

$$\epsilon_{MU} = -\left(\frac{\Delta \lambda}{\Delta Y} \right)\left(\frac{Y}{\lambda} \right) = 1 - \alpha.$$

Now consider the ratio of the change in utility resulting from a small increase in income at income level Y_i to the change resulting from the same increase at some base level of income Yb:

$$\frac{\Delta U_i}{\Delta U_b} = \frac{Y_i^{\alpha-1}}{Y_b^{\alpha-1}}$$

The value $(Y_i/Y_b)^{(\alpha-1)}$ is the ratio of the changes in utility resulting from the small change in income at the two income levels. If all individuals had this utility function, the value $(Y_i/Y_b)^{(\alpha-1)}$ would express the extra utility gained by an individual at income level i from a small increase in income as a multiple of the extra utility which would be gained by an individual at the base level of income from the same small increase. If $Y_i > Y_b$ ($Y_i < Y_b$), the multiple is less (greater) than unity.

Following the above line of reasoning, the appropriate income distribution weight can be expressed in algebraic form as follows (note that we are inverting the ratio discussed in the previous paragraph):

$$d_i = \left(\frac{\overline{Y}}{Y_i} \right)^n$$

where

d_i = the distribution weight for income group i
$\bar{Y}$ = the average level of income for the economy
Y_i = the average income level of group i
n = *the elasticity (responsiveness)* of marginal utility with respect to an increase in income, expressed as the ratio of the percentage fall in marginal utility to the percentage rise in income (note that $n = \epsilon_{MU}$ is expressed as a positive number and that $n = (1-\alpha)$ for the utility function discussed above).

Recall that we have already encountered the elasticity of the marginal utility of income in the analysis of the consumption rate of interest (Technical note 5.2) and in the analysis of risk aversion (Technical note 9.1).

As an example of the use of such weights, suppose that a net beneficiary of a project is in an income group which has a level of income equal to, say, $750 per annum ($Y_i$), and that the national average income is $1500 ($\bar{Y}$); the distribution weight for that individual would then be:

$$d_i = \left(\frac{1500}{750}\right)^n = 2^n.$$

While the value of d_i finally depends on the value of n, it can be seen that if n takes a value greater than zero, which it must since $0 < a < 1$ for the utility function under consideration, then when the project beneficiary in question enjoys a level of income below the national average, the distribution weight applied to her net benefits will be greater than one; and if the level of consumption is above the national average the weight will be less than one. If n = 0.8, then an individual at consumption level $750 per annum will have her benefits weighted by a factor of 1.74 (i.e. $(1500/750)^{0.8}$). Someone at an income level of $2500 per annum will have his benefits weighted by a factor of: 0.66 (i.e. $(1500/2500)^{0.8}$). An additional $1.00 going, for example, to someone earning an income of $4250 per annum would be valued at only 43% of the value of an additional $1.00 going to someone at the average ($1500 per annum) income level. These income distribution weights can be illustrated graphically as in Figure 11.4.

To this point, based on our simple utility function, we have assumed that n in the formula $d_i = (\bar{Y}/Y_i)^n$ has a value between zero and one. As noted above, n indicates the responsiveness (or 'elasticity') of marginal utility of income with respect to a one unit increase in income. In other words, it indicates how rapidly utility declines as the income level increases. The higher the value of n, the faster the rate at which marginal utility falls. While n must be positive, to be consistent with diminishing marginal utility, the utility function need not take the particular form we have assumed and there is no theoretical reason why n must be less than unity. Indeed, it was pointed out in the discussion of Technical note 5.2 that a survey of 50 countries concluded that n averages around 1.25. Let us now recalculate the values of d_i for various values of n as shown in Table 11.8.

As the value of n increases, the spread of values of the distribution weights increases quite dramatically. For someone in the lowest income group ($250 per annum), the weight increases to 36 when n = 2 and to 216 when n = 3. For someone in the highest income group

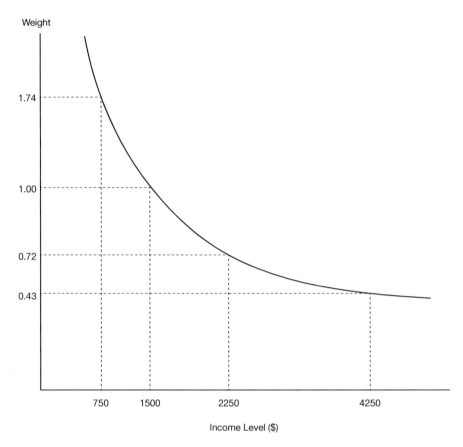

Figure 11.4 Weighting factors for extra income. Assuming n=0.8, an extra dollar of income to a person at income level $750 is valued at $1.74, whereas an extra dollar of income to a person at income level $4250 is valued at $0.43 in the appraisal of the income distribution effects of a project.

Table 11.8 Responsiveness of distribution weights to changes in value of n

$/Annum	Distribution Weight			
	n = 0	n = 1	n = 2	n = 3
250	1.00	6.00	36.00	216.00
750	1.00	2.00	4.00	8.00
1250	1.00	1.20	1.44	1.73
1500	1.00	1.00	1.00	1.00
1750	1.00	0.71	0.73	0.63
2250	1.00	0.67	0.44	0.30
2750	1.00	0.55	0.30	0.16
3250	1.00	0.46	0.21	0.10
3750	1.00	0.40	0.16	0.06
4250	1.00	0.35	0.12	0.04

($4250 per annum), the weight falls to 0.12 when $n = 2$, and to 0.04 when $n = 3$. These calculations are not intended to be prescriptive but rather to illustrate how sensitive the distribution weight is to a change in the assumed value of n. How, then, do we determine the appropriate value of n for a particular economy?

While values of n have been calculated for a range of countries,[1] the estimated values for similar countries (typically in the 1.5 to 2.5 range for developed economies) vary too much to be more than a guide to policy-makers. In practice, the choice of value of n represents a value judgement that someone at the policy-making level will have to make if distributional weighting is to be used explicitly. In other words, to express an opinion as to the appropriate value of n is to express an *opinion* as to the value of an additional dollar in the hands of a rich person, as opposed to a poor person.

The formula $d_i = (\bar{Y}/Y_i)^n$ simply expresses, in algebraic form, the point that the distribution weights we use are determined by *two* factors:

i   the relative income/consumption level of the project beneficiaries ($\bar{Y}/Y_i$); and,
ii   the value judgement that is made about the utility or satisfaction that is gained by project beneficiaries at different income levels; the value of n.

Can the policy-maker or project analyst escape the awkward task of making such a value judgement, in the case of any given project, by simply refraining from using distribution weights at all? Unfortunately, no; as long as we do not apply distribution weights *explicitly*, we are *implicitly* assigning a value $d_i = 1.00$ to the net benefits accruing to each and every project beneficiary, irrespective of how much income he receives. And assigning a value $d_i = 1$ to all income groups implies, as we have seen in Table 11.8, that we are using a value $n = 0$. In other words, we are implicitly assuming that the distribution of the project's benefits does not matter at all. This is as much of a value judgement as the situation in which we assume that distribution *does* matter and we choose to give n a value that is greater than 0.

Some analysts have tried to avoid making a value judgement by turning the question of distributional weighting on its head. For example, we saw from Table 11.7 that a distribution weight of approximately 3 justified the selection of project E over project D. If we observe that project E is actually chosen in preference to project D, the weight implicitly used by the decision-maker to value net benefits to low-income groups must be at least 3. In other words, in some circumstances we can infer from the choices the decision-maker makes what are regarded as the appropriate or "threshold" distribution weights. This type of threshold analysis can be useful in eliminating patently absurd trade-offs, but it does not provide any independent information about the appropriate weights. As outlined in Example 11.2, this approach has been formalised by Burton Weisbrod.

11.7 Distributional weighting in practice

In the ideal situation, distributional weighting of project benefits would be undertaken across all public sector projects being appraised and evaluated as well as, perhaps, being extended to private sector projects under review by governments for the award of incentives such as subsidies, company tax holidays, relief from import duties, etc. In such a comprehensive system information would be required on two levels.

First, at the national or central level, information on the government's objectives with respect to income distribution (among individuals, regions, or sectors) would have to be translated into a set of explicit distribution weights as discussed above. These weights would then be made available as *national parameters* for project analysts at all levels. This could be described as a *top-down* approach, as opposed to the *bottom-up* approach mentioned briefly at the end of the previous section and discussed in more detail in the next section. Second, at the level of the project itself, information on the income distribution implications of the potential investment would have to be compiled. For instance, in order to apply the distribution weights that are given to us as national parameters we would need to know the following about the project in question:

i   identification of the project's gainers and losers within the Referent Group;
ii classification of the Referent Group's gainers and losers; i.e. to which particular income category they belong; and,
iii (iii) quantification of gains and losses to members of the Referent Group, i.e. by how much do the net incomes of the gainers and losers increase or decrease?

If information were available to us at all of these levels, we could proceed with the task of distributional weighting. If the analyst has followed the approach recommended in this book it should be evident that the relevant project level information about (i) and (iii) already exists, for this is, in effect, what is contained in the disaggregated Referent Group Analysis discussed in Chapter 6. We need *only* identify the income group to which each Referent Group stakeholder belongs and then apply the relevant distribution weight assuming this is available from the decision-makers.

Regarding the provision of national parameters, however, most governments appear neither able nor willing to translate their stated commitment to income redistribution into a system of explicit weights, and, in most instances, the departments within the various government bodies and other agencies vested with responsibility for appraising and evaluating projects do not possess the authority or power to decide for themselves the relative importance that should be given to the distribution objective, as opposed to other important objectives such as economic efficiency, especially where these might be in conflict. In the real world, therefore, the project analyst is most likely to be operating in a vacuum as far as such national parameters are concerned.

The bottom-up approach

We prefer to consider CBA and distributional weighting from the perspective of it serving as a potentially powerful means of *identifying* the possible distributional implications of government project selection, and thereby sharpening our understanding of the value judgements often implicit in government decisions. In other words, through CBA we can make ourselves, as project analysts, and the relevant policy-makers more aware of the distributional implications of decisions concerning project selection. This is essentially the procedure advocated by UNIDO[2] in its *Guidelines for Project Evaluation*, in which the immediate objective is to get the policy-makers to confront the income distribution objectives implicit in project choice and to make explicit the value judgments underlying what are essentially political decisions with respect to choice among projects. In this context, distribution weights can be interpreted as the focus of a dialogue between the analyst and the policy-maker, the main purpose of which is to induce policy-makers to make their assumptions and judgements explicit and consistent.

e.g.

EXAMPLE 11.1 Taking income distribution into account in the analysis of an irrigation project

Two alternative designs of an irrigation project have been prepared. They are mutually exclusive and differ chiefly with respect to location. Alternative A would be located in the Central Region where a well-developed infrastructure would keep investment and certain operating costs relatively low. Alternative B would be located in the Southern Region which is both poor and underdeveloped, though equally well-endowed with suitable natural resources for such a project. The chief differences between the alternatives are:

1   The capital costs of Project B are significantly higher because of the additional investment required in roads and other infrastructure, and
2   Project B would bring additional income and employment to one of the poorer regions of the country, whereas Project A would cost less initially, but would contribute to further inequality of income distribution in the country.

The opportunity cost of unskilled labour in the Southern Region is estimated to be much lower than in the Central Region, but transport costs will be much higher, even after the infrastructure has been established, as distances to markets and sources of input supply are much greater. In addition, supervisory personnel would have to come from the Central Region for some time to come, and they require a wage supplement to offset the harsher living conditions in the Southern Region. The net result is that operating costs, calculated in terms of efficiency prices, would be very similar. As the two projects produce essentially identical outputs and the gross operating benefits are equal, the question is which alternative will the decision-maker prefer?

Table 11.9 Threshold distribution weights

Discount rate (%)	NPV A ($)	NPV B ($)	Threshold distributional weight (NPV[A] = NPV[B])
10	360	200	1.8
15	315	150	2.1
20	270	100	2.7

The results of the CBA of the projects in Example 11.1 are summarised in Table 11.9. The decision-maker's initial response may be to choose the project with the higher aggregate Referent Group net benefit: Project A. However, once the relative income distribution effects of the two projects have been drawn to the decision-maker's attention, the analyst may be asked to apply income distributional weights in the comparison of the projects. However, if no national guidelines exist on the appropriate weights, the analyst is not in a position to supply this information since it is the preserve of the political sphere (which the decision-maker occupies) rather than a technical matter for the analyst to deal with.

Suppose that the analyst calculated the aggregate Referent Group net benefits of the two projects at three different discount rates, 10%, 15% and 20%, and then worked out what the distribution weight on additional income accruing to the Southern Region would have to be for the weighted NPV of Project B to be at least as much as the unweighted NPV of Project A. The threshold distribution weights for this example are shown in the right-hand column of Table 11.9.

If the appropriate discount rate is 10%, then a distribution weight of at least 1.8 on net benefits of Project B will be required for it to be preferred to Project A; at a discount rate of 15%, the weight would have to be at least 2.1, and at 20%, it would have to be at least 2.7. (The reason that the required weight rises as the discount rate increases is that the more capital-intensive project (B) becomes relatively less attractive as the cost of capital rises. This issue is discussed further in the following section of this chapter.) The decision-maker might respond that it seems reasonable to value net benefits generated in the Southern Region at roughly two to three times as much as those generated in the Central Region, and might, therefore, favour alternative B even if the discount rate were as high as 20%.

e.g.

EXAMPLE 11.2 Calculating implicit income distribution weights

An early attempt to work out implicit distribution weights was undertaken in the United States in the 1960s by Weisbrod (1968) who compared three similar water resource projects (Projects 2 to 4) that had been selected in preference to a project with a higher benefit/cost ratio (Project 1). In each case he was able to disaggregate the project's net benefits among four income/racial groups. On the initial assumption that the sum of the net weighted benefits of each of the three chosen projects per dollar of project cost was equal to the sum of the net weighted benefits of the rejected project, he was able to find the values of the implicit distribution weights from the coefficients derived by solving the set of simultaneous equations:

Project 1: $0.36a + 0.63b + 0.007c + 0.003d = 1$
Project 2: $0.172a + 0.479b + 0.025c + 0.039d = 1$
Project 3: $0.161a + 0.294b + 0.066c + 0.33d = 1$
Project 4: $0.023a + 0.470b + 0.007c + 0.041d = 1$

where the coefficients indicate the proportion of each dollar of a project's benefits accruing to each group, and the variables a, b, c and d are the distribution weights for the four groups of project beneficiaries. The equations were solved simultaneously, yielding implicit distribution weights: $a = -1.3$; $b = 2.2$; $c = 9.3$; and, $d = -2.0$. (Who were the groups? The weights a, b, c and d referred to Poor Whites, Non-Poor Whites, Poor Non-Whites and Non-Poor Non-Whites respectively.) As a form of sensitivity analysis the exercise was then repeated allowing the net weighted benefits of the three preferred projects (Projects 2 to 4) to vary from 5% to 20% more than the net weighted benefits of Project 1. In Exercise 5, at the end of this chapter, the reader is invited to set up and solve a set of equations for the implicit distribution weights used by decision-makers.

In the way illustrated by Example 11.1, the project analyst has clearly delineated the political component of project formulation and selection, and has facilitated the final choice by spelling out the consequences of the alternatives. Ideally, this approach or dialogue could be repeated for all potential projects. Where inconsistencies appear, the project analyst can draw them to the attention of the minister or policy-maker in question and they can be resolved to a degree at which the range of implied income distribution parameters has been sufficiently narrowed for some concrete conclusions to be reached as to their (inferred) appropriate values. This process can be formalised as described in Example 11.2.

In the absence of the kind of analysis described in Example 11.2, the project analyst can perhaps do no more than try to establish, among alternative projects, who the gainers and losers are, how much they gain or lose, and what the efficiency and income distribution implications would be of choosing one rather than another project; in other words, prepare a disaggregated Referent Group Analysis along the lines developed in this book.

11.8 Worked example: incorporating income distribution effects in the NFG project

In Chapter 6 it was found that the NFG project generated negative aggregate Referent Group net benefits (–$201.9 thousand at a 5% discount rate). The results of the disaggregated Referent Group Analysis are reproduced in the first column of Table 11.10. (For the purposes of this example, it will be assumed that the relevant discount rate is 5%.) The second column provides a set of distribution weights (d_i) which are used to calculate the adjusted or weighted net benefits as shown in the last column. The values for d_i were chosen deliberately so that a decision reversal results, as compared with the result based on unweighted net benefits. With net benefits to wage-earners (gainers) being given a relatively high weight of 4, and those to downstream users (losers) a relatively low weight of 0.5, the weighted aggregate net benefit becomes marginally positive ($1.97 thousand).

In the absence of information about distribution weights, we could have considered the threshold values of d_i; i.e. by asking what combination of values for d_i would result in a positive aggregate net benefit. Let us assume, for example, that the policy-makers are satisfied that the d_i value for government revenue/expenditure should be 1, and for landowners it should be 0.6, but that there is less clarity as to the values of d_i for wage-earners and downstream users. We could then construct a trade-off matrix showing what possible combinations of these two d_i values would produce a positive value for aggregate Referent Group net benefits. An example of such a matrix is shown in Table 11.11.

What this shows is that for any d_i value of less than 4 for wage-earners, the d_i value for downstream users would need to be extremely low for the aggregate (weighted) net

Table 11.10 Distributional weighting in the NFG project

	Net Benefit (5%) ($000s)	d_i	Weighted Net Benefit ($000s)
Government	−173.2	1.00	−173.2
Landowners	33.2	0.6	19.92
Wage earners	53.2	4.0	212.8
Downstream users	−115.1	0.5	−57.55
Aggregate	−201.9		1.97

Table 11.11 Threshold combinations of distribution weights

Combinations of d_i	1	2	3
Wage earners	3.5	3.0	2.5
Downstream users	0.3	0.05	<0

benefit to be positive; this would imply that the costs the project imposes on these users are regarded as being relatively unimportant. When wage earners' d_i is set at 3, the d_i for downstream users would need to be 0.05, and if it is set at 2.5, the d_i for downstream users would need to be negative, which is an unlikely value since it implies that imposing costs on this category of users is actually regarded as an advantage of the project. It is through this type of sensitivity testing and threshold analysis that the analyst can engage in a useful elicitation process with decision-makers.

11.9 Inter-temporal distribution considerations

Until now in this chapter it has been assumed that the marginal value of a dollar is the same regardless of whether that dollar is consumed or saved; this assumption is based on the supposition that, from an individual viewpoint, income is allocated optimally between consumption and savings. Furthermore, it may be the case that the aggregate effect of individual savings decisions, based on the market rate of interest, is to generate the optimal level of savings in the economy as a whole. However, in some cases it is legitimate to ask whether, from the viewpoint of society as a whole, an extra dollar that is saved is considered to have the same value as an extra dollar that is consumed, and whether the level of aggregate savings determined by market processes is sufficient.

The answer to that question is related to the question of whether the social time preference rate is different from the market rate of interest, as discussed in Chapter 5. As noted above, individuals use the market rate of interest to make decisions about consumption and saving, and they maximise their individual welfare by saving up to the point at which the present value of extra future consumption resulting from extra saving equals the value of the extra consumption forgone in the present. In other words, at the market rate of interest the present value of an extra dollar consumed is the same as the present value of an extra dollar saved. If, however, the social time preference rate is considered to be lower than the market rate, because, for instance, the preferences of future generations are believed to be ignored in market transactions, then the present value of a dollar saved will be greater than the present value of a dollar consumed. In that event, in determining the relative value of a dollar of benefit to the poor, as opposed to the rich, we need to take into account the fact that the rich save a higher proportion of an additional dollar than do the poor; in other words, while an extra dollar of benefit to the rich may worsen the *atemporal* distribution of income, it may improve the *inter-temporal* distribution. Before developing this argument, however, we can review the reasons outlined in Chapter 5 as to why the relative value of savings and consumption might differ.

The argument runs as follows: the portion of current income which is not consumed by us today is saved. Savings finance investment, and investment today generates consumable output in the future. Therefore, the decision we make today, regarding how much of our income is spent now on immediate consumption and how much is saved now for future consumption, is essentially a decision about how consumption should be distributed among those living today and those living in the future – *inter-temporal* distribution. Other things being equal, the more that is consumed by us today, the less that is left for future generations to consume, and vice versa. If the social time preference rate is lower than the market rate of interest, the present value of the additional benefit that could be generated by one extra dollar of savings is greater than the additional benefit generated by one extra dollar of consumption.

Using a social discount rate lower than the market rate of interest will result in raising the NPV of all investment projects, and raising by more the NPV of projects which are more capital-intensive relative to that of projects which are less capital-intensive. (This effect of changing the discount rate was discussed in detail in Chapter 3.) However, the lower social discount rate makes no allowance for the different effects of projects on saving and reinvestment of project net benefits, as discussed by Marglin (1963). This issue was not addressed in the Efficiency Analysis because it is generally of concern only in a limited number of developing economies which experience an acute shortage of capital. In these circumstances the analyst may wish to take projects' indirect contributions to capital formation into account in project selection by establishing what part of the Referent Group's net benefits are saved and what parts are consumed, and by attaching a premium to that part which is saved.

It cannot be assumed that all members of the Referent Group save the same proportion of any income gained or lost. It is commonly observed that individuals at higher-income levels save a higher percentage of income than individuals at lower-income levels. One reason for this is obvious: at some low level of income the individual needs to spend total income on consumption goods in order to survive; at higher-income levels the luxury of saving to provide for the future can be considered. It follows that a project which largely benefits high-income groups will result in more saving than one which benefits low-income groups (assuming the net benefits are positive). If an extra dollar of saving is worth more to the economy than an extra dollar of consumption, it is possible that the net benefits to a relatively better-off group could be weighted *more* than those to a relatively poorer group, taking into account both the *atemporal* and *inter-temporal* income distribution effects.

Recalling our discussion of distributional weighting in the previous section, we learned that one extra unit of consumption may not convey the same benefit to all income groups; the higher the income group, the lower the relative value of an additional dollar of consumption, and vice versa. This was the rationale for the use of income distribution weights to assign greater value to net benefits accruing to the poor than to the rich. Once we introduce a premium (or shadow-price) on savings, we need to make an additional adjustment to the raw estimates of Referent Group net benefits. As noted above, this adjustment will tend to favour projects that generate relatively more net benefit for the rich, since the rich generally save a higher proportion of any additional dollar of income than do the poor. It is possible that the decision-maker will be faced with a trade-off between a better *atemporal* and a better *inter-temporal* distribution of income. Such a case is illustrated by Example 11.3.

The problem in applying this approach is how to estimate the value of savings relative to consumption: the premium on savings. One method is to consider the implications of the social time preference rate as discussed in Chapter 5. Suppose that a dollar of saving results in a dollar of investment yielding the market rate of interest, r, in perpetuity. At the social rate of time preference, r_s, the present value of the investment project's net benefit stream is (r/r_s); for example if r = 6% and r_s = 5%, the present value generated by a dollar of saving, matched by investment, is \$1.2. In other words, the premium on saving is 20% and the weight attached to the portion of net benefits saved would be 1.2.

e.g.

EXAMPLE 11.3 Incorporating atemporal and inter-temporal distribution effects

Suppose that the effects of two projects can be summarised as follows:

Project A generates a net Referent Group benefit of $100. $80 is saved and $20 is consumed by a group with above average income. Assuming that $1.00 saved is worth the same to the economy as $1.20 consumed (i.e.. the shadow-price of savings is 1.2), and that the distribution weight for this group is 0.75 then: Net Benefit (in terms of dollars of consumption)

= $80(1.2) + $20(0.75)

= $96 + $15

= $111

Project B also has a net benefit of $100. Of this $40 is saved and $60 is consumed by members of the Referent Group who enjoy an average level of income. As the same shadow-price of saving (1.2) applies, and the sub-group's distribution weight will have a value of 1.0, then:

Net Benefit (in terms of Dollars of consumption)

= $50(1.2) + $50(1.0)

= $60 + $50

= $110

In this instance, the combined effect of introducing distribution weights and the shadow-price of saving is to favour the project that benefits the relatively richer group, Project A, whereas in the absence of a shadow-price of saving, Project B would have been favoured.

Another method is to try to elicit from the decision-makers their implicit premium on savings through the choices they make. For example, we could confront the decision-makers, perhaps by means of a choice experiment, with a number of project options where a choice would have to be made between an extra dollar saved by any member of society versus an extra dollar consumed by individuals in various income categories. As a result of this process it might be inferred that the decision-maker would be indifferent as to whether the extra dollar was saved, or consumed by an individual in a particular income group, Y_i. Since there is a premium on saving there must also be a premium on consumption by individuals in chosen income category Y_i; in other words, the *atemporal* distributional weight attached to net benefits to this group must exceed unity, reflecting the group's low-income status. Ignoring the small amount of saving that might be undertaken by individuals in this low income group, we can infer that the *atemporal* distributional weight (d_i) for that group must be equal to the premium on saving. Consider Example 11.4.

e.g.

EXAMPLE 11.4 Deriving the premium on savings indirectly

The hypothetical example in Table 11.12 reports the consumption levels of the different income groups in the first column, and the *atemporal* distributional weights are given in the second column. In this example, the mean level of consumption is $1500. The decision-maker is asked, through a choice experiment, to specify the critical consumption level (C_c) at which the value of an extra dollar of consumption to an individual at the critical level is equal to the value of an extra dollar saved.

Table 11.12 Composite distribution weights

Consumption $/annum ($C_i$)	Distributional weight (d_i)
250	6.00
750	2.00
C_c = 1250	1.20
$\bar{C}$ = 1500	1.00
1750	0.71
2250	0.67
2750	0.55

Since a dollar saved is assumed to be more valuable than a dollar consumed by the average income earner, the critical consumption level nominated by the decision-maker will turn out to be less than the average level. Assume that through a process of iteration the decision-makers identify consumption level, C_c = $1250 as the critical level at which they place the same value on an additional dollar of an individual's consumption as on a dollar of additional saving. From Table 11.12 it can be seen that the distribution weight for this critical consumption level is 1.2. In other words, a dollar of extra consumption enjoyed by individuals at the critical consumption level is 1.2 times as valuable as each extra dollar consumed by an individual at the average level, and, we now know, is worth the same as an extra dollar of saving. We can therefore conclude that the implicit premium on saving is 1.2. Each extra dollar of saving has to be weighted by a factor of 1.2 in the analysis of inter-temporal distribution.

If the decision-maker's income distribution objectives are to be accommodated by a system of *atemporal* and *inter-temporal* weights, it will be necessary for the analyst to disaggregate net benefit for each Referent Group gainer or loser into its consumption and savings components, and then weight the consumption component by the atemporal distributional weight and the savings component by the inter-temporal distribution weight (or savings premium). If we were to follow this procedure in the context of the NFG project discussed previously (see Table 11.10), it would be necessary to disaggregate each stakeholder group's net benefits into their consumption and savings components, and then apply the respective d_i to the consumption net benefit and the savings premium to the savings portion of the net benefit of each Referent Group beneficiary or loser.

Finally it should be noted that once Referent Group net benefits have been adjusted by a set of distribution weights the adding-up property of the spreadsheet framework will no longer hold: weighted Referent Group net benefits need not be equal to Efficiency less non-Referent Group net benefits. For this reason we prefer that, where distributional weighting is employed, the weighted net benefits are presented in a separate part of the spreadsheet model, and that the decision-maker's attention is drawn to the value judgments implicit in both the weighted and unweighted net benefit summaries.

11.10 Further reading

The issues of *atemporal* and *inter-temporal* distribution of income are dealt with at length in I.M.D. Little and J.A. Mirrlees, *Manual of Industrial Project Analysis in Developing Countries* (Paris: OECD, 1968) and P. Dasgupta, A. Sen and S. Marglin, *Guidelines for Project Evaluation* (New York: United Nations, 1972). For an early study showing how distributional weights can be derived from government's project choices, see B.A. Weisbrod, "Income Redistribution Effects and Benefit-Cost Analysis", in S.B. Chase (ed.), *Problems in Public Expenditure Analysis* (Washington, DC: The Brookings Institution, 1968). On the reinvestment of project net benefits, see S.A. Marglin, "The Opportunity Costs of Public Investment", *Quarterly Journal of Economics*, 77(2) (1963): 274–289.

Exercises

1   In both countries A and B two groups of income recipients are identified. Within each group income is evenly distributed, but the upper income group receives a disproportionate share of national income:

Country A: the lower 40% of income recipients receive 30% of total income and the upper 60% of income recipients receive 70% of total income;

Country B: the lower 50% of income recipients receive 40% of total income and the upper 50% of income recipients receive 60% of total income.

i   Draw a Lorenz Curve for each country (remember to label the axes);
ii   Calculate Gini Coefficients for countries A and B;
iii   Explain briefly what the Gini Coefficients tell us about the comparison of the degree of inequality of income distribution in the two countries.

2   From the data in the table on the distribution of income (by quintile) between two states, A and B, you are required to compare the degree of income inequality between the two states using Lorenz Curves and Gini Coefficients.

	Percentage of Total Disposable Income	
Quintile	State A	State B
1st	50	35
2nd	20	30
3rd	15	25
4th	10	6
5th	5	4
Total	100	100

3   A decision-maker believes that each member of the community derives utility from their income according to the following formula:

$$U = \alpha Y^\beta$$

where

    U = annual level of utility

    Y = annual income level

    $\alpha = 1$

    $\beta = 0.5$.

The average level of income in the community is $1800 per annum. Given the above information, what income distributional weights should be placed on net benefits accruing to individuals at the following income levels (in each case briefly explain your answer):

i   annual income = $1800;
ii   annual income = $200;
iii annual income = $7200.

4   The data in the table below show the breakdown of the Referent Group (RG) benefits for two projects (A and B), information about each sub-group's income level and marginal propensity to save (mps). In addition, you have ascertained that the mean income level is $30,000 per annum, that an appropriate value for 'n' (the elasticity of marginal utility with respect to an increase in income) is 2, and that the premium on savings relative to consumption is 10%.

	RG1	RG2	RG3
Income p.a. ($)	20,000	40,000	90,000
Mps	0.05	0.20	0.60
Net Benefits – A ($)	10	20	90
Net Benefits – B ($)	20	30	40

With this information you are required to calculate and compare the projects' net benefits in terms of:

i   unweighted aggregate Referent Group net benefits;
ii   aggregate net benefits weighted for *atemporal* distributional objectives (assuming consumption and savings have the same value at the margin); and,
iii aggregate net benefits weighted for both *atemporal* and *inter-temporal* distributional objectives.

5   You have been provided with the following information about three public sector projects with a breakdown of the Referent Group net benefits (unweighted) among the three Referent Group (RG) categories (in $):

	RG 1	RG 2	RG 3
Project A	30	30	60
Project B	40	20	40
Project C	40	30	10

A decision-maker considers that all three projects have equivalent weighted net benefits, and equal to the net benefit of a fourth project, after weighting for its distributional effects. If the weighted net benefit of the latter project is calculated to be $100, find the implicit values of the three distribution weights for RG1, RG2, and RG3, on the assumption that these weights have an average value of 1.

Notes

1   See the paper by Layard *et al.* referred to in Chapter 5. Also note that an estimate of *n* is a byproduct of demand analysis by means of a linear expenditure system. See for example Alan Powell, "A complete system of consumer demand equations for the Australian economy fitted by a model of additive preferences", *Econometrica*, 34 (1966): 661–675.
2   United Nations International Development Organization, *Guidelines for Project Evaluation* (New York: United Nations, 1972), Chapter 17, pp. 248–258.

12 Economic impact analysis

12.1 Introduction

A large project, such as construction of a major highway or development of a large mine will have a significant impact on the economy. The spending in the construction and operating phases will generate income and employment, and public sector decision-makers often take these effects into account in deciding whether or not to undertake the project. An economic impact analysis is a different procedure from a cost-benefit analysis in that it attempts to predict, but not evaluate, the effects of a project. Keynes is reputed to have observed that digging a purposeless hole in the ground or building a hospital might have the same economic impact but very different levels of net benefit. Since the data assembled in the course of a CBA are often used as inputs to an economic impact analysis, the two types of analyses tend to be related in the minds of decision-makers and may be undertaken by the same group of analysts.

In this chapter we survey briefly three approaches to economic impact analysis: (1) the income multiplier approach; (2) the inter-industry model; and (3) the computable general equilibrium model. Use of the latter two approaches involves a degree of technical expertise and would generally not be undertaken by the non-specialist. In discussing economic impacts we reiterate that these are not the same as the costs or benefits measured by a CBA. However, there may be costs and benefits associated with the project's economic impact and the decision-maker may wish to take these into account.

The decision whether or not to take the multiplier or flow-on effects into account in the evaluation should be based on an assessment of the extent to which similar such effects would or would *not* occur in the absence of the project in question. When choosing between alternative projects, this would depend on the extent to which the multiplier effects can be expected to vary significantly between the alternatives; when faced with an accept vs reject decision for a discrete project the analyst would need to assess whether the same investment in the alternative, next best use could be expected to generate multiplier effects of the same or similar magnitude. These points are taken up again in the discussion of the multiplier effects.

DOI: 10.4324/9781003312758-12

12.2 Multiplier analysis

12.2.1 *The closed economy*

For simplicity, the concept of the national income multiplier can be developed in the context of a closed economy – one that does not engage in foreign trade. This simplifying assumption will be dropped later in the discussion.

The value of the gross product of an economy can be measured in two equivalent ways: (1) as the value of all the goods and services produced by the economy in a given time period, usually a year; or (2) as the value of the incomes to factors of production – labour, capital and land – generated in the course of producing those goods and services. The two measures are equivalent because of the concept of the circular flow of income. Figure 12.1 shows the flow of income from firms (the sector which demands the services of factors of production, and supplies goods and services) to the household sector (the sector which demands goods and services, and supplies the services of factors of production), and the flow of expenditures from households to firms. It also shows the activities of government which collects tax revenues and allocates them to the funding of purchases of goods and services from firms for the use of the household sector.

While the concept of the circular flow of income implies that it is always identically true that the flow of income equals the flow of expenditure, there is only one level of income which can be sustained by any given level of demand for goods and services. This equilibrium level of income can be calculated by means of the equilibrium condition:

$$Y = C + I + G$$

In this expression, Y is the equilibrium level of income and C, I and G are expenditures on the three types of goods and services produced by the economy – consumption goods and services, investment goods, and government goods and services.

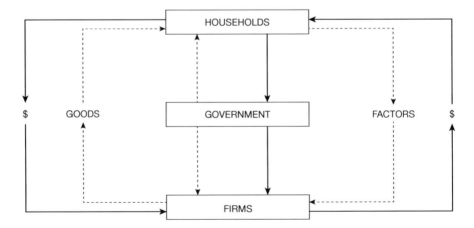

Figure 12.1 The circular flow of income. Households supply factors of production to firms in exchange for income, and firms supply goods and services to households in exchange for sales revenue. Government levies taxes on both households and firms, and supplies goods and services to households and factors of production to firms. Solid lines represent money flows and dotted lines represent flows of goods and services.

In applying the equilibrium condition we need to distinguish between those variables that are explained by this simple model of national income determination, and those that are not. The former variables, Y and C in this model, are termed *endogenous*, i.e. their values are determined by the model, and the latter, I and G, are *exogenous*, or autonomous, i.e. their values are determined in a manner not explained by the model.

Consumption expenditure can be related to income by means of a consumption function:

C = C* + bY.

In this expression C* is the level of autonomous consumption expenditure, and b is the marginal propensity to consume – the proportion of an additional dollar of income that is consumed, as opposed to saved. Since income must be either saved or consumed, the savings function can be written as:

S = (1 – b)Y – C*,

where (1 – b) is the marginal propensity to save – the proportion of an extra dollar of income that households will save. When the income level Y > C*/(1 – b), savings are positive, but when Y < C*/(1 – b) savings are negative (i.e. being run down) because households are financing a portion of their consumption expenditure out of past savings. In either event, according to the consumption function, a proportion (1 – b) of any additional dollar of income earned will be saved.

When the above information is substituted into the equilibrium condition, it becomes:

Y=C*+bY+I*+G*,

where the starred values indicate exogenous variables. This expression can be simplified by collecting the Y terms on the LHS of the equation and solving for the equilibrium level of income:

$$Y = \left(\frac{1}{(1-b)}\right)(C* + I* + G*).$$

If we know the level of expenditure chosen by government, the level of investment expenditure chosen by firms, and the level of autonomous consumption, we can work out the equilibrium level of national income.

The term 1/(1 – b) is termed the national income multiplier because it determines the equilibrium level of income as a multiple of the sum of the autonomous expenditures. It also relates changes in national income to changes in the level of autonomous expenditure. For example, suppose that a firm decided to undertake an additional investment project, the change in the level of national income could be calculated as:

$$\Delta Y = \left(\frac{1}{(1-b)}\right)\Delta I*.$$

To consider the implications of this result, suppose that b = 0.9 and that ΔI* = $100, then ΔY = $1000. In other words, an additional expenditure of $100 on autonomous investment results in a $1000 increase in the equilibrium level of national income.

The intuition behind this striking result of the model can be seen by following the money trail left by the increase in the level of autonomous investment. When an extra $100 is spent, it represents $100 of extra income to households, of which $100b is spent, and this amount in turn represents extra income to households, of which $100b^2$ is spent, and so on. The total increase in expenditure (or income) is given by:

$$\Delta Y = 100 + 100b + 100b^2 + 100b^3 + 100b^4 + 100b^5 + \cdots$$

The sum of this geometric progression can be calculated in the usual way to give:

$$\Delta Y = \left(\frac{100}{(1-b)} \right).$$

It will be evident from this example that, in theory, the only thing that prevents nominal income rising without limit in response to an increase in autonomous expenditure is the fact that a proportion $(1-b)$ leaks out of the expenditure cycle at each round. The lower the value of b, the less the proportion of any income increase that is passed on in the form of expenditure to generate further increases in income.

Now consider an economy in which there are two types of leakages – into household savings and into tax revenues. In this model, consumption expenditure can be related to disposable income:

$$C = C* + b(Y - T*),$$

where $T*$ represents taxes paid by households to government. Now the equilibrium level of income is given by:

$$Y = \left(\frac{1}{(1-b)} \right)(C* + I* + G* - bT*).$$

Suppose that the government decided to increase expenditure by some amount $\Delta G*$, and to finance it by printing money rather than by raising taxes. The resulting increase in national income would be:

$$\Delta Y = \left(\frac{1}{(1-b)} \right)\Delta G*.$$

If, on the other hand, the government decided to raise taxes in order to finance the additional expenditure, the increase in national income would be:

$$\Delta Y \left(\frac{1}{(1-b)} \right)(\Delta G* - b\Delta T*).$$

If the additional expenditure was fully funded by the tax increase, $\Delta G^* = \Delta T^*$, the increase in national income would be:

$$\Delta Y = \left(\frac{1}{(1-b)}\right)(1-b)\,\Delta G^* = \Delta G^*.$$

This latter expression demonstrates that the "balanced budget multiplier" – the national income multiplier to be applied to an increase in government expenditure fully funded by an increase in taxes – is unity when the level of taxation is autonomous. The reason for this result is that, in the round after the initial expenditure, the increase in household consumption which would have occurred in the absence of a tax increase, $b\Delta G^*$, is offset by the reduction in household expenditure caused by the tax increase, $b\Delta T^*$, and these effects cancel out in the case of a balanced budget, so that only the initial expenditure adds to national income.

To this point we have established that an additional public or private project, which represents an increase in the level of autonomous expenditure, will have a multiplied effect on the equilibrium level of national income. In the simple example presented above, a $100 project raises national income by $1000, which seems to imply that in addition to the $100 worth of project output, there is an additional $900 worth of benefit from the project. However, we have to be sceptical of this interpretation of the result of the model because it fails to distinguish between real and nominal increases in national income.

If the economy is at full employment, no increase in real national income is possible; in other words, no net increase in the volume of goods and services produced is possible because there are no additional factor inputs which can be allocated to production. In that event the increase in the equilibrium level of national income (or product) would be purely nominal, consisting simply of the same volume of goods valued at higher prices resulting from inflation. If the national product in real terms is denoted by the flow of goods, Q, and the price level by P, we can express national income or product by PQ, and the increase in national income by:

$$\Delta Y = P\Delta Q + Q\Delta P$$

where the term $P\Delta Q$ represents a real benefit to the economy in the form of higher output, and the term $Q\Delta P$ represents the increase in the valuation of the original level of production as a result of inflation. Clearly in a social CBA, we would be interested only in the former outcome since individuals benefit from increased availability of goods and services, but not from higher prices even though these are matched by higher incomes. At full employment, $\Delta Q = 0$ and there is no benefit to be measured.

Now suppose that the economy is operating at less than full employment. In that case, some fraction of the increase in national income resulting from the multiplier effect represents an increase in the value of output of goods and services. However, before we could count this as a benefit of the project, we would need to determine what that fraction is, and how much of it represents a net benefit. In an economy beset by structural unemployment, where the skills of the unemployed do not match the skills required by business, it may not be possible to increase real output and the multiplier effect may be simply to raise the price level. Determining the fraction of the multiplier effect which represents an increase in output is clearly not a simple matter. Furthermore, as argued in Chapter 5, otherwise unemployed resources may have an opportunity cost. Unemployed individuals may value their leisure time, or value goods and services, such as food or home and car repairs, that they produce outside the market economy. The net gain to the unemployed is the wage less the value of such items. This was recognised in Chapter 5 by shadow-pricing the labour used in a project at less than the wage rate, but at more than zero. A similar shadow-price should also be used in assessing the net benefit of any induced changes in real output level as a result of the project.

It must be emphasised that the income and employment multiplier effects will normally last only for the duration of the activity in question. For instance, if the investment and construction phase of a project is limited to one or two years, the associated flow-on impact on output and employment will be limited to that two-year period. It is often the case that policy-makers/politicians overlook this relatively short duration when referring to the income and employment impacts of a project.

Finally, it must be recognised that any increase in autonomous expenditure will generate a multiplier effect, so that in comparisons among projects the multiplier effects can generally be ignored, unless, of course, it is evident that the magnitude of the multiplier effect varies significantly among the alternatives being considered.

12.2.2 The open economy

For an economy which imports and exports goods and services, the national income equilibrium condition is given by:

$$Y = C + I + G + X - M.$$

This expanded form of the condition recognises that only domestic production of goods and services generates income in the economy. The imported components of C, I, G and X, totalling M in aggregate, are therefore subtracted from the total value of goods and services produced. Exports can be regarded as exogenous – determined by foreign demand – while expenditure on imported goods and services will be positively related to income.

The multiplier for this open economy can be determined as before by substituting the behavioural relations into the equilibrium condition and solving for equilibrium national income. The behavioural relations are the consumption function:

$$C = C^* + b(Y - T),$$

the import function:

$$M = M^* + mY,$$

and a tax function that allows tax collections to rise as income rises:

$$T = tY.$$

Incorporating this relationship between T and Y recognises that the tax system imposes a brake on the expansion of the economy by increasing the tax leakage as income rises. The equilibrium level of income in this model is given by:

$$Y = \left(\frac{1}{(1-b(1-t)+m)}\right)(C^* + I^* + G^* + X^* - M^*).$$

Some plausible values for b, t and m can be selected as follows. For a mixed economy in which the government sector produces 30% of total production of goods and services, the value of t will be around 0.3; for an open economy for which 25% of the value of goods and services consumed is imported, the value of m will be around 0.25; and household saving out of disposable income can be assumed to be 10% as before, giving a value of 0.9 for b. Using these values a multiplier of 1.61 is obtained. This is considerably lower than the value obtained for the closed economy model because of the addition of the tax and import leakages. The implication of this result is that an additional $100 of government expenditure, not matched by any increase in taxes, would increase national income by $161, consisting of the $100 expenditure plus second and additional round effects totalling $61.

This more realistic model suggests that multiplier effects are in reality much smaller than in our closed economy model. Again, as noted above, we need to recognise that some of the increase in equilibrium income may simply reflect price changes, and that any increase in real income – the volume of goods and services produced in the economy – will come at some opportunity cost so that the net benefit is lower than the real increase in income.

As noted above, the size of the multiplier falls as the size of the leakages from the economy increases. Since the smaller the region, the larger the import leakage it can be concluded that regional multipliers will be smaller than state multipliers, and state multipliers smaller than national multipliers. A relatively small region might easily import half of the value of the goods and services it consumes. Substituting the value $m = 0.5$ into the above multiplier model reduces the value of the multiplier to 1.15. Furthermore, a relatively small region may not get the full benefit of the initial round of expenditure. For example, suppose an additional tourist spends $100, represented as $\Delta X^* = \$100$ in the national income multiplier model. However, a portion of this expenditure may immediately leak out of the region without generating any first-round income increase at all. Suppose that half of the expenditure simply goes to the national head office of a hotel chain so that the amount of additional income generated in the region – income to service providers such as hotel, restaurant and transport staff – is only $50. Now if we apply the multiplier we find that the rise in regional income is only $57.

If multiplier effects are relatively small, and have to be qualified by distinguishing real from nominal effects, and netting out opportunity costs, it might be asked why various client groups are often keen to have them estimated. One industry which often emphasises these effects is the tourism industry, since it usually operates at below capacity. For a hotel, or a bus tour which has vacant places, the marginal cost of serving an additional customer is very low. This means that almost the whole of the induced rise in regional income constitutes a net benefit to business. This case is similar to the case of an economy with significant unemployment where multiplier-induced increases in income are real, as opposed to nominal, and the net benefit is similar to the gross benefit because of the low opportunity cost of additional factor services.

12.2.3 Crowding out

In the analysis of the open economy multiplier several key variables were held constant while government expenditure was increased. However, if additional expenditure is financed by borrowing, as opposed to tax increases, the increase in supply of government bonds can be expected to result in a rise in the rate of interest, which in turn can affect the level of private consumption and investment. Furthermore a rise in the interest rate may attract foreign lenders and this will result in an appreciation of the exchange rate, thereby encouraging imports and making the country's exports more expensive for foreigners to purchase. In effect, the financing of the increase in government expenditure "crowds out" private sector activity, such as consumption, investment and exports, and encourages the substitution of imports for domestic production.

We can now rewrite the equation for the equilibrium level of income in the open economy as:

$$Y = \left(\frac{1}{\left(1 - b(1 - t) + m\right)} \right) \left(C(r) + I(r) + G^* + X(r) - M(r) \right)$$

where, instead of being autonomous variables treated as constants, $C(r)$, $I(r)$, $X(r)$ and $M(r)$ are now endogenous variables affected by the increase in the interest rate, r, resulting from financing increased government expenditure, with C, I and X responding negatively, and M positively, the latter two effects being transmitted via the exchange rate. The rise in the rate of interest causes a negative wealth effect for consumers, thereby reducing C, and an increase in the cost of capital for investors, thereby reducing I.

When these adjustments are made to the model, the increase in national income resulting from increased government expenditure is at least partially offset by a decline in consumption, investment and exports and substitution of imports for domestic production. Taking the displacement of private activity into account can result in a dramatic downward revision to the value of the multiplier. Table 12.1 reports estimates for various countries.

Table 12.1 Range of fiscal multipliers for OECD countries

Range	Countries
> 1	Poland
0.6–0.8	Australia
0.4–0.6	USA, Spain, Canada, Japan, Denmark, Portugal, France
0.2–0.4	Korea, Luxembourg, New Zealand, Germany, UK, Sweden, Finland, Austria, Switzerland
0–0.2	Czech Republic, Slovak Republic, Netherlands

Note: Dollar increase in value of real economic output per dollar of additional government spending.

(Source: OECD Economic Outlook, Interim Report 2009, Figure 3.4)

12.2.4 Cost-benefit analysis of fiscal stimulus

During the Global Financial Crisis (GFC) in the first decade of the twenty-first century, governments of many advanced countries adopted programmes of government spending in an attempt to ward off recession. While these measures were largely successful in limiting the impact of the GFC on GDP, they had to be redoubled at the end of the second decade to counter the more significant impact of the Covid-19 pandemic. Suppose that the government borrows a dollar to hire otherwise unemployed factors of production to produce a consumption good valued at \$B. Notwithstanding our discussion of shadow-pricing in Chapter 5, let us assume that the opportunity cost of the factors of production involved in producing the consumption good, and in subsequent rounds of expenditure through the multiplier effect, is zero. The gross benefit of the project consists of the value placed on the consumption good plus the value of the real increase in Gross Domestic Product (GDP) resulting from the \$1 expenditure, which, in the case of a \$1 increase in government expenditure, is measured by the multiplier, μ. The opportunity cost of the project is the value of the private activity forgone in order to finance the purchase of the \$1 bond. As we saw in Chapter 5, this latter value is the marginal cost of public funds (MCF), which we denote here by λ. Once we introduce the MCF, we need to distinguish between the after-tax increase in real GDP, $(1 - t)\mu$, and the tax revenues, $t\mu$, which should be shadow-priced at the MCF. In summary, the benefits of the \$1 stimulus expenditure are, in dollar terms: B, $(1 - t)\mu$, and $t\mu\lambda$ and the opportunity cost is μ. For the project to break even, with a net benefit of zero, the value of B must be at least equal to: $\lambda (1 - t\mu) - (1 - t)\mu$. Plausible coefficient values are: $\mu = 0.8$, $t = 0.3$, and $\lambda = 1.2$, which yields the value B = \$0.35. Given the chosen coefficient values, this result is interpreted as follows: if the government borrows a dollar to hire factors of production, with zero opportunity cost, to produce a consumption good, the value consumers place on that good has to be \$0.35 or higher for the policy to pass the cost-benefit test. (For further discussion of this model, and an extension to the case of a productivity-enhancing government expenditure, see Dahlby (2009).)

12.2.5 The employment multiplier

Governments often emphasise job creation as a consequence of increased public expenditure. Additional jobs are created to the extent that the increase in national income is real as opposed to nominal. A real increase in income is obtained by drawing otherwise unemployed resources into market activity, whereas a nominal increase simply involves shifting the existing labour force around from one job to another. A "jobs multiplier" can be constructed by using the ratio of the number of jobs to the national income in conjunction with the national income multiplier. For example, suppose that the ratio of employment to income, L/Y, is denoted by k. Using the relations L = kY, and ΔL = k ΔY, the multiplier relationship can be expanded to predict the number of jobs created, providing that the increase in income is real as opposed to nominal:

$$\Delta L = k\Delta Y = k \cdot \left(\frac{1}{\left(1 - b\left(1 - t\right) + m\right)} \right) \Delta G^*$$

where $(k/[1 - b(1 - t) + m])$ is the employment multiplier. For example, if the average wage, Y/L, is $50,000, the value of k is 0.00002 and, if the national income multiplier is 1.45, the value of the employment multiplier in the national income model is 0.000029. This means that, according to the model, an extra $1 million of government expenditure would create 29 new jobs.

12.3 Inter-industry analysis

More detailed estimates of the economic impact of a proposed project can be obtained on an industry basis by using an input–output model which takes into account inter-industry sales and purchases of intermediate inputs. The input–output model measures the impact of any increase in final demand expenditure (an increase in the value of an exogenous variable) on the level of output of each industry. As in the case of the multiplier model, these impacts are not predictions or forecasts, because other events may occur to counter or magnify the industry impacts, but they indicate the size and direction of the effects of the increase in final demand. Governments are interested in this information because of concern for the welfare of particular industries which may exhibit high unemployment or be located in economically depressed regions.

Inter-industry analysis recognises that each industry uses two types of inputs to produce its output: factors of production such as land, labour and capital; and intermediate inputs consisting of goods and services produced by itself or by other industries. The value of the gross output of each industry equals the value of output used as intermediate inputs plus the value of output used in final demand. Inter-industry transactions net out of national income transactions since a revenue for one industry represents a cost to another, or, in terms of the national income model, because all inter-industry transactions occur within the *Firms* sector of Figure 12.1 and do not form part of the circular flow of income. Each industry's contribution to national income is measured by the value of the final demand goods it supplies, or, equivalently by the value of the factor incomes it generates. This contribution is referred to as "value added" since it measures the difference between the value of the goods and services the industry sells and the value of the goods and services it buys.

In modelling inter-industry transactions we distinguish between the value of goods produced by industry i, x_i, and the value of its production that enters final demand, y_i. If the coefficient aij is used to represent the sales of industry i to industry j per dollar's worth of output of industry j, then total sales of industry i to industry j can be represented by:

$$x_{ij} = a_{ij} x_j$$

And total value of output of industry i can be written as:

$$x_i = \sum_j a_{ij} x_j + y_i$$

A disadvantage of the input–output model as described here is the assumption of a set of fixed relationships between inputs and output. A more general approach would allow the input–output coefficients to vary with the level of output, or over time with technological change.

In order to illustrate how the input–output model is constructed, consider an economy consisting of three industries. Using the above relation we can write a set of equations determining the gross output of each industry:

$$a_{11}x_1 + a_{12}x_2 + a_{13}x_3 + y_1 = x_1$$

$$a_{21}x_1 + a_{22}x_2 + a_{23}x_3 + y_2 = x_2$$

$$a_{31}x_1 + a_{32}x_2 + a_{33}x_3 + y_3 = x_3$$

This set of equations can be simplified to:

$$(1-a_{11})x_1 - a_{12}x_2 - a_{13}x_3 = y_1$$

$$-a_{21}x_1 + (1-a_{22})x_2 - a_{23}x_3 = y_2$$

$$-a_{31}x_1 - a_{32}x_2 + (1-a_{33})x_3 = y_3$$

which can be written in matrix form as:

$$(I-A)X = Y$$

where I is the identity matrix (a matrix with '1' on the diagonal and '0' elsewhere), A is the matrix of inter-industry coefficients, a_{ij}, X is the vector of industry gross output values, and Y is the vector of values of final demands for industry outputs. Solving this equation in the usual way (by pre-multiplying both sides by $(I-A)^{-1}$) an equation relating gross industry outputs to final demands can be obtained:

$$X = (I-A)^{-1}Y$$

This equation can be used to predict the effect on industry output of any change in final demand. For example, if a private investment project were to involve specified increases in final demands for the outputs of the three industries, the effects on gross industry outputs could be calculated.

Table 12.2 illustrates the inter-industry structure of a three-industry closed economy. Note that the sum of inter-industry sales plus value added equals the value of the gross output of each industry. Similarly the sum of inter-industry purchases plus final demand sales also equals the value of gross output. The inter-industry coefficients a_{ij} are obtained as the ratios of inter-industry purchases to the value of the gross output of each industry. Expressed in the form of the matrix A, these are:

$$A = \begin{bmatrix} 0.1 & 0.2 & 0.1 \\ 0 & 0.2 & 0.3 \\ 0 & 0 & 0.2 \end{bmatrix}$$

Table 12.2 Inter-industry structure of a small closed economy

Purchases	Sales			Final demand	Gross output
	1	2	3		
1	100	400	300	200	1000
2	0	400	900	700	2000
3	0	0	600	2400	3000
Value added	900	1200	1200	3300	
Gross output	1000	2000	3000		

We now construct the matrix $(I - A)$:

$$(I-A) = \begin{bmatrix} 0.9 & -0.2 & -0.1 \\ 0 & 0.8 & -0.3 \\ 0 & 0 & 0.8 \end{bmatrix}$$

and, using the *Minverse* function in *Excel©*, we obtain the inverse matrix:

$$(I-A)^{-1} = \begin{bmatrix} 1.11 & 0.28 & 0.24 \\ 0 & 1.25 & 0.47 \\ 0 & 0 & 1.25 \end{bmatrix}$$

We can now write down the set of relations $X = (I - A)^{-1} Y$:

$x_1 = 1.11\ y_1 + 0.28\ y_2 + 0.24\ y3$

$x_2 = 1.25\ y_2 + 0.47\ y_3$

$x_3 = 1.25\ y_3$

which relates the value of gross output of each industry to the value of final demand for the products. This set of equations can also be written in the form of changes in values of gross output in response to changes in values of final demand:

$\Delta x_1 = 1.11\ \Delta y_1 + 0.28\ \Delta y_2 + 0.24\ \Delta y_3$

$\Delta x_2 = 1.25\ \Delta y_2 + 0.47\ \Delta y_3$

$\Delta x_3 = 1.25\ \Delta y_3.$

Suppose that a public or private project costs $100, consisting of expenditures of $20, $30 and $50 on the goods and services produced by industries 1, 2 and 3 respectively. Using the above set of equations we can calculate the effects on the values of gross output of industries 1, 2 and 3:

$\Delta x_1 = 42.6,\ \Delta x_2 = 61.0$ and $\Delta x_3 = 62.5$

Note that the total increase in the value of gross output exceeds the $100 increase in the value of final demand because each industry needs to produce enough output to satisfy inter-industry flows as well as final demand.

12.3.1 Inter-industry analysis and the national income multiplier

The estimates of increases in values of industry output obtained above are first-round effects only. In increasing their output levels industries will hire additional services of factors of production and so the incomes paid to the owners of these factors – land, labour and capital – will rise. As predicted by the multiplier model, some of this additional income will be spent, thereby generating second and subsequent rounds of increases in final demands, leading to an eventual increase in national income in excess of the original $100 increase in the exogenously determined level of final demand.

The input–output model can be modified to incorporate multiplier effects by adding a set of equations which play the same role as the consumption function in the closed economy multiplier model. Suppose that the final demands for the outputs of the three industries are given by:

$$y_1 = 0.2 \, y + g_1$$

$$y_2 = 0.3 \, y + g_2$$

$$y_3 = 0.4y + g_3$$

where y is the level of national income and g_i are the autonomous levels of demand for the output of each industry. Recall that it was changes in these levels of autonomous demand, resulting from the expenditures associated with the public or private project, that set off the first round effects on the values of gross output of the three industries. Each of the above equations can be thought of as an industry-specific consumption function, where the coefficient on y plays the role of the coefficient b in the aggregate consumption function of Section 12.2.1, and g_i plays the role of C*. By summing the equations we can see that:

$$\sum_i y_i = 0.9y + \sum_i g_i$$

where 0.9 is the marginal propensity to consume, as in our initial illustrative model of national income determination.

Now consider the system of inter-industry equations Y = (I – A)X, and replace Y by the set of industry-specific expenditure functions and rearrange to get:

$$0.9 \, x_1 - 0.2 \, x_2 - 0.1 \, x_3 - 0.2 \, y = g_1$$

$$0 \, x_1 - 0.8 \, x_2 - 0.3 \, x_3 - 0.3 \, y = g_2$$

$$0 \, x_1 - 0 \, x_2 - 0.8 \, x_3 - 0.4 \, y = g_3$$

We now have a set of three equations in four unknowns, the x_i and y. In order to close the system, we need to add the condition, derived from the concept of circular flow of national income, which determines equilibrium: the value of final demand output should equal the value added in producing this output – the value of factor incomes paid by the industries (see Table 12.2). The values added can be expressed as proportions of the values of the outputs of the three industries:

$$y = 0.9\, x_1 + 0.6\, x_2 + 0.4\, x_3$$

When this equation is added to the other three and the system solved (using Cramer's Rule and the MDETERM function in *Excel©*) for national income, y, the following result is obtained:

$$y = 10(g_1 + g_2 + g_3)$$

In other words, the equilibrium level of national income is ten times the sum of the levels of autonomous demands, i.e. the multiplier is 10, as we saw earlier in our illustrative closed-economy national income determination model. On the basis of the hypothetical values chosen for the coefficients of the model, a change in the level of autonomous demand for any one of the goods would produce a ten-fold change in the level of equilibrium national income.

12.3.2 Inter-industry analysis and employment

In the above discussion the input–output model was used to calculate the effect on national income of an increase in autonomous demand for the three goods produced in the economy. The model can be extended to calculate the effect of the demand increase on employment of factors of production.

   Suppose that the level of input of factor of production i to industry j is given by:

$$v_{ij} = b_{ij}\, x_j$$

where b_{ij} is a coefficient and x_j is gross output of industry j as before. Supposing that there are two inputs, labour and capital for example, total input levels are given by:

$$v_1 = b_{11}x_1 + b_{12}x_2 + b_{13}x_3$$
$$v_2 = b_{21}x_1 + b_{22}x_2 + b_{23}x_3$$

or, in matrix notation, V = BX, where V is the vector of factor inputs, B the matrix of employment coefficients, and X the vector of industry outputs. However, we already know that $X = (I - A)^{-1}Y$ and so we can write $V = B\,(I - A)^{-1}Y$, a set of equations which relates the levels of inputs of the two factors to the levels of final demand for the three goods. Any change in the level of autonomous demand for any of the three goods can be traced back through this system of equations to calculate the effects on the levels of the factor inputs.

   For example, let the matrix of factor input coefficients be:

$$B = \begin{bmatrix} 0.6 & 2 & 1.2 \\ 0.5 & 0.4 & 0.3 \end{bmatrix}$$

We can now solve for V, using the *MMULT* operation in *Excel©*:

$$V = \begin{bmatrix} v_1 \\ v_2 \end{bmatrix} = \begin{bmatrix} 0.6 & 2 & 1.2 \\ 0.5 & 0.4 & 0.3 \end{bmatrix} \begin{bmatrix} 1.11 & 0.28 & 0.24 \\ 0 & 1.25 & 0.47 \\ 0 & 0 & 1.25 \end{bmatrix} \begin{bmatrix} y_1 \\ y_2 \\ y_3 \end{bmatrix}$$

to get the following result:

$$v_1 = 0.67y_1 + 2.67y_2 + 2.58y_3$$

$$v_2 = 0.56y_1 + 0.64y_2 + 0.68y_3$$

Now suppose that input v_1 represents labour, and that the increases in the levels of y_i are $20, $30 and $50 respectively, then the increase in employment – the number of jobs created in the first round – is given by:

$$\Delta v_1 = 0.67*20 + 2.67*30 + 2.58*50 = 222.5$$

The above calculation takes account of first round effects only. If second and subsequent round effects are also to be considered, this can be done by solving the input–output multiplier model for the levels of final demand, y_i, for both the original and the new levels of autonomous expenditures, g_i. The two sets of y_i estimates can then be used to calculate two sets of estimates of employment levels of the two factors, v_i, and the difference between these values would indicate the extra employment resulting from the original change in the levels of autonomous expenditures and the subsequent multiplier effects in the economy.

12.4 General equilibrium analysis

A general equilibrium in an economy is a situation in which all markets clear in the sense that, at the prevailing prices, the quantity supplied of each good or factor of production equals the quantity demanded. This implies that, at the equilibrium prices, no economic agent has any incentive to change the quantity they supply or demand of any good or factor. In principle, a situation of equilibrium will continue until there is a change in the value of some exogenous variable, such as a tax rate or level of autonomous government expenditure. A computable general equilibrium (CGE) model can be constructed to determine the equilibrium values of prices and quantities traded in the economy, and to calculate the changes in these values which would result from some change in an exogenous variable. For example, a large project could involve a significant change in the exogenously determined level of investment, which, in turn, could cause changes in the values of variables of interest to the policy-maker – employment, the wage level, the exchange rate, and so on. A CGE model could be used in this way to assess the economic impact of a large project.

It is beyond the scope of this discussion to consider in detail the structure of CGE models. However, simple CGE models can be constructed from information contained in the input–output model and it is useful to consider in principle how the model described in the previous section of this chapter could be developed into a CGE model. Our simple input–output model was of a closed economy producing three goods and using two factors of production. The model could be solved for the equilibrium values of the two inputs and the three outputs. In a general equilibrium model values are calculated as the product of two variables – price and quantity. Equilibrium in such a model with three goods and two factors consists of five equilibrium prices and five equilibrium quantities. In order to solve for the values of these 10 variables, a system of 10 equations is required.

Three of the required equations can be obtained from the input–output coefficients by assuming that competition in the economy ensures that total revenue equals total cost; this means that for each industry the value of its sales equals the sum of the values of the intermediate inputs and factors of production used to produce the quantity of the good sold. Two more equations are obtained from the factor markets where factor prices have to be set at levels at which the quantities of factors supplied by households equal the quantities demanded by industries; this means, for example, that the market wage has to be high enough to ensure that the total quantity of labour supplied is sufficient to meet the demands of the three industries. Three equations are obtained from the markets for goods in final demand; goods prices must be at levels at which the quantities demanded by households are equal to the final demand quantities supplied by industries. An income equation is required to ensure that the income households receive from the supply of factors of production to industries is equal to their expenditure on final demand goods. Lastly, since the CGE model determines relative prices only, the price of either one good or one factor must be set at an arbitrary level: for example, the wage could be set equal to 1, or the price level, as measured by some price index, could be set equal to an arbitrary value. In either event, this normalization adds one more equation to the model and completes the system of 10 equations required to solve for the 10 variables.

It will be evident that the model described above is very limited in that it contains no government or foreign sector. However, our purpose here is simply to convey the flavour of such models rather than to provide a recipe for constructing one. The solution to a CGE model is often expressed in terms of a set of proportionate changes in the values of variables of interest (solving for rates of change is a way of linearising a non-linear model). For example, in the analysis of the impact of a significant exogenously driven expansion of an export sector of the economy the model might predict that the exchange rate would appreciate (the value of the domestic currency would rise in terms of foreign currency), the domestic price level would fall, and the wage rate would fall. The first step in interpreting these results is to ask whether they are consistent with economic theory: the exchange rate rises because of an increase in exports; prices of imported consumer goods fall in domestic currency terms, leading to consumers substituting imports for domestically produced goods; the wage falls because of a net decline in employment owing to the fact that the capital-intensive exporting sector is unable to absorb all the labour released by the shrinking domestic consumer goods sector. Since the direction of changes predicted by the CGE model is consistent with economic theory, we can have some confidence in the results and turn our attention to the magnitudes of the changes predicted.

12.5 Case study: the impact of the ICP Project on the economy

An economic impact analysis would estimate the effect of the proposed ICP project on Thailand's gross national product. As noted earlier, the gross impact of the project may be larger than its net impact to the extent that project outlays reflect expenditures that are simply displaced from other areas of the economy. We will consider the gross impact first, and then the net impact of the project.

It will be recalled from the discussion of the national income multiplier that money which leaks out of the circular flow of income as savings and taxes has no further impact on the level of national income. A large proportion of the ICP project expenditures is in this category since it represents imports of capital inputs, such as machinery, and inter mediate inputs, such as cotton. Only expenditures on inputs sourced from within the economy will have multiplier effects on national income through their effects on consumers' incomes and expenditures. For example, receipt of wages will stimulate spending by workers; receipt of rent will stimulate spending by landlords; and receipt of service fees by banking and insurance companies, and water and power utilities, will, eventually, stimulate spending by shareholders of these institutions. Applying the multiplier to the total of these categories of expenditure in each year will provide an estimate of the gross impact of the project in that year.

The net impact of the project on GNP will be lower than the gross impact because some of the above expenditures are simply displaced from other sectors of the economy. For example, the analysis of the opportunity cost of the project inputs established that the wages paid to skilled labour and the rent paid to the landlord would have been paid in the absence of the project. Any gross multiplier effects attributed to these expenditures cancel out in the analysis of net impact. The most significant net impact of the project on the economy probably results from the wages paid to unskilled labour. The unskilled wage bill could be adjusted by the national income multiplier to provide an estimate of this impact.

As discussed earlier, it is important not to confuse "net impact" with "net benefit". The first round net benefits resulting from employment of each unit of unskilled labour have already been included in the estimate of Referent Group benefits as the difference between the wage and the shadow-price. There could be additional net benefits to the extent that additional spending by unskilled labour generates additional demand for the services of unskilled labour, additional wages, and hence further spending. However, as far as the case study is concerned, these additional net benefits are likely to be relatively small and it is unlikely that the results of this kind of impact analysis would affect the decision about the project.

Because of the importance of the distinction between net benefit and net impact, the reporting of impact effects derived from multiplier, input–output or CGE models should not be integrated with the cash flow analysis and social CBA results. An appropriate format is a separate statement of impact assessment results, including an assessment of both their expected longevity/duration and their net impact relative to the scenario without the project.

12.6 Further reading

A good discussion of national income determination and the interindustry model is contained in T.F. Dernburg and J.D. Dernburg, *Macroeconomic Analysis: An Introduction to Comparative Statics and Dynamics* (Reading, MA: Addison-Wesley, 1969). For more on fiscal stimulus, see B. Dahlby, "Once on the Lips, Forever on the Hips", C.D. Howe Institute Backgrounder No. 121, December 2009.

Appendix to Chapter 12

The annual economic impact of the Virginia Creeper Trail

In the appendix to Chapter 8 a study by Bowker et al. (2007) was described in which the annual benefits of the Virginia Creeper Trail (VCT) were calculated in the form of an area of consumer surplus under a demand curve estimated from sample data obtained from users of the trail. Non-primary purpose trail users were omitted from this analysis in order to avoid attributing to the VCT consumer surplus generated by their primary purpose recreation facility. In the estimation of economic impact, however, all expenditures by tourists, whether primary purpose or not, contribute to local economic activity.

In the consumer surplus calculations we identified the Referent Group as residents of Grayson and Washington counties and this focus will be maintained in the analysis of economic impact. In consequence, expenditures by local residents using the trail will not be considered in the impact analysis on the grounds that, in the absence of the trail, these sums would have been spent on accessing some other recreational facility in the locality, with the result that the existence of the VCT has no net effect on the expenditures of residents. It should be noted that in the absence of the VCT, tourists (non-residents) could also be assumed to make comparable expenditures, with corresponding economic impact, elsewhere, perhaps in their home jurisdictions, and it could be concluded that the VCT makes no net contribution to national output. However, from the viewpoint of the Referent Group, it is the local impact which is relevant.

The sample survey of users provided information on tourist expenditures on accommodation, food, transport, equipment rentals, guides, use fees and miscellaneous items in the area within 25 miles of the trail, which we take to be the region of interest to the Referent Group. The average expenditure values for primary purpose day users (PPDU), non-primary purpose day users (NPDU), primary purpose overnight users (PPON), and non-primary purpose overnight users (NPON) are reported in Table A12.1. To calculate expenditures associated with use of the VCT by non-primary purpose users (NPDU and NPON), a proportion of total trip expenditure was assigned: for NPDU the proportion was the time spent on the trail as a proportion of the day, and for NPON it was the time spent on the trail as a proportion of time spent in the region.

In order to calculate regional economic impact, the expenditures reported in Table A12.1 could be treated as changes in autonomous consumption expenditure, ΔC^*, in the multiplier models described in Section 12.2 of this chapter. Alternatively, they could be treated as changes in final demand expenditures on a range of goods and services, Δy_i, in the inter-industry model of Section 12.3, and this is the approach adopted by Bowker and his colleagues.

As discussed in Section 12.3, changes in final demand (termed *direct* impacts by Bowker et al.) result in changes in industry outputs (termed *indirect* impacts); for example, additional spending in restaurants is matched by additional demand for the output of food and other industries, as well as the additional services provided by the restaurant sector itself. In order to produce additional output these industries need to hire additional factors of production, such as labour and business premises, as discussed in Section 12.3. The additional value-added, as computed by the input–output model, takes the form of higher incomes to local labour, property owners and tax revenues to government, and additional spending of this additional income causes further rounds of output increases (termed *induced effects*) as

Table A12.1 Average tourist expenditures in the VCT region per person per trip (2003 US$)

Category of user	PPDU	NPDU	PPON	NPON
Expenditure/person/trip	17.16	12.31	82.10	7.02

described in Section 12.2. The overall effect is summarised by the regional multiplier, as described in Section 12.3.1.

When the per person-trip expenditures reported in Table A12.1 are multiplied by the corresponding use data reported in Table A8.1, an estimate of the total increase in direct expenditure in the region of $1.17 million is obtained. When this figure is adjusted by the output multiplier of 1.34 obtained from the input–output model, the annual regional economic impact of the VCT (in 2003 US dollars) is estimated to be $1.56 million. As can be seen from Section 12.3.2, the input–output model can also provide estimates of the changes in the value added (incomes) of the factors of production used to produce the additional industry outputs. Bowker et al. report the following estimates of additional regional factor incomes: labour, $610,372; property, $126,098; and indirect business taxes, $105,103.

As discussed in Section 12.3.2, the additional factor income to labour can be converted, by means of the employment multiplier, to an estimate of the additional number of jobs generated by the additional direct expenditures. In the case of the VCT the estimate is 27 positions. These "jobs" could represent existing employees working longer hours, otherwise unemployed workers becoming employed, or labour migrating to the region and, perforce, becoming members of the Referent Group.

In the discussion of economic impact we were careful to distinguish impact from the net benefit measured by Efficiency or Referent Group Analysis. What, if any, are the net benefits associated with the economic impacts described above? First, as already noted, the impacts are local and are likely to be offset by equivalent reductions in other regions. Second, there will be opportunity costs of inputs associated with the provision of additional outputs. For example, the extra working hours required to earn the additional $610,372 in wages come with an opportunity cost in the form of forgone leisure, if in no other form. If pressed for an answer, we might make the assumption that half of the additional labour income is offset by opportunity cost, that extra property income represents pure rent, and that additional local tax revenues are a net benefit to the region. On that basis we might add $0.54 million to the $1.4–2.4 million already identified in Chapter 8 as the annual gross benefits to members of the Referent Group who use the Virginia Creeper Trail.

Exercises

1   The annual Gross Regional Product (GRP) of the Northern Territory consists of private consumption and investment goods, government goods and services and exports of goods and services. The Territory imports significant quantities of goods and services.

   i   Develop a simple model that can be used to determine the equilibrium level of regional income in the Territory.
   ii   Use your model of regional income determination to work out a formula for the regional income multiplier in the Territory.
   iii How would the value of this regional multiplier compare with the size of the national income multiplier? Explain.

iv   Explain how you would use your estimate of the regional income multiplier to calculate the impact of an advertising campaign aimed at attracting tourists to the Territory.

2   An advertising campaign is expected to increase expenditures of foreign tourists in Far North Queensland (FNQ) by $100 million per annum. You have been asked to estimate the economic impact of this increase on the FNQ economy. You are told that out of any extra dollar earned by FNQ residents, they pay 30 cents in taxes. Out of any dollar increase in disposable (after tax) income, they save 20 cents and spend 30 cents on goods produced outside of FNQ. It is estimated that 30 cents of every dollar spent by foreign tourists accrues as income to non-residents of FNQ, such as national hotel chain headquarters.

i   Develop and explain a simple model of the FNQ economy which can be used to estimate the impact of the advertising campaign on FNQ's Gross Regional Product (GRP). (Hint: follow the procedure described on page 339 to derive a multiplier formula which you will find differs slightly from the formula derived there.)
ii   Use the data provided above to estimate the FNQ multiplier and calculate the size of the increase in FNQ's GRP.
iii   If the FNQ GRP per capita is $50,000, what is the likely size of the employment effects in FNQ of the advertising campaign?

3   Given the structure of inter-industry relations reported in Table 12.2 use *Excel*© to cal-culate the effect on gross industry outputs of $100 of public expenditure, consisting of expenditures of $20, $30 and $50 on the products of industries 1, 2 and 3 respectively.
4   Suppose that the inter-industry structure of a three-commodity, two-factor closed economy is described by the following table:

Inter-industry structure of a small closed economy

	Sales			*Final demand*	*Gross output*
	1	*2*	*3*		
Purchases					
1	300	200	200	300	1000
2	0	300	700	1000	2000
3	0	0	600	2400	3000
Value added	700	1500	1500	3700	
Gross output	1000	2000	3000		

Work out the first-round effects on industry outputs of an exogenous $100 increase in final demand for the output of industry 2.

13 Writing the cost-benefit analysis report

13.1 Introduction

If cost-benefit analysis is to assist in the decision-making process the analyst must be able to convey and interpret the main findings of the CBA in a style that is user-friendly and meaningful to the decision-makers. The analyst should never lose sight of the fact that the findings of a CBA are intended to inform the decision-making process. CBA is a *decision-support* tool, not a *decision-making* tool. To this end, it is imperative that the report provides the decision-maker with information about the project and the analysis which is directly relevant to the decision that has to be made, and to the context in which it is to be used.

There is no blueprint for report writing, as every project or policy decision will be different in various respects, as will the decision-making context and framework in which the analyst will be operating. Good report writing is essentially an art that can be developed and refined through practical experience. The purpose of this final chapter is to identify what we consider to be some of the key principles the analyst should follow when preparing a report and to illustrate how these might be applied in the drafting of a report on the ICP Project we have used in previous chapters as a case study.

As an illustration, we have selected the report on the ICP Project prepared by one of our postgraduate students at the University of Queensland, Australia, who has kindly given us permission to use it in this way. In the sections below we explain why we consider this study to be an example of good CBA reporting. The report, reproduced as an appendix to this chapter, deals with all seven scenarios described in Chapter 4 regarding the location and tax treatment of the ICP Project. The complete report included, in the form of an appendix, spreadsheet solutions to all seven scenarios but, to save space, we present only the spreadsheet dealing with the base case; spreadsheets dealing with the remaining scenarios can be found on the Support Material website. The report is reproduced in its original form, with the exception of a few minor editorial corrections and the spreadsheets referred to above. It consists of an *executive summary*, four main sections – *introduction*, *methodology*, *analysis* and *conclusion* – and the *appendix* containing the summaries of variables and results and the printouts of the spreadsheet tables.

Before commenting on this report, we should note that the starting point for the exercise, the ICP case study problem appended to Chapter 4, is already some way along the path which the analyst must follow. As in any area of investigation, framing the questions to be answered is an important first step. The ICP case study asked for an analysis of seven different scenarios involving two different locations and various tax regimes. These scenarios would have been established in discussions between the analyst and the client about the terms

DOI: 10.4324/9781003312758-13

of reference for the study. It must not be assumed that the analyst plays a passive role in developing the terms of reference. As noted in Chapter 1, the cost-benefit analyst may play an important role in project design by asking pertinent questions about the options which are available. This significant contribution to the decision-making process is sometimes referred to as *scoping* the project.

13.2 Contents of the report

13.2.1 The executive summary

The executive summary should not be written until the analyst has completed a draft of the rest of the report. The executive summary is, as its title indicates, written for the decision-makers' benefit. These are usually senior civil servants and/or politicians who are, most likely, not technical experts in the field, and who will, most probably, delegate reading the rest of the report to their staff. It is reasonable to assume that they will not want to read more than a page or two. This abbreviated report manages to highlight the main issues, summarise the main findings, and make a recommendation in one page. The executive summary begins with a succinct description of the project, its purpose, the players, and the decision to be made. It then describes the options, discusses the form of analysis undertaken and the main findings, and makes some recommendations for the Minister's consideration. In this instance, the writer has chosen not to include a summary table, but has instead reported some of the results in the text. Given that this report will constitute a major input into a negotiation process between the government of Thailand and ICP, the report could have incorporated a summary table showing the trade-off between ICP's private net benefit and the Referent Group net benefits over the seven project options, perhaps at a single dis-count rate. As the subsequent analysis shows, in the ICP case, this trade-off summary is a clear example of a zero-sum-game where each additional dollar gained by ICP represents an equivalent loss in net benefit by the Referent Group.

13.2.2 The introduction

Bearing in mind that the main body of the report is written independently of the execu-tive summary, the introduction provides a brief summary of the report, the players, and the options to be analysed and compared. This section should not contain too much detail about the project and the information used to undertake the analysis, but should rather aim to provide the reader with a fairly general description of what the project is about and what decision the analysis intends to inform.

13.2.3 The methodology

While it is imperative that any report of this sort should provide an account of the meth-odology used in the analysis, it is not necessary for the report to contain a detailed account of what CBA is about. It should be assumed that the reader is already familiar with the concept and principles of CBA. However, as conventions, terms and approaches can vary quite considerably, the writer of our sample report has been careful to point out in sub-section 2.1 exactly what is meant by *Project (or Market)*, *Private*, *Efficiency*, and *Referent Group Analysis*, in anticipation that the reader will need to have this information to follow the rest of the report. Sub-section 2.2 explains which decision criteria are used and over what range the discount rates are allowed to vary in the sensitivity analysis. The author is also mindful to refer the reader (sub-section 2.3) to the *Key Variables* table in the spreadsheet (attached as Appendix Table A13.6) where all the input data and assumptions are contained.

The main assumptions on which the findings of the analysis rest are summarised in sub-section 2.4. It should be noted that this section does not attempt to summarise each and every piece of input information but instead focuses on those assumptions that are perhaps most open to the analyst's interpretation and discretion. This includes assumptions underlying the estimation of shadow-prices for the Efficiency Analysis, as well as other unknowns such as the projected exchange rate, possible future tax changes, and so on.

Since the analysis of the ICP project was an exercise there was no need to comment on the quality of the data, but in a real-world situation it is worth detailing the sources of the data used in the analysis and reporting any verification process that was followed. This information could be included in a data appendix.

13.2.4 The analysis

This section presents the findings of the analysis. It does not discuss in detail how the cash flows were derived, as this detail is contained in the spreadsheet tables in the appendix. Tables A13.1 to A13.4 summarise the NPVs and IRR for each project option, in terms of the four types of CBA we undertake: Market (Project), Private, Efficiency, and Referent Group. In terms of the decision to be made in this case, the two analyses that are really important are the Private CBA which shows how much ICP stands to gain or lose under each option, and the Referent Group CBA, which is what matters from the Government of Thailand's perspective. The writer has chosen, for sake of completeness, to summarise the results for the Project (Market) and the Efficiency Analyses as well. These are not really important to the decision in this case and could well have been omitted, or perhaps mentioned briefly in passing. However, in general, their relevance will depend on the context. If, for instance, we were reporting on a project that is being funded by an international agency such as the World Bank or a multilateral regional bank such as the Asian or African Development Bank, the agency could well be interested in knowing how the project performed in global efficiency terms, in which case reporting also on the Efficiency CBA would be required.

It should also be noted that a sensitivity analysis has been undertaken under each option, and it has been established, and reported by means of informative charts (see Figures A13.1 and A13.2), that the ranking of options from both ICP and the government's perspective is not unambiguous. At a real discount rate of around 20% there is a switching of the ranking of options, which could be significant for ICP's decision as 20% is not outside the possible bounds of a private investor's required rate of return.

13.2.5 The conclusion

Note that the conclusion does not merely summarise what has already been said elsewhere in the report. It synthesises the main findings, discusses their relevance and explores their possible implications for the government's negotiations and decision-making. Attention is also drawn to areas where further, more detailed work might be undertaken before a final decision is reached. In this instance the writer has also made a fairly concrete recommendation in favour of Option 2, although in the context of her other recommendations, it is qualified and should not be interpreted as the final word.

13.3 Other issues

Because the ICP case study is designed as an exercise to be undertaken on the basis of Chapters 1–6, the report does not deal with many of the issues covered in Chapters 7–12: the possibility of changes in output or input prices as a result of the project; non-market valuation; risk analysis; the shadow-price of foreign exchange; income distribution effects; and economic impact analysis. These issues are dealt with in the appendices to Chapters 7–12. In a real-world situation, all these issues would need to be explored with the client in the scoping process and sources of information identified if they were to be included in the analysis.

Appendix to Chapter 13

Report on International Cloth Products Ltd: Spinning Mill proposal prepared by Angela Mcintosh[1]

Executive summary

International Cloth Products Ltd (ICP), a foreign-owned textile firm, proposes to invest in a spinning mill in Thailand at the end of 1999. The mill will have a capacity of 15 million pounds (lbs) of yarn, which will satisfy ICP's current demand for 10 million pounds, as well as demand from other Thai textile firms of approximately five million pounds.

ICP is seeking a concession from the Government to induce it to invest in Thailand. ICP has suggested that it is likely to invest in the Bangkok area if it receives either duty-free entry of cotton *or* a ten-year exemption from profits tax. The Government would prefer the mill to be located in Southern Thailand, due to the policy to disperse industries to "deprived areas". ICP will consider locating the mill in Southern Thailand if it is given duty-free entry of cotton *and* a ten-year exemption from profits tax.

The report considers the following options:

- Option 1: Bangkok – No Concessions
- Option 2: Bangkok – No Duties
- Option 3: Bangkok – No Profits Tax
- Option 4: Southern Thailand – No Concessions
- Option 5: Southern Thailand – No Duties
- Option 6: Southern Thailand – No Profits Tax
- Option 7: Southern Thailand – No Duties, No Profits Tax.

Using social cost-benefit analysis methodology, each option can be ranked according to performance from: the market perspective; ICP's perspective; the economy as a whole; and from Thailand's perspective. Based on the data from the analyses, Options 4 and 7 can clearly be rejected. Option 4 offers the best outcome for Thailand, but offers ICP no incentive and is ranked last from ICP's perspective. Conversely, Option 7 (ICP's highest ranked Option) will be disastrous for Thailand, with negative net present values for all discount rates.

Option 2 warrants serious consideration. This Option (Bangkok – No Duties on Imported Cotton) satisfies ICP's request to receive one concession if it invests in the Bangkok area.

It is ranked fourth from both ICP's and Thailand's perspectives and, though in terms of the economy as a whole, it is not as efficient as any of the Southern Thailand options, the differences between Options 1–3 and Options 4–7 in efficiency terms are relatively small (IRR 22% compared to IRR 23%).

Unless ICP can be convinced to accept Option 5 (ranked fifth from ICP's perspective and third from Thailand's perspective), none of the Southern Thailand options seems likely, unless a different set of alternatives, including subsidization of any investment in Southern Thailand, is considered. However, to determine whether or not other options are viable would require further, detailed analysis and direction from the Minister. Therefore, Option 2 is the recommended Option.

1 Introduction

This report examines a proposal by International Cloth Products Ltd (ICP) to invest in a spinning mill in Thailand at the end of 1999. ICP is a foreign-owned textile firm that manufactures cloth products (textile prints, finished and unfinished cloth) in the Bangkok area of central Thailand. It currently imports 10 million pounds (lbs) of cotton yarn each year. However, if the proposed spinning mill (with a capacity of 15 million lbs of yarn) goes ahead, ICP will be able to supply its own demand, as well as approximately five million lbs of yarn to other textile firms in Thailand. There appears to be an existing market for the excess supply, as Thailand's textile industry currently imports around 16.5 million lbs of cotton yarn per annum.

ICP is seeking a concession from the Government to induce it to invest in Thailand. ICP has suggested that it is likely to invest in the Bangkok area if it receives either duty-free entry of cotton *or* a ten-year exemption from profits tax. Although ICP has indicated its preference to locate the mill adjacent to its existing plant in the Bangkok area, it will consider locating the mill in Southern Thailand (consistent with the Government's policy to disperse industries to "deprived areas"), if it is given both concessions – that is, duty-free entry of cotton *and* a ten-year exemption from profits tax. To accommodate the variety of alternatives open to the Government and ICP regarding the proposed spinning mill, the following options have been considered:

- Option 1: Bangkok – No Concessions (duties and profits tax paid by ICP)
- Option 2: Bangkok – No Duties (i.e. duty free entry of cotton)
- Option 3: Bangkok – No Profits Tax
- Option 4: Southern Thailand – No Concessions
- Option 5: Southern Thailand – No Duties
- Option 6: Southern Thailand – No Profits Tax
- Option 7: Southern Thailand – No Duties, No Profits Tax.

2 Methodology

2.1 SOCIAL COST-BENEFIT ANALYSIS

The social cost-benefit analysis methodology is designed to provide information about the level and distribution of the costs and benefits of each identified alternative. Four different points of view are taken into account:

1 *Project (Market) Analysis:* The project cost-benefit analysis values all project inputs and outputs at private market prices (does not take account of tax, interest on loans etc.) and determines whether the project is efficient from a market perspective.

2   *Private Analysis*: The private cost-benefit analysis examines the proposed project from the private firm's perspective by taking the Project Analysis and netting out tax, interest and debt flows.

3   *Efficiency Analysis*: The Efficiency Analysis is similar to the Project Analysis, except that "shadow-prices" are used where applicable. Shadow-prices are used to adjust the observed, imperfect market price to measure marginal benefit or marginal cost. They are not the prices at which project inputs or outputs are actually traded. For example, minimum wages represent a labour market imperfection. If the labour input is additional to the current supply (for example, taken from the existing, unskilled labour pool), then it is priced at the "shadow wage". Alternatively, if the labour input is reallocated from another market (for example, skilled labour), then the market wage is used. Another example is taxes, which also distort the market, and need careful treatment in the Efficiency Analysis. The Efficiency Analysis determines whether the project is an efficient allocation of resources across all groups impacted by the proposal – if the money measure of the benefits of the proposal exceeds the amount required to theoretically compensate those who are adversely affected by the project, then the project is deemed to be an efficient allocation of resources (a potential Pareto improvement).

4   *Referent Group Analysis*: The Referent Group is comprised of the stakeholder(s) deemed to be relevant. For this project the Referent Group can be broadly categorised as Thailand. The Referent Group Analysis also examines the distributional effects of the proposal on the members of the Referent Group. The Referent Group for the proposal at hand can be broken down into the following sub-groups: the domestic bank, the Thai Government, domestic labour likely to be engaged on the project, the Electricity Corporation of Thailand, and the domestic insurance company. There are two methods of determining Referent Group impacts. Method A involves identifying the costs and benefits for each member of the Referent Group, while method B involves subtracting the non-Referent Group net benefits from the Efficiency Analysis. If the results of method A equal those for method B, then the analysis is shown to be internally consistent.

2.2 DISCOUNTED CASH FLOWS, INTERNAL RATE OF RETURN

The four components of the analysis (Project (Market), Private, Efficiency and Referent Group) have been assessed using discounted cash flow techniques: see Appendices. Benefits and costs, measured as cash flows, have been projected over each year of the project (from 1999–2009) and then discounted back to present value (1999) amounts using real discount rates of 4%, 8%, 12%, 16% and 20%. These different discount rates allow a sensitivity analysis to be undertaken of the net present values for each discount rate. As the discount rates increase, the NPVs decrease. If the net present value (NPV) is greater than zero, then for the relevant discount rate, the project is worth undertaking.

The internal rate of return (IRR) has also been determined for each component of the analysis. The IRR represents the discount rate that reduces the NPV to zero. If the IRR exceeds the discount rate, then the project should usually be accepted. This issue is discussed further in Section 3.2 (Private Analysis).

2.3 KEY VARIABLES

Key variables for each option are set out in the Appendices and these variables have been used to analyse each option. Every item identified in these Appendices may be varied and may, indeed, change over time (see Section 2.4 Assumptions).

Options 1 and 4 set out the base cases for Bangkok and Southern Thailand, respectively, in that they have all duties and taxes included. Options 2 and 5 have the duty on imported raw cotton reduced to 0% (compared to 10%) in Options 1, 3, 4 and 6. Options 3 and 6 have the tax on profits reduced from 50% to 0%. Option 7 examines the costs and benefits for each component of the analysis should ICP locate in Southern Thailand with concessions on both duties and profits tax granted (cotton import duty 0% and profits tax 0%).

2.4 ASSUMPTIONS

For any cost-benefit analysis, a number of assumptions need to be made. For this proposal, the following key assumptions have been made.

Operating costs   Operating costs have been assumed to remain constant over the life of the project. This may or may not be the case. For example, the cost of imported cotton may change with changes in world prices and other market forces. Similarly, the cost of labour (skilled and unskilled), electricity, water, spare parts, insurance and rent may all change over time. However, undertaking a sensitivity analysis helps to mitigate this potential problem.

Shadow-prices   The shadow wage for labour has been assumed by the Government to be 50% of the market wage in Bangkok and zero in Southern Thailand. This may change over time, especially if the Government's "deprived area" policy is successful (e.g. shadow wages in deprived areas could increase).

The variable cost to the Electricity Corporation of Thailand of supplying power to ICP for the proposed mill is estimated to be 50% of the standard rate. In the absence of information on the supply of electricity to Southern Thailand, this shadow rate has also been used to cost electricity in the efficiency analyses for Options 4–7 (the Bangkok standard rate of electricity charge has also been used for the other components of the analysis). Further information on alternative electricity providers could be found through a company search, or Government contacts. In terms of supply costs, the relevant supplier could be approached to obtain the information or, alternatively, annual reports may be a useful source of information.

Insurance has been shadow-priced at zero, as the insurance risk has been assumed to be widely spread. This may change over time. An examination of workplace health and safety incidents could be undertaken to find data on accidents in spinning mills to determine whether employee insurance premiums are likely to change. Additionally, insurance costs and shadow-prices have been applied equally across all options. However, Southern Thailand may have peculiarities that warrant different rates being applied. Further discussions with insurance companies could be undertaken to determine whether insurance premiums in Southern Thailand are higher or lower than those in Bangkok. For example, Southern Thailand may have different rates of crime and weather conditions, factors which may affect an insurance premium.

Interest on loan   ICP intends to finance 75% of imported spindles and equipment under a foreign credit. The interest component of this loan (8%) has not been calculated on a reducing balance.

Profits tax, import duties   Like profits tax, import duties on spindles, ancillary equipment, working capital and imported cotton are assumed to be fixed for the length of the project. However, these rates may change over time. It would be prudent to check with the

Government's Treasury to determine whether increases or decreases in taxes/duties are planned or likely over the coming decade, including the introduction of new taxes/duties. Of course if the Government changes, any forecasted rates may also change.

Depreciation The rules for depreciation are assumed to be fixed (at a rate of 10%). However, it may be useful to check with the appropriate accounting body whether any review of accounting standards is planned for the near future and/or if different depreciation methods are planned to be introduced during the life of the project.

Foreign exchange rate The exchange rate is assumed to be fixed at Bt.44 = US$1 for the life of the project. This assumption is correct if exchange rates are fixed in Thailand. If, on the other hand, Thailand has a floating exchange rate, then depending on monetary policy and global impacts, the rate may or may not fluctuate. Again, the sensitivity analysis deals with this potential problem.

Sunk costs ICP currently incur 7.5 million Baht per annum in overhead cost. This can be considered a "sunk cost" and has not been included in the analysis.

Production capacity It is assumed that ICP will operate at 50% capacity for the first year and 100% for every year after that. An examination of ICP's business plan and project implementation plan would be prudent to verify these assumptions.

3 Analysis

Each of the four components of the social cost-benefit analysis is examined separately below, for each option. All values are in millions of Baht. The information in this section is taken from the Appendix Tables. Summary information on all components and options is contained in Appendix Table A13.5, Appendix Table A13.6 and Appendix Table A13.7.

3.1 PROJECT (MARKET) ANALYSIS

From the market perspective (see Table A13.1), Option 2 is preferred over the other options, with Options 5 and 7 ranked equal second. This is the case for both the NPV and IRR calculations. NPVs for all options are positive, except for Options 4 and 6 when a discount rate of 20% is used. This is because the IRR for those Options is 17%.

Table A13.1 Market Analysis

Discount Rates (for NPV):		4%	8%	12%	16%	20%	RANK	IRR	RANK
Bangkok	Option 1	405.3	259.8	152.0	70.7	8.3	3	21%	3
	Option 2	568.5	393.4	263.3	164.8	89.0	1	26%	1
	Option 3	405.3	259.8	152.0	70.7	8.3	3	21%	3
Southern	Option 4	341.4	200.9	96.9	18.5	−41.7	4	17%	4
Thailand	Option 5	504.5	334.5	208.2	112.6	39.1	2	23%	2
	Option 6	341.4	200.9	96.9	18.5	−41.7	4	17%	4
	Option 7	504.5	334.5	208.2	112.6	39.1	2	23%	2

3.2 PRIVATE ANALYSIS

Not surprisingly, Option 7 is the most favoured option for ICP (see Table A13.2). This is because the costs of relocating to Southern Thailand are outweighed by the benefits of not having to pay profits tax or import duty on imported cotton. Option 3 is the next best option for ICP. Deciding on the third ranking is different when comparing the NPV and IRR results for Options 2 and 6. This is due to a phenomenon called "switching", which occurs because the NPV curves of these mutually exclusive projects cross one another, as illustrated in Figure A13.1.

It is not possible to determine precisely whether ICP would prefer Option 2 over Option 6 and vice versa because information on the cost of capital has not been supplied. If the cost of capital is greater than 20.27%, then Option 2 would be preferred over Option 6; and if the cost of capital is less than 20.27%, then Option 6 would be preferred to Option 2. The easiest way of obtaining this information would be to approach ICP directly. However, in the absence of that information, the NPV decision rule is preferred over the IRR rule and so Option 6 is preferred over Option 2. Just as Option 7 was not a surprising first choice, neither is Option 4 for last place, with additional costs, no concessionary benefits and negative NPVs at discount rates of 16% and 20%.

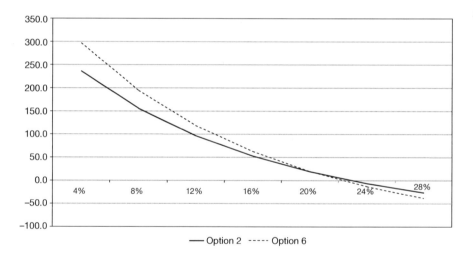

Figure A13.1 Switching Options 2 and 6: NPV Schedules.

Table A13.2 Private Analysis

Discount Rates (for NPV):		4%	8%	12%	16%	20%	RANK	IRR	RANK
Bangkok	Option 1	154.5	88.7	40.8	5.4	−21.1	6	17%	6
	Option 2	236.1	155.4	96.4	52.5	19.2	4	23%	3
	Option 3	361.2	252.9	173.3	113.8	68.6	2	29%	2
Southern	Option 4	119.1	53.7	6.2	−28.9	−55.2	7	13%	7
Thailand	Option 5	200.7	120.5	61.8	18.2	−14.8	5	18%	5
	Option 6	296.7	194.0	118.6	62.4	19.8	3	22%	4
	Option 7	459.9	327.5	229.9	156.5	100.5	1	31%	1

3.3 EFFICIENCY ANALYSIS

All Options involve efficient allocation of resources across all groups impacted by the proposal, with the Southern Thailand Options ranking slightly higher than the Bangkok Options (see Table A13.3). Of course, this analysis does not take into account any distributional effects.

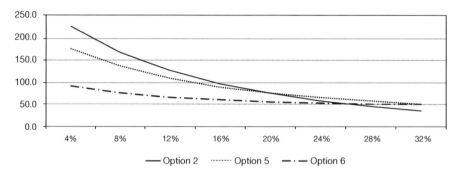

Figure A13.2 NPV schedules.

Table A13.3 Efficiency Analysis

Discount Rates (for NPV):		4%	8%	12%	16%	20%	RANK	IRR	RANK
Bangkok	Option 1	408.9	269.8	166.6	88.7	28.9	2	22%	2
	Option 2	408.9	269.8	166.6	88.7	28.9		22%	
	Option 3	408.9	269.8	166.6	88.7	28.9		22%	
Southern	Option 4	408.1	271.4	170.1	93.6	39.4	1	23%	1
Thailand	Option 5	408.1	271.4	170.1	93.6	39.4		23%	
	Option 6	408.1	271.4	170.1	93.6	39.4		23%	
	Option 7	408.1	271.4	170.1	93.6	39.4		23%	

Table A13.4 Referent Group Analysis

Discount Rates (for NPV):		4%	8%	12%	16%	20%	RANK	IRR
Bangkok	Option 1	236.0	181.1	141.4	112.1	90.2	2	N/A
	Option 2	154.5	114.3	85.7	65.0	49.8	4	N/A
	Option 3	29.3	16.9	8.9	3.7	0.5	6	N/A
Southern	Option 4	270.1	217.7	179.9	152.1	131.4	1	N/A
Thailand	Option 5	188.5	150.9	124.3	105.0	91.0	3	N/A
	Option 6	92.5	77.5	67.5	60.9	56.4	5	N/A
	Option 7	−70.7	−56.1	−43.8	−33.3	−24.3	7	N/A

As would be expected, the Option that ranked lowest in the Private Analysis (Option 4) is ranked first in the Referent Group Analysis (see Table A13.4): not only will the Government enjoy receiving cotton import duties and profits tax under this Option, but many unskilled workers will earn significantly more than they currently do. Conversely, Option 7 (ICP's highest ranked Option) will be disadvantageous for the Referent Group, with negative NPVs for all discount rates.

It is noted that the NPV rankings are consistent for all discount rates except 20%. Again, by plotting the NPV curves, it is possible to observe in Figure 13.2 that they cross at different points, explaining this anomaly.

IRRs are not applicable because the net benefit streams for the Referent Group Analyses of all Options have more than one change of sign (i.e. positive, negative, positive), leading to an inaccurate result. The reason for this sign change is due to negative cash flows for the domestic bank and the Government in 2000 (all other cash flows are positive) for the Referent Group members.

4 Conclusion

Based on the data from the analyses, Options 4 and 7 can clearly be rejected. Option 4 offers the best outcome for the Referent Group, but offers ICP no incentive and is ranked last in terms of the Private Analysis. Conversely Option 7 is clearly unacceptable to the Referent Group. Option 2, however, warrants serious consideration. This Option: Bangkok – No Duties on Imported Cotton satisfies ICP's request to receive one concession if it invests in the Bangkok area. It is ranked fourth in both the Private and Referent Group Analyses and, though in terms of the Efficiency Analysis it is not as efficient as any of the Southern Thailand options, the differences between Options 1–3 and Options 4–7 in efficiency terms are relatively small (IRR 22% compared to IRR 23%). Option 2 is also preferred over Option 3 (which also satisfies ICP's request for a concession). Option 3 is ranked sixth in the Referent Group ranking.

Unless ICP can be convinced to accept Option 5 (ranked fifth from ICP's perspective and third from Thailand's perspective), none of the Southern Thailand options seems likely. A further alternative that could be considered is subsidizing ICP for some or all of the costs to move the spinning mill to Southern Thailand. This could potentially be coupled with a reduction in import duties on raw materials. Moving to Southern Thailand involves an additional injection of up-front funds in 1999 which are not discounted. It can be seen from the Private Cash Flows in the Attachments that the costs to ICP between 2001 and 2004 are less in Southern Thailand than in Bangkok. However, to determine whether or not other options are viable, further detailed analysis and direction from the Minister would be required.

In conclusion, Option 2 is the recommended option for Thailand.

Note

1   This report was written as an assignment while the author was studying as a post-graduate student at the University of Queensland (2001).

Table A13.5 Summary of ICP Project results

Summary Information

Net Present Values (millions of Baht)

		Project					Private					Efficiency					Referent			
Discount Rates	4%	8%	12%	16%	20%	4%	8%	12%	16%	20%	4%	8%	12%	16%	20%	4%	8%	12%	16%	20%
Bangkok																				
Option 1	405.3	259.8	152.0	70.7	8.3	154.5	88.7	40.8	5.4	−21.1	408.9	269.8	166.6	88.7	28.9	236.0	181.1	141.4	112.1	90.2
Option 2	568.5	393.4	263.3	164.8	89.0	236.1	155.4	96.4	52.5	19.2	408.9	269.8	166.6	88.7	28.9	154.4	114.3	85.7	65.0	49.8
Option 3	405.3	259.8	152.0	70.7	8.3	361.2	252.9	173.3	113.8	68.6	408.9	269.8	166.6	88.7	28.9	29.3	16.9	8.9	3.7	0.5
Southern Thailand																				
Option 4	341.4	200.9	96.9	18.5	−41.7	119.1	53.7	6.2	−28.9	−55.2	408.1	271.4	170.1	93.6	34.9	270.1	217.7	179.9	152.1	131.4
Option 5	504.5	334.5	208.2	112.6	39.1	200.7	120.5	61.8	18.2	−14.8	408.1	271.4	170.1	93.6	34.9	188.5	150.9	124.3	105.0	91.0
Option 6	341.4	200.9	96.9	18.5	−41.7	296.7	194.0	118.6	62.4	19.8	408.1	271.4	170.1	93.6	34.9	92.5	77.5	67.5	60.9	56.4
Option 7	504.5	334.5	208.2	112.6	39.1	459.9	327.5	229.9	156.5	100.5	408.1	271.4	170.1	93.6	34.9	−70.7	−56.1	−43.8	−33.3	−24.3

Internal Rates of Return

	Project	Private	Efficiency	Referent
Bangkok				
Option 1	21%	17%	22%	N/A
Option 2	26%	23%	22%	N/A
Option 3	21%	29%	22%	N/A
Southern Thailand				
Option 4	17%	13%	23%	N/A
Option 5	23%	18%	23%	N/A
Option 6	17%	22%	23%	N/A
Option 7	23%	31%	23%	N/A

Table A13.6 ICP Option 1 Key Variables

OPTION 1 (BANGKOK – NO CONCESSIONS): KEY VARIABLES

Exchange Rate Bt/1US$	44	
ICP Capacity (lbs of yarn)	15,000,000	
ICP demand for yarn (lbs)	10,000,000	
ICP yarn for sale (lbs)	5,000,000	

Capacity Output		**Discount Rates**	
1999	Nil	1	4%
2000	50%	2	8%
2001–2009	100%	3	12%
		4	16%
		5	20%

REVENUES	Rates	Cost Savings	Yarn Sales
CIF price of yarn/lb ($US)	0.58		
CIF price of yarn/lb (Bt)	25.52	255,200,000	127,600,000
Import Duty/lb yarn (Bt)	2.10	21,000,000	10,500,000
Customs Charges/lb yarn (Bt)	0.30	3,000,000	1,500,000
TOTAL		279,200,000	139,600,000

INVESTMENT COSTS: BANGKOK

	No. (units)	Cost (US$)	Cost (Bt)	Import Duty	Total Cost (Bt)	Shadow Price
Fixed Investment	[1999]					
Spindles	40,000	4,000,000	176,000,000	5%	184,800,000	176,000,000
Ancillary equipment		850,000	37,400,000	5%	39,270,000	37,400,000
Installation (skilled)		100,000	4,400,000		4,400,000	4,400,000
Building (unskilled)		818,182	36,000,000		36,000,000	18,000,000
Construction						
Start-up Costs (skilled)		409,091	18,000,000		18,000,000	18,000,000
TOTAL					282,470,000	253,800,000
Working Capital	[2000]					
3 months' raw materials		1,200,000	52,800,000	10%	58,080,000	52,800,000
1 year's spare parts		243,000	10,692,000	5%	11,226,600	10,692,000
TOTAL					69,306,600	63,492,000

OPERATING COSTS: BANGKOK

Note: water, power and raw materials cost 50% in 2000, with all other costs 100%.

	No.	Cost (US$)	Cost (Bt)	Import Duty	Total Cost (Bt)
Raw Materials [2000 on]					
Imported Cotton (lbs)	16,200,000	4,860,000	213,840,000	10%	235,224,000

	No. (employees)	Cost p.a. (Bt) (/employee)	Total Cost (Bt)	Bangkok Shadow Price 50%
Labour [2000 on]				
Supervisors and Technicians	100	150,000	15,000,000	15,000,000
Other Skilled Workers	1000	22,500	22,500,000	22,500,000
Unskilled Workers	500	12,000	6,000,000	3,000,000
TOTAL	1600		43,500,000	40,500,000

		(lbs of yarn)	Cost p.a. (Bt) (/million lbs)	Total Cost (Bt)	Bangkok Shadow Price 50%
Fuel, Water, Spare Parts	[2000 on]				
Water		15,000,000	480,000	7,200,000	7,200,000
Electricity		15,000,000	480,000	7,200,000	3,600,000
TOTAL				14,400,000	10,800,000
Spare Parts (% of equipment + 5% duty)	[2000 on]		10%	22,407,000	21,340,000
TOTAL				22,407,000	21,340,000

(*continued*)

Table A13.6 Cont.

Insurance and Rent					
Insurance	*[2000 on]*			3,000,000	0
Rent	*[2000 on]*			3,000,000	3,000,000
TOTAL				6,000,000	3,000,000
FINANCING		*Amount (Bt)*	*Interest (p/a)*	*Term (yrs)*	*Interest on O'Draft*
Loan for Spindles and Equipment *foreign*		160,050,000	8%	5	
Working Capital Overdraft *local*		60,000,000	10%	9	−6,000,000
TAXES AND DEPRECIATION		*Rate (%)*	*Initial Fixed Inv Costs*	*Dep'n Expense*	
Profits Tax		50%			
Depreciation	*[2000 on]*	10%	264,470,000	26,447,000	

Table A13.7 Option 1: Bangkok – No Concessions

OPTION 1: BANGKOK – NO CONCESSIONS (MILLIONS OF BAHT)

Project Cash Flow	1999	2000	2001	2002	2003	2004	2005	2006	2007	2008	2009
Investment Costs											
Fixed Investment	−282.5										
Working Capital		−69.3									69.3
TOTAL	−282.5	−69.3									69.3
Operating Costs											
Raw Materials		−117.6	−235.2	−235.2	−235.2	−235.2	−235.2	−235.2	−235.2	−235.2	−235.2
Labour		−43.5	−43.5	−43.5	−43.5	−43.5	−43.5	−43.5	−43.5	−43.5	−43.5
Fuel, Water		−7.2	−14.4	−14.4	−14.4	−14.4	−14.4	−14.4	−14.4	−14.4	−14.4
Spare Parts		−22.4	−22.4	−22.4	−22.4	−22.4	−22.4	−22.4	−22.4	−22.4	−22.4
Insurance and Rent		−6.0	−6.0	−6.0	−6.0	−6.0	−6.0	−6.0	−6.0	−6.0	−6.0
TOTAL		−196.7	−321.5	−321.5	−321.5	−321.5	−321.5	−321.5	−321.5	−321.5	−321.5
Revenues											
Cost Savings		139.6	279.2	279.2	279.2	279.2	279.2	279.2	279.2	279.2	279.2
Yarn Sales		69.8	139.6	139.6	139.6	139.6	139.6	139.6	139.6	139.6	139.6
TOTAL		209.4	418.8	418.8	418.8	418.8	418.8	418.8	418.8	418.8	418.8
NET CASH FLOW	−282.5	−56.6	97.3	97.3	97.3	97.3	97.3	97.3	97.3	97.3	166.6

Discount Rates	4%	8%	12%	16%	20%
NPV	405.3	259.8	152.0	70.7	8.3
IRR	21%				

Private Cash Flow	1	2	3	4	5	6	7	8	9	10
	1999	2000	2001	2002	2003	2004	2005	2006	2007	2008	2009
FINANCING											
Principal											
Loan	160.1	−27.3	−29.5	−31.8	−34.4	−37.1					
Overdraft		60.0									−60.0
Interest											
Loan		−12.8	−10.6	−8.3	−5.7	−3.0					
Overdraft			−6.0	−6.0	−6.0	−6.0	−6.0	−6.0	−6.0	−6.0	−6.0
NET FINANCING FLOW	160.1	19.9	−46.1	−46.1	−46.1	−46.1	−6.0	−6.0	−6.0	−6.0	−66.0
NCF (Equity Pre-Tax)	−122.4	−36.7	51.2	51.2	51.2	51.2	91.3	91.3	91.3	91.3	100.6
TAXES											
Revenues		209.4	418.8	418.8	418.8	418.8	418.8	418.8	418.8	418.8	418.8
Operating Costs		−196.7	−321.5	−321.5	−321.5	−321.5	−321.5	−321.5	−321.5	−321.5	−321.5
Depreciation		−26.4	−26.4	−26.4	−26.4	−26.4	−26.4	−26.4	−26.4	−26.4	−26.4
Interest on loans		−12.8	−16.6	−14.3	−11.7	−9.0	−6.0	−6.0	−6.0	−6.0	−6.0
PROFITS (BEFORE TAX)		−26.6	52.2	56.6	59.1	61.9	64.8	64.8	64.8	64.8	64.8
TAX CASH FLOW		13.3	−27.1	−28.3	−29.6	−30.9	−32.4	−32.4	−32.4	−32.4	−32.4
EQUITY (AFTER TAX) PRIVATE CASH FLOW	−122.4	−23.4	24.1	22.9	21.6	20.3	58.9	58.9	58.9	58.9	68.2

Private Cash Flow	1	2	3	4	5	6	7	8	9	10

Discount Rates	4%	8%	12%	16%	20%
NPV	154.5	88.7	40.8	5.4	−21.1
IRR	17%				

(continued)

Efficiency Analysis	1999	2000	2001	2002	2003	2004	2005	2006	2007	2008	2009
INPUTS											
Investment Costs											
Fixed Investment	−253.8										
Working Capital		−63.5									63.5
TOTAL	−253.8	−63.5	0.0	0.0	0.0	0.0	0.0	0.0	0.0	0.0	63.5
Operating Costs											
Raw Materials		106.9	−213.8	−213.8	−213.8	−213.8	−213.8	−213.8	−213.8	−213.8	−213.8
Labour		−40.5	−40.5	−40.5	−40.5	−40.5	−40.5	−40.5	−40.5	−40.5	−40.5
Fuel, Water		−5.4	−10.8	−10.8	−10.8	−10.8	−10.8	−10.8	−10.8	−10.8	−10.8
Spare Parts		−21.3	−21.3	−21.3	−21.3	−21.3	−21.3	−21.3	−21.3	−21.3	−21.3
Insurance and Rent		−3.0	−3.0	−3.0	−3.0	−3.0	−3.0	−3.0	−3.0	−3.0	−3.0
TOTAL		−177.2	−289.5	−289.5	−289.5	−289.5	−289.5	−289.5	−289.5	−289.5	−289.5
Outputs											
Revenues											
Cost Savings		127.6	255.2	255.2	255.2	255.2	255.2	255.2	255.2	255.2	279.2
Yarn Sales		63.8	127.6	127.6	127.6	127.6	127.6	127.6	127.6	127.6	127.6
TOTAL		191.4	382.8	382.8	382.8	382.8	382.8	382.8	382.8	382.8	382.8
EFFICIENCY ANALYSIS		253.8	−49.3	93.3	93.3	93.3	93.3	93.3	93.3	93.3	156.8
Discount Rates	4%	8%	12%	16%	20%						
NPV	408.9	269.8	166.6	88.7	28.9						
IRR	22%										

Referent Group Analysis

METHOD A:	1999	2000	2001	2002	2003	2004	2005	2006	2007	2008	2009
Domestic Bank											
Overdraft		−60.0									60.0
Interest from Overdraft			6.0	6.0	6.0	6.0	6.0	6.0	6.0	6.0	6.0
Thai Government											
Lost Import Duties		−18.0	−36.0	−36.0	−36.0	−36.0	−36.0	−36.0	−36.0	−36.0	−36.0
Spindles I.Duty	8.8										
Equipment I.Duty	1.9										
W/Cap raw materials I.Duty		5.3									−5.3
W/Cap Spare parts I.Duty		0.5									−0.5
Imported Cotton I.Duty		10.7	21.4	21.4	21.4	21.4	21.4	21.4	21.4	21.4	21.4

METHOD A:	1999	2000	2001	2002	2003	2004	2005	2006	2007	2008	2009
Spare Parts Duty		1.1	1.1	1.1	1.1	1.1	1.1	1.1	1.1	1.1	1.1
Profits Tax		−13.3	27.1	28.3	29.6	30.9	32.4	32.4	32.4	32.4	32.4
Domestic Labour Initial Labour	18.0										
Ongoing Labour		3.0	3.0	3.0	3.0	3.0	3.0	3.0	3.0	3.0	3.0
Domestic Electricity Corp Electricity		1.8	3.6	3.6	3.6	3.6	3.6	3.6	3.6	3.6	3.6
Domestic Insurance Co Insurance		3.0	3.0	3.0	3.0	3.0	3.0	3.0	3.0	3.0	3.0
TOTAL	28.7	−65.9	29.2	30.3	31.6	33.0	34.5	34.5	34.5	34.5	88.6

Table A13.7 Cont.

METHOD B:	1999	2000	2001	2002	2003	2004	2005	2006	2007	2008	2009
Efficiency Analysis	−253.8	−49.3	93.3	93.3	93.3	93.3	93.3	93.3	93.3	93.3	156.8
Less Non-Referent Grp											
ICP	−122.4	−23.4	24.1	22.9	21.6	20.3	58.9	58.9	58.9	58.9	68.2
Foreign Credit – Principal	−160.1	27.3	29.5	31.8	34.4	37.1					
Foreign Credit – Interest		12.8	10.6	8.3	5.7	3.0					
TOTAL	28.7	−65.9	29.2	30.3	31.6	33.0	34.5	34.5	34.5	34.5	88.6
Discount Rates	4%	8%	12%	16%	20%						
NPV	236.0	181.1	141.4	112.1	90.2						
IRR	N/A										

Appendix 1
Case study assignments

As discussed in the Preface, the National Fruit Growers (NFG) and International Cloth Products (ICP) Case Studies can be used as student assignments. While the solution spreadsheets are reported in the text, these are in the form of *values* only. An assignment requiring the student to perform a sensitivity analysis (or for advanced students, a risk analysis) on selected project variables necessitates the construction of a solution spreadsheet based on *formulae*. The values reported in the text provide a convenient way for the student to check the accuracy of their spreadsheet formulae. (The Excel files for the NFG and ICP Case Studies are available to instructors from the Support Material website.)

For more demanding assignments, the following case studies are presented in approximate order of complexity, with the simpler ones first. While most of these studies are loosely based on actual projects, the case studies should be regarded as hypothetical and they are not intended to represent the activities of any particular firm or organization. Spreadsheet solutions are available to Instructors on the Support Material website.

A1.1 South Australian Olive Oil Project*

Introduction

In 2011, a foreign investor, Virgin Olives Incorporated (VOI), is considering establishment of an irrigated olive-growing and oil-processing project in the Adelaide Plains of South Australia. This region has an ideal climate for growing olive trees, with mild winters, low humidity and a dry summer. It has a history of successful olive growing dating back to the 1860s. Australia currently imports about 95% of its olive consumption. Domestic demand is increasing at around 10% per annum, and despite recent plantings of olive trees maturing over the next few years, it is not expected that Australia will become self-sufficient in olive production in the foreseeable future. VOI is seeking financial assistance from the State Government of South Australia in the form of a concessional loan and exemption from Federal Government profits taxes. You are required to undertake a comprehensive CBA with a view to advising the government on the desirability of the project from a South Australian perspective. Assume the project has a life of 30 years (to end 2041).

Fixed investment costs

VOI is intending to purchase 135 hectares of land in the Adelaide Plains, with access to underground water for irrigation, at a cost of $2000 per hectare (Ha). It should be assumed that 15% of the land cannot be cultivated to allow for on-farm infrastructure such as buildings, dams, roads, etc. Irrigation infrastructure, including locally manufactured pumps, motors, and sprinkler system, will cost $23,000 per Ha. Major repairs at 70% of initial cost will be required after 15 years. VOI will also need to purchase a water allocation licence for drawing 10 megalitres (ML) of water per Ha per annum. The licence will cost $300,000. To prepare the land for planting, a local contractor will be employed at a cost of $150 per Ha, using 50% skilled labour and 50% unskilled labour. VOI will plant and water 200 olive cuttings per Ha at $25 per cutting, and $1,500 per Ha for planting and watering, using 70% unskilled labour, 30% skilled labour. VOI will also require imported vehicles and farm equipment costing $235,000. All vehicles and equipment have a 10-year life, and are replaced at the same initial cost. A machinery shed and office (with a 30-year life) is to be constructed at a cost of $54,000, consisting of locally produced materials (30%) and local labour (70%), of which 50% is skilled and 50% unskilled labour.

Assume all the above investment costs occur in 2011.

In 2015, the oil extraction factory will be established and will begin operations in the same year. This will consist of a building costing $75,000 (assume the same cost components as with the shed), and an imported extraction centrifuge and harvester costing $1,000,000. Assume these have a life to the end of the project.

Working capital

VOI will need to establish a stock of spare parts, fuel, chemicals and office supplies in 2012 amounting to $57,000. Assume 50% imported; 50% locally produced goods.

Operating costs (from 2015 onwards)

Fixed costs

    i   Pruning: $70,000 per annum (all unskilled labour)
    ii Water charges: $400 per megalitre (ML) payable to the State water authority, a public corporation
    iii Fertiliser and Herbicides: $1,100 per Ha/annum (50% local materials; 50% unskilled labour)
    iv Management, communications, office supplies etc.: $135,000/annum (90% skilled labour, 10% locally produced materials)

Variable costs

These are directly related to production level from 2015 onwards; all figures relate to full capacity output.

    i   Harvesting labour (unskilled): $650/Ha
    ii Harvesting fuel: $170/Ha
    iii Factory labour: $457/Ha (50% skilled; 50% unskilled)
    iv Factory electrical power: $100/Ha payable to the Electricity Trust of SA (ETSA), a public corporation
    v   Repairs and maintenance: $39,000/annum (25% imported spares; 25% local spares; 50% skilled labour)
    vi Bottling: $15/Ha (25% unskilled labour; 75% local materials)
    vii Freight: $100/Ha (paid to an interstate contractor)

Output

The olive trees are not expected to bear fruit until 2015, when the oil extraction plant is expected to operate at 40% of capacity, with production increasing to 50% in 2016, 70% in 2017, and 95% in 2018. At full capacity (2019) the orchard will produce 22kg of fruit per tree per annum. The expected oil extraction rate is 0.3 litre/kg fruit. Olive oil currently sells in the world market at EUR 9 per litre and imported olive oil is subject to an import duty of 12%. The current exchange rate is EUR/AUD = 1.45.

Financing

VOI plans to finance the project by borrowing $2.5 million from the Bank of South Australia in 2011. If VOI borrowed at commercial rates they would expect to repay the loan as an annuity over 15 years at a real interest rate of 8.5% per annum. VOI intends to apply for a concessional loan from the state government at a real interest rate of 5% per annum, repayable over 15 years. The balance is to be financed from VOI's equity.

Taxes and duties

Costs of imported inputs to VOI include a 7% import duty, paid to the Federal Government. Imported olives carry an import duty of 12%. The Government of South Australia also imposes an agricultural levy equal to $0.50 per litre of olive oil produced, payable by VOI. The prices of all imported inputs reported above are *inclusive of duties*, and the domestic market price of olive oil is the world price plus import duty. Company (profits) tax of 35% is payable to the Australian Federal Treasury, with any losses being off-settable against VOI's other operations in Australia in the same year. (This does not apply during a period of taxation exemption, if applicable. You should treat depreciation and interest on debt as deductible against profits tax. The agricultural levy is *not* deductible against profits tax.)

Environmental costs

There are two major problems associated with clearing land for agricultural use such as the VOI operation. These are:

i *declining water quality* in rivers, streams and creeks which can potentially lead to degradation of estuarine and inshore marine ecosystems. For example, the loss of seagrass beds and mangrove forests in Nepean Bay, Boston Bay and Spencer Gulf coastal waters has been attributed to a combination of nutrient enrichment, industrial pollution and discharges of agricultural drainage water;

ii *habitat fragmentation and loss*, which is a major threat to biodiversity conservation. One way of mitigating the negative effects of fragmentation is to improve habitat connectivity. Habitat corridors have been shown to be valuable for the conservation of various groups of wildlife and in various situations (e.g. urban, agricultural, and production forest landscapes). Retained areas of native forest within plantations are beneficial for wildlife conservation and form an important component of the conservation programme in agricultural landscapes.

In general, nature or habitat corridors can be located along environmental contours to ensure habitat continuity, especially along riparian corridors. The riparian corridors are also important for the protection of water catchments, tend to be species-rich and structurally diverse relative to surrounding areas. It has been estimated that if VOI set aside a further 15% of the land in the form of riparian buffer zones and nature corridors, the negative environmental impacts of the project would be negligible. VOI's output would be reduced proportionately. If no buffer zone is created on the land, the non-marketed environmental cost to the wider community of South Australia would amount to an estimated $200,000 per annum. Assume that the avoided environmental cost is proportional to the amount of land set aside (from 0% to 15%) as a riparian buffer.

Other

i The shadow-price for unskilled labour is equal to 70% of the market wage: assume that when otherwise unemployed labour is employed on the project there is no net change to Federal Government taxes and social security payments; all labour employed on the project comes from South Australia;

ii   Electrical power produced by the Electricity Trust of South Australia (ETSA) is sold to VOI at a price equal to the marginal internal cost of production. The external CO_2 emission cost, assumed to be borne by the community of South Australia, is estimated at 30% of the selling price of power. ETSA pays a carbon tax of this amount to the State government;

iii The opportunity cost of land is 120% of the market price; the land will be sold back to the South Australia Government at the end of the project's life at its original (real) purchase price.

iv   The opportunity cost of water to the South Australian community as a whole is estimated to be 120% of the market price on a per ML basis (the water licence itself has a zero opportunity cost and it will be sold back to the South Australian government at the original (real) purchase price by the end of the project's life);

v   The domestic price of olive oil is equal to world price plus 12% duty;

vi   Assume straight line depreciation over the life of each asset: with irrigation infrastructure depreciate the entire initial cost over the first 15 years, and the major repair costs over the remaining years;

vii Assume a 30-year life for the project (ending in 2041) with zero salvage values for any assets (except for the land and water allocation licence which will be sold back to the South Australian government at original (real) cost);

viii Assume VOI repatriates all after tax profits to Europe.

Note

*   The invaluable contributions of Kim Nguyen and Prabha Prayaga are duly acknowledged. The names and information used in this case study are hypothetical.

A1.2 Walnuts Tasmania Project

Introduction

Walnuts Tasmania Limited (WALT) is considering establishment of an irrigated walnut growing project in eastern Tasmania. This region has an ideal climate for growing walnut trees, with cold winters and mild summers. The walnut industry in Australia is rapidly growing and is expected to continue growing in the coming years. Australia currently still imports 4000 tonnes of kernels (9000 tonnes of in-shell equivalent) each year. WALT has applied to the Australian government for a subsidised loan, under the Rural Reconstruction Program, to establish a walnut plantation in eastern Tasmania in 2013. In support of its application the company argues that the project will reduce Australia's dependence on imported walnuts, generate jobs in an area of high unemployment, and help to reduce soil salinity in the local area.

Industry sources suggest that WALT requires a real rate of return of 10% on its equity capital. The Australian government has commissioned a cost-benefit analysis of the project and the Tasmanian government has asked for a separate analysis of the effects on Tasmania. As part of its application, WALT has supplied the following information about the project.

Investment costs

The investment costs for the project are given in Table App_1.1.

Table App_1.1 Investment costs (constant 2013 prices)

Costs	Number	Price ($/unit)	Life (years)
Fixed Investment (2013)			
Farm equipment (units)	10	250,000	13
Vehicles (units)	8	75,000	7
Buildings (m²)	625	2500	20
Working capital (2014)			
Fertiliser stocks (tons)	5	1250	
Insecticide stocks (litres)	6250	75	
Spare parts (units)	25	2500	
Fuel stocks (litres)	1250	1.75	

Assume the project has a life of 20 years (to end 2033). The lives of fixed investment are book lives for tax purposes and none of the capital will have to be replaced during the life of the project. At the end of the project the cost of land rehabilitation is expected to be

$125,000 and a salvage value of 10% of fixed investment can be assumed. Farm equipment and spare parts have an import duty of 10% and vehicles have an import duty of 25%.

Operating costs

The operating costs of the project are given in Table App_1.2.

Table App_1.2 Operating costs (constant 2013 prices)

Operating costs	Units/yr	Price/unit
Rent on land (Ha)	250	75
Fuel (litres)	6250	1.75
Seeds (Kg)	625	50
Fertilisers (tonnes)	7.5	1250
Insecticides (litres)	7500	75
Water (ML)	2250	50
Spare parts (units)	30	2500
Casual labour (days)	250	150
Administration (units)	30	2500
Insurance (units)	2.5	20,663
Management (units)	30	7500
Miscellaneous (units)	2.5	19,250

A tax on fuel of 10% is levied to discourage use. Assume that this tax is used to offset any negative health effects. In addition, the prices of various inputs employed by WALT are subsidised: seeds 25%, fertiliser 30%, insecticide 12% and water 15%. The opportunity costs of land and labour are 60% of market values. For the purposes of this project, assume that all operating costs are fixed for the life of the project.

Financing

WALT has applied to the Australian government for a subsidised loan, under the Rural Reconstruction Program, of $1.5 million, at a nominal interest rate of 4% per annum, repayable over 15 years. WALT will also take out a bank overdraft from an Australian Bank in 2014 for $102,500, at 8% (nominal), which it intends to repay after 6 years. The balance of the required funding will be met from WALT's own funds (equity). The current business income tax on profits is 30%, payable annually. WALT has revenue from other operations in Australia against which it can deduct any losses for tax purposes. The market rate of interest (nominal) is expected to be 5% and inflation is forecast to be 2.5% per annum over the life of the project.

Revenues

The project is expected to produce at 25% capacity in 2014, with production increasing to 50% in 2015 and 75% in 2016. At full capacity (2017 onwards) the project is expected to produce 250 tonnes of walnuts per annum, which has a market price of $8,750 per tonne. At the end of the project it is expected that the plantation would produce 250 tonnes of timber at a market price of $2,500 per tonne. Imported walnuts are subject to an import duty of 10%.

Environmental costs

The Tasmanian government is broadly supportive but has expressed concern about the use of fertilisers and insecticides causing stream pollution which will adversely affect coastal oyster farms. This could be offset by establishing riparian buffer zones within the plantation. The Tasmanian government has recommended a riparian buffer of 5% to prevent the loss to oyster farms downstream; however, WALT's output would be reduced proportionately with no corresponding reduction in costs. If no buffer zone is created on the land, the loss to oyster farms downstream amounts to an estimated $250,000 per annum.

An environmental benefit of the project is the reduction in soil salinity. But the establishment of a 5% riparian zone would also proportionately reduce the benefits from soil salinity reduction. If no riparian zone was established, the benefit from reduced soil salinity amounts to $12,500 per annum.

Assume that the environmental costs and benefits are proportional to the amount of land set aside (from 0% to 5%) as a riparian buffer.

Referent Group

For the purpose of this analysis, assume that all stakeholders are part of the Referent Group. However, for the purposes of negotiations with WALT, the Australian and Tasmanian governments also wish to know the net benefits to labour and the Australian Banks.

Note: you should set the accuracy check in this section at four (4) decimal places.

Assignment instructions

You are required to undertake a comprehensive CBA (Project (Market), Private, Efficiency and Referent Group analysis), with a view to advising the government on the desirability of the project from the perspectives of the Australian government, Tasmania and WALT. While the Australian government uses a 4% discount rate (real) for public sector investment decision-making, you should also undertake a sensitivity analysis at 2% and 6%. It is understood that WALT requires a minimum return of 10% (real) on its equity. Your report should advise the government on whether the project is worthwhile, giving reasons. You should also discuss its desirability and profitability from the perspectives of the Tasmanian state government and WALT respectively, commenting on whether WALT should be offered any further inducements to invest in the project (WALT says that it would be more likely to undertake the project if it were given relief from import duties on the initial investment in equipment and vehicles).

Sensitivity Analysis

You should also conduct and report the results of a sensitivity analysis in which you calculate and comment on, at least, the sensitivity of the results to:

1   the riparian buffer zone:
    a   If the required riparian buffer to achieve the full environmental benefit to oyster farmers varies between 4% and 6%.

    b   Calculate the threshold size of the riparian buffer zone at which (i) the aggregate Referent Group NPV becomes positive at 4% discount rate; and (ii) the IRR to WALT is at its minimum acceptable level. Assume the recommended buffer is at the base case (5%) level.

2   the opportunity costs of land and labour (30% variation each side of the base-case estimate).

3   Identify one other main variable which when varied/changed has a strong impact on the Referent Group net benefits and discuss the implications for the various stakeholders.

Initially undertake a sensitivity analysis showing the effects of variability of these inputs individually and jointly on the net benefits to the Referent Group. You should also undertake a risk analysis assuming a triangular probability distribution for the three variables in 2 and 3 above. In your report you should also indicate if there are any omitted costs and benefits that could be of potential significance to the decision-maker and might warrant further investigation.

A1.3 A tuna cannery in Papua New Guinea

Introduction

Western Pacific Tuna Products (WPTP), a company registered in the Philippines, proposes to set up a tuna harvesting and processing operation on the north coast of Papua New Guinea (PNG). It plans to import 12 medium-sized purse seiners, together with support vessels, to catch 32,000 metric tonnes (mt) of skipjack and yellowfin tuna per annum in PNG's Exclusive Economic Zone (EEZ). A processing plant will be constructed to can the tuna for export to the European Union (EU), and fishmeal will be produced as a by-product and sold locally. The project will take one year to establish and will run for a further 20 years, operating at 30% capacity in year 1, 70% in year 2 and at full capacity thereafter. All costs are estimated in 2007 US dollars (see Table App_1.3).

Table App_1.3 Capital costs

Capital costs*	$ millions	Salvage value (%)
Vessels	18,000,000	20
Land and improvements	2,000,000	80
Buildings and facilities	6,000,000	50
Equipment	4,250,000	15
Working capital	3,600,000	100

Note: *includes 5% import duty.

The firm plans to borrow 60% of the cost of vessels, land and improvements, buildings and facilities, equipment and working capital at a real interest rate of 12% with the loan repayable over 10 years. It will be able to depreciate vessels over 15 years, buildings and facilities over 20 years and equipment over 10 years for tax purposes. Replacement costs will be incurred in year 11 of the project: 25% of the initial cost of vessels and 50% of the initial cost of equipment, and this investment can also be depreciated for tax purposes at the annual rates indicated above. Salvage values are market value of assets sold in PNG and receipts are treated as income for tax purposes. Company tax is levied on taxable income, defined as revenue net of export tax and EU duty, less operating costs, access fees, and interest and depreciation expenses. Resident company tax rate is 25%, and non-resident company tax rate is 48%. If taxable income is negative in any year the company receives a tax refund calculated as the tax rate times the amount of the loss.

Table App_1.4 Operating costs

Operating costs (at full capacity) $ millions pa	
Labour	8,250,000.00
Materials*	5,000,000.00
Maintenance	2,500,000.00
Fuel**	24,000,000.00
Electricity	1,000,000.00
Insurance	100,000.00
Miscellaneous***	4,500,000.00

Notes:
*   includes 5% import duty.
**   includes 10% fuel tax.
*** includes 10% Value Added Tax (VAT).

In Years 1 and 2 all operating costs, except insurance, will be incurred at 30% and 70% of capacity costs respectively. In subsequent years operating costs are at full capacity level irrespective of the volume of catch (see Table App_1.4).

The proportion of the wage bill accounted for by local labour is 80%. Local labour pays an average income tax rate of 3% and foreign labour 15%.

The processing plant will produce 50 cases of canned tuna and 0.15 metric tons (mt) of fishmeal per mt of raw tuna processed. Canned tuna sells for $35 per case in Europe and fishmeal sells for $430 per mt in PNG. The company will have to pay EU import duty and PNG export tax levied on the value of canned tuna sold in the EU: the EU import duty rate for this class of product is 10% and the PNG export duty rate is 5%. Under the Lomé Convention, EU duties on imports from developing counties can be waived and WPTP wants the PNG government to apply for this exemption for the project.

Tuna purse seiners operating in PNG's EEZ pay an annual fee for access to the tuna stocks. Distant Water Fishing Nation (DWFN) vessels pay an annual royalty of 6% of the anticipated value of their catch, based on the Bangkok price of tuna. Domestic vessels pay an annual fee of $3000 per vessel. PNG's tuna stocks are judged to be close to fully exploited and it is thought that fishing activity by DWFN vessels will have to be curtailed to accommodate the catch of the proposed project. WPTP has asked for its vessels to be treated as domestic vessels.

There is significant unemployment in the coastal region of PNG and a study has estimated that the opportunity cost of local labour is 50% of the pre-tax wage. A further study has concluded that the fishing and canning operation will cause significant water, air and noise pollution. While an exact figure could not be placed on the cost of pollution, one expert suggested that it might be around $20 per mt of tuna processed per annum.

WPTP requires a 20% real rate of return on equity capital for projects in the Pacific Islands Region. It emphasises the employment benefits the project will bring to PNG, and claims that it will require a range of concessions to make the project viable from the viewpoint of its equity holders. It has asked for exemption from duties on imports of vessels, equipment, working capital and materials; exemption from export tax; exemption from fuel tax, and exemption from VAT on miscellaneous items. It wishes to be treated as a resident company for income tax purposes, and for its vessels to be domestic flagged for the purposes of determining access fees. It requests that PNG apply to have its sales in the EU classified as developing country in origin and exempt from EU duty under the Lomé Convention.

Assignment instructions

On behalf of the Government of Papua New Guinea, you are required to undertake and report the findings of a cost-benefit analysis of the WPTP proposal. The analysis is to be conducted in millions of 2007 US$, to two decimal places. You should calculate IRRs for the Project (Market), Private and Efficiency Analyses and report NPVs for the Referent Group on an aggregated and disaggregated basis for the year 2007 (the present year.) The PNG government uses a 5% discount rate (real) for investment decision-making, but is interested in knowing the sensitivity of the results at 3% and 8% discount rates. The Government of PNG also wishes advice on which, if any, of the concessions requested by WPTP should be granted.

You should also conduct and report the results of a sensitivity analysis in which you analyse and comment on, at least, the sensitivity of the results to: (i) the expected annual catch of the fleet, and (ii) the price of canned tuna in Europe. You are encouraged to explore the sensitivity of the results to other variables with a view to identifying those that have most impact on the results and that would warrant further investigation. In your report you should also indicate if there are any omitted costs and benefits that could be of potential significance to the decision-maker.

Your written report should not be more than 12 pages in length, excluding tables, and a one-page executive summary. Printouts of base-case scenario spreadsheets only should be included as appendixes, and each CBA (i.e. Project (Market), Private, etc.) to be printed on one A4 size page. Results of the sensitivity analyses should be reported in tables included in the text. You must submit an electronic copy of all your spreadsheets with your printed assignment.

A1.4 Urban water supply in South-East Queensland*

Introduction

The Department of Water Resources (DWR) is planning to increase South-East Queensland's urban water supply to meet the anticipated extra demand over the 24-year period from 2011 to 2034 inclusive. Demand is expected to rise from the anticipated 2010 level of 500,000 megalitres per annum (ML/a) in equal annual increments of 6000 ML/a to 644,000 ML/a in 2034.

DWR is considering a proposal to construct a dam on the Elizabeth River in the years 2009 and 2010 which would provide a yield of up to 90,000 ML/a. This would meet the anticipated additional demand in the years 2011 to 2024 inclusive. In 2024 the dam wall would be raised, thereby providing a further yield of up to 54,000 ML/a which would meet the anticipated extra demand until 2034.

DWR does not have desalination technology or expertise, but a French company, Aqua Vite (AV), has suggested an alternative way of meeting the anticipated extra demand for water. It proposes to build a series of four desalination plants, each with a capacity of 36,000 ML/a. Plants 1 and 3 would be built with extra tunnel and pipeline capacity which would be utilised by Plants 2 and 4. The plants would each take two years to construct and would be built so as to become operational in 2011 (Plant 1), 2017 (Plant 2), 2023 (Plant 3) and 2029 (Plant 4). AV would operate as a commercial company and would sell the water it produced to DWR. At the end of 2034 AV would sell the desalination infrastructure to DWR at an agreed price.

In order to maintain capacity and the additional 144,000 ML/a water supply provided by each option, further expenditures by DWR would be required in the period 2035–58 inclusive. The dam and its extension would require a capital refurbishment programme, the equipment comprising the interconnection network and the water treatment plant would have to be replaced at some stage and the annual fixed and variable operating costs would be incurred. If the desalination option were chosen, DWR would bear the costs of replacement of plant and equipment and the annual fixed and variable costs of the plants. At the end of 2058, the economic life of both projects will be over.

You have been engaged as a consultant by DWR to evaluate the AV proposal from the viewpoint of the State of Queensland: DWR wishes to know which of the two proposals is the least-cost method of supplying the additional demand in the period 2011–34, and maintaining that supply in the subsequent period 2035–58. Your recommendation will be either to accept the AV proposal, or to reject it and proceed with the Elizabeth River dam projects. Note that the State Government expects to receive from the Commonwealth 20% of any additional GST revenues generated in Queensland as a result of either project. Present values are to be estimated at 4%, 6% and 8% real rates of interest.

The estimated costs of the two projects are detailed below. All costs reported are in 2009 dollars.

The Elizabeth River Dam Project

Capital costs

It is estimated that 2% of all capital costs reported in Tables App_1.5 and App_1.6 consists of Goods and Services Tax (GST) payments.

Table App_1.5 Initial dam capital costs

Initial capital costs ($ millions)	2009	2010
Dam	850	850
Interconnection		
Pipelines and pump stations	355	355
Water treatment plant		
Plant	12	12
Equipment	50	50
Storages	5	5

Table App_1.6 Dam refurbishment costs

Year	2013	2018	2023	2028	2033	2035–58
Capital refurbishment ($ millions)	0.14	0.845	0.595	2.195	0.33	20 p.a.

Operating costs

Annual fixed and variable costs in the year 2011 together with their composition and tax components in that year are shown in Table App_1.7. (The labour, energy and materials proportions apply to both the Fixed Annual Cost and the Variable Cost.)

Table App_1.7 Dam operating costs

Operating costs	
Fixed annual cost ($ million/pa)	18
Variable cost ($/ML)	225
Labour (proportion)*	0.4
Energy (proportion)**	0.1
Materials (proportion)**	0.5

Notes: * Includes 3% payroll tax.
** Includes 10% GST.

Cost of energy

A consultant estimates that the energy price will rise at a rate 1% above the general rate of price inflation, starting in 2012, over the period until 2034. From 2035 onwards the energy price is expected to follow the general rate of price inflation.

The desalination project

Capital costs

It is estimated that 2% of all capital costs reported in Table App_1.8 consists of Goods and Services Tax (GST) payments.

Table App_1.8 Desalination project capital costs

Capital costs ($ millions)	Plant 1		Plant 2		Plant 3		Plant 4	
Year	2009	2010	2015	2016	2021	2022	2027	2028
Plant								
Tunnels and marine infrastructure	160	160	0	0	160	160	0	0
Plant, buildings and equipment	250	250	176	176	250	250	176	176
Land acquisition	5	0	0	0	5	0	0	0
Interconnection								
Pipelines, pumps and tanks	95	95	0	0	95	95	0	0

In addition to the capital costs reported above, capital expenditures of $20 million per annum will be required in the years 2035–58 inclusive for refurbishment of the plant and related infrastructure.

Operating costs

Fixed annual costs and variable costs for each Plant, together with the composition of costs (calculated at 2009 prices) and the tax components are shown in Table App_1.9.

Table App_1.9 Desalination project operating costs

Operating costs of each plant	Plant 1	Plant 2	Plant 3	Plant 4
Fixed annual cost ($ million/pa)	22	10	22	10
Variable cost ($/ML)	350	350	350	350
Labour (proportion)*	0.4	0.4	0.4	0.4
Energy (proportion)**	0.3	0.3	0.3	0.3
Materials (proportion)**	0.3	0.3	0.3	0.3

Notes: * Includes 3% payroll tax.
** Includes 10% GST.

Cost of energy

As noted above, a consultant estimates that the energy price will rise at a rate 1% above the general rate of price inflation, starting in 2012, over the period until 2034. From 2035 onwards the energy price is expected to follow the general rate of price inflation.

The Aqua Vite venture: financial and tax flows

Aqua Vite (AV) will borrow $500 million at an 8% real rate of interest from an overseas bank in 2009. The loan, together with interest, will be repaid in equal annual amounts, in the form of an annuity, over the 15-year period starting in 2010.

AV will sell to DWR the water it produces at a price of $5250 (2009 dollars) per ML. It will pay 30% business income tax on its earnings net of operating costs, interest payments and depreciation allowances (assume that AV can deduct any losses against taxable income from its other Australian projects):

Depreciation allowances as shown in Table App_1.10 can be claimed starting in the year 2011 (Plant 1), 2017 (Plant 2), 2023 (Plant 3) and 2029 (Plant 4). At the end of 2034 AV will sell its desalination plants to DWR for a surrender value of $100 million (2009 dollars), which will be subject to business income tax. AV says that it requires a real rate of return of 9% on the project if it is to proceed.

Table App_1.10 Desalination project depreciation

Depreciation (years)	
Tunnels and marine infrastructure	40
Plant, buildings and equipment	30
Pipelines, pumps and tanks	25

Labour market

There is very low unemployment in Queensland, especially in the construction industry, but it is expected that job vacancies can be filled by migrants from other states.

External costs

Both projects are thought to involve significant external costs. The Dam Project involves flooding the Elizabeth River Valley with consequent loss of recreation facilities and wildlife habitat. The Elizabeth River is home to a rare species of lungfish, together with other animals such as platypus, crayfish and frogs. A contingent valuation study undertaken by a consultant estimated the annual cost of the inundation of the valley to users and nonusers at $40 million (2009 dollars) starting in 2011.

The Desalination Project produces highly saline water as a waste product and it is proposed to release this as a brine trail in the ocean with consequent damage to marine ecosystems in the area of release. The volume of waste product is proportional to the volume of water produced and the amount of damage is proportional to the volume of waste. A consultant has estimated an environmental cost of $400 per ML of water produced (2009 dollars) starting in 2011. Concerns have also been voiced about the loss in visual and recreational amenity in the vicinity of the desalination plants.

The Dam Project may provide some benefits in the form of flood mitigation, but the Desalination Project may be a more reliable source of supply in the immediate future. No attempt has been made to quantify these effects.

Assignment instructions

On behalf of the Government of Queensland, you are required to undertake and report the findings of cost analyses of the Elizabeth River Dam and AV Desalination proposals. The analysis is to be reported in *millions* of 2009 Australian dollars, to two decimal places, with present values in year 2009 to be calculated.

Since the government of Queensland is mainly interested in the costs to Queensland of the two proposals for meeting SE Queensland's additional water supply needs into the middle of the century, the costs reported in the Project (Market) and Efficiency Analyses, and in the Department of Water Resources (DWR) section of the Private Analysis, are to be entered as positive numbers. In the Aqua Vite (AV) section of the Private Analysis, the usual convention is to be followed, with revenues entered as positive numbers and costs as negative numbers. In the Referent Group Analysis costs to Queensland are to be entered as positive numbers (with any incidental benefits to Queensland entered as negative costs), and net benefits to other groups are to be entered as positive numbers. These conventions must be borne in mind in summing the overall effects of the two projects.

You should summarise the results of the Project (Market), Private, Efficiency and Referent Group Analyses by calculating and reporting NPVs, with the NPV for the Referent Group reported on an aggregated and disaggregated basis. The costs calculated in the Project (Market) and Efficiency Analysis should also be expressed on a per ML basis in the Summary Table of Results. The IRR for the Private analysis of the AV Project should also be reported, and the impact of the proposals on DWR's budget estimated. The government generally uses a 6% discount rate (real) for investment decision-making, but is interested in knowing the sensitivity of the results at 4% and 8% discount rates. The government also wishes to be advised about the viability of the AV proposal from AV's perspective.

You should also conduct and report the results of a sensitivity analysis in which you calculate and comment on, at least, (i) the sensitivity of the results to the expected increase in the real price of energy, and (ii) the likely external costs of the projects. You are encouraged to explore the sensitivity of the results to a small number of other variables with a view to identifying those that have most impact on the results and that would warrant further investigation. In your report you should also indicate if there are any omitted costs and benefits that could be of potential significance to the decision-maker and might warrant further investigation. You must submit an electronic copy of all your spreadsheets with your printed assignment.

Note

* We would like to thank Peter Jacob and the late Tony Hand of Marsden Jacob Associates for help in obtaining cost data used in this exercise. All responsibility for the exercise and the results rests with the course instructors.

A1.5 The Scottish Highlands and Islands remote dental care programme

Introduction

There are currently 40 General Dental Practices (GDP) in the Scottish Highlands and Islands serving a population of around 250,000. The GDPs deal with all routine dental care, but each year a few hundred patients have to travel a considerable distance to hospital in Aberdeen for specialist advice, diagnosis and a treatment plan to manage complex restorative, prosthetic or periodontal problems.

The National Health Service (NHS) is considering two alternative proposals, starting in 2006, to reduce the number of hospital visits required and instead to treat a significant number of these patients at the GDPs:

1 *The Outreach Project*, proposed by the NHS Scottish Division, would provide for specialists to travel from Aberdeen to provide treatment at eight selected GDPs located in regional centres; or
2 *The Teledentistry Project*, proposed by a Canadian company, would provide, maintain and operate videoconferencing equipment which would enable local dentists to provide specialist care at the local GDPs with the guidance of specialists located in Aberdeen.

It is estimated that under either proposal there will be 400 patients who can be treated locally or regionally in 2007 instead of having to travel to Aberdeen. The amount of specialist time involved in treating patients under either proposal will be the same as under the existing system, but the Outreach Project will involve extra specialist time in the form of travel. Because of changes in population size and structure, the number of specialist patients eligible for local treatment is estimated to rise by 10 each year starting in 2008.

The Outreach Project

The NHS Scottish Division proposes to fund an upgrade of space and equipment at eight GDPs located in regional centres to enable specialists travelling from Aberdeen to treat cases at these GDPs. Specialist treatment is provided at no cost to the patient and GDP costs are reimbursed by the NHS.

The Scottish Division proposes to up-grade the eight regional GDPs in 2006 and to commence the Outreach Project in 2007. It is estimated that 50% of the eligible 400 patients per annum can be treated in 2007 and 100% thereafter. The project will run for 15 years from 2007.

Capital costs (at 2006 prices)

The NHS will incur a cost of £4250 in up-grading each of the eight Regional GDPs in 2006. Of this sum, 40% will be spent on specialist equipment, which is imported and subject to a 20% tariff. The equipment has a life of 5 years and a salvage value of 10%. The remainder of the funds will be spent on refurbishing GDP work space; the refurbishments also have a life of 5 years, but no salvage value.

Operating costs (at 2006 prices)

These costs are reported in Table App_1.12.

The Teledentistry project

A Canadian company, Telemedicine Services Incorporated (TSI), proposes to facilitate the treatment of specialist cases using its teledentistry service: PC-based videoconference hardware and software coupled with medium band width ISDN lines provide live audio and video transmission between a referring dentist and patient in a remote location and a specialist at the base hospital. The referring dentist provides the treatment at the local GDP under the supervision of the specialist in Aberdeen. Each GDP will require an upgrade of equipment costing £1000 (in addition to capital costs incurred by TSI) in 2006, and funded by the NHS. Specialist treatment is provided at no cost to the patient, though under the Teledentistry project there is a fee for the initial consultation with the local dentist, and GDP operating costs are reimbursed by the NHS.

TSI proposes to set up the system in 2006 with a capacity to treat 600 patients per year (to allow for the projected increase in patient numbers) and operate it from 2007–21 (15 years). The company proposes to facilitate treatment of 50% of eligible patients in 2007 and then 100% of eligible patients from 2008 onwards.

TSI revenues and costs

Capital costs (see Table App_1.11)

Table App_1.11 TSI's capital costs

Capital costs for teledentistry (at 2006 prices)	Cost per GDP unit (£)	Cost per hospital unit (£)
Videoconferencing unit	860	1,660
Codec and software	1,396	1,396
Imaging equipment	1,176	–
ISDN connection	99	99
Total capital cost	3,531	3,155

Note: 40 GDP units and three hospital units will be required. All items, except the ISDN connection, will have to be up-graded in years 2011 and 2016 at a cost of 7.5% of initial capital cost. Salvage value at the end of the project is 10% of initial cost (except for the ISDN connection). Costs of equipment and software include 20% import duties. Equipment and software (including the 5 yearly up-grade) can be depreciated at 20% of initial cost per annum over 5 years for tax purposes. Company tax rate is 30% and TSI has profits from other UK operations against which any losses on the Teledentistry Project can be deducted. GDP dentists and specialists will need to be trained in videoconferencing procedures: training costs of £1475 will be incurred in 2006.

Operating costs (at 2006 prices)

- Annual videoconferencing training: £40 p.a. per GDP and hospital unit starting i 2007
- Telephone charges: line rental £152 p.a. per GDP
- Equipment operator in Aberdeen: 1 Technician at an annual salary of £20,000
- Repairs and maintenance cost: £30 p.a. per unit (GDP and hospital).

Financing

TSI will receive a £10,000 Grant from Scottish Highlands and Islands Development Board (SHIDB) in 2007. TSI will borrow £80,000 from the Scottish Development Bank (SDB), a state bank, in year 2006 repayable over 5 years at a concessional rate of interest (5% real rate of interest per annum). TSI estimates its cost of capital at a real rate of 12% p.a.

Revenues (at 2006 prices)

The NHS envisages that it will pay TSI a fee of £140 per patient treated under the teledentistry programme. TSI hopes to negotiate a higher fee.

NHS costs

As noted earlier, the NHS will incur a capital cost of a £1000 equipment up-grade per GDP in 2006. Equipment cost is at 2006 prices and is inclusive of 20% import duty.

Variable costs of the two proposals and the current hospital programme

Table App_1.12 Variable costs per patient incurred by patients and NHS (at 2006 prices)

	Teledentistry (£)	Outreach visits (£)	Hospital visits (£)
Costs to patient			
Diagnosis and treatment plan (i)	29.50	–	–
Travel costs (ii)	4.39	42.5	–
Travel and consultation time (iii)	129.21	173.43	844.47
NHS costs			
TSI fee per patient	140		
Pre-consultation costs	46.42	9.17	9.17
Consultation costs	112.72	54.50	54.50
Post-consultation costs	29.72	18.17	18.17
Nurse and administration costs	0.20	3.69	3.69
Diagnostic images	7.38	0.88	0.88
Patient travel & accommodation (iv)	–	–	250.64
Specialist travel & accommodation (v)	–	130	–
Specialist travel time (vi)	–	22.50	–

Notes: (i) fee paid to NHS. (ii) cost of return car or bus travel from home to local or regional GDP. (iii) patient time valued at the market wage. (iv) cost of return car, bus, ferry or air travel from home to Aberdeen. (v) based on £650 return travel and accommodation costs per 5 patients treated. (vi) based on one day's travel time per 5 patients. Travel time valued at the specialist wage of £112.50 per day.

NHS hospital capital cost

Under both the Outreach and Teledentistry Programmes there will be between 400 and 600 fewer patients per annum treated in Aberdeen. The NHS estimates that this will free up one fully equipped $75m^2$ consulting facility from 2008 for other uses. NHS estimates its opportunity cost of consulting facilities at £3500/m^2, at 2006 prices, with a life of 20 years. It uses the equivalent annual cost method to work out an annual cost to estimate any long-run cost saving.

Assignment instructions

You are required to conduct a cost-benefit analysis of the Outreach and Teledentistry proposal, each of which proposes remote treatment of a given number of patients instead of treatment in hospital in Aberdeen. Your spreadsheets should contain a Variables Table plus Project (Market), Private, Efficiency and Referent Group Analyses for both proposals. The Project (Market) Analysis should detail the cost of hospital treatment, provide a breakdown of the Teledentistry proposal into its TSI and NHS components, and analyse the Outreach proposal, but it should ignore patient costs. The Private Analysis concerns TSI only. The Referent Group consists of the residents and government of the United Kingdom. Net Referent Group benefits are to be disaggregated into those accruing to the NHS, the tax office, financial institutions, and patients.

The proposals should be assessed against the costs of the current treatment regime using real rates of discount of 6%, 9% and 12%. The NHS wishes to know which if either of the two proposals should be implemented, and in the case of the Teledentistry programme, the level at which TSI's fee should be set. The sensitivity of the results to values of key variables, including the TSI fee, the value of patient time, and any other variables you consider to be important, is to be assessed. Your recommendation to the NHS should include some discussion of factors not covered by the analysis, such as changes in waiting times for treatment, and suggestions for further work on the analysis.

While the Teledentistry and Outreach proposals involve the same amount of specialist treatment time as the current hospital treatment system, the Outreach Project involves additional specialist time in the form of travel. The NHS is concerned about the increasing shortage of dental specialists, which is likely to result in a significant rise in the cost of specialist time. It has been estimated that the cost of specialist time is likely to rise at an annual rate of 8–10% above the rate of inflation starting in 2007. Perform a sensitivity analysis of this scenario and indicate how it would affect the relative net benefits of the two proposals.

The format of the report

The report is to be presented in a format similar to that of the ICP case study reported in Chapter 13 of the text. It should begin with an executive summary of approximately one page in length. The main body of the report should not be more than 10 pages in length (1.5 spacing, font size 12) including your summary tables of results. The printouts of the spreadsheets, including the Input Data should be in a separate appendix. You should print only the spreadsheets for the base-case scenario. Try to condense these so all years fit on one A4 page (landscape). You must submit an electronic copy of all your spreadsheets with your printed assignment.

A1.6 The Defarian Early Childhood Intervention Program (DECIP)*

Introduction

The DECIP was designed by the Department of Education in North Carolina as an experimental early childhood intervention programme to provide intensive pre-school services to children in very low-income households and classified as "high-risk" in relation to expected intellectual and social development. The programme was designed to provide participant families with intensive, specialist day care for their children from infancy to 5 years of age. The programme began in 1985 with a randomly selected group of 200 children born in 1985 into low-income households and considered "high-risk". An equal number of children were assigned to the DECIP and to a control group who could, if their parents chose, attend a regular pre-school.

Under DECIP all 100 children followed a specially designed pre-school programme for the first five years of their lives, at a centre which operated from 7.30 am to 5.30 pm, for five days per week for 50 weeks per annum. The specially designed curriculum emphasised language development but also provided for other developmental needs, including medical and nutritional services.

The children began by attending a nursery in their first year. Each of the 10 nursery groups accommodated 10 infants and was staffed by three trained carers. In years two and three of the programme, the children were organised into groups of eight with two teachers/carers per group. In years four and five, each group contained 12 children with two staff.

Researchers then traced: (i) the educational attainment of participants in both groups to age 21 (i.e. year 2006); (ii) the educational attainment and employment performance of mothers of both groups over the same 21-year period; and (iii) the incidence of smoking and criminal records of both groups by age 21 years.

The main costs and benefits of DECIP can be grouped into two categories: (i) those that impact directly on the project's participants and families (namely, the Department of Education, the child participants in DECIP, their mothers, and future generations born to DECIP participants); and (ii) those that are external to the project or are indirect effects. The benefits that need to be quantified for the cost-benefit analysis are:

i   lifetime effects on educational attainment and earnings of DECIP participants in comparison with the control group (direct effect);
ii lifetime effects on earnings of the mothers of DECIP participants compared with the control group (direct effect);
iii lifetime effects on earnings of descendants of DECIP participants compared with the control group (direct effect);
iv reduced need for special education while at school-going age (direct effect);

v   health and longevity (life expectancy) benefits to participants (both direct and indirect effect);

vi   reduced crime costs to state government and society (external effect);

vii   reduction in welfare payments to DECIP participants by Federal government (indirect effect).

The costs of DECIP can be classified into two categories:

i   the additional costs to the State's Education Department of the provision of child-care under DECIP in comparison with public pre-schooling and/or parental care;

ii   the extra educational costs to the State's Education Department to provide DECIP participants with additional post-secondary education, given their higher retention rates and performance in secondary education.

On the basis of the information provided in the following sections you are required to undertake a cost-benefit analysis of DECIP on a *per child* basis. For the purpose of the study the Project (Market) Analysis should include only the direct costs and benefits to the project's participants and families as defined above. Any indirect effects and externalities should be included in the Efficiency Analysis. The Referent Group includes all those affected by the project at the level of the State of North Carolina, i.e. excluding the loan, import duties, indirect taxes, income taxes and welfare payments paid to/by the Federal Treasury Department. Although the project's historical cash flow begins in 1985, net present values for the Referent Group should be calculated for the present year (2006). All $ amounts provided below are expressed in constant 2006 prices unless otherwise indicated.

Programme costs and financing

Child-care costs

To operate the DECIP centre required the rental of appropriate space in the locality, the cost of equipment, the employment of suitably qualified carers and pre-school teachers, as well as volunteers, and material inputs including those required for medical and nutritional services provided by the programme. It should be assumed that each child attends the centre for 40 hours per week and that the property in which the centre is located is reallocated from existing spare capacity from the properties owned by the State Government. Assume that all classrooms are being used each year. School equipment includes 40% imported components subject to a 10% ad valorem tax (included in cost). Details of these child-care inputs and costs are provided in Table App_1.13.

Table App_1.13 DECIP child-care inputs and costs

	Year 0	Years 1 to 2	Years 3 to 4
Centre rental (per group) ($)	1800	2000	2400
Equipment, supplies, etc. (per group) ($)	1000	1200	1500
No. of staff/group	3	2	2
No. of children/group	10	8	12
No. of volunteers/group	2	2	2
Salaries/staff/annum ($)	30,000	40,000	45,000

For the control group, it was found that participation rates in regular public pre-schools over the first 5 years of their lives were: 20%, 30%, 70%, 75% and 80% respectively. The marginal costs for a child attending a regular public pre-school for 40 hours a week are estimated at $2000 per annum, of which 70% is staff salaries and 30% material inputs. (Assume 40% imports subject to 10% tariff.)

It should be assumed that a control group child not in a pre-school was cared for by paid carers or relatives, or by unpaid family members. For the purpose of the analysis you should assume that each hour of paid care costs $3.00. The opportunity cost, given the surplus of carers available in the informal sector, is estimated at $2.00 per hour. Assume also that total hours of parental care are split evenly between paid carers and unpaid family labour. The opportunity cost of family labour in low-income households is estimated at $4 per hour. Volunteers at DECIP are not paid a wage for their assistance, but it has been estimated that the opportunity cost of their time (forgone leisure) is $1.50 per hour.

Educational costs

One of the main benefits of DECIP is the higher educational attainment of participants. School histories were kept for all participants in DECIP and in the control group. The major difference between the two groups was in the percentage of school years a child spent in special education. For the control group 60% of children spent 30% of their 12 years of schooling in special education. For the DECIP group 20% of the group spent 10% of their schooling in special education. It has been estimated that each year of special education has an additional marginal cost of $12,000 per child per annum compared with the marginal cost of a regular public school. Teachers' salaries account for 60% of the cost in special education. However, there is a shortage of teachers qualified to provide special education, and though they are paid at the same rate as regular teachers, it has been estimated that their opportunity cost is around 25% more than their wage. Assume that the additional special education costs are spread evenly over the child's life from 6 to 16 years of age.

The higher educational attainment of the DECIP participants by age 21 implies that there is an increase in the numbers of students attending higher educational institutions after completion of grade 12. The costs of providing additional higher educational places must therefore be counted as a cost to the programme. It was found that the probability of being enrolled in a 3-year post-secondary educational programme increased from 15% for the control group to 40% for the DECIP group. Each additional post-secondary place is estimated to cost $10,000 per annum for 3 years, from student age 17 to 19 years. Students are charged fees of $2000 per annum, paid to the State's Education Department.

Financing

The cost of implementing the project was financed partially through a special loan from the Federal Government in 1985 equal to $200,000 (2006 prices) for the experimental programme, at an interest rate (real) of 2% per annum, repayable over 5 years from 1986. The balance was financed under the State Government's regular budgetary allocation to the Education Department.

Programme benefits

Lifetime earnings of participants

Econometric studies have found that an individual's lifetime earnings can be reasonably predicted on the basis of educational attainment by age 21. Educational attainment levels were recorded for all 200 participants in 2006, i.e. at age 21. Table App_1.14 shows the distribution of each group level across a range of educational level categories and Table App_1.15 shows the probabilities for each educational category falling into each income group. From these data the mean expected earnings for the two groups can be estimated for the year 2006. It should be assumed that thereafter the earnings gap between the DECIP and control group widens by 2% per annum from age 21 (2006) to age 65 (2051) and that income tax, payable to Federal Treasury, is 30% of earnings. Wages are regarded a reasonable indicator of opportunity cost.

Table App_1.14 Educational level attained by age 21 years (2006) (%)

	Category 1 (<T9 years)	Category 2 (9–11 years)	Category 3 (12 years)	Category 4 (2–3 years post-secondary)	Category 5 (Completed university degree)
Control group	35	22	30	10	3
DECIP Group	10	23	32	25	10

Table App_1.15 Probability of starting income level by educational category (2006) (%)

	$8–10K	$10 > 15K	$15–30K	$30–50K	$50–70K
Cat. 1	80	15	4	1	0
Cat. 2	75	18	5	2	0
Cat. 3	50	25	15	6	4
Cat. 4	20	20	30	25	5
Cat. 5	10	10	30	40	10

Lifetime earnings of participants' mothers

The provision through DECIP of five years of high-quality, full-time care and education of programme children increased the opportunities for their mothers (mostly single parents with very low educational backgrounds) to obtain employment, training and other productivity-enhancing activities. These opportunities resulted in increased annual earnings for the mothers of the DECIP group compared with the control group equivalent to $2000 per annum (2006 prices, undiscounted) over the 21-year period 1985 to 2006. Assume that mothers work another 20 years beyond 2006 and that this earnings differential increases by 1% per annum in real terms from 2006 to 2025, when the mothers are assumed to retire. Assume that income tax is levied at 30% on mothers' earnings, and is payable to Federal Treasury, and that the market wage reflects opportunity cost.

Lifetime earnings of participants' descendants

There is substantial evidence of a positive relationship between parental education and income, and the educational attainment and income of their children. Estimates have been derived for the elasticity of child income with respect to the income of parent(s). This allows us to calculate the impact of the higher education and income levels of the DECIP group in comparison with the control group, on the earnings of future generations, making some simplifying assumptions about the numbers of offspring produced and years spent in the labour force. It is estimated that on average the future generations' annual earnings for participants in the DECIP group are between $1000 and $3000 more than the amount for those in the control group. In the base-case scenario, a mid-point estimate should be used, and it should be assumed that this difference applies from year 2035 and continues in perpetuity. Assume that marginal income taxes on earnings, payable to Federal Treasury, are 30% and that the market wage reflects opportunity cost.

Reduced health costs

There are numerous health benefits associated with higher levels of educational attainment, but estimating these is extremely complex. For the purpose of this study it is to be assumed that the main health-related benefits can be estimated through the lower incidence of cigarette and marijuana smoking among the better-educated DECIP group in comparison with the control group. The main effects are the improvement in health and longevity and the reduced cost of health care. Participants in the two groups were surveyed in 2005 (aged 20 years) and it was found that the rates of smoking for the control group and DECIP programme group were about 55% and 40% respectively. It has been estimated that being a smoker at age 20 costs the state's health sector an additional $750 per annum more than a non-smoker. Assume this applies from age 21 onwards for the rest of the individual's life. Assume also that the patient pays one-third of the health cost in fees to the state health provider.

From available data it is also estimated that being a non-smoker at age 20 increases longevity by 7 years. For the purpose of the analysis, assume that this implies that the average age at death of a non-smoker is 77 years compared with 70 for a smoker. The value of human life is the subject of much debate, and is a highly contentious issue generally. Values from the various studies vary between $150,000 and $300,000 per year of life (2006 prices). For the purpose of the base-case scenario, use a mid-point value per year. (Assume this is net of any additional costs to the state.)

Reduced crime rates

One important effect of the DECIP is the reduction in crimes committed by the programme children. Investigation of official records of the state police and courts indicated that programme participants had significantly lower juvenile delinquency and crime rates than the control group. The incidence of arrest by age 21 was 20% in the case of DECIP participants compared with 40% for the control group. It has been estimated that each arrest costs the state criminal justice system $2000 (2006 prices) on average. For the purpose of the analysis, assume that the cost saving occurs when the child is aged 18 years. In addition, from other studies it is believed that the likelihood of the child becoming a career criminal in adulthood is also significantly reduced if the child had not committed a crime by age 21.

Among the target population for DECIP, there is a 40% chance that the child has a criminal career in adulthood. It has been estimated that a career criminal costs the justice system approximately $30,000 in present value terms (2006 prices). If participation in DECIP reduces the probability of the child committing a crime by age 21 from 40% to 20%, it is reasonable to assume that the probability of becoming a career criminal in adulthood is also reduced by the same proportion. Apart from the decreased justice system costs, there are also the reduced costs to the victims of crime. These are estimated to be 120% of the state's justice system costs. For the purpose of this study it should be assumed that the cost savings occur in the year the child turns 30 years of age.

Reduced welfare payments

As the DECIP increases the probability of a child attaining a higher level of education and becoming employed, it is to be expected that there will be a reduction in the level of social security payments from the Federal Government. It was found that at age 21, 10% of the participants in DECIP were dependent on welfare payments as opposed to 25% of the control group. The average level of welfare payments from the Federal Government to each social security-dependent household has been estimated at $20,000 per annum, and the marginal costs to administer the programme at $2000 per recipient household per annum. For the purpose of the analysis it should be assumed that the reduced welfare payments occur over the age 21 to 65 years for the DECIP participant and that *no* income tax is paid on this income.

Assignment instructions

On behalf of the Education Department in the State of North Carolina, you are required to undertake and report the findings of a cost-benefit analysis of the DECIP programme on a per-participant basis over the expected lifetime of the participants. You should calculate IRRs for the Project (Market), Private, Efficiency and Referent Group Analyses and report NPVs for the Referent Group on an aggregated and disaggregated basis for the year 2006 (the present year.) The state (and Federal) governments use a 5% discount rate (real) for investment decision-making, but are interested in knowing the sensitivity of the results at 3% and 8% discount rates.

You should also conduct and report the results of a sensitivity analysis in which you analyse and comment on, at least, the sensitivity of the results to: (i) the expected earnings of the participant families, and (ii) the assumed value of life in estimating the health benefits. You are encouraged to explore sensitivity to other variables with a view to identifying those that have most impact on the results and that would warrant further investigation. In your report you should also indicate other omitted costs and benefits that could be of potential significance to the decision-maker.

Although the Federal Government is not part of the Referent Group, it is important for the State Government to know what the overall impact of the project is on Federal Treasury finances as they might want to build a case for further concessional loans should it be decided to repeat the DECIP project in the future. Your report should offer advice to the State Government in relation to this issue.

Your written report should not be more than 12 pages in length, excluding tables and a one-page executive summary. You must submit an electronic copy of all your spreadsheets with your printed assignment.

Note

* This is a hypothetical project based on a programme described in the references provided.

References

Here are some useful references (not provided to students before undertaking this case study).

Barnett, W. Steven (1993) "Benefit-Cost Analysis of Preschool Education: Findings from a 25-Year Follow-Up", *American Journal of Orthopsychiatrics*, 63(4): 500–508.

Barnett, W. Steven (2000) "Economics of Early Childhood Intervention", in J.P. Shankhoff and S.J. Meisels (eds), *Handbook of Early Childhood Intervention* (2nd edition), Cambridge: Cambridge University Press, pp. 589–610.

Diefendorf, M. and Goode, S. (2005) *The Long Term Economic Benefits of High Quality Early Childhood Intervention Programs: Minibibliography*, NECTC Clearinghouse on Early Intervention and Early Childhood Special Education, Chapel Hill, NC: UNC-CH. Available at: www.nectac.org/~pdfs/pubs/econbene.pdf

Masse, L.N. and Barnett, W. Steven (2002) "A Benefit-Cost Analysis of the Abecedarian Early Childhood Intervention". Available at: http://nieer.org/resources/research/AbecedarianStudy.pdf

A1.7 A pulp mill for Tasmania?*

Introduction

A Scandinavian company, Nordic Forest Products Ltd (NFP), is proposing to build and operate a pulp mill at Devil River on the north coast of Tasmania. Construction will start in 2011 and the mill will take two years to build (Years 0 and 1). In its first year of operation (2013, Year 2 of the project) it will convert 3.2 million tonnes of fibre obtained from eucalypt logs into pulp for the export market (see Table App_1.16). The quantity of fibre processed is scheduled to rise by 80,000 tonnes per annum until a throughput of 4 million tonnes of fibre per annum is achieved in Year 12 of the project (see Table App_1.16). The Year 12 level of throughput will be maintained until Year 31 of the project, after which the mill will cease to operate. In the long run the primary source of fibre will be privately owned eucalypt plantations, with 33% of private plantation fibre coming from plantations established under Managed Investment Schemes (MIS). However, these plantations will not be fully mature at the start of operations and the shortfall in the initial years of the project will be made up of logs from the state's native forests and plantations operated by Forestry Services Tasmania (FST). For the purpose of your analysis, assume that all the timber logged for this project would otherwise have been processed into wood chips for export.

Table App_1.16 Sources of eucalypt fibre (tonnes)

	Project year 2	Annual change years 3–12
FST native forest	2,560,000	−171,000
FST plantations	150,000	0
Private plantations	490,000	+251,000

The mill will cost A$2 billion to construct and a further A$200 million will be required to construct pipelines for water supply and effluent disposal (see Table App_1.17). In addition, Highways Tasmania (70%) and the Federal Government (30%) are proposing to spend A$200 million on road upgrades to cope with the transport of logs. No scrap value or decommissioning cost of the mill has been assessed.

Table App_1.17 Capital costs (millions)

	Project year 0	Project year 1
Plant and equipment	600	1400
Pipelines	0	200
Road upgrades	100	100

The ratio of fibre input to pulp output is 4.1:1 for native forest logs and 3.6:1 for plantation logs. Pulp is expected to sell for US$590 per tonne in the initial year of the mill's operation. However, over Project Years 3–11 the world pulp price is expected to fall by 1.4% per annum and then remain constant in real terms. The exchange rate is expected to be US$0.85 = A$1. The project expects to earn A$20 per tonne of pulp produced from sales of surplus energy, and it expects to receive a further A$22 per tonne of pulp from sale of renewable energy certificates (RECs) earned under the Federal Government's renewable energy subsidy programme. Although the project receives the proceeds from the sale of the RECs, the external benefits from the reduced reliance on fossil fuels (and output of carbon) are to be treated as a benefit to the global population, also valued at A$22 per tonne of pulp output.

Operating costs

NFP will pay stumpage fees (i.e. royalties on logs) to the suppliers of logs, a road levy to Forest Services Tasmania to pay for the maintenance of logging roads, and the harvesting and transport costs required to deliver the logs to the mill. The stumpage rate on logs supplied from native forests is set at 2% of the world pulp price. Stumpage rates on plantation timber are negotiated with suppliers at the values shown in Table App_1.18.

Table App_1.18 Cost of fibre ($/tonne fibre)

	FST native forest	FST plantations	Private plantations
Stumpage	13.18*	27.00	33.00
FST road costs	7	7	4
Harvesting cost	20	15	15
Transport cost	24	15	10

Note: * Calculated as 2% of the pulp price reported above converted to A$ at the exchange rate reported above.

Other operating costs including costs of labour, maintenance, chemicals and energy, and ocean freight are shown in Table App_1.19.

Table App_1.19 Operating costs ($/tonne of pulp)

Operating cost	Price in A$ and US$
Labour (including 6.1% state payroll tax)	42
Maintenance	38
Chemicals and energy	US$ 50
Ocean freight	US$ 70

The project will also use 26,000 megalitres (ML) of water per annum, which will be supplied by Tasmania's Hydro Electric Commission (HEC) at a price of A$35 per ML.

Financing

Shares in 20% of the proposed project will be owned by Tasmanian shareholders with the remaining shares held by Scandinavian interests. In Year 1, NFP will borrow 70% of the capital cost it incurs to undertake the project through an export credit facility provided by a consortium of Scandinavian banks at a real interest rate of 3.5% over a 20-year term, with interest and principal repayments starting in Year 2, on an annuity basis. The balance of the capital costs will be borne by the Tasmanian and Scandinavian shareholders. The capital cost of the mill and the pipelines can be depreciated on a straight line basis over the operating life of the project for tax purposes. The business income tax rate is 30%. NFP requires a 10% real rate of return on the project. NFP has income from other ventures in Australia against which any losses from this project can be offset. The Scandinavian shareholders will repatriate their after-tax profits.

Opportunity costs

- *Timber*: Independent analysis of world pulpwood markets suggests that the stumpage value of native forest timber supplied to the mill is actually 4% of the world pulp price. In addition, logs constituting 5% of the volume of timber supplied from native forests are thought to come from high conservation value areas where the non-timber value of stands exceeds the stumpage value by 10%. It has been calculated that the stumpage fees for logs supplied from private plantations under the MIS are subsidised by 30% through federal government business income tax concessions.
- *Labour*: While the mill will generate employment in its construction and operation phases, it is estimated that construction workers will be diverted from other projects and that only a small proportion of the operation's labour force would otherwise have been unemployed. It is estimated that 5.6% of the gross operations wage bill constitutes employment benefits which are divided in the following amounts: workers 1.2%; State 0.4% in the form of payroll tax; and Commonwealth 4% in the form of personal income tax, GST and reduced social security payments.
- *Water*: While the water supplied to the mill by Hydro Tasmania will not involve any reduction in the volume of power generated, it will be diverted from irrigation. Irrigators in the region are currently paying A$46 per megalitre.
- *Transport*: It has been estimated that supplying the mill with fibre will involve an additional 4.2 million kilometres of log-truck travel per annum. Currently Tasmanian roads serve an estimated 2,900 million vehicle kilometres per annum, and the annual cost of traffic accidents, including injury and loss of life, is estimated by the Bureau of Transport Economics to be A$310 million.
- *Air and water quality*: It has been estimated that operation of the mill will increase the ambient concentration of ultra-fine particles in the northern Tasmanian airshed, resulting in a 0.75% increase in the incidence of respiratory disease in the region. This will result in three additional deaths, 300 additional hospital admissions and an additional 300 working days lost per annum. A study has suggested that the cost of each death is A$1.026 million, hospital admissions cost A$3,870 each, and days of work lost cost A$150 per day.
- The mill will dispose of 64,000 tonnes of effluent per day through a pipeline into Bass Strait. While the limits set for dioxins and furans per litre of waste discharged equal or improve on levels set by the US EPA, Environment Canada and the EU, as well as meeting various best practice guidelines, there is concern for the long-run effect of the effluent on Bass Strait seal colonies and fisheries, but no estimate of the cost is available.
- The plantations supplying the mill will reduce stream run-off by absorbing rainfall and releasing it into the atmosphere through evapo-transpiration. The reduction in

stream-flow may affect the availability of irrigation water in northern Tasmania, but there is no estimate of the impact.

- Proponents of the mill argue that sustainable forestry is carbon-neutral and that exporting pulp, as opposed to wood-chips, will reduce greenhouse gas emissions associated with ocean transport. Opponents of the mill argue that cutting mature forests reduces the amount of carbon stored in the trees and the soil.
- Tasmanian farmers, fishermen, winegrowers and tourism operators have expressed concern that the pulp mill will affect Tasmania's "clean green" image. Mill supporters point to the example of New Zealand, which has several pulp mills, but enjoys a positive environmental image.

Assignment instructions

On behalf of the government of Tasmania, you are required to undertake and report the findings of a cost-benefit analysis of the NFP proposal. The analysis is to be conducted in millions of 2011 A\$, to two decimal places. You should calculate NPVs and IRRs for the Project (Market), Private and Efficiency Analyses and report NPVs in 2011 for the Referent Group (Tasmanian Government and community) on an aggregated and disaggregated basis. The Tasmanian government uses a 5% discount rate (real) for investment decision-making, but is interested in knowing the sensitivity of the results at 3% and 8% discount rates.

The government also requires an analysis of the impact of the project on the Commonwealth government's revenue and expenditure flows, as well as NFP, the foreign banks and any other non-Referent Group stakeholders affected by the project.

You should also conduct and report the results of a sensitivity analysis in which you calculate and comment on, at least, the sensitivity of the results to: (i) the world price of pulp, and (ii) the exchange rate. You are encouraged to explore the sensitivity of the results to a small number of other variables with a view to identifying those that have most impact on the results and that would warrant further investigation. You should also consider and discuss the implications of the external, environmental benefits to the global community from reduced reliance on fossil fuels. Should these be treated as benefits in the Referent Group analysis? Discuss also the threshold levels of any other environmental benefits or costs in terms of the Referent Group net benefit at least breaking even. You should also indicate if there are any omitted costs and benefits that could be of potential significance to the decision-maker. You must submit an electronic copy of all your spreadsheets with your printed assignment.

Note

* Although based on an actual project proposal using publicly available information, some of the numbers and assumptions have been modified or simplified to make the project more suitable for teaching purposes.

A1.8 Qingcheng Water Project[1]

Introduction

Water supplies for agriculture, domestic and industrial uses in China's Shandong Province have long been sourced from the Yellow River (the Yellow River region was one of the four initial major ancient irrigation societies, alongside Mesopotamia, the Nile valley, and the Indus Basin).[2] However, silt loading has become a major environmental and public health concern over the past century, as usage of Yellow River water for irrigation, domestic and industrial water supplies continues without many alternatives.[3] Although some infrastructure has been developed to combat the adverse effects of silt loading (which include increased salinity, erosion and the reduced health of the river system in general), this sort of response has largely been confined to the lower half of the Yellow River. Hence erosion has continued to impact the Loess Plateau and the North China Plain.[4] Indeed, the majority of erosion in the North China Plain is caused by increasing salinity and inefficient water usage.[5]

Under these circumstances, the International Water Company (IWC) – an internationally renowned water company that processes and supplies water in many countries around the world – has expressed its interested in the province. IWC and Qingcheng Municipality Government (QMG) propose to establish a new company in 2009, named China Water Company (CWC) which will operate over a period of 30 years. It will be responsible only for the water treatment project via a new desalination plant (not transmission pipelines, and other existing infrastructure). According to the proposal, some facets of the "BOT" (Build-Operate-Turnover) model will be adopted for this project, implying that IWC will receive all the revenue in the first 15 years of operation (from 2009 to 2023), and then turn over the entire project to QMG in 2024 for a determined purchase price of ¥100 million.

Variables

Consumption

Water is currently pumped from the Yellow River, with the resulting price of raw water set at ¥0.94 per cubic metre. The final price for drinkable water (set in 2000 by the government) was ¥2 per cubic metre. CWC, on the other hand, will not source its water from the Yellow River but from the nearby East China Sea and other salt-water deposits which have infiltrated the region. It will be able to source water for ¥0.90 per cubic metre, and implement a variable pricing scheme based on usage and demand in order to encourage consumers to use water more efficiently. The move from the current government-set price to the new variable scheme will be gradual in order to avoid welfare losses due to switching costs for water users or over-consumption in response to lower prices. Thus, for the first 10 years of the project (including the first nine years of operation), the average price for water will be ¥2.00 per cubic metre. From years 2019 through 2028, the average water price will be ¥1.85 per cubic metre of water. Finally, for the last ten years of the project, the average price will be ¥1.80 per cubic metre.

A study of the city's water industry shows that the residents of Qingcheng demand 185,044,126 cubic metres of drinkable water per annum. It is estimated that CWC will initially satisfy 80% of this demand. The sourcing arrangement for the remaining proportion will remain unchanged. It is assumed that water consumption behaviour becomes more efficient over the life of the project. In consequence, more people and projects can benefit from the same water output. It is expected that access to clean water will rise to 85% of Qingcheng's urban population of 2.4677 million for the period from 2018 through 2027. This number is then predicted to increase again to 90% in the last ten years of the project.

Investment costs

IWC proposes to invest ¥112.5 million in 2009 to purchase 25 units of water machinery (equipment), 4 vehicles, and a 30,000 sq m plant, all of which are expected to last for 30 years (not including the start-up year). The estimated investment costs are reported in Table App_1.20, including the salvage value on the initial capital inputs once their usage-life is complete.

Table App_1.20 Investment costs: Qingcheng Water Project

Investment costs	Units	Price (¥)	Cost (¥)
Fixed investment			
Water equipment (units)	25	520,000 [1]	13,000,000
Vehicles (units)	4	125,000 [2]	500,000
Buildings (m²)	30,000	3,300	99,000,000
Total			112,500,000
Salvage value	10%		
Depreciation			
	(Life years)		
Equipment	30		
Vehicles	30		
Buil dings	30		

Notes: 1 Including import duty at 10% of c.i.f. value.
2 Including import duty at 20% of c.i.f. value.

In addition, initial investments have to be made in the first operational year, 2010, in the form of working capital to ensure the smooth operation of the project. In terms of quantities needed, it was determined that 25% of the quantities of raw water (cubic metre), chlorine (tons), the flocculent Polyaluminium Chloride (PAC) (tons) and spare parts for maintenance (rounded to the nearest whole part) needed for full annual operation would be required as working capital. Prices per unit and tax/subsidy constraints are the same for working capital investments as for operating costs.

IWC hopes to produce at 82% of its full capacity in its first operational year, and at full capacity from the second year. Assume all operating costs are incurred in proportion to output levels.

Operating costs

Approximately 47.8 million kV of electricity, 2000 units of chlorine, 1800 tons of PAC and 25 units of spare parts for water equipment maintenance are required annually to operate the plant. The corresponding price, government subsidy, and import duty at c.i.f. values are listed in Table App_1.21.

Table App_1.21 Raw materials: Qingcheng Water Project

	Units	Price
Raw water(cubic metre)	151,056,430	0.90
Power (kilo volt p.a.)	47,785,798	0.70 [1]
Chlorine (tons p.a.)	2000	1910 [2]
PAC (tons p.a.)	1800	3696 [3]
Maintenance (p.a.)	25	576,000 [4]
Insurance (p.a.)	1	230,000
Management (per month)	12	2,998,000
Miscellaneous (p.a.)	1	36,553,100

Notes:
1 Containing government subsidy at 10% of price.
2 Including import duty at 8.6% of c.i.f. value.
3 Including import duty at 5.5% of c.i.f. value.
4 Including import duty at 7.0% of c.i.f. value.

CWC is expected to employ 326 workers in its operations. All workers will be recruited from Qingcheng and paid the annual market rate of ¥8,028 for their work. There is considerable unemployment in the city and the shadow-price of unskilled labour recruited in Qingcheng is believed by the government to be 50% of the market price. All skilled management is assumed to be valued efficiently in the labour market regardless of its place of employment.

Financing

IWC intends to raise US$ 8.01 million (equivalent to 62.5 million Yuan as per the spot exchange rate) from foreign banks in 2009. The credit carries a real interest rate of 5.88% p.a., repayable from 2010 in ten equal annual instalments (including principal and interest). Interest payable on debt is allowable against corporate tax. The remainder of the investment will be financed from IWC's own sources, but the company expects to be able to repatriate all profits as and when they are made.

Tax and incentives

China's current tax policies encourage Foreign Direct Investment (FDI). Thus, CWC will be exempt from taxation for the first two years of operation, and will pay a reduced tax rate equivalent to 50% of the standard corporate tax rate (which is 24% of taxable profits) for the following three years. Starting from the sixth year of the project the company will pay the full tax rate.

Corporate taxes[6] are levied on the profits calculated after allowing for depreciation and interest. For accounting purposes, equipment, vehicles and buildings are depreciated using the straight-line method over the lives indicated previously. The depreciation charge can be levied from 2010 onwards.

External costs and benefits

Agricultural production

A serious side-effect of the old method of sourcing water from the Yellow River was severe soil erosion which has steadily reduced the amount of productive agricultural land in the region over the past 50 years. Were the Qingcheng region to continue pumping water from the Yellow River, this erosion would continue at a steady rate.

For this reason it is required that any municipalities which pump water from the Yellow River must contribute an "erosion" tax to the central government, which in part helps to pay aid packages to farmers affected by erosion. The erosion tax is set at 3% of the value of water pumped (the value is calculated at the market prices which consumers pay for the pumped-in water). It is estimated that prior to this water project, all water was pumped and transported from the Yellow River, and after the project inception (starting year 2010) all water will be provided either by CWC or the smaller water companies (no water will be pumped from the Yellow River). This erosion tax is paid by the local government and is not necessarily passed down to consumers.

This water production project will combat the erosion problem, and the residents of Qingcheng in Shandong Province will save 5 square kilometres of agricultural land per year which would have been lost to erosion otherwise. Each 5 km² of productive land has a net value of approximately ¥1 million per year to the farming population (including jobs and agricultural products that would have been lost otherwise), and 5% of this value would have been paid to the local government in the form of small business farming revenue taxes.

Since use of this land is assumed to continue to produce in perpetuity, in the final project year its present value should be considered as a lump sum. QMG believes that a 5% discount rate would be appropriate for this exercise.

Population health

It is widely known that unhealthy drinking water can induce diseases such as cancer, liver disease, lithiasis, heart disease, dementia and ossification. According to a report by the United Nations, unhealthy drinkable water and other poor sanitary conditions such as foul air, waste and noise can induce 80% of chronic diseases. It is estimated that the per capita medical expense for Qingcheng's urban resident on such chronic diseases is 85 Yuan per year. Authorities believe that: (i) unhealthy water accounts for 25% of that cost figure for chronic diseases; and (ii) the health benefit to the population is valued in terms of avoidable health costs only, such that gains in productivity due to improved health are not considered.

Pollution

Per capita atmospheric contaminant emissions in China stand at 1.8 tons p.a. at the moment. The construction and operation of the project will cause an initial 0.018 ton increase in per capita emissions in the Qingcheng region (this equates to a 1% increase in emissions initially in the region).

Using the classic Gaussian air pollutant dispersion equation, research has indicated that air pollution is likely to stay within the environmental regulation zones within the region. It is also estimated from this dispersion equation that this concentration of air pollutants will slowly, but steadily, build up in the area, with ambient air pollution increasing incrementally by 0.001% per year over the course of the project per year.

Studies have shown that in the local region, each 1% increase in emissions costs the government ¥29 million in health and well-being costs. It has been assumed for the purpose of this cost-benefit analysis that this cost would remain uniform through different increments of pollution emissions (e.g. 0.1% change in emissions is associated with a ¥2.9 million change in associated costs; 0.05% change in emissions is associated with a ¥1.45 million change in associated costs, etc.).

In addition, it is expected that there will be approximately ¥20 million in other direct environmental damages associated with the operation of the full-production plant. This amount will accrue each year. The government can choose to mitigate these costs by requiring IWC to install a pollution scrubber and disposal system, which will offset direct environmental costs by ¥10 million per year, and reduce the starting increase in emissions from 0.018 to 0.01 tons per capita. The pollution scrubber and disposal system would be installed in 2009 and would cost ¥20 million. It is expected that there will be no additional operating and maintenance costs, and that its market price is a reasonable indication of opportunity cost.

Assignment instructions

On behalf of the Qingcheng Municipal Government (QMG), you are required to undertake and report on the findings of a cost-benefit analysis of the CWC water project over the expected lifetime of the operation. You should calculate IRRs for the Project (Market), Private, Efficiency and Referent Group Analyses and report NPVs for the Referent Group on an aggregated and disaggregated basis for the year 2009 (the present year). Although the Central Government should be included as a stakeholder in the Referent Group Analysis, QMG is also seeking advice on the extent to which it should seek assistance from Central Government if the project goes ahead, especially in relation to negotiations with the non-Referent Group stakeholders, IWC and the foreign bank. The Central and Local governments use a 5% discount rate (real) for investment decision-making, but are interested in knowing the sensitivity of the results at 10% and 15% discount rates.

Your report should offer advice to QMG in relation to this issue, by addressing the following questions:

1 Is IWC likely to be interested in participating in this water project?
2 Should QMG support the proposed project by granting IWC the proposed tax concessions and/or other concessions, and if so, should QMG request assistance from the Central Government?
3 What conditions, if any, should be imposed on the project if approval is given?

It is acknowledged that there is a level of uncertainty involved in answering the three questions posed above. You are asked to conduct a sensitivity analysis to provide a more rigorous understanding of the issue at hand.

Sensitivity analysis and risk analysis

You should conduct and report the results of a sensitivity analysis in which you analyse and comment on, at least, the sensitivity of the results to: (i) the expected water usage (%) for the region over the life of the project; (ii) the assumed value of the benefits of clean water (health); and (iii) the expected growth in pollution over the life of the project. You are also encouraged to explore sensitivity to other variables with a view to identifying those that have most impact on the results and that would warrant further investigation. In your report you should also indicate other omitted costs and benefits that could be of potential significance to the decision-maker.

Written report

Your written report should not be more than 12 pages in length, excluding tables and a one-page executive summary. Printouts of cash-flows (to be included as appendixes) should be cut-off at year 21 and each part of the CBA (i.e. Project (Market), Private, Efficiency and Referent Group Analyses) printed on one A4 size page to save paper, but you must submit an electronic copy of all your spreadsheets with your printed assignment. (Only the base-case scenario spreadsheets should be printed.)

Notes

1 The project described in this study is not a case study of an actual project. It draws heavily on a case study developed for teaching purposes by Professor Qi Jian Hong, School of Economics, Shandong University, PRC. Acknowledgments are also due to Laura Davidoff and Kim Nguyen who adapted the original case study for teaching purposes at the School of Economics, University of Queensland.
2 Chengrui, Mei and Dregne, Harold (2001) "Review Article: silt and the future development of China's Yellow River", *The Geographic Journal*, 167(1) (2001): 7–22.
3 Ibid.
4 Ibid.
5 McVicar, Tim (2002) "Overview", in T.R. McVicar, Li Rui, J. Walker, R.W. Fitzpatrick, and Liu Changming (eds), *Regional Water and Soil Assessment for Managing Sustainable Agriculture in China and Australia*. ACIAR Monograph No. 84 [online]. Canberra, Australia: ACIAR. Available at: www.eoc.csiro.au/aciar/book/index.html (accessed June 2009).
6 Under China's financial federalism, corporate tax falls into the category of local government revenue, while import tariffs go to the Central Government.

A1.9 Highway Project 2012*

Introduction

The Government of Jambalaya Island (GOJ) is considering a plan to construct a four-lane high speed, limited access highway network of 233 km, linking Queens Town to Hope Bay and Black Town. The project is to be undertaken in two stages. Stage 1 will begin in 2013 and will complete the section between Queens Town and Williamsburg (85 km) by the end of 2016. Stage 2 will complete the two stretches of highway between Williamsburg and Hope Bay (85 km west), and to Black Town (63 km north) to be constructed between 2016 and 2018. The routes are illustrated in Figure App_1.1.

It is proposed that this project will be undertaken as a public-private partnership (PPP) in terms of which the private contractor (the "concessionaire") will operate the project under a build-operate-transfer (BOT) scheme. Under this scheme the concessionaire will be responsible for building, operating and maintaining the highway. It is proposed that the new highway will be operated as a toll road through which the concessionaire is expected to recoup its share of the capital investment and operating and maintenance costs. The concessionaire will operate the highway for 15 years after completion of Stage 2 of the project (2019 to 2033) after which it will hand over the project to the GOJ, at no charge. Thereafter, the GOJ assumes responsibility for operations and maintenance until the end of 2058, the assumed end of the project's life.

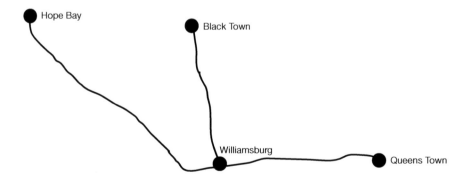

Figure A1.1 Map showing route of planned highway.

You are required to undertake a comprehensive cost-benefit analysis of the project from the perspective of GOJ and the concessionaire. All calculations should be done in millions of US dollars, rounded to two decimal places and expressed in constant 2013 prices. Note that all values reported in this project summary are at 2013 prices and assume an exchange rate of J$80 = US$1.

Investment costs

The investment costs are given in Table App_1.22 which shows the year-by-year break-down, the allocation between the private and public sectors, and, the composition of inputs.

Table App_1.22 Composition of capital costs

Year	Concessionaire	GOJ	Composition of construction costs					
			Labour		Materials		Equipment	
			Local	Foreign	Local	Foreign	Local	Foreign
2013	4000	4000	640	640	480	960	320	1760
2014	4000	4000	960	1120	800	2240	640	2240
2015	4000	4000	960	1120	800	2240	640	2240
2016	8000	8000	2000	800	800	3200	1200	1600
2017	8000	8000	4000	2000	800	4800	1600	2800
2018	–	12000	4800	800	2400	2400	800	800

Note: All values in J$ millions.

Highway construction

Land is purchased at the start of each stage of construction. Included in GOJ capital costs are land purchase costs. Assume that in 2013 J$3,200 million (of the J$4,000 million) and in 2016 J$6,400 million (of the J$8,000 million) is for land. The land purchased for the highway has an opportunity cost estimated at approximately 50% of the cost paid to the previous occupants.

Assume local labour consists of 50% unskilled and 50% skilled/managerial, and foreign labour is 100% skilled/managerial. The shadow-price of unskilled labour is 20% of the wage and the opportunity cost of skilled labour (local and foreign) is 100% of their wage.

Assume that all local materials and equipment prices are inclusive of a sales tax of 10% paid to GOJ, and all imported materials and equipment costs include a 5% import duty. Note that no sales tax is levied on imports.

Toll system infrastructure

The toll system will require the construction of six toll plazas to be installed as new sections of the road are completed. The capital cost of each plaza is estimated at J$150 million (2013 prices). The first two plazas will be installed in 2016, another two in 2017 and two more in 2018. The composition of this cost is: 40% imported materials, including 5% import duty; 40% local materials including 10% sales tax; 10% skilled labour; and 10% unskilled labour.

Operating and maintenance costs and salvage value

Highway

It should be assumed that expenditure on highway maintenance (excluding the toll facilities) will amount to J$300 million per annum, beginning in 2016. Assume the same composition as with total capital construction costs over the period 2013–18 as shown in Table App_1.22.

Toll system

Operation of the toll system will require regular maintenance and periodic rehabilitation equal to 5% of the total initial capital cost of the toll utilities, beginning in 2019. For the purpose of efficiency pricing assume the same composition as for the initial capital costs of the toll plazas.

Salvage values

The salvage value of the highway and toll utilities will be approximately 50% of the initial capital cost. For the Efficiency Analysis assume that the benefits of all salvage values are equal to their value at market prices. If at the end of the 15-year concession GOJ decide to scrap the toll, allow for the 50% salvage value of the toll plazas at the end of 2033, otherwise for toll plazas and highway at the end of 2058. The salvage value of the toll plazas and highway will accrue to GOJ.

Benefits from the highway

The main categories of project benefits that should be considered in the project appraisal are:

i   reduction in vehicle operating costs to road users due to improved road surface;
ii value of time saved by passengers and drivers;
iii reduction in road maintenance costs on existing roads due to lower traffic volumes;
iv reduction of accident costs due to improved safety;
v   reduced pollution due to efficiency gains in vehicle use;
vi toll revenues received by the concessionaire and GOJ.

Table App_1.23 shows the forecast traffic volume on the new highway network. The forecast total includes vehicles travelling in both directions on existing roads and the new highway combined.

Table App_1.23 Forecast of total annual road usage from 2017

Route	Distance (km)	Total forecast (vehicles)
Queens Town–Williamsburg	85	10,000
Williamsburg–Hope Bay	85	4000
Williamsburg–Black Town	63	2000

It has been estimated that with the toll, 60% of the forecast traffic between the towns along the route of the highway will use the highway, with the other 40% continuing to use the existing road network. Based on recent trends it should be assumed that this forecast traffic volume will increase by 4% each year, starting with year 2018, until the end of the project's life. Assume that in 2017 and 2018 the usage of the highway is equal to 36% of the forecast total volume of traffic for the complete highway. In 2019, when Stage 2 opens, the highway will operate at 100% of the forecast volume of highway traffic from the first year onwards.

The following sub-sections outline the details necessary for calculation of each component of project benefit.

Reduced vehicle operating costs (VOCs)

For the purpose of this study it will be assumed that under the "without-project" scenario the existing road network will remain unchanged. It will also be assumed that the total vehicle and passenger kilometres are the same with and without the new highway. Of this traffic volume, approximately 60% is expected to be passenger vehicles (private cars, motorbikes, taxis, minibuses and larger buses) and 40% trucks. The expected VOC saving (in 2013 prices) is J$3.50 per km for passenger vehicles and J$20 per km for trucks. For all components of VOC assume that the efficiency price is the same as the market price. Treat VOC savings as external benefits for the purposes of the Project and Private Analyses.

Value of time saving

Each passenger vehicle is expected to carry, on average, 16 passengers including the driver, and each truck carries on average 1.5 passengers including the driver. Passengers travel for different reasons; work, commuting and leisure activities. It has been estimated that time spent travelling on GOJ's highways can be broken down as follows: 15% work; 52% commuting, and 33% leisure (these proportions apply to passenger vehicles only). Assume trucks' drivers and passengers travel exclusively for work purposes. Estimates of the time opportunity cost per person-kilometre on existing roads for the three categories of passenger are: J$6, J$5, and J$1 respectively (2013 prices). All travel time saved should be treated as external non-financial benefits (i.e. should be omitted from the Project and Private Analyses) and should be priced at the appropriate shadow prices in the efficiency and Referent Group non-financial benefits accounts. It has been estimated that the new highway will reduce travel time by 70% for passenger vehicles and 45% for trucks.

Reduced maintenance costs on existing road network

Since 60% of the traffic moving between the places to be connected by the new highway is expected to shift from the existing roads onto the toll highway, the lower volume of traffic remaining on the existing roads implies both the need for less maintenance and the opportunity to defer scheduled rehabilitation works. It has been estimated that, in the absence of the new highway, annual maintenance costs on these sections of existing roads would be J$4,000 million per annum (in 2013 prices) with effect from 2017. These consist of 95% capital works and 5% operating expenses. It is expected that the lower traffic load will reduce annual capital works by 10% and annual operating expenses by 20%. For the purpose of the Efficiency Analysis it should be assumed that the efficiency prices of reduced capital and operating costs have the same composition as new highway construction and operating costs. Treat all reduced maintenance costs as benefits to GOJ in both the Private and Efficiency Analyses.

Reduced accident costs

It has been estimated that total annual costs associated with road accidents will amount to approximately J$4,000 million per annum in 2013 and that these costs can be expected to rise proportionately with the increase in traffic volume, *ceteris paribus*. If these costs are apportioned over the entire country's road network, it can be estimated that 20% of the total traffic accident costs are incurred on the sections of the road network where the new highway is to be located. With 60% of the traffic expected to shift to the safer highway, it has been estimated that accidents over this section of the network will be reduced by 40% (i.e. 40% of 20% of the annual forecast total cost of traffic accidents). For the purpose of the Efficiency Analysis it should be assumed that the shadow-price of the cost of accidents is the same as the market price. Treat all benefits from reduced accidents as external to the Project and Private Analyses.

Reduced pollution costs

The main form of pollution caused by traffic is air pollution, the main pollutants being: carbon dioxide (CO_2), carbon monoxide (CO), hydrocarbons (HC), nitrous oxide (NO), sulphur oxide (SO) and particulates (PM). Drawing on estimates of the costs of these pollutants from studies undertaken in other countries, and, given the composition of the entire country's traffic fleet by type of vehicle, it has been estimated that the pollution cost savings from the new highway will amount to, in efficiency prices, J$0.20 per vehicle kilometre travelled by passenger vehicles on the new highway and J$0.40 per vehicle kilometre travelled by trucks on the new highway. For the purpose of the Efficiency and Referent Group Analyses it should be assumed that the full cost of this externality is borne by the residents of the country. Treat all benefits from reduced pollution as external to the Project and Private Analyses.

Revenues from tolls

Toll charges are expected to be set at J$10 per kilometre for passenger vehicles and J$20 per vehicle kilometre for trucks, in 2013 prices, and indexed for inflation. There is no sales tax on the toll, the concessionaire retains all the proceeds from the tolls until the end of its concession (at the end of 2033).

Tax and financing arrangements

The concessionaire is required to pay 25% company tax on its earnings. While interest payments on loans can be treated as a tax-deductible cost, no provision is made for depreciation allowances as a deduction against income for tax purposes.

   The concessionaire will finance its share of the initial capital cost with loans from international banks of US$200 million drawn in 2016 and repayable over the next 15 years at a 3% (real) interest rate. The balance of its investment is financed from its own funds (equity) held by its parent company in France. The other investor is the GOJ which borrows the equivalent of US$600 million domestically at 4.5% (real) interest rate in 2016, repayable over 40 years, and finances the balance of its expenditure from its own funds.

Arrangements on termination of the concession

On termination of the concession at the end of 2033 the highway and its maintenance are handed over to and will become the full responsibility of GOJ. There will be no payment to the concessionaire. Assume that the road is operated for a further 25 years (i.e. until the end of year 2058) and that the salvage value is 50% of the initial highway capital cost. A decision has to be made as to whether the highway should be managed with or without a toll system from 2034 onwards. It has been predicted that removal of the toll would result in a 30% increase of the total forecast traffic flow moving to the highway from the existing road network.

The toll utilities have a salvage value of 50% of initial cost, and will be scrapped either at the end of 2033, if GOJ decides not to continue the toll, or at the end of 2058 if they do. (In either case the salvage value will accrue to GOJ.)

It should be assumed that if the toll is removed in 2034 all benefits, with the exception of the reduced maintenance costs on existing roads, increase proportionately with the increase in traffic volume; i.e. by 30%, including reduced accident costs as described above.

Referent Group definition

For the purpose of the analysis assume that all stakeholders with the exception of the concessionaire, foreign labour and the foreign lender are part of the Referent Group. However, for the purposes of negotiations with the concessionaire GOJ also wishes to know what its net benefits are under the alternative scenarios. (The Private CBA should show the returns on equity of both the private concessionaire and GOJ.)

Assignment instructions

Your task is to undertake a complete cost-benefit analysis of the proposed project as detailed above, with and without the toll after 2033. The results of the Project, Private, Efficiency and Referent Group Analyses should be calculated and discussed.

While GOJ uses a 6% discount rate (real) for public sector investment decision-making, you should also undertake a sensitivity analysis at 9% and 12%. It is understood that the concessionaire requires a minimum return of 12% (real) on its equity. Your report should advise the government on whether the project is worthwhile and which scenario (tolls or no tolls after 2033) you consider the best, giving reasons.

You should also conduct and report the results of a sensitivity analysis in which you calculate and comment on, at least, the sensitivity of the results to:

i   traffic forecasts reported in Table App_1.23 (range between low 2%, base case 4%, and high 5%);
ii vehicle operating costs (20% variation each side of the base-case estimate);
iii opportunity costs of travel for the different categories of commuters (30% variation each side of the base case estimate).

Initially undertake a sensitivity analysis showing the effects of variability of these inputs individually and jointly on the net benefits to the Referent Group. You should also undertake ExcelSim or @RISK analysis assuming a triangular probability distribution for the same three inputs. In your report you should also indicate if there are any omitted costs and benefits that could be of potential significance to the decision-maker and might warrant further investigation.

In your discussion of your findings you should identify (in 250 words or less) what other variables if any, should be selected for further sensitivity/risk analysis, and explain why. (Please note: it is assumed that import duties and sales tax are fixed and are not subject to change. Hence these variables are not to be used in sensitivity/risk analysis.)

The format of the report

Your written report should be not more than 12 pages in length, excluding tables. It should be on A4 size pages (portrait orientation only) in PDF format, 12-point Times New Roman font, double line spacing, and 2.5 cm margins on all sides.

The report should begin with an executive summary of no more than one page in length. Results of the sensitivity analyses should be reported in summary tables included in the text, and where necessary, in more detailed tables in an Appendix (not included in the 12-page limit). Do not attach copies of spreadsheets (e.g. in PDF) to your main report, though sections showing summary results can be cut and pasted into the report.

Excel files for all scenarios and sensitivity analysis should be submitted electronically, formatted in landscape in normal view, and left unlocked so calculations can be checked. Each Excel file (and/or sheet in a workbook) should be clearly and logically labelled with your student number and scenario for reader-friendly identification. For ease of reading you need only show the cash flows for the first 21 years of the project (i.e. up to and including 2034) and the last two years (2057 and 2058), hiding all the years in between so the entire sheet can fit on one screen.

Note

* This case study, which was developed for teaching purposes at the University of Queensland, is based on a report prepared by Dessau Soprin International Inc. for the Development Bank of Jamaica ("Highway 2000 Project: Economic Cost-Benefit Analysis", July 2000). The details of the project and its financing as specified here are hypothetical and do not necessarily correspond to those of the original report.

A1.10 International Mining Corporation (IMC) Copper Mining Project[*]

Introduction

The Government of Indonesia is considering a proposed joint venture with a foreign investor, International Mining Corporation (IMC), to develop a new copper mine in the mountainous and remote Eastern Province (EP) where recent geological surveys have revealed significant copper deposits. Under the proposal, Eastern Province Mining Limited (EPML), in which the Government of Indonesia (GOI) will hold 30% of the shares, will mine and mill the copper ore on site. The concentrate will then be transported (in slurry form) by pipeline to a dedicated port facility at the mouth of the Eastern Province River, from where it will be shipped to Japan for refining and sale on the world market. Although the main product is copper, the concentrate will also contain some quantities of gold and silver that will also be extracted from the concentrate at the refining stage and sold on the world market.

Eastern Province is a low-income region with a population of 5 million and a per capita income of around US$300 per annum, considered the least developed part of the country. The local population living in the area rely mainly on subsistence agriculture and fishing in the Eastern Province for their livelihoods. At present there is very little economic or social infrastructure in the area, which means that apart from on-site investment in the mine, mill and tailings dam, EPML will need to make substantial off-site investments in infrastructure and logistics, such as transport equipment, the construction of roads, bridges, wharfs, an airstrip, storage facilities, housing, power generation and supply, as well as the establishment of a school, hospital, shops, recreational facilities and other amenities for the locally engaged and expatriate employees and families. The GOI is eager for the project to proceed as it is expected to provide a significant injection of investment in the region, and opportunities for training and employment of local workers as well as some interstate, migrant workers from other underdeveloped areas of the EP and elsewhere in Indonesia.

It is also expected that the project will generate some backward linkages into the local economy. EPML will be required to sub-contract certain services to locally based contractors and to buy some of its supplies locally, such as food. Under the proposal the local land-owners are to be compensated for the use of their land for mining-related activities, and priority is to be given to the local population for employment and training by EPML. Part of the compensation payments are to be paid into the Eastern Province Development Trust Fund that will be used to finance development projects in the EP region once the fund is established. EPML will also pay royalties to GOI, based on the value of its mineral sales net of transportation, treatment and refining costs.

Although IMC has a strong interest in the project, it is concerned that the additional off-site costs associated with the establishment and operation of the extensive economic and social infrastructure will reduce the profitability of the project significantly. It is also felt that much of EPMLs off-site investment will be of significant benefit to the local population. In its proposal to GOI it has argued that it will participate in the project only if it is granted concessions in the form of:

- exemption from import duties on all imported goods;
- a tax holiday in the form of exemption from company taxes over the first 10 years of the mine's operation;
- lower royalties;
- a smaller share of equity for GOI.
- The Treasury Department of Indonesia is not in favour of making these concessions, especially as IMC has no obligation to reinvest its after-tax profits in Indonesia and there are no restrictions on the remittance of profits to its overseas shareholders.

You have been contracted by GOI to advise on the project. You are required to prepare a report based on a cost-benefit analysis of the project in which you estimate the net benefits of the project from the perspective of the people of Indonesia – the Referent Group. As this report is also to be used to inform GOI in its negotiations with IMC you are required to consider a number of scenarios in which you show the net benefits to both the Referent Group and IMC.

You should assume that the initial investment begins during 2003 and that the project will come to an end – the mine will be closed down completely – at the end of 2020. GOI uses a discount rate of 6% (real) for public sector investment appraisal, and requires sensitivity analysis over a range of discount rates from 5–10% (real).

All available details of the project are provided in the following sections. Where information is missing or ambiguous you are required to make what you consider the most reasonable assumption, which should be discussed explicitly in the text of your report. All values should be reported in thousands of New Rupiah, our assumed "new" currency of Indonesia, which exchanges at Rp2.5 to US$1.0 (to simplify conversions and calculations). All prices are in constant 2003 prices. Assume that relative prices are unaffected by inflation.

The report should contain an executive summary of approximately one page in length, and should be no more than 12 pages in total, including tables and charts. Printouts of the detailed spreadsheets should be attached in an appendix, and should also be provided in electronic format on disk. For an example of a completed report, see the appendix to Chapter 13 in the text.

Investment costs

The project is an open-cut mining operation with most material drilled and blasted. EPML will begin the construction stage of the project in 2003. Mining, milling and the overseas refining operations are expected to begin three years later, in 2006.

Total initial investment in the project amounts to approximately US$1.25 billion, consisting of both on-site and off-site expenditure and involving a combination of imported and locally produced goods, local and expatriate labour, and is spread over three years: 25% in 2003; 45% in 2004; and, 30% in 2005. A detailed breakdown is provided in Table App_1.24.

Table App_1.24 Details of initial capital expenditure, 2003–05

All amounts in Rp 000s unless otherwise stated.

Item	Composition Imports US$000s (c.i.f.)[1]	Local Materials[2]	Labour[3]
Roads	11,800	48,750	81,250
Buildings	3800	17,000	14,875
Wharf	3275	6750	6750
Airstrip	950	2656	5313
Power Supplies	73,000	28,750	57,500
Facilities	18,000	75,000	125,000
Mining Equip.	470,000	71,875	71,875
Milling Equip.	83,000	27,000	13,500
Logistics Equip.	92,500	37,500	75,000
Other Equip.	5500	7500	7500
Construction	5000	30,000	105,000
Excavation		16,500	88,000

Notes:
1. Import duties are applied to c.i.f. prices at a rate of 10% *ad valorem*.
2. Consisting exclusively of locally manufactured materials (65%), contractors' margins (15%) local skilled labour (5%) and local unskilled labour (15%).
3. Consisting of a mixture of expatriate salaries (40%), local managerial and professional wages (10%), local skilled (30%), and local unskilled (20%).

It should be noted that there are both direct labour and indirect labour inputs to be considered in the Efficiency Analysis.

During the operation of the mine certain items of infrastructure and equipment will need to be replaced as shown in Table App_1.25.

Table App_1.25 Details of replacement capital expenditure, 2009–17

Item	2009	2013	2017
Imported equipment (US$000s c.i.f.)[1]	6,612	8,264	4,959
Local materials[2]	12,500	25,000	12,500
Installation[3]	12,500	18,750	12,500

Notes: 1. Import duties are 10% ad valorem. 2. Composition as for Table App_1.25. 3. Consisting exclusively of labour – composition as for Table App_1.24. 4. All amounts in Rp '000's unless otherwise stated.

Working capital consisting of a mixture of imported equipment spares (60%), fuel supplies (35%) and locally produced materials (5%) is to be built up from a level of approximately Rp25 million in 2004 to Rp75 million in 2005, and maintained at that level over the remainder of the project's life until 2020 when it is to be completely run down (or sold off at cost). See Table App_1.26 for details.

Table App_1.26 Composition of additions to working capital

Item	2004	2005
Imported equipment (US$000s c.i.f.)[1]	6,600	10,909
Fuel (US$000s c.i.f.)[2]	3,182	6,364
Local materials[3]	1,250	2,500

Notes: 1. Import duties are 10% *ad valorem*. 2. Composition as for Table App_1.24. All amounts in Rp '000's unless otherwise stated.

Salvage values, depreciation rates and provision for rehabilitation

For the purpose of this study it should be assumed that the mine and mill equipment has a salvage value at the end of the project life amounting to 10% of the initial cost. Offsite infrastructure (including construction but excluding excavation costs) has an end value amounting to 30% of its initial cost; logistic and other equipment has an end-value equal to 20% of its initial cost. Under Indonesian tax legislation IMC is permitted to depreciate all initial investment over the 15 year operating life of the project, starting in 2006, using the straight-line method. Replacement investment is depreciated over the 3 years following the year of the investment, also using the straight-line method, and is assumed to have no salvage value.

It is also understood that EPML will need to rehabilitate the mine site and the Eastern Province Catchment Area on closure of the mine. It is expected that this will cost Rp625 million, in year 2020. EPML is permitted to include an annual provision for rehabilitation as an operating expense for tax purposes. You should treat this as a sinking fund with an annual real interest rate of 7%.

The output of the mine

In the first year of operations, 2006, it is expected that 25% of full capacity will be reached; in the second year, 50%; in the third year 75%; and, in the fourth year 100%. Based on the geological information available, EPML expects to operate the mine at full capacity until the end of 2020.

When operating at full capacity the mine is expected to extract 85 million tons of material per annum. It is believed that this will consist of 40% ore and 60% waste. When milled, a concentrate equal to 2% of the ore tonnage is produced for treatment at a refinery. EPML will contract out the treatment to a refinery in Japan, and will sell the refined minerals on the world markets. Refined copper produced can be expected to amount to 30% of the weight of the concentrate, and currently sells on the world market at US$1.25 per pound (lb). Refined silver equivalent to 0.004% and gold equivalent to 0.0025% of the weight of the concentrate are also extracted during the refining process. In world markets the current price of gold averages around US$290 per ounce, while the price of silver averages around US$5 per ounce.

As noted earlier, the company will invest in substantial off-site infrastructure including roads, utilities and community amenities such as a school, clinic and recreational facilities for its staff. It is understood that the extended families of IMC employees and local land-owners will also have access to and benefit from most of these. In the opinion of an expert in the area the value of the additional benefits to the local communities from these amenities could reasonably estimated as the equivalent of 30% of the annual overhead expenditure on 'Utilities' and 'Community Services' (Rp110 million). The use of these amenities by the wider (non-employee) community should be treated as an additional 'output' of the project and its value added to the mineral output in the Efficiency Analysis, and as an equivalent gain in the Referent Group Analysis.

Operating costs

Overheads, insurance and compensation

Total overheads, including all transportation of inputs and outputs, but excluding compensation payments and provision for rehabilitation, are expected to total about Rp410 million per annum, with effect from 2004. The full details of these are provided in Table App_1.27. As IMC has another operation in Indonesia and will be running the project administration from the same head office in Jakarta, a significant part of the overhead costs itemised in Table App_1.27 (50%) are already being incurred.

Table App_1.27 Composition of overhead costs

	Expatriate wages	Local skilled	Local unskilled	Local materials[1]
Administration	45,500	6,500	3,250	9,750
Management	49,500	5,500		
Transport & Engineering	54,000	27,000	13,500	40,500
Maintenance	18,000	7,500	1,500	3,000
Utilities	15,000	30,000	15,000	15,000
Community Services	3,500	7,000	14,000	10,500
Other		3,000	3,000	9,000

Notes:
1. Composition as for Table 24.
2. All amounts in Rp000s unless otherwise stated.

The company will have to increase its existing insurance premium from US$100,000 to US$200,000 per annum. This is paid to an overseas company. No taxes or duties apply.

It has been agreed with GOI and representatives of the local clans inhabiting the catchment area that EPML will make annual compensation payments into a development trust fund amounting to Rp30 million per annum (in real terms) once operations begin in 2006. This is to be treated as compensation for use of communally held land by the mine and its associated activities. A study has estimated that this level of compensation represents approximately twice the opportunity cost of the land resources affected by the mine, based on their present uses.

Mining and milling

Mining and milling costs consist of a number of on-site activities including operations, maintenance, engineering, training, metallurgy and port operations. They consist of a combination of fixed and variable cost and can be disaggregated by type of input as shown in Table App_1.28.

Table App_1.28 Composition of mining and milling costs

(All amounts in Rp000s unless otherwise stated)					
	Imports US$000s (c.i.f.)	Local materials	Expatriate labour	Skilled labour	Unskilled labour
Mining – fixed cost	10,455	12,575	5,925	12,000	8,400
Mining variable cost /ton material	0.1230	0.1479	0.0700	0.1424	0.1000
Milling – fixed cost	7,710	8,900	3,900	9,250	5,700
Milling variable cost /ton ore	0.6235	0.2618	0.1150	0.2735	0.1675

Freight

In addition to the costs of handling and transportation of the concentrate by pipeline in slurry form to the port (included under milling costs), the concentrate needs to be shipped to Japan. EPML will pay US$25 per ton of concentrate for freighting it to the refinery gate.

Treatment

The concentrate is treated before being refined at a charge by the Japanese refinery of US$95 per ton of concentrate.

Refining

Once treated the copper is refined at US$0.10 per lb of refined copper produced. The refining of gold and silver is charged at US$5 and US$3 respectively per ounce produced.

Royalties

EPML pays royalties to GOI. These are 2% of the fob value of sales, i.e. the gross value of sales less the cost of freight to Japan, and the cost of treatment and refining in Japan.

Taxation

EPML will pay corporate taxes of 30% on net profits. For purposes of calculating taxes, losses incurred in one year may be offset against IMC's other operations in Indonesia. Royalties, interest payments, depreciation and provision for rehabilitation may also be treated as tax deductible expenses.

Finance

The initial investment (2003 to 2005) will be financed partly by debt (US$500 million) and the remainder through IMC's own funds. The US$500 million foreign loan (repayable in US dollars) is to be raised in 2003 on the international capital market at a fixed interest rate of 7% per annum (in real terms), repayable as an annuity over 15 years, beginning in 2006. IMC has negotiated an interest free period of grace until the beginning of 2006. In the preliminary negotiations GOI has indicated that it expects to be allocated a 30% share of the equity in EPML, without making any contribution to the equity capital. Dividends are to be calculated on the basis of the net cash after debt service, royalties, tax and provision for depreciation and rehabilitation, and will be paid only in those years in which this balance is positive.

Referent Group stakeholders

The Referent Group consists of all parties engaged in the project except the following: IMC, the Japanese shipping and refining company, the foreign lender, the overseas insurance company and expatriate labour. The Referent Group Analysis should show the net benefits on a disaggregated basis (i.e. by stakeholder group) as the GOI is concerned to know how the benefits of the project will be distributed. You are advised to set up separate working tables to calculate the net benefits for some stakeholder groups (i.e. GOI, unskilled labour, and local contractors) as each has numerous sources of net benefit, making the calculation rather complex.

Efficiency pricing

Labour

Various categories of labour cost enter the analysis: the direct costs of those employed on the project and the indirect costs of labour inputs in other components of project cost. The details of these are provided in the preceding sections. For the purpose of efficiency pricing, it should be noted that for only one category of labour can the opportunity cost be considered significantly different from the market price, namely, unskilled labour. A recent study by a labour economist at the local university estimates the opportunity cost of unskilled labour in the Eastern Province at 20% of the legal minimum wage (which is the wage paid by the company and associated operations). In the Efficiency Analysis you should price all unskilled labour accordingly.

Local contractors

As part of its policy of promoting small-scale local enterprises, the GOI has stipulated that local contractors should be engaged by the project wherever possible. It is envisaged that they will be engaged primarily for the supply of locally produced and non-tradeable goods and services. It has been estimated that 25% of the income they earn as "contractors' margin" (i.e. as a mark-up or commission from these activities) is a rent (a payment in excess of opportunity cost) attributable to the market power they have. In the Efficiency Analysis this needs to be taken into account.

HINT: As unskilled labour and/or contractors' margins also enter as cost components of "Local Materials" which is, in turn, a major input in a number of expenditure items, it is important that all the items of expenditure of which unskilled labour and local materials are components are entered in the Variables Table of the spreadsheet on a disaggregated basis. This should simplify the subsequent revaluation of these items in terms of efficiency prices. This revaluation can be undertaken as part of Table App_1.24 to simplify the remainder of the spreadsheet.

Conversions and assumptions

For the purpose of this study you should use the following conversions:

- 2204.62 pounds (lbs) per ton
- 32.1507 ounces (oz.) per kilogram
- 1000 kilograms (Kg.) per ton.

Any other information requirements will need to be based on your own assumptions. It is essential that these are made explicit in your report.

Scenarios and sensitivity analysis

Once you have undertaken a complete CBA for the Base Case Scenario (Project (Market), Private, Efficiency and Referent Group Analysis) you need to calculate the net benefits to IMC and to the Referent Group under a number of alternative policy scenarios, as you are also required to provide GOI with information and advice for its negotiations with IMC. The policy variables IMC wishes to discuss in the negotiations are: company taxes; import duties; royalties; and, GOI's share of equity (dividends payable).

You should consider, at least, the following scenarios in addition to the base case:

1   exemption from all duties;
2   exemption from company taxes;
3   exemption from duties and taxes;
4   exemption from royalties;
5   reducing GOI's equity to 10%;
6   both (4) and (5);
7   all concessions (1) to (5).

Your final report should comment on the relative attractiveness (ranking) of these from GOI and IMC's perspective, and should offer advice to GOI on how you expect the negotiations to proceed, based on these calculations.

You should also identify those variables (other than the above) to which the project's net benefits are most sensitive and discuss the possible implications for the project if their actual values moved within a band of, say, 10–20% around the estimated "best guess" values.

The sample report included as the Appendix to Chapter 13 of the text is a guide to the appropriate style of presentation of your report. You must submit an electronic copy of all your spreadsheets with your printed assignment.

Note

*   This assignment case study is hypothetical and is not based on the activities of an actual company. Any similarity with regard to the activities or name of an actual company is purely coincidental.

A1.11 Comparative levelised cost of electricity: renewables vs coal

Introduction: the task at hand

A private mining company, Rio Blanco Corporation (RBC), which is a major electricity user, needs to decide whether to enter into a new contract with one of the State's main electricity retailers for the supply of its electricity needs for the next 20 years. The dilemma it faces is that the State's power grid is sourcing its electricity exclusively from existing, black coal-fired power stations at a supply price, equal to the marginal cost of electricity generation, of $60/MWh (constant 2020 prices).

It is widely acknowledged that the external costs of carbon emissions from coal-fired electricity generation are extremely high relative to other technologies, especially renewables. It is estimated that for every MWh of electricity generated by a black coal-fired power station, 0.8 tonnes of carbon is emitted. There is growing recognition of the need for electricity regulatory authorities to factor in the Social Cost of Carbon (SCC) in comparative cost calculations. To internalise such external costs the State Government is planning to introduce a carbon price/tax. For coal-fired electricity generation this additional cost, over and above the current contracted supply price, could be significant. Although there is some uncertainty as to how much and when the carbon charge will be implemented, the prevailing view among most stakeholders is that this will happen 'sooner rather than later'. For the purpose of this study it should be assumed that a carbon tax, initially set at $20/tCO$_2$ (constant 2020 prices) will be introduced at the beginning of 2024 and will increase by $2/tCO$_2$ per annum until it reaches a level of $40/tCO$_2$ (constant 2020 prices), and remains at $40/tCO$_2$ to the end of the project life-pan of 20 years (i.e. to 2040).

The alternative to sourcing its electricity from the grid is for RBC to invest itself as producer/consumer (a '*prosumer*') in a new electricity generation plant, using a renewable technology. The two options under consideration are onshore wind turbines *versus* solar photovoltaic (PV). The construction and management of the preferred option is to be contracted to a private supplier: for the solar option, Sunshine Solar (SS), a local State-based company, and for the onshore wind option, Deutsche Onshore Wind (DOW), a Germany-based multinational company. Both companies have existing operations in the State, are considered world leaders in renewable electricity generation, and will be paid a project management fee (equal to 10% of, and included in, the estimated fixed operating and maintenance costs in Table App_1.29). In the case of DOW the full amount of this fee will be remitted to its parent company in Germany.

If RBC switches to become a *prosumer* with one of the renewable options, it will avoid having to purchase any electricity from the grid at the current price and avoid paying the additional carbon charge from 2024 onwards. This will also reduce the amount of electricity generated by the coal-fired power stations by an amount equivalent to the MWh generated by RBC.

You are required to undertake a cost-benefit analysis and prepare a report for RBC, comparing the costs and benefits to RBC of sourcing electricity over the next 20 years from either onshore wind- or solar-powered generators, vs continuing with supply from the grid using coal-fired generation sources. Your comparison of the two options should be based on their Levelised Cost of Electricity (LCOE) as well as the usual decision-making criteria.

As the proposed investment has wider social and economic implications, your analysis and report should also compare the costs and benefits from the social perspective of the State, using your estimates of the appropriate Social Cost of Carbon (SCC) and any other external costs or benefits, to be included in the Efficiency CBA.

You will also need to show in the Referent Group CBA, the distribution of all costs and benefits among the various stakeholders affected, and discuss other social and environmental costs *not* quantified in the CBA, which could impact on RBCs final decision. Your conclusions and recommendations should also include a discussion of the policy implications your analysis raises from the State Government's perspective.

Details of renewable options

Both proposed plants, if adopted, are to be constructed in 2020 and will begin production in 2021. They both have an economic life of 20 years, with an end value equal to 10% of the initial capital cost.[1]

Your analysis should consider the options both with and without the carbon tax in the Market and Private CBA, and with the Social Cost of Carbon (SCC) in the Efficiency and Referent Group CBA. You are required to research and justify the range of values used in your analysis for the SCC, and the level at which the Government sets its carbon charge.

Due to gradual wear-and-tear of the installed capacity, it should also be assumed that the efficiency of the onshore wind option will decline by 0.5% per annum and the solar option by 0.75% per annum. This means that the calculated annual MWh sent out per annum will decline at this rate.

In the Private CBA you will be analysing the projects from the viewpoint of RBC. Your recommendation to RBC will be either to accept one of the renewable *prosumer* options, or to reject them in favour of continued supply from the grid. Net Present Values and Levelised Cost of Electricity (LCOE) in the Market and Private CBA are to be estimated at a discount rate of 10% (real), with sensitivity testing at 8% and 12% (real). For the Private CBA you should report the IRRs.

For the Efficiency and Referent Group CBAs you are required to report Net Present Values (NPVs) using a real social discount rate of 5% with sensitivity testing at 3% and 7%, given that the public sector traditionally uses a 'social' discount rate somewhat lower than that used by private investors.

The estimated costs and other input data are detailed in Table App_1.29. (All costs reported are in A$2020 prices.)

Table App_1.29 Available input data on costs and benefits of renewable electricity generation options

Plant Specific Input Data	Onshore wind	Solar PV
Installed (Nameplate) Capacity (MW)	150	200
Capacity Factor	40%	30%
Initial Capital Costs ($mn/MW)	$2.00	$1.75
Fixed O&M Costs ($/MW of nameplate capacity, per annum)*	$30,000	$20,000
Variable O&M Costs ($/MWh sent out per annum)*	$12.00	$2.00
Productivity Drop-Off (% output pa)	0.50%	0.75%

* Including GST of 10%

Financing

To finance the investment RBC will need to take out a commercial loan in 2020 of $200mn at a 6% real rate of interest with a consortium of overseas financial institutions, to be repaid in equal annual amounts as an annuity, over the life of the project starting in 2021. (In the Private CBA enter the loan repayments from RBCs perspective but keep the interest cost as a positive amount.)

Private companies pay 30% business company tax on profits net of operating costs, interest payments and depreciation allowances. RBC can deduct any losses against taxable profits from their other Australian projects.

Depreciation allowances can be claimed as tax deductible costs using the straight-line method over a period of 10 years starting in the year 2021; i.e. no depreciation beyond year 2030.

External costs and benefits

Both prosumer projects can be expected to involve significant external benefits relative to electricity sourced from coal-fired generation in terms of reduced carbon emissions. The reductions in carbon emissions are expected to amount to 0.8 tons of carbon per MWh of electricity generated. However, both the renewable project options are expected to result in other external social and environmental costs to the local communities which have not been quantified. These need to be identified and discussed in your report.

Sensitivity and Scenario Analysis

You are required to undertake sensitivity testing on at least four of the variables individually (apart from the discount rates) that can be considered most uncertain and with significant implications for the results and recommendations. These should include the Social Cost of Carbon, and the Capacity Factor of the two plants, plus two other key variables selected and justified. Your report should also explain and substantiate the range of values selected for sensitivity testing. (For the purpose of this study it should be assumed that the nameplate capacities of the two plants *cannot* be varied.)

Your sensitivity analysis should also include derivation of the threshold values for:

(i) the carbon tax at which the wind and solar options become at least as cost-effective as coal-fired generation from the grid in the Private CBA; and

(ii) the SCC at which the wind and solar options become at least as cost-effective as coal-fired generation from the grid in the Efficiency CBA.

(You are encouraged to derive threshold values for other variables such as the Capacity Factors.)

You should also report the results of a scenario analysis based on the most optimistic, best guess, and most pessimistic, sets of values for the combination of variables included in your sensitivity analysis

Instructions

On behalf of RBC you are required to undertake and report the findings of a cost-benefit analysis of the solar and onshore wind proposals as possible alternatives to continued supply from the national grid using coal-fired electricity generation.

The analysis is to be reported in millions of 2020 Australian dollars, to two decimal places, with PV of costs and LCOEs in year 2020 to be calculated. *It is highly recommended that all costs and revenues/benefits are entered as positive values in the spreadsheets and the net values are derived by subtracting costs from revenues/benefits in the corresponding formulae. The loan payments receipts should be calculated from RBCs perspective.*

In your report, which should be approximately twelve pages in length, you should also indicate if there are any omitted costs and benefits that could be of potential significance to the decision-maker and might warrant further investigation.

Note

1 The same end-value calculated for the Market and Private CBA should be used in the Efficiency CBA. The end-value should not be deducted from capital costs when calculating depreciation.

A1.12 Cost-benefit analysis of the proposed repeal of Water Saving Regulations*

Background

Water Saving Regulations

Queensland had in place regulations under which new buildings in most areas must achieve certain water savings targets, and rainwater tanks (RWTs) are generally the means selected to achieve these targets. Compliance costs were considered by various stakeholders in the community to be excessive and in September 2012, the State Government ordered an analysis of a proposed repeal of water savings regulations. A rudimentary CBA was undertaken by a government department, followed-up by a more comprehensive CBA by a local consulting company, of maintaining existing regulations on water savings targets through the installation of RWTs in new property developments. This case study is based on their analysis, and much of this document draws directly from it.

You are now required to undertake a more comprehensive analysis in which you are to introduce two additional dimensions into the analysis: (i) the use of shadow (efficiency) prices where relevant; and, (ii) undertaking a Referent Group analysis showing how the costs and benefits of maintaining the existing regulations are distributed across the various stakeholder groups. More details as to what is required follow in the rest of this document.

Stormwater Quality Regulations

The Queensland State Government also has in place a State Planning Policy governing the quality of stormwater run-off, which requires a reduction of pollutants from untreated stormwater before it enters the waterways. A range of actions can be used to meet these requirements, including evaporation, reuse and infiltration to native soils, or filtration through a soil and plant stormwater 'bio-retention' treatment system. These actions add to the costs of urban developments, through capital, operating and land costs, and are calibrated to limit the levels of pollutants to the required levels. Bioretention systems can be placed on both public and private land, but tend to be placed on areas already

* Prepared by Richard Brown, School of Economics, University of Queensland. This case study is based on the report *Assessment of Proposed Repeal of Water Saving Regulations* prepared for Queensland Competition Authority by Marsden Jacob Associates (MJA, 2012), some sections of which have been reproduced in this document. This study has been prepared and amended for training purposes and differs from the actual MJA study in a number of respects. MJA and QCA are acknowledged for allowing us to use their report and spreadsheets.

designated as public land (road reserve, drainage reserve or park land). Once completed, bioretention areas arguably have little aesthetic difference to other areas of public open space.

BOX 1 Bioretention systems

Bioretention systems are an aspect of Water Sensitive Urban Design (WSUD) which seek to maintain near-to natural flow levels and pollutant loads of stormwater into receiving waters (creeks and streams) in urbanised areas. The objective is to minimise the impacts of urbanisation on receiving waters. Urbanisation increases the proportion of impervious surfaces (such as roofs, roads, footpaths), which increases runoff from rainfall events, depositing pollutants into nearby waterways and affecting waterway health.

Bioretention systems operate by filtering stormwater runoff through densely planted surface vegetation and then percolating runoff through a filter media. During percolation, pollutants are retained through fine filtration, adsorption and some biological uptake.

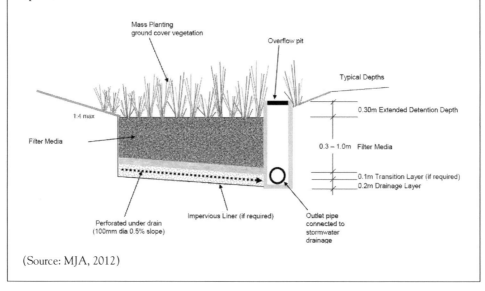

(Source: MJA, 2012)

Bioretention systems combine various WSUD treatment types in one 'treatment train'. The system is designed to carry out primary and/or secondary treatment processes of stormwaters and retard flows. This retention or retardation can enable sediments to precipitate out of the water taking along with it some pollutants. The use of biological processes to 'treat' stormwater while facilitating conveyance and retention gives rise to the title of bioretention.

Identified components of costs and benefits

MJA (2012) identified the components of cost and benefits shown in Table App_1.30 for inclusion in the CBA, acknowledging that there could well be others not included here.

Table App_1.30 Types of costs and benefits included in the MJA CBA model

Benefits / Avoided Costs	
Deferred augmentation capital costs	The benefit to the State Government (Department of Energy and Water Supply, DEWS) from deferring the need to augment bulk water infrastructure (e.g. a desalination plant or a dam) due to the use of water from RWTs. This is the benefit estimated by the Queensland Water Commission (QWC). Similar to the QWC, we consider only the deferral of one augmentation within the analysis.
Avoided augmentation fixed OPEX	The fixed component of operating water supply infrastructure, borne by DEWS, that is deferred and potentially avoided by the use of RWTs.
Avoided variable OPEX	This is the avoided variable OPEX associated with the lower required volume of water for distribution to users by BCC once the RWTs become operative.
Bioretention CAPEX savings	The savings to property developers from reducing the area of construction of bioretention areas for residential developments due to the installation of RWTs.
Bioretention OPEX savings	This is the avoided cost of maintaining bioretention areas in residential developments (e.g. weeding, keeping vegetation healthy, etc) over and above the cost of maintaining public open space; assumed to be borne 50% by Brisbane City Council, and 50% by property owners.
Costs	
Capital cost of tanks	The cost to property developers of the RWT including installation cost.
Operating costs	Energy and maintenance costs incurred by property owners associated with RWTs.
Abatement costs if tanks not replaced	This represents the additional abatement costs to Brisbane City Council for the biological nutrient removal through wastewater treatment that may be incurred to mitigate any degradation in waterway health due to stormwater that is no longer mitigated by RWTs that are not replaced after the end of their lives.

(Source: Adapted from MJA, 2012)

One essential task is to provide a realistic estimate of how many additional RWTs would be installed *with* the regulations in place versus a situation where the regulations are repealed (*without* regulation), bearing in mind that some developers and property owners will elect to install RWTs on a voluntary basis. The alternative scenarios are depicted in Figure App_1.2.

Costs and benefits at market prices[1]

Table App_1.31 provides the estimates of the quantities and prices for the various items of costs and benefit.

Taxes, subsidies and externalities[2]

Taxes and subsidies

Table App_1.32 provides relevant details of import duties, taxes and subsidies included in market prices, and the relevant level of government affected.

Number of Tanks

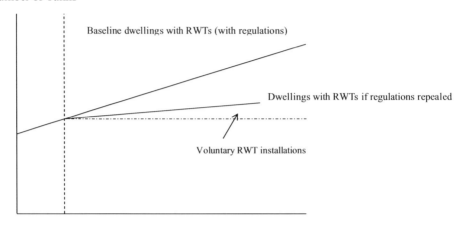

Year

Figure A1.2 Policy change relative to baseline.
(Source: MJA, 2012)

Table App_1.31 Input data for rain water tank CBA

Assumed length of retained regulation	25 years (2012=Yr 0)
Discount rate (real)	4%
Tanks installed with regulation (units p.a.)	25,000
Tanks installed without regulation (units p.a.)	5000
Tank yield (KL p.a.)	50
Cost of tank and pump	$4500
Energy and maintenance costs per tank/annum (from yr.1)	$40
Bioretention Capital Costs avoided per new tank	$820
Bioretention Operating Costs avoided per new tank (from yr.1)	$15
Abatement cost if tanks fail (p.a. from year 10 onwards)	$1,500,000
Delayed augmentation capital cost (from 2034 to 2037)	$1,000,000,000
Avoided annual augmentation operating costs (2035–37)	$30,000,000
Variable operating costs for water supplied ($/KL)	$0.60

Table App_1.32 Cost/benefit item

	Import Duties/Taxes/Subsidies		
	%	Type	Stakeholder
Cost of tank and pump	15%	import duties	Fed Treasury
Tank energy and maintenance costs	20%	subsidy	State Govt
Bioretention Capital costs /tank	40%	subsidy	Fed Treasury
Bioretention Operating costs /tank	40%	subsidy	Fed Treasury
Abatement cost if tanks fail	10%	taxes	Fed Treasury

External costs

In addition, there are various external, environmental costs that will be avoided if the water saving regulations and RWTs are to be retained. Those that have been identified and quantified for this study, using appropriate non-market valuation methods include the following:

A. The future construction of the dam in 2034 to augment the bulk water supply is estimated to generate significant environmental costs in terms of losses of both agricultural output and ecosystem services produced by the land that will be inundated by the dam, and the corridor running through private and state-owned land along the 75 km route of the pipeline from the dam to the water treatment plant. This includes sections of valuable rainforests and pristine sections of the river catchment. Moreover, the operation and maintenance of the pipeline and pumping stations to transport the water will generate further externalities through both the maintenance of the pipeline, and the high energy requirements to operate the pumping stations. A team of environmental economists from the university has estimated that the external costs of the dam and pipeline construction will amount to a further 75% of the dam's construction costs at market prices, and the external costs associated with the annual operating and maintenance costs of the dam and pipeline will be approximately 50% of the augmentation operating costs at market prices.

B. The treatment and delivery to consumers of the additional water from the augmentation dam are also expected to generate substantial external environmental costs, primarily in the form of air pollution from the additional energy required for the water treatment plant and delivery of the treated, potable water, and the safe disposal of the chemicals and other toxic residuals produced at the water treatment plant. The same team of environmental economists estimates that these external costs will amount to a further 125% of the variable operating costs of supplying water to the consumers.

Your task

You are required to undertake a comprehensive BCA (Project/Private,[3] Efficiency and Referent Group analysis), with a view to advising the State government on the desirability or not of retaining the existing water saving regulations. (For the purpose of the analysis and report, you should undertake a CBA of *retaining* the existing regulations rather than *removing* them.) You are encouraged to consider and comment on the distribution of the regulation's costs and benefits by stakeholder, and any related potential policy implications that could be considered at all levels of government.

Referent groups

For the purpose of this analysis assume that the stakeholders of interest to your client are all those located in Queensland, including the costs and benefits to Queensland State and Local Government. However, as the Regulatory Impact Statement of which this CBA forms part is to be submitted to the Australian Competition and Consumer Commission (ACCC), and, as the regulation has budgetary implications for Federal Treasury, you should also consider the implications for Federal Government. Furthermore, as Brisbane City Council (BCC) is a stakeholder with major responsibility for the supply of water and the health of the city's waterways and the adjoining bay, your analysis should also consider the regulation from BCC's perspective. The disaggregated Referent Group analysis should also include the costs and benefits to property developers and subsequent property owners, as well as the Queensland public at large.

Sensitivity testing

As part of your case study you are required to undertake and report on the following:

A. A detailed sensitivity analysis and scenario analysis in which you test the robustness of your results and conclusions to variations in the value of selective variables/parameters. Your sensitivity analysis should test each variable listed below, individually and then jointly with a view to describing three scenarios: most pessimistic; best guess; and, most optimistic. This should include the following variables:

  (i).   discount rate for the base case: 2% and 6%
  (ii).   tank yield 30 and 90 kl/annum
  (iii). import duties/taxes on tank cost 0% and 30%
  (iv).   externalities varied by 25 percentage points above and below their estimated non-market values; i.e. if the given estimate of an externality is equivalent to 50% of the cost at market prices, you should vary this between 25% and 75% market cost, and so on.

B. Discuss whether you believe any other variable is of sufficient relevance and uncertainty for inclusion in a sensitivity analysis, with a clear and concise justification for your opinion.

C. You are also required to identify and discuss any other efficiency costs or benefits not included in the analysis, which you believe should be taken into account, with a clear statement justifying your position.

D. As part of your sensitivity analysis you should also identify the respective *threshold values* for each individual variable selected for sensitivity testing, at which the NPV of the efficiency net benefit is zero. This should also include calculation of the threshold value for the *Other Unrecorded Efficiency Benefits* identified by you under C above. You should also comment on the likelihood of each threshold value being within a feasible range of values for that variable.

Notes

1 All values are in constant 2012 prices. These are not the same as the estimates on which the MJA (2012) study was based.
2 This information is purely hypothetical and intended only for illustrative, training purposes and is not intended to reflect in any way the actual situation with respect to taxes, subsidies and externalities, nor the division of costs and benefits among the various stakeholders identified.
3 As this is not an investment project the Project and Private CBAs can be merged, provided it is possible to show the net benefit stream for each major stakeholder separately.

A1.13 Cost-benefit analysis of the proposed extension to the Mount Beno Walking Trail

Introduction and project description

The Mount Beno Walking Trail (MBWT) located outside Brisbane, Queensland is currently a six kilometre paved walking trail with additional unpaved sections that attract many types of users, mainly recreational walkers and nature-lovers. The MBWT was established and operated by the National Parks Authority of Brisbane (NPAB). As part of a larger effort to promote out-door recreation and appreciation of the local natural environment among Brisbane's residents and visitors, the NPAB is proposing an extension of the MBWT that would add 12 kilometres to the total paved track currently available to users. This extension, to be constructed over the period 2018 to 2023, would connect the village of Mount Beno (the starting and finishing point of the existing trail) with the village of Mount Illustrious, the location of another popular recreational area operated by the NPAB. With the extended track the NPAB will also be offering a series of guided walks, for a nominal fee to be paid to the NPAB, led by their rangers and/or suitably qualified volunteer ornithologists, ecologists and botanists.

One of the project options under consideration is for the State Government of Queensland (GOQ) to use the MBWT extension as part of an on-the-job training programme for at-risk, unemployed youth from the greater Brisbane region. In view of high rates of youth unemployment and, according to local authorities, an associated increase in the crime rate, the GOQ recently introduced a Youth Skill Development Program (YSDP) to train otherwise unemployed, at-risk school leavers for possible careers in the State's rapidly expanding eco-tourism sector. Under the YSDP they will be instructed in skills required both to construct and maintain hiking trails, and to become rangers and guides for subsequent employment by the NPAB or other similar eco-tourism parks. The GOQ will be requesting financial support from the Federal Government to support the YSDP, possibly in the form of tax exemptions.

You have been invited by GOQ and NPAB to undertake a cost-benefit analysis (CBA) of the MBWT extension and to prepare a report in which you present and discuss the results of your CBA and make recommendations to NPAB and GOQ.

Project Costs at Market Prices[1]

Investment costs

The main investment will be the costs involved in the construction of the additional 12 km track, including some picnic spots and public toilets, and additional car and bicycle parking facilities at each end of the completed trail. It is envisaged that the construction will occur in three 2-year stages, i.e. 2 km per annum over six years, beginning in 2018 (Project Year 0). On the completion of each 4 km stage that section of the track will be opened to the public, including the guided walks.

The estimated total investment cost for the construction of the MBWT (excluding land acquisition) would be between $8 to $9.5 million, spread over the period 2018 to 2023 as detailed below. In addition, acquisition of privately-owned land to provide an adequate corridor along sections of the trail adjacent to private residential and agricultural lots is expected to amount to a total of 30 Ha at an average cost of $100,000 per Ha, spread evenly in three stages, in years 2018, 2020 and 2022.

Details of the other investment costs are as follows:

i.   Equipment costing $5 million will be purchased in 2018;
ii. Materials costing $40,000 per annum from 2018 to 2023;
iii. Other costs of $10,000 per annum from 2018 to 2023;
iv. Two options for the construction component of the investment are under consideration.

Option 1: Out-sourcing completely to a private company. A private company will be paid a total of $0.7 million for each of the 4 km sections, 50% at the start of each section and the balance on completion of each section.

Option 2: Employing local youth as part of the YSDP where the recruits are trained on a 4-year, in-service programme to develop a number of relevant trades and skills, including, initially, carpentry, masonry, landscaping, etc. for the construction and maintenance of walking trails, and, as wild-life rangers and guides in the State's extensive national parks-based eco-tourism facilities. It is envisaged that an initial group of 10 youth trainees will be recruited through the YSDP by the NPAD at the beginning of 2018 to undergo training and to work on the construction of the new trails. Two trainers/supervisors will be recruited by NPAB for the period 2018–2023, each at a cost of $150,000 per annum. Each trainee on the YSDP will receive a GOQ allowance of $36,000 per annum for the 4-year duration of each award. At the end of each award period a new intake of 10 YSDP trainees will be assigned to the MBWT project. Participants in the YSDP will no longer qualify for the usual welfare payments from the Federal Government amounting to an estimated $20,000 per annum. (The YSDP will continue over the full life of the project as discussed in Section 2.3.)

It should be assumed that the extension to the MBWT has a life of 25 years (to the end of 2043) when it would need to be almost completely re-constructed. For the purpose of the CBA, assume that at the end of the project's life the acquired land cannot be sold, and that the equipment component of initial investment costs has a salvage value equivalent to 20% of its cost (with no GST payable).

Under both options, the NPAB will also need to invest in working capital consisting of stocks of materials for maintenance and operations, totalling $600,000 in 2023 to be built-up evenly from 2018.

Loan

NPAB intends to take out an $8 million loan (in Australian dollars) in 2018 from the International Environment Foundation (IEF, a UN concessional funding agency for projects promoting environmental awareness and education, based in Geneva) at an interest rate of 3% per annum (real), to be repaid over 20 years as an annuity.

Operating and maintenance costs

Once in operation, the additional tracks will need to be maintained on a regular basis. It is estimated that the materials required for maintenance will cost $24,000 per annum, starting in 2024. In addition, the usage of the tracks by the general public will need to be supervised, including the production and distribution of information materials including maps, wild-life guides, etc., and the operation and maintenance of the new rangers' information and first-aid hut to be constructed at the Mount Illustrious end of the new trail. These are expected to cost another $6000 per annum from 2024 onwards. Under *both* options an estimated 10 hours per week of volunteers' time will be required for additional ranger/guide duties.

Under <u>Option 1</u>, maintenance tasks will be sub-contracted out to a local company at an annual fee of $50,000 per annum from 2024 onwards. Suitably qualified additional staff (both regular paid employees and of NPAB and local volunteers) will be recruited as rangers and guides and for track supervision. Two full-time rangers will be employed from 2024 onwards at an annual cost of $100,000 each.

Under <u>Option 2</u>, the locally engaged YSDP trainees will undertake all maintenance and ranger/guide tasks, (along with the additional volunteers), under the supervision of a specifically recruited NPAB ranger at a cost of $100,000 per annum from 2024 onwards. The YSDP trainees (totalling 10 each year) will continue to be recruited from 2024 onwards on a 4-year basis and paid an annual allowance of $36,000 from GOQ until the final year of the project. (Assume that the last intake recruited in 2038 is retained for 6 years, to the end of the project's life in 2043. This implies that a total of 60 youth will be trained through the YSDP over the MBWT project's life.)

Opportunity cost information

i.   GST of 10% (assume it is a distortionary tax) is paid on all materials, machinery, spares and other material inputs included under investment (including working capital) and O&M costs
ii   All material inputs are sourced through additional supply
iii. Land has an opportunity cost estimated at 150% of its market value
iv. Sub-contractors' fees, supervisors' and trainers' costs consist entirely of skilled labour with an opportunity cost the same as market costs
v.   The social opportunity cost of the youth trainees' time is estimated at 20% of their YSDP allowance, not taking into account the welfare benefits they will cease receiving from Federal Government.
vi. Volunteers' time is valued at $25 per hour.
vii. GOQ uses an opportunity cost of capital of 6% per annum (real) for discounting, but requires sensitivity testing at 4% and 8%.

Project benefits

The NPAB and GOQ have asked you to take into account both tangible and intangible costs and benefits to the various stakeholders. At this stage NPAB has identified the additional recreational benefits accruing to the users of the trail as the main component of project benefit. The only financial benefit accruing to NPAB will be the additional revenue from the guided eco-walks, although you are also required to estimate the implications of introducing a general $10 entry fee for all MBWT users, including those on the guided walks.

Recent survey data show that the existing MBWT can be expected to attract about 30,000 visits per annum. (Note that this refers to the number of *visits* with each of the estimated 4800 visitors undertaking 6.25 visits per annum, at an average travel cost of $30 per visit.) It is estimated that the effect of adding the extension will increase the number of visitors to MBWT by 20% in each of the three discrete years immediately after the completion of each stage; i.e. in 2020, 2022 and 2024. It is also estimated that as each new stage of the trail is opened, another group of visitors will be attracted to the MBWT to participate in one of the proposed guided walks for which they will need to pay a fee of $10 per walk to NPAB. This group of visitors is expected to increase the total number of MBWT visitors by 800 from 2020, increasing by 20% in 2022 and 2024 as each additional stage is completed. Each visitor participating in guided walks is expected to undertake 6 visits per annum, also at an average travel cost of $30 per visit.

It is expected that in line with the projected increase in general eco-tourism visits to Queensland's national parks the numbers of new visitors using the MBWT for both guided and un-guided walks will increase by 5% per annum from 2026 onwards. Without the MBWT extension the number of visitors is projected to increase by 2% per annum from 2026 (i.e. it is assumed that without the extension there will be zero growth in visitors between 2018 and 2025).

To estimate the dollar value of the non-market benefits to visitors attributable to the MBWT extension, the NPAB engaged a team of environmental economists from one of the local universities to undertake a non-market valuation study using the Travel Cost Method (TCM).[2] From this an estimate was made of the track users' Willingness to Pay (WTP) and Consumer Surplus (CS) from their access to and usage of the existing trail. The average travel cost of getting to and from the MBWT was estimated at $30 per visit and the additional average entry cost for a guided walk is $10 per person. For each category of user a linear demand curve was estimated from the TCM studies.

For un-guided walks: $W_u = a - bP_u = 10 - 0.125P_u$
For guided walks: $W_g = c - d\,P_g = 12 - 0.15P_g$
where, W_u = number of visits *per visitor* on unguided walks
W_g = number of visits *per visitor* on guided walks
P_u = average price *per visit* for an unguided walk
P_g = average price *per visit* for a guided walk (where price is the average travel cost per visit plus the fee).

For the purpose of the analysis it should be assumed that these estimated demand functions for visits per visitor apply to the use of the MBWT with the extension. The main effect of the extension will be through the increase in the numbers of *visitors* who use the track, as indicated above. It should be assumed that any increase in user benefits from visits to the MBWT are additional to benefits from other recreational activities. The relevant CBA input data for numbers of visits and consumer surplus can be derived from the TCM model estimates and data given above. To assist you with your initial calculations the following data are provided for initial entry into your Variables table. However, you will need to eventually derive and link these values from the TCM model to the benefits section of your spreadsheet.

TRAIL VISITOR INPUT DATA	Visits/annum	Consumer Surplus ($/visit)
Unguided, no entry fee	30,000	
Additional Guided, no entry fee	4800	
Unguided, with entry fee	24,000	
Additional Guided, with entry fee	3600	
CS/visit (unguided, without entry fee)		$25
CS/visit (guided, without entry fee)		$20
CS/visit (unguided, with entry fee)		$20
CS/visit (guided, with entry fee)		$15

Another important social benefit will be the training and employment benefits accruing over the longer term to the youth engaged through the YSDP. You should assume that in the absence of the MBWT extension project the 60 youths recruited to the MBWT project would not otherwise have participated in the YSDP.

The *Referent Group* should be treated as the population of Queensland, with GOQ the main public sector stakeholder. However, as the project has implications for other non-Referent Group Stakeholders, you should also estimate and include in your report the implications for them, especially the Federal Government who GOQ is intending to approach for possible supplementary funding for the project in light of the gains it is expected to make through the additional GST revenues and reduction in welfare payments to the trainees on the YSDP.

Alternative scenarios and Sensitivity Analysis

Your CBA should consider four alternative scenarios:

Options 1 and 2 as discussed above (without and with the YSDP component respectively) and under each Option: with no general entry fee for visitors to the MBWT (Options 1a and 2a); and, with a $10 general entry fee for both groups participating on the guided and unguided walks (Options 1b and 2b). The entry fees will be collected by NPAB and volunteers within their usual working hours. Assume no other fee collection costs and that the general entry fee can only be introduced under the extension of MBWT, and is therefore not an option on the existing trail.

You are also required to undertake a *Sensitivity Analysis* in which you estimate and report on the implications of variations in the values attached to the most critical variables in the CBA, including, where appropriate, their *Threshold Values*; e.g. the threshold value at which an option breaks even, and/or, at which their ranking switches.

These should include at least:

i.   The assumed number of initial visitors and the growth in visitor numbers;
ii. Full exemption from the 10% GST;

iii. Other possible benefits generated by the training of the 60 youths recruited through the YSDP. (These can be included in your spreadsheet under Option 2 as '*Other Unaccounted Benefits/Costs*' over the project's life.)

You are also encouraged to discuss (but not quantify) the possible implications of other unaccounted costs and/or benefits not included in your CBA but which you believe could impact potentially on the project options, especially if they could affect their ranking.

Notes

1 All dollar amounts are in constant 2018 prices unless otherwise stated.
2 See Chapter 8.6.1 for an explanation and example of the TCM approach.

A1.14 Cost-benefit analysis of a proposed Drug Court programme in the State of Euphoria, Federal Republic of Oz

Background

In the state of Euphoria in the Federal Republic of Oz consideration is being given to the establishment of a Drug Court. A Drug Court is responsible for sentencing and supervising the treatment of participants with drug or alcohol problems, who have committed an offence under the influence of drugs or alcohol. It excludes offenders who have committed sex offences or where an act of violence occasioning actual bodily harm is involved. Drug Courts have been operating for some years in Australia, Canada, Ireland, Scotland and England. Evaluations of drug courts elsewhere have found that they can be both less costly and more effective than the alternative of imprisonment.

The impact of drug abuse on the Euphorian community and criminal justice system increased throughout the last decade. Studies associate the incidence of drug dependency with various community challenges such as ill health, unemployment and homelessness. When poor education, high unemployment and housing crises arise in combination, prospects for drug affected people are severely limited relative to others not so affected. Key factors and actions that resulted in this programme being initiated were: by early 2015, more than 80% of prisoners had a history of drug and alcohol abuse, with backgrounds of social disadvantage, low education, high unemployment, significant health issues (including mental health) and poor family and social links; and, about 44% of Euphoria's prisoners were returning to prison within two years of being released and this rate was increasing at about 3% per annum.

The proposed Euphorian Drug Court Program (EDCP) represents a fundamental shift in the way in which the courts will deal with drug offenders. It has been developed in line with the principles of therapeutic jurisprudence. These principles recognise that the law can operate as a therapeutic agent, intervening effectively in the lives of 'appropriately motivated individuals' to resolve issues in their lives that contribute to their criminal activities. It seeks to protect the community by focusing on the rehabilitation of offenders from drug or alcohol addiction with the ultimate goal of bringing stability to participants' chaotic lifestyles and reintegrating them into the community.

Under the EDCP proposal an offender in the Drug Court will be sentenced to a Drug Treatment Order (DTO). The DTO consists of two parts, a custodial part and a treatment and supervision part. The custodial sentence is suspended to allow for the treatment of the offender. The treatment and supervision will involve conditions being imposed, which are intended to address the offender's drug and alcohol dependency. Sanctions and rewards are used to reward compliant behaviour and sanction non-compliant behaviour. The DTO objectives are:

- to reduce drug/alcohol usage and re-offending and achieve greater stability and integration into the community for participants;
- to link transitional housing and homelessness assistance for people experiencing homelessness or housing crises to the Drug Court. Housing support is a unique feature of the EDCP, provided in collaboration with the Euphorian Office of Housing (EOH).

Under the proposal, the Drug Court Magistrate will have the responsibility for the supervision of participants placed in the Drug Court programme. A multi-disciplinary team consisting of a case manager, clinician, specialist community correction officers and a dedicated police prosecutor and defence lawyer will assist the Drug Court Magistrate. This team will work with the Drug Court Magistrate in managing and supervising participants on the DTO. If participants breach the DTO, it can be cancelled and participants sentenced to serve the unexpired portion of their sentence. The EDCP initiative is proposed as an experimental response to the failure of current custodial sanctions to adequately address drug use and related offending. If implemented, the first Euphorian Drug Court will commence operations in January 2016 and will be located in Euphoria's capital city, Utopia. It would then be trialled as a six-year pilot programme. The DTO is planned as a three-phase programme. For the purpose of this study it should be assumed each DTO spans a period of two years. Successful graduates are released on completion of the DTO with any remaining part of their sentence suspended. Under the base case, a fixed number of 50 new DTO participants per year is assumed. The first intake will enter the EDCP 1 January 2016 and the last on 1 January 2021. In the base case scenario it is assumed, conservatively, that 50% of those sentenced to a DTO drop out of the treatment program at the end of the first year.

Program costs

The Department of Justice (DOJ) and the Department of Human Services (DHS) costs are detailed in Tables App_1.33 to App_1.35. Table App_1.33 provides details of the start-up costs incurred in 2015, including details of any GST and import duties. The start-up costs of DOJ are to be financed partly through a loan from the Community Services Bank of Oz (CSBO).

Table App_1.33a Start-up costs (2015) ($ thousands)

	DoJ	DHS
Construction*	$7500	$1000
Materials**	$4500	$2000
Training***	$18,000	$7000

* including GST of 10%
** including import duties of 15%
*** Magistrate's (DoJ) and counsellors' (DHS) time

Table App_1.33b Loan from CSBO

Disbursement (2015)	$20,000
Term (yrs)	6
Interest rate (real)	3%

Table App_1.34 provides details of the estimated one-off costs incurred for each new DTO entering the EDCP.

Table App_1.34 New DTO entrant one-off costs

DOJ	$000's per DTO Entrant
Magistrate's Court Costs*	$40
Magistrate	$30
Court Services*	$25
Euphorian Legal Aid (Solicitors)	$15
Police (Prosecutors)	$25
Prison Costs*	$15

* Including GST of 10%

Table App_1.35 provides estimates of the ongoing costs incurred each year per active DTO participant over the two years of the DTO.

Table App_1.35 On-going active DTO costs

DOJ	Market Price $000's per annum per active DTO
Corrections (Case Managers)	$75
Accommodation Service	$40
Drug Testing Service*	$10
DHS	
Drug Treatment (Clinicians)	$45
Counselling (excl. Salvation Army volunteers)	$15
Housing Services	$12
Housing- Property Mgt/Maintenance*	$4
Housing- Lease Costs	$10
Volunteers (Salvos time)	$0

* Including GST of 10%

Market distortions

Apart from GST there are several other market distortions for which a set of shadow prices is needed for undertaking the Efficiency CBA.

- The accommodation service used by the DOJ for the DTOs is in government-owned apartments subject to rent control. It has been estimated that the opportunity cost is about 25% higher than the market price.
- Housing Services and Housing Lease Costs have an opportunity cost 10% higher than the market price.
- Professional labour including magistrates, solicitors and prosecutors (see Tables App_1.33 and App_1.34) and Case Managers and Counsellors (see Table App_1.35) are in short supply in Euphoria. Economists at the local university estimate their opportunity cost at 10% above the market wage.

- In the case of suitably qualified drug treatment clinicians (see Table App_1.35) there is an extreme shortage; the opportunity cost is estimated at 50% above the regulated market wage.
- In addition to the paid counsellors, the EDCP also relies on the voluntary counselling services offered by the Salvation Army; one volunteer-year per DTO entrant. The opportunity cost is estimated to be the equivalent of the average wage in Euphoria for a semi-skilled worker ($25,000 per annum before tax).

Programme benefits

The benefits of the programme depend critically on reducing the incidence of re-offending (recidivism) among the program's participants. Based on studies from other countries the estimated differences between EDCP participants' and a control group of similar drug-affected offenders' recidivism rates are given in Table App_1.36.

Table App_1.36 Estimated recidivism rates for control vs DTO groups

	Control Group	DTO Group	
		Drop-outs	Graduates
Re-offending Rates (%)	60%	40%	10%
Offences per re-offender pa	6	4	1

These data in conjunction with the information provided in Table App_1.37a can be used to estimate the reduction in the number of re-offences attributable to the EDCP, and the associated reduction in costs.

Table App_1.37a Additional data for reduced costs to DOJ and DHS

	Units/Market Prices ($000s)
Prison days per offence (DOJ)	30
Prison Costs per day ($) (DOJ)*	$0.10
Court Costs per offence (DOJ)*	$0.20
Reduced Drug treatment costs ($/graduate/annum) (DHS)	$1.75
Reduced Drug-related health costs ($/graduate/annum) (DHS)	$1.00
Reduced Public Housing costs ($/graduate/annum) (DHS)**	See table note
Reduced hospitalisation(bed days/DTO/annum) (DHS)	10
Hospitalisation costs per bed/day (DHS)	$1.00

* Including GST
** equal to the sum of 'Housing – Property Mgt/Maintenance' and 'Housing Lease Costs' in Table 35

As shown in Table App_1.37b, some of the cost savings occur while the DTOs are active in the programme, while others continue over the full duration of the extended period to 2035. The cost savings *during* the DTO occur for the duration of the EDCP only for those who remain active, while the cost savings *from the successful completion of the programme* continue for each graduate until 2035.

Table App_1.37b Categories of reduced costs attributable to DTOs

---

Reduced Costs *During* DTO (per active DTO)
- Reduced Prison Costs
- Reduced Court Costs
- Reduced Drug Treatment Costs
- Reduced Drug-related Health Costs
- Reduced Hospitalisation Costs

Reduced Costs *From* graduating DTOs (per graduate)
- Reduced Court Costs
- Reduced Prison Costs
- Reduced Drug Treatment Costs
- Reduced Drug-related Health Costs
- Reduced Public Housing Costs
- Reduced Hospitalisation Costs
- Reduced Costs to Families*
- Reduced Costs to Victims*
- Reduced Costs to Society*
- Reduced Unemployment Benefits (Federal Government)
- Increased Earnings of Graduates (before income tax)*
- Income Tax (Federal Government)

---

* assume total expenditure of graduates and GST remain constant

Quality of life and reduced fatalities

In addition to the reduced costs to the various stakeholder groups there will be other benefits to the wider community from the reduction in drug-related crime. Families of treated addicts, victims of drug-related crimes, the EDCP graduates themselves and others in Euphorian society will benefit in several ways:

> First, they will incur less expenditure in relation to the previous addicts' behaviour.
> Second, they will enjoy a better quality of life.
> Third, there will a reduction in the number of fatalities attributable to successful graduation of the DTOs; among both (i) victims of drug-related crimes and (ii) the DTO graduates themselves.

Details of the cost savings and Quality-Adjusted Life Years (QALYs) gained and fatalities avoided per graduating DTO are given in Table App_1.38, along with the estimated Value of a Statistical Life (VSL) for the people of the Federated Republic of Oz, which was estimated on the basis of Benefit Transfer from other international studies. To simplify the analysis it should be assumed that:

- the cost reductions continue beyond the duration of the EDCP to the year 2035
- the full value of the QALYs gained and fatalities avoided is realised in the same year as the DTOs graduation from the EDCP[1]

Table App_1.38 Benefits to families, victims, graduates and society

	Market Prices ($000s)/units	QALYs/Graduate
Reduced costs to families ($ & QALY)*	$5	4
Reduced costs to victims ($ & QALY)*	$10	5
Reduced costs to society ($ & QALY)*	$15	1
Fatalities avoided (no. of victims/graduate)**	0.25	
Fatalities avoided (no. of graduates/graduate)**	0.5	
Value of Statistical Life		
VSL ($000's/person)	$3500	
Statistical Life (years)	40	

* equivalent to one VSL-year
** based on $ estimate of VSL

Productivity gains and earnings of graduates

Another important benefit expected from the EDCP will be the increased productivity attributable to the successful treatment of the graduating DTOs. It should be assumed that each graduate is employed for the remainder of the extended project life (to 2035) earning an income equivalent to the wage for the average semi-skilled worker in Euphoria ($25,000 per annum before tax). This will save the Federal Government unemployment benefits totalling $12,000 per annum. The average rate of income tax payable to the Federal Government for that income category is 20%. The gross wage is considered a good indicator of opportunity cost.

Benefit decay

It can be expected that not all graduates will remain drug- and crime-free throughout the extended project life. In this case there will be some decay in the value of benefits each year. In the base case, assume an annual 'benefit decay' rate of 10% per annum, of the gross value of all benefits calculated from Table App_1.37, starting one year after graduation. For those benefits calculated from Table App_1.38, assume a drop of 10% for each year in which these benefits occur.

Instructions

You have been engaged by the State of Euphoria (SOE) to complete a benefit-cost analysis (BCA) of the proposed EDCP. This includes an assessment of the financial, economic and social impacts of establishing the programme and its operation over the six-year period 2016–2021. (As each DTO stays in the programme for 2 years, the active DTOs from the final 2021 intake will continue in the EDCP through 2022.) As some of the benefits discussed above are of a longer-term duration, you should consider all of these, where relevant, over the period 2015–2035. The analysis should be conducted in thousands of 2015 dollars and the required discount rate is 4% (real), but SOE normally requires sensitivity testing around this rate.

Although your main objective of the report is to make recommendations to the SOE based on the net benefits to the Referent Group, you are also required to consider the costs and benefits to: (i) the various sub-categories of stakeholders within the Government of the State of Euphoria; the Department of justice (DOJ) and Department of Human Services (DHS), and, (ii) 'other' stakeholders in the SOE, aggregated as one stakeholder group, which includes the families, victims, the offenders and the rest of Euphorian society at large; and, (iii) other non-Referent Group stakeholders in the Federal Republic of Oz; viz. the Federal Government of Oz and Community Services Bank of Oz (CSBO).

Sensitivity analysis

In addition to presenting and discussing the results for your 'Base Case' scenario, you should also undertake extensive sensitivity testing to check the robustness of your findings and recommendations. One of your roles is to identify the variables that give rise to the greatest variability in the results. Among these should be: (i) the drop-out rate among EDCP participants; (ii) the comparable recidivism rates between the DTOs and the control group; (iii) the benefit decay rate; and, (iv) the Value of a Statistical life (VSL).

As part of your sensitivity analysis you should:

(i)   show the results also for the *most pessimistic* and *most optimistic* scenarios;
(ii) discuss whether you believe any other variable is of sufficient relevance and uncertainty for inclusion in a sensitivity analysis, with a clear and concise justification for your opinion;
(iii) identify the respective *threshold values* for each individual variable selected for sensitivity testing, at which the NPV of the aggregate Referent Group net benefit becomes zero. You should also comment on the likelihood of each threshold value being within the feasible range of actual values for that variable;
(iv) identify and discuss any other costs or benefits not included in the analysis, which you believe should be taken into account, with a clear statement justifying your position. (To guide you, two relevant papers on the topic are suggested; Cohen (2000) and Dossetor (2011). These should be cited in your report if/when used, together with any other reference material from your own research.)[2]

Notes

1 For a basic explanation of a QALY see Phillips (2009) 'What is a QALY?" www.researchgate.net/publication/255655713_What_is_a_QALY
    and for a discussion of the Value of a Statistical Life (VSL), see Campbell and Brown Chapter 8.9. As the VSL is a present value, to convert a VSL to a QALY you need to take discounting into account.
2 M.A. Cohen (2000) Measuring the Costs and Benefits of Crime and Justice, in. *Measurement and Analysis of Crime and Justice*, edited by D. Joffee, pp. 263–315, Vol. 4 Criminal Justice 2000, Washington DC: National Institute of Justice, US Department of Justice.
    K. Dossetor (2011) Cost-benefit analysis and its application to crime prevention and criminal justice research, AIC Technical and Background Paper, No. 42, Canberra: Australian Institute of Criminology.

Appendix 2
Discount and annuity factors

Discount factors

	1%	2%	3%	4%	5%	6%	7%	8%	9%	10%	11%	12%	13%	14%	15%	16%	17%	18%	19%	20%	25%	30%
1	0.990	0.980	0.971	0.962	0.952	0.943	0.935	0.926	0.917	0.909	0.901	0.893	0.885	0.877	0.870	0.862	0.855	0.847	0.840	0.833	0.800	0.769
2	0.980	0.961	0.943	0.925	0.907	0.890	0.873	0.857	0.842	0.826	0.812	0.797	0.783	0.769	0.756	0.743	0.731	0.718	0.706	0.694	0.640	0.592
3	0.971	0.942	0.915	0.889	0.864	0.840	0.816	0.794	0.772	0.751	0.731	0.712	0.693	0.675	0.658	0.641	0.624	0.609	0.593	0.579	0.512	0.455
4	0.961	0.924	0.888	0.855	0.823	0.792	0.763	0.735	0.708	0.683	0.659	0.636	0.613	0.592	0.572	0.552	0.534	0.516	0.499	0.482	0.410	0.350
5	0.951	0.906	0.863	0.822	0.784	0.747	0.713	0.681	0.650	0.621	0.593	0.567	0.543	0.519	0.497	0.476	0.456	0.437	0.419	0.402	0.328	0.269
6	0.942	0.888	0.837	0.790	0.746	0.705	0.666	0.630	0.596	0.564	0.535	0.507	0.480	0.456	0.432	0.410	0.390	0.370	0.352	0.335	0.262	0.207
7	0.933	0.871	0.813	0.760	0.711	0.665	0.623	0.583	0.547	0.513	0.482	0.452	0.425	0.400	0.376	0.354	0.333	0.314	0.296	0.279	0.210	0.159
8	0.923	0.853	0.789	0.731	0.677	0.627	0.582	0.540	0.502	0.467	0.434	0.404	0.376	0.351	0.327	0.305	0.285	0.266	0.249	0.233	0.168	0.123
9	0.914	0.837	0.766	0.703	0.645	0.592	0.544	0.500	0.460	0.424	0.391	0.361	0.333	0.308	0.284	0.263	0.243	0.225	0.209	0.194	0.134	0.094
10	0.905	0.820	0.744	0.676	0.614	0.558	0.508	0.463	0.422	0.386	0.352	0.322	0.295	0.270	0.247	0.227	0.208	0.191	0.176	0.162	0.107	0.073
11	0.896	0.804	0.722	0.650	0.585	0.527	0.475	0.429	0.388	0.350	0.317	0.287	0.261	0.237	0.215	0.195	0.178	0.162	0.148	0.135	0.086	0.056
12	0.887	0.788	0.701	0.625	0.557	0.497	0.444	0.397	0.356	0.319	0.286	0.257	0.231	0.208	0.187	0.168	0.152	0.137	0.124	0.112	0.069	0.043
13	0.879	0.773	0.681	0.601	0.530	0.469	0.415	0.368	0.326	0.290	0.258	0.229	0.204	0.182	0.163	0.145	0.130	0.116	0.104	0.093	0.055	0.033
14	0.870	0.758	0.661	0.577	0.505	0.442	0.388	0.340	0.299	0.263	0.232	0.205	0.181	0.160	0.141	0.125	0.111	0.099	0.088	0.078	0.044	0.025
15	0.861	0.743	0.642	0.555	0.481	0.417	0.362	0.315	0.275	0.239	0.209	0.183	0.160	0.140	0.123	0.108	0.095	0.084	0.074	0.065	0.035	0.020
16	0.853	0.728	0.623	0.534	0.458	0.394	0.339	0.292	0.252	0.218	0.188	0.163	0.141	0.123	0.107	0.093	0.081	0.071	0.062	0.054	0.028	0.015
17	0.844	0.714	0.605	0.513	0.436	0.371	0.317	0.270	0.231	0.198	0.170	0.146	0.125	0.108	0.093	0.080	0.069	0.060	0.052	0.045	0.023	0.012
18	0.836	0.700	0.587	0.494	0.416	0.350	0.296	0.250	0.212	0.180	0.153	0.130	0.111	0.095	0.081	0.069	0.059	0.051	0.044	0.038	0.018	0.009
19	0.828	0.686	0.570	0.475	0.396	0.331	0.277	0.232	0.194	0.164	0.138	0.116	0.098	0.083	0.070	0.060	0.051	0.043	0.037	0.031	0.014	0.007
20	0.820	0.673	0.554	0.456	0.377	0.312	0.258	0.215	0.178	0.149	0.124	0.104	0.087	0.073	0.061	0.051	0.043	0.037	0.031	0.026	0.012	0.005
21	0.811	0.660	0.538	0.439	0.359	0.294	0.242	0.199	0.164	0.135	0.112	0.093	0.077	0.064	0.053	0.044	0.037	0.031	0.026	0.022	0.009	0.004
22	0.803	0.647	0.522	0.422	0.342	0.278	0.226	0.184	0.150	0.123	0.101	0.083	0.068	0.056	0.046	0.038	0.032	0.026	0.022	0.018	0.007	0.003
23	0.795	0.634	0.507	0.406	0.326	0.262	0.211	0.170	0.138	0.112	0.091	0.074	0.060	0.049	0.040	0.033	0.027	0.022	0.018	0.015	0.006	0.002
24	0.788	0.622	0.492	0.390	0.310	0.247	0.197	0.158	0.126	0.102	0.082	0.066	0.053	0.043	0.035	0.028	0.023	0.019	0.015	0.013	0.005	0.002
25	0.780	0.610	0.478	0.375	0.295	0.233	0.184	0.146	0.116	0.092	0.074	0.059	0.047	0.038	0.030	0.024	0.020	0.016	0.013	0.010	0.004	0.001
26	0.772	0.598	0.464	0.361	0.281	0.220	0.172	0.135	0.106	0.084	0.066	0.053	0.042	0.033	0.026	0.021	0.017	0.014	0.011	0.009	0.003	0.001
27	0.764	0.586	0.450	0.347	0.268	0.207	0.161	0.125	0.098	0.076	0.060	0.047	0.037	0.029	0.023	0.018	0.014	0.011	0.009	0.007	0.002	0.001
28	0.757	0.574	0.437	0.333	0.255	0.196	0.150	0.116	0.090	0.069	0.054	0.042	0.033	0.026	0.020	0.016	0.012	0.010	0.008	0.006	0.002	0.001
29	0.749	0.563	0.424	0.321	0.243	0.185	0.141	0.107	0.082	0.063	0.048	0.037	0.029	0.022	0.017	0.014	0.011	0.008	0.006	0.005	0.002	0.001
30	0.742	0.552	0.412	0.308	0.231	0.174	0.131	0.099	0.075	0.057	0.044	0.033	0.026	0.020	0.015	0.012	0.009	0.007	0.005	0.004	0.001	0.000
40	0.672	0.453	0.307	0.208	0.142	0.097	0.067	0.046	0.032	0.022	0.015	0.011	0.008	0.005	0.004	0.003	0.002	0.001	0.001	0.001	0.000	0.000
50	0.608	0.372	0.228	0.141	0.087	0.054	0.034	0.021	0.013	0.009	0.005	0.003	0.002	0.001	0.001	0.001	0.000	0.000	0.000	0.000	0.000	0.000

Annuity factors

	1%	2%	3%	4%	5%	6%	7%	8%	9%	10%	11%	12%	13%	14%	15%	16%	17%	18%	19%	20%	25%	30%
1	0.990	0.980	0.971	0.962	0.952	0.943	0.935	0.926	0.917	0.909	0.901	0.893	0.885	0.877	0.870	0.862	0.855	0.847	0.840	0.833	0.800	0.769
2	1.970	1.942	1.913	1.886	1.859	1.833	1.808	1.783	1.759	1.736	1.713	1.690	1.668	1.647	1.626	1.605	1.585	1.566	1.547	1.528	1.440	1.361
3	2.941	2.884	2.829	2.775	2.723	2.673	2.624	2.577	2.531	2.487	2.444	2.402	2.361	2.322	2.283	2.246	2.210	2.174	2.140	2.106	1.952	1.816
4	3.902	3.808	3.717	3.630	3.546	3.465	3.387	3.312	3.240	3.170	3.102	3.037	2.974	2.914	2.855	2.798	2.743	2.690	2.639	2.589	2.362	2.166
5	4.853	4.713	4.580	4.452	4.329	4.212	4.100	3.993	3.890	3.791	3.696	3.605	3.517	3.433	3.352	3.274	3.199	3.127	3.058	2.991	2.689	2.436
6	5.795	5.601	5.417	5.242	5.076	4.917	4.767	4.623	4.486	4.355	4.231	4.111	3.998	3.889	3.784	3.685	3.589	3.498	3.410	3.326	2.951	2.643
7	6.728	6.472	6.230	6.002	5.786	5.582	5.389	5.206	5.033	4.868	4.712	4.564	4.423	4.288	4.160	4.039	3.922	3.812	3.706	3.605	3.161	2.802
8	7.652	7.325	7.020	6.733	6.463	6.210	5.971	5.747	5.535	5.335	5.146	4.968	4.799	4.639	4.487	4.344	4.207	4.078	3.954	3.837	3.329	2.925
9	8.566	8.162	7.786	7.435	7.108	6.802	6.515	6.247	5.995	5.759	5.537	5.328	5.132	4.946	4.772	4.607	4.451	4.303	4.163	4.031	3.463	3.019
10	9.471	8.983	8.530	8.111	7.722	7.360	7.024	6.710	6.418	6.145	5.889	5.650	5.426	5.216	5.019	4.833	4.659	4.494	4.339	4.192	3.571	3.092
11	10.368	9.787	9.253	8.760	8.306	7.887	7.499	7.139	6.805	6.494	6.207	5.938	5.687	5.453	5.234	5.029	4.836	4.656	4.486	4.327	3.656	3.147
12	11.255	10.575	9.954	9.385	8.863	8.384	7.943	7.536	7.161	6.814	6.494	6.194	5.918	5.660	5.421	5.197	4.988	4.793	4.611	4.439	3.725	3.190
13	12.134	11.348	10.635	9.986	9.394	8.853	8.358	7.904	7.487	7.103	6.750	6.424	6.122	5.842	5.583	5.342	5.118	4.910	4.715	4.533	3.780	3.223
14	13.004	12.106	11.296	10.563	9.899	9.295	8.745	8.244	7.786	7.367	6.982	6.628	6.302	6.002	5.724	5.468	5.229	5.008	4.802	4.611	3.824	3.249
15	13.865	12.849	11.938	11.118	10.380	9.712	9.108	8.559	8.061	7.606	7.191	6.811	6.462	6.142	5.847	5.575	5.324	5.092	4.876	4.675	3.859	3.268
16	14.718	13.578	12.561	11.652	10.838	10.106	9.447	8.851	8.313	7.824	7.379	6.974	6.604	6.265	5.954	5.668	5.404	5.162	4.938	4.730	3.887	3.283
17	15.562	14.292	13.166	12.166	11.274	10.477	9.763	9.122	8.544	8.022	7.549	7.120	6.729	6.373	6.047	5.749	5.475	5.222	4.990	4.775	3.910	3.295
18	16.398	14.992	13.754	12.659	11.690	10.828	10.059	9.372	8.756	8.201	7.702	7.250	6.840	6.467	6.128	5.818	5.534	5.273	5.033	4.812	3.928	3.304
19	17.226	15.678	14.324	13.134	12.085	11.158	10.336	9.604	8.950	8.365	7.839	7.366	6.938	6.550	6.198	5.877	5.584	5.316	5.070	4.843	3.942	3.311
20	18.046	16.351	14.877	13.590	12.462	11.470	10.594	9.818	9.129	8.514	7.963	7.469	7.025	6.623	6.259	5.929	5.628	5.353	5.101	4.870	3.954	3.316
21	18.857	17.011	15.415	14.029	12.821	11.764	10.836	10.017	9.292	8.649	8.075	7.562	7.102	6.687	6.312	5.973	5.665	5.384	5.127	4.891	3.963	3.320
22	19.660	17.658	15.937	14.451	13.163	12.042	11.061	10.201	9.442	8.772	8.176	7.645	7.170	6.743	6.359	6.011	5.696	5.410	5.149	4.909	3.970	3.323
23	20.456	18.292	16.444	14.857	13.489	12.303	11.272	10.371	9.580	8.883	8.266	7.718	7.230	6.792	6.399	6.044	5.723	5.432	5.167	4.925	3.976	3.325
24	21.243	18.914	16.936	15.247	13.799	12.550	11.469	10.529	9.707	8.985	8.348	7.784	7.283	6.835	6.434	6.073	5.746	5.451	5.182	4.937	3.981	3.327
25	22.023	19.523	17.413	15.622	14.094	12.783	11.654	10.675	9.823	9.077	8.422	7.843	7.330	6.873	6.464	6.097	5.766	5.467	5.195	4.948	3.985	3.329
26	22.795	20.121	17.877	15.983	14.375	13.003	11.826	10.810	9.929	9.161	8.488	7.896	7.372	6.906	6.491	6.118	5.783	5.480	5.206	4.956	3.988	3.330
27	23.560	20.707	18.327	16.330	14.643	13.211	11.987	10.935	10.027	9.237	8.548	7.943	7.409	6.935	6.514	6.136	5.798	5.492	5.215	4.964	3.990	3.331
28	24.316	21.281	18.764	16.663	14.898	13.406	12.137	11.051	10.116	9.307	8.602	7.984	7.441	6.961	6.534	6.152	5.810	5.502	5.223	4.970	3.992	3.331
29	25.066	21.844	19.188	16.984	15.141	13.591	12.278	11.158	10.198	9.370	8.650	8.022	7.470	6.983	6.551	6.166	5.820	5.510	5.229	4.975	3.994	3.332
30	25.808	22.396	19.600	17.292	15.372	13.765	12.409	11.258	10.274	9.427	8.694	8.055	7.496	7.003	6.566	6.177	5.829	5.517	5.235	4.979	3.995	3.332
40	32.835	27.355	23.115	19.793	17.159	15.046	13.332	11.925	10.757	9.779	8.951	8.244	7.634	7.105	6.642	6.233	5.871	5.548	5.258	4.997	3.999	3.333
50	39.196	31.424	25.730	21.482	18.256	15.762	13.801	12.233	10.962	9.915	9.042	8.304	7.675	7.133	6.661	6.246	5.880	5.554	5.262	4.999	4.000	3.333

Glossary

Decision-maker: the individual or organization (the client) which commissions the cost-benefit analysis as an aid to deciding whether or not to support the project.

Efficiency Analysis: the calculation of the overall net benefits of a project, irrespective of to which groups they accrue, or of whether or not the effects of the project are correctly measured by market prices.

Impact Analysis: calculation of the effect of a project on national or regional Gross Domestic (or Regional) Product (GDP or GRP).

Market Analysis: calculation of the net benefits of a project in which all inputs and outputs are valued at market prices; commodities which are not traded in markets are priced at zero.

Pareto Improvement: the result of a change in the allocation of resources which leaves at least one agent in the economy better off without making any other agent worse off.

Potential Pareto Improvement: a situation in which a project could result in a Pareto Improvement if its benefits and/or costs could be redistributed without cost among the relevant economic agents, even though such a redistribution is not included as part of the project; the gainers from the project could compensate the losers and still be better off.

Private Analysis: calculation of the net benefit of a project to its proponent, which may be the equity holders of a private firm, a public-private partnership (PPP), a government department, a government foreign aid agency, a non-government organization (NGO), an international organization or similar body.

Project: any proposed action which will change the allocation of resources.

Project Analysis: this term is sometimes used to refer to the Market Analysis as described above.

Referent Group: the group of economic agents the decision-maker identifies as the stakeholders to be considered in appraising the project; these may be residents of a region, members of a social or ethnic group or any other identifiable group determined by the decision-maker.

Referent Group Analysis: calculation of the net benefits of the project to the Referent Group.

Resources: the scarce factors of production – land, labour, capital, materials and management – which are involved in undertaking a project.

Social Cost-Benefit Analysis: calculation of all the benefits and costs of a project, whether or not the relevant inputs or outputs are traded in markets, and irrespective of which groups gain or lose as a result of the project – also termed the Efficiency Analysis.

Index

Printed in Great Britain
by Amazon

34007505R00271